Differer
Equati

and

Calculus of Variations

for

BSc and B Tech Students of All Indian Universities

Differential Equations

and

Calculus of Variations

for BSc and B Tech Students of All Indian Universities

Rakesh Kumar
Department of Mathematics
Hindu College, Moradabad, UP

Nagendra Kumar
Department of Mathematics
MMH College, Ghaziabad, UP

CBS Publishers & Distributors Pvt Ltd

New Delhi • Bengaluru • Chennai • Kochi • Pune
Hyderabad • Kolkata • Manipal • Mumbai • Nagpur • Patna

Differential Equations and Calculus of Variations

ISBN: 978-81-239-2204-1

First Edition: 2013

Published by Satish Kumar Jain for
CBS Publishers & Distributors Pvt Ltd
4819/XI Prahlad Street, 24 Ansari Road, Daryaganj, New Delhi 110 002, India.
Ph: 23289259, 23266861, 23266867 Fax: 011-23243014 Website: www.cbspd.com
e-mail: delhi@cbspd.com; cbspubs@airtelmail.in

Corporate Office: 204 FIE, Industrial Area, Patparganj, Delhi 110 092
Ph: 4934 4934 Fax: 4934 4935 e-mail: publishing@cbspd.com; publicity@cbspd.com

Branches

- **Bengaluru:** Seema House 2975, 17th Cross, K.R. Road, Banasankari 2nd Stage, Bengaluru 560 070, Karnataka
  Ph: +91-80-26771678/79 Fax: +91-80-26771680 e-mail: bangalore@cbspd.com
- **Chennai:** 20, West Park Road, Shenoy Nagar, Chennai 600 030, Tamil Nadu
  Ph: +91-44-26260666, 26208620 Fax: +91-44-42032115 e-mail: chennai@cbspd.com
- **Kochi:** 36/14 Kalluvilakam, Lissie Hospital Road, Kochi 682 018, Kerala
  Ph: +91-484-4059061-65 Fax: +91-484-4059065 e-mail: kochi@cbspd.com
- **Pune:** Bhuruk Prestige, Sr. No. 52/12/2+1+3/2 Narhe, Haveli (Near Katraj-Dehu Road Bypass), Pune 411 041, Maharashtra
  Ph: +91-20-64704058, 64704059, 32342277 Fax: +91-20-24300160 e-mail: pune@cbspd.com

Representatives

- **Hyderabad** 0-9885175004
- **Kolkata** 0-9831437309
- **Manipal** 0-9742022075
- **Mumbai** 0-9833017933
- **Nagpur** 0-9021734563
- **Patna** 0-9334159340

Printed at India Binding House, Noida, UP

Preface

The topic *Differential Equations and Calculus of Variations* is an essential part of the mathematics course for the undergraduate students studying in the fields of sciences and engineering. Keeping this in mind, this book has been developed according to the UGC unified syllabus implemented at undergraduate level in all the Indian universities. This book, divided into two parts **A** and **B**, consists of four sections and fourteen chapters. Part **A** is a major part of the book dealing with the study of differential equations, partial differential equations and Laplace transform, covered in ten chapters. Part **B** is devoted to the calculus of variations and contains four chapters.

Chapter 1 is introductory and Chapters 2 and 3 are concerned with the linear differential equations of nth order with constant coefficients and of 2nd order with variable coefficients. Series solutions of second order differential equations have been described in chapter 4. In this chapter the solutions of Legendre's equation and Bessel's equation, orthogonal properties and reccurrence relations are discussed. In chapter 5, hypergeometric functions, confluent hypergeometric functions, their differentiation and integral representation have been described. Chapter 6 deals with the study of orthogonality of functions and Sturm-Liouville problem. Chapters 7 and 8 are devoted to partial differential equations. Chapter 7 is concerned with the partial differential equations of first order and their solutions whereas Chapter 8 is concerned with the partial differential equations of second order and their solutions. Laplace and inverse Laplace transform with their properties have been described in Chapter 9. Applications of Laplace transform to solve the differential equations, partial differential equations and integral equations have been discussed in Chapter 10.

Chapter 11 is the first chapter of calculus of variation that introduces variational problems with fixed boundaries and Chapter 12 describes variational method of boundary value problems. Chapter 13 is concerned with variational problems with moving boundaries and Chapter 14 is concerned with sufficient conditions for an extremum.

This book provides theoretical background of the subject with well graded set of detailed solved examples. Each chapter consists of unsolved problems in the form of exercises. For better understanding of the subject, a large number of objective type questions are also given at the end of each section—differential equations, partial differential equations, Laplace transforms and calculus of variations.

We would like to thank to our colleagues and undergraduate students who inspired us to write this book. We would like to express our gratitude to our family members for their moral support, encouragement and inspiration during the writing of this book.

Finally, we are thankful to Mr YN Arjuna, Senior Director—Publishing, Editorial and Publicity, CBS Publishers & Distributors Pvt Ltd, New Delhi, who took all the pains to bring out this book.

Rakesh Kumar | Nagendra Kumar

Contents

PART A
Differential Equations

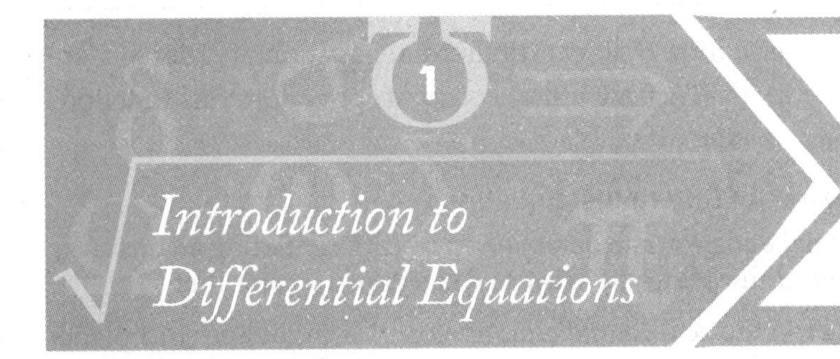

1.1 INTRODUCTION

An equation which involves independent variables, dependent variables and their derivatives, is called a differential equation.

If an equation which involves only one independent variable, one or more dependent variables and their derivatives, is called an ordinary differential equation, e.g.

$$\text{(a)}\ \frac{dy}{dx}=\frac{1+x^2}{1-y^2} \qquad \text{(b)}\ \begin{cases}\dfrac{dy}{dt}+2y+3x=t\\ \dfrac{dx}{dt}+3y+2x=t^2\end{cases} \qquad \text{(c)}\ \left[1+\left(\frac{dy}{dx}\right)^2\right]^{\frac{3}{2}}=K\frac{d^2y}{dx^2}$$

If an equation which involves more than one independent variable, dependent variable and its partial derivatives, is called a partial differential equation, e.g.

$$\text{(a)}\ \frac{\partial z}{\partial x}+\frac{\partial z}{\partial y}=xy \quad \text{(b)}\ \frac{\partial^2 z}{\partial x^2}+2\frac{\partial^2 z}{\partial y^2}+\frac{\partial z}{\partial y}\frac{\partial z}{\partial x}+z=e^{x+y}$$

Now, first we shall discuss the ordinary differential equations.

Order. The order of a differential equation is the order of the highest differential coefficient which occurs in the differential equation.

Degree. The degree of a differential equation is the power of the highest order differential coefficient which occurs in the differential equation. For example, the differential equation

$$f(x,y)\left(\frac{d^m y}{dx^m}\right)^n+\phi(x,y)\left(\frac{d^{m-1}y}{dx^{m-1}}\right)^n+\ldots=0 \text{ is of order } m \text{ and degree } n.$$

Linear and non-linear differential equations. A differential equation is said to be linear differential equation if the dependent variable and all its derivatives are occurring in the first degree, otherwise the differential equation is said to be non-linear. The differential equation

$$a_0(x)\frac{d^n y}{dx^n}+a_1(x)\frac{d^{n-1}y}{dx^{n-1}}+a_2(x)\frac{d^{n-2}y}{dx^{n-2}}+\ldots+a_{n-1}(x)\frac{dy}{dx}+a_n(x)y=F(x)$$

is the most general form of linear differential equation of order n. Here $a_0(x), a_1(x), ... a_n(x), F(x)$ are either constants or continuous functions of x only, on any interval (a, b).

1.1.1 Formation of Differential Equations

The differential equation can be obtained by differentiating an ordinary equation and eliminating the arbitrary constants among them.

Example 1: Find the differential equation of all straight lines in a plane.

Solution: Consider the equation of a plane is $ax+by+cz=d$.

Thus the straight line in xy-plane is given by $ax+by=d$ ...(1)
where a, b, d are arbitrary constants.

On differentiating (1) with respect to 'x', we get

$$a+b\frac{dy}{dx}=0$$

Again differentiating with respect to 'x', we get

$$b\frac{d^2y}{dx^2}=0 \Rightarrow \frac{d^2y}{dx^2}=0,$$

which is a differential equation of order 2 and degree 1.

Example 2: Find the differential equation of all circles of radius 'a' whose centres lie on the y-axis.

Solution: Let the centre of any such circle is (0, k), then the equation of the circle is

$$(x-0)^2+(y+k)^2=a^2 \Rightarrow x^2+(y-k)^2=a^2 \quad ...(1)$$

where k is the arbitrary constant. Differentiating (1) with respect to 'x', we have

$$2x+2(y-k)\frac{dy}{dx}=0 \Rightarrow (y-k)=-x\Big/\left(\frac{dy}{dx}\right)$$

Substituting the value of $(y-k)$ in (1), we obtain

$$x^2+\frac{x^2}{\left(\frac{dy}{dx}\right)^2}=a^2 \Rightarrow x^2\left[1+\left(\frac{dx}{dy}\right)^2\right]=a^2$$

which is the required differential equation.

Example 3: Find the differential equation of all parabolas whose vertex is (h, k).

Solution: The equation of the parabola is $(y-k)^2 = 4a(x-h)$ ...(1)

where a is a constant. Differentiating (1) with respect to x, we get

$$2(y-k)\frac{dy}{dx} = 4a$$

Hence from (1), we have

$$(y-k)^2 = 2(y-k)\frac{dy}{dx}.(x-h) \Rightarrow y-k = 2(x-h)\frac{dy}{dx}$$

which is required differential equation.

1.1.2 Solution of Differential Equation

Any relation between dependent and independent variables not containing the derivatives of dependent variable with respect to the independent variable, which on substitution in the differential equation reduces it to an identity, is called a solution of differential equation. A solution or integral is also known as a primitive because the differential equation can be derived from it.

The *general* or *complete solution* of a differential equation is the solution which contains as many numbers of arbitrary constants as its order. For example, $y = Ae^{3x} + Be^{5x}$ is the general solution of the differential equation $\frac{d^2y}{dx^2} - 8\frac{dy}{dx} + 15y = 0$.

A solution which is obtained from the general solution of a differential equation by putting the particular values for the arbitrary constants is called a particular solution. So for A =1 and B = 2 the solution $y = e^{3x} + 2e^{5x}$ is a particular solution of the differential equation $\frac{d^2y}{dx^2} - 8\frac{dy}{dx} + 15y = 0$.

1.2 DIFFERENTIAL EQUATIONS OF FIRST ORDER AND FIRST DEGREE

The differential equation of the type $\frac{dy}{dx} = f(x,y)$ is known as a differential equation of first order and first degree. We shall now discuss Various types of differential equations of first order and first degree by providing some examples.

1.2.1 Type 1. Variables Separable Form

A differential equation of the form

$$\frac{dy}{dx} = f(x,y) \qquad ...(1)$$

is separable if we may write $f(x, y) = M(x).N(y)$ or $f(x, y) = \frac{M(x)}{N(y)}$, where $M(x)$ and $N(y)$ are real valued functions of x and y respectively. Thus equation (1) is equivalent to $N(y)dy = M(x)dx$. The solution of this equation is obtained on integrating it.

Example 4: Solve $3e^x \tan y dx + (1-e^x)\sec^2 y dy = 0$.

Solution: Separating the variables, we get $\frac{3e^x}{1-e^x}dx + \frac{\sec^2 y}{\tan y}dy = 0$.

On integrating, we obtain $3\int \frac{e^x}{1-e^x}dx + \int \frac{\sec^2 y}{\tan y}dy = \text{constant}$

$\Rightarrow -3\log(1-e^x) + \log \tan y = \log C \Rightarrow \tan y = C(1-e^x)^3$.

Example 5: Solve $y - x\frac{dy}{dx} = a^2\left(y + \frac{dy}{dx}\right)$.

Solution: Given equation can be written as $y(1-a^2) = (a^2+x)\frac{dy}{dx}$.

Separating the variable, we get $(1-a^2)\frac{dx}{(a^2+x)} = \frac{dy}{y}$

On integrating, we obtain

$$(1-a^2)\log(a^2+x) = \log y + \log C \Rightarrow Cy = (a^2+x)^{(1-a^2)}.$$

1.2.2 Type 2. Homogeneous Equations

A function with each term of same degree n is called a homogeneous function of degree n. For example,

$$f(x,y) = a_0x^n + a_1x^{n-1}y + a_2x^{x-2}y^2 + ... + a_ny^n.$$

Test: Any function $f(x, y)$ is said to be a homogeneous function of degree n in x and y, if $f(tx, ty) = t^n f(x, y)$

The differential equation $\frac{dy}{dx} = f(x, y)$ ...(1)

is said to be homogeneous differential equation if $f(x, y)$ is homogeneous function of degree 0, i.e., if $f(x, y)$ can be written in the form $g\left(\frac{y}{x}\right)$ or $h\left(\frac{x}{y}\right)$. Such equations can be solved by substituting $y = vx$ or $x = vy$. If differential equation is homogeneous then it is equivalent to

$$\frac{dy}{dx} = g(y/x) \qquad ...(2)$$

Substituting $y = vx$ in equation (2), we have

$$v + x\left(\frac{dv}{dx}\right) = g(v) \Rightarrow x\left(\frac{dv}{dx}\right) = g(v) - v,$$

which is now separable equation and can be solved by variables separable method. The solution is deduced by replacing v from (y/x).

Example 6: Solve $(1 + e^{x/y})\, dx + e^{x/y}(1 - x/y)\, dy = 0$.

Solution: Given equation can be written as

$$e^{x/y}\left[1 - \frac{x}{y}\right] + \left[1 + e^{x/y}\right]\frac{dx}{dy} = 0 \qquad ...(1)$$

Putting $x = vy$, and $\frac{dx}{dy} = v + y\frac{dv}{dy}$, the equation (1) becomes

$$e^{v}(1 - v) + (1 + e^{v})\left[v + y\frac{dv}{dy}\right] = 0$$

$$\Rightarrow \quad e^{v} - ve^{v} + v + ve^{v} + y\frac{dv}{dy}(1 + e^{v}) = 0$$

$$\Rightarrow \quad (v + e^{v}) + (1 + e^{v})\, y\frac{dv}{dy} = 0$$

$$\Rightarrow \quad \left(\frac{1 + e^{v}}{v + e^{v}}\right)dv + \frac{1}{y}dy = 0.$$

On integrating, we get $\log\left[(v + e^{v})y\right] = \log c$

$$\Rightarrow \quad (v + e^{v})y = c$$

$$\Rightarrow \quad \left[\frac{x}{y} + e^{x/y}\right]y = c \Rightarrow x + y\, e^{x/y} = c.$$

Example 7: Solve $\left(x\cos\frac{y}{x} + y\sin\frac{y}{x}\right)y - \left(y\sin\frac{y}{x} - x\cos\frac{y}{x}\right)x\frac{dy}{dx} = 0.$

Solution: Given equation can be written as

$$\frac{dy}{dx} = \frac{\left(x\cos\frac{y}{x} + y\sin\frac{y}{x}\right)y}{\left(y\sin\frac{y}{x} - x\cos\frac{y}{x}\right)x} \qquad ...(1)$$

Putting $y = vx$, and $\frac{dy}{dx} = v + x\frac{dv}{dx}$ the equation (1) reduces to

$$v + x\frac{dv}{dx} = \left(\frac{x\cos v + vx\sin v}{vx\sin v - x\cos v}\right)\frac{vx}{x} = \left(\frac{\cos v + v\sin v}{v\sin v - \cos v}\right)v$$

$$\Rightarrow \qquad x\frac{dv}{dx} = \frac{2v\cos v}{v\sin v - \cos v}$$

On integrating, we get

$$\frac{1}{2}\int \frac{v\sin v - \cos v}{v\cos v}\,dv = \int \frac{1}{x}\,dx + c$$

$$\Rightarrow \qquad -\frac{1}{2}\log(v\cos v) = \log x + \frac{1}{2}\log k$$

$$\Rightarrow \qquad \frac{k}{x^2} = v\cos v \Rightarrow k = xy\cos\left(\frac{y}{x}\right).$$

1.2.3 Type 3. Equations Reducible to Homogeneous

If equation $\frac{dy}{dx} = f(x, y)$ is of the form

$$\frac{dy}{dx} = \frac{Ax + By + C}{A'x + B'y + C'} \qquad ...(1)$$

It can be reduced to the homogeneous form by substituting $x = X + h$, $y = Y + k$. Therefore

$$\frac{dy}{dx} = \frac{dY}{dX} \qquad ...(2)$$

and the equation (1) reduces to

$$\frac{dY}{dX} = \frac{AX + BY + Ah + Bk + C}{A'X + B'Y + A'h + B'k + C'}$$

Now, choose h and k so that $Ah + Bk + C = 0$ and $A'h + B'k + C' = 0$. which gives

$$\frac{h}{BC' - CB'} = \frac{k}{CA' - AC'} = \frac{1}{AB' - BA'}$$

$$\Rightarrow \quad h=\frac{BC'-CB'}{AB'-BA'},\ k=\frac{CA'-AC'}{AB'-BA'};\ \text{(Does } h, k \text{ always exist?)}$$

By substituting these values in (2), we get

$$\frac{dY}{dX}=\frac{AX+BY}{A'X+B'Y}$$

which is a homogeneous differential equation and can be solved by the method discussed in Type-2.

Case of failure. In this case, the differential equation reduces to variables separable type.

If $\frac{A}{A'}=\frac{B}{B'}$ then the value of h and k will not exist. In this case, let

$$\frac{A}{A'}=\frac{B}{B'}=\frac{1}{m} \Rightarrow A'=Am \text{ and } B'=Bm.$$

Then the given equation (4) becomes $\frac{dy}{dx}=\frac{Ax+By+C}{m(Ax+By)+C'}$

To solve it first we put Ax + By = z and than apply the method of variables separable to solve the transformed differential equation.

Example 8: Solve $(6x+2y-10)\frac{dx}{dy}-2x-9y+20=0.$

Solution: Given equation can be written as $\frac{dy}{dx}=\frac{2x+9y-20}{6x+2y-10}$ ...(1)

Now putting $x=X+h, y=Y+k$,

then $dx=dX, dy=dY$ and $\frac{dy}{dx}=\frac{dY}{dX}$

Therefore

$$\frac{dY}{dX}=\frac{2(X+h)+9(Y+k)-20}{6(X+h)+2(Y+k)-10} \Rightarrow \frac{dY}{dX}=\frac{2X+2h+9Y+9k-20}{6X+6h+2Y+2k-10} \quad ...(2)$$

Now choose h and k,

such that $2h+9k-20=0$ and $6h+2k-10=0.$

On solving these equations, we get

$$\frac{h}{-90+40}=\frac{k}{-120+20}=\frac{1}{4-54} \Rightarrow h=1,\ k=2.$$

Putting $h = 1, k = 2$ in equation (2), we obtain $\frac{dY}{dX}=\frac{2X+9Y}{6X+2Y}$, which is an homogeneous equation.

Putting $Y = vX$ and $\dfrac{dY}{dX} = v + X\dfrac{dv}{dX}$,

we get $v + X\dfrac{dv}{dX} = \dfrac{2X + 9v\,X}{6X + 2v\,X}$

$$\Rightarrow X\frac{dv}{dX} = \frac{2 + 9v - 6v - 2v^2}{6 + 2v} \quad \Rightarrow \quad \frac{2(3+v)}{(3v - 2v^2 + 2)}dv = \frac{dX}{X}$$

$$\Rightarrow \frac{-2}{v-2}dv + \frac{2}{2v+1}dv = \frac{dX}{X}$$

On integrating, we get $x + 2y - 5 = c(2x - y)^2$.

Example 9: Solve $\dfrac{(4x + 6y + 5)dy}{(3y + 2x + 4)dx} = 1$

Solution: We can write $\dfrac{dy}{dx} = \dfrac{3y + 2x + 4}{4x + 6y + 5} = \dfrac{2x + 3y + 4}{2(2x + 3y) + 5}$.

Now putting $2x + 3y = v$ and $2 + 3\dfrac{dy}{dx} = \dfrac{dv}{dx}$, we get

$$\frac{1}{3}\left[\frac{dv}{dx} - 2\right] = \frac{v + 4}{2v + 5}$$

$$\Rightarrow \quad \frac{dv}{dx} - 2 = \frac{3v + 12}{2v + 5}$$

$$\Rightarrow \quad \frac{dv}{dx} = 2 + \frac{3v + 12}{2v + 5} = \frac{7v + 22}{2v + 5}$$

$$\Rightarrow \quad \left[\frac{2v + 5}{7v + 22}\right]dv = dx$$

$$\Rightarrow \quad \frac{2}{7}\left[\frac{7v + \frac{35}{2}}{7v + 22}\right]dv = dx$$

$$\Rightarrow \quad \frac{2}{7}\left[1 - \frac{9}{2(7v + 22)}\right]dv = dx.$$

On integrating, we get

$$\frac{2}{7}v - \frac{9}{49}\log(7v+22) = x + c$$

$$\Rightarrow \quad \frac{2}{7}(2x+3y) - \frac{9}{49}\log(14x+21y+22) = x + c.$$

1.2.4 Type 4. Exact Differential Equations

Suppose the equation $\frac{dy}{dx} = f(x, y)$ is of the form

$$M(x, y)dx + N(x, y)dy = 0 \qquad \text{...(1)}$$

The equation (1) is said to be exact if there exists a function $u(x, y)$ such that

$$Mdx + Ndy = du = \frac{\partial u}{\partial x}dx + \frac{\partial u}{\partial y}dy$$

i.e., if there exists a function $u(x, y)$ such that $M = \frac{\partial u}{\partial x}$ and $N = \frac{\partial u}{\partial y}$.

Obviously $\frac{\partial M}{\partial y} = \frac{\partial^2 u}{\partial y \partial x} = \frac{\partial^2 u}{\partial x \partial y} = \frac{\partial N}{\partial x}$

Therefore the necessary and sufficient condition for the ordinary differential equation $Mdx + Ndy = 0$ to be exact is that $\frac{\partial M}{\partial y} = \frac{\partial N}{\partial x}$.

Steps for Solving Exact Differential Equations

Step 1: Integrate M with respect to x keeping y constant.

Step 2: Integrate only those terms of N which do not contain x, with respect to y.

Step 3: Result of Step 1+ Result of Step 2 =Constant, is the solution of the given differential equation.

1.2.5 Type 5. Equations Reducible to Exact Differential Form

Sometimes a differential equation which is not exact can be made exact on multiplication by a suitable function. This suitable function is known as "integrating factor" (I. F.). Various rules for obtaining integrating factors are discussed below.

Rule 1: **Inspection Method.** Sometimes it is easy to find the integrating factor (I.F.) by inspection. We re-arrange the terms by dividing or multiplying some terms and separate the variables so that the given equation becomes exact.

Rule 2: If M is of the form $M = yf_1(x.y)$ and N is of the form $N = xf_2(x.y)$, and $Mx - Ny \neq 0$, then $(Mx - Ny)^{-1}$ be an integrating factor (I.F.).

Remark: If $Mx - Ny = 0$ *i.e.* $\frac{M}{N} = \frac{y}{x}$, then on substituting it in equation (1), we get $\frac{y}{x}.dx + dy = 0 \Rightarrow y.dx + xdy = 0$

On integrating, we get the required solution $xy = c$ *(always in this case)*.

Rule 3: If the given equation $Mdx + Ndy = 0$ is homogeneous equation and $Mx + Ny \neq 0$, then $(Mx + Ny)^{-1}$ is an I. F.

Remark: If $Mx + Ny = 0$ *i.e.*, $\frac{M}{N} = -\frac{y}{x}$, then on substituting it in equation (1), we get $-\frac{y}{x}dx + dy = 0 \Rightarrow \frac{dx}{x} = \frac{dy}{y}$

On integrating, we get the required solution $x = cy$ *(always in this case)*.

Rule 4: If $\left(\frac{\partial M}{\partial y} - \frac{\partial N}{\partial x}\right)\Big/N$ is a function of x alone, say *f(x)*, then I.F. is equal to $e^{\int f(x)dx}$.

Rule 5: If $\left(\frac{\partial M}{\partial x} - \frac{\partial N}{\partial y}\right)\Big/M$ is a function of y alone, say *f(y)*, then I.F. is equal to $e^{\int f(y)dy}$.

Rule 6: If the equation $\frac{dy}{dx} = f(x, y)$ is of the form

$x^a y^b [My\,dx + Nx\,dy] + x^r y^s [pydx + qxdy] = 0$,

where a, b, M, N, r, s, p and q are all constants, then $I.F. = x^h y^k$, where h and k are chosen such that after multiplying the given differential equation by I.F. it becomes exact. This exact differential equation can be solved by the above described method.

Example 10: Solve $(x^2 - ay)dx = (ax - y^2)dy$.

Solution: Given equation can be written as

$$(x^2 - ay)dx + (y^2 - ax)dy = 0 \qquad ...(1)$$

Comparing with $Mdx + Ndy = 0$, we get

$M = x^2 - ay$ and $N = y^2 - ax$.

Now $\frac{\partial M}{\partial y} = -a,\ \frac{\partial N}{\partial x} = -a \Rightarrow \frac{\partial M}{\partial y} = \frac{\partial N}{\partial x}$.

Therefore, equation (1) is exact.

Hence, the solution is

$$\int_{y(const.)} M\,dx + \int (\text{terms in N not containing } x)\, dy = c,$$

where c is an arbitrary constant of integration.

$$\Rightarrow \int_{y(const.)} (x^2 - ay)\,dx + \int y^2\,dy = c \Rightarrow \frac{x^3}{3} - axy + \frac{y^3}{3} = c$$

$$\Rightarrow x^3 - 3axy + y^3 = 3c$$

$$\Rightarrow x^3 - 3axy + y^3 = A.$$

Example 11: Solve $x\,dx + y\,dy + \frac{x\,dy - y\,dx}{x^2 + y^2} = 0$.

Solution: Given equation can be written as

$$\frac{1}{2}(2x\,dx + 2y\,dy) + \frac{\left(\frac{x\,dy - y\,dx}{x^2}\right)}{1 + \left(\frac{y}{x}\right)^2} = 0$$

$$\Rightarrow \quad d\left(\frac{x^2 + y^2}{2}\right) + \frac{d(y/x)}{1 + (y/x)^2} = 0$$

$$\Rightarrow \quad d\left(\frac{x^2 + y^2}{2}\right) + d\left\{\tan^{-1}\left(\frac{y}{x}\right)\right\} = 0$$

Hence the required solution is $\frac{x^2 + y^2}{2} + \tan^{-1}\left(\frac{y}{x}\right) = c$.

Example 12: Solve $y \sin 2x\,dx - (1 + y^2 + \cos^2 x)\,dy = 0$

Solution: Here $M = y\sin 2x,\ N = -(1 + y^2 + \cos^2 x)$.

Implies that $\frac{\partial M}{\partial y} = \sin 2x,\ \frac{\partial N}{\partial x} = 2\sin x \cos x = \sin 2x$.

Now solution is

$$\int_{y(const.)} M\,dx + \int (\text{terms in N not containing } x)\,dy = c,$$

$$\Rightarrow \quad \int_{y(const.)} y\sin 2x\,dx + \int -(1+y^2)dy = c$$

$$\Rightarrow \quad -\frac{1}{2}y\cos 2x - y - \frac{1}{3}y^3 = c$$

$$\Rightarrow \quad 3y\cos 2x + 6y + 2y^3 = -6c.$$

Example 13: Solve $y\,dx - x\,dy + (1+x^2)dx + x^2\sin y\,dy = 0$.

Solution: Now dividing by x^2, we get

$$\frac{y\,dx - x\,dy}{x^2} + \left(\frac{1}{x^2}+1\right)dx + \sin y\,dy = 0.$$

$$\Rightarrow \quad d\left(\frac{-y}{x}\right) + d\left(x - \frac{1}{x}\right) - d(\cos y) = 0$$

On integrating it, we obtain

$$\frac{-y}{x} + \left(x - \frac{1}{x}\right) - \cos y = c$$

$$\Rightarrow \quad x^2 - y - 1 - x\cos y = xc.$$

Example 14: Solve

$(xy\sin xy + \cos xy)y\,dx + (xy\sin xy - \cos xy)x\,dy = 0$.

Solution: We have

$$(xy\sin xy + \cos xy)y\,dx + (xy\sin xy - \cos xy)x\,dy = 0 \quad ...(1)$$

Here $M = (xy\sin xy + \cos xy)y$ and $N = (xy\sin xy - \cos xy)x$

Clearly $\frac{\partial M}{\partial y} \neq \frac{\partial N}{\partial x}$. The equation is not exact. Now

$$Mx - Ny = (xy\sin xy + \cos xy)xy - xy(xy\sin xy - \cos xy) = 2xy\cos xy.$$

Hence, Integrating Factor $= \frac{1}{Mx - Ny} = \frac{1}{2xy\cos xy}$.

Now multiplying the equation (1) by $\frac{1}{2xy\cos xy}$ the equation becomes $\frac{1}{2}\left\{\tan xy + \frac{1}{xy}\right\}y\,dx + \frac{1}{2}\left\{\tan xy - \frac{1}{xy}\right\}x\,dy = 0$

$$\Rightarrow \quad \left(y\tan xy + \frac{1}{x}\right)dx + \left(x\tan xy - \frac{1}{y}\right)dy = 0 \quad ...(2)$$

Now, the equation (2) is exact, so that solution is given by

$$\int_{y(const.)}\left(y\tan xy+\frac{1}{x}\right)dx+\int\left(-\frac{1}{y}\right)dy=c$$

$\Rightarrow\quad -\log\cos xy+\log x-\log y=\log c$

$\Rightarrow\quad x=cy\cos(xy).$

Example 15: Solve $\left(y+\frac{1}{3}y^3+\frac{1}{2}x^2\right)dx+\frac{1}{4}(1+y^2)x\,dy=0.$

Solution: Here $M=y+\frac{1}{3}y^3+\frac{1}{2}x^2,\ N=\frac{1}{4}(1+y^2)x$

Clearly $\frac{\partial M}{\partial y}\neq\frac{\partial N}{\partial x}$, so equation is not exact.

Now $\frac{1}{N}\left(\frac{\partial M}{\partial y}-\frac{\partial N}{\partial x}\right)=\frac{4}{(1+y^2)x}\left[(1+y^2)-\frac{1}{4}(1+y^2)\right]$

$=\frac{4}{(1+y^2)x}\cdot\frac{3}{4}(1+y^2)=\frac{3}{x}$, which is a function of x only ...(1)

Therefore integrating factor = $e^{\int\frac{3}{x}dx}=e^{3\log x}=x^3.$

Now, multiplying the equation by integrating factor x^3, we get

$$\left(x^3y+\frac{1}{3}x^3y^3+\frac{1}{2}x^5\right)dx+\frac{1}{4}(1+y^2)x^4\,dy=0 \qquad ...(2)$$

Now equation (2) is exact. Hence the solution is

$$\int_{y(const.)}\left(x^3y+\frac{1}{3}x^3y^3+\frac{x^5}{2}\right)dx=c\Rightarrow\frac{x^4}{4}y+\frac{1}{12}x^4y^3+\frac{x^6}{12}=c$$

$\Rightarrow\quad 3x^4y+x^4y^3+x^6=12c$

Example 16: Solve $(xy^3+y)dx+2(x^2y^2+x+y^4)dy=0$.

Solution: Here $\frac{\partial M}{\partial y}\neq\frac{\partial N}{\partial x}$, so given equation is not exact.

Now $\frac{1}{M}\left(\frac{\partial N}{\partial x}-\frac{\partial M}{\partial y}\right)=\frac{1}{y(xy^2+1)}\left\{4xy^2+2-3xy^2-1\right\}$

$=\quad\frac{1}{y(xy^2+1)}(xy^2+1)=\frac{1}{y}$, which is a function of y alone ...(1)

Therefore integrating factor $=e^{\int\frac{1}{y}dy}=e^{\log y}=y.$

Now, multiplying the equation by integrating factor y, we obtain

$$y^2(xy^2+1)\,dx+2y(x^2y^2+x+y^4)\,dy=0 \qquad ...(2)$$

This equation (2) is exact. Hence the solution is

$$\int y^2\left(xy^2+1\right)dx+\int 2y^5dy=C$$

$$\Rightarrow \quad y^4\frac{x^2}{2}+xy^2+2\frac{y^6}{6}=c$$

$$\Rightarrow \quad 3x^2y^4+6xy^2+2y^6=6c.$$

Example 17: Solve $x(3y\,dx+2x\,dy)+8y^4(y\,dx+3x\,dy)=0.$

Solution: We can write $(3xy+8y^5)\,dx+(2x^2+24xy^4)\,dy=0 \qquad ...(1)$

Here $\dfrac{\partial M}{\partial y}\neq\dfrac{\partial N}{\partial x}$, so equation is not exact.

Let the integrating factor be x^hy^k. Therefore, multiplying by x^hy^k, we get

$$(3x^{h+1}y^{k+1}+8y^{k+5}x^h)\,dx+(2x^{h+2}y^k+24x^{h+1}y^{k+4})\,dy=0. \qquad ...(2)$$

Now $\dfrac{\partial M}{\partial y}=\dfrac{\partial N}{\partial x}$ gives

$$3(k+1)x^{h+1}y^k+8(k+5)y^{k+4}x^h=2(h+2)x^{h+1}y^k+24(h+1)x^hy^{k+1}$$

On equating coefficients of same powers, we obtain

$$3(k+1)=2(h+2) \qquad ...(3)$$

and

$$8(k+5)=24(h+1). \qquad ...(4)$$

On solving (3) and (4) for h and k, we get $h = 1$, $k = 1$.

Thus from (2), we get

$$(3x^2y^2+8xy^6)\,dx+(2x^3y+24x^2y^5)\,dy=0$$

which is exact. Thus solution is $\int(3x^2y^2+8xy^6)dx=c$

$$\Rightarrow \quad x^3y^2+4x^2y^6=c.$$

1.2.6 Type 6. Standard Linear Differential Equations

A differential equation of the form $\dfrac{dy}{dx}+Py=Q$, where P and Q are the functions of x alone, is called a linear differential equation.

Solution of linear equation. To solve such type of differential equation we multiply both sides by $I.F. = e^{\int P\,dx}$

We have $e^{\int P\,dx}.dy + e^{\int P\,dx}.Py\,dx = e^{\int P\,dx}Q\,dx$

Hence on integrating both sides, we get $ye^{\int Pdx} = \int\left[Q.e^{\int Pdx}\right]dx + C$

which is the required solution of the given linear differential equation.

Example 18: Solve $(1+x^2)\dfrac{dy}{dx} + 2xy - 4x^2 = 0.$

Solution: We can write $\dfrac{dy}{dx} + \dfrac{2x}{1+x^2}y = \dfrac{4x^2}{1+x^2}$, which is linear differential equation.

Here $P = \dfrac{2x}{1+x^2}$, $Q = \dfrac{4x^2}{1+x^2}$.

Hence, integration factor $(I.F.) = e^{\int Pdx} = e^{\int \frac{2x}{1+x^2}dx} = e^{\log(1+x^2)} = 1 + x^2.$

Hence, the solution is given by $I.F. \times y = \int I.F. \times Qdx + c$, which gives

$$(1+x^2)y = \int (1+x^2)\frac{4x^2}{(1+x^2)}dx + c$$

$$\Rightarrow \quad (1+x^2)y = \frac{4x^3}{3} + c.$$

Example 19: Solve $(1+y^2)dx = (\tan^{-1} y - x)dy.$

Solution: We can write $\dfrac{dx}{dy} + \dfrac{x}{1+y^2} = \dfrac{\tan^{-1} y}{1+y^2}$, which is linear equation in x.

Thus $I.F. = e^{\int \frac{dy}{1+y^2}}.dy = e^{\tan^{-1} y}.$

and the solution is $e^{\tan^{-1} y}x = \int \dfrac{\tan^{-1} y}{1+y^2}e^{\tan^{-1} y}dy + c$

Let $\tan^{-1} y = t$, then $\dfrac{dy}{1+y^2} = dt$. So we have

$$e^{\tan^{-1} y}x = \int te^t dt + c = te^t - e^t + c$$

$$\Rightarrow \quad xe^{\tan^{-1} y} = \tan^{-1} y e^{\tan^{-1} y} - e^{\tan^{-1} y} + c$$

$\Rightarrow \quad x = \tan^{-1} y + ce^{-\tan^{-1} y} - 1$

Example 20: Solve $\dfrac{dy}{dx} = y\tan x - 2\sin x.$

Solution: Given differential equation is $\dfrac{dy}{dx} - y\tan x = -2\sin x$

Here $I.F. = e^{-\int \tan x dx} = e^{\log\cos x} = \cos x.$

Hence, the solution is $\cos x . y = \int \cos x . (-2\sin x) dx + c$

$\Rightarrow \quad y . \cos x = -\int 2\sin x \cos x \, dx + c$

$\Rightarrow \quad y\cos x = \dfrac{\cos 2x}{2} + c.$

Example 21: Solve $x\log x \dfrac{dy}{dx} + y = 2\log x.$

Solution: We can write $\dfrac{dy}{dx} + \dfrac{1}{x\log x} y = \dfrac{2}{x}$, which is a linear differential equation of first order.

So $I.F. = e^{\int \frac{1}{x\log x} dx} = e^{\log(\log x)} = \log x.$

Hence, the solution is $y . \log x = \int \log x . \dfrac{2}{x} dx + c.$ ...(1)

Now, $I = \int \dfrac{2\log x}{x} dx = \int 2t dt$ (*if* $\log x = t$, $\dfrac{1}{x} dx = dt$)

$= t^2 = (\log x)^2.$

Thus by (1), we have $y(\log x) = (\log x)^2 + c.$

1.2.7 Type 7. Equations Reducible to Linear Differential Equations

Sometimes equations which are not linear can be reduced to linear form by suitable transformation.

Consider the equation $\dfrac{dy}{dx} + Py = Qy^n$

On dividing y^n, we get $y^{-n} \dfrac{dy}{dx} + Py^{-n+1} = Q$

This equation can be reduced to the linear form by the substituting $v=y^{-n+1}$. This type of equations is also known as Bernoulli's equation.

Example 22: Solve $\frac{dy}{dx}+\frac{y}{x}=y^2$.

Solution: Given that $\frac{dy}{dx}+\frac{y}{x}=y^2$. ...(1)

On dividing by y^2, we have $y^{-2}\frac{dy}{dx}+y^{-1}\frac{1}{x}=1$ ...(2)

Let $y^{-1}=v$, then $-y^{-2}\frac{dy}{dx}=\frac{dv}{dx}$

$\Rightarrow \quad y^{-2}\frac{dy}{dx}=-\frac{dv}{dx}$.

Thus, equation (2) becomes $-\frac{dv}{dx}+\frac{v}{x}=1$

$\Rightarrow \quad \frac{dv}{dx}-\frac{v}{x}=-1$

which is a linear equation.

Now $I.F.=e^{\int -\frac{1}{x}dx}=e^{-\log x}=\frac{1}{x}$

Hence solution is $\frac{1}{x}v=\int \frac{1}{x}\times(-1)\,dx+c$

$\Rightarrow \quad \frac{y^{-1}}{x}=-\log x+c$

$\Rightarrow \quad 1=xy\,(c-\log x)$.

1.2.8 Type 8. Change of Variables

A suitable substitution often reduces a given differential equation which does not directly come under any of the forms discussed so far, to one of these forms. This device is known as the change of the independent or the dependent variable (as case may be).

Example 23: Solve $x\frac{dy}{dx}+y\log y=xye^x$. ...(1)

Solution: On dividing by xy, we get $\frac{1}{y}\frac{dy}{dx}+\frac{\log y}{x}=e^x$ ...(2)

Let $\log y=t$ and $\frac{1}{y}\frac{dy}{dx}=\frac{dt}{dx}$, then (2) becomes $\frac{dt}{dx}+\frac{t}{x}=e^x$ ...(3)

Now $I.F. = e^{\int \frac{1}{x} dx} = e^{\log x} = x.$

Hence, the solution is $x.\log y = \int xe^x dx + c$

$$\Rightarrow \quad x \log y = xe^x - e^x + c.$$

Example 24: Solve $\dfrac{dy}{dx} - \dfrac{\tan y}{1+x} = (1+x)\, e^x \sec y.$ ...(1)

Solution: On dividing by sec y, we get $\cos y \dfrac{dy}{dx} - \dfrac{\sin y}{1+x} = (1-x)e^x$ ...(2)

Let $\sin y = t \;\Rightarrow\; \cos y \dfrac{dy}{dx} = \dfrac{dt}{dx}$, then (2) becomes

$$\frac{dt}{dx} - \frac{t}{1+x} = (1+x)\, e^x.$$

Now $I.F. = e^{\int -\frac{1}{1+x} dx} = e^{-\log(1+x)} = \dfrac{1}{1+x}.$

Hence, the solution is $\dfrac{1}{1+x} \sin y = \int \dfrac{1}{1+x}(1+x)e^x \, dx + c$

$$\Rightarrow \quad \frac{\sin y}{1+x} = e^x + c.$$

Example 25: Solve $\dfrac{x\,dx + y\,dy}{x\,dy - y\,dx} = \sqrt{\dfrac{a^2 - x^2 - y^2}{x^2 + y^2}}$ ...(1)

Solution: Let $x = r\cos\theta$, $y = r\sin\theta$, then we have

$$\frac{\partial x}{\partial r} = \cos\theta, \quad \frac{\partial x}{\partial \theta} = -r\sin\theta, \quad \frac{\partial y}{\partial r} = \sin\theta, \quad \frac{\partial y}{\partial \theta} = r\cos\theta$$

By advanced calculus $dx\,dy = J\,dr\,d\theta$

$$= \begin{vmatrix} \dfrac{\partial x}{\partial r} & \dfrac{\partial x}{\partial \theta} \\ \dfrac{\partial y}{\partial r} & \dfrac{\partial y}{\partial \theta} \end{vmatrix} dr\,d\theta$$

$$= \begin{vmatrix} \cos\theta & -r\sin\theta \\ \sin\theta & r\cos\theta \end{vmatrix} dr\,d\theta$$

$$= \quad r(\cos^2\theta + \sin^2\theta)\,dr\,d\theta = r\,dr\,d\theta.$$

Also $x^2 + y^2 = r^2$ and $\theta = \tan^{-1}\dfrac{y}{x}$

which implies $2r\,dr = 2x\,dx + 2y\,dy \Rightarrow r\,dr = x\,dx + y\,dy$

and $d\theta = \dfrac{1}{1+\dfrac{y^2}{x^2}}\left[\dfrac{y\,dx - x\,dy}{x^2}\right] \Rightarrow y\,dx - x\,dy = -r^2\,d\theta$

Thus, the equation (1) becomes $-\dfrac{r\,dr}{r^2\,d\theta} = \sqrt{\dfrac{a^2 - r^2}{r^2}}$

$$\Rightarrow \quad -\frac{dr}{d\theta} = \sqrt{a^2 - r^2}$$

$$\Rightarrow \quad \int \frac{dr}{\sqrt{a^2 - r^2}} = -\int d\theta + c$$

$$\Rightarrow \quad \sin^{-1}\frac{r}{a} = -\theta + c$$

$$\Rightarrow \quad r = a\sin(-\theta + c)$$

$$\Rightarrow \quad \sqrt{x^2 + y^2} = -a\sin\left\{\tan^{-1}\left(\frac{y}{x}\right) - c\right\}.$$

EXERCISE 1.1

1. Find the differential equations for the family of curves by eliminating arbitrary constants.

   (i) $y = Ae^x + Be^{-x} + C$ (ii) $y = Ax^3 + Bx^2$

   (iii) $\dfrac{x^2}{a^2} + \dfrac{y^2}{b^2} = 1$ (iv) $e^{2y} + 2axe^y + a^2 = 0$

2. Find the differential equations of the following curves.

   (i) All circles of radius a (ii) All circles.

   (iii) All circles touching the y-axis and centres are on the x-axis.

Solve the following differential equations

3. (a) $(1 + x)\,y\,dx + (1 - y)\,x\,dy = 0$. (b) $\dfrac{dy}{dx} + \sqrt{\dfrac{(1 - y^2)}{(1 - x^2)}} = 0$.

   (c) $3e^x \tan y\,dx + (1 - e^x)\sec^2 y\,dy = 0$.

4. $(x^2 + 2xy)dy + (2xy + y^2 + 3x^2)dx = 0$.

5. $(x^3 - 3xy^2)\,dx = (y^2 - 3x^2y)\,dy$.

6. $x\,dy - y\,dx = \sqrt{(x^2 + y^2)}\,dx.$

7. $x\,dy - y\,dx = 2\sqrt{y^2 - x^2}.$

8. $(2x + 3y - 5)\dfrac{dy}{dx} + 3x + 2y - 5 = 0.$

9. $\dfrac{dy}{dx} = \dfrac{(x - y) + 3}{2(x - y) + 5}$

10. $(x + 1)\dfrac{dy}{dx} - xy = e^x (x + 1)^{n+1}.$

11. $\dfrac{dy}{dx} + \dfrac{y}{(1 - x^2)^{3/2}} = \dfrac{x + \sqrt{(1 - x^2)}}{(1 - x^2)^2}$

12. $(2x - 10y^3)\dfrac{dy}{dx} + y = 0.$

13. $\dfrac{dy}{dx} + \dfrac{2}{x}y = 3x^2 y^{4/3}$

14. $\dfrac{dy}{dx} + y\cos x = y^n \sin 2x.$

15. $\dfrac{dy}{dx}(x^2y^3 + xy) = 1.$

16. $(1 + e^{x/y})dx + e^{x/y}\left(1 - \dfrac{x}{y}\right)dy = 0.$

17. $xy^3 (y\,dx + 2x\,dy) + (3y\,dx + 5x\,dy) = 0.$

18. $x(3y\,dx + 2x\,dy) + 2xy(3y\,dx + 4x\,dy) = 0.$

19. $y(axy + e^x)\,dx - e^x\,dy = 0.$

20. $\sin y\dfrac{dy}{dx} = \cos y(1 - x\cos y).$

21 $3\dfrac{dy}{dx} + \dfrac{2}{x + 1}y = \dfrac{x^3}{y^2}.$

ANSWERS 1.1

1. (i) $\dfrac{d^3y}{dx^3} - \dfrac{dy}{dx} = 0$ (ii) $x^2\dfrac{d^2y}{dx^2} - 4x\dfrac{dy}{dx} + 6y = 0$

(iii) $xy\frac{d^2y}{dx^2} + x\left(\frac{dy}{dx}\right)^2 - y\frac{dy}{dx} = 0$

(iv) $(1-x^2)\left(\frac{dy}{dx}\right)^2 + 1 = 0$

2. (i) $\left[1+\left(\frac{dy}{dx}\right)^2\right]^3 = a^2\left(\frac{d^2y}{dx^2}\right)^2$

(ii) $\left[1+\left(\frac{dy}{dx}\right)^2\right]\frac{d^3y}{dx^2} - 3\frac{dy}{dx}\left(\frac{d^2y}{dx^2}\right)^2 = 0$

(iii) $2xy\frac{dy}{dx} + x^2 - y^2 = 0$

3. (a) $xy = ce^{y-x}$.

(b) $y\sqrt{(1-x^2)} + x\sqrt{(1-y^2} = c.$

(c) $\tan y = c\,(1-e^x)^3$.

4. $x(x^2 + y^2 + xy) = c$

5. $y^2 - x^2 = c\,(x^2 + y^2)^2$

6. $y + \sqrt{(x^2 + y^2)} = cx^2$.

7. $y + \sqrt{(y^2 - x^2)} = cx^3$.

8. $\frac{3}{2}(x^2 + y^2) + 2xy - 5(x+y) = c.$

9. $x - 2y + \log(x - y + 2) = 0.$

10. $y = (e^x + c)\,(x+1)^n$.

11. $y = \frac{ce^{-x}}{\sqrt{(1-x^2)}} + \frac{x}{\sqrt{(1-x^2)}}$.

12. $(x - 2y^3)y^2 = c.$

13. $7y^{-1/3} = cx^{2/3} - 3x^2$

14. $y^{1-n} = ce^{(n-1)\sin x} + 2\sin x + \frac{2}{n-1}$.

15. $1 + x(y^2 + ce^{-y^2} + 2) = 0.$

16. $x + ye^{x/y} = c.$

17. $x^4y^8 + 4x^3y^5 = c.$

18. $x^3y^2 + 4x^2y^6 = c$

19. $ax^2y - ay + 2e^x = 0$

20. $\sec y = x + 1 + ce^x$.

21. $(x+1)^2 y = \frac{x^6}{6} + \frac{2x^5}{5} + \frac{x^4}{4} + c.$

2
Linear Differential Equations of nth Order with Constant Coefficients

2.1 INTRODUCTION

A differential equation of the form $Ly = F(x)$, that is

$$\left(a_0(x)\frac{d^n}{dx^n} + a_1(x)\frac{d^{n-1}}{dx^{n-1}} + a_2(x)\frac{d^{n-2}}{dx^{n-2}} + ... + a_n(x)\right)y = F(x) \quad ...(1)$$

where $a_0(x) \neq 0$, $a_1(x), a_2(x), ..., a_n(x)$ and $F(x)$ are continuous functions on real interval $a \leq x \leq b$; is called a linear differential equation of nth order. If F(x) is identically zero $\forall\, x \in [a, b]$, then equation (1) reduces to

$$a_0(x)\frac{d^n y}{dx^n} + a_1(x)\frac{d^{n-1}y}{dx^{n-1}} + a_2(x)\frac{d^{n-2}y}{dx^{n-2}} + ... + a_n(x)y = 0 \quad ...(2)$$

which is called homogeneous linear differential equation of nth order.

Definition: If in equation (1) we consider the coefficient $a_0(x)$, $a_1(x)$,..., $a_n(x)$, are constants. Then it can be written as

$$\frac{d^n y}{dx^n} + p_1\frac{d^{n-1}y}{dx^{n-1}} + p_2\frac{d^{n-2}y}{dx^{n-2}} + ... + p_n y = Q(x) \quad ...(3)$$

where $p_1, p_2, ..., p_n$ are constants and $Q(x)$ is a function of x only. The differential equation (3) is called a linear differential equation of nth order with constant coefficients.

If $Q(x) = 0, \forall \in [a, b]$, then equation (3) reduces to

$$\frac{d^n y}{dx^n} + p_1\frac{d^{n-1}y}{dx^{n-1}} + p_2\frac{d^{n-2}y}{dx^{n-2}} + ... + p_n y = 0 \quad ...(4)$$

This is known as homogeneous linear differential equation of order n with constant coefficients.

The general solution of equation (4) is called complementary function (C.F.) and contains n arbitrary constant. A solution of equation (3) free from arbitrary constants is called a particular integral (P.I.)

2.1.1 The Operator D

For conveniences denote the differential operator $\frac{d}{dx}$ by D, so that the operators $\frac{d^2}{dx^2}, \frac{d^3}{dx^3}, \ldots \frac{d^n}{dx^n}$ are denoted by $D^2, D^3, D^4, \ldots, D^n$, respectively. Thus the equation (3) can be written as

$$(D^n + p_1 D^{n-1} + p_2 D^{n-2} + \ldots + p_{n-1} D + p_n) y = Q$$

In short we can write it as follows $f(D)y = Q$ ...(5)

where $f(D) = D^n + p_1 D^{n-1} + p_2 D^{n-2} + \ldots + D^n$, *i.e.* $f(D)$ is an nth degree polynomial in D.

2.2 DEPENDENT AND INDEPENDENT SOLUTIONS OF LINEAR DIFFERENTIAL EQUATIONS

2.2.1 Linearly Dependent Solutions

Any n solutions $y_1(x), y_2(x), \ldots y_n(x)$ are said to be linearly dependent solutions of (2) if there exist n constants $c_1, c_2, c_3 \ldots c_n$ not all zero, such that

$$c_1 y_1 + c_2 y_2 + \ldots + c_n y_n = 0 \;\forall\; x \in (a, b).$$

2.2.2 Linearly Independent Solutions

The n solutions $y_1, y_2, \ldots y_n$ are said to be linearly independent if whenever $c_1 y_1 + c_2 y_2 + \ldots + c_n y_n = 0, \forall\; x \in (a,b) \Rightarrow c_1 = c_2 = \ldots = c_n = 0$.

2.2.3 Fundamental Set of Solutions and Fundamental Matrix

Any collection $y_1, y_2, \ldots y_n;\; x \in (a, b)$ of n-solutions of the equation (2) is said to be fundamental set of solution of (2) if $t \geq 0$, are linearly independent on (a, b).

The matrix

$$\Phi = \begin{bmatrix} y_1 & y_2 & \cdots & y_n \\ y_1' & y_2' & \cdots & y_n' \\ y_1'' & y_2'' & \cdots & y_n'' \\ \vdots & \vdots & & \vdots \\ y_1^{(n-1)} & y_2^{(n-1)} & \cdots & y_n^{(n-1)} \end{bmatrix}$$

is called the fundamental matrix of (2) if $y_1, y_2, \ldots. y_n$ are linearly independent.

2.2.4 The Wronskian

If $y_1, y_2, \ldots. y_n$ are n-linearly independent solutions of equation (2), then the scalar function

$$W(y_1 y_2 y_3, \ldots y_n) = \begin{bmatrix} y_1 & y_2 & \cdots & y_n \\ y_1' & y_2' & \cdots & y_n' \\ y_1'' & y_2'' & \cdots & y_n'' \\ \vdots & \vdots & & \vdots \\ y_1^{(n-1)} & y_2^{(n-1)} & \cdots & y_n^{(n-1)} \end{bmatrix}$$

is called the Wronskian of $y_1, y_2, y_3 \ldots y_n$ for $x \in (a, b)$ If $y_1(x)$ and $y_2(x)$ be two linearly independent solution of linear homogeneous differential equation of second order

$$a_0(x)\frac{d^2y}{dx^2} + a_1(x)\frac{dy}{dx} + a_2(x)y = 0 \quad \ldots(3)$$

then the Wronskian of $y_1(x)$ and $y_2(x)$ is defined as

$$W(y_1, y_2)(x) = \begin{vmatrix} y_1(x) & y_2(x) \\ y_1'(x) & y_2'(x) \end{vmatrix}$$

Theorem 2.1: Any linear combination of m solution of nth order homogeneous linear differential equation

$$a_0(x)\frac{d^ny}{dx^n} + a_1(x)\frac{d^{n-1}y}{dx^{n-1}} + \ldots + a_n(x)y = 0 \quad \ldots(1)$$

is also a solution of this equation (1).

Proof: Let $y_1, y_2, \ldots y_m$ are m-solutions of (1) then we have

$$a_0(x)\frac{d^ny_i}{dx^n} + a_1(x)\frac{d^{n-1}y_i}{dx^{n-1}} + \ldots + a_{n-1}(x)\frac{dy_i}{dx} + a_n(x)\,y_i = 0, \ \forall\, i = 1, 2, \ldots m \quad \ldots(2)$$

Any linear combination of these m-solution is

$$y = c_1y_1 + c_2y_2 + \ldots + c_my_m = \sum_{i=1}^{m} c_iy_i \quad \ldots(3)$$

Then if $y^{(j)}$ denote the jth derivative of y, we have $y^{(j)} = \sum_{i=1}^{m} c_iy_i^{(j)}$

Now

$$a_0(x)\frac{d^ny}{dx^n} + a_1(x)\frac{d^{n-1}y}{dx^{n-1}} + \ldots + a_n(x)y$$

$$= a_0(x)\sum_{i=1}^{m} c_iy_i^{(n)} + a_1(x)\sum_{i=1}^{m} c_i\, y_i^{(n-1)} + \ldots + a_n(x)\sum_{i=1}^{m} c_iy_i$$

$$= \sum_{i=1}^{m} c_i\left[a_0(x)y_i^{(n)} + a_1(x)y_i^{(n-1)} + \ldots + a_n(x)y_i\right]$$

$$= \sum_{i=1}^{m} c_i 0 = 0, \text{ [using equation (2)].}$$

This shows that the linear combination

$y = c_1y_1 + c_2y_2 + c_3y_3 + ... + c_my_m$ is a solution of (1).

Theorem 2.2: The n-solutions $y_1, y_2, ... y_n$ of a differential equation

$$a_0(x)\frac{d^ny}{dx^n} + a_1(x)\frac{d^{n-1}y}{dx^n} + ... + a_n(x)\, y = 0 \qquad ...(1)$$

on [a, b] are linearly independent iff their Wronskian

$$W(y_1, y_2, y_3, ..., y_n) \neq 0.$$

or the n solutions $y_1, y_2, ... y_n$ of a differential equation (1) are linearly dependent iff their Wronskian is identically zero.

Proof: Suppose $y_1, y_2, ... y_n$ are linearly dependent on [a, b], then there exist $c_1, c_2, ... c_n$, not all zero such that

$$\sum_{i=1}^{n} c_iy_i(x) = 0 \text{ for all } x \in [a, b].$$

It follows that $\sum_{i=1}^{n} c_iy_i^j(x) = 0$ for all $i = 1, 2, 3, ..., (n-1)$.

Since all $c_1, c_2, ... c_n$ are not zero, we must have

$$\begin{vmatrix} y_1 & y_2 & \cdots & y_n \\ y_1' & y_2' & \cdots & y_n' \\ y_1'' & y_2'' & \cdots & y_n'' \\ \vdots & \vdots & & \vdots \\ y_1^{(n-1)} & y_2^{(n-1)} & \cdots & y_n^{(n-1)} \end{vmatrix} = 0 \quad \Rightarrow W(y_1, y_2, ... y_n)(x_0) = 0.$$

But $x_0 \in [a, b]$ is an arbitrary point, so we have

$$W(y_1, y_2, ... y_n)(x) = 0, \forall\, x \in [a, b].$$

Conversely let $W(y_1, y_2, ..., y_n)(x) = 0 \;\forall\, x \in [a, b]$, hence if $x_0 \in [a, b]$, we have

$$W(y_1, y_2, ... y_n)(x_0) = 0 \Rightarrow \begin{vmatrix} y_1 & y_2 & \cdots & y_n \\ y_1' & y_2' & \cdots & y_n' \\ y_1'' & y_2'' & \cdots & y_n'' \\ \vdots & \vdots & & \vdots \\ y_1^{(n-1)} & y_2^{(n-1)} & \cdots & y_n^{(n-1)} \end{vmatrix} = 0$$

This gives a system of n linear equations in n unknowns $k_1, k_2, ..., k_n$, not all zero, as

$$k_1y_1(x_0)+k_2y_2(x_0)+...+k_ny_n(x_0)=0$$

$$k_1y_1'(x_o)+k_2y'(x_o)+...+k_ny_n(x_o)=0$$

....

....

$$k_1y_1^{(n-1)}(x_0)+k_2y_2^{(n-1)}(x_0)+...+k_ny_n^{(n-1)}(x_0)=0$$

Since x_0 is an arbitrary point, we have

$c_1\,y_1(x)+c_2\,y_2(x)+...+c_n\,y_n(x)=0, \forall\, x\in[a,b]$, where $c_1, c_2, ... c_n$ are not all zero simultaneously.

$\Rightarrow \quad y_1(x), y_2(x), y_3(x), ... y_n(x)$ are linearly dependent.

2.2.5 Derivation of a Differential Equation

If $y_1, y_2, ... y_n$ are n linearly independent functions on $[a, b]$ then the corresponding differential equation, whose solutions are these n linearly independent functions, is given by

$$\begin{vmatrix} y & y_1 & y_2 & \cdots & y_n \\ y' & y_1' & y_2' & \cdots & y_n' \\ y'' & y_1'' & y_2'' & \cdots & y_n'' \\ \vdots & \vdots & \vdots & & \vdots \\ y^{(n)} & y_1^{(n)} & y_2^{(n)} & \cdots & y_n^{(n)} \end{vmatrix} = 0$$

Example 1: Are the functions $y_1(x)=\sin x$ and $y_2(x)=\sin x-\cos x$ linearly independent solutions of $y''+y=0$? Further if $\sin x + 3\cos x = c_1y_1(x)+c_2y_2(x)$ then determine the constants c_1 and c_2.

Solution: Given $y_1(x)=\sin x$, then we have

$$y_1'(x)=\cos x \Rightarrow y_1''(x)=-\sin x \Rightarrow y_1''(x)+y_1=0$$

Hence $y_1(x)$ is a solution of $y''+y=0$.

Similarly, we can show that $y_2(x)$ is a solution of $y''+y=0$.

Now, Wronskian $W(y_1, y_2)(x)$ is given by

$$W(y_1,y_2)(x)=\begin{vmatrix} y_1(x) & y_2(x) \\ y_1'(x) & y_2'(x) \end{vmatrix} = \begin{vmatrix} \sin x & \sin x-\cos x \\ \cos x & \cos x+\sin x \end{vmatrix}$$

$$\equiv \quad \sin x\cos x+\sin^2 x-\sin x\cos x+\cos^2 x=1\neq 0$$

This shows that $y_1(x)$ and $y_2(x)$ are linearly independent.

Now $\sin x+3\cos x=c_1\,y_1(x)+c_2\,y_2(x)$

$\Rightarrow \quad \sin x+3\cos x=c_1\sin x+c_2(\sin x-\cos x)$

$=(c_1+c_2)\sin x-c_2\cos x$

On comparing the coefficient of $\sin x$ and $\cos x$ on both sides, we get

$$c_1+c_2=1, -c_2=3 \quad \Rightarrow c_1=4, c_2=-3$$

Example 2: Determine the differential equation whose linearly independent solution set is given by $\{e^x, xe^x, x^2e^x\}$.

Solution: If y is a general solution of the differential equation whose linearly independent solutions are e^x, xe^x, x^2e^x, then the differential equation is given by

$$W(y, y_1, y_2, y_3) = 0 \Rightarrow \begin{vmatrix} y(x) & y_1(x) & y_2(x) & y_3(x) \\ y'(x) & y_1'(x) & y_2'(x) & y_3'(x) \\ y''(x) & y_1''(x) & y_2''(x) & y_3''(x) \\ y'''(x) & y_1'''(x) & y_2'''(x) & y_3'''(x) \end{vmatrix} = 0$$

$$\Rightarrow \begin{vmatrix} y(x) & e^x & xe^x & x^2e^x \\ y'(x) & e^x & e^x + xe^x & 2xe^x + x^2e^x \\ y''(x) & e^x & 2e^x + xe^x & 2e^x + 4xe^x + x^2e^x \\ y'''(x) & e^x & 3e^x + xe^x & 6e^x + 6xe^x + x^2e^x \end{vmatrix} = 0$$

$$\Rightarrow e^{3x} \begin{vmatrix} y & 1 & x & x^2 \\ y' & 1 & 1+x & 2x + x^2 \\ y'' & 1 & 2+x & 2 + 4x + x^2 \\ y''' & 1 & 3+x & 6 + 6x + x^2 \end{vmatrix} = 0$$

$$\Rightarrow \begin{vmatrix} y & 1 & x & x^2 \\ y' - y & 0 & 1 & 2x \\ y'' - y' & 0 & 1 & 2x + 2 \\ y''' - y'' & 0 & 1 & 2x + 4 \end{vmatrix} = 0, \text{ since } e^{3x} \neq 0$$

$$\Rightarrow \begin{vmatrix} y' - y & 1 & 2x \\ y'' - y' & 1 & 2x+2 \\ y''' - y'' & 1 & 2x+4 \end{vmatrix} = 0 \Rightarrow \begin{vmatrix} y' - y & 1 & 2x \\ y'' - 2y' + y & 0 & 2 \\ y''' - 2y'' + y' & 0 & 2 \end{vmatrix} = 0$$

$$\Rightarrow \begin{vmatrix} y'' - 2y' + y & 2 \\ y''' - 2y'' + y' & 2 \end{vmatrix} = 0$$

$$\Rightarrow 2[y'' - 2y' + y - y''' + 2y'' - y'] = 0$$

$$\Rightarrow y''' - 3y'' + 3y' - y = 0$$

which is the required differential equation.

Example 3: Use Wronskian to show that the functions e^t, e^{-t}, e^{2t} are linearly independent on any interval $[a, b]$. Determine the differential equation with these functions as independent solutions.

Solution: Let $\phi_1(t) = e^t, \phi_2(t) = e^{-t}, \phi_3(t) = e^{2t}$, then

$$W(\phi_1,\phi_2,\phi_3)(t) = \begin{vmatrix} \phi_1 & \phi_2 & \phi_3 \\ \phi_1' & \phi_2' & \phi_3' \\ \phi_1'' & \phi_2'' & \phi_3'' \end{vmatrix} = \begin{vmatrix} e^t & e^{-t} & e^{2t} \\ e^t & -e^t & 2e^{2t} \\ e^t & e^{-t} & 4e^{2t} \end{vmatrix}$$

$$= e^{2t}\begin{vmatrix} 1 & 1 & 1 \\ 1 & -1 & 2 \\ 1 & 1 & 4 \end{vmatrix} = -6e^{2t} \neq 0$$

for all $t \in \mathbb{R}$. Hence the functions e^t, e^{-t}, e^{2t} are L.I. on any interval of $\mathbb{R}$.

If u is a general solution of the differential equation whose linearly independent solutions are e^t, e^{-t}, e^{2t}, then the differential equation is given by

$$W(u,\phi_1,\phi_2,\phi_3) = 0 \Rightarrow \begin{vmatrix} u & \phi_1 & \phi_2 & \phi_3 \\ u' & \phi_1' & \phi_2' & \phi_3' \\ u'' & \phi_1'' & \phi_2'' & \phi_3'' \\ u''' & \phi_1''' & \phi_2''' & \phi_3''' \end{vmatrix} = 0$$

$$\Rightarrow \begin{vmatrix} u & e^t & e^{-t} & e^{2t} \\ u' & e^t & -e^{-t} & 2e^{2t} \\ u'' & e^t & e^{-t} & 4e^{2t} \\ u''' & e^t & -e^{-t} & 8e^{2t} \end{vmatrix} = 0$$

$$\Rightarrow e^{2t}\begin{vmatrix} u & 1 & 1 & 1 \\ u' & 1 & -1 & 2 \\ u'' & 1 & 1 & 4 \\ u''' & 1 & -1 & 8 \end{vmatrix} = 0$$

$$\Rightarrow u'''\begin{vmatrix} 1 & 1 & 1 \\ 1 & -1 & 2 \\ 1 & 1 & 4 \end{vmatrix} + (-1)^1 u''\begin{vmatrix} 1 & 1 & 1 \\ 1 & -1 & 2 \\ 1 & -1 & 8 \end{vmatrix} + (-1)^2 u'\begin{vmatrix} 1 & 1 & 1 \\ 1 & 1 & 4 \\ 1 & -1 & 8 \end{vmatrix} + (-1)^3 u\begin{vmatrix} 1 & -1 & 2 \\ 1 & 1 & 4 \\ 1 & 1 & 8 \end{vmatrix} = 0$$

$$\Rightarrow -6u''' + 12u'' + 6u' - 12u = 0$$

$$\Rightarrow \frac{d^3u}{dt^3} - 2\frac{d^2u}{dt^2} - \frac{du}{dt} + 2u = 0.$$

which is the required differential equation on every closed interval $[a,b]$ where $-\infty < a < b < \infty$.

Example 4: Test the linear independence of the functions t^2, e^t, e^{-t}. Find the differential equation with these functions as independent solutions.

Solution: Let $\phi_1(t)=t^2, \phi_2(t)=e^t, \phi_3(t)=e^{-t}$, then

$$W(\phi_1,\phi_2,\phi_3)(t) = \begin{vmatrix} t^2 & e^t & e^{-t} \\ 2t & e^t & -e^{-t} \\ 2 & e^t & e^{-t} \end{vmatrix} = e^t e^t \begin{vmatrix} t^2 & 1 & 1 \\ 2t & 1 & -1 \\ 2 & 1 & 1 \end{vmatrix} = 2(t^2-2)$$

Thus $W(\phi_1,\phi_2,\phi_3)(t) \neq 0$ identically $\Rightarrow$ These function ϕ_1,ϕ_2,ϕ_3 are L.I.

If u be the general solution of the differential equation whose linearly independent solutions are t^2, e^t, e^{-t}, then the differential equation is given by $W(u,\phi_1,\phi_2,\phi_3)(t)=0$

$$\Rightarrow \begin{vmatrix} u & t^2 & e^t & e^{-t} \\ u' & 2t & e^t & -e^{-t} \\ u'' & 2 & e^t & e^{-t} \\ u''' & 0 & e^t & -e^{-t} \end{vmatrix} = 0 \Rightarrow e^t e^{-t} \begin{vmatrix} u & t^2 & 1 & 1 \\ u' & 2t & 1 & -1 \\ u'' & 2 & 1 & 1 \\ u''' & 0 & 1 & -1 \end{vmatrix} = 0$$

$$\Rightarrow u''' \begin{vmatrix} t^2 & 1 & 1 \\ 2t & 1 & -1 \\ 2 & 1 & 1 \end{vmatrix} + (-1)u'' \begin{vmatrix} t^2 & 1 & 1 \\ 2t & 1 & -1 \\ 0 & 1 & -1 \end{vmatrix} + (-1)^2 u' \begin{vmatrix} t^2 & 1 & 1 \\ 2 & 1 & 1 \\ 0 & 1 & -1 \end{vmatrix} + (-1)^3 u \begin{vmatrix} 2t & 1 & -1 \\ 2 & 1 & 1 \\ 0 & 1 & -1 \end{vmatrix} = 0$$

$$\Rightarrow \quad 2(t^2-2)u''' - 4t\,u'' - 2(t^2-2)u' + 4t\,u = 0$$

$$\Rightarrow \quad \frac{d^3u}{dt^3} - \frac{2t}{t^2-2}\frac{d^2u}{dt^2} - \frac{du}{dt} + \frac{2t}{(t^2-2)}u = 0$$

This is the required differential equation on any closed interval $[a, b] \subset R_0 = R - \{-\sqrt{2}, \sqrt{2}\}$.

Example 5: If $\phi_1(t)$ and $\phi_2(t)$ are two distinct solutions of a differential equation

$$\frac{d^2u}{dt^2} + p(t)\frac{du}{dt} + q(t)u = 0 \qquad ...(1)$$

on $[a, b]$ such that $\phi_2(t) \neq 0$ for all $t \in [a, b]$ and $p(t), q(t)$ are continuous on $[a, b]$.

(i) Then prove that $\dfrac{d}{dt}\left[\dfrac{\phi_1(t)}{\phi_2(t)}\right] = -\dfrac{W(\phi_1,\phi_2)(t)}{[\phi_2(t)]^2}$ for all $t \in [a, b]$.

(ii) Deduce that if $W(\phi_1, \phi_2)(t) = 0$ for all $t \in [a, b]$ then two solutions $\phi_1(t)$ and $\phi_2(t)$ are linearly dependent on $[a, b]$.

(iii) If two solutions $\phi_1(t)$ and $\phi_2(t)$ are linearly independent then the function $\phi(t) = \dfrac{\phi_1(t)}{\phi_2(t)}$ is a monotonic function on $[a, b]$.

(iv) If two solutions $\phi_1(t)$ and $\phi_2(t)$ of differential equation (1), are linearly independent on $[a, b]$ and a_1, a_2, b_1, b_2 are constants such that $a_1b_2 - b_1a_2 \neq 0$, then prove that the two solutions $a_1\phi_1 + a_2\phi_2$ and $b_1\phi_1 + b_2\phi_2$ are also linearly independent on $[a, b]$.

Solution:

(i) Given ϕ_1, ϕ_2 are two solutions of the differential equation (1), so that

$$\frac{d}{dt}\left[\frac{\phi_1(t)}{\phi_2(t)}\right] = \frac{-\phi_1(t)\phi_2'(t) + \phi_1'(t)\phi_2(t)}{[\phi_2(t)]^2} = \frac{-[\phi_1(t)\phi_2'(t) - \phi_1'(t)\phi_2(t)]}{[\phi_2(t)]^2}$$

$$= \frac{-\begin{vmatrix} \phi_1(t) & \phi_2(t) \\ \phi_1'(t) & \phi_2(t) \end{vmatrix}}{[\phi_2(t)]^2} = \frac{-W(\phi_1, \phi_2)(t)}{[\phi_2(t)]^2}.$$

(ii) Now if $W(\phi, \phi_2)(t) = 0$ for all $t \in [a, b]$,

then $\dfrac{d}{dt}\left[\dfrac{\phi_1(t)}{\phi_2(t)}\right] = 0 \Rightarrow \dfrac{\phi_1(t)}{\phi_2(t)} =$ constant, [on integrating]

$\Rightarrow \phi_1(t) - c\phi_2(t) = 0 \quad \Rightarrow \phi_1(t)$ and $\phi_2(t)$ are linearly dependent on $[a, b]$.

(iii) To check the monotonicity of $\phi(t)$, let us differentiate $\phi(t)$ with respect to 't', we get

$$\phi'(t) = -\frac{W(\phi_1, \phi_2)(t)}{[\phi_1(t)]^2} \qquad \text{...(2)}$$

Given that $\phi_1(t)$ and $\phi_2(t)$ are linearly independent on $[a, b]$

$\Rightarrow W(\phi_1, \phi_2)(t) \neq 0$, for any $t \in [a, b]$.

$\Rightarrow$ either $W(\phi_1, \phi_2)(t) < 0$ or $W(\phi_1\ \phi_2)(t) > 0$ for all $t \in [a, b]$.

Thus from (2) clearly either

$\phi'(t) > 0$ or $\phi'(t) < 0$ for all $t \in [a, b] \Rightarrow \phi(t)$ is a monotonic on $[a, b]$.

(iv) Let $\psi_1(t) = a_1\phi_1(t) + a_2\phi_2(t), \psi_2(t) = b_1\phi_1(t) + b_2\phi_2(t)$, then

$$W(\psi_1, \psi_2)(t) = \begin{vmatrix} \psi_1 & \psi_2 \\ \psi_1' & \psi_2' \end{vmatrix} = \begin{vmatrix} a_1\phi_1 + a_2\phi_2 & b_1\phi_1 + b_2\phi_2 \\ a\phi_1' + a_2\phi_2' & b_1\phi_1' + b_2\phi_2' \end{vmatrix}$$

$$= \begin{vmatrix} a_1\phi_1 & b_1\phi_1 \\ a_1\phi_1' & b_1\phi_1' \end{vmatrix} + \begin{vmatrix} a_1\phi_1 & b_2\phi_2 \\ a_1\phi_1' & b_2\phi_2' \end{vmatrix} + \begin{vmatrix} a_2\phi_2 & b_1\phi_1 \\ a_2\phi_2' & b_1\phi_1' \end{vmatrix} + \begin{vmatrix} a_2\phi_2 & b_2\phi_2 \\ a_2\phi_2' & b_2\phi_2' \end{vmatrix}$$

$$= a_1b_1 \begin{vmatrix} \phi_1 & \phi_1 \\ \phi_1' & \phi_1' \end{vmatrix} + a_1b_2 \begin{vmatrix} \phi_1 & \phi_2 \\ \phi_1' & \phi_2' \end{vmatrix} + a_2b_1 \begin{vmatrix} \phi_2 & \phi_1 \\ \phi_2' & \phi_1' \end{vmatrix} + a_2b_2 \begin{vmatrix} \phi_2 & \phi_2 \\ \phi_2' & \phi_2' \end{vmatrix}$$

$$= a_1b_2\, W(\phi_1, \phi_2)(t) - a_2b_1\, W(\phi_1, \phi_2)(t) = (a_1b_2 - a_2b_1) W(\phi_1, \phi_2)(t) \quad ...(3)$$

As ϕ_1, ϕ_2 are L.I. on $[a, b] \Rightarrow W(\phi_1, \phi_2)(t) \neq 0$, so that from (3) it is clear that $W(\psi_1, \psi_2)(t) \neq 0$, for all $t \in [a, b] \Rightarrow \psi_1, \psi_2$ are linearly independent on $[a, b]$.

Example 6: Let $\{\phi_1, \phi_2\}$ be a set of two linearly independent solutions of equation

$$a_0 \frac{d^2u}{dt^2} + a_1(t)\frac{du}{dt} + a_2(t)u = 0 \quad ...(1)$$

where $a_0(t), a_1(t), a_2(t)$ are continuous on $[a, b]$ and $a_0(t) \neq 0$ for all $t \varepsilon [a, b]$. Let $\{\psi_1, \psi_2\}$ be another set of linearly independent solutions of (1) on $[a, b]$. If $W(\phi_1, \phi_2)(t)$ and $W(\psi_1, \psi_2)(t)$ has usual meaning then prove that there exists a constant $c \neq 0$ such that

$$W(\psi_1, \psi_2)(t) = c\, W(\phi_1, \phi_2)(t), \text{ for all } t \in [a, b].$$

Solution: As $\{\phi_1, \phi_2\}$ be a set of two linearly independent solutions of (1) on $[a, b]$ and $\{\psi_1, \psi_2\}$ be another set of two linearly independent solution of same differential equation. So there exist constants a_1, a_2, b_1, b_2 such that $\psi_1(t) = a_1\phi_1(t) + a_2\phi_2(t)$ and $\psi_2(t) = b_1\phi_1(t) + b_2\phi_2(t)$.

Now $W(\psi_1, \psi_2)(t) = \begin{vmatrix} \psi_1 & \psi_2 \\ \psi_1' & \psi_2' \end{vmatrix}$

$$= \begin{vmatrix} a_1\phi_1(t) + a_2\phi_2(t) & b_1\phi_1(t) + b_2\phi_2(t) \\ a_1\phi_1'(t) + a_2\phi_2'(t) & b_1\phi_1'(t) + b_2\phi_2'(t) \end{vmatrix}$$

$$= \begin{vmatrix} a_1\phi_1(t) & b_1\phi_1(t) \\ a_1\phi_1'(t) & b_1\phi_1'(t) \end{vmatrix} + \begin{vmatrix} a_1\phi_1(t) & b_2\phi_2(t) \\ a_1\phi_1'(t) & b_2\phi_2'(t) \end{vmatrix} + \begin{vmatrix} a_2\phi_2(t) & b_1\phi_1(t) \\ a_2\phi_2'(t) & b_1\phi_1'(t) \end{vmatrix} +$$

$$\begin{vmatrix} a_2\phi_2(t) & b_2\phi_2(t) \\ a_2\phi_2'(t) & b_2\phi_2'(t) \end{vmatrix}$$

$$= \quad a_1b_1.0 + a_1b_2\begin{vmatrix}\phi_1(t) & \phi_2(t)\\ \phi_1'(t) & \phi_2'(t)\end{vmatrix} + a_2b_1\begin{vmatrix}\phi_2(t) & \phi_1(t)\\ \phi_2'(t) & \phi_1'(t)\end{vmatrix} + a_2b_2\,.0$$

$$= \quad a_1b_2\, W(\phi_1,\phi_2)(t) - a_2b_1\, W(\phi_1,\phi_2)(t) = (a_1b_2 - b_1a_2)W(\phi_1,\phi_2)(t)$$

Thus if $a_1b_2 - b_1a_2 = c$, then we have $W(\psi_1,\psi_2)(t) = c\,W(\phi_1,\phi_2)(t)$ ...(2)

But $\{\psi_1,\psi_2\}$ and $\{\phi_1,\phi_2\}$ are both linearly independent sets of solutions of (1) so neither $W(\psi_1,\psi_2)(t)=0$ nor $W(\phi_1,\phi_2)(t)=0$ for any $t\in[a,b]$. Thus by (2) it is clear that $c \neq 0$ for all $t\in[a,b]$. Hence the result follows.

Example 7: Find the Wronskian of the set $\{1-x, 1+x, 1-3x\}$. Determine whether this set L.I. on $\mathbb{R}$.

Solution: Here

$$W(1-x, 1+x, 1-3x) = \begin{vmatrix} 1-x & 1+x & 1-3x \\ \frac{d}{dx}(1-x) & \frac{d}{dx}(1+x) & \frac{d}{dx}(1-3x) \\ \frac{d^2}{dx^2}(1-x) & \frac{d^2}{dx^2}(1+x) & \frac{d^2}{dx^2}(1-3x) \end{vmatrix}$$

$$= \begin{vmatrix} 1-x & 1+x & 1-3x \\ -1 & 1 & -3 \\ 0 & 0 & 0 \end{vmatrix} = 0$$

For linear independence consider the equation

$c_1(1-x) + c_2(1+x) + c_3(1-3x)$

$\Rightarrow 0(-c_1 + c_2 - 3c_3)\,x + (c_1 + c_2 + c_3) = 0$

which is satisfied for all x if both coefficient are zero. Thus

$-c_1 + c_2 - 3c_3 = 0$ and $c_1 + c_2 + c_3 = 0$

On solving simultaneously, we get $c_1 = -2c_3, c_2 = c_3$ with c_3 is arbitrary. Choose $c_3 = 2$ then

$c_1 = -4,\ c_2 = 2,\ c_3 = 2$ satisfies (1). Thus the set $\{1-x, 1+x, 1-3x\}$ is linearly dependent.

Example 8:

(i) Determine whether the set $\{t^3, |t^3|\}$ is L.I. on $[-1, 1]$.

(ii) Find Wronskian $W(t^3, |t^3|)$.

(iii) Do the results of (i) and (ii) contradict the theorem that the solutions u_1 and u_2 of the linear homogeneous equation

$$Lu \equiv p_0(t)\frac{d^2u}{dt^2} + p_1(t)\frac{du}{dt} + p_2(t)\,u = 0$$

on $[a, b]$ to be L.I. iff $W(u_1, u_2) \neq 0$ for any $t \in [a,b]$

(iv) The two solutions of $\frac{d^2u}{dt^2} - \frac{2}{t}\frac{du}{dt} = 0$ on $[-1, 1]$ are $u_1 = t^3$ and $u_2 = |t^3|$. Does this result contradict the solution of (iii).

Solution:

(i) Consider the equation $c_1t^3 + c_2|t^3| = 0$ ...(1)

But we know that $|t^3| = \begin{cases} t^3, & if\ t \geq 0 \\ -t^3, & if\ t < 0 \end{cases}$

Hence for $t \geq 0$, the equation (1) becomes $c_1t^3 + c_2t^3 = 0$ ...(2)

and for $t < 0$, the equation (1) becomes $c_1t^3 - c_2t^3 = 0$ ...(3)

Solving (2) and (3) simultaneously for c_1 and c_2, then the only solution is $c_1 = 0 = c_2$

Hence the given set is L.I. on $[-1, 1]$.

(ii) We have $|t^3| = \begin{cases} t^3 & if\ t > 0 \\ 0 & if\ t = 0 \\ -t^3 & if\ t < 0 \end{cases}$,

which implies that $\frac{d}{dt}|t^3| = \begin{cases} 3t^3 & if\ t > 0 \\ 0 & if\ t = 0 \\ -3t^2 & if\ t < 0 \end{cases}$.

Then for $t > 0$, $W\left(t^3, |t^3|\right) = \begin{vmatrix} t^3 & t^3 \\ 3t^2 & 3t^2 \end{vmatrix} = 0$

For $t = 0$, $W(t^3, |t^3|) = \begin{vmatrix} t^3 & 0 \\ 3t^2 & 0 \end{vmatrix} = 0$

and for $t < 0$, $W(t^3, |t^3|) = \begin{vmatrix} t^3 & -t^3 \\ 3t^2 & -3t^2 \end{vmatrix} = 0$.

Thus $W(t^3, |t^3|) = 0$ for each $t \in [-1, 1]$.

(iii) No. Although the Wronskian of $t^3, |t^3|$ is identically vanishes but two functions $t^3, |t^3|$ are not solutions of the same linear homogeneous equations of the form $Lu = 0$, which has no singular point in $[-1, 1]$.

(iv) No. Although $W(t^3, |t^3|) = 0$ for all $t \in [-1, 1]$ and both $u_1 = t^3$ and $u = |t^3|$ are L.I. solutions of the same linear homogeneous equation $\frac{d^2u}{dt^2} - \frac{2}{t}\frac{du}{dt} = 0$.

But this differential equation is not of the form $Lu = 0$ because the coefficient $-\frac{2}{t}$ is not continuous at $t = 0$ on [-1, 1].

EXERCISE 2.1

1. Prove that the Wronskian of $e^{at}\sin\beta t$ and $e^{at}\cos\beta t$, $\beta \neq 0$ is $\beta\, e^{2at}$.
2. Find Wronskian of $e^{m_1 t}, e^{m_2 t}, e^{m_3 t}$. Show that the set $\{e^{m_1 t}, e^{m_2 t}, e^{m_3 t}\}$ is linearly independent on $\mathbb{R}$ if m_1, m_2, m_3 are distinct.
3. Show that the set of functions $\{t, t^2, t^3\}$ is linearly independent on $\mathbb{R}$. Determine the differential equation with these functions as linearly independent solution.
4. Show that the functions $t^2, t^2 \log t$ are linearly independent on $\mathbb{R}$. Determine the differential equation with these functions as linearly independent solutions.
5. If $\{\phi_1, \phi_2\}$ be a set of two linearly independent solutions of the second order homogeneous linear differential equation

$$a_0(t)\frac{d^2u}{dt^2} + a_1(t)\frac{du}{dt} + a_2(t)u = 0,$$

where $a_0(t)$, $a_1(t)$, $a_2(t)$ are continuous on $[a, b]$

If $\phi_1''(t_0) = \phi_2''(t_0) = 0$ at some point to $t_0 \in [a, b]$, then prove that $a_1(t_0) = a_2(t_0) = 0$.

6. Test the linear independence or linear dependence of the following sets of functions.

(i) $\{\sin 3t, \cos 3t\}$. (ii) $\{te^t, e^t, \sin t\}$.

(iii) $\{t, 2t-7, 1\}$ (iv) $\{\sin t, 2\cos t, 3\sin t + \cos t\}$.

(v) $\{\sin 3t, \sin t, \sin^3 t\}$. (vi) $\{t, t^2, t^4\}$.

7. Determine the differential equation corresponding to linearly independent sets in problem (6) as its solutions.

8. Show that the functions $\{t, |t|\}$ are linearly independent on $\mathbb{R}$ although the Wronskian vanishes.

9. Prove that t and t^2 are linearly independent on $-1<t<1$ although $W(t, t^2)=0$ at $t=0$. From this what you conclude about the possibility of t and t^2 being solution of a second order linear equation? Show that these are solutions of $t^2u'' - 2tu' + 2u = 0$. Does it contradicts your conclusion?

10. If ϕ_1, ϕ_2 are linearly independent solution of

$$\frac{d^2u}{dt^2} + a_1(t)\frac{du}{dt} + a_2(t)\, u = 0,$$

where $a_1(t)$ and $a_2(t)$ are continuous on $[a, b]$, then prove that $c_1\phi_1$ and $c_2\phi_2$ are also be linearly independent provided c_1 and c_2 are both not equal to zero.

11. Show that the Wronskian of the functions $x^\alpha, x^\beta, x^\gamma$ $(x>0)$ is equal to $(\alpha-\beta)(\beta-\gamma)(\gamma-\alpha)x^{\alpha+\beta+\gamma-3}$.

ANSWERS 2.1

2. $(m_1-m_2)(m_2-m_3)(m_3-m_1)e^{t(m_1 t m_2 t m_3)}$

3. $t^3u''' - 3t^2u'' + 6tu' - 6u = 0$

4. $t^2u'' - 3t\,u' + 4u = 0$

6. (i) L.I. (ii) L.I.
   (iii) L.D. (iv) L.D.
   (v) L.D. (vi) L.I.

7. (i) $u'' + 9u = 0$.
   (ii) $u''' + (\tan t - 2)u'' + (\tan t + 1)u' + \tan t\; u = 0$
   (vi) $t^3u''' - 4t^2u'' + 8t\,u' - 8u = 0$.

2.3 SOLUTION OF LINEAR DIFFERENTIAL EQUATIONS

First of all we discuss some theorems regarding solutions of linear differential equations.

Theorem 2.3: If $y=y_1, y=y_2, \ldots$ and $y=y_n$ are n linearly independent solutions of the differential equation

$$(D^n + p_1D^{n-1} + p_2D^{n-1} + \ldots + p_n)y = 0 \qquad \ldots(1)$$

Then $u = c_1y_1 + c_2y_2 + \ldots c_ny_n$ also be its solution, where $c_1, c_2, \ldots, c_n$ are arbitrary constants.

Proof: Since $y = y_1, y = y_2, ... y = y_n$ are solution of equation (1), we have

$$D^n y_1 + p_1 D^{n-1} y_1 + p_2 D^{n-2} y_1 + ... + p_n y_1 = 0$$
$$D^n y_2 + p_1 D^{n-1} y_2 + p_2 D^{n-2} y_2 + ... + p_n y_2 = 0$$
$$..$$
$$D^n y_n + p_1 D^{n-1} y_n + p_2 D^{n-2} y_n + ... + p_n y_n = 0. \quad ...(2)$$

Now $D^n u + p_1 D^{n-1} u + p_2 D^{n-2} u + ... + p_n u = D^n (c_1 y_1 + c_2 y_2 + ... + c_n y_n)$

$$+ p_1 D^{n-1} (c_1 y_1 + c_2 y_2 + ... + c_n y_n) + p_2 D^{n-2} (c_1 y_1 + c_2 y_2 + ... + c_n y_n)$$
$$+ ... + p_n (c_1 y_1 + c_2 y_2 + ... + c_n y_n)$$
$$= c_1 (D^n y_1 + p_1 D^{n-1} y_1 + p_2 D^{n-2} y_1 + ... + p_n y_1)$$
$$+ c_2 (D^n y_2 + p_1 D^{n-1} y_2 + p_2 d^{n-2} y_2 + ... + p_n y_2) + ...$$
$$+ c_n (D^n y_n + p_1 D^{n-1} y_n + p_2 D^{n-1} y_n + ... + p_n y_n)$$
$$= c_1 (0) + c_2 (0) + ... + c_n (0) = 0 \quad \text{[By using (2)].}$$

which show that $u = c_1 y_1 + c_2 y_2 + ... + c_n y_n$ is a solution of equation (1).

Note: The solution $u = c_1 y_1 + c_2 y_2 + ... + c_n y_n$ of the differential equation (1) is known as general or complete solution.

Theorem 2.4: If $y = u$ is the complete solution of the equation $f(D)y = 0$ and $y = v$ is a particular solution (containing no arbitrary constants) of the equation $f(D)y = Q$. Then the complete solution of the equation $f(D)y = Q$ be $y = u + v$.

Proof: Since $y = u$ is the complete solution of the equation $f(D)y = 0$ ...(1)

So we have $f(D)u = 0$. ...(2)

Also $y = v$ is a particular solution of the equation $f(D)y = Q$ ...(3)

Then $f(D)v = Q$. ...(4)

Adding (2) and (4), we obtain $f(D)(u + v) = Q$.

Thus $y = u + v$ satisfies the equation (3), hence it is the complete solution (C.S.) because it contains n arbitrary constants.

The part $y = u$ is called the complementary function (C.F.) and the part $y = v$ is called the particular integral (P.I.) of the equation (3).

Hence the complete solution of equation (3) is $y = C.F. + P.I.$

Thus to solve the equation (3), we first find the C.F., i.e. the C.S. of equation (1) and then the P.I., i.e. a particular solution of equation (3).

2.3.1 Complementary Function

Consider the differential equation $f(D)y = Q$.

The complementary function is actually the solution of the given differential equation (1) when it's right hand side member Q is replaced by zero. To find C.F. we first form the auxiliary equation.

Note 1: C.F. has no connection with P.I. If $Q = 0$, then P.I. is zero and the C.F. becomes the C.S.

Note 2: Solution of differential equation $f(D)y = Q$ should always be written as $y = C.F. + P.I.$ no matter P.I. is zero or not.

2.3.2 Auxiliary Equation

Consider the differential equation

$$(D^n + p_1 D^{n-1} + p_2 D^{n-2} + ... p_n)y = 0. \quad ...(1)$$

Let $y = e^{mx}$ be a solution of (1), then

$$Dy = m^{mx}, D^2y = m^2e^{mx}, ..., D^{n-2}y = m^{n-2}e^{mx}, D^{n-1}y = m^{n-1}e^{mx}, D^n y = m^n e^{mx}.$$

Substituting the value of $y, Dy, D^2y, ... D^n y$ in (1), we get

$$(m^n + p_1 m^{n-1} + p_2 m^{n-2} + ... + p_n)e^{mx} = 0$$

$$\Rightarrow \quad m^n + p_1 m^{n-1} + p_2 m^{n-2} + ... + p_n = 0, \text{ since } e^{mx} \neq 0. \quad ...(2)$$

Thus $y = e^{mx}$ will be a solution of equation (1) if m satisfies the polynomial equation (2). This equation (2) is called the auxiliary equation for the differential equation (1).

Definition: The equation obtained by equating to zero the symbolic coefficient of y is called the auxiliary equation, briefly written as A.E.

Steps for Finding Auxiliary Equation

Step 1: Replace y by 1.

Step 2: Replace $\frac{d}{dx}$ by m, $\frac{d^2}{dx^2}$ by $m^2, ...,$ $\frac{d^n}{dx^n}$ by m^n.

Step 3: By doing so, we get an algebraic equation in m of degree n, which is the required auxiliary equation.

Note: In order to find out the auxiliary equation, we assume Q as zero (if it is not so) in the given equation $f(D)y = Q$.

2.3.3 Rules for Finding the Complementary Function

Case I: *Complementary functions when all the roots of the auxiliary equation are real and distinct.*

Consider the equation $\left(D^n + a_1 D^{n-1} + a_2 D^{n-2} + ... + a_n\right)y = 0 \quad ...(1)$

where all the a_i's are constants.

Equation (1) is equivalent to $(D-m_1)(D-m_2)...(D-m_n)y=0.$ ...(2)

Equation (2) will be satisfied by the solutions of the equations

$$(D-m_1)y=0, (D-m_2)y=0,...,(D-m_n)y=0.$$

Now consider the equation $(D-m_1)y=0$ *i.e.* $\frac{dy}{dx}-m_1 y=0.$

It is a linear equation, so I.F. $=e^{\int -m_1 dx}=e^{-m_1 x}$

Therefore, its solution is $y.e^{-m_1 x}=\int 0.e^{-m_1 x}dx+c_1 \Rightarrow y=c_1 e^{m_1 x}.$

Similarly, the solution of $(D-m_2)y=0$ is $y=c_2 e^{m_2 x}.$

The solution of $(D-m_n)y=0$ *is* $y=c_n e^{m_n x}.$

Hence, the general solution of (2) is given by

$$C.F.=c_1 e^{m_1 x}+c_2 e^{m_2 x}+...+c_n e^{m_n x}.$$

Example 1: Solve $\frac{d^2y}{dx^2}+3\frac{dy}{dx}+2y=0.$

Solution: The given equation is $\frac{d^2y}{dx^2}+3\frac{dy}{dx}+2y=0.$

Here $Q=0 \Rightarrow P.I.=0$ and the auxiliary equation is

$$m^2+3m+2=0 \Rightarrow m=-1,-2.$$

The roots are real and distinct. Hence the C.F. $=c_1 e^{-x}+c_2 e^{-2x}$

Hence the complete solution is $y=C.F.+P.I.=c_1 e^{-x}+c_2 e^{-2x}$, where c_1, c_2 are arbitrary constants of integration.

Case II: *Complementary functions when the roots of auxiliary equation are equal.*

(a) When two roots of auxiliary equation are equal, i.e. let $m_1=m_2$. Solution of equation (1) is (as in Case I)

$$Y=C.F.+P.I=c_1 e^{m_1 x}+c_2 e^{m_2 x}+...+c_n e^{m_n x}+0,$$

because $P.I.=0$ as $Q=0$

$$=(c_1+c_2)e^{m_1 x}+c_3 e^{m_3 x}+...+c_n e^{m_n x}, \text{ since } m_1=m_2$$

$$=ce^{m_1 x}+c_3 e^{m_3 x}+...+c_n e^{m_n x}.$$

It contains $(n-1)$ arbitrary constants and is, therefore, not the complete solution of equation (1).

The part of C.F. corresponding to the repeated root is the complete solution of

$(D-m_1)(D-m_1)=0 \Rightarrow (D-m_1)v=0$ [putting $(D-m_1)y=v$]

$\Rightarrow \quad \dfrac{dv}{dx}-m_1 v = 0.$

Its solution is $v = c_2 e^{m_1 x}$. Hence $(D-m_1)y = c_2 e^{m_1 x}$

$\Rightarrow \quad \dfrac{dy}{dx} - m_1 y = c_2 e^{m_1 x}$, which is a linear equation.

Its solution is $ye^{-mx} = \int c_2 e^{m_1 x} e^{m_1 x} dx + c_1 \Rightarrow y = (c_2 x + c_1)e^{m_1 x}$.

Hence part of complementary function $(C.F.) = (c_1 + c_2 x)e^{m_1 x}$.

Hence complete C.F. $= (c_1 + c_2 x)e^{m_1 x} + c_3 e^{m_2 x} + \ldots c_n e^{m_n x}$.

(b) If however, three roots of the auxiliary equation are equal say $m_1 = m_2 = m_3$, then proceeding as above, we get

$$\text{C.F.} = (c_1 + c_2 x + c_3 x^3)e^{m_1 x} + c_4 e^{m_4 x} + \ldots + c_n e^{m_n x}.$$

In general, if there are r roots of the auxiliary equation are equal, then we have

$$\text{C.F.} = (c_1 + c_2 x + c_3 x^2 + \ldots + c_r x^{r-1})e^{m_1 x} + c_{r+1} e^{m_{r+1} x} + \ldots + c_n e^{m_n x}.$$

Example 2: Solve $\dfrac{d^3 y}{dx^3} - 3\dfrac{d^2 y}{dx^2} + 4y = 0.$

Solution: The auxiliary equation is $m^3 - 3m^2 + 4 = 0$

$\Rightarrow \quad (m+1)(m-2)^2 = 0 \Rightarrow m = -1, 2, 2.$

The roots are real and repeated, so C.F. $= c_1 e^{-x} + (c_2 + c_3 x)e^{2x}$ and $P.I. = 0$.

Hence the complete solution is $y = C.F. + P.I. = c_1 e^{-x} + (c_2 + c_3 x)e^{2x}$, where c_1, c_2, c_3 are constants of integration.

Case III: *Complementary functions when two roots of auxiliary equation are imaginary.*

Let $m = \alpha \pm i\beta$, then C.F. $= c_1 e^{(\alpha+i\beta)x} + c_2 e^{(\alpha-i\beta)x} = e^{\alpha x}(c_1 e^{i\beta x} + c_2 e^{-i\beta x})$

$= e^{\alpha x}[c_1(\cos\beta x + i\sin\beta x) + c_2(\cos\beta x - i\sin\beta x)]$

$= e^{\alpha x}(A_1 \cos\beta x + A_2 \sin\beta x)$, where $A_1 = c_1 + c_2, A_2 = i(c_1 - c_2)$

$= Ae^{\alpha x}\cos(\beta x + B)$ or $Ae^{\alpha x}\sin(\beta x + B)$.

From the above discussion we infer that if $\alpha + i\beta$ is the roots of the A.E. then C.F. can be written in any one of the above form.

Then C.F. $= Ae^{\alpha x}\cos(\beta x + B) + c_3 e^{m_3 x} + c_4 e^{m_4 x} + \ldots + c_n e^{m_n x}$

Example 3: Solve $\frac{d^2y}{dx^2} + 2p\frac{dy}{dx} + (p^2 + q^2)\,y = 0.$

Solution: The A.E. is $m^2 + 2pm + (p^2 + q^2) = 0$

$\Rightarrow \quad m = -\,p \pm iq.$

The roots are imaginary, so $C.F. = c_1 e^{-px}\cos(qx + c_2)$ and P.I. = 0.

Hence complete solution is $y = C.F. + P.I. = c_1 e^{-px}\cos(qx + c_2)$.

Example 4: Solve $(D^4 - n^4)y = 0$ *where* $D \equiv \frac{d}{dx}$.

Solution: The auxiliary equation is

$$m^4 - n^4 = 0 \Rightarrow (m^2 - n^2)(m^2 + n^2) = 0 \Rightarrow m = \pm n, \pm\, in.$$

Now $C.F. = c_1 e^{nx} + c_2 e^{-nx} + e^{0x}(c_3 \cos nx + c_4 \sin nx)$

$= c_1 e^{nx} + c_2 e^{-nx} + c_3 \cos nx + c_4 \sin nx$ and $P.I. = 0$.

Hence the solution is

$y = C.F. + P.I. = c_1 e^{nx} + c_2 e^{-nx} + c_3 \cos nx + c_4 \sin nx$,

where c_1, c_2, c_3 and c_4 are arbitrary constants of integration.

Example 5: Solve $\left(D^4 + m^4\right)y = 0$

Solution: Auxiliary equation is

$$M^4 + m^4 = 0 \Rightarrow M^4 + m^4 + 2M^2m^2 = 2M^2m^2$$

$$\Rightarrow \quad (M^2 + m^2)^2 = 2M^2m^2 \Rightarrow M^2 + m^2 = \pm\sqrt{2}\,Mm.$$

Case I: Taking (+ve) sign: $M^2 + m^2 - \sqrt{2}\,Mn = 0$, then we have

$$M = \frac{\sqrt{2}\,m \pm \sqrt{2m^2 - 4m^2}}{2} = \frac{\sqrt{2}\,m \pm i\sqrt{2}\,m}{2} = \frac{m}{\sqrt{2}} \pm \frac{im}{\sqrt{2}}.$$

Case II: Taking (*-ve*) sign: $M^2 + m^2 + \sqrt{2}\,Mm = 0$, then we have

$$M = \frac{-\sqrt{2}m \pm \sqrt{2m^2 - 4m^2}}{2} = \frac{-\sqrt{2}m \pm i\sqrt{2}m}{2} = -\frac{m}{\sqrt{2}} \pm i\frac{m}{\sqrt{2}}$$

Here, $P.I. = 0$, and

$$C.F. = e^{\frac{m}{\sqrt{2}}x}\left(c_1 \cos\frac{m}{\sqrt{2}}x + c_2 \sin\frac{m}{\sqrt{2}}x\right) + e^{-\frac{m}{\sqrt{2}}x}\left(c_3 \cos\frac{m}{\sqrt{2}}x + c_4 \sin\frac{m}{\sqrt{2}}x\right).$$

Hence complete solution is

$$y = e^{\frac{m}{\sqrt{2}}x}\left(c_1 \cos\frac{m}{\sqrt{2}}x + c_2 \sin\frac{m}{\sqrt{2}}x\right) + e^{-\frac{m}{\sqrt{2}}x}\left(c_3 \cos\frac{m}{\sqrt{2}}x + c_4 \sin\frac{m}{\sqrt{2}}x\right),$$

where c_1, c_2, c_3 and c_4 are arbitrary constants of integration.

Case III: *Complementary functions when roots of auxiliary equation are repeated imaginary.*

Let $m_1 = m_2 = \alpha + i\beta$ *and* $m_3 = m_4 = \alpha - i\beta$, then by Case IInd

$$\text{C.F.} = e^{\alpha x}\,[(c_1 + c_2 x)\cos\beta x + (c_3 + c_4 x)\sin\beta x] + c_5 e^{m_5 x} + \ldots + c_n e^{m_n x}.$$

Example 6: Solve $\dfrac{d^4y}{dx^4} - 4\dfrac{d^3y}{dx^3} + 8\dfrac{d^2y}{dx^2} - 8\dfrac{dy}{dx} + 4y = 0.$

Solution: Auxiliary equation is

$$m^4 - 4m^3 + 8m^2 - 8m + 4 = 0 \Rightarrow (m^2 - 2m + 2)^2 = 0$$

$$\Rightarrow \quad m = \frac{2 \pm \sqrt{4-8}}{2}\,(\textit{twice}) = \frac{2 \pm 2i}{2}\,(\textit{twice}) = 1 \pm i,\ 1 \pm i$$

Thus C.F. $= e^x\,[(c_1 + c_2 x)\cos x + (c_3 + c_4 x)\sin x]$ and P.I. = 0.

The complete solution is $y = e^x\,[(c_1 + c_2 x)\cos x + (c_3 + c_4 x)\sin x]$,

where c_1, c_2, c_3 and c_4 are arbitrary constants of integration.

Example 7: Solve the differential equation

$$(D^2 + 1)^3 (D^2 + D + 1)^2 y = 0, \text{ where } D \equiv \frac{d}{dx}.$$

Solution: The auxiliary equation is $(m^2 + 1)^3 (m^2 + m + 1)^2 = 0$

$$\Rightarrow (m^2 + 1)^3 = 0 \text{ and } (m^2 + m + 1)^2 = 0$$

$$\Rightarrow m = \pm i, \pm i, \pm i, \ \frac{-1 \pm \sqrt{3}i}{2}, \frac{-1 \pm \sqrt{3}i}{2}.$$

Here $P.I. = 0$ and

$$C.F. = e^{0x}\,[(c_1 + c_2 x + c_3 x^2)\cos x + (c_4 + c_5 x + c_6 x^2)\sin x]$$
$$+ e^{-\frac{x}{2}}\left[(c_7 + c_8 x)\cos\frac{\sqrt{3}}{2}x + (c_9 + c_{10}x)\sin\frac{\sqrt{3}}{2}x\right]$$

$$= (c_1 + c_2 x + c_3 x^3)\cos x + (c_4 + c_5 + c_6 x^2)\sin x$$
$$+ e^{-x/2}\left\{(c_7 + c_8 x)\cos\frac{\sqrt{3}}{2}x + (c_9 + c_{10}x)\sin\frac{\sqrt{3}}{2}x\right\}$$

Therefore, the solution is

$$y = (c_1 + c_2 x + c_3 x^2)\cos x + (c_4 + c_5 x + c_6 x^2)\sin x$$
$$+ e^{-x/2}\left\{(c_7 + c_8 x)\cos\frac{\sqrt{3}}{2}x + (c_9 + c_{10}x)\sin\frac{\sqrt{3}}{2}x\right\},$$

where $c_1, c_2, c_3, c_4, c_5, c_6, c_7, c_8, c_9$ *and* c_{10} are arbitrary constants of integration.

Case IV: *Complementary functions when roots of auxiliary equation are irrational.*

Let $m_1 = \alpha + \sqrt{\beta}$ and $m_2 = \alpha - \sqrt{\beta}$, then

$$\text{C.F.} = c_1 e^{(\alpha+\sqrt{\beta})x} + c_2 e^{(\alpha-\sqrt{\beta})x}$$

$$= e^{\alpha x}(c_1 . e^{\sqrt{\beta}x} + c_2 e^{-\sqrt{\beta}x}) = e^{\alpha x}(A_1 \cos h\sqrt{\beta}x + A_2 \sin h\sqrt{\beta}x)$$

$$= Ae^{\alpha x} \cos h(\sqrt{\beta}\,x + B) \text{ or}$$

$$Ae^{\alpha x} \sin h(\sqrt{\beta}x + B) \text{ or } c_1 \cos h(\alpha + \sqrt{\beta})x + c_2 \sin h(\alpha + \sqrt{\beta})x$$

Hence complete

$$\text{C.F.} = c_1 \cos h\left(\alpha + \sqrt{\beta}\right)x + c_2 \sin h\left(\alpha + \sqrt{\beta}\right)x + c_3 e^{m_3 x} + c_4 e^{m_4 x} + \ldots + c_n e^{m_n x}$$

Case V: *Complementary functions when roots of auxiliary equation are repeated irrational.*

Let $m_1 = m_2 = \alpha + \sqrt{\beta}$ and $m_3 = m_4 = \alpha - \sqrt{\beta}$, then by case II,

$$\text{Complete C.F.} = e^{\alpha x}[(c_1 + c_2 x)\cos h\sqrt{\beta}\,x + (c_3 + c_4 x)\sin h\sqrt{\beta}\,x] + c_5 e^{m_5 x} + c_6 e^{m_6 x} + \ldots + c_n e^{m_n x}.$$

Example 8: Solve the differential equation

$$(D^4 - 6D^2 + 9)y = 0, \text{ where } D \equiv \frac{d}{dx}.$$

Solution: The auxiliary equation is

$$m^4 - 6m^2 + 9 = 0 \quad \Rightarrow (m^2 - 3)^2 = 0 \quad \Rightarrow m = \pm\sqrt{3}.$$

Here $P.I. = 0$ and $C.F. = (c_1 + c_2 x)\cosh\sqrt{3}\,x + (c_3 + c_4 x)\sinh\sqrt{3}\,x$

Therefore, the solution is

$$y = (c_1 + c_2 x)\cosh\sqrt{3}\,x + (c_3 + c_4 x)\sinh\sqrt{3}\,x.$$

Example 9: Solve the differential equation

$$(D^2 - 6D + 7)(D^2 - 6D + 10)y = 0.$$

Solution: The auxiliary equation is

$$(m^2 - 6m + 7)(m^2 - 6m + 10) = 0 \Rightarrow m = 3 \pm \sqrt{2},\ 3 \pm i.$$

Here $P.I. = 0$ and

$$C.F. = e^{3x}(c_1 \cosh\sqrt{2}\,x + c_2 \sinh\sqrt{2}\,x) + e^{3x}(c_3 \cos x + c_4 \sin x)$$

Therefore, the solution is

$$y = e^{3x}(c_1 \cosh\sqrt{2}\,x + c_2 \sinh\sqrt{2}\,x) + e^{3x}(c_3 \cos x + c_4 \sin x).$$

EXERCISE 2.2

Solve the following differential equations

1. $\frac{d^2y}{dx^2} - 7\frac{dy}{dx} + 12y = 0.$
2. $\frac{d^3y}{dx^3} - 3\frac{d^2y}{dx^2} + 3\frac{dy}{dx} - y = 0.$
3. $\frac{d^2y}{dx^2} + (a+b)\frac{dy}{dx} + aby = 0.$
4. $\frac{d^3y}{dx^3} - 2\frac{d^2y}{dx^2} + 4\frac{dy}{dx} - 8y = 0.$
5. $\frac{d^4y}{dx^4} + 8\frac{d^2y}{dx^2} + 16y = 0.$
6. $\frac{d^3y}{dx^3} - 9\frac{d^2y}{dx^2} + 23\frac{dy}{dx} + 15y = 0.$
7. $\frac{d^3y}{dx^3} - \frac{d^2y}{dx^2} - \frac{dy}{dx} - 2y = 0.$
8. $\frac{d^4y}{dx^4} - 2\frac{d^3y}{dx^3} + 2\frac{d^2y}{dx^2} - 2\frac{dy}{dx} + y = 0.$
9. $\frac{d^4y}{dx^4} + 13\frac{d^2y}{dx^2} + 36y = 0.$
10. $\frac{d^3y}{dx^3} - 6\frac{d^2y}{dx^2} + 11\frac{dy}{dx} - 6y = 0.$
11. Solve $\frac{d^2y}{dx^2} + y = 0$ given $y = 2$ for $x = 0$, $y = 2$ for $x = \frac{\pi}{2}$.

ANSWERS 2.2

1. $y = c_1e^{3x} + c_2e^{4x}$
2. $y = (c_1 + c_2x + c_3x^2)e^x.$
3. $y = c_1e^{-ax} + c_2e^{-bx}$
4. $y = c_1e^{2x} + c_2\cos(c_2x + c_3).$
5. $y = (c_1 + c_2x)\cos 2x + (c_3 + c_4x)\sin 2x.$
6. $y = c_1e^x + c_2e^{3x} + c_3e^{5x}$
7. $y = c_1e^{2x} + c_2e^{-x/2}\cos\left((\sqrt{3}x/2) + c_3\right).$
8. $y = c_1\cos(x + c_2) + (c_3 + c_4x)e^x.$
9. $y = c_1\cos 3x + c_2\sin 3x + c_3\cos 2x + c_4\sin 2x.$

10. $y = c_1 e^x + c_2 e^{2x} + c_3 e^{3x}$

11. $y = 2(\cos x - \sin x)$.

2.4 RULES FOR FINDING THE PARTICULAR INTEGRAL (P.I.)

Consider the differential equation

$$(D^n + p_1 D^{n-1} + p_2 D^{n-2} + \dots + p_{n-1} D + p_n) y = Q$$

We can write it as $f(D)y = Q$, then

$$P.I. = \frac{1}{f(D)} Q.$$

Rules for particular integral:

(i) $\frac{1}{f(D)} Q$ is the particular integral of $f(D)y = Q$.

(ii) $\frac{1}{D} Q = \int Q\,dx$

(iii) $\frac{1}{D-a} Q = e^{ax} \int Q e^{-ax}\,dx$

(iv) $\frac{1}{D+a} Q = e^{-ax} \int Q e^{ax}\,dx$.

2.4.1 Case I: Particular integral when $Q = e^{ax}$:

Since

$$De^{ax} = ae^{ax}$$
$$D^2 e^{ax} = a^2 e^{ax}$$
$$\dots\dots\dots\dots\dots$$
$$D^{n-1} e^{ax} = a^{n-1} e^{ax}$$
$$D^n e^{ax} = a^n e^{ax}$$

Therefore

$$(D^n + p_1 D^{n-1} + \dots + p_{n-1} D + p_n) e^{ax} = (a^n + p_1 a^{n-1} + p_2 a^{n-2} + \dots + p_{n-1} a + p_n) e^{ax}$$

$$\Rightarrow \quad f(D) e^{ax} = f(a) e^{ax}.$$

Operating both side by $\frac{1}{f(D)}$, we obtain $\frac{1}{f(D)} f(D) e^{ax} = \frac{1}{f(D)} f(a) e^{ax}$

$$\Rightarrow \quad \frac{1}{f(D)} e^{ax} = \frac{1}{f(a)} e^{ax} \quad \text{provided } f(a) \neq 0.$$

Case of failure: If $f(a) = 0$. Then the above method fails since $f(a) = 0$. Here $D = a$ is a root of $f(D) = 0$.

Therefore $(D - a)$ is a factor of $f(D)$. Let

$$f(D) = (D - a)\phi(D) \text{ where } \phi(a) \neq 0$$

Then $\frac{1}{f(D)}e^{ax} = \frac{1}{(D-a)\phi(D)}e^{ax} = \frac{1}{D-a}.\frac{1}{\phi(D)}e^{ax} = \frac{1}{D-a}\frac{1}{\phi(a)}e^{ax}$

$= \frac{1}{\phi(a)}e^{ax}\int e^{ax}.e^{-ax}dx = \frac{1}{\phi(a)}e^{ax}.x.$

Hence $\frac{1}{f(D)}e^{ax} = x\left[\frac{1}{\phi(a)}e^{ax}\right].$

If a be a double root of $f(D)=0$, then suppose that $f(D)=(D-a)^2\psi(D)$.

Now

$$P.I. = \frac{1}{(D-a)^2\psi(D)}e^{ax} = \frac{1}{(D-a)^2\psi(a)}.e^{ax}$$

$$= \frac{1}{\psi(a)}\frac{1}{D-a}.\frac{1}{D-a}e^{ax} = \frac{1}{\psi(a)}\frac{1}{D-a}xe^{ax}$$

$$= \frac{1}{\psi(a)}e^{ax}\int e^{-ax}xe^{ax}dx = \frac{1}{\psi(a)}e^{ax}.\frac{x^2}{2!} = \frac{x^2e^{ax}}{2\psi(a)}.$$

In general, if the root 'a' of $f(D)=0$, repeats r times then suppose that $f(D)=(D-a)^r F(D)$.

Thus $P.I. = \frac{x^r}{r!}\frac{e^{ax}}{F(a)}$

This result can also be written as

$$\frac{1}{f(D)}e^{ax} = x^r\frac{1}{\left[f^{(r)}(D)\right]_{D=a}}e^{ax}, \text{ if } f(a)=f'(a)=f^{(r-1)}(a)=0.$$

Example 1: Solve $(4D^2+4D-3)y = e^{2x}$.

Solution: A.E. is $4m^2+4m-3=0 \Rightarrow m = \frac{1}{2}, -\frac{3}{2}$.

Therefore, $C.F. = C_1e^{x/2} + C_2e^{-3x/2}$

and $P.I. = \frac{1}{4D^2+4D-3}e^{2x} = \frac{1}{4(2)^2+4(2)-3}e^{2x} = \frac{1}{21}e^{2x}.$

Complete solution is $y = C_1e^{x/2} + C_2e^{-3x/2} + \frac{1}{21}e^{2x}$.

Example 2: Solve $(D^2-a^2)\,y = e^{ax} - e^{-ax}$.

Solution: A.E. is $m^2 - a^2 = 0 \Rightarrow m = \pm a$.

Therefore $C.F. = c_1 e^{ax} + c_2 e^{-ax}$

and $P.I. = \frac{1}{D^2 - a^2}(e^{ax} - e^{-ax}) = \frac{1}{D^2 - a^2}(e^{ax}) - \frac{1}{D^2 - a^2}(e^{-ax})$

$$= x.\frac{1}{2D}e^{ax} - x\frac{1}{2D}e^{-ax} = \frac{x}{2}\int e^{ax}dx - \frac{x}{2}\int e^{-ax}dx = \frac{x}{2}\frac{e^{ax}}{a} - x\left(\frac{e^{-ax}}{-2a}\right)$$

$$= \frac{x}{2}\frac{(e^{ax} + e^{-ax})}{a} = \frac{x}{a}\cos hax$$

Hence complete solution is

$$y = C.F. + P.I. = c_1 e^{ax} + c_2 e^{-ax} + \frac{x}{a}\cos hax$$

where c_1 and c_2 are arbitrary constants of integration.

Example 3: Solve $(D+2)(D-1)^2\, y = e^{-2x} + 2\sin hx.$

Solution: A.E. is $(m+2)(m-1)^2 = 0 \Rightarrow m = 1, 1, -2$

Therefore $C.F. = c_1 e^{-2x} + (c_2 + c_3 x)e^x$

and P.I. $= \frac{1}{(D+2)(D-1)^2}[e^{-2x} + e^x - e^{-x}]$

Now

$$\frac{1}{(D+2)(D-1)^2}e^{-2x} = \frac{1}{(D+2)}\left[\frac{1}{(D-1)^2}e^{-2X}\right] = \frac{1}{9}\frac{1}{(D+2)}e^{-2x} = \frac{1}{9}xe^{-2x} \quad ...(1)$$

$$\frac{1}{(D+2)(D-1)^2}e^x = \frac{1}{(D-1)^2}\left[\frac{1}{(D+2)}e^x\right] = \frac{1}{(D-1)^2}\left(\frac{1}{1+2}e^x\right)$$

$$= \frac{1}{3}\frac{1}{(D-1)^2}e^x = \frac{1}{3}x^2\frac{1}{2}e^x = \frac{x^2}{6}e^x \quad ...(2)$$

$$\frac{1}{(D+2)(D-1)^2}e^{-x} = \frac{1}{(-1+2)(-1-1)^2}e^{-x} = \frac{1}{4}e^{-x}. \quad ...(3)$$

Hence from (1), (2) and (3), we have $P.I. = \frac{x}{9}e^{-2x} + \frac{x^2}{6}e^x + \frac{1}{4}e^{-x}.$

Hence complete solution is

$$y = c_1 e^{-2x} + (c_2 + c_3 x) + \frac{x}{9}e^{-2x} + \frac{x^2}{6}e^x + \frac{1}{4}e^{-x},$$

where c_1, c_2 *and* c_3 are arbitrary constants.

Example 4: Solve $\left(D^3 - 3D^2 + 3D - 1\right)y = e^x + 2.$

Solution: Auxiliary equation is $m^3 - 3m^2 + 3m - 1 = 0$

$\Rightarrow \quad (m-1)^3 = 0 \Rightarrow m = 1, 1, 1$

Therefore $C.F. = (c_1 + c_2 x + c_3 x^2)e^x$

and $$P.I. = \frac{1}{D^3 - 3D^2 + 3D - 1}(e^x + 2) = \frac{1}{(D-1)^3}e^x + \frac{1}{(D-1)^3}2e^{0x}$$

$$= x^3 \frac{1}{3.2.1}e^x - 2 = \frac{x^3}{6}e^x - 2.$$

Hence solution is $y = (c_1 + c_2x + c_3x^2)e^x + \frac{x^2}{6}e^x - 2$, where c_1, c_2, c_3 are arbitrary constants.

2.4.2 Case II: Particular integral when $Q = \sin(ax+b)$ or $\cos(ax+b)$:

We have

$$D\sin(ax+b) = a\cos(ax+b)$$

$$D^2 \sin(ax+b) = (-a^2)\sin(ax+b)$$

$$D^3 \sin(ax+b) = (-a^3)\cos(ax+b)$$

$$D^4 \sin(ax+b) = +a^4 \sin(ax+b)$$

....

$$(D^2)^n \sin(ax+b) = (-a^2)^n \sin(ax+b)$$

Gives $f(D^2)\sin(ax+b) = f(-a^2)\sin(ax+b)$

Therefore $$\frac{1}{f(D^2)}\sin(ax+b) = \frac{1}{f(-a^2)}\sin(ax+b)$$

Similarly $$\frac{1}{f(D^2)}\cos(ax+b) = \frac{1}{f(-a^2)}\cos(ax+b).$$

Case of failure: If $f(-a^2) = 0$ then the above method fails. Now, we proceed as follows:

$$\frac{1}{f(D^2)}\cos(ax+b) = x\frac{1}{f'(D^2)}\cos(ax+b)$$

$$\frac{1}{f(D^2)}\sin(ax+b) = x\frac{1}{f'(D^2)}\sin(ax+b).$$

If $f'(-a^2) = 0$, then

$$\frac{1}{f(D^2)}\sin(ax+b) = x^2\frac{1}{f''(D^2)}\sin(ax+b)$$

$$\frac{1}{f(D^2)}\cos(ax+b) = x^2\frac{1}{f''(D^2)}\cos(ax+b).$$

Similarly $$\frac{1}{D^2+a^2}\sin(ax+b) = -\frac{x}{2a}\cos(ax+b)$$

$$\frac{1}{D^2+a^2}\cos(ax+b) = \frac{x}{2a}\sin(ax+b).$$

Example 5: Solve $\frac{d^3y}{dx^3} - 3\frac{d^2y}{dx^2} + 4\frac{dy}{dx} - 2y = e^x + \cos x.$

Solution: A.E. is $m^3 - 3m^2 + 4m - 2 = 0 \Rightarrow m = 1, 1 \pm i$

Therefore $C.F. = c_1 e^x + e^x (c_2 \cos x + c_3 \sin x)$

and $$P.I. = \frac{1}{(D^3 - 3D^2 + 4D - 2}e^x + \frac{1}{(D^3 - 3D^2 + 4D - 2)}\cos x$$

$$= x\frac{1}{(3D^2 - 6D + 4)}e^x + \frac{1}{(-D + 3 + 4D - 2)}\cos x$$

$$= x.\frac{1}{(7-6)}e^x + \frac{1}{(3D+1)}\cos x = xe^x + \frac{(3D-1)}{9D^2 - 1}\cos x$$

$$= xe^x - \frac{1}{10}(-3\sin x - \cos x).$$

Hence complete solution is

$$y = c_1 e^x + e^x (c_2 \cos x + c_3 \sin x) + xe^x + \frac{1}{10}(3\sin x + \cos x).$$

Example 6: Solve $\frac{d^3y}{dx^3} + a^2\frac{dy}{dx} = \sin ax.$

Solution: A.E. is $m^3 + a^2 m = 0 \Rightarrow m = 0, \pm ai$

Therefore $C.F. = c_1 + c_2 \cos ax + c_3 \sin ax$

and $$P.I. = \frac{1}{D^3 + a^2 D}\sin ax = \frac{D^{-1}\sin ax}{D^2 + a^2} = \frac{1}{D^2 + a^2}\left(\frac{-\cos ax}{a}\right)$$

$$= -\frac{x}{a}\frac{1}{2D}\cos ax = \frac{-x}{2a^2}\sin ax.$$

Hence complete solution is $y = c_1 + c_2 \cos ax + c_3 \sin ax + \frac{x}{2}\sin ax,$ where c_1, c_2, c_3 are arbitrary constants.

Example 7: Solve $(D^2 - 4)y = \cos^2 x.$

Solution: A.E. is $m^2 - 4 = 0 \Rightarrow m = \pm 2$

Therefore $C.F. = c_1 e^{2x} + c_2 e^{-2x}$

and $$P.I. = \frac{1}{D^2 - 4}\cos^2 x = \frac{1}{D^2 - 4}\left(\frac{\cos 2x + 1}{2}\right) = -\left(\frac{1}{16}\cos 2x + \frac{1}{8}\right).$$

Hence complete solution is $y = c_1 e^{2x} + c_2 e^{-2x} - \left(\frac{1}{16}\cos 2x + \frac{1}{8}\right).$

Example 8: Solve $\frac{d^2y}{dx^2} - 8\frac{dy}{dx} + 9y = 40\sin 5x.$

Solution: A.E. is $m^2 - 8m + 9 = 0 \Rightarrow m = \dfrac{8 \pm 2\sqrt{7}}{2} = 4 \pm \sqrt{7}$

Therefore $C.F. = e^{4x}(c_1 \cos h\sqrt{7}x + c_2 \sin h\sqrt{7}\,x)$

and $P.I. = \dfrac{1}{D^2 - 8D + 9} 40 \sin 5x = \dfrac{1}{-25 - 8D + 9} 40 \sin 5x$

$$= \frac{(D-2)}{-8(D+2)(D-2)} 40 \sin 5x = \frac{(+25\cos 5x - 10\sin 5x)'}{-(-25-4)}$$

$$= \frac{1}{29}(25\cos 5x - 10 \sin 5x).$$

Hence complete solution is

$$y = e^{4x}(c_1 \cosh\sqrt{7}x + c_2 \sin\sqrt{7}\,x) + \frac{1}{29}(25\cos 5x - 10\sin 5x).$$

2.4.3 Case III: When $Q = x^m$, where m being a positive integer then $P.I. = \dfrac{1}{f(D)} x^m$:

Steps for solving $f(D)$

(1) The $f(D)$ must be reduced in the form of $\{1 + \phi(D)\}$ *or* $\{1 - \phi(D)\}$ on taking lowest degree term common from $f(D)$ so that I[st] term become unity.

(2) Now written the function $f(D)$ in numerator as $\{1-\phi(D)\}^{-1}$ or $\{1+\phi(D)\}^{-1}$.

(3) Expand $\{1 + \phi(D)\}^{-1}$ and $\{1-\phi(D)\}^{-1}$ in the ascending power of D as far as the term containing D^m operating on x^m term-by-term.

Example 9: Solve $\dfrac{d^2y}{dx^2} - 4y = x^2$.

Solution: A.E. is $m^2 - 4 = 0 \Rightarrow m = \pm 2$.

Therefore $C.F. = c_1 e^{2x} + c_2 e^{-2x}$

$$P.I. = \frac{1}{D^2 - 4} x^2 = -\frac{1}{4}\left(1 - \frac{D^2}{4}\right)^{-1} x^2$$

$$= -\frac{1}{4}\left(1 + \frac{D^2}{4} \ldots\right) x^2 = -\frac{1}{4}\left(x^2 + \frac{1}{2}\right) = -\frac{x^2}{4} - \frac{1}{8}.$$

Hence complete solution is $y = c_1 e^{2x} + c_2 e^{-2x} - \left(\dfrac{x^2}{4} + \dfrac{1}{8}\right)$.

Example 10: Solve $\frac{d^3y}{dx^3}+2\frac{d^2y}{dx^2}+\frac{dy}{dx}=c^{2x}+x^2+x.$

Solution: A.E. is $m^3+2m^2+m=0 \Rightarrow m\left(m^2+2m+1\right)=0 \Rightarrow m=0,-1,-1$

Therefore $C.F.=c_1+\left(c_2+c_2x\right)e^{-x}$

and $P.I.=\frac{1}{D^3+2D^2+D}\left(e^{2x}+x^2+x\right)$

$$=\frac{1}{8+8+2}e^{2x}+D^{-1}\left\{1+\left(D^2+2D\right)\right\}^{-1}\left(x^2+x\right)$$

$$=\frac{1}{18}e^{2x}+D^{-1}\left[1-\left(D^2+2D\right)+\left(D^4+4D^2+4D^3\right)-\ldots\right]\left(x^2+x\right)$$

$$=\frac{1}{18}e^{2x}+\left\{D^{-1}-(D+2)+\left(D^3+4D+4D^2\right)-\ldots\right\}\left(x^2+x\right)$$

$$=\frac{1}{18}e^{2x}+\left(\frac{x^3}{3}+\frac{x^2}{2}-\left\{2x+1+2\left(x^2+x\right)\right\}+(8x+4+8)\right)$$

$$=\frac{1}{18}e^{2x}+\left(\frac{x^3}{3}-\frac{3}{2}x^2+4x+11\right)$$

Hence complete solution is

$$y=c_1+\left(c_2+c_3x\right)e^{-x}+\frac{1}{18}e^{2x}+\left(\frac{x^3}{3}-\frac{3}{2}x^2+4x+11\right).$$

2.4.4 Case IV: Particular integral when $Q=e^{ax}V$, where V is a function of x

By successive differentiation we see that

$$De^{ax}V=e^{ax}DV+ae^{ax}V=e^{ax}(D+a)V,$$

$$D^2\left(e^{ax}V\right)=e^{ax}(D+a)^2V,$$

$$D^3\left(e^{ax}V\right)=e^{ax}(D+a)^3V,$$

....

$$D^n\left(e^{ax}V\right)=e^{ax}(D+a)^nV.$$

Therefore, $f(D)\left(e^{ax}V\right)=e^{ax}f(D+a)V.$

Hence, we have $\frac{1}{f(D)}\left(e^{ax}V\right)=e^{ax}\left[\frac{1}{f(D+a)}V\right].$

Thus e^{ax} which is on the right of $\frac{1}{f(D)}$ may be taken out to the left and D is replaced by $(D+a)$.

Example 11: Solve $(D^2+D-2)y=e^x$.

Solution: A.E. is $(m+2)(m-1)=0 \Rightarrow m=1,-2$

Therefore $C.F.=c_1e^x+c_2e^{-2x}$

and $$P.I.=\frac{1}{D^2+D-2}e^x=e^x\frac{1}{(D+1)^2+(D+1)-2}1$$

$$=e^x\frac{1}{D^2+2D+1+D+1-2}1=\frac{e^x}{(D^2+3D)}1$$

$$=\frac{e^x}{3}D^{-1}\left(1+\frac{D}{3}\right)^{-1}=\frac{1}{3}xe^x-\frac{1}{9}e^x.$$

Hence complete solution is $y=c_1e^x+c_2e^{-2x}+\frac{1}{3}xe^x-\frac{1}{9}e^x$.

Example 12: Solve $\frac{d^2y}{dx^2}-4y=x\sin hx$.

Solution: A.E. is $m^2-4=0 \Rightarrow m=\pm 2$

Therefore $C.F.=c_1e^{2x}+c_2e^{-2x}$

and $$P.I.=\frac{1}{D^2-4}x\sin hx=\frac{1}{D^2-4}x\left(\frac{e^x-e^{-x}}{2}\right)$$

$$=\frac{1}{2}\left[e^x\left(\frac{1}{D^2+1+2D-4}x\right)-\left(e^{-x}\frac{1}{D^2+1-2D-4}\right)x\right]$$

$$=\frac{1}{2}\left[e^x\frac{1}{-3\left\{1-\left(\frac{D^2}{3}+\frac{2}{3}D\right)\right\}}x+e^{-x}\frac{1}{3\left\{1-\left(\frac{D^2}{3}-\frac{2}{3}D\right)\right\}}x\right]$$

$$=\frac{1}{6}\left[-e^x\left\{1+\left(\frac{D^2}{3}+\frac{2}{3}D\right)\right\}x+e^{-x}\left\{1+\left(\frac{D^2}{3}-\frac{2}{3}D\right)\right\}x\right]$$

$$=\frac{1}{6}\left[-e^x\left(x+\frac{2}{3}\right)+e^{-x}\left(x-\frac{2}{3}\right)\right]=\frac{1}{6}\left[-x(e^x-e^{-x})-\frac{2}{3}(e^x+e^{-x})\right]$$

$$=\frac{1}{3}\left[-x\sin hx-\frac{2}{3}\cos hx\right]=-\frac{x}{3}\sin hx-\frac{2}{9}\cos hx.$$

Hence complete solution is $y=c_1e^{2x}+c_2e^{-2x}-\frac{x}{3}\sin hx-\frac{2}{9}\cos hx$.

2.4.5 Case V: When $Q = xV$ where V is the function of x

Then

$$P.I. = \frac{1}{f(D)}xV = x\frac{1}{f(D)}V + \left[\frac{d}{dD}\frac{1}{f(D)}\right]V.$$

In general,

$$\frac{1}{f(D)}x^m V = x^m \frac{1}{f(D)}V + mx^{m-1}\left[\frac{d}{dD}\frac{1}{f(D)}\right]V + \cdot$$

$$\frac{m(m-1)}{2!}x^{m-2}\left[\frac{d^2}{dD^2}\frac{1}{f(D)}\right]V + \ldots$$

2.4.6 Case VI: When Q is any other function of x, which does not discussed so far

Now first we factorizes $f(D)$ into linear factors as follows:

$$f(D) = (D-m_1)(D-m_2)(D-m_3)\ldots\ldots(D-m_n)$$

Thus $P.I. = \frac{1}{f(D)}Q = \frac{1}{(D-m_1)(D-m_2)(D-m_3)\ldots\ldots(D-m_n)}Q$

$$= \left(\frac{A_1}{D-m_1} + \frac{A_2}{D-m_2} + \ldots\ldots + \frac{A_n}{D-m_n}\right)Q \qquad \text{(by partial fraction)}$$

$$= A_1\frac{1}{D-m_1}Q + A_2\frac{1}{D-m_2}Q + \ldots\ldots A_n\frac{1}{D-m_n}Q$$

$$= A_1 e^{m_1 x}\int Qe^{-m_1 x}\,dx + A_2 e^{m_2 x}\int Qe^{-m_2 x}\,dx + \ldots + A_n e^{m_n x}\int Qe^{-m_n x}\,dx.$$

Example 13: Solve $(D^2+2D+1)y = x\cos x$.

Solution: Auxiliary equation is

$$m^2 + 2m + 1 = 0 \Rightarrow (m+1)^2 = 0 \Rightarrow m = -1, -1$$

Therefore $C.F. = (c_1 + c_2 x)e^{-x}$

and $P.I. = \frac{1}{D^2+2D+1}x\cos x = x\frac{1}{D^2+2D+1}\cos x - \frac{2D+2}{(D^2+2D+1)^2}\cos x$

$$= x\frac{1}{(-1+2D+1)}\cos x - \frac{2(D+1)}{4D^2}\cos x = \frac{x}{2}\sin x - \frac{1}{2}\left(D^{-1}+D^{-2}\right)\cos x$$

$$= \frac{x}{2}\sin x - \frac{1}{2}\sin x + \frac{1}{2}\cos x$$

Hence complete solution is $y = (c_1 + c_2 x)e^{-x} + \frac{\pi}{2}\sin x - \frac{1}{2}\sin x + \frac{1}{2}\cos x.$

Example 14: Solve $\frac{d^2y}{dx^2} + y = x^2\sin 2x.$

Solution: Auxiliary equation is $m^2 + 1 = 0 \Rightarrow m^2 = -1 \Rightarrow m = \pm i$

Therefore $C.F. = (c_1 \cos x + c_2 \sin x)$

$$P.I. = \frac{1}{D^2+1} x^2 \sin 2x = \frac{1}{D^2+1} x^2 \left(\frac{e^{2ix} - e^{-2ix}}{2i} \right)$$

$$= \frac{1}{2i}\left[e^{2ix} \left(\frac{1}{(D+2i)^2+1} x^2 \right) - e^{-2ix} \left(\frac{1}{(D^2-2i)^2+1} x^2 \right) \right]$$

$$= \frac{1}{2i}\left[e^{2ix} \left(\frac{1}{D^2-4+4iD+1} x^2 \right) - e^{-2ix} \left(\frac{1}{D^2-4-4iD+1} x^2 \right) \right]$$

$$= \frac{1}{2i}\left[\frac{e^{2ix}}{-3} \left\{ 1 - \left(\frac{D^2}{3} + \frac{4}{3} iD \right) \right\}^{-1} x^2 + \frac{e^{-2ix}}{3} \left\{ 1 - \left(\frac{D^2}{3} - \frac{4}{3} iD \right) \right\}^{-1} x^2 \right]$$

$$= \frac{1}{2i}\left[\frac{e^{2ix}}{-3} \left\{ 1 + \left(\frac{D^2}{3} + \frac{4iD}{3} \right) + \left(\frac{D^2}{3} + \frac{4iD}{3} \right)^2 \ldots \right\} x^2 + \frac{e^{-2ix}}{3} \left\{ 1 \right. \right.$$

$$\left. \left. + \left(\frac{D^2}{3} - \frac{4}{3} iD \right) + \left(\frac{D^2}{3} - \frac{4iD}{3} \right)^2 \ldots \right\} x^2 \right]$$

$$= \frac{1}{2i}\left[\frac{e^{2ix}}{-3} \left(x^2 + \frac{2}{3} + \frac{8ix}{3} - \frac{16.2}{9} \right) + \frac{e^{-2ix}}{3} \left(x^2 + \frac{2}{3} - \frac{8ix}{3} - \frac{16.2}{9} \right) \right]$$

$$= -\frac{e^{2ix} - e^{-2ix}}{2i} \frac{x^2}{3} - \frac{4x}{9} (e^{2ix} + e^{2ix}) + \frac{26}{27} \left(\frac{e^{2ix} - e^{-2ix}}{2i} \right)$$

$$= -\frac{x^2}{3} \sin 2x - \frac{8}{9} x \cos 2x + \frac{26}{27} \sin 2x.$$

Hence complete solution is

$$y = c_1 \cos x + c_2 \sin x - \frac{x^3}{3} \sin 2x - \frac{16}{9} x \cos 2x + \frac{26}{27} \sin 2x.$$

Example 15: Solve $(D^2 + a^2)\, y = \sec ax$.

Solution: A.E. is $m^2 + a^2 = 0 \Rightarrow m = \pm ai$

Therefore $C.F. = c_1 \cos ax + c_2 \sin ax$

Now P.I. $= \dfrac{1}{D^2 + a^2} \sec ax = \dfrac{1}{(D+ia)(D-ia)} \sec ax$

$$=\frac{1}{2ai}\left[\frac{1}{D-ai}-\frac{1}{D+ai}\right]\sec ax$$

$$=\frac{1}{2ai}\left[\frac{1}{D-ai}\sec ax-\frac{1}{D-ai}\sec ax\right]=\frac{1}{2ai}(A_1-A_2)$$

Where $A_1=\frac{1}{D-ai}\sec ax=e^{iax}\int e^{-iax}\sec ax\,dx$

$$=e^{iax}\int(\cos ax-i\sin ax)\sec ax\,dx=e^{iax}\int(1-i\tan ax)dx$$

$$=e^{iax}\left(x+\frac{i}{a}\log\cos ax\right)$$

and $A_2=\frac{1}{(D+ai)}\sec ax=e^{-iax}\int e^{iax}\sec ax\,dx$

$$=e^{-iax}\int(\cos ax+i\sin ax)\sec ax\,dx$$

$$=e^{-iax}\int(1+i\tan ax)dx=e^{-iax}\left(x-\frac{i}{a}\log\cos ax\right)$$

Now $P.I.=\frac{1}{2ia}[A_1-A_2]$

$$=\frac{1}{2ia}\left[e^{iax}\left(x+\frac{i}{a}\log\cos ax\right)-e^{-iax}\left(x-\frac{i}{a}\log\cos ax\right)\right]$$

$$=\frac{1}{2ia}\left[x(e^{iax}-e^{-iax})+\frac{i}{a}\log(\cos ax)(e^{iax}+e^{-iax})\right]$$

$$=\frac{1}{a}\left[x\sin ax+\frac{1}{a}\cos ax\log\cos ax\right].$$

Complete solution is

$$y=c_1\cos ax+c_2\sin ax+\frac{1}{a}\left[x\sin ax+\frac{1}{a}\cos ax\log\cos ax\right].$$

EXERCISE 2.3

Solve the following differential equation:

1. $\frac{d^2y}{dx^2}-3\frac{dy}{dx}+2y=e^{5x}.$

2. $\frac{d^2y}{dx^2}+31\frac{dy}{dx}+240y=27e^{-x}$

3. $\frac{d^2y}{dx^2}+4y=\tan 2x.$

4. $\frac{d^2y}{dx^2}+y=\operatorname{cosec} x.$

5. $\dfrac{d^2y}{dx^2}+5\dfrac{dy}{dx} = e^{2x}$.

6. $\dfrac{d^2y}{dx^2}+y = \cos 2x$.

7. $\dfrac{d^3y}{dx^3}-3\dfrac{d^2y}{dx^2}+4\dfrac{dy}{dx}-2y=e^x \cos x$.

8. $\dfrac{d^2y}{dx^2}-4y = e^x+\sin 2x$.

9. Find the integral of the equation $\dfrac{d^2y}{dx^2}-2n\cos\alpha\dfrac{dx}{dt}+n^2x=a\cos nt$, where $x=0$ and $\dfrac{dx}{dt}=0$, when $t=0$.

10. $(D^4+2D^3-3D^2)y = 3e^{2x}+4\sin x$.

11. $y''-2y'+2y= x+e^x \cos x$.

12. $\dfrac{d^2y}{dx^2}-2\dfrac{dy}{dx}+2y = xe^{3x}+\sin 2x$.

13. $(D^2-4D+4)y= e^{2x}+x^2+\cos 2x$.

14. $\dfrac{d^2y}{dx^2}+y = \sin 3x \cos 2x$.

15. $\dfrac{d^2y}{dx^2}-2\dfrac{dy}{dx}-3y = 2e^{2x}+10\sin 3x$ given that $y(0)= 2, y'(0)=4$.

16. $(D^2-4D+3)y = 2xe^{3x}+3e^x \cos 2x$.

17. $(D^2-4D+13)y=e^{2x}\cos 3x+(x^2+x+9)$.

18. $\dfrac{d^2y}{dx^2}+2\dfrac{dy}{dx}+y = \dfrac{e^{-x}}{x^2}$.

19. $\dfrac{d^2y}{dx^2}-\dfrac{dy}{dx}-6y=e^x \cosh x$.

20. $(D-1)(D+1)^2 y = e^{2x}+x^2$.

21. $\dfrac{d^2y}{dx^2}+6\dfrac{dy}{dx}+9y = 5e^{3x}$.

22. $(D^2-3D+2)y = 2x^2e^{4x}+5e^{3x}$.

23. $\dfrac{d^2y}{dx^2}-2\dfrac{dy}{dx}+y = xe^x \sin x$.

24. $(D^2-4D+4)y=8x^2 e^{2x} \sin 2x$.

25. $(D-1)^2(D^2+1)^2 y = \sin^2\frac{1}{2}x + e^x.$

26. $(D^2-1)y = \sinh x \cosh x \cos x.$

27. $\frac{d^4y}{dx^4} + 2\frac{d^2y}{dx^2} + y = x^2 \cos x.$

28. $\frac{d^2y}{dx^2} + 2\frac{dy}{dx} + 2y = e^{-x}\sec^3 x.$

29. $\frac{d^2y}{dx^2} + y = x - \cot x.$

30. Find the solution of the equation $\frac{d^2y}{dx^2} - y = 1$, which vanishes when $x = 0$ and tends to a finite limit as $x \to \infty$.

31. $(D^3 - 6D^2 + 11D - 6)y = e^{-2x} + e^{-3x}.$

ANSWERS 2.3

1. $y = c_1 e^x + c_2 e^{2x} + \frac{1}{12}e^{5x}.$

2. $y = c_1 e^{-15x} + c_2 e^{-16x} + \frac{136}{105}e^{-x}.$

3. $y = c_1 \cos 2x + c_2 \sin 2x - \frac{1}{4}\cos 2x \log \tan\left(x + \frac{\pi}{4}\right).$

4. $y = c_1 \cos x + c_2 \sin x + \sin \log \sin x - x \cos x.$

5. $y = c_1 e^{-2x} + c_2 e^{-3x} + \frac{1}{20}e^{2x}.$

6. $y = c_1 \cos(x + c_2) - \frac{1}{3}\cos 2x.$

7. $y = c_1 e^x + e^x(c_2 \cos x + c_3 \sin x) + xe^x + \frac{1}{10}(3\sin x + \cos x).$

8. $y = c_1 e^{2x} + c_2 e^{-2x} - \frac{1}{3}e^x - \frac{1}{8}\sin 2x.$

9. $x = -ae^{nt\cos\alpha}\frac{\sin(nt\sin\alpha)}{n^2\sin 2\alpha} + \frac{a\sin nt}{2n^2\cos\alpha}.$

10. $y = c_1 + c_2 x + c_3 e^x + c_4 e^{-3x} + \frac{3}{20}e^{2x} + \frac{2}{5}(\cos x - 2\sin x).$

11. $y = e^x(c_1 \cos x + c_2 \sin x) + \frac{1}{2}(x+1) + \frac{1}{2}xe^x \sin x.$

12. $y = c_1 e^x + c_2 e^{2x} + \frac{e^{3x}}{4}(2x-3) + \frac{3}{20}\cos x - \frac{1}{20}\sin 2x.$

13. $y=(c_1+c_2x)e^{2x}+\frac{1}{2}x^2e^{2x}+\frac{1}{8}(2x^3+6x^2+9x+6)-\frac{1}{8}\sin 2x.$

14. $y=c_1\cos x+c_2\sin x+\frac{1}{48}(-\sin 5x-12x\cos x).$

15. $y=\frac{29}{12}e^{3x}-\frac{1}{12}e^{-x}-\frac{2}{3}e^{2x}+\frac{1}{3}(\cos 3x-2\sin 3x).$

16. $y=c_1e^x+c_2e^{3x}+e^{3x}\left(\frac{x^2}{2}-\frac{x}{2}\right)-\frac{3}{8}e^x(\cos 2x+\sin 2x).$

17. $y=e^{2x}(c_1\cos 3x+c_2\sin 3x)+\frac{xe^{2x}\sin 3x}{6}+\frac{1}{13}\left(x^2+\frac{21}{13}x+\frac{1579}{169}\right).$

18. $y=(c_1+c_2x)e^{-x}-e^{-x}\log x.$

19. $y=(c_1+c_2+c_3x^2)e^{-x}+\frac{1}{6}x^3e^{-x}.$

20. $y=c_1e^{2x}+(c_1+c_3x)e^{-x}+\frac{\pi}{9}e^{2x}-\frac{e^x}{4}$

21. $y=(c_1+c_2x)+\frac{5}{36}e^{3x}.$

22. $y=c_1e^x+c_xe^{2x}+\frac{e^{3x}}{5}\left[x^2+\frac{12}{5}x-\frac{62}{25}\right].$

23. $y=(c_1+c_2x)e^x-e^x\left[x\sin x+\cos x\right].$

24. $y=c_1+c_2x+3\sin 2x-2x^2\sin 2x-4xe^{2x}\cos 2x.$

25. $y=\left(c_1+c_2x+\frac{1}{8}x^2\right)e^x+(c_3+c_4x)\sin x+(c_5+c_6x)\cos x.$

26. $y=c_1e^x+c_2e^{-x}+\frac{1}{20}(2\sin x\cosh 2x+\cos x\sinh 2x)-\frac{1}{32}x^2\sin x+\frac{1}{2}$

27. $y=(c_1+c_2x)\sin x+(c_3+c_4x)\cos x+\frac{1}{12}x^3\sin x+\frac{1}{48}(9x^2-x^4)\cos x.$

28. $y=e^{-x}\left[c_1\cos x+c_2\sin x+\frac{\sin x\tan x}{2}\right].$

29. $y=c_1\cos x+c_2\sin x-x\cos^2 x-\sin x\log(\operatorname{cosec} x-\cot x).$

30. $y=e^x-1.$

31. $y=c_1e^x+c_2e^{2x}+c_3e^{3x}-\frac{1}{120}(2e^{-2x}+e^{-3x})$

2.5 HOMOGENEOUS LINEAR DIFFERENTIAL EQUATIONS (EULER-CAUCHY EQUATIONS)

A differential equation of the form

$$x^n \frac{d^n y}{dx^n} + a_1 x^{n-1} \frac{d^{n-1} y}{dx^{n-1}} + \ldots + a_{n-1} x \frac{dy}{dx} + a_n y = Q,$$

where $a_1, a_2, a_3, \ldots a_n$ are constant and Q is a function of x, is called a homogeneous linear differential equation or Euler-Cauchy linear equation.

Now reduce the equation (1) into linear differential equation with constant coefficients by replacing the independent variable x, as follows:

Set $z = \log x$ or $x = e^z$, so that $\frac{dy}{dx} = \frac{dy}{dz}\frac{dz}{dx} = \frac{dy}{dx}\frac{1}{x}$

or $\quad x\frac{dy}{dx} = \frac{dy}{dz} = Dy$, where $D = \frac{d}{dz} = x\frac{d}{dx} \Rightarrow \frac{d}{dx} = \frac{1}{x}\frac{d}{dz}$

Now $\frac{d^2y}{dx^2} = \frac{1}{x}\frac{d}{dz}\left(\frac{dy}{dz}\cdot\frac{dz}{dx}\right) = \frac{1}{x}\frac{d}{dz}\left(\frac{1}{x}\frac{dy}{dz}\right) = \frac{1}{x^2}\left(\frac{d^2y}{dz^2} - \frac{dy}{dz}\right)$

$\Rightarrow \quad x^2\frac{d^2y}{dx^2} = D^2y - Dy = D(D-1)y.$

Similarly, we can find $x^3\frac{d^3y}{dx^3} = D(D-1)(D-2)y$ and so on.

On substituting these values in equation (1), we get the linear differential equation with constant coefficients.

Steps for Solution

(i) Put $x = e^z$ so that $z = \log x$, $D \equiv \frac{d}{dz}$

(ii) Replace

$x\frac{d}{dz}$ by D, $x^2\frac{d^2}{dx^2}$ by $D(D-1)$, $x^3\frac{d^3}{dx^3}$ by $D(D-1)(D-2)$ and so on.

(iii) On substituting all these values in equation (1), the equation becomes linear differential equation with constant co-efficient. Now solve this equation by the methods, discussed earlier.

2.6 LEGENDRE'S LINEAR DIFFERENTIAL EQUATIONS

A differential equation of the form

$$(a+bx)^n \frac{d^n y}{dx^n} + a_1 (a+bx)^{n-1} \frac{d^{n-1} y}{dx^{n-1}} + \ldots + a_{n-1}(a+bx)\frac{dy}{dx} + a_n y = Q \quad \ldots(1)$$

where $a_1, a_2, \ldots, a_n$ are constants and Q is a function of x, is called Legendre's linear differential equation.

Now, we reduce the equation (1) into the linear differential equation with constant coefficients by substituting $a + bx = e^z$ i.e. $z = \log(a+bx)$.

So that $\dfrac{dy}{dx} = \dfrac{dy}{dz} \cdot \dfrac{dz}{dx} = \dfrac{b}{a+bx} \dfrac{dy}{dz}$

$$\Rightarrow \quad (a+bx)\frac{dy}{dx} = b\frac{dy}{dz} = bDy, \text{ where } D = \frac{d}{dz}$$

$$\frac{d^2y}{dx^2} = \frac{d}{dx}\left(\frac{dy}{dx}\right) = \frac{d}{dx}\left(\frac{b}{a+bx}\frac{dy}{dz}\right) = \frac{-b^2}{(a+bx)^2}\frac{dy}{dz} + \frac{b}{(a+bx)}\frac{d^2y}{dz^2}\frac{dz}{dx}$$

$$= \frac{-b^2}{(a+bx)^2}\frac{dy}{dz} + \frac{b^2}{(a+bx)^2}\frac{d^2y}{dz^2}$$

$$(a+bx)^2\frac{d^2y}{dx^2} = b^2(D^2y - Dy) = b^2D(D-1)y.$$

Similarly, we can find

$$(a+bx)^3\frac{d^3y}{dx^3} = b^3D(D-1)(D-2)y, \text{ and so on.}$$

Substituting these values in equation (1), we get the linear differential equation with constant coefficient.

Steps for Solution

(i) Put $(a+bx) = e^z$ *i.e.* $z = \log(a+bx)$

(ii) Replace in the equation

$$(a+bx)\frac{d}{dx} \quad by \quad bD,$$

$$(a+bx)^2\frac{d}{dx^2} \quad by \quad b^2D(D-1),$$

$$(a+bx)^3\frac{d}{dx^3} \quad by \quad b^3D(D-1)(D-2) \text{ and so on.}$$

(iii) Substituting these values in equation (1), the equation reduced to linear differential equation with constant coefficients. Now solve this equation by the methods, discussed earlier.

Remarks: One can solve Legendre's linear differential equation first reducing in the homogenous equation and then transforming in the linear differential equation with constant coefficients. So, these equations are also known as reducible to homogeneous linear differential equations.

Example 1: Solve $x^3\dfrac{d^3y}{dx^3} + 2x^2\dfrac{d^2y}{dx^2} + 3x\dfrac{dy}{dx} - 3y = x^2 + x.$

Solution: Let $x = e^z$ *i.e.*, $z = \log x$, then we have

$$x\frac{dy}{dx} = Dy, \; x^2\frac{d^2y}{dx^2} = D(D-1)y, \; x^3\frac{d^3y}{dx^3} = D(D-1)(D-2)y.$$

Thus equation given equation becomes

$$D(D-1)(D-2)y+2D(D-1)y+3Dy-3y=e^{2z}+e^{z}$$

$\Rightarrow \quad (D^3-D^2+3D-3)y=e^{2z}+e^{z}$, which is a L.D.E. with constant coefficients.

A.E. is $m^3-m^2+3m-3=0 \Rightarrow m=1, \pm\sqrt{3}\,i$

Now $C.F.=c_1e^{z}+c_2(\cos\sqrt{3}\,z+c_3\sin\sqrt{3}\,z)$

and $P.I.=\dfrac{1}{D^3-D^2+3D-3}e^{2z}+\dfrac{1}{D^3-D^2+3D-3}e^{z}$

$$=\frac{1}{8-4+6-3}e^{2z}+e^{z}\frac{1}{(D+1)^3-(D-1)^2+3(D+1)-3}1$$

$$=\frac{1}{7}e^{2z}+e^{z}\frac{1}{(D^3+2D^2+4D}.1=\frac{1}{7}e^{2z}+\frac{e^{z}}{4D}\left\{1+\left(\frac{D^2}{4}+\frac{1}{2}D\right)\right\}^{-1}.1$$

$$=\frac{1}{7}e^{2z}+\frac{e^{z}}{4}D^{-1}\left(1-\frac{D^2}{4}-\frac{1}{2}D...\right).1=\frac{1}{7}e^{2z}+\frac{e^{z}}{4}\left(D^{-1}(1)-\frac{1}{4}D-\frac{1}{2}\right).1$$

$$=\frac{1}{7}e^{2z}+\frac{e^{z}}{4}\left(z-\frac{1}{2}\right).$$

Complete solution is

$$y=c_1e^{z}+(c_2\cos\sqrt{3}z+c_3\sin\sqrt{3}\,z)+\frac{1}{7}e^{2z}+\frac{e^{z}}{4}\left(z-\frac{1}{2}\right)$$

$$=c_1x+\left[c_2\cos\left(\sqrt{3}\log x\right)+c_3\sin\left(\sqrt{3}\log x\right)\right]+\frac{1}{7}x^2+\frac{x}{4}\left(\log x-\frac{1}{2}\right).$$

Example 2: Solve $(x^2D^2+3xD+1)y=\dfrac{1}{(1-x)^2}, D=\dfrac{d}{dx}$

Solution: Putting $x=e^{z}$ so that $z=\log x$ and $D_1=\dfrac{d}{dz}=xD$, we get

$$[D_1(D_1-1)+3D_1+1]y=\frac{1}{(1-e^{Z})^2} \Rightarrow \qquad (D_1+1)^2y=\frac{1}{(1-e^{z})^2}$$

A.E. is $(m+1)^2=0 \Rightarrow m=-1,-1$ Thus $C.F.=(c_1+c_2z)e^{-z}$

and $P.I.=\dfrac{1}{(D_1+1)^2}\dfrac{1}{(1-e^{z})^2}=\dfrac{1}{D_1+1}e^{-z}\displaystyle\int\frac{e^{z}}{(1-e^{z})^2}dz$

Let $\quad 1-e^{z}=t \Rightarrow e^{z}\,dz=-dt$, we obtain

$$P.I.=\frac{1}{D_1+1}e^{-z}\int\frac{-dt}{t^2}=\frac{1}{D_1+1}\frac{e^{-z}}{1-e^{z}}=e^{-z}\int\frac{e^{z}.e^{-z}}{(1-e^{z})}dz=e^{-z}\int\frac{e^{-z}}{(e^{-z}-1)}dz$$

$$=e^{-z}\left[-\log(e^{-z}-1)\right]=e^{-z}\log\left(\frac{e^{z}}{1-e^{z}}\right).$$

Hence complete solution is $y=(c_1+c_2 z)e^{-z}+e^{-z}\log\left(\dfrac{e^z}{1-e^z}\right)$

$$=(c_1+c_2\log x)x^{-1}+x^{-1}\log\frac{x}{1-x}$$

where c_1 and c_2 are arbitrary constants of integration.

Example 3: Solve $(3x+2)^2\dfrac{d^2y}{dx^2}+3(3x+2)\dfrac{dy}{dx}-36y=3x^2+4x+1.$

Solution: Let $3x+2=e^z$ *i.e.*, $z=\log(3x+2)$, so that

$$(3x+2)\frac{dy}{dx}=3Dy,\ (3x+2)^2\frac{d^2y}{dx^2}=3^2D(D-1)y,\text{ where } D=\frac{d}{dz}.$$

Thus given equation reduces to

$$\left[3^2D(D-1)+3.3D-36\right]y=3\left(\frac{e^z-2}{3}\right)^2+4\left(\frac{e^z-2}{3}\right)+1$$

$$\Rightarrow\quad 9\left(D^2-4\right)y=\frac{1}{3}e^{2z}-\frac{1}{3}\Rightarrow\left(D^2-4\right)y=\frac{1}{27}\left(e^{2z}-1\right)$$

which is a linear equation with constant coefficient.

Its auxiliary equation is $m^2-4=0\Rightarrow m=\pm 2$.

Therefore $C.F.=c_1e^{2z}+c_2e^{-2z}=c_1(3x+2)^2+c_2(3x+2)^{-2}$

and $P.I.=\dfrac{1}{27}\dfrac{1}{D^2-4}\left(e^{2z}-1\right)=\dfrac{1}{27}\left[\dfrac{1}{D^2-4}e^{2z}-\dfrac{1}{D^2-4}e^{0z}\right]$

$$=\frac{1}{27}\left[z.\frac{1}{2D}e^{2z}-\frac{1}{0-4}e^{0z}\right]=\frac{1}{27}\left[\frac{z}{2}\int e^{2z}dz+\frac{1}{4}\right]$$

$$=\frac{1}{27}\left[\frac{z}{4}e^{2z}+\frac{1}{4}\right]=\frac{1}{108}\left[(3x+2)^2\log(3x+2)+1\right].$$

Hence complete solution is

$$y=c_1(3x+2)^2+c_2(3x+2)^{-2}+\frac{1}{108}\left[(3x+2)^2\log(3x+2)+1\right].$$

EXERCISE 2.4

1. $x^2\dfrac{d^2y}{dx^2}-4x\dfrac{dy}{dx}+6y=\dfrac{42}{x^4}.$

2. $\dfrac{d^2y}{dx^2}+\dfrac{1}{x}\dfrac{dy}{dx}=\dfrac{12\log x}{x^2}.$

3. $x^2\dfrac{d^2y}{dx^2}+2x\dfrac{dy}{dx}-2y=(x+1)^2.$

4. $x^4 \frac{d^4y}{dx^4} + 6x^3 \frac{d^3y}{dx^3} + 9x^2 \frac{d^2y}{dx^2} + 3x\frac{dy}{dx} + y = (1+\log x)^2.$

5. $x^2 \frac{d^3y}{dx^3} - 4x\frac{d^2y}{dx^2} + 6\frac{dy}{dx} = 4.$

6. $(x^3D^3 + 2xD - 2)y = x^2 \log x + 3x.$

7. $x^2 \frac{d^2y}{dx^2} - 2x\frac{dy}{dx} + 2y = x + x^2 \log x + x^3.$

8. $x^2 \frac{d^2y}{dx^2} + 9x\frac{dy}{dx} + 25y = 50.$

9. $(x+1)^2 \frac{d^2y}{dx^2} - 3(x+1)\frac{dy}{dx} + 4y = x^2.$

10. $(1+x)^2 \frac{d^2y}{dx^2} + (1+x)\frac{dy}{dx} + y = 2\sin\log(1+x).$

11. $(2x+3)^2 \frac{d^2y}{dx^2} - 2(2x+3)\frac{dy}{dx} - 12y = 6x.$

12. $16(x+1)^4 \frac{d^4y}{dx^4} + 96(x+1)^3 \frac{d^3y}{dx^3} + 104\frac{d^2y}{dx^2} + 8(x+1)\frac{dy}{dx} + y = x^2 + 4x + 3.$

13. Obtain general solution of the differential equation
$x^2y'' + xy' - y = x^2e^x.$

14. $x^2 \frac{d^2y}{dx^2} + x\frac{dy}{dx} + y = \sin(\log x^2).$

15. $x^3 \frac{d^3y}{dx^3} + 2x^2 \frac{d^2y}{dx^2} + 2y = 10\left(x + \frac{1}{x}\right).$

ANSWERS 2.4

1. $y = c_1x^2 + c_2x^3 + \frac{1}{x^4}$

2. $y = c_1 + c_2 \log x + 2(\log x)^2.$

3. $y = c_1x^{-5} + c_5x^{-4} - \frac{x^2}{14} - \frac{x}{9} - \frac{1}{20}.$

4. $y = (c_1 + c_2 \log x)\cos(\log x) + (c_3 + c_4 \log x)\sin(\log x) + (\log x)^2$
$+ 2\log x - 3.$

5. $y = c_1 + c_2x^3 + c_3x^4 + \frac{2}{3}x.$

6. $y = c_1 x + c_2 x \cos(\log x + c_3) + \frac{1}{2}x^2(\log x - 2) + 3x\log x.$

7. $y = (c_1 - \log x)x + \left[c_2 - \log x + \frac{1}{2}(\log x)^2\right]x^2 + \frac{1}{2}x^3.$

8. $y = c_1 x^{-4}\cos(3\log x + c_2) + 2.$

9. $y = c_1 + c_2\log(1+x) + \frac{1}{2}\left[[\log(1+x)]^2(x+1)^2 - 2(x+1) + \frac{1}{4}\right].$

10. $y = c_1\cos\log(1+x) + c_2\sin\log(1+x) - \log(1+x)\log(1+x).$

11. $y = c_1(2x+3)^{-1} + c_2(2x+3)^3 - \frac{3}{4}(2x+3) + 3.$

12. $y = (c_1 + c_2 z)e^{z/2} + (c_3 + c_4 z)e^{-z/2} + \frac{2}{9}e^z + \frac{1}{225}e^{2z},$

where $z = \log(1+x).$

13. $y = c_1 x + c_2 x^{-1} + \left[x - 3 + \frac{3}{x}\right]e^x.$

14. $y = c_1\cos(\log x) + c_2\sin(\log x) - \frac{1}{3}\sin(\log x^2).$

15. $y = \frac{c_1}{x} + x\left[c_2\cos(\log x) + c_3\sin(\log x)\right] + 5x + \frac{2}{x}\log x.$

2.7 SIMULTANEOUS LINEAR DIFFERENTIAL EQUATIONS WITH CONSTANT COEFFICIENTS

Such linear differential equations in which there are one independent variable and two or more than two dependent variables are called simultaneous linear differential equations. To solve such equations, we must have as many simultaneous equations as the number of dependent variable. Here we shall consider equations with constant coefficients only.

Let x, y be the two dependent variables and t the independent variable. Consider the simultaneous equations

$$f_1(D)x + f_2(D)y = T_1 \qquad ...(1)$$

and $$\phi_1(D)x + \phi_2(D)y = T_2 \qquad ...(2)$$

where $D = \frac{d}{dt}$ and T_1, T_2 are function of t. To eliminate y, we operate $\phi_2(D)$ on both sides of (1) and $f_2(D)$ on both side of (2) and then, subtract to get

$$\left[f_1(D)\phi_2(D)-\phi_2(D)f_2(D)\right]x=\phi_2(D)T_1-f_2(D)T_2$$

which is a linear equation in x and t and can be solved by the methods already discussed.

Substituting the value of x in either (1) or (2), we get the value of y.

Example 1: Solve $\frac{dx}{dt}+7x-y=0,\ \frac{dy}{dt}+2x+5y=0.$

Solution. Let $D\equiv\frac{d}{dt}$, then the given system of equation reduces to

$$(D+7)x-y=0 \qquad ...(1)$$
$$2x+(D+5)y=0. \qquad ...(2)$$

Now operating equation (1) by $(D+5)$ and adding in (2), we get

$$(D+7)(D+5)x+2x=0 \Rightarrow \left(D^2+12D+37\right)x=0$$

A.E. is $m^2+12m+37=0 \Rightarrow m=\frac{-12\pm\sqrt{144-148}}{2} \Rightarrow m=-6\pm i$

Therefore C.F. $=e^{-6t}\left(c_1\cos t+c_2\sin t\right)$.

This gives $x=e^{-6t}\left(c_1\cos t+c_2\sin t\right)$. ...(3)

From (1), we have

$$y=\frac{d}{dt}\left[e^{-6t}\left(c_1\cos t+c_2\sin t\right)\right]+7\left[e^{-6t}\left(c_1\cos t+c_2\sin t\right)\right]$$

$$=-6e^{-6t}\left(c_1\cos t+c_2\sin t\right)+e^{-6t}\left(-c_1\sin t+c_2\cos t\right)$$
$$+7\left[e^{-6t}\left(c_1\cos t+c_2\sin t\right)\right]$$

$$=c_1e^{-6t}\cos t+c_2e^{-6t}\cos t+c_2e^{-6t}\cos t-c_2e^{-6t}\sin t$$

$$=\left(c_1+c_2\right)e^{-6t}\cos t+\left(c_2-c_1\right)e^{-6t}\sin t$$

$$\Rightarrow \quad y=Ae^{-6t}\cos t+Be^{-6t}\sin t \qquad ...(4)$$

where $A=c_1+c_2$ and $B=c_2-c_1$.

Equations (3) and (4) give the complete solution.

Example 2: Solve $\frac{dx}{dt}+2x-3y=t,\ \frac{dy}{dt}-3x+2y=e^{2t}.$

Solution: Substituting $D=\frac{d}{dt}$, equations reduce to

$$(D+2)x-3y=t \qquad ...(1)$$
$$-3x+(D+2)y=e^{2t}. \qquad ...(2)$$

Operating equation (1) by $D+2$ and multiplying (2) by 3 and adding, we get

$$\{(D+2)^2-9\}x=(D+2)t+3e^{2t}$$
$$(D^2+4D-5)x=(2t+1)+3e^{2t}.$$

A.E. is $m^2+4m-5=0 \Rightarrow m=1,-5$

Therefore $C.F.=c_1e^t+c_2e^{-5t}$

$$P.I.=\frac{1}{D^2+4D-5}(3e^{2t}+2t+1)=\frac{1}{D^2+4D-5}3e^{2t}+\frac{1}{D^2+4D-5}(2t+1)$$

$$=\frac{1}{7}3e^{2t}-\frac{1}{5}\left[1-\left(\frac{D^2+4D}{5}\right)\right]^{-1}(2t+1)=\frac{3}{7}e^{2t}-\frac{1}{5}(2t+1)-\frac{4}{25}D(2t+1)$$

$$=\frac{3}{7}e^{2t}-\frac{1}{5}(2t+1)-\frac{8}{25}$$

Hence $x=c_1e^t+c_2e^{-5t}+\frac{3}{7}e^{2t}-\frac{2}{5}t-\frac{13}{25}.$ ...(3)

Substituting the value of x in (1), we have

$$3y=\frac{d}{dt}\left(c_1e^t+c_2e^{-5t}+\frac{3}{7}e^{2t}-\frac{2}{5}t-\frac{13}{25}\right)+2\left(c_1e^t+c_2e^{-5t}+\frac{3}{7}e^{2t}-\frac{2}{5}t-\frac{13}{25}\right)-t$$

$$=\left(c_1e^t-5c_2e^{-5t}+\frac{6}{7}e^{2t}-\frac{2}{5}\right)+2\left(c_1e^t+c_2e^{-5t}+\frac{3}{7}e^{2t}-\frac{2}{5}t-\frac{13}{25}\right)-t$$

$$=\left[3c_1e^t-3c_2e^{-5t}+\frac{12}{7}e^{2t}-\frac{9}{5}t-\frac{36}{25}\right].$$ ...(4)

Equations (3) and (4) give the complete solution.

Example 3: Solve $\frac{d^2x}{dt^2}+4x+5y=t^2$, $\frac{d^2y}{dt^2}+5x+4y=t+1.$

Solution: Putting $D=\frac{d}{dt}$, the given equations become

$(D^2+4)x+5y=t^2$ ...(1)

$5x+(D^2+4)y=t+1.$ ...(2)

To eliminate y, multiplying both side of (1) by (D^2+4) and both side of (2) by 5 and subtracting, we get

$$\left[(D^2+4)^2-25\right]x=(D^2+4)t^2-5(t+1)$$

$$\Rightarrow \quad (D^4+8D^2-9)x=4t^2-5t-3.$$

A.E. is

$$m^4+8m^2-9=0 \Rightarrow (m^2+9)(m^2-1)=0 \Rightarrow m=\pm 3i,\ \pm 1.$$

Hence $C.F.=c_1e^t+c_2e^{-t}+c_3\cos 3t+c_4\sin 3t$

$$P.I. = \frac{1}{D^4+8D^2-9}\left(4t^2-5t-3\right)$$

$$= \frac{1}{-9\left[1-\left(\frac{D^4+8D^2}{9}\right)\right]}\left(4t^2-5t-3\right)$$

$$= -\frac{1}{9}\left[1-\left(\frac{D^4+8D^2}{9}\right)\right]^{-1}\left(4t^2-5t-3\right) = -\frac{1}{9}\left[1+\frac{D^4+8D^2}{9}+\ldots\right]\left(4t^2-5t-3\right)$$

$$= -\frac{1}{9}\left[4t^2-5t-3+\frac{8}{9}(8)\right] = -\frac{1}{9}\left[4t^2-5t+\frac{37}{9}\right].$$

Therefore

$$x = c_1e^t + c_2e^{-t} + c_3\cos 3t + c_4\sin 3t - \frac{4t^2}{9} + \frac{5t}{9} - \frac{37}{81}. \qquad \ldots(3)$$

Now $\dfrac{dx}{dt} = c_1e^t - c_2e^{-t} - 3c_2\sin 3t + 3c_4\cos 3t - \dfrac{8}{9}t + \dfrac{5}{9}$

and $\dfrac{d^2x}{d+2} = c_1e^t + c_2e^{-t} - 9c_3\cos 3t - 9c_4\sin 3t - \dfrac{8}{9}$

Substituting the value of x and $\dfrac{d^2x}{dt^2}$ in (1), we obtain

$$5y = t^2 - 4c_1e^t - 4c_2e^{-t} - 4c_3\cos 3t - 4c_4\sin 3t + \frac{16t^2}{9} - \frac{20}{9}t + \frac{148}{81}$$

$$-c_1e^t - c_2e^{-t} + 9c_3\cos 3t + 9c_4\sin 3t + \frac{8}{9}$$

$$\Rightarrow \quad y = \frac{1}{5}\left[-5c_1e^t - 5c_2e^{-t} + 5c_3\cos 3t + 5c_4\sin 3t + \frac{25}{9}t^2 - \frac{20}{9}t + \frac{220}{81}\right].$$

$$= -c_1e^t - c_2e^{-t} + c_3\cos 3t + c_4\sin 3t + \frac{1}{9}\left(5t^2 - 4t + \frac{44}{9}\right). \qquad \ldots(4)$$

Equations (3) and (4) give the complete solution.

Example 4: Solve the following simultaneous equations

$$\frac{dx}{dt} = 2y,\ \frac{dy}{dt} = 2z,\ \frac{dz}{dt} = 2x.$$

Solution: The given equations are

$$\frac{dx}{dt} = 2y \qquad \ldots(1)$$

$$\frac{dy}{dt} = 2z \qquad \ldots(2)$$

$$\frac{dz}{dt} = 2x. \qquad \ldots(3)$$

Differentiating (1) w. r. t. 't' we get $\frac{d^2x}{dt^2} = 4\frac{dy}{dt} = 2(2z)$.

Differentiating again w. r. t. 't', we obtain

$$\frac{d^3x}{dt^3} = 4\frac{dz}{dt} = 4(2x)$$

Or $\quad (D^3 - 8)x = 0$ where $D = \frac{d}{dt}$.

A.E. is $m^3 - 8 = 0 \Rightarrow (m-2)(m^2 + 2m + 4) = 0$.

Hence $m = 2, \frac{-2 \pm 2i\sqrt{3}}{2} \Rightarrow m = 2, -1 \pm i\sqrt{3}$.

Therefore $x = c_1 e^{2t} + c_2 e^{-t}\cos(\sqrt{3}t - c_3)$. ...(4)

From (1), we get

$$y = \frac{1}{2}\frac{dx}{dt} = \frac{1}{2}\left[2c_1 e^{2t} - c_2 e^{-t}\cos(\sqrt{3}\,t - c_3) - c_2\sqrt{3}e^{-t}\sin(\sqrt{3}t - c_3)\right].$$

$$= c_1 e^{2t} + c_2 e^{-t}\left[\cos\frac{2\pi}{3}\cos(\sqrt{3}t - c_3) - \sin\frac{2\pi}{3}\sin(\sqrt{3}t - c_3)\right]$$

$$= c_1 e^{2t} + c_2 e^{-t}\cos\left(t\sqrt{3} - c_3 + \frac{2\pi}{3}\right). \quad ...(5)$$

From (3), we get

$$z = \frac{1}{2}\frac{dy}{dt}$$

$$= \frac{1}{2}\left[2c_1 e^{2t} - c_2 e^{-t}\cos\left(\sqrt{3}\,t - c_3 + \frac{2\pi}{3}\right) - c_2\sqrt{3}e^{-t}\sin\left(\sqrt{3}t - c_3 + \frac{2\pi}{3}\right)\right]$$

$$= c_1 e^{2t} + c_2 e^{-t}\left[\cos\frac{2\pi}{3}\cos\left(\sqrt{3}\,t - c_3 + \frac{2\pi}{3}\right) - \sin\frac{2\pi}{3}\sin\left(\sqrt{3}t - c_3 + \frac{2\pi}{3}\right)\right]$$

$$= c_1 e^{2t} + c_2 e^{-t}\cos\left(\sqrt{3}\,t - c_3 + \frac{4\pi}{3}\right). \quad ...(6)$$

Equations (4), (5) and (6) give the complete solution.

Example 5: Solve

$$\frac{dx}{dt} + \frac{dy}{dt} - 2y = 2\cos t - 7\sin t,$$

$$\frac{dx}{dt} - \frac{dy}{dt} + 2x = 4\cos t - 3\sin t.$$

Solution: Putting $D = \frac{d}{dt}$, equations reduces to

$$Dx + (D-2)y = 2\cos t - 7\sin t. \quad ...(1)$$

$$(D+2)x - Dy = 4\cos t - 3\sin t. \qquad ...(2)$$

Multiplying (1) by D and (2) by $(D - 2)$, we get

$$D^2x + D(D-2)y = -2\sin t - 7\cos t$$

$$(D^2-4)x - D(D-2)y = -4\sin t - 8\cos t - 3\cos t + 6\sin t.$$

On adding these equations, we get

$$(2D^2-4)x = -18\cos t \Rightarrow (D^2-2)x = -9\cos t$$

A.E. is $m^2 - 2 = 0 \Rightarrow m = \pm\sqrt{2}$

Therefore $C.F. = c_1 e^{\sqrt{2}t} + c_2 e^{-\sqrt{2}t}$

$$P.I. = \frac{1}{D^2-2}(-9\cos t) = 3\cos t$$

Hence $x = c_1 e^{\sqrt{2}t} + c_2 e^{-\sqrt{2}t} + 3\cos t.$ ...(3)

Putting the value of x in (2), we get

$$\frac{dy}{dt} = \sqrt{2}\,c_1 e^{\sqrt{2}t} - \sqrt{2}\,c_2 e^{-\sqrt{2}t} - 3\sin t + 2c_1 e^{\sqrt{2}t} + 2c_2 e^{-\sqrt{2}t} + 6\cos t - 4\cos t + 3\sin t$$

$$\Rightarrow \quad \frac{dy}{dt} = (2+\sqrt{2})c_1 e^{\sqrt{2}t} + (2-\sqrt{2})c_2 e^{-\sqrt{2}t} + 2\cos t.$$

Integrating w. r. t. 't', we get

$$y = (\sqrt{2}+1)c_1 e^{\sqrt{2}t} - (\sqrt{2}-1)c_2 e^{-\sqrt{2}t} + 2\sin t + c_3. \qquad ...(4)$$

Equation (3) and (4), give the complete solution.

Example 6: Solve the simultaneous differential equations

$$\frac{dx}{dt} + \frac{dy}{dt} + 3x = \sin t, \quad \frac{dx}{dt} + y - x = \cos t.$$

Solution: Let $\frac{d}{dt} \equiv D$ then given system of equation reduces to

$$(D+3)x + Dy = \sin t \qquad ...(1)$$

$$(D-1)x + y = \cos t. \qquad ...(2)$$

Multiplying (2) by D and subtracting from (1), we get

$$[(D+3) - D(D-1)]x = \sin t - D\cos t$$

$$\Rightarrow \quad (-D^2+2D+3)x = 2\sin t.$$

A.E. is $-m^2 + 2m + 3 = 0 \Rightarrow m = -1, 3$

Therefore $C.F. = c_1 e^{-t} + c_2 e^{3t}$

$$P.I. = \frac{1}{-D^2+2D+3} 2\sin t = \frac{1}{2D+4} 2\sin t = \frac{1}{(D+2)}\sin t$$

$$= \frac{D-2}{D^2-4}\sin t = \frac{\cos t - 2\sin t}{-5}$$

Hence $x = c_1 e^{-t} + c_2 e^{3t} - \frac{1}{5}(\cos t - 2\sin t)$. ...(3)

From (2), we have

$$y = x + \cos t - \frac{d}{dt}\left[c_1 e^{-t} + c_2 e^{3t} - \frac{1}{5}(\cos t - 2\sin t)\right]$$

$$= c_1 e^{-t} + c_2 e^{3t} - \frac{1}{5}(\cos t - 2\sin t) + \cos t + c_1 e^{-t} - 3c_2 e^{3t} - \frac{1}{5}(\sin t + 2\cos t)$$

$$\Rightarrow \quad y = 2c_1 e^{-t} - 2c_2 e^{3t} + \frac{1}{5}\sin t + \frac{2}{5}\cos t. \qquad ...(4)$$

Equation (3) and (4) give the complete solution.

Example 7: Solve the simultaneous equations

$$t\frac{dx}{dt} + 2(x - y) = t, \quad t\frac{dy}{dt} + x + 5y = t^2.$$

Solution: Let $t = e^z$ or $z = \log t$ and $D_1 = \frac{d}{dz}$.

Then the equations become

$$(D_1 + 2)x - 2y = e^z \qquad ...(1)$$

$$x + (D_1 + 5)y = e^{2z}. \qquad ...(2)$$

Now multiplying (1) by $(D_1 + 5)$ and (2) by 2 and adding, we get

$$[(D_1 + 2)(D_1 + 5) + 2]x = (D_1 + 5)e^z + 2e^{2z} \Rightarrow (D_1^2 + 7D_1 + 12)x = 6e^z + 2e^{2z}.$$

A.E. is $m^2 + 7m + 12 = 0 \Rightarrow m = -3, -4$

Therefore $C.F. = c_1 e^{-3z} + c_2 e^{-4z}$

and $P.I. = \frac{1}{D^2 + 7D + 12}(6e^z + 2e^{2z}) = \frac{1}{1 + 7 + 12}6e^z + \frac{2e^{2z}}{4 + 14 + 12}$

$$= \frac{6e^z}{20} + \frac{2e^{2z}}{30} = \frac{3e^z}{10} + \frac{e^{2z}}{15}$$

Hence $x = c_1 e^{-3z} + c_2 e^{-4z} + \frac{3e^z}{10} + \frac{e^{2z}}{15}$ ...(3)

and $D_1 x = -3c_1 e^{-3z} - 4c_2 e^{-4z} + \frac{3e^z}{10} + \frac{2e^{2z}}{15}$ ...(4)

Now from (1) and (4), we get

$$y = -\frac{1}{2}c_1 e^{-3z} - c_2 e^{-4z} - \frac{1}{20}e^z + \frac{2}{15}e^{2z}. \qquad ...(5)$$

Putting $t = e^z$, (3) and (5) give the required general solution

$$x = c_1 t^{-3} + c_2 t^{-4} + \frac{3t}{10} + \frac{t^2}{15}$$

and $y=-\frac{1}{2}c_1t^{-3}-c_2t^{-4}+\frac{2t^2}{15}-\frac{t}{20}$.

Example 8: Solve the simultaneous differential equations

$$D^2x+m^2y=0, D^2y-m^2x=0 \text{ , where } D=\frac{d}{dt}.$$

Solution: The given equations can be written as

$$D^2x+m^2y=0 \qquad ...(1)$$

$$-m^2x+D^2y=0. \qquad ...(2)$$

Now operating D^2 on (1) and m^2 on (2) and subtracting, we get

$$\left(D^4+m^4\right)x=0.$$

A.E. is $A^4+m^4=0 \Rightarrow \left(A^4+2A^2m^2+m^4\right)-2A^2m^2=0$

$$\Rightarrow \quad \left(A^2+m^2\right)^2-\left(\sqrt{2}Am\right)^2=0$$

$$\Rightarrow \quad \left(A^2+m^2+\sqrt{2}Am\right)\left(A^2+m^2-\sqrt{2}Am\right)=0$$

$$\Rightarrow \quad A=\frac{-m\sqrt{2}\pm\sqrt{2m^2-4m^2}}{2} \text{ or } A=\frac{m\sqrt{2}\pm\sqrt{2m^2-4m^2}}{2}$$

$$\Rightarrow \quad A=-\frac{m}{\sqrt{2}}\pm\frac{im}{\sqrt{2}} \text{ or } \frac{m}{\sqrt{2}}\pm\frac{m}{\sqrt{2}}i.$$

Hence

$$x=e^{-mt/\sqrt{2}}\left[c_1\cos\left(\frac{mt}{\sqrt{2}}\right)+c_2\sin\left(\frac{mt}{\sqrt{2}}\right)\right]+e^{mt/\sqrt{2}}\left[c_3\cos\left(\frac{mt}{\sqrt{2}}\right)+c_4\sin\left(\frac{mt}{\sqrt{2}}\right)\right]$$

$$\frac{dx}{dt}=e^{-mt/\sqrt{2}}\left[-\frac{c_1m}{\sqrt{2}}\sin\left(\frac{mt}{\sqrt{2}}\right)+\frac{c_2m}{\sqrt{2}}\cos\left(\frac{mt}{\sqrt{2}}\right)\right]-\frac{me^{-mt/\sqrt{2}}}{\sqrt{2}}\left[c_1\cos\left(\frac{mt}{\sqrt{2}}\right)\right.$$

$$\left.+c_2\sin\left(\frac{mt}{\sqrt{2}}\right)\right]+e^{mt/\sqrt{2}}\left[-\frac{m}{\sqrt{2}}c_3\sin\left(\frac{mt}{\sqrt{2}}\right)+\frac{m}{\sqrt{2}}c_4\cos\left(\frac{mt}{\sqrt{2}}\right)\right]$$

$$+\frac{m}{\sqrt{2}}e^{mt/\sqrt{2}}\left[c_3\cos\left(\frac{mt}{\sqrt{2}}\right)+c_4\sin\left(\frac{mt}{\sqrt{2}}\right)\right]$$

$$=\frac{m}{\sqrt{2}}e^{-mt/\sqrt{2}}\left[-(c_1+c_2)\sin\left(\frac{mt}{\sqrt{2}}\right)+(c_2-c_1)\cos\left(\frac{mt}{\sqrt{2}}\right)\right]$$

$$+\frac{me^{mt/\sqrt{2}}}{\sqrt{2}}\left[(c_3+c_4)\cos\frac{mt}{\sqrt{2}}+(c_4-c_3)\sin\left(\frac{mt}{\sqrt{2}}\right)\right]$$

And $\frac{d^2x}{dt^2}=m^2e^{-mt/\sqrt{2}}\left[c_1\sin\left(\frac{mt}{\sqrt{2}}\right)-c_2\cos\left(\frac{mt}{\sqrt{2}}\right)\right]+m^2e^{mt/\sqrt{2}}\left[c_4\cos\left(\frac{mt}{\sqrt{2}}\right)\right.$

$$\left.-c_3\sin\left(\frac{mt}{\sqrt{2}}\right)\right]. \qquad ...(3)$$

Now from equation (1), we obtain

$$y = -\frac{1}{m^2}\frac{d^2x}{dt^2} = e^{-mt/\sqrt{2}}\left[c_2\cos\left(\frac{mt}{\sqrt{2}}\right) - c_1\sin\left(\frac{mt}{\sqrt{2}}\right)\right]$$

$$+e^{mt/\sqrt{2}}\left[c_3\sin\left(\frac{mt}{\sqrt{2}}\right) - c_4\cos\left(\frac{mt}{\sqrt{2}}\right)\right] \quad ...(4)$$

Hence the required general solution is given by (3) and (4).

EXERCISE 2.5

Solve the following simultaneous equations

1. $\frac{dx}{dt} + 5x + y = e^t, \ \frac{dy}{dt} - x + 3y = e^{2t}.$
2. $\frac{dx}{dt} - y = e^{-t}, \ \frac{dy}{dt} + x = e^t.$
3. $\frac{dy}{dt} - 17y + 2\frac{dx}{dt} - 8x = 0; \ 13\frac{dy}{dt} - 53y - 2x = 0.$
4. $3\frac{dx}{dt} + 2\frac{dy}{dt} - 4x + 3y = 8e^{-3t}; \ 4\frac{dx}{dt} + \frac{dy}{dt} + 3x + 4y = 8e^{-3t}$ given that $x = \frac{1}{5}, \ y = 0$ when $t = 0$.
5. $\frac{d^2x}{dt^2} + 4x + y = te^{3t}; \ \frac{d^2y}{dt^2} + y - 2x = \cos^2 t.$
6. $t^2\frac{d^2x}{dt^2} + t\frac{dx}{dt} + 2y = 0; \ t^2\frac{d^2y}{dt^2} + t\frac{dy}{dt} - 2x = 0.$
7. $\frac{d^2x}{dt^2} - 2\frac{dy}{dt} - x = e^t\cos t; \ \frac{d^2y}{dt^2} + 2\frac{dx}{dt} - y = e^t\sin t.$
8. $\frac{dx}{dt} = ax + by; \ \frac{dy}{dt} = a'x + b'y.$
9. $(D-1)x + Dy = 2t+1; \ (2D+1)x + 2Dy = t.$
10. $tdx = (t-2x)dt; \ tdy = (tx + ty + 2x - t)dt.$
11. $t\frac{dx}{dt} + y = 0; \ t\frac{dy}{dt} + x = 0.$
12. $\frac{d^2x}{dt^2} + 16x - 6\frac{dy}{dt} = 0; \ \frac{dx}{dt} + \frac{d^2y}{dt^2} + 16y = 0.$
13. $4\frac{dx}{dt} + 9\frac{dy}{dt} + 11x + 31y = e^t; \ 3\frac{dx}{dt} + 7\frac{dy}{dt} + 8x + 24y = e^{2t}.$

ANSWERS 2.5

1. $x = -(c_1 + c_2t)e^{-4t} + c_2e^{-4t} + \frac{4e^t}{25} - \frac{e^{2t}}{36}; \ y = (c_1 + c_2t)e^{-4t} + \frac{e^t}{25} + \frac{7e^{2t}}{36}.$

2. $x = c_1 \cos t + c_2 \sin t + \frac{1}{2}\left(e^t - e^{-t}\right); \quad y = c_2 \cos t - c_1 \sin t + \frac{1}{2}\left(e^t - e^{-t}\right).$

3. $x = 6c_2 e^{5t} - 7c_1 e^{3t}; \; y = c_1 e^{3t} + c_2 e^{5t}.$

4. $x = e^{-t}\left[\cos 2t - \frac{1}{10}\sin 2t\right] - \frac{4}{5}e^{-2t}; \; y = e^{-t}\left[-\frac{4}{5}\cos 2t - \frac{21}{20}\sin 2t\right] + \frac{4}{5}e^{-3t}.$

5. $x = c_1 \cos\sqrt{3}t + c_2 \sin\sqrt{3}t + c_3 \cos\sqrt{2}t + c_4 \sin\sqrt{2}t + \frac{1}{22}e^{3t}$

$$+\frac{5}{66}e^{3t}\left(t - \frac{3}{2}\right) - \frac{1}{12} - \frac{1}{4}\cos 2t$$

$$y = -c_1 \cos\sqrt{3}t - c_2 \sin\sqrt{3}\,t - 2c_3 \cos\sqrt{2}t - 2c_4 \sin\sqrt{2}t$$

$$+\frac{t}{66}e^{3t} - \frac{23}{2452}e^{3t} + \frac{1}{3}.$$

6. $x = t\left[c_1 \cos(\log t) + c_2 \sin(\log t)\right] + \frac{1}{t}\left[c_3 \cos(\log t) + c_4 \sin(\log t)\right];$

$$y = t\left[c_1 \sin(\log t) - c_2 \cos(\log t)\right] + \frac{1}{t}\left[c_4 \cos(\log t) - c_3 \sin(\log t)\right].$$

7. $x = (c_1 + c_2 t)\cos t + (c_3 + c_4 t)\sin t + \frac{1}{25}e^t(4\sin t - 3\cos t);$

$$y = (c_3 + c_4 t)\cos t - (c_1 - c_2 t)\sin t + \frac{1}{25}e^t(4\cos t - 3\sin t).$$

8. $x = c_1 e^{\alpha_1 t} + c_2 e^{\alpha_2 t}; \; y = \frac{1}{b}\left[c_1(\alpha_1 - a)e^{\alpha_1 t} + c_2(\alpha_2 - a)e^{\alpha_2 t}\right]$

where $\alpha_1 = \frac{1}{2}\left[a + b' + \sqrt{(a+b)^2 - 4(a'b)}\right],$

$$\alpha_2 = \frac{1}{2}\left[a + b' - \sqrt{(a-b)^2 + 4(a'b)}\right].$$

9. $x = -t - \frac{2}{3}; \; y = \frac{1}{2}t^2 + \frac{4}{3}t + c.$

10. $x = c_1 t^{-2} + \frac{1}{3}t; \; y = c_2 e^t - c_1 t^{-2} - \frac{1}{3}t.$

11. $x = c_1 t + c_2 / t; \; y = c_2 / t - c_1 t.$

12. $x = c_1 \cos 2t + c_2 \sin 2t + c_3 \cos 8t + c_4 \cos 8t;$

$$y = c_1 \sin 2t + c_2 \cos 2t - c_3 \sin 8t + c_4 \cos 8t.$$

13. $x = (c_1 + c_2 t)e^{-4t} + \frac{32}{25}e^t - \frac{49}{36}e^{2t};$

$$y = -(c_2 + c_1 + c_2 t)e^{-4t} + \frac{19}{36}e^{2t} - \frac{11}{25}e^t.$$

3

Linear Differential Equations of Second Order with Variable Coefficients

3.1 INTRODUCTION

The differential equations of the form

$$\frac{d^2y}{dx^2} + P\frac{dy}{dx} + Qy = R,$$

where P, Q and R are functions of x alone, are called second order linear differential equations with variable coefficients. There is no general method to solve such type of equations, but we shall consider few methods in which the integral can be found.

3.2 COMPLETE SOLUTION OF $y'' + Py' + Qy = R$ IN TERMS OF ONE KNOWN INTEGRAL BELONGING TO THE COMPLEMENTARY FUNCTION

Given that $$\frac{d^2y}{dx^2} + P\frac{dy}{dx} + Qy = R. \qquad ...(1)$$

Let $y = u$ be a known integral of complementary function so u is a solution if (1) when its right hand side is taken to be zero. Thus $y = u$ is solution of $y'' + Py' + Qy = 0$.

So that $$\frac{d^2u}{dx^2} + P\frac{du}{dx} + Qu = 0. \qquad ...(2)$$

Now let the complete solution of (1) is $y = uv$, ...(3)

where v is a function of x, will now be determined from (3). Therefore we have

$$\frac{dy}{dx} = v\frac{du}{dx} + u\frac{dv}{dx},\ \frac{d^2y}{dx^2} = v\frac{d^2u}{dx^2} + 2\frac{du}{dx}\frac{dv}{dx} + u\frac{d^2v}{dx^2} \qquad ...(4)$$

Now using (3) and (4) in (1), we obtain

$$\left(v\frac{d^2u}{dx^2}+2\frac{du}{dx}\frac{dv}{dx}+u\frac{d^2v}{dx^2}\right)+P\left(v\frac{du}{dx}+u\frac{dv}{dx}\right)+Quv=R$$

$$\Rightarrow \quad v\left(\frac{d^2u}{dx^2}+P\frac{du}{dx}+Qu\right)+u\left(\frac{d^2v}{dx^2}+P\frac{dv}{dx}\right)+2\frac{du}{dx}\frac{dv}{dx}=R.$$

$$\Rightarrow \quad u\frac{d^2v}{dx^2}+\frac{dv}{dx}\left(2\frac{du}{dx}+Pu\right)=R, \text{ by using (2)}$$

$$\Rightarrow \quad \frac{d^2v}{dx^2}+\frac{dv}{dx}\left(\frac{2}{u}\frac{du}{dx}+P\right)=\frac{R}{u}. \qquad \text{...(5)}$$

Now putting $\frac{dv}{dx}=p, \frac{d^2v}{dx^2}=\frac{dp}{dx}$, equation (5) becomes

$$\frac{dp}{dx}+p\left(P+\frac{2}{u}\frac{du}{dx}\right)=\frac{R}{u}. \qquad \text{...(6)}$$

which is a linear equation in p and x. Its integrating factor is

$$e^{\int\left(P+2/u\frac{du}{dx}\right)dx}=e^{\int Pdx+2\log u}=u^2e^{\int Pdx}$$

Therefore the solution of equation (6) is given by

$$pu^2e^{\int Pdx}=\int\frac{R}{u}u^2e^{\int Pdx}dx+C_1 \Rightarrow \frac{dv}{dx}=\frac{1}{u^2e^{\int Pdx}}\left\{\int Rue^{\int Pdx}dx+C_1\right\}.$$

On integrating, we obtain

$$v=\int\left[\frac{1}{u^2e^{\int Pdx}}\left\{\int Rue^{\int Pdx}dx+C_1\right\}\right]dx+C_2.$$

Now putting the value of v in equation (3), we get

$$y=u\left(\int\left\{\frac{1}{u^2e^{\int Pdx}}\int Rue^{\int Pdx}dx\right\}dx+C_1\int\frac{e^{-\int Pdx}}{u^2}dx+C_2\right). \qquad \text{...(7)}$$

This solution includes the given solution $y = u$. Since (7) contains two arbitrary constants, so it is the required complete solution of the given differential equation.

3.2.1 Rules for Getting an Integral u of Complementary Function (C.F.)

Let the second order differential equation be $y''+Py'+Qy=R$.

Rule 1: If $a^2+Pa+Q=0$, then $y=e^{ax}$ is a solution

Proof: If $y=e^{ax}$ then $\frac{dy}{dx}=ae^{ax}$ and $\frac{d^2y}{dx^2}=a^2e^{ax}$.

Putting these values in (1), we get

$$\left(a^2 + Pa + Q\right)e^{ax} = 0 \Rightarrow a^2 + Pa + Q = 0.$$

Particular cases:

(i) Let $a = 1$, then $1 + P + Q = 0$ and $y = e^x$ is a solution.

(ii) Let $a = -1$, then $1 - P + Q = 0$ and $y = e^{-x}$ is a solution.

Rule 2: If $m(m-1) + Pmx + Qx^2 = 0$, then $y = x^m$ is a solution

Proof: If $y = x^m$ then $\dfrac{dy}{dx} = mx^{m-1}$ and $\dfrac{d^2y}{dx^2} = m(m-1)x^{m-2}$.

Putting these values in (1), we get

$$\left[m(m-1) + Pmx + Qx^2\right]x^{m-2} = 0 \Rightarrow m(m-1) + Pmx + Qx^2 = 0.$$

Particular cases:

(i) Let $m = 1$, then $P + Qx = 0$ and $y = x$ is a solution.

(ii) Let $m = 2$, then $2 + 2Px + Qx^2 = 0$ and $y = x^2$ is a solution.

3.2.2 Working Rule for Finding Complete Primitive Solution

If an Integral u of a C.F. is known or can be obtained by the above rules then we find complete primitive solution as follows:

Step 1: Put the equation in standard form $y'' + Py' + Qy = R$ in which the coefficient of $\dfrac{d^2y}{dx^2}$ is unity.

Step 2: Find an integral u of C.F. by using the above rules. If a solution (or integral) u is given in a problem then the given integral is the C.F. and there is no need to find u.

Step 3: Suppose that the complete solution of given equation is $y = uv$ where u has been obtained in step 2, then the given equation reduces to

$$\frac{d^2v}{dx^2} + \left(P + \frac{2}{u}\frac{du}{dx}\right)\frac{dv}{dx} = \frac{R}{u}. \qquad \text{...(1)}$$

Step 4: Setting $\dfrac{dv}{dx} = p, \dfrac{d^2v}{dx^2} = \dfrac{dp}{dx}$, then (1) becomes linear equation in x and p. If $R \neq 0$, solve it as usual and if $R = 0$, then solve it using variables separable method because in this case variables p and x will be separable.

Step 5: First replace p by $\frac{dv}{dx}$ and separate the variables of v and x, then integrate. Now determine v. Put this value of v in the assumed solution $y = uv$. This will give us the desired complete solution of the given equation.

Example 1: Solve $\left(1-x^2\right)\frac{d^2y}{dx^2} + x\frac{dy}{dx} - y = x\left(1-x^2\right)^{3/2}$.

Solution: Comparing it with $\frac{d^2y}{dx^2} + P\frac{dy}{dx} + Qy = R$, we get

$$P = \frac{x}{1-x^2},\ Q = -\frac{1}{1-x^2},\ R = x\sqrt{1-x^2}$$

This gives $P + Qx = 0$. Hence, a part of C.F. = x

Let $y = vx$ be the complete solution of the given differential equation, then

$$\frac{dy}{dx} = v + x\frac{dv}{dx} \text{ and } \frac{d^2y}{dx^2} = x\frac{d^2v}{dx^2} + 2\frac{dv}{dx}.$$

Substituting these values in the given equation, we get

$$\frac{d^2v}{dx^2} + \frac{2-x^2}{x\left(1-x^2\right)}\frac{dv}{dx} = \left(1-x^2\right)^{1/2}$$

$$\Rightarrow \quad \frac{dp}{dx} + \frac{\left(2-x^2\right)}{x\left(1-x^2\right)}p = \left(1-x^2\right)^{1/2}, \text{ where } p = \frac{dv}{dx} \qquad \ldots(1)$$

Here $I.F. = e^{\int \frac{2-x^2}{x\left(1-x^2\right)}dx} = e^{\int\left(\frac{2}{x} + \frac{x}{1-x^2}\right)dx} = \frac{x^2}{\sqrt{1-x^2}}$.

Hence, the solution of (1) is

$$p\frac{x^2}{\sqrt{1-x^2}} = \int \sqrt{1-x^2}\,\frac{x^2}{\sqrt{1-x^2}}\,dx + C_1$$

$$\Rightarrow \quad p\frac{x^2}{\sqrt{1-x^2}} = \frac{x^3}{3} + C_1, \text{ where } C_1 \text{ is a constant of integration.}$$

$$\Rightarrow \quad p = \frac{x}{3}\sqrt{1-x^2} + C_1\frac{\sqrt{1-x^2}}{x^2}$$

$$\Rightarrow \quad \frac{dv}{dx} = \frac{x}{3}\sqrt{1-x^2} + C_1\frac{\sqrt{1-x^2}}{x^2}.$$

On integrating, we get

$$v = \frac{1}{3}\int x\sqrt{1-x^2}\,dx + C_1 \int \frac{\sqrt{1-x^2}}{x^2}\,dx + C_2$$

$$= -\frac{1}{9}\left(1-x^2\right)^{3/2} + C_1\left[-\frac{\sqrt{1-x^2}}{x} - \sin^{-1} x\right] + C_2$$

where C_2 is also a constant of integration.

Hence the complete solution is given by

$$y = vx = -\frac{x\left(1-x^2\right)^{3/2}}{9} - C_1\left[\sqrt{1-x^2} + x\sin^{-1}x\right] + C_2 x.$$

Example 2: Solve $\frac{d^2y}{dx^2} - \cot x\frac{dy}{dx} - (1-\cot x)y = e^x \sin x.$

Solution: Comparing with the standard form, we get

$$P = -\cot x, Q = -(1-\cot x), R = e^x \sin x$$

This implies that $1 + P + Q = 1 - 1 + \cot x - \cot x = 0.$ Hence a part of $C.F. = e^x$.

Let $y = ve^x$ be the complete solution of given equation, then

$$\frac{dy}{dx} = ve^x + e^x\frac{dv}{dx}, \quad \frac{d^2y}{dx^2} = ve^x + 2e^x\frac{dv}{dx} + e^x\frac{d^2v}{dx^2}.$$

Substituting these values in the given equation, we get

$$\frac{d^2v}{dx^2} + (2-\cot x)\frac{dv}{dx} = \sin x$$

$$\Rightarrow \quad \frac{dp}{dx} + (2-\cot x)p = \sin x, \text{ where } p = \frac{dv}{dx} \qquad ...(1)$$

This is a linear differential equation of first order in p and x. Hence

$$I.F. = e^{\int(2-\cot x)dx} = \frac{e^{2x}}{\sin x}.$$

Thus solution of (1) is given by

$$p\frac{e^{2x}}{\sin x} = \int \sin x\frac{e^{2x}}{\sin x}\,dx + C_1 = \frac{e^{2x}}{2} + C_1,$$

where C_1 is an arbitrary constant of integration.

$$\Rightarrow \quad p = \frac{1}{2}\sin x + C_1 e^{-2x}\sin x \Rightarrow \frac{dv}{dx} = \frac{1}{2}\sin x + C_1 e^{-2x}\sin x.$$

On integrating, we get

$$v = -\frac{1}{2}\cos x - \frac{1}{5}C_1 e^{-2x}(\cos x + 2\sin x) + C_2.$$

Hence the complete solution is

$$y = ve^x = \left[-\frac{1}{2}\cos x - \frac{1}{5}C_1 e^{-2x}(\cos x + 2\sin x + C_2)\right]e^x.$$

Example 3: Solve $x\frac{d^2y}{dx^2} - (2x-1)\frac{dy}{dx} + (x-1)y = 0$ given that $y = e^x$ is an integral included in the complementary function.

Solution: Given equation can be written as

$$\frac{d^2y}{dx^2} - \frac{(2x-1)}{x}\frac{dy}{dx} + \frac{(x-1)}{x}y = 0.$$

Putting $y = ve^x$, we get $\frac{d^2v}{dx^2} + \left[P + \frac{2}{u}\frac{du}{dx}\right]\frac{dv}{dx} = 0.$

$$\Rightarrow \quad \frac{dz}{dx} + \left[-\frac{2x-1}{x} + \frac{2}{e^x}e^x\right]z = 0, \text{ where } \frac{dv}{dx} = z$$

$$\Rightarrow \quad \frac{dz}{dx} + \frac{z}{x} = 0 \Rightarrow \frac{dz}{z} + \frac{dx}{x} = 0$$

On integrating, we get $\log z + \log x = \log c_1$

$$\Rightarrow \quad z = \frac{C_1}{x} \Rightarrow \frac{dv}{dx} = \frac{C_1}{x}$$

Again integrating, we get $v = C_1 \log x + C_2$

Hence the complete solution is $y = ve^x = e^x(C_1 \log x + C_2)$.

Example 4: Solve $x^2y'' - (x^2 + 2x)y' + (x+2)y = x^3e^x$ given that $y = x$ is a solution.

Solution: Given equation can be written as

$$y'' - \frac{x^2 + 2x}{x^2}y' + \frac{x+2}{x^2}y = xe^x.$$

On putting $y = vx$, we get $\frac{d^2v}{dx^2} + \left[P + \frac{2}{u}\frac{du}{dx}\right]\frac{dv}{dx} = \frac{R}{u}$

$$\Rightarrow \frac{d^2v}{dx^2} + \left[-\frac{x^2 + 2x}{x^2} + \frac{2}{x}.1\right]\frac{dv}{dx} = \frac{xe^x}{x}$$

$$\Rightarrow \quad \frac{d^2v}{dx^2}-\frac{dv}{dx}=e^x \Rightarrow \frac{dz}{dx}-z=e^x, \text{ where } z=\frac{dv}{dx} \quad ...(1)$$

which is a linear differential equation. Hence, we have $I.F.=e^{-\int dx}=e^{-x}$ and the solution is $ze^{-x}=\int e^x e^{-x}dx+C_1=x+C_1$

$$\Rightarrow \quad z=e^x\, x+C_1e^x \Rightarrow \frac{dv}{dx}=e^x\, x+C_1\, e^x.$$

Again integrating, we obtain

$$v=xe^x-e^x+C_1e^x+C_2=(x-1)e^x+C_1e^x+C_2$$

Hence the complete solution is $y=vx=\left(x^2-x+C_1x\right)e^x+C_2x.$

Example 5: Solve $(x+2)\frac{d^2y}{dx^2}-(2x+5)\frac{dy}{dx}+2y=(x+1)e^x.$

Solution: We can write $\frac{d^2y}{dx^2}-\frac{2x+5}{x+2}\frac{dy}{dx}+\frac{2y}{x+2}=\frac{(x+1)e^x}{x+2}$ ...(1)

Applying $1+\frac{P}{a}+\frac{Q}{a^2}=0,$ with $a=2,$ we have

$$1-\frac{2x+5}{2x+4}+\frac{2}{4x+8}=0.$$

Hence $y=e^{2x}$ is a part of C.F.

Putting $y=e^{2x}v$ in (1), we obtain $\frac{d^2v}{dx^2}+\left[P+\frac{2}{u}\frac{du}{dx}\right]\frac{dv}{dx}=\frac{R}{u}$

$$\Rightarrow \quad \frac{d^2v}{dx^2}+\left[-\frac{2x+5}{x+2}+\frac{2}{e^{2x}}2e^{2x}\right]\frac{dv}{dx}=\frac{(x+1)e^x}{e^{2x}(x+2)}$$

$$\Rightarrow \quad \frac{d^2v}{dx^2}+\frac{2x+3}{x+2}\frac{dv}{dx}=\frac{(x+1)e^{-x}}{x+2}$$

$$\Rightarrow \quad \frac{dz}{dx}+\frac{2x+3}{x+2}z=\frac{x+1}{x+2}e^{-x}, \text{ where } \frac{dv}{dx}=z$$

which is a linear differential equation. x

Hence $I.F.=e^{\int\frac{2x+3}{x+2}dx}=e^{\int\left(2-\frac{1}{x+2}\right)dx}=e^{2x-\log(x+2)}=\frac{e^{2x}}{x+2}.$

and its solution is given by

$$z\frac{e^{2x}}{x+2}=\int\frac{e^{2x}}{(x+2)}\frac{x+1}{(x+2)}e^{-x}dx+C=\int\frac{e^x(x+1)}{(x+2)^2}dx+C$$

$$= \int e^x \left[\frac{1}{x+2} - \frac{1}{(x+2)^2}\right] dx + C = \int \frac{e^x dx}{x+2} - \int \frac{e^x dx}{(x+2)^2} + C$$

$$= \frac{e^x}{(x+2)} + \int \frac{e^x}{(x+2)^2} - \int \frac{e^x dx}{(x+2)^2} + C = \frac{e^x}{(x+2)} + C$$

$$\Rightarrow \quad z = e^{-x} + C(x+2)e^{-2x} \Rightarrow \frac{dv}{dx} = e^{-x} + C(x+2)e^{-2x}$$

On integrating, we obtain

$$v = \int e^{-x}\, dx + C \int (x+2)e^{-2x}\, dx + C_1$$

$$= -e^{-x} + C\left[\frac{(x+2)e^{-2x}}{-2} - \int \frac{e^{-2x}}{-2} dx\right] + C_1 = -e^{-x} - \frac{Ce^{-2x}}{4}[2x+5] + C_1$$

Hence the complete solution is

$$y = uv = e^{2x}\left[-e^{-x} - \frac{Ce^{-2x}}{4}(2x+5) + C_1\right] = -e^x - \frac{C}{4}(2x+5) + C_1 e^{2x}.$$

Example 6: Solve the differential equation

$$(x\sin x + \cos x)\frac{d^2y}{dx^2} - x\cos x\frac{dy}{dx} + y\cos x = 0,$$

in which $y = x$ is a solution.

Solution: We can write

$$\frac{d^2y}{dx^2} - \left(\frac{x\cos x}{x\sin x + \cos x}\right)\frac{dy}{dx} + \left(\frac{\cos x}{x\sin x + \cos x}\right)y = 0. \qquad ...(1)$$

Here a part of C.F. $= x$. Let $y = vx$ be the complete solution of equation (1), then we have

$$\frac{d^2v}{dx^2} + \left[\frac{2}{x} - \frac{x\cos x}{x\sin x + \cos x}\right]\frac{dv}{dx} = 0$$

$$\Rightarrow \quad \frac{dp}{dx} + \left[\frac{2}{x} - \frac{x\cos x}{x\sin x + \cos x}\right]p = 0 \text{ where } p = \frac{dv}{dx}$$

$$\Rightarrow \quad \frac{dp}{p} + \left[\frac{2}{x} - \frac{x\cos x}{x\sin x + \cos x}\right]dx = 0.$$

Integration yields, $px^2 = C_1(x\sin x + \cos x)$

$$\Rightarrow \quad \frac{dv}{dx} = C_1\left[\frac{\sin x}{x} + \frac{\cos x}{x^2}\right].$$

Again integrating, we get $v = -C_1 \dfrac{\cos x}{x} + C_2.$

Hence the complete solution is given by $y = vx$

$= -C_1 \cos x + C_2 x$,

where C_1 and C_2 are the arbitrary constants of integration.

Example 7: Solve $x\dfrac{dy}{dx} - y = (x-1)\left[\dfrac{d^2y}{dx^2} - x + 1\right].$

Solution: This equation can be written as

$$\frac{d^2y}{dx^2} - \frac{x}{x-1}\frac{dy}{dx} + \frac{y}{x-1} = (x-1).$$

Here $P + Qx = 0$, so $y = x$ is a part of the complementary function.

Putting $y = vx$, we get

$$\frac{d^2y}{dx^2}\left(-\frac{x}{x-1} + \frac{2}{x}\right)\frac{dv}{dx} = \frac{(x-1)}{x}$$

$$\Rightarrow \quad \frac{dp}{dx} + \left(\frac{2}{x} - \frac{x}{x-1}\right)p = \frac{x-1}{x}, \text{ where } p = \frac{dv}{dx}$$

This is a linear equation in p and x, so we have

$$I.F. = e^{\int\left(\frac{2}{x} - \frac{x-1}{x-1} - \frac{1}{x-1}\right)dx} = e^{2\log x - \log(x-1) - x} = \frac{x^2}{x-1}e^{-x}.$$

$$\text{Therefore } p\frac{x^2}{x-1}e^{-x} = \int xe^{-x}\,dx + C_1 = -xe^{-x} - e^{-x} + C_1$$

$$\Rightarrow \quad p = -\frac{x-1}{x} - \frac{x-1}{x^2} + \frac{C_1(x-1)e^x}{x^2}$$

$$\Rightarrow \quad \frac{dv}{dx} = -1 + \frac{1}{x^2} + C_1\left(\frac{1}{x} - \frac{1}{x^2}\right)e^x.$$

Integration yields, $v = -x - \dfrac{1}{x} + C_1\dfrac{e^x}{x} + C_2.$

Hence the complete solution is

$$y = vx = -x^2 - 1 + C_1 e^x + C_2 x.$$

Example 8: Solve $\dfrac{d^2y}{dx^2} + \left(1 + \dfrac{2}{x}\cot x - \dfrac{2}{x^2}\right)y = x\cos x$, given that $\dfrac{\sin x}{x}$ is an integral included in the C.F.

Solution: Putting $y = v.\dfrac{\sin x}{x}$ in the given equation, we get

$$\frac{d^2v}{dx^2} + 2\left(\cot x - \frac{1}{x}\right)\frac{dv}{dx} = x^2 \cot x.$$

$$\Rightarrow \quad \frac{dp}{dx} + 2\left(\cot x - \frac{1}{x}\right)p = x^2 \cot x, \text{ where } p = \frac{dv}{dx}.$$

which is a linear equation in p.

Thus $I.F. = e^{\int 2\left(\cot x - \frac{1}{x}\right)dx} = e^{2(\log \sin x - \log x)} = \dfrac{\sin^2 x}{x^2}.$

Hence $p\dfrac{\sin^2 x}{x^2} = \int x^2 \cot x \dfrac{\sin^2 x}{x^2}\, dx + C_1 = \dfrac{1}{2}\int \sin 2x \, dx + C_1$

$$= -\frac{1}{4}\cos 2x + C_1 = -\frac{1}{4} + \frac{1}{2}\sin^2 x + C_1$$

$$\Rightarrow \quad p = \frac{dv}{dx} = -\frac{1}{4}x^2 \operatorname{cosec}^2 x + \frac{1}{2}x^2 + C_1 x^2 \operatorname{cosec}^2 x.$$

On integrating, we obtain

$$v = \left(C_1 - \frac{1}{4}\right)\left[-x^2 \cot x + 2x \log \sin x - 2\int \log \sin x \, dx\right] + \frac{x^3}{6} + C_2.$$

Hence the required solution is

$$y = v\frac{\sin x}{x} = \left(C_1 - \frac{1}{4}\right)\left[-x\cot x + 2\sin x \log \sin x - \frac{2\sin x}{x}\int \log \sin x \, dx\right] + \frac{x^2 \sin x}{6} + C_2\frac{\sin x}{x}.$$

Example 9: Solve

$$x\frac{d^2y}{dx^2}(x\cos x - 2\sin x) + (x^2+2)\frac{dy}{dx}\sin x - 2y(x\sin x + \cos x) = 0.$$

Solution: The given equation can be written as

$$\frac{d^2y}{dx^2} + \frac{-(x^2+2)\sin x}{x(x\cos x - 2\sin x)}\frac{dy}{dx} - \frac{2(x\sin x + \cos x)}{x(x\cos x - 2\sin x)}y = 0, \quad ...(1)$$

where $P = \dfrac{(x^2+2)\sin x}{x(x\cos x - 2\sin x)}, Q = -\dfrac{2(x\sin x + \cos x)}{x(x\cos x - 2\sin x)}$

This gives $2 + 2Px + Qx^2 = 0$, so $y = x^2$ is a solution of the equation (1). On putting $y = vx^2$, we get

$$\frac{d^2v}{dx^2}+\left[\frac{4}{x}+\frac{(x^2+2)\sin x}{x(x\cos x-2\sin x)}\right]\frac{dv}{dx}=0.$$

$$\Rightarrow \quad \frac{dp}{dx}+\left[\frac{4}{x}+\frac{(x^2+2)\sin x}{x(x\cos x-2\sin x)}\right]p=0, \text{ where } p=\frac{dv}{dx}.$$

$$\Rightarrow \quad \frac{dp}{p}+\left[\frac{4}{x}-\frac{(-x^2\sin x+2x\cos x-2x\cos x-2\sin x)}{x(x\cos x-2x\sin x)}\right]dx=0$$

On integrating, we get

$$\log p+4\log x-\log\left[x(x\cos x-2\sin x)\right]=\log C_1$$

$$\Rightarrow \quad p=\frac{dv}{dx}=C_1\frac{(\cos x-2\sin x)}{x^4}$$

Again integrating, we get

$$v=C_1\left[\int\frac{\cos x}{x^2}dx-\int\frac{2\sin x}{x^3}dx\right]+C_2$$

$$=C_1\left[\frac{\sin x}{x^2}-\int-\frac{2}{x^3}\sin x\,dx-\int\frac{2\sin x}{x^3}dx\right]+C_2$$

$$=C_1\frac{\sin x}{x^2}+C_2.$$

Hence the complete solution is $y=vx^2=C_1\sin x+C_2x^2$.

EXERCISE 3.1

Solve the following differential equations:

1. $\frac{d^2y}{dx^2}-x^2\frac{dy}{dx}+xy=x.$

2. $x\frac{d^2y}{dx^2}-(3+x)\frac{dy}{dx}+3y=0.$

3. $(2x^3-a)\frac{d^2y}{dx^2}-6x^2\frac{dy}{dx}+6xy=0$, when $y=x$ is a solution.

4. $x^2\frac{d^2y}{dx^2}+x\frac{dy}{dx}-y=0$, when $x+\frac{1}{x}$ is a one integral.

5. $xy''-y'-4x^3y=-4x^5$ given that $y=e^{x^2}$ is a solution of the left hand side equated to zero.

6. $\left(1-x^2\right)\frac{d^2y}{dx^2}-x\frac{dy}{dx}-a^2y=0$ given that $y=ce^{\alpha\sin^{-1}x}$ is an integral.

7. $\sin^2 x\frac{d^2y}{dx^2}=2y$ given that $y=\cot x$ is a solution.

8. $(3-x)\frac{d^2y}{dx^2}-(9-4x)\frac{dy}{dx}+(6-3x)y=0.$

9. $\frac{d^2y}{dx^2}-2(x+1)\frac{dy}{dx}+(x+2)y=(x-2)e^x.$

ANSWERS 3.1

1. $y=1+C_1 x\int x^{-2}e^{(x^3)/3}\,dx+C_2x.$

2. $y=-C_1\left(x^3+3x^2+6x+6\right)+C_2e^x.$

3. $y=C_1\left(x^3+a\right)+C_2x.$

4. $y=-\frac{C_1}{2x}+C_2\left(x+\frac{1}{x}\right).$

5. $y=-\frac{1}{4}C_1e^{-x^2}+C_2\,e^{x^2}+x^2.$

6. $y=\left[-\frac{C_1}{2a}e^{-2a\sin^{-1}x}+C_2\right]ce^{a\sin^{-1}x}.$

7. $Cy=1+(C_1-x)\cot x.$

8. $y=C_1e^{3x}\left(-4x^3-42x^2+150x-183\right)+C_2e^x.$

9. $y=C_1x^3e^x+C_2e^x-\frac{\left(x^2-2x\right)}{2}e^x.$

3.3 NORMAL FORM (REMOVAL OF FIRST ORDER DERIVATIVE)

Consider the differential equation

$$\frac{d^2y}{dx^2}+P\frac{dy}{dx}+Qy=R. \qquad ...(1)$$

Set $y=uv$, where u is not an integral solution of C.F., we have

$$\frac{dy}{dx}=v\frac{du}{dx}+u\frac{dv}{dx} \text{ and } \frac{d^2y}{dx^2}=u\frac{d^2v}{dx^2}+2\frac{du}{dx}\frac{dv}{dx}+v\frac{d^2u}{dx^2}.$$

On putting these values of y, $\frac{dy}{dx}$, $\frac{d^2y}{dx^2}$ in (1), we get

$$\left[u\frac{d^2v}{dx^2}+2\frac{dv}{dx}\frac{du}{dx}+v\frac{d^2u}{dx^2}\right]+P\left[u\frac{dv}{dx}+v\frac{du}{dx}\right]+Q.uv=R$$

$$\Rightarrow \quad u\frac{d^2v}{dx^2}+\frac{dv}{dx}\left[Pu+2\frac{du}{dx}\right]+v\left[\frac{d^2u}{dx^2}+P\frac{du}{dx}+Qu\right]=R$$

$$\Rightarrow \quad \frac{d^2v}{dx^2}+\frac{dv}{dx}\left[P+\frac{2}{u}\frac{du}{dx}\right]+\frac{v}{u}\left[\frac{d^2u}{dx^2}+P\frac{du}{dx}+Qu\right]=\frac{R}{u} \qquad ...(2)$$

Here the expression in last bracket on L.H.S. is not zero as $y=u$ is not a part of C.F.

Now to remove the first derivative, we shall choose $P+\frac{2}{u}\frac{du}{dx}=0$.

This gives

$$\frac{du}{u}=-\frac{1}{2}Pdx \Rightarrow \log u=-\frac{1}{2}\int P\,dx \Rightarrow u=e^{-\frac{1}{2}\int P\,dx} \qquad ...(3)$$

Now we first have to find out $\frac{d^2u}{dx^2}+P\frac{du}{dx}+Qu$, for this value of u.

From (3), we have

$$\frac{du}{dx}=-\frac{P}{2}e^{-\frac{1}{2}\int Pdx}=-\frac{1}{2}Pu$$

$$\frac{d^2u}{dx^2}=-\frac{1}{2}\frac{dP}{dx}u-\frac{P}{2}\frac{du}{dx}=-\frac{1}{2}\frac{dP}{dx}u-\frac{P}{2}\left(-\frac{1}{2}Pu\right)$$

$$=-\frac{1}{2}\frac{dP}{dx}u+\frac{1}{4}P^2u$$

Therefore, we have

$$\frac{d^2u}{dx^2}+P\frac{du}{dx}+Qu=-\frac{1}{2}\frac{dP}{dx}u+\frac{1}{4}P^2u+P\left(-\frac{1}{2}Pu\right)+Qu$$

$$=u\left[Q-\frac{1}{2}\frac{dp}{dx}-\frac{1}{4}P^2\right]$$

Thus, equation (2) is transformed to

$$\frac{d^2v}{dx^2}+\frac{v}{u}u\left[Q-\frac{1}{2}\frac{dP}{dx}-\frac{P^2}{4}\right]=\frac{R}{u}.$$

$$\Rightarrow \quad \frac{d^2v}{dx^2}+v\left[Q-\frac{1}{2}\frac{dP}{dx}-\frac{P^2}{4}\right]=\frac{R}{u}$$

$$\Rightarrow \quad \frac{d^2v}{dx^2}+Q_1v=R_1,$$

where $Q_1 = \left[Q - \frac{1}{2}\frac{dP}{dx} - \frac{P^2}{4}\right]$, $R_1 = \frac{R}{u}$, $u = e^{-\int P/2dx}$.

Hence the complete solution of the equation is given by $y = uv$.

Remark: This method works only if value of Q_1 is either constant or $\frac{\text{constant}}{x^2}$.

Example 1: Solve $\frac{d^2y}{dx^2} + \frac{1}{x^{1/3}}\frac{dy}{dx} + \left[\frac{1}{4x^{2/3}} - \frac{1}{4x^{4/3}} - \frac{6}{x^2}\right]y = 0.$

Solution: Here $P = x^{-1/3}$, so we have $u = e^{-(1/2)\int P dx} = e^{-(1/2)\int x^{-1/3} dx} = e^{-\frac{3x^{2/3}}{4}}$

We have

$$Q - \frac{P^2}{4} - \frac{1}{2}\frac{dP}{dx} = \frac{e^{-2/3}}{4} - \frac{x^{-4/3}}{6} - \frac{6}{x^2} - \frac{x^{-2/3}}{4} - \frac{1}{2}\left(-\frac{1}{3}\right)x^{-4/3} = \frac{6}{x^2}$$

Therefore on substituting $y = uv$ the equation reduces to

$$\frac{d^2v}{dx^2} - \frac{6v}{x^2} = 0 \Rightarrow x^2\frac{d^2v}{dx^2} - 6v = 0.$$

Putting $x = e^z$ and denoting $\frac{d}{dz}$ by D, we get

$$[D(D-1)-6]v = 0 \Rightarrow [D^2 - D - 6]v = 0. \qquad ...(1)$$

Now auxiliary equation is $m^2 - m - 6 = 0 \Rightarrow m = 3, -2$.

Hence the solution of (1) is $C_1 e^{3z} + C_2 e^{-2z}$

Now the solution of the given equation is

$$y = uv = [C_1 x^3 + C_2 x^{-2}]e^{-\frac{3x^{2/3}}{4}}.$$

Example 2: Solve $\frac{d}{dx}\left[\cos^2 x \frac{dy}{dx}\right] + (\cos^2 x)y = 0$

Solution: Given equation can be written as

$$\frac{d^2y}{dx^2}\cos^2 x - 2\cos x \sin x \frac{dy}{dx} + (\cos^2 x)y = 0$$

$$\Rightarrow \quad \frac{d^2y}{dx^2} - 2\tan x \frac{dy}{dx} + y = 0.$$

Here $P = -2\tan x, Q = 1, R = 0$, so we have

$$Q_1 = Q - \frac{1}{2}\frac{dP}{dx} - \frac{P^2}{4} = 1 - \frac{1}{2}(-2\sec^2 x) - \frac{4\tan^2 x}{4}$$

$$= 1 + \sec^2 x - \tan^2 x = 1 + 1 = 2,$$

$$R_1 = \text{Re}^{1/2\int P\,dx} = 0,$$

$$u = e^{-(1/2)\int(-2\tan x)dx} = e^{\int \tan x\,dx} = e^{\log \sec x} = \sec x.$$

Now the normal form of the equation be $\frac{d^2v}{dx^2} + 2v = 0$ ...(1)

Here A.E. is $(m^2 + 2)u = 0 \Rightarrow m = \pm i\sqrt{2}$.

Now the solution of (1) is $v = C_1 \cos\sqrt{2}x + C_2 \sin\sqrt{2}x$

Hence the complete solution is

$$y = uv = \left[C_1 \cos\sqrt{2}x + C_2 \sin\sqrt{2}x\right]\sec x.$$

Example 3: Solve $\frac{d^2y}{dx^2} - 4x\frac{dy}{dx} + (4x^2 - 1)y = -3e^x \sin 2x.$

Solution: Here $P = -4x$, $Q = (4x^2 - 1)$, $R = -3e^{x^2} \sin 2x$.

So we have $u = e^{-\frac{1}{2}\int P\,dx} = e^{-\frac{1}{2}\int -4x\,dx} = e^{x^2}$

$$Q_1 = Q - \frac{1}{2}\frac{dP}{dx} - \frac{1}{4}P^2 = 4x^2 - 1 - \frac{1}{2}(-4) - \frac{1}{4}16x^2 = 1,$$

$$R_1 = -\frac{3e^{x^2}\sin 2x}{e^{x^2}} = -3\sin 2x$$

Setting $y = uv$, the equation transforms to normal form

$$\frac{d^2v}{dx^2} + v = -3\sin 2x$$

which is a linear equation with constant coefficients.

A.E. is $m^2 + 1 = 0 \Rightarrow m = \pm i$

Thus, the complementary function (C.F.) $= C_1 \cos x + C_2 \sin x$ and the particular integral $= \frac{1}{D^2 + 1}(-3\sin 2x) = \sin 2x$.

Therefore $v = C_1 \cos x + C_2 \sin x + \sin 2x$.

Now the required solution is

$$y = uv = e^{x^2}(C_1 \cos x + C_2 \sin x + \sin 2x).$$

Example 4: Solve $\left(\frac{d^2y}{dx^2}+y\right)\cot x+2\left(\frac{dy}{dx}+y\tan x\right)=\sec x.$

Solution: Given equation can be written as

$$\frac{d^2y}{dx^2}+2\tan x\frac{dy}{dx}+y\left(2\tan^2 x+1\right)=\sec x\tan x.$$

Here we have

$$u=\exp\left(-\frac{1}{2}\int P\,dx\right)=\exp\left(-\frac{1}{2}\int 2\tan x\,dx\right)=\cos x$$

Setting $y=uv,$ the equation reduces to

$$\frac{d^2v}{dx^2}+v\left(Q-\frac{1}{2}\frac{dP}{dx}-\frac{P^2}{4}\right)=\frac{R}{u}$$

$$\Rightarrow\quad \frac{d^2v}{dx^2}+v\left(2\tan^2 x+1-\frac{2}{2}\sec^2 x-\frac{4}{4}\tan^2 x\right)=\sec^2 x\tan x$$

$$\Rightarrow\quad \frac{d^2v}{dx^2}=\tan x\sec^2 x=\tan x\,d(\tan x).$$

On integrating, we get $\frac{dv}{dx}=\frac{\tan^2 x}{2}+C_1=\frac{1}{2}\left(\sec^2 x-1\right)+C_1.$

Further integration yields, $v=\frac{1}{2}(\tan x-x)+C_1x+C_2.$

Hence the required solution is

$$y=uv=\cos x\left[\left(\frac{\tan x-x}{2}\right)+C_1x+C_2\right].$$

Example 5: Solve

$$x^2(\log x)^2\frac{d^2y}{dx^2}-2x\log x\frac{dy}{dx}+\left[2+\log x-2(\log x)^2\right]y=x^2(\log x)^3.$$

Solution: Given equation can be written as

$$\frac{d^2y}{dx^2}-\frac{2}{x(\log x)}\frac{dy}{dx}+\left[\frac{2}{x^2(\log x)^2}+\frac{1}{x^2(\log x)}-\frac{2}{x^2}\right]y=\log x.$$

Taking $u=\exp\left(-\frac{1}{2}\int Pdx\right)=\exp\left(-\frac{1}{2}\int-\frac{2}{x\log x}dx\right)=e^{\log\log x}=\log x$

and putting $y=uv,$ the equation reduces to

$$\frac{d^2v}{dx^2}+v\left[Q-\frac{1}{2}\frac{dP}{dx}-\frac{P^2}{4}\right]=\frac{R}{u}$$

$$\Rightarrow \quad \frac{d^2v}{dx^2}+v\left[\frac{2}{x^2(\log x)^2}+\frac{1}{x^2(\log x)}-\frac{2}{x^2}-\frac{1}{2}\frac{2}{x^2(\log x)}\right.$$

$$\left.+\frac{2}{x^2(\log x)^2}-\frac{4}{4x^2(\log x)^2}\right]=\frac{\log x}{\log x}$$

$$\Rightarrow \quad x^2\frac{d^2v}{dx^2}-2v=x^2.$$

Putting $x=e^z$ and denoting $\frac{d}{dz}$ by D, we have

$$[D(D-1)-2]v=e^{2z} \Rightarrow (D^2-D-2)v=e^{2z}.$$

The complementary function $=C_1\, e^{2z}+C_2e^{-z}$

and the particular integral $=\dfrac{1}{(D-2)(D+1)}e^{2z}=\dfrac{ze^{2z}}{3}$

Hence the solution of the transformed equation is

$$v=(C_1x^2+C_2\, x^{-1})+\frac{(\log x)x^2}{3}$$

Now the required solution is

$$y=uv=\left[(C_1x^2+C_2x^{-1})+\frac{(\log x)x^2}{3}\right]\log x.$$

Example 6: Solve $\dfrac{d^2y}{dx^2}-2\tan x\dfrac{dy}{dx}+5y=\sec x\,.\,e^x.$

Solution: Here $P=-2\tan x, Q=5$ and $R=\sec x\,.\,e^x$, so that

$$u=\exp\left(-\frac{1}{2}\int -2\tan x\,dx\right)=e^{\log sex\, x}=\sec x.$$

Now $Q_1=Q-\dfrac{P^2}{4}-\dfrac{1}{2}\dfrac{dP}{dx}=5-\tan^2x-\dfrac{1}{2}(-2\sec^2x)=6,\ R_1=\dfrac{R}{u}=e^x.$

Substituting $y=uv$, the given equation becomes

$$\frac{d^2v}{dx^2}+6v=e^x \qquad ...(1)$$

The auxiliary equation is $(m^2+6)=0 \Rightarrow m=\pm i\sqrt{6}.$

Hence the complementary function $=c_1\cos\sqrt{6}\,x+c_2\sin\sqrt{6}x$

and particular integral $=\dfrac{1}{D^2+6}e^x=\dfrac{e^x}{7}.$

Thus the solution of (1) is $v=\left[\frac{e^x}{7}+c_1\cos\sqrt{6}x+c_2\sin\sqrt{6}\,x\right]$.

Hence the required solution is

$$y=uv=\sec x\left[\frac{e^x}{7}+c_1\cos\sqrt{6}x+c_2\sin\sqrt{6}\,x\right].$$

EXERCISE 3.2

1. $\frac{d^2y}{dx^2}+\frac{2}{x}\frac{dy}{dx}+x^2y=0.$

2 $\frac{d^2y}{dx^2}-4x\frac{dy}{dx}(4x^2-3)+y=e^{x^2}.$

3. $\frac{d^2y}{dx^2}+2x\frac{dy}{dx}+(x^2+5)y=xe^{-1/2x^3}.$

4. $x\frac{d}{dx}\left(x\frac{dy}{dx}-y\right)-2x\frac{dy}{dx}+2y+x^2y=0.$

5. $\frac{d^2y}{dx^2}-2x\frac{dy}{dx}+(x^2+2)y=e^{\frac{1}{2}(x^2+2x)}.$

6. $\frac{d^2y}{dx^2}-2bx\frac{dy}{dx}+b^2x^2y=0.$

7. $\frac{d^2y}{dx^2}-4x\frac{dy}{dx}+(4x^2-1)y=0.$

8. $x\frac{d^2y}{dx^2}+2\frac{dy}{dx}=n^2xy.$

9. Solve the following differential equation by removing the first derivative $(1+2x)^2\frac{d^2y}{dx^2}-6(1+2x)\frac{dy}{dx}+16y=8(1+2x)^2$, given that $y(0)=y'(0)=2$.

ANSWERS 3.2

1. $y=\frac{C_1(\sin nx+C_2)}{x}$

2. $e^{-x^2}y=C_1e^x+C_2e^{-x}-1.$

3. $y=e^{-\frac{x^2}{2}}\left(C_1\cos(2x+C_2)+\frac{1}{4}x\right).$

4. $y=x(C_1\cos x+C_2\sin x).$

5. $y=\left(C_1 \cos\sqrt{3}+C_2 \sin\sqrt{3}x\right)e^{x^2/2}+\frac{1}{4}e^x e^{x^2/2}$.

6. $y=e^{bx^2/2}\,C_1 \sin\left(x\sqrt{b}+C_2\right)$.

7. $y=e^{x^2}\left(C_1 \cos x+C_2 \sin x\right)$.

8. $y=\frac{1}{x}(c_1e^{nx}+c_2e^{-nx})$.

9. $y=\frac{1}{4}(1+2x)^2+(1+2x)-(1+2x)^2.\log(1+2x)+\frac{3}{4}$.

3.4 CHANGE OF INDEPENDENT VARIABLE

Here we discuss the method to find the complete solution of

$$\frac{d^2y}{dx^2}+P\frac{dy}{dx}+Qy=R \qquad ...(1)$$

by changing the independent variable x to z so that the equation becomes with constant coefficients.

Method: To change the independent variable x to z, we take

$$\frac{dy}{dx}=\frac{dy}{dz}\times\frac{dz}{dx} \qquad ...(2)$$

Then $\frac{d^2y}{dx^2}=\frac{d}{dx}\left(\frac{dy}{dx}\right)=\frac{d}{dx}\left[\frac{dy}{dz}\cdot\frac{dz}{dx}\right]=\frac{dy}{dz}\cdot\frac{d^2z}{dx^2}+\frac{dz}{dx}\cdot\frac{d}{dz}\left(\frac{dy}{dz}\right)\frac{dz}{dx}$

$$\Rightarrow \quad \frac{d^2y}{dx^2}=\frac{dy}{dz}\cdot\frac{d^2z}{dx^2}+\left(\frac{dz}{dx}\right)^2\frac{d^2y}{dz^2}. \qquad ...(3)$$

Substituting these values in (1), we get

$$\frac{dy}{dz}\cdot\frac{d^2z}{dx^2}+\left(\frac{dz}{dx}\right)^2\frac{d^2y}{dz^2}+P\frac{dy}{dz}\frac{dz}{dx}+Qy=R$$

$$\Rightarrow \quad \frac{d^2y}{dz^2}+P_1\frac{dy}{dz}+Q_1y=R_1 \qquad ...(4)$$

where $P_1=\dfrac{\frac{d^2z}{dx^2}+P\frac{dz}{dx}}{\left(\frac{dz}{dx}\right)^2}$, $Q_1=\dfrac{Q}{\left(\frac{dz}{dx}\right)^2}$, $R_1=\dfrac{R}{\left(\frac{dz}{dx}\right)^2}$.

Here P_1, Q_1 and R_1 are functions of x which can be transformed into function of z using the relation $z=f(x)$. After getting equation (4) we choose z such that the equation (4) can be easily integrated.

For this purpose we choose z according to following two cases:

Case I: Choose z such that Q_1 = constant = a^2 (say).

$$\Rightarrow \quad \frac{Q}{\left(\frac{dz}{dx}\right)^2} = a^2 \Rightarrow a\frac{dz}{dx} = \sqrt{Q} \Rightarrow dz = \frac{\sqrt{Q}}{a}dx.$$

Integration yields, $z = \int \frac{\sqrt{Q}}{a} dx.$

Thus equation (4) becomes $\frac{d^2y}{dz^2} + P_1\frac{dy}{dz} + a^2y = R_1.$ ...(5)

If this choice of z makes P_1 constant then equation (5) can be solved easily.

Case II: Choose z such that $P_1 = 0$ *i.e.*, $\frac{d^2z}{dx^2} + P\frac{dz}{dx} = 0$, then

$$\frac{d}{dx}\left(\frac{dz}{dx}\right) + P\frac{dz}{dx} = 0$$

$$\Rightarrow \quad \frac{dz}{dx} = e^{-\int Pdx} \Rightarrow z = \int \left(e^{-\int Pdx}\right)dx.$$

In this case the equation (4) becomes $\frac{d^2y}{dz^2} + Q_1y = R_1.$

This equation can be solved if Q_1 is constant or of the form $\frac{\lambda}{z^2}$, for some constant λ.

Steps for Solution

1. Make the coefficient of $\frac{d^2y}{dx^2}$ as 1 if it is not so.
2. Compare the equation with standard form $y'' + Py' + Qy = R$ and get P, Q and R.
3. We choose z so that $\left(\frac{dz}{dx}\right)^2 = Q$. For this purpose Q is taken in such a way that it remains the whole square of a function without surd and its negative sign is ignored.
4. Obtain z (on integration) and $\frac{d^2z}{dx^2}$ (on differentiation).
5. Find P_1, Q_1 and R_1 by the formulae

$$P_1 = \frac{\frac{d^2z}{dx^2} + P\frac{dz}{dx}}{\left(\frac{dz}{dx}\right)^2}, Q_1 = \frac{Q}{\left(\frac{dz}{dx}\right)^2}, R_1 = \frac{R}{\left(\frac{dz}{dx}\right)^2}.$$

6. Find reduced equation $\frac{d^2y}{dz^2} + P_1\frac{dy}{dz} + Q_1 y = R_1$ which we solve for y in terms of z.
7. We write the complete solution as y in terms of x by replacing the value of z in terms of x.

Example 1: Solve $\frac{d^2y}{dx^2} + \frac{2}{x}\frac{dy}{dx} + \frac{a^2}{x^4}y = 0.$

Solution: Choose z such that $Q_1 = 1$, then

$$\left(\frac{dz}{dx}\right)^2 = Q = \frac{a^2}{x^4} \Rightarrow \frac{dz}{dx} = \pm\frac{a}{x^2} \Rightarrow z = \pm\frac{a}{x}$$

and changing the independent variable from x to z given by the relation $z = \frac{a}{x}$, we get

$$P_1 = \left(\frac{d^2z}{dx^2} + P\frac{dz}{dx}\right) \Big/ \left(\frac{dz}{dx}\right)^2 = 0.$$

Hence, the equation becomes $\frac{d^2y}{dz^2} + y = 0.$

Its solution is $y = c_1 \cos z + c_2 \sin z.$

Now the required solution is $y = c_1 \cos\frac{a}{x} + c_2 \sin\frac{a}{x}.$

Example 2: Solve $\frac{d^2y}{dx^2} + \cot x\frac{dy}{dx} + 4y\,\text{cosec}^2\, x = 0.$

Solution: If $z = f(x)$, the equation is transformed into

$$\frac{d^2y}{dz^2} + P_1\frac{dy}{dz} + Q_1\, y = 0$$

where $P_1 = \left(\frac{d^2z}{dx^2} + P\frac{dz}{dx}\right) \Big/ \left(\frac{dz}{dx}\right)^2$ and $Q_1 = Q \Big/ \left(\frac{dz}{dx}\right)^2.$

Now, choose z such that $Q_1 = 1$, then $\left(\frac{dz}{dx}\right)^2 = Q = 4\,\text{cosec}^2\, x$

$$\Rightarrow \quad \left(\frac{dz}{dx}\right) = 2\,\text{cosec}\, x$$

Integration yields, $z = 2\log\tan\frac{x}{2}$.

Here $P_1 = \frac{-2\operatorname{cosec} x\cot x + 2\operatorname{cosec} x\cot x}{4\operatorname{cosec}^2 x} = 0.$

Therefore by the change of the independent variable from x to z the equation transforms into

$$\frac{d^2y}{dz^2} + y = 0.$$

Its solution is $y = c_1\cos z + c_2\sin z$.

Hence required solution is

$$y = c_1\cos\left(2\log\tan\frac{x}{2}\right) + c_2\sin\left(2\log\tan\frac{x}{2}\right).$$

Example 3: Solve $x\frac{d^2y}{dx^2} - \frac{dy}{dx} + 4x^3 y = x^5$.

Solution: The given equation can be written as $\frac{d^2y}{dx^2} - \frac{1}{x}\frac{dy}{dx} + 4x^2 y = x^4$.

Here $Q = 4x^2$, $P = -\frac{1}{x}$ and $R = x^4$. Now, choose z such that $Q_1 = 1$,

then $\left(\frac{dz}{dx}\right)^2 = 4x^2 \Rightarrow \frac{dz}{dx} = 2x \Rightarrow z = x^2$

Now $P_1 = \left[\frac{d^2z}{dx^2} + P\left(\frac{dz}{dx}\right)\right] \Big/ \left(\frac{dz}{dx}\right)^2 = \left[2 + \left(-\frac{1}{x}\right).2x\right] \Big/ 4x^2 = 0,$

$$R_1 = \frac{x^4}{4x^2} = \frac{x^2}{4} = \frac{z}{4}.$$

Hence the equation transforms into $\frac{d^2y}{dz^2} + y = \frac{z}{4}$.

Clearly complementary function $= c_1\cos z + c_2\sin z$

and the particular integral $= \frac{1}{D^2+1}\left(\frac{z}{4}\right) = (1 - D^2 + \ldots)\frac{z}{4} = \frac{z}{4}$.

So its solution is $y = c_1\cos z + c_2\sin z + \frac{z}{4}$.

Hence the required solution is $y = c_1\cos x^2 + c_2\sin x^2 + \frac{x^2}{4}$.

Example 4: Solve $\frac{d^2y}{dx^2} + (3\sin x - \cot x)\frac{dy}{dx} + 2y\sin^2 x = e^{-\cos x}\sin^2 x$.

Solution: Here $P = 3\sin x - \cot x$, $Q = 2\sin^2 x$, $R = e^{-\cos x}\sin^2 x$.

Choosing z such that $\frac{d^2z}{dx^2}+P\frac{dz}{dx}=0$, we get

$$\frac{d^2z}{dx^2}+(3\sin x-\cot x)\frac{dz}{dx}=0$$

$$\Rightarrow \quad z=\int\left[\exp\left(-\int P\,dx\right)\right]dx=\int\exp\left[-\int\left\{3\sin x-\left(\frac{\cos x}{\sin x}\right)\right\}\right]dx$$

$$=\int\exp(3\cos x+\log\sin x)dx=\int\sin x\,.\,e^{3\cos x}dx=\frac{e^{3\cos x}}{-3},$$

$$Q_1=\frac{Q}{\left(\frac{dz}{dx}\right)^2}=\frac{2\sin^2 x}{\sin^2 x\,.\,e^{6\cos x}}=\frac{2}{9z^2},$$

$$R_1=\frac{R}{\left(\frac{dz}{dx}\right)^2}=\frac{e^{-\cos x}\sin^2 x}{\sin^2 x e^{6\cos x}}=\frac{1}{-(3z)^{1/3}(9z^2)}.$$

Hence the equation transforms into

$$\frac{d^2y}{dz^2}+\frac{2y}{9z^2}=\frac{1}{-(3z)^{1/3}(9z^2)}\Rightarrow 9z^2\frac{d^2y}{dz^2}+2y=-\frac{1}{(3z)^{1/3}}$$

Putting $z=e^t$, denote $\frac{d}{dt}$ by D, we get

$$[9D(D-1)+2]y=-\frac{1}{3^{1/3}}e^{-t/3}$$

Therefore, the complementary function $=c_1e^{(2/3)t}+c_2e^{(1/3)t}$

and the particular integral

$$=-\frac{1}{(3)^{1/3}}\cdot\frac{1}{9D^2-9D+2}e^{-t/3}=-\frac{1}{(3)^{1/3}}\cdot\frac{e^{-t/3}}{6}$$

So its solution is

$$y=c_1e^{(2/3)t}+c_2e^{(1/3)t}-\frac{e^{-(1/3)t}}{6.(3)^{1/3}}=c_1\,z^{2/3}+c_2z^{1/3}-\frac{(z)^{-1/3}}{6.(3)^{1/3}}$$

$$=c_1\frac{e^{2\cos x}}{(9)^{1/3}}+c_2\frac{e^{\cos x}}{(-3)^{1/3}}-\frac{(-3)^{1/3}e^{-\cos x}}{6.(3)^{1/3}}$$

Hence the required solution is $y=k_1\,e^{2\cos x}+k_2\,e^{\cos x}+\frac{e^{-\cos x}}{6}$.

Example 5: Solve $x\dfrac{d^2y}{dx^2}-\dfrac{dy}{dx}-4x^3y=8x^3\sin x^2$.

Solution: The given equation can be written as

$$\frac{d^2y}{dx^2}-\frac{1}{x}\frac{dy}{dx}-4x^2y=8x^2\sin x^2.$$

Here $P=-\dfrac{1}{x}, Q=-4x^2$ and $R=8x^2\sin x^2$. Now choosing z such that $\dfrac{d^2z}{dx^2}+P\dfrac{dz}{dx}=0$

$$\Rightarrow \quad z=\int \exp\left(-\int Pdx\right)dx=\int \exp\left\{-\int -\frac{1}{x}\right\}dx=\int x dx=\frac{x^2}{2}.$$

Therefore $Q_1=Q\Big/\left(\dfrac{dz}{dx}\right)^2=\dfrac{-4x^2}{x^2}=-4$

and $\quad R_1=R\Big/\left(\dfrac{dz}{dx}\right)^2=\dfrac{8x^2\sin x^2}{x^2}=8\sin x^2$.

On changing independent variable x to z, the equation transforms into $\dfrac{d^2y}{dz^2}-4y=8\sin 2z$.

Hence the complementary function $=c_1e^{2z}+c_2e^{-2z}$.

and the particular integral $=8.\dfrac{\sin 2z}{D^2-4}=-\sin 2z$

Therefore, the solution is

$$y=c_1e^{2z}+c_2e^{-2z}-\sin 2z=c_1e^{x^2}+c_2e^{-x^2}-\sin x^2.$$

Example 6: Solve $(1+x)^2\dfrac{d^2y}{dx^2}+(1+x)\dfrac{dy}{dx}+y=4\cos\log(1+x)$.

Solution: We have

$$\frac{d^2y}{dx^2}+\frac{1}{(x+1)}\frac{dy}{dx}+\frac{y}{(1+x)^2}=\frac{4}{(1+x)^2}\cos\log(1+x) \qquad ...(1)$$

Choose z such that $Q_1=1$, then

$$\left(\frac{dz}{dx}\right)^2=\frac{1}{(1+x)^2}\Rightarrow \frac{dz}{dx}=\frac{1}{1+x} \qquad ...(2)$$

Integration yields $z=\log(1+x)$.

From (2), we have $\frac{d^2z}{dx^2} = -\frac{1}{(1+x)^2}$.

Hence $P_1 = \frac{-\frac{1}{(1+x)^2} + \frac{1}{(1+x)} \cdot \frac{1}{(1+x)}}{\frac{1}{(1+x)^2}} = 0$

and $R_1 = \frac{R}{\left(\frac{dz}{dx}\right)^2} = 4\cos\log(1+x) = 4\cos z$.

Hence (1) becomes

$$\frac{d^2y}{dz^2} + y = 4\cos z.$$

Auxiliary equation is $m^2 + 1 = 0 \Rightarrow m = \pm i$

Hence $C.F. = c_1 \cos z + c_2 \sin z$

and $P.I. = \frac{1}{D^2+1}(4\cos z) = 4 \cdot \frac{z}{2}\sin z = 2z\sin z$.

Now complete solution is

$$y = c_1 \cos z + c_2 \sin z + 2z\sin z$$

$$= c_1 \cos\log(1+x) + c_2 \sin\log(1+x) + 2\log(1+x)\sin\log(1+x).$$

Example 7: Solve $\frac{d^2y}{dx^2}\cos x + \frac{dy}{dx}\sin x - 2y\cos^3 x = 2\cos^5 x$.

Solution: The given equation can be written as

$$\frac{d^2y}{dx^2} + \frac{\sin x}{\cos x}\frac{dy}{dx} - 2y\cos^2 x = 2\cos^4 x.$$

Here $P = \tan x$, $Q = -2\cos^2 x$ and $R = 2\cos^4 x$.

Now taking z such that $\frac{d^2z}{dx^2} + P\frac{dz}{dx} = 0$, then

$$z = \int \left\{\exp\left(-\int P dx\right)\right\} dx = \int \left\{\exp\left(-\int \tan x\, dx\right)\right\} dx = \int \cos x\, dx = \sin x,$$

$$Q_1 = \frac{Q}{\left(\frac{dz}{dx}\right)^2} = -\frac{2\cos^2 x}{\cos^2 x} = -2$$

$$R_1 = \frac{R}{\left(\frac{dz}{dx}\right)^2} = \frac{+2\cos^4 x}{\cos^2 x} = 2\cos^2 x.$$

Changing the independent variable from x to z, the equation becomes

$$\frac{d^2y}{dz^2} - 2y = 2\cos^2 x = 2 - 2\sin^2 x = 2 - 2z^2 .$$

Here the complementary function $= c_1 e^{\sqrt{2}z} + c_2 e^{-\sqrt{2}z}$
and the particular integral

$$= \frac{1}{-2\left(1 - \frac{D^2}{2}\right)} . 2\left(1 - z^2\right) = -\left[1 + \frac{D^2}{2} + \frac{D^2}{4} + ...\right]\left(1 - z^2\right)$$

$$= -1\left(1 - z^2\right) + 1 = z^2$$

Hence the required solution is

$$y = c_1 e^{\sqrt{2}z} + c_2 e^{-\sqrt{2}z} + z^2 = c_1 e^{\sqrt{2}\sin x} + c_2 e^{-\sqrt{2}\sin x} + \sin^2 x .$$

Example 8: Solve $\frac{d^2y}{dx^2} + \left(1 - \frac{1}{x}\right)\frac{dy}{dx} + 4x^2 y e^{-2x} = 4\left(x^2 + x^3\right)e^{-3x}$.

Solution: Here $P = \left(1 - \frac{1}{x}\right)$, $Q = 4x^2 e^{-2x}$ and $R = 4\left(x^2 + x^3\right)e^{-3x}$.

Choosing z such that $\frac{d^2z}{dx^2} + P\left(\frac{dz}{dx}\right) = 0$, then

$$z = \int \exp\left(-\int P dx\right)dx = \int \exp\left[-\int\left\{\left(1 - \frac{1}{x}\right)dx\right\}\right]dx$$

$$= \int e^{-x+\log x} dx = \int x e^{-x}\, dx = -xe^{-x} - e^{-x} = -e^{-x}(1 + x) ,$$

$$Q_1 = \frac{Q}{\left(\frac{dz}{dx}\right)^2} = \frac{4x^2 e^{-2x}}{x^2 e^{-2x}} = 4 ,$$

$$R_1 = \frac{R}{\left(\frac{dz}{dx}\right)^2} = \frac{4x^3(1 + x)e^{-3x}}{x^2 e^{-2x}} = 4(1 + x)e^{-x} = -4z .$$

Changing independent variable from x to z, the equation becomes

$$\frac{d^2y}{dz^2} + 4y = -4z.$$

Clearly the complementary function $= c_1 \cos 2z + c_2 \sin 2z.$

The particular integral $= \frac{1}{4 + D^2}(-4z) = \frac{1}{4}\left(1 - \frac{D^2}{4} + ...\right)(-4z) = -z.$

Therefore $y = c_1 \cos 2z + c_2 \sin 2z - z$

Hence, the solution of the given equation is

$$y = c_1 \cos\left[-2(1+x)e^{-x}\right] + c_2 \sin\left[-2(1+x)e^{-x}\right] + (1+x)e^{-x}$$

$$= c_1 \cos\left[2(1+x)e^{-x}\right] - c_2 \sin\left[2(1+x)e^{-x}\right] + (1+x)e^{-x}$$

EXERCISE 3.3

Solve the following differential equations by changing the independent variable.

1. $\left(a^2 - x^2\right)\dfrac{d^2y}{dx^2} - \dfrac{a^2}{x}\dfrac{dy}{dx} + \dfrac{x^2}{a}y = 0.$

2. $x^6 \dfrac{d^2y}{dx^2} + 3x^5 \dfrac{dy}{dx} + a^2 y = \dfrac{1}{x^2}.$

3. $\left(1+x^2\right)\dfrac{d^2y}{dx^2} + 2x\left(1+x^2\right)\dfrac{dy}{dx} + 4y = 0.$

4. $\dfrac{d^2y}{dx^2} - \cot x \dfrac{dy}{dx} - \left(\sin^2 x\right) y = \left(\cos x - \cos^2 x\right).$

5. $\dfrac{d^2y}{dx^2} + (\tan x - 1)^2 \dfrac{dy}{dx} - x(x-1)y \sec^4 x = 0.$

6. $\dfrac{d^2y}{dx^2} - \left(1 + 4e^x\right)\dfrac{dy}{dx} + 3ye^{2x} = \exp\left\{2\left(x + e^x\right)\right\}$

7. $\dfrac{d^2y}{dx^2} + \tan x\left[\dfrac{dy}{dx} + y\cot^2 x\right] = 0.$

ANSWERS 3.3

1. $y = c_1 \sin\left[\sqrt{\left(\dfrac{a^2 - x^2}{a^2}\right)} + c_2\right].$

2. $y = c_1 \cos\left(\dfrac{a}{2x^2}\right) + c_2 \sin\left(\dfrac{a}{2x^2}\right) + \dfrac{1}{a^2x^2}.$

3. $y\left(1+x^2\right) = c_1\left(1-x^2\right) + 2c_2 x.$

4. $y = c_1 e^{-\cos x} + c_2 e^{\cos x} - \cos x.$

5. $y = c_1 e^{-x \tan x} + c_2 e^{(x-1)\tan x}.$

6. $y = c_1 e^{3x} + c_2 e^{x} - e^{2e^x}.$

7. $y = c_1 \sin(\sin x) + c_2 \cos(\sin x).$

3.5 METHOD OF VARIATION OF PARAMETERS

Here we discuss the method to find the solution of $y'' + Py' + Qy = R$ by method of variation of parameters.

Method: Consider the differential equation

$$\frac{d^2y}{dx^2} + P\frac{dy}{dx} + Qy = R. \qquad ...(1)$$

Let the complementary function of (1) be

$$y = Au + Bv \qquad ...(2)$$

where u and v are the functions of x and A, B are constants. As u and v are parts of C.F., so

$$\frac{d^2u}{dx^2} + P\frac{du}{dx} + Qu = 0 \qquad ...(3)$$

and $$\frac{d^2v}{dx^2} + P\frac{dv}{dx} + Qv = 0. \qquad ...(4)$$

Let the complete solution of (1) be

$$y = Au + Bv \qquad ...(5)$$

where A and B are not constants, but suitable functions of x to be chosen such that (5) satisfies (1). Now from (5), we have

$$y' = Au' + Bv' + (A'u + B'v). \qquad ...(6)$$

Let us choose A and B such that $A'u + B'v = 0$. ...(7)

Then (6) becomes $y' = Au' + Bv'$ ...(8)

It gives $$y'' = A'u' + Au'' + B'v' + Bv''. \qquad ...(9)$$

Substituting the values of y, y', y'' in (1), we get

$$(A'u' + Au'' + B'v' + Bv'') + P(Au' + Bv') + Q(Au + Bv) = R$$

$$\Rightarrow \quad A'u' + B'v' + A(u'' + Pu' + Qu) + B(v'' + Pv' + Qv) = R$$

$$\Rightarrow \quad A'u' + B'v' = R, \text{ using (3) and (4)} \qquad ...(10)$$

Solving (7) and (10) for A' and B', we get

$$\frac{A'}{-Rv} = \frac{B'}{Ru} = \frac{1}{uv' - u'v}$$

This implies that

$$A' = \frac{-Rv}{uv' - u'v} = \phi(x)(say) \qquad ...(11)$$

$$B' = \frac{Ru}{uv' - u'v} = \psi(x)(say) \qquad ...(12)$$

Integrating (11), we get $A = \int \phi(x)dx + a$, where a is a constant of integration. ...(13)

Integrating (12), we get $B = \int \psi(x)dx + b$ where b is a constant of integration. ...(14)

Putting the above values in (5), we get

$$y = \left[\int \phi(x)dx + a\right]u + \left[\int \psi(x)dx + b\right]v$$

$$\Rightarrow \quad y = u\int \phi(x) + v\int \psi(x)dx + au + bv.$$

This gives the complete solution of (1).

Steps for Solution

1. First find the parts of C.F.
2. Write them by u and v.
3. Consider $y = Au + Bv$ as the complete solution
4. A and B are determined by the formulae

   $$A = \int \frac{-Rv}{uv' - u'v}dx + c_1, \quad B = \int \frac{Ru}{uv' - u'v}dx + c_2$$

   where c_1 and c_2 are the arbitrary constants of integration.
5. Thus the complete solution is given by $y = Au + Bv$.

   Note: We can also use the formulae

   $$A = \int -\frac{Rv}{W}dx + a \text{ and } B = \int \frac{Ru}{W}dx + b$$

   to find A and B respectively, where $W = \begin{vmatrix} u & v \\ u' & v' \end{vmatrix} = uv' - u'v$ is called Wronskian of u, v which are parts of complementary function.

Example 1: Solve $\dfrac{d^2y}{dx^2} + a^2y = \sec ax$ by the method of variation of parameters.

Solution: The complementary function is

$$y = A\cos ax + B\sin ax. \qquad ...(1)$$

Now let A and B be the functions of x and let (1) satisfies the given equation, then

$$\frac{dy}{dx} = -Aa\sin ax + Ba\cos ax + \cos ax\frac{dA}{dx} + \sin ax\frac{dB}{dx}.$$

Now if $\cos ax \dfrac{dA}{dx} + \sin ax \dfrac{dB}{dx} = 0,$ ...(2)

Then $\dfrac{d^2y}{dx^2} = - Aa^2 \cos ax - Ba^2 \sin ax - \dfrac{dA}{dx} a \sin x + \dfrac{dB}{dx} a \cos ax.$

If (1) satisfies the given equation, when A and B are the function of x, we have

$$-\frac{dA}{dx} . a \sin ax + a \cos ax \frac{dB}{dx} = \sec ax . \quad ...(3)$$

Solving (2) and (3), we get

$$a\frac{dB}{dx} = 1 \text{ and } a\frac{dA}{dx} = - \tan x$$

On integrating, we get

$$A = \frac{1}{a^2} \log \cos ax + c_1 \text{ and } B = \frac{x}{a} + c_2 .$$

Hence the solution is

$$y = c_1 \cos ax + c_2 \sin ax + \frac{1}{a^2}(\log \cos ax)\cos ax + \frac{x}{a} \sin ax .$$

Example 2: Solve by the method of variation of parameters

$$\frac{d^2y}{dx^2} - y = \frac{2}{1+e^x} .$$

Solution: Here $u = e^x$, $v = e^{-x}$ which are parts of C.F., also $R = \dfrac{2}{1+e^x}$.

Let $y = Ae^x + Be^{-x}$ be the complete solution of given equation where A and B are the suitable functions of x, then A and B are determined as follows:

$$A = \int \frac{-Rv}{uv' - u'v} dx + c_1 = \int \frac{-2.e^{-x}}{(1+e^x)\{e^x(-e^{-x}) - e^x . e^{-x}\}} dx + c_1$$

$$= - 2\int \frac{e^{-x}}{(1+e^x)(-2)} dx + c_1 = \int \frac{e^{-x}}{1+e^x} dx + c_1$$

$$= \int \frac{e^{-x}}{1+e^{-x}} dx + c_1 = \log\left[\frac{1+e^x}{e^x}\right] - e^{-x} + c_1$$

$$B = \int \frac{Ru}{uv' - u'v} dx + c_2 = \int \frac{2e^x}{(1+e^x)\{e^x(-e^{-x} - e^x.e^{-x})\}} dx + c_2$$

$$= - \int \frac{e^x}{1+e^x} dx + c_2 = - \log(1+e^x) + c_2 .$$

Hence the complete solution is

$$y = \left[\log\left(\frac{1+e^x}{e^x}\right) - e^{-x} + c_1\right]e^x + \left[-\log\left(1+e^x\right) + c_2\right]e^{-x}$$

where c_1 and c_2 are arbitrary constants of integration.

Example 3: Solve $x^2\dfrac{d^2y}{dx^2} - 2x(1+x)\dfrac{dy}{dx} + 2(x+1)y = x^3$ by the method of variation of parameters.

Solution: The given equation can be written as

$$\frac{d^2y}{dx^2} - \frac{2}{x}(1+x)\frac{dy}{dx} + \frac{2}{x^2}(x+1)y = x.$$

Let us first solve

$$\frac{d^2y}{dx^2} - \frac{2}{x}(1+x)\frac{dy}{dx} + \frac{2}{x^2}(1+x)y = 0. \qquad ...(1)$$

Since $P + Qx = 0$, so $y = x$ is a part of its solution. Putting $y = vx$, we get

$$\frac{d^2v}{dx^2} - 2\frac{dv}{dx} = 0$$

Its solution is

$$v = Ae^{2x} + B$$

Hence $y = Axe^{2x} + Bx$ is a solution of (1) which is also the complementary function of the given equation. Now let A and B are functions of x, then

$$\frac{dy}{dx} = 2\,Axe^{2x} + Ae^{2x} + B + xe^{2x}\frac{dA}{dx} + x\frac{dB}{dx}$$

Now, if

$$xe^{2x}\frac{dA}{dx} + x\frac{dB}{dx} = 0 \qquad ...(2)$$

then $\dfrac{d^2y}{dx^2} = 4Axe^{2x} + 2Ae^{2x} + \dfrac{dB}{dx} + (2x+1)e^{2x}\dfrac{dA}{dx}.$

Substituting these values in the given equation, we get

$$\frac{dB}{dx} + e^{2x}\frac{dA}{dx}(2x+1) = x. \qquad ...(3)$$

Solving (2) and (3) by cross multiplication, we get

$$\frac{dB}{dx} = -\frac{1}{2} \Rightarrow B = -\frac{x}{2} + c_1$$

and $\dfrac{dA}{dx}=\dfrac{1}{2}e^{-2x} \Rightarrow A=-\dfrac{1}{4}e^{-2x}+c_2$.

Hence the required solution is

$$y=\left[-\frac{1}{4}e^{-2x}+c_2\right]xe^{2x}+\left(-\frac{x}{2}+c_1\right)x=-\frac{x}{4}-\frac{x^2}{2}+c_1\,x+c_2x\,e^{2x}.$$

Example 4: Solve $\dfrac{d^2y}{dx^2}+(1-\cot x)\dfrac{dy}{dx}-y\cot x=\sin^2 x$ by the method of variation of parameters.

Solution: Let us first solve

$$\frac{d^2y}{dx^2}+(1-\cot x)\frac{dy}{dx}-y\cot x=0 \qquad ...(1)$$

Here $1-P+Q=1-(1-\cot x)-\cot x=0$, so $y=e^{-x}$ is a solution of (1).

Putting $y=ve^{-x}$, the equation (1) reduces to

$$\frac{d^2v}{dx^2}+\left(1-\cot x-\frac{2}{e^{-x}}.e^{-x}\right)\frac{dv}{dx}=0$$

$$\Rightarrow \quad \frac{dp}{dx}=(1+\cot x)p\text{, where }\frac{dv}{dx}=p$$

$$\Rightarrow \quad \frac{dp}{p}=(1+\cot x)dx.$$

On integrating both sides, we get + $\log c_1+\log p=x+\log\sin x$

$$\Rightarrow \quad \frac{dv}{dx}=c_1\sin x\,.\,e^{x}$$

Again integrating, we get

$$v=\int e^{x}\sin x\,dx=-c_1\frac{e^{x}}{2}(\cos x-\sin x)+c_2$$

Therefore, we have $y=ve^{-x}=A(\cos x-\sin x)+Be^{-x}$.

which is the complementary function of the given equation.

Now let A and B are functions of x, then

$$\frac{dy}{dx}=A(-\sin x-\cos x)-Be^{-x}+(\cos x-\sin x)\frac{dA}{dx}+e^{-x}\frac{dB}{dx}$$

If $\dfrac{dA}{dx}(\cos x-\sin x)+e^{-x}\dfrac{dB}{dx}=0$, ...(2)

then $\frac{d^2y}{dx^2} = A(-\cos x + \sin x) + Be^{-x} + (-\sin x - \cos x)\frac{dA}{dx} - e^{-x}\frac{dB}{dx}$.

Putting these values in the given equation, we get

$$(-\sin x - \cos x)\frac{dA}{dx} - e^{-x}\frac{dB}{dx} = \sin^2 x. \quad ...(3)$$

Solving (2) and (3) by cross multiplication, we get

$$\frac{dA}{dx} = \frac{-\sin x}{2} \Rightarrow A = \frac{\cos x}{2} + c_1$$

and $\frac{dB}{dx} = \frac{e^x \sin 2x}{4} - \frac{e^x(1-\cos 2x)}{4}$

$$\Rightarrow \quad B = -\frac{e^x}{20}(2\cos 2x - \sin 2x) + c_2 - \frac{e^x}{4} - \frac{e^x}{20}(-2\sin 2x - \cos 2x).$$

Hence the solution is

$$y = \left(\frac{\cos x}{2} + c_1\right)(\cos x - \sin x) - \left[\frac{e^x}{20}(\cos 2x - 3\sin 2x) - \frac{e^x}{4} + c_2\right]e^{-x}$$

$$= c_1(\cos x - \sin x) - \frac{1}{10}(\sin 2x - 2\cos 2x) + c_2 e^{-x} + \frac{1}{4}.$$

Example 5: Solve $\frac{d^2y}{dx^2} + y = \operatorname{cosec} x$ by the method of variation of parameters.

Solution: A.E. $D^2 + 1 = 0 \Rightarrow D = \pm i$

Therefore $C.F. = A\cos x + B\sin x$

Here $u = \cos x$, $v = \sin x$ so that $uv' - u'v = \sin^2 x + \cos^2 x = 1$. Let $y = A\cos x + B\sin x$ be the complete solution of given equation where A and B are suitable functions of x, then A and B are determined as follows:

$$u = \int \frac{-Rv}{uv' - u'v}dx + c_1 = \int \frac{-\sin x . \operatorname{cosec} x \, dx}{1} + c_1 = -\int dx + c_1 = -x + c.$$

$$v = \int \frac{Ru}{uv' - u'v}dx + c_2 = \int \frac{\cos x . \operatorname{cosec} x \, dx}{1} + c_2 = \log \sin x + c_2$$

Hence the required solution is

$$y = c_1 \cos x + c_2 \sin x - x\cos x + \sin x \log \sin x.$$

Example 6: Solve $\frac{d^2y}{dx^2} + 4y = 4\tan 2x$ by the method of variation of parameters.

Solution: Here A.E. is $m^2 + 4 = 0 \Rightarrow m = \pm 2i$.

So the complementary function $= A\cos 2x + B\sin 2x$.

Let $y = A\cos 2x + B\sin 2x$ be the complete solution of given equation where A and B are functions of x, then

$$\frac{dy}{dx} = -2A\sin 2x + 2B\cos 2x + \cos 2x\frac{dA}{dx} + \sin 2x\frac{dB}{dx}.$$

If $$\cos 2x\frac{dA}{dx} + \sin 2x\frac{dB}{dx} = 0. \qquad ...(1)$$

Then $$\frac{d^2y}{dx^2} = -4A\cos 2x - 4B\sin 2x - 2\sin 2x\frac{dA}{dx} + 2\cos 2x\frac{dB}{dx}.$$

Substituting these values in the given equation, we have

$$\cos 2x\frac{dB}{dx} - \sin 2x\frac{dA}{dx} = 2\tan 2x. \qquad ...(2)$$

Solving (1) and (2) by cross multiplication, we get

$$\frac{dB}{dx} = 2\sin 2x \Rightarrow B = -\cos 2x + c_1$$

and $$\frac{dA}{dx} = \frac{2\sin^2 2x}{\cos 2x} = 2\cos 2x - 2\sec 2x.$$

$$\Rightarrow \quad A = \sin 2x - \log(\sec 2x + \tan 2x) + c_2.$$

Hence the solution is

$$y = [\sin 2x - \log(\sec 2x + \tan 2x) + c_2]\cos 2x + (c_1 - \cos 2x)\sin 2x$$

$$= c_1\sin 2x + c_2\cos 2x - \log[\sec 2x + \tan 2x]\cos 2x.$$

Example 7: Solve $(x+2)y_2 - (2x+5)y_1 + 2y = (x+1)e^x$ by the method of variation of parameters.

Solution: This equation can be written as

$$\frac{d^2y}{dx^2} - \frac{(2x+5)}{(x+2)}\frac{dy}{dx} + \frac{2}{(x+2)}y = \frac{(x+1)e^x}{(x+2)}$$

Here $2^2 + 2P + Q = 4 - \frac{2(2x+5)}{(x+2)} + \frac{2}{(x+2)} = 0$,

so e^{2x} is a part of C.F.

Now putting $y = e^{2x}.v$ in $\frac{d^2y}{dx^2} - \frac{(2x+5)}{(x+2)}\frac{dy}{dx} + \frac{2}{(x+2)}y = 0$, we get

$$\frac{d^2v}{dx^2} + \frac{(2x+3)}{(x+2)}\frac{dv}{dx} = 0$$

$$\Rightarrow \quad \frac{dp}{dx} + \frac{(2x+3)}{(x+2)}p = 0, \text{ where } p = \frac{dv}{dx}$$

$$\Rightarrow \quad p\frac{e^{2x}}{(x+2)} = c_1 \Rightarrow \frac{dv}{dx} = c_1(x+2)e^{-2x}$$

Integration yields

$$v = -c_1\frac{(x+2)}{2}.e^{-2x} - c_1\frac{e^{-2x}}{4} + c_2 = -c_1\frac{(2x+5)}{4}.e^{-2x} + c_2.$$

Hence the complementary function of the given equation is

$$y = A(2x+5) + Be^{2x}$$

Now let A and B are function of x, then

$$\frac{dy}{dx} = (2x+5)\frac{dA}{dx} + e^{2x}\frac{dB}{dx} + 2A + 2Be^{2x}$$

If $\quad (2x+5)\frac{dA}{dx} + e^{2x}\frac{dB}{dx} = 0. \qquad ...(1)$

Then $\frac{d^2y}{dx^2} = 2\frac{dA}{dx} + 2\frac{dB}{dx}e^{2x} + 4Be^{2x}.$

Substituting these values in the given equation, we get

$$2(x+2)\frac{dA}{dx} + 2(x+2)\frac{dB}{dx}e^{2x} = (x+1)e^{x}. \qquad ...(2)$$

Solving (2) and (3), we get

$$\frac{dA}{dx} = -\frac{(x+1)e^x}{4(x+2)^2} = -\frac{e^x}{4}\left[\frac{1}{(x+2)} - \frac{1}{(x+2)^2}\right]$$

$$\Rightarrow \quad A = -\frac{e^x}{4(x+2)} + c_1$$

and $\quad \frac{dB}{dx} = \frac{(2x+5)(x+1)e^{-x}}{4(x-2)^2} = \frac{e^{-x}}{4}\left[2 - \frac{1}{(x+2)} - \frac{1}{(x+2)^2}\right]$

$$\Rightarrow \quad B = \frac{e^{-x}}{4}\left[\frac{1}{(x+2)} - 2\right] + c_2 .$$

Hence the solution is

$$y = (2x+5)\left[-\frac{e^x}{4(x+2)} + c_1\right] + \frac{e^x}{4}\left[\frac{1}{(x+2)} - 2\right] + c_2 e^{2x}$$

$$= c_1(2x+5) + c_2 e^{2x} - e^x .$$

EXERCISE 3.4

Solve the following differential equations by the method of variation of parameters.

1. $(1-x)\frac{d^2x}{dy^2} + x\frac{dy}{dx} - y = (1-x)^2.$

2. $\frac{d^2y}{dx^2} + (\tan x - 3\cos x)\frac{dy}{dx} + 2\cos^2 x . y = \cos^4 x.$

3. $\frac{d^2y}{dx^2} + y = \sec x \tan x .$

4. $\frac{d^3y}{dx^3} - 6\frac{d^2y}{dx^2} + 11\frac{dy}{dx} - 6y = e^{2x}.$

5. $\frac{d^2y}{dx^2} - 3\frac{dy}{dx} + 2y = \frac{e^x}{1+e^x}.$

6. $(D^2+1)y = \tan^2 x .$

7. $\frac{d^2y}{dx^2} + y = \tan x$

8. $(1-x^2)\frac{d^2y}{dx^2} + x\frac{dy}{dx} - y = x(1-x^2)^{3/2} .$

9. $\frac{d^2y}{dx^2} - 2y = e^x \sin x .$

ANSWERS 3.4

1. $y = c_1 e^x + c_2 x + (x^2 + x + 1).$

2. $y = c_1 e^{\sin x} + c_2 e^{2\sin x} - \frac{5}{4} - \frac{3}{2}\sin x - \frac{1}{2}\sin^2 x .$

3. $y = c_1 \cos x + c_2 \sin x + x\cos x + \sin x \log \sec x - \sin x .$

4. $y = -xe^{2x} + c_1 e^x + c_2 e^{2x} + c_2 e^{3x} .$

5. $y=(e^x+e^{2x})\log(1+e^x)+(c_2-x)e^{2x}+(c_1-1-x)e^x$.

6. $y=c_1\cos x+c_2\sin x-\cos(\sec x-\cos x)+\sin x\log(\sec x+\tan x)+\sin^2 x$.

7. $y=c_1\cos x+c_2\sin x-\cos x\log(\sec x+\tan x)$.

8. $y=c_1(\sqrt{1-x^2}+x\sin^{-1}x)+c_2x-\frac{x}{9}(1-x^2)^{3/2}$.

9. $y=c_1+c_2e^{2x}-\frac{e^x}{2}\sin x$.

3.6 DIFFERENTIAL EQUATIONS OF OTHER TYPES

3.6.1 Equations of the Form $y''=f(y)$

Steps for solving differential equation of the form

$$\frac{d^2y}{dx^2}=f(y), \quad ...(1)$$

are as given below:

Step 1: Multiply both sides by $2\frac{dy}{dx}$ to get

$$2\frac{dy}{dx}\cdot\frac{d^2y}{dx^2}=2\frac{dy}{dx}f(y), \text{ which is now integrable.} \quad ...(2)$$

Step 2: Integrate (2) w. r. t. 'x' as

$$\left(\frac{dy}{dx}\right)^2=2\int f(y)dy=\psi(y)+c_1, \; c_1 \text{ is a constant of integration.}$$

$$\Rightarrow \quad \frac{dy}{dx}=\sqrt{\psi(y)+c_1}, \text{ which is again integrable} \quad ...(3)$$

Step 3: Integrate (3) w. r. t. 'x' as $\int\frac{dy}{\sqrt{\psi(y)+c_1}}=x+c_2$, where c is a constant of integration.

Example 1: Solve $\sin^3 y\frac{d^2y}{dx^2}=\cos y$.

Solution: The given equation may be written as

$$\frac{d^2y}{dx^2}=\operatorname{cosec}^2 y\cot y \quad ...(1)$$

On multiplying both sides by $2\dfrac{dy}{dx}$, we get

$$2\frac{dy}{dx}\frac{d^2y}{dx^2} = 2\operatorname{cosec}^2 y \cot y \frac{dy}{dx} \qquad ...(2)$$

On integrating, we get

$$\left(\frac{dy}{dx}\right)^2 = -\cot^2 y + c_1$$

$$\Rightarrow \quad \frac{dy}{dx} = \sqrt{(c_1 - \cot^2 y)} \Rightarrow \frac{dy}{\sqrt{c_1 - \cot^2 y}} = dx$$

$$\Rightarrow \quad \frac{\sin y\, dy}{\sqrt{c_1 \sin^2 y - \cos^2 y}} = dx \Rightarrow \frac{\sin y\, dy}{\sqrt{c_1 - (1 + c_1)\cos^2 y}} = dx$$

Again integrating, we get $-\dfrac{1}{\sqrt{1+c_1}} \sin^{-1}\left(\dfrac{\sqrt{(1+c_1)}}{c_1} \cos y\right) = x + c_2$

$$\Rightarrow \quad \sin\left[(x + c_2)\sqrt{(1+c_1)}\right] + \frac{\sqrt{(1+c_1)}}{c_1}\cos y = 0.$$

Example 2: Solve $\dfrac{d^2y}{dx^2} = \sec^2 y \tan y$ under the conditions

$y = 0$ and $\dfrac{dy}{dx} = 1$ when $x = 0$.

Solution: On multiplying both sides by $2\dfrac{dy}{dx}$, we get

$$2\frac{dy}{dx}\frac{d^2y}{dx^2} = 2\sec^2 y \tan y \frac{dy}{dx}.$$

Integration yields

$$\left(\frac{dy}{dx}\right)^2 = \tan^2 y + c_1 \Rightarrow \frac{dy}{dx} = \sqrt{\tan^2 y + c_1}. \qquad ...(1)$$

On putting $y = 0$ and $\dfrac{dy}{dx} = 1$ at $x = 0$, we get $c_1 = 1$. Therefore by (1) we have

$$\frac{dy}{dx} = \sqrt{\tan^2 y + 1} = \sqrt{\sec^2 y} = \sec y \Rightarrow \cos y\, dy = dx.$$

On integrating, we get

$$\sin y = x + c_2 \qquad ...(2)$$

On putting $y=0$, $x=0$, we get $c_2=0$

Hence from (2), we get $\sin y = x \Rightarrow y = \sin^{-1} x$.

Example 3: A motion is governed by $\dfrac{d^2x}{dt^2}=\dfrac{36}{x^2}$ given that at $t=0$, $x=8$ and $\dfrac{dx}{dt}=0$, find the displacement at any time t.

Solution: On multiplying both sides by $2\dfrac{dx}{dt}$, we get

$$2\frac{dx}{dt}\frac{d^2x}{dt^2}=\frac{36}{x^2}2\frac{dx}{dt}.$$

Integration yields $\left(\dfrac{dx}{dt}\right)^2=-\dfrac{72}{x}+c_1$ ...(1)

But at $t=0$, $\dfrac{dx}{dt}=0$, $x=8$, so we have $0=-\dfrac{72}{8}+c_1 \Rightarrow c_1=9$.

Thus from equation (1), we have

$$\left(\frac{dx}{dt}\right)^2=-\frac{72}{x}+9=\frac{-72+9x}{x} \Rightarrow \frac{dx}{dt}=3\sqrt{\frac{x-8}{x}}$$

On integrating, we get

$$\int\frac{\sqrt{x}\,dx}{\sqrt{x-8}}=3\int dt+c_2 \Rightarrow \int\frac{x}{\sqrt{x^2-8x}}dx=3t+c_2$$

$$\Rightarrow \quad \frac{1}{2}\int\left[\frac{2x-8}{\sqrt{x^2-8x}}+\frac{8}{\sqrt{(x-4)^2-(4)^2}}\right]dx=3t+c_2$$

$$\Rightarrow \quad \sqrt{x^2-8x}+4\cosh^{-1}\frac{x-4}{4}=3t+c_2.$$

At $t=0$, $x=8$, we get $c_2=0$. Hence the required solution is

$$\sqrt{x^2-8x}+4\cosh^{-1}\frac{x-4}{4}=3t.$$

3.6.2 Equations of the Form $\dfrac{d^2y}{dx^2}=f(x)$

To solve the equations of this type, we first integrate to get

$$\frac{dy}{dx}=\int f(x)dx+c=F(x)(say).$$

Then again integrating, we get $y = \int F(x)dx + c'$ as the required solution.

3.6.3 Equations of the Form $\dfrac{d^n y}{dx^n} = f(x)$

On integrating, again and again n times, we get the required solution.

Example 4: Solve $\dfrac{d^2y}{dx^2} = xe^x$.

Solution: Integration yields $\dfrac{dy}{dx} = xe^x - e^x + c_1 = e^x(x-1) + c_1$

Again integrating, we get $y = (x-1)e^x - e^x + c_1 x + c_2$

$$= e^x(x-2) + c_1 x + c_2 .$$

Example 5: Solve $\dfrac{d^2y}{dx^2} = x^2 \sin x$.

Solution: On integrating, we get

$$\frac{dy}{dx} = -x^2 \cos x + \int 2x \cos x \, dx + c_1$$

$$= -x^2 \cos x + 2x \sin x - \int 2 \sin x \, dx + c_1$$

$$= -x^2 \cos x + 2x \sin x + 2 \cos x + c_1$$

Again integrating, we get

$$y = -x^2 \sin x + \int 2x \sin x \, dx - 2x \cos x + \int 2 \cos x \, dx + 2 \sin x + c_1 x + c_2$$

$$= -x^2 \sin x - 2x \cos x + 2 \sin x - 2x \cos x - 2 \sin x + 2 \sin x + c_1 x + c_2$$

$$= -x^2 \sin x - 4x \cos x + 2 \sin x + c_1 x + c_2 .$$

3.6.4 Equations of the Form which does not Contain *y* Directly

In this case, the order of equation may be expressed by assuming as a dependent variable, the lowest derivative which presents itself in the given equation. Thus when we write $\dfrac{dy}{dx} = p, \dfrac{d^2y}{dx^2} = \dfrac{dp}{dx}, \ldots, \dfrac{d^n y}{dx^n} = \dfrac{d^{n-1}p}{dx^{n-1}}$ in the general equation

$$f\left(\frac{d^n y}{dx^n} \cdots \frac{dy}{dx}, x\right) = 0,$$

the order of the equation is lowered by one and it takes the form

$$f\left(\frac{d^{n-1}p}{dx^{n-1}}, \dots, p, x\right) = 0.$$

This may be possible to solve for p.

3.6.5 Equations of the Form which does not Contain *x* Directly

The equations that do not contain x directly are of the form

$$f\left(\frac{d^n y}{dx^n}, \frac{d^{n-1}y}{dx^{n-1}}, \dots \frac{dy}{dx}, y\right) = 0. \qquad \dots(1)$$

On substituting $\frac{dy}{dx} = p$, $\frac{d^2y}{dx^2} = \frac{dp}{dy} \cdot \frac{dy}{dx} = \frac{dp}{dy} p$ in equation (1), we get

$$\left[\frac{dp^{n-1}}{dx^{n-1}} \dots p, y\right] = 0. \qquad \dots(2)$$

Now equation (2) is solved for p. Let after integration, we obtain

$$p = f_1(y) \Rightarrow \frac{dy}{dx} = f_1(y) \Rightarrow \frac{dy}{f_1(y)} = dx$$

Integration yields $\int \frac{dy}{f_1(y)} = x + c$.

3.6.6 Equations of the form $f\left(\frac{d^n y}{dx^n}, \frac{d^{n-1}y}{dx^{n-1}}, x\right) = 0$, where differential coefficients differ in order by 1

In such case, we write $q = \frac{d^{n-1}y}{dx^{n-1}}$ so that the equation reduces to

$$f\left(\frac{dq}{dx}, q, x\right) = 0$$

which is an equation of I[st] order between q and x and may be integrated for q.

Thus if $q = \frac{d^{n-1}y}{dx^{n-1}} = F(x)$ then we can find y by successive integration.

3.3.7 Equations of the form $f\left(\frac{d^n y}{dx^n}, \frac{d^{n-2}y}{dx^{n-2}}, x\right) = 0$, where differential coefficients differ in order by 2

In such cases if we write $q = \frac{d^{n-2}y}{dx^{n-2}}$, then equation reduces to

$$f\left(\frac{d^2q}{dx^2}, q, x\right) = 0$$

from which q may be found as it is a second order differential equation in q and x.

Thus if $q = \frac{d^{n-2}y}{dx^{n-2}} = F(x)$, then we can find value of y by successive integration.

Example 6: Solve $\frac{d^2y}{dx^2} = \left[1 - \left(\frac{dy}{dx}\right)^2\right]^{1/2}$.

Solution: Given equation does not contain x directly, so set $\frac{dy}{dx} = p$

then $\frac{d^2y}{dx^2} = \frac{dp}{dx}$.

Thus the equation reduces to $\frac{dp}{dx} = \sqrt{1-p^2} \Rightarrow \frac{dp}{\sqrt{1-p^2}} = dx.$

On integrating, we get

$$\sin^{-1} p = x + c_1 \Rightarrow p = \sin(x + c_1) \Rightarrow \frac{dy}{dx} = \sin(x + c_1)$$

Integration yields $y = -\cos(x + c_1) + c_2$.

Example 7: Solve $y\frac{d^2y}{dx^2} + \left(\frac{dy}{dx}\right)^2 = y^2$. ...(1)

Solution: Given equation does not contain x directly, so set $\frac{dy}{dx} = p$,

then

$$\frac{d^2y}{dx^2} = \frac{dp}{dx} = \frac{dp}{dy} \cdot \frac{dy}{dx} = p\frac{dp}{dy}$$

Using these in (1), we get

$$yp\frac{dp}{dy} + p^2 = y^2 \Rightarrow p\frac{dp}{dy} + \frac{p^2}{y} = y \qquad \text{...(2)}$$

Let $p^2 = z$ so that $2p\frac{dp}{dy} = \frac{dz}{dy}$, then (2) becomes

$$\frac{1}{2}\frac{dz}{dy} + \frac{z}{y} = y \Rightarrow \frac{dz}{dy} + \frac{2z}{y} = 2y$$

which is linear equation in z. Now $I.F. = e^{\int 2/y\, dy} = e^{2\log y} = y^2$.
Hence the solution is given by

$$y^2 z = \int 2y.(y)^2\, dy + c \Rightarrow p^2 y^2 = \frac{y^4}{2} + c$$

$$\Rightarrow \quad 2p^2y^2 = y^4 + k \quad \Rightarrow \sqrt{2}\, yp = \sqrt{y^4 + k}$$

$$\Rightarrow \quad \sqrt{2}\, y\frac{dy}{dx} = \sqrt{y^4 + k} \quad \Rightarrow \frac{\sqrt{2}ydy}{\sqrt{y^4 + k}} = dx$$

$$\Rightarrow \quad \frac{1}{\sqrt{2}}\frac{dt}{\sqrt{t^2 + k}} = dx \text{, using } y^2 = t \text{ and } 2y\, dy = dt.$$

On integrating, we obtain

$$\frac{1}{\sqrt{2}}\sin h^{-1}\frac{t}{k} = x + c \Rightarrow y^2 = k \sin h\left(\sqrt{2}x + c\right).$$

Example 8: Solve $y(1 - \log y)\frac{d^2y}{dx^2} + (1 + \log y)\left(\frac{dy}{dx}\right)^2 = 0$. ...(1)

Solution: Given equation does not contain x directly. So set $\frac{dy}{dx} = p$, then

$$\frac{d^2y}{dx^2} = \frac{dp}{dx} = \frac{dp}{dy}.\frac{dy}{dx} = p.\frac{dp}{dy}$$

Thus equation (1) becomes $y(1 - \log y)\frac{dp}{dy}p + (1 + \log y)p^2 = 0$

$$\Rightarrow \quad y(1 - \log y)\frac{dp}{dy} + (1 + \log y)p = 0$$

$$\Rightarrow \quad \frac{dp}{p} = -\frac{(1 + \log y)}{(1 - \log y)}\frac{dy}{y}$$

$$\Rightarrow \quad \frac{dp}{p} = -\frac{(1 + t)}{(1 - t)}dt = \left(1 + \frac{2}{t - 1}\right)dt \text{ using } \log y = \text{t and } \frac{dy}{y} = dt.$$

On integrating, we get

$$\log p = t + 2\log(t - 1) + \log c$$

$$\Rightarrow \quad \log p = \log e^t + \log c(t - 1)^2 \Rightarrow p = ce^t(t - 1)^2$$

$$\Rightarrow \quad \frac{dy}{dx} = cy(\log y - 1)^2$$

$$\Rightarrow \quad \frac{dy}{y(\log y - 1)^2} = c\, dx.$$

On integrating, we get $-\dfrac{1}{(\log y-1)}=cx+d \Rightarrow (1-\log y)=\dfrac{1}{cx+d}$.

Example 9: Solve $y\dfrac{d^2y}{dx^2}-\left(\dfrac{dy}{dx}\right)^2=y^2\log y$. ...(1)

Solution: The given equation does not contain x directly. So let $\dfrac{dy}{dx}=p$,

then $\dfrac{d^2y}{dx^2}=p\dfrac{dp}{dy}$. Thus equation (1) becomes

$$yp\frac{dp}{dy}-p^2=y^2\log y$$

$$\Rightarrow \quad p\frac{dp}{dy}-\frac{1}{y}p^2=y\log y.$$

Put $p^2=v$ so that $2p\dfrac{dp}{dy}=\dfrac{dv}{dy}$. Hence the above equation becomes

$$\frac{1}{2}\frac{dv}{dy}-\frac{1}{y}v=y\log y \Rightarrow \frac{dv}{dy}-\frac{2}{y}v=2y\log y \qquad ...(2)$$

which is linear in v and y. Now $I.F.=e^{-\int(2/y)dy}=e^{-2\log y}=\dfrac{1}{y^2}$.

Hence solution of (2) is given by

$$v\frac{1}{y^2}=\int 2y\log y.\frac{1}{y^2}dy+c_1$$

$$\Rightarrow \quad \frac{p^2}{y^2}=2\int\log y.\frac{dy}{y}+c_1 \Rightarrow \left(\frac{dy}{dx}\right)^2\frac{1}{y^2}=\frac{2(\log y)^2}{2}+c_1$$

$$\Rightarrow \quad dx=\frac{dy}{y\sqrt{(\log y)^2+c_1}}$$

Let $\log y=t$, so that $\dfrac{dy}{y}=dt$. Then equation (4) becomes

$$dx=\frac{dt}{\sqrt{t^2+c_1}}$$

Integration yields,

$$\sinh^{-1}\frac{t}{\sqrt{c_1}}=x+c_2 \Rightarrow t=\sqrt{c_1}\sinh(x+c_2).$$

$$\Rightarrow \quad \log y=\sqrt{c_1}\sinh(x+c_2).$$

Example 10: Solve $\left(\frac{dy}{dx}\right)^2 - y\frac{d^2y}{dx^2} = n\left\{\left(\frac{dy}{dx}\right)^2 + a^2\left(\frac{d^2y}{dx^2}\right)\right\}^{1/2}$ ...(1)

Solution: The given equation does not contain x directly. So let $\frac{dy}{dx} = p$,

then $\frac{d^2y}{dx^2} = p\frac{dp}{dy}$.

Thus equation (1) becomes $p^2 - yp\frac{dp}{dy} = n\left[p^2 + a^2p^2\left(\frac{dp}{dy}\right)^2\right]^{1/2}$

$\Rightarrow \quad p = yp_1 + n\left(1 + a^2p_1^2\right)^{1/2}$, taking $\frac{dp}{dy} = p_1$.

which is Clairaut's form (in p and y as variables), so its solution is

$$p = yc + n\left(1 + a^2c^2\right)^{1/2} \Rightarrow \frac{dy}{dx} = yc + n\sqrt{1 + a^2c^2}$$

$$\Rightarrow \quad \frac{c\,dy}{yc + n\sqrt{(1 + a^2c^2)}} = c\,dx.$$

Integration yields $\log\left[cy + n\sqrt{1 + a^2c^2}\right] = cx + \log c'$

$$\Rightarrow \quad yc + n\sqrt{1 + a^2c^2} = c'e^{cx}.$$

Example 11: Solve $\frac{d^4y}{dx^4} - \cot x\frac{d^3y}{dx^3} = 0$. ...(1)

Solution: Let $\frac{d^3y}{dx^3} = z$, then $\frac{d^4y}{dx^4} = \frac{dz}{dx}$

Thus equation (1) becomes $\frac{dz}{dx} - z\cot x = 0 \Rightarrow \frac{dz}{z} = \cot x\,dx$

Integration yields $\log z = \log\sin x + \log c_1 \Rightarrow z = c_1\sin x$

$$\Rightarrow \quad \frac{d^3y}{dx^3} = c_1\sin x$$

On integrating, we get $\frac{d^2y}{dx^2} = -c_1\cos x + c_2$.

Successive integration two times yields

$$y = c_1\cos x + c_2\frac{x^2}{2} + c_3x + c_4.$$

Example 12: Solve $\frac{d^4y}{dx^4} - a^2\frac{d^2y}{dx^2} = 0$.

Solution: Let $\frac{d^2y}{dx^2} = q$, then given equation reduces to $\frac{d^2q}{dx^2} - a^2q = 0$.

Its solution is $q = c_1 \cos hax + c_2 \sin hax$

$$\Rightarrow \quad \frac{d^2y}{dx^2} = c_1 \cos hax + c_2 \sin hax.$$

Successive integration two times yields

$$y = \frac{c_1}{a^2}\cos hax + \frac{c_2}{a^2}\sin hax + c_3x + c_4.$$

EXERCISE 3.5

Solve the following differential equations

1. $\frac{d^2y}{dx^2} = \left[1 + \left(\frac{dy}{dx}\right)^2\right]^{1/2}$

2. $x\frac{d^2y}{dx^2} + x\left(\frac{dy}{dx}\right)^2 - \frac{dy}{dx} = 0.$

3. $2x\frac{d^2y}{dx^3}\cdot\frac{d^2y}{dx^2} = \left(\frac{d^2y}{dx^2}\right)^2 - a^2.$

4. $e^{x^2/2}\left[x\frac{d^2y}{dx^2} - \frac{dy}{dx}\right] = x^3.$

5. $\left(1+x^2\right)\frac{d^2y}{dx^2} + x\frac{dy}{dx} + ax = 0.$

6. $\frac{d^2y}{dx^2} - \frac{a^2}{x\left(a^2 - x^2\right)}\frac{dy}{dx} = \frac{x^2}{a\left(a^2 - x^2\right)}.$

7. $\left(1+x^2\right)\frac{d^2y}{dx^2} + 1 + \left(\frac{dy}{dx}\right)^2 = 0.$

8. $\frac{d^5y}{dx^5} - n^2\frac{d^3y}{dx^3} = e^{ax}.$

9. $\frac{d^2y}{dx^2} + \left(\frac{dy}{dx}\right)^2 + 1 = 0.$

10. $\frac{d^2y}{dx^2} = e^{-2y}.$

11. $\frac{d^2y}{dx^2} = 2(y^3 + y)$, when $y(0) = 0$, $y'(0) = 1$.

12. $y'' = \sqrt{y}$, when $y(0) = 1$, $y'(0) = \frac{2}{3}$.

13. $y'' = \sqrt{\left[1 - (y')^2\right]}$.

ANSWERS 3.5

1. $y = -\cos h(x + c_1) + c_2$.

2. $y = \log\left(x^2 + 2c_1\right) + c_2$.

3. $15c_1^2y = 4\left(c_1x + a^2\right)^{5/2} + c_2x + c_3$.

4. $y = e^{-x^2/2} + c_1\frac{x^2}{2} + c_2$.

5. $y = c_2 - ax + c_1 \log\left(x + \sqrt{1+x^2}\right)$.

6. $2ay + x^2 = c_1 \sqrt{a^2 - x^2} + c_2$. 7. $y = -cx + (1+c^2)\log(x-c) + c_1$.

8. $y = c_1 e^{nx} + c_2 e^{-nx} + c_3 + c_4 x + c_5 x^2 + e^{ax}/a^3(a^2 - x^2)$.

9. $y = \log \sin(x - c_1) + c$. 10. $\cos h(c_1 x + c_2) = c_1 e^y$.

11. $y = \tan x$. 12. $y^{1/4} = \frac{2}{\sqrt{3}} x + 1$.

13. $y = -\cos(x + c_1) + c_2$.

3.7 METHOD OF OPERATIONAL FACTORS

If operator is factorable then this method can be applied.

Consider a linear equation of second order

$$\frac{d^2y}{dx^2} + P\frac{dy}{dx} + Qy = R \quad ...(1)$$

Suppose it can be written as

$$f_1(D) f_2(D) y = R \quad ...(2)$$

where $f_1(D) f_2(D)$ are linear factors. Then by taking $f_2(D)y = v$, the equation (2) becomes

$$f_1(D)v = R.$$

Now it is a first order differential equation which can be solved by the method discussed earlier. This process will be clear from the following examples.

Example 1: Solve $x\frac{d^2y}{dx^2} + (x-2)\frac{dy}{dx} - 2y = x^3$.

Solution: This equation can be written as $\left[xD^2 + (x-2)D - 2\right]y = x^3$

$$\Rightarrow \quad (xD-2)(D+1)y = x^3.$$

Let $(D+1)y = v$, then we have

$$(xD-2)v = x^3 \Rightarrow \frac{dv}{dx} - \frac{2}{x}v = x^2.$$

which is a linear equation of first order.

Now $I.F. = e^{\int 1\frac{x}{2}dx} = e^{-2\log x} = \frac{1}{x^2}$.

Therefore its solution is

$$v.\frac{1}{x^2} = \int \frac{1}{x^2}.x^2\, dx + c_1 = x + c_1$$

$$\Rightarrow \quad v = x^3 + c_1 x^2.$$

Substituting the value of v in $(D+1)y = v$, we get

$$(D+1)y = x^3 + c_1\, x^2.$$

which is a linear equation of first order, so we have $I.F. = e^{\int 1.dx} = e^x$.
Therefore, the required solution is given by

$$y.e^x = \int \left(x^3 + c_1 x^2\right) e^x \, dx + c_2$$

$$= \left(x^3 + c_1 x^2\right) e^x - \left(3x^2 + 2c_1 x\right) e^x + \left(6x + 2c_1\right) e^x - 6e^x + c_2$$

$$\Rightarrow \quad y = x^3 + c_1 x^2 - 3x^2 - 2c_1 x + 6x + 2c_1 - 6 + c_2 e^{-x}$$

$$= x^3 + \left(c_1 - 3\right) x^2 - 2\left(c_1 - 3\right) x + 2\left(c_1 - 3\right) + c_2 e^{-x}.$$

Example 2:. Solve $\dfrac{d^3y}{dx^3} - 6\dfrac{d^2y}{dx^2} + 11\dfrac{dy}{dx} - 6y = e^{2x}$.

Solution: This equation can be written as $\left[D^3 - 6D^2 + 11D - 6\right] y = e^{2x}$.

$$\Rightarrow \quad (D-1)(D-2)(D-3)y = e^{2x}.$$

Let $(D-3)y = v$, then $(D-1)(D-2)\,v = e^{2x}$.

Again let $(D-2)v = z$, we have $(D-1)z = e^{2x}$

which is a linear equation of first order. Here $I.F. = e^{\int -1.dx} = e^{-x}$.

So that $z.e^{-x} = \int e^{2x}.e^{-x} \, dx + c_1 = e^x + c_1 \Rightarrow z = e^{2x} + c_1 e^x$.

Substituting value of z in $(D-2)v = z$, we obtain

$$(D-2)v = e^{2x} + c_1 e^x, \text{ a linear equation of first order.}$$

Now $I.F. = e^{\int -2dx} = e^{-2x}$.

Therefore $v.e^{-2x} = \int \left(e^{2x} + c_1 e^x\right) e^{-2x} \, dx + c_2 = x - c_1 e^{-x} + c_2$

$$\Rightarrow \quad v = xe^{2x} - c_1 e^x + c_2 e^{2x}.$$

On substituting value of v in $(D-3)y = v$, we obtain

$(D-3)y = xe^{2x} - c_1 e^x + c_2 e^{2x}$, a linear equation of first order.

Again $I.F. = e^{\int -3dx} = e^{-3x}$.

Therefore, the required solution is given by

$$y.e^{-3x} = \int \left(xe^{2x} - c_1 e^x + c_2 e^{2x}\right) e^{-3x} \, dx + c_3$$

$$= \int \left(xe^{-x} - c_1 e^{-2x} + c_2 e^{-x}\right) dx + c_3$$

$$= (-x-1)e^{-x} + \frac{c_1}{2} e^{-2x} - c_2 e^{-x} + c_3$$

$$\Rightarrow \quad y = (-x-1-c_2)e^{2x} + \frac{c_1}{2} e^x + c_3 e^{3x}.$$

EXERCISE 3.6

Solve the following differential equations by factorization of the operator.

1. $x\dfrac{d^2y}{dx^2}+(1-x)\dfrac{dy}{dx}-y=e^x.$

2. $x\dfrac{d^2y}{dx^2}-(1-x)\dfrac{dy}{dx}-y=x^2.$

3. $3x^2\dfrac{d^2y}{dx^2}+(2-6x^2)\dfrac{dy}{dx}-4y=0.$

ANSWERS 3.6

1. $y=e^x\log x+c_1 e^x\int\dfrac{e^{-x}}{x}dx+c_2e^x.$

2. $y=c_1(x-1)+c_2 e^{-x}+x^2.$

3. $y=c_2e^{2x}+c_1 e^{2x}\int e^{-\frac{2}{3x}-2x}.dx$

4

Series Solution of Second Order ODEs: Legendre and Bessel Functions

4.1 INTRODUCTION

We have discussed ordinary differential equations of second order with variable and constant coefficients in previous chapters, which have solutions in terms of elementary functions namely the algebraic functions and elementary transcendental functions.

In general, the higher order differential equations have no solutions in terms of elementary functions in a simple manner. In such cases one of the standard methods is to derive a pair of linearly independent solutions in the term of infinite series. The power series method is the standard method for solving linear differential equations with variable coefficients which gives solution in the form of power series.

An expression of the form

$$a_0 + a_1(x - x_0) + a_2(x - x_0)^2 + \ldots = \sum_{r=0}^{\infty} a_r (x - x_0)^r$$

is called power series in $(x - x_0)$ or centered at x_0.

In particular, the series $a_0 + a_1 x + a_2 x^2 + a_3 x^3 + \ldots = \sum_{r=0}^{\infty} a_r x^r$ is called a power series in x or centered at origin.

Examples of the power series are:

(a) $e^x = \sum_{r=0}^{\infty} \frac{x^r}{r!} = 1 + x + \frac{x^2}{2!} + \frac{x^3}{3!} + \ldots$

(b) $\sin x = \sum_{r=0}^{\infty} \frac{(-1).2x^{2r+1}}{(2r+1)!} = x - \frac{x^3}{3!} + \frac{x^5}{5!} - \ldots$

(c) $\cos x = \sum_{r=0}^{\infty} \frac{(-1)^r .x^{2r}}{(2r)!} = 1 - \frac{x^2}{2!} + \frac{x^4}{4!} - \ldots.$

(d) $\frac{1}{(1-x)} = \sum_{r=0}^{\infty} x^r = 1 + x + x^2 + x^3 + \ldots (|x| < 1)$

4.2 SERIES SOLUTION OF DIFFERENTIAL EQUATIONS

Consider the equation

$$P_0(x)\frac{d^2y}{dx^2} + P_1(x)\frac{dy}{dx} + P_2(x)y = 0. \qquad ...(1)$$

where $P_0(x), P_1(x), P_2(x)$ are polynomials in x. We have to find its solution in terms of infinite convergent series.

4.2.1 Ordinary Point

A point $x = x_0$ is said to be an ordinary point of the equation (1) if the functions $\frac{P_1(x)}{P_0(x)}$ and $\frac{P_2(x)}{P_0(x)}$ are both analytic at $x = x_0$. Simply, in other words, we can say that the point $x = 0$ is ordinary point of (1) if $P_0(0)$ does not vanish, i.e. $P_0(0) \neq 0$.

If $x=0$ is an ordinary point of (1), we consider its solution as

$$y = a_0 + a_1 x + a_2 x^2 + ... = \sum_{r=0}^{\infty} a_r x^r.$$

4.2.2 Linearly Dependent and Linearly Independent Series

Two power series are said to be linearly dependent if one is constant multiple of other, otherwise are said to linearly independent.

If the general solution of (1) in series form contains two linearly independent series, say y_1 and y_2, the complete solution of (1) is given by $y = c_1 y_1 + c_2 y_2$, where c_1 and c_2 are constants.

4.2.3 Method to find the series solution when $x = 0$ is an ordinary point of the differential equation

In this case, let series solution of (1) be

$$y = a_0 + a_1 x^1 + a_2 x^2 + a_3 x^3 + ... = \sum_{r=0}^{\infty} a_r x^r. \qquad ...(2)$$

Step 1: Find $\frac{dy}{dx}, \frac{d^2y}{dx^2}$ etc.

Step 2: Substitute the values of y, $\frac{dy}{dx}$ and $\frac{d^2y}{dx^2}$ in the given differential equation.

Step 3: Equating to zero the coefficients of x^0, x, we find easily a_2, a_3 in terms of a_0 and a_1.

Step 4: Equating to zero the coefficient of x^n, we get recurrence relations in terms of n.

Step 5: Determine the values of $a_4, a_5, a_6, \ldots$ from the recurrence relations in terms of a_0 and a_1 by putting $n = 2, 3, 4, \ldots$.

Step 6: Substituting the values of $a_1, a_2, a_4, \ldots$ in (2), we get required series solution.

Example 1: Solve $\dfrac{d^2y}{dx^2} + y = 0.$ ...(1)

Solution: Since $x = 0$ is an ordinary point of (1), let its series solution be

$$y = a_0 + a_1 x + a_2 x^2 + a_3 x^3 + \ldots = \sum_{r=0}^{\infty} a_r x^r. \quad \ldots(2)$$

From (2), we have $\dfrac{dy}{dx} = \sum_{r=1}^{\infty} r\, a_r\, x^{r-1}$, $\dfrac{d^2y}{dx^2} = \sum_{r=2}^{\infty} r(r-1) a_r\, x^{r-2}$.

Substituting the values of $y, \dfrac{dy}{dx}$ *and* $\dfrac{d^2y}{dx^2}$ in the given differential equation, we get

$$\sum_{r=2}^{\infty} r(r-1) a_r\, x^{r-2} + \sum_{r=0}^{\infty} a_r x^r = 0$$

$$\Rightarrow \left[2.1a_2 + 3.2\, a_3\, x + \ldots + (n+2)(n+1) a_{n+2}\, x^n + \ldots\right]$$

$$+\left[a_0 + a_1\, x + a_2\, x^2 + \ldots + a_n x^n + \ldots\right] = 0$$

$$\Rightarrow (2a_2 + a_0) + (6a_3 + a_1)x + \ldots + \left[(n+2)(n+1)a_{n+2} + a_n\right]x^n + \ldots = 0$$

Equating to zero the coefficients of x^0 and x^1, we obtain

$$2a_2 + a_0 = 0, \text{ i.e. } a_2 = -\frac{a_0}{2!}$$

$$6a_3 + a_1 = 0, \text{ i.e. } a_3 = -\frac{a_1}{3!}$$

Equating to zero the coefficient of x^n, we have

$$(n+2)(n+1)a_{n+2} + a_n = 0 \Rightarrow a_{n+2} = -\frac{a_n}{(n+2)(n+1)}$$

which is called the recurrence relation.

Putting $n = 2, 3, 4, 5, \ldots$, in the recurrence relation

$$a_4 = -\frac{a_2}{4.3} = +\frac{a_0}{4.3.2.1} = \frac{a_0}{4!}$$

$$a_5 = \frac{a_3}{5!} \text{ and so on.}$$

Substituting these values in (2), we get

$$y = a_0 + a_1 x - \frac{a_0}{2!} x^2 - \frac{a_1}{3!} x^3 + \frac{a_0}{4!} x^4 + \frac{a_1}{5!} x^5 - \ldots$$

$$\Rightarrow \quad y = \left(a_0 - \frac{a_0}{2!} x^2 + \frac{a_0}{4!} x^4 \ldots \right) + \left(a_1 x - \frac{a_1}{3!} x^3 + \frac{a_1}{5!} x^5 - \ldots \right)$$

$$\Rightarrow \quad y = a_0 \left(1 - \frac{x^2}{2!} + \frac{x^4}{4!} - \ldots \right) + a_1 \left(x - \frac{x^3}{3!} + \frac{x^5}{5!} - \ldots \right)$$

$\Rightarrow \quad y = a_0 \cos x + a_1 \sin x$, which is required solution.

Example 2: $\dfrac{d^2y}{dx^2} - y = 0.$ ...(1)

Solution: Since $x = 0$ is an ordinary point of (1), let its series solution be

$$y = a_0 + a_1 x + a_2 x^2 + \ldots = \sum_{r=0}^{\infty} a_r x^r. \qquad \ldots(2)$$

Then $\dfrac{dy}{dx} = \sum\limits_{r=1}^{\infty} r\, a_r x^{x-1} \quad \dfrac{d^2y}{dx^2} = \sum\limits_{r=2}^{\infty} r(r-1) a_r x^{r-2}.$

Substituting the values of $y, \dfrac{dy}{dx}$ and $\dfrac{d^2y}{dx^2}$ in the given differential equation, we obtain

$$\sum_{r=2}^{\infty} r(r-1) a_r x^{r-2} - \sum_{r=0}^{\infty} a_r x^r = 0$$

$$\Rightarrow \quad \left[2.1.a_2 + 3.2a_3 x + 4.3a_4 x^2 + \ldots + (r+2)(r+1) a_{n+2} x^n + \ldots \right]$$
$$- \left[a_0 + a_1 x + a_2 x^2 + \ldots + a_n x^n + \ldots \right] = 0.$$

Equating to zero the lowest power of x i.e., x^0, we get

$$2a_2 - a_0 = 0 \; i.e., a_2 = \frac{a_0}{2!}$$

Equate to zero the coefficient of x, we get $6a_3 - a_1 = 0 \; i.e. \; a_3 = \dfrac{a_1}{3!}$ etc

With these coefficients the series (2) becomes

$$y = a_0 + a_1 x + \frac{a_0}{2!} x^2 + \frac{a_1}{3!} x^3 + \frac{a_0}{4!} x^4 + \ldots$$

$$\Rightarrow \quad y = a_0\left(1+\frac{x^2}{2!}+\frac{x^4}{4!}+...\right)+a_1\left(x+\frac{x^3}{3!}+\frac{x^5}{5!}+...\right),$$ which is required solution.

Example 3: Solve in series $(1+x^2)\frac{d^2y}{dx^2}+x\frac{dy}{dx}-y=0$ ...(1)

Solution: Since $x=0$ is the ordinary point of the given equation, let its series solution be

$$y = a_0 + a_1x + a_2x^2 + ... = \sum_{r=0}^{\infty} a_r x^r. \quad ...(2)$$

Then $\frac{dy}{dx} = \sum_{r=1}^{\infty} r\, a_r\, x^{r-1}$, $\frac{d^2y}{dx^2} = \sum_{r=2}^{\infty} r(r-1)a_r\, x^{r-2}$.

Substituting the values of y, $\frac{dy}{dx}$ and $\frac{d^2y}{dx^2}$ in the given differential equation, we have

$$(1+x^2)\sum_{r=2}^{\infty} r(r-1)a_r x^{r-2} + x\sum_{r=1}^{\infty} r a_r x^{r-1} - \sum_{r=0}^{\infty} a_r x^r = 0$$

$$\Rightarrow \quad (1+x^2)(2a_2+6a_3x+12\,a_4x^2+...)+x(a_1+2a_2x+3a_3x^3+...)$$
$$-(a_0+a_1x+a_2x^2+...)=0$$

$$\Rightarrow (2a_2-a_0)+(6a_3+a_1-a_1)x+(2a_2+12a_4+2a_2-a_2)x^2+......=0.$$

Equating to zero the coefficients of x^0 and x^1, we obtain

$2a_2-a_0=0$ or $a_2=\frac{a_0}{2}$,

$6a_3=0$ or $a_3=0$,

$12a_4+3a_2=0$ or $a_4=-\frac{a_0}{8}$.

The complete solution of the differential equation is given by

$$y = a_0\left[1+\frac{x^2}{2}-\frac{x^4}{8}...\right]+a_1\,x$$

Example 4: Find the series solution about of the $x = 0$ differential equation about $x=0$ $(1-x^2)y^2+2xy'-y=0$ ...(1)

Solution: Since $P_0(0)=1-0^2=1\neq 0$, so that point $x=0$ is an ordinary point of (1). Let its solution be

$$y = a_0 + a_1 x + a_2 x^2 + \ldots = \sum_{r=0}^{\infty} a_r x^r \qquad \ldots(1)$$

Then $\dfrac{dy}{dx} = \sum_{r=1}^{\infty} r\, a_r\, x^{r-1}, \; \dfrac{d^2y}{dx^2} = \sum_{r=2}^{\infty} r(r-1) a_r\, x^{r-2}.$

Substituting the values of y, $\dfrac{dy}{dx}$ and $\dfrac{d^2y}{dx^2}$ in the given differential equation, we obtain

$$(1-x^2)\sum_{r=2}^{\infty} r(r-1)a_r\, x^{r-2} + 2x\sum_{r=1}^{\infty} ra_r\, x^{r-1} - \sum_{r=0}^{\infty} a_r x^r = 0$$

$$\Rightarrow \quad \sum_{r=2}^{\infty} r(r-1)a_r\, x^{r-2} - \sum_{r=2}^{\infty} r(r-1)a_r\, x^r + \sum_{r=1}^{\infty} ra_r\, x^r - \sum_{r=0}^{\infty} a_r x^r = 0$$

$$\Rightarrow \quad \sum_{r=0}^{\infty} (r+2)(r+1)a_{r+2}\, x^r - \sum_{r=0}^{\infty} (r+2)(r+1)a_{r+2} x^{r+2} + 2\sum_{r=0}^{\infty} (r+1)a_{r+1} x^{r+1} - \sum_{r=0}^{\infty} a_r x^r = 0$$

$$\Rightarrow \quad \sum_{r=0}^{\infty} \left[\left\{(r+2)(r+1)a_{r+2} - a_r\right\} x^r - (r+2)(r+1)a_{r+2} x^{r+2} + 2(r+1)a_{r+1} x^{r+1}\right] = 0.$$

Equating to zero the lowest power of x *i.e.* x^0, we get

$$2a_2 - a_0 = 0 \Rightarrow a_2 = \frac{a_0}{2!}.$$

Equating to zero the coefficient of x, we get

$$6a_3 - a_1 + 2a_1 = 0 \Rightarrow a_3 = -\frac{a_1}{6} = -\frac{a_1}{3!}.$$

Equating to zero the coefficient of x^r, we get

$$(r+2)(r+1)a_{r+2} - a_r - r(r-1)a_r + 2ra_r = 0$$

$$\Rightarrow \quad (r+2)(r+1)a_{r+2} - \left[1 + r^2 - r - 2r\right]a_r = 0 \Rightarrow$$

$$a_{r+2} = \frac{r^2 - 3r + 1}{(r+2)(r+1)} a_r.$$

Putting $r = 2, 4, 6, \ldots$, we get

$$a_4 = -\frac{1}{6} a_2 = -\frac{1}{6} \cdot \frac{a_0}{2} = -\frac{4}{2(4)!} a_0$$

$$a_6 = \frac{5}{6.5} a_4 = -\frac{5}{6.5} \cdot \frac{4}{2(4)!} a_0 = -\frac{20}{2(6)!} a_0$$

$$a_8 = \frac{19}{8.7} a_6 = -\frac{4.5.19}{8.7.2(6)!} a_0 = \frac{4.5.19}{2(8)!} a_0$$

Putting $r = 3, 5, 7, \ldots$ we get $a_5 = \frac{1}{5.4} a_3 = \frac{1}{5.4} \cdot \frac{a_1}{3!} = -\frac{a_1}{(5)!}$

$$a_7 = \frac{11}{7.6} a_5 = -\frac{11}{7!} a_1 \text{ etc.}$$

With these coefficients series (2) becomes

$$y = a_0 + a_1 x + \frac{a_0}{2!} x^2 - \frac{a_1}{3!} x^3 - \frac{4a_0}{2(4)!} a_0 x^4 - \frac{a_1}{5!} x^5$$

$$- \frac{20}{2(6)!} a_0 x^6 - \frac{11}{7!} a_1 x^7 - \frac{380}{2(8)!} x^8 \ldots.$$

$$= \frac{a_0}{2}\left[2 + x^2 - \frac{4}{4!} x^4 - \frac{20}{6!} x^6 - \frac{380}{8!} x^8 - ..\right]$$

$$+ a_1 \left[x - \frac{1}{3!} x^3 - \frac{1}{5!} x^5 - \frac{11}{7!} x^7 - \ldots\right].$$

EXERCISE 4.1

Solve the following differential equations by power series:

1. $\frac{d^2y}{dx^2} + xy = 0.$

2. $\frac{d^2y}{dx^2} + 4y = 0.$

3. $\frac{d^2y}{dx^2} - x\frac{dy}{dx} - y = 0.$

4. $\frac{d^2y}{dx^2} + x^2 y = 0$

5. $\left(x^2 - 1\right)\frac{d^2y}{dx^2} + x\frac{dy}{dx} - y = 0.$

6. $\left(x^2 - 1\right)\frac{d^2y}{dx^2} + 4x\frac{dy}{dx} + 2y = 0.$

7. $\left(x^2 - 1\right)\frac{d^2y}{dx^2} + 2y = 0$, near $x = 0$ given that $y(0) = 4, y'(0) = 5.$

8. $\left(x^2 - 1\right)\frac{d^2y}{dx^2} + 3x\frac{dy}{dx} + xy = 0; y(0), y'(0) = 3.$

ANSWERS 4.1

1. $y = a_0\left[1 - \frac{x^3}{3!} + \frac{4 \cdot 6^{x^6}}{6!} - \frac{7 \cdot 4\, x^9}{9!} + \ldots\right] + a_1\left[x - \frac{2x^4}{4!} + \frac{5 \cdot 2\, x^7}{7!} \ldots\right].$

2. $y = a_0\left[1 - 2x^2 + \frac{2}{3}x^4 - \ldots\right] + a_1\left[x - \frac{2}{3}x^3 + \frac{4}{15}x^5 - \ldots\right].$

3. $y = a_0\left[1 + \frac{x^2}{2} + \frac{x^4}{2.4} + \frac{x^6}{2.4.6} + ..\right] + a_1\left[x + \frac{x^3}{3} + \frac{x^5}{3.5} + \frac{x^7}{3.5.7} + \ldots\right]$

4. $y = a_0\left[1 - \frac{x^4}{12} + \frac{x^8}{672} - \ldots\right] + a_1\left[x - \frac{x^5}{20} + \frac{x^9}{1440} - \ldots\right].$

5. $y = a_0\left[1 + \frac{x^2}{2} + \frac{x^4}{4} + \ldots\right] + a_1 x.$

6. $y = a_0\left[1 + x^2 + x^4 + \ldots\right] + a_1\left[x + x^3 + x^5 + \ldots\right].$

7. $y = 4\left(1 - x^4\right) + 5\left[x - \frac{x^3}{3} - \frac{x^5}{15} - \ldots\right]$

8. $y = 2 + 3x + \frac{11}{6}x^3 + \frac{1}{4}x^4 + \ldots.$

4.3 SINGULAR POINTS

Consider the second order homogeneous linear differential equation

$$P_0(x)\frac{d^2y}{dx^2} + P_1(x)\frac{dy}{dx} + P_2(x)y = 0 \qquad \ldots(1)$$

where $P_0(x), P_1(x), P_2(x)$ are polynomials in x.

Singular point: If the point $x = x_0$ is not an ordinary point of the equation (1), it is called a singular point. Singular points are of two types:

(i) Regular singular point: A singular point $x = x_0$ of the equation (1) is called a regular singular point if the functions

$$(x - x_0)\frac{P_1(x)}{P_0(x)} \text{ and } (x - x_0)^2\frac{P_2(x)}{P_0(x)} \text{ are both analytic at } x = x_0.$$

(ii) Irregular singular point: A singular point which is not regular is called irregular singular point.

4.4 METHOD OF SERIES SOLUTION WHEN $x = 0$ IS A REGULAR SINGULAR POINT OF THE EQUATION (METHOD OF FROBENIUS)

$$P_0(x)\frac{d^2y}{dx^2} + P_1(x)\frac{dy}{dx} + P_2(x)y = 0. \qquad ...(1)$$

Since $x = 0$ is a regular singular point of (1), so let its series solution be

$$y = x^m\left(a_0 + a_1 x + a_2 x^2 + ...\right) = \sum_{r=0}^{\infty} a_r x^{m+r} \qquad ...(2)$$

Then $\dfrac{dy}{dx} = \sum_{r=0}^{\infty} (m+r)a_r x^{m+r-1}$,

$$\frac{d^2y}{dx^2} = \sum_{r=0}^{\infty} (m+r)(m+r-1)a_r x^{m+r-2}.$$

Step 1: Substitute the value of y, $\dfrac{dy}{dx}$ and $\dfrac{d^2y}{dx^2}$ in the given differential equation.

Step 2: Equating to zero the coefficient of lowest power of x, we get a quadratic equation which is called indicial equation.

Step 3: Equating to zero the coefficients of other powers of x, we obtain the values of $a_1, a_2, a_3, a_4, ...$ in terms of a_0.

Step 4: Substituting the values of $a_1, a_2, a_3, a_4, ...$ in equation (2), we get series solution of given differential equation. But this is not complete solution of the differential equation. The complete solution of the given differential equation depends on the nature of roots of the quadratic equation. For different cases, the solutions of the differential equations are given as follows:

Case I: If the roots m_1, m_2 of the indicial equation are distinct and not differing by an integer, then the complete solution is

$$y = c_1(y)_{m_1} + c_2(y)_{m_2}.$$

Case II: If the roots $y = c_1y_1 + c_2y_2$ of the indicial equation are equal i.e., $m_1 = m_2$, then complete solution is

$$y = c_1(y)_{m_1} + c_2\left(\frac{\partial y}{\partial m}\right)_{m=m_1}$$

Case III: When the roots of the indicial equation m_1, m_2 are distinct and differ by an integer, we have following two different cases:

(i) If some of the coefficients of series of y become infinite at $m_1 < m_2$, we replace a_0 by $b_0(m - m_1)$. The complete solution is

$$y = c_1(y)_{m_2} + c_2\left(\frac{\partial y}{\partial m}\right)_{m=m_1}$$

(ii) If a coefficient of y-series is indeterminate at $m_1 < m_2$, the complete primitive is given by putting $m = m_2$ in y which contains two arbitrary constants.

By putting $m = m_1$ in y, we get a series which is merely a constant multiple of one of the series contained in the first solution. So we reject the solution obtained by putting $m = m_1$.

Note: All the problems having ordinary points can also be solved by Frobenius method which is clear from example 5.

Example 1: Solve in series $3x\dfrac{d^2y}{dx^2} + 2\dfrac{dy}{dx} + y = 0.$ ...(1)

Solution: Since $x = 0$ is a regular singular point of (1), let its series solution be

$$y = x^m\left(a_0 + a_1 x + a_2x^2 + \dots\right) = \sum_{r=0}^{\infty} a_r x^{m+r} \quad \text{...(2)}$$

Then $\dfrac{dy}{dx} = \sum_{r=0}^{\infty}(m+r)a_r\, x^{m+r-1}$

and $\dfrac{d^2y}{dx^2} = \sum_{r=0}^{\infty}(m+r)(m+r-1)a_r\, x^{m+r-2}.$

Substituting the values of y, $\dfrac{dy}{dx}$ and $\dfrac{d^2y}{dx^2}$ in (1), we have

$$3x\sum_{r=0}^{\infty}(m+r)(m+r-1)a_r\, x^{m+r-2} + 2\sum_{r=0}^{\infty}(m+r)a_r x^{m+r-1} + \sum_{r=0}^{\infty}a_r x^{m+r} = 0$$

$$\Rightarrow \quad 3\sum_{r=0}^{\infty}(m+r)(m+r-1)a_r\, x^{m+r-1} + 2\sum_{r=0}^{\infty}(m+r)a_r x^{m+r-1} + \sum_{r=0}^{\infty}a_r x^{m+r} = 0$$

$$\Rightarrow \quad \sum_{r=0}^{\infty}\left[3(m+r)(m+r-1) + 2(m+r)\right]a_r\, x^{m+r-1} + \sum_{r=0}^{\infty}a_r x^{m+r} = 0.$$

Equate to zero the coefficient of lowest power of $x\left(i.e., x^{m-1}\right)$ corresponding to $r = 0$, we get

$$\left[3m(m-1)+2m\right]a_0=0 \Rightarrow m(3m-1)a_0=0 \Rightarrow m=0,\, m=\frac{1}{3}$$

since $a_0 \neq 0$.

Equate to zero the coefficient of x^m, we get

$$\left[3(m+1)m+2(m+1)\right]a_1+a_0=0 \Rightarrow a_1=-\frac{a_0}{(m+1)(3m+2)}$$

Equate to zero the coefficient of x^{m+1}, we have

$$\left[3(m+2)(m+1)+2(m+2)\right]a_2+a_1=0 \Rightarrow (m+2)(3m+5)a_2=-a_1$$

$$\Rightarrow a_2=\frac{-a_1}{(m+2)(3m+5)} \Rightarrow a_2=\frac{a_0}{(m+1)(m+2)(3m+2)(3m+5)}$$

Similarly, we get

$$a_3=\frac{-a_0}{(m+1)(m+2)(m+3)(3m+2)(3m+5)(3m+8)} \text{ etc.}$$

When $m=0$, we get $a_1=-\frac{1}{2}a_0,\, a_2=\frac{1}{20}a_0,\, a_3=-\frac{1}{480}a_0$ *etc.*

Hence the first solution is $y_1=a_0\left(1-\frac{x}{2}+\frac{1}{20}x^2-\frac{1}{480}x^3+\ldots\right)$

When $m=\frac{1}{3}$, we have $a_1=-\frac{1}{4}a_0,\, a_2=\frac{1}{56}a_0,\, a_3=-\frac{1}{1680}a_0$

Hence the second solution is

$$y_2=a_0\left(x^{1/3}-\frac{1}{4}x^{4/3}+\frac{1}{56}x^{7/3}-\frac{1}{1680}x^{10/3}+\ldots\right).$$

The complete solution of equation (1) is

$$y=c_1a_0\left(1-\frac{x}{2}+\frac{x^2}{20}-\frac{x^3}{480}+\ldots\right)+c_2a_0x^{1/3}\left(1-\frac{x}{4}+\frac{x^2}{56}-\frac{x^3}{1680}+\ldots\right).$$

Example 2: Find the general solution of $\frac{d^2y}{dx^2}+(x-3)\frac{dy}{dx}+y=0$ near to $x=2$.

Solution: We have $\frac{d^2y}{dx^2}+(x-3)\frac{dy}{dx}+y=0$ ...(1)

On comparing the given equation with

$$P_0(x)\frac{d^2y}{dx^2}+P_1(x)\frac{dy}{dx}+P_2(x)y=0.$$

We have $P_0(x)=1, P_1(x)=(x-3), P_2(x)=1$.

Obviously $\dfrac{P_1(x)}{P_0(x)}$ and $\dfrac{P_2(x)}{P_0(x)}$ are both analytic at $x=2$, so $x=2$ is an ordinary point of the given equation. To find the solution near $x=2$, we shall find series solution in powers of $(x-2)$. Let

$$y=a_0+a_1(x-2)+a_2(x-2)^2+\ldots=\sum_{r=0}^{\infty}a_r(x-2)^r. \qquad \ldots(2)$$

Then we have

$$\frac{dy}{dx}=\sum_{r=1}^{\infty}r\,a_r(x-2)^{r-1} \text{ and } \frac{d^2y}{dx^2}=\sum_{r=2}^{\infty}r(r-1)a_r(x-2)^{r-2}.$$

Substituting the values of $y, \dfrac{dy}{dx}$ and $\dfrac{d^2y}{dx^2}$ in equation (1), we get

$$\sum_{r=2}^{\infty}r(r-1)a_r(x-2)^{r-2}+(x-3)\sum_{r=1}^{\infty}ra_r(x-2)^{r-1}+\sum_{r=0}^{\infty}a_r(x-2)^r=0$$

$$\Rightarrow \sum_{r=2}^{\infty}r(r-1)a_r(x-2)^{r-2}+(x-2-1)\sum_{r=1}^{\infty}ra_r(x-2)^{r-1}+\sum_{r=0}^{\infty}a_r(x-2)^r=0$$

$$\Rightarrow \quad \sum_{r=2}^{\infty}r(r-1)a_r(x-2)^{r-2}+\sum_{r=1}^{\infty}ra_r(x-2)^{r}-\sum_{r=1}^{\infty}ra_r(x-2)^{r-1}$$

$$+\sum_{r=0}^{\infty}a_r(x-2)^r=0$$

$$\Rightarrow \quad \sum_{r=0}^{\infty}(r+2)(r+1)a_{r+2}(x-2)^r+\sum_{r=0}^{\infty}ra_r(x-2)^r$$

$$-\sum_{r=0}^{\infty}(r+1)a_{r+1}(x-2)^r+\sum_{r=0}^{\infty}a_r(x-2)^r=0$$

$$\Rightarrow \quad \sum_{r=0}^{\infty}\left[(r+2)(r+1)a_{r+2}-(r+1)a_{r+1}+(r+1)a_r\right](x-2)^r=0.$$

Equating to zero the coefficients of various powers of $(x-2)$, we get

$$2a_2-a_1+a_0=0 \Rightarrow a_2=\frac{a_1-a_0}{2} \qquad \ldots(3)$$

and $\quad (n+2)(n+1)a_{n+2}-(n+1)a_{n+1}+(n+1)a_n=0,\ n\geq 1$

$$\Rightarrow \quad a_{n+2}=\frac{(a_{n+1}-a_n)}{n+2} \text{ for } n\geq 1. \qquad \ldots(4)$$

Putting $n = 1, 2, 3, \ldots$ in (4), we get

$$a_3 = \frac{a_2 - a_1}{3} = \frac{1}{3}\left[\frac{a_1 - a_0}{2} - a_1\right] = -\frac{a_0 + a_1}{6}$$

$$a_4 = \frac{a_3 - a_2}{4} = \frac{1}{4}\left[-\frac{a_0 + a_1}{6} - \frac{a_1 - a_0}{2}\right] = \frac{1}{12}a_0 - \frac{1}{6}a_1.$$

Putting these values in (2), we get

$$y = a_0 + a_1(x-2) + \frac{a_1 - a_0}{2}(x-2)^2 - \frac{a_0 + a_1}{6}(x-2)^2 + \left(\frac{a_0}{12} - \frac{a_1}{6}\right)(x-2)^4 + \ldots$$

$$= a_0\left[1 - \frac{1}{2}(x-2)^2 - \frac{1}{6}(x-2)^3 + \frac{1}{12}(x-2)^4 + \ldots\right] + a_1\left[(x-2) + \frac{1}{2}(x-2)^2 - \frac{1}{6}(x-2)^3 - \frac{1}{6}(x-2)^4 + \ldots\right].$$

Example 3: Solve in series the equation

$$2x^2\frac{d^2y}{dx^2} + (2x^2 - x)\frac{dy}{dx} + y = 0. \qquad \ldots(1)$$

Solution: Since $x = 0$ is a singular point of (1), let its series solution be

$$y = x^m(a_0 + a_1x + a_2x^2 + \ldots) = \sum_{r=0}^{\infty} a_r x^{m+r} \qquad \ldots(2)$$

Then $\dfrac{dy}{dx} = \sum_{r=0}^{\infty}(m+r)a_r x^{m+r-1}$

and $\dfrac{d^2y}{dx^2} = \sum_{r=0}^{\infty}(m+r)(m+r-1)a_r x^{m+r-2}.$

Substituting the values of $y, \dfrac{dy}{dx}$ and $\dfrac{d^2y}{dx^2}$ in equation (1), we get

$$2x^2\sum_{r=0}^{\infty}(m+r)(m+r-1)a_r x^{m+r-2} + (2x^2 - x)\sum_{r=0}^{\infty}(m+r)a_r x^{m+r-1} + \sum_{r=0}^{\infty}a_r x^{m+r} = 0$$

$$\Rightarrow \quad 2x^2\left[m(m-1)a_0x^{m-2} + (m+1)ma_1x^{m-1} + (m+2)(m+1)a_2x^m + (m+3)(m+2)a_3x^{m+1} + \ldots\right]$$
$$+ (2x^2 - x)\left[ma_0x^{m-1} + (m+1)a_1x^m + (m+2)a_2x^{m+1} + \ldots\right]$$
$$+ \left[a_0x^m + a_1x^{m+1} + a_2x^{m+2} + a_3x^{m+3} + \ldots\right] = 0$$

Equating to zero the coefficient of lowest power of x, we get indicial equation

$$2m(m-1)a_0 - ma_0 + a_0 = 0 \Rightarrow m = 1, \frac{1}{2} \text{ since } a_0 \neq 0$$

Roots are $1, \frac{1}{2}$, which are distinct and not differing by an integer.

Equating to zero the coefficient of x^{m+1}, we get

$$2m(m+1)a_1 + 2ma_0 - (m+1)a_1 + a_1 = 0$$

$$\Rightarrow \quad (2m^2 + m)a_1 + 2ma_0 = 0 \Rightarrow a_1 = -\frac{2}{2m+1}a_0$$

Equating to zero the co-efficient of x^{m+2}, we have

$$2(m+2)(m+1)a_2 + 2(m+1)a_1 - (m+2)a_2 + a_2 = 0$$

$$\Rightarrow \quad (2m+3)(m+1)a_2 + 2(m+1)a_1 = 0$$

$$\Rightarrow \quad a_2 = -\frac{2}{(2m+3)}a_1 \Rightarrow a_2 = \frac{4}{(2m+1)(2m+3)}a_0$$

Similarly, we get

$$a_3 = -\frac{8}{(2m+1)(2m+3)(2m+5)}a_0 \text{ and so on.}$$

Substitute these values in equation (2), we get

$$y = a_0 x^m \left[1 - \frac{2}{2m+1}x + \frac{4}{(2m+1)(2m+3)}x^2 - \frac{8}{(2m+1)(2m+3)(2m+5)}x^3 + \ldots\right]. \quad \ldots(3)$$

Putting $m = \frac{1}{2}$, the first solution is

$$y_1 = a_0\, x^{1/2}\left[1 - x + \frac{1}{2}x^2 - \frac{1}{6}x^3 \ldots\right].$$

Putting $m = 1$ the second solution is

$$y_2 = a_0 x\left(1 - \frac{2}{3}x + \frac{4}{3.5}x^2 - \frac{8}{3.5.7}x^3 - \ldots\right).$$

The complete solution $y = c_1y_1 + c_2y_2$ gives

$$y = c_1a_0x^{1/2}\left(1 - x + \frac{x^2}{2} - \frac{x^3}{6} + \ldots\right).$$

$$+ c_2a_0\left(1 - \frac{2}{3}x^2 + \frac{4}{3.5}x^2 - \frac{8}{3.5.7}x^3 + \ldots\right.$$

Example 4: Solve in series $x^2 \frac{d^2y}{dx^2} + 5x\frac{dy}{dx} + x^2y = 0.$ ...(1)

Solution: Since $x = 0$ is a regular singular point of (1), let its series solution be

$$y = a_0^m \left(a_0 + a_1x + a_2x^2 + a_3x^3 + ...\right) = \sum_{r=0}^{\infty} a_r\, x^{m+r} \quad ...(2)$$

Then $\frac{dy}{dx} = \sum_{r=0}^{\infty} (m+r)a_r x^{m+r-1}$

and $\frac{d^2y}{dx^2} = \sum_{r=0}^{\infty} (m+r)(m+r-1)a_r x^{m+r-2}.$

Substituting the values of $y, \frac{dy}{dx}$ and $\frac{d^2y}{dx^2}$ in (1), we have

$$x^2 \sum_{r=0}^{\infty} (m+r)(m+r-1)a_r x^{m+r-2} + 5x \sum_{r=0}^{\infty} (m+r)a_r x^{m+r-1} + x^2 \sum_{r=0}^{\infty} a_r x^{m+r} = 0$$

$$\Rightarrow \quad \sum_{r=0}^{\infty} (m+r)(m+r-1)a_r\, x^{m+r} + 5\sum_{r=0}^{\infty} (m+r)a_r\, x^{m+r} + \sum_{r=0}^{\infty} a_r x^{m+r+2} = 0$$

$$\Rightarrow \quad \sum_{r=0}^{\infty} \left[(m+r)(m+r-1) + 5(m+r)\right]a_r\, x^{m+r} + \sum_{r=0}^{\infty} a_r\, x^{m+r+2} = 0. \quad ...(3)$$

Equating to zero the coefficient of lowest power of x corresponding to $r = 0$, we get $\left[m(m+1) + 5m\right]a_0 = 0 \Rightarrow m(m+4)a_0 = 0$, which is called indicial equation.

Since $a_0 \neq 0$, we have $m = 0, -4$.

The roots are distinct and differing by an integer. Equate to zero the coefficient of x^{m+1}, we get

$$\left[(m+1)m + 5(m+1)\right]a_1 = 0 \Rightarrow (m+5)(m+1)a_1 = 0$$

$$\Rightarrow \quad a_1 = 0 \left[\because m \neq 5, m \neq -1\right]$$

Equating the coefficient of x^{m+2} equal to 0, we get

$$\left[(m+2)(m+1) + 5(m+2)\right]a_2 + a_0 = 0$$

$$\Rightarrow \quad (m+2)(m+6)a_2 + a_0 = 0$$

$$\Rightarrow \quad a_2 = -\frac{a_0}{(m+2)(m+6)}$$

Equating coefficient of x^{m+3} equal to 0, we get

$$(m+3)(m+2)a_3 + 5(m+3)a_3 + a_1 = 0$$

$$\Rightarrow \quad a_3 = -\frac{a_1}{(m+3)(m+7)} \Rightarrow a_3 = 0$$

Equating coefficient of $x^{m+4} = 0$, we get

$$a_4 = \frac{a_0}{(m+2)(m+4)(m+6)(m+8)}$$

Substitute these values in equation (2), we get

$$y = a_0 x^m \left[1 - \frac{x^2}{(m+2)(m+6)} + \frac{x^4}{(m+2)(m+4)(m+6)(m+8)} - \dots\right]. \quad \dots(4)$$

Putting $m = 0$ in equation (4), we have

$$y_1 = a_0\left(1 - \frac{x^2}{2.6} + \frac{x^4}{2.4.6.8} - \dots\right). \quad \dots(5)$$

Putting $m = -4$ in equation (4), the coefficients of x^4 and higher powers of x become infinite. Then replacing a_0 by $b_0(m+4)$, we obtain

$$y = b_0 x^m \left[(m+4) - \frac{(m+4)}{(m+2)(m+6)}x^2 + \frac{x^4}{(m+2)(m+6)(m+8)} - \dots\dots\right]. \quad \dots(6)$$

Differentiating both sides of equation (6), w. r. t.'m', we get

$$\frac{\partial y}{\partial m} = y \log x + b_0\, x^m \left[1 + \frac{m^2 + 8m + 20}{(m^2 + 8m + 12)^2}\right.$$

$$\left. - \frac{(3m^2 + 32m + 76)}{(m^3 + 16m^2 + 76m + 96)^2}x^4 + \dots\dots\right]. \quad \dots(7)$$

Putting $m = -4$ in the above equation, we have

$$y_2 = \left(\frac{\partial y}{\partial m}\right)_{m=-4} = (y)_{m=-4}\log x + b_0 x^{-4}\left(1 + \frac{x^2}{4} - \frac{x^4}{64} + \dots\right)$$

$$y_2 = b_0 x^{-4} \log x\left[-\frac{x^4}{16} - \frac{x^6}{96} - \dots\right] + b_0\, x^{-4}\left[1 + \frac{x^2}{4} - \frac{x^4}{64} + \dots\right].$$

The complete solution $y = c_1 y_1 + c_2 y_2$ of the given differential equation is

$$y = c_1\, a_0 \left[1 - \frac{x^2}{12} + \frac{x^2}{384} - \ldots\right] + c_2 b_0 x^{-4} \log x \left(-\frac{x^2}{16} - \frac{x^6}{96} - \ldots\right)$$

$$+ c_2 b_0\, x^{-4}\left(\frac{x^2}{4} - \frac{x^4}{64} + \ldots\right).$$

Example 5: Solve $(2+x^2)\dfrac{d^2y}{dx^2} + x\dfrac{dy}{dx} + (1+x)y = 0.$ ...(1)

Solution: Here $x = 0$ is an ordinary point because

$$\frac{P_1(x)}{P_0(x)} = \frac{x}{x^2+2}, \frac{P_2(x)}{P_0(x)} = \frac{1+x}{2+x^2}$$

are analytic at $x = 0$. Suppose solution of (1) is $y = \sum_{r=0}^{\infty} a_r x^{m+r}$, so that

$$\frac{dy}{dx} = \sum_{r=0}^{\infty} a_r (m+r) x^{m+r-1}$$

$$\frac{d^2y}{dx^2} = \sum_{r=0}^{\infty} a_r (m+r)(m+r-1).\, x^{m+r-2}.$$

Substituting the values of $y, \dfrac{dy}{dx}$ and $\dfrac{d^2y}{dx^2}$ in (1), we get

$$\sum_{r=0}^{\infty} a_r \Big[(2+x^2)(m+r)(m+r-1)x^{m+r-2}$$

$$+ x(m+r)x^{m+r-1} + (1+x)x^{m+r}\Big] = 0$$

$$\Rightarrow \quad \sum_{r=0}^{\infty} a_r \left[2(m+r)(m+r-1)x^{m+r-2} + \{(m+r)^2 + 1\}x^{m+r} + x^{m+r+1}\right] = 0$$

...(2)

Equating to zero the coefficient of lowest power of x *i.e.*, x^{m-2}, we get $2m(m-1)a_0 = 0 \Rightarrow m = 0, 1$; as $a_0 \neq 0$ being the coefficient of first term.

Equating to zero the coefficient of the next higher power of x *i.e.*, x^{m-1}, we get

$$2(m+1).\, m.\, a_1 = 0. \qquad \text{...(3)}$$

If $m = 0$, a_1 is indeterminate. Equating to zero the coefficients of general terms x^{m+r}

$$a_{r+2} 2(m+r+2)(m+r+1) + \left\{(m+r)^2 + 1\right\} a_r + a_{r-1} = 0$$

$$\Rightarrow \qquad a_{r+2} = -\frac{a_r\left\{(m+r)^2+1\right\}+a_{r-1}}{2(m+r+1)(m+r+2)}. \qquad \text{...(4)}$$

Putting $r = 0, 1, 2, 3, \ldots$ we have

$$a_2 = -\frac{(m^2+1)a_0}{2(m+1)(m+2)} \text{ (as the term } a_{-1} \text{ does not exist)}$$

$$a_3 = -\frac{\left\{(m+1)^2+1\right\}a_1+a_0}{2(m+2)(m+3)} = -\frac{(m^2+2m+2)a_1+a_0}{2(m+2)(m+3)}$$

$$a_4 = -\frac{\left\{(m^2+2)^2+1\right\}a_2+a_1}{2(m+3)(m+4)}$$

$$= +\frac{(m^2+4m+5)(m^2+1)a_0}{2^2(m+1)(m+2)(m+3)(m+4)} - \frac{a_1}{2(m+3)(m+4)}.$$

Therefore, the solution is given by

$$y = a_0x^m + a_1x^{m+1} - \frac{(m^2+1)a_0}{2(m+1)(m+2)}x^{m+2} - \left[\frac{(m^2+2m+2)a_1+a_0}{2(m+2)(m+3)}\right]x^{m+3}$$

$$+\left[\frac{(m^2+4m+5)(m^2+1)a_0}{2^2(m+1)(m+2)(m+3)(m+4)} - \frac{a_1}{2(m+3)(m+4)}\right]x^{m+4} + \ldots$$

$$= a_0x^m\left[1 - \frac{(m^2+1)}{2(m+1)(m+2)}x^2 - \frac{1}{2(m+2)(m+3)}x^3\right.$$

$$\left. + \frac{(m^2+4m+5)(m^2+1)}{2^2(m+1)(m+2)(m+3)(m+4)}x^4 + \ldots\right]$$

$$+ a_1x^{m+1}\left[1 - \frac{(m^2+2m+2)}{2(m+2)(m+3)}x^2 - \frac{1}{2(m+3)(m+4)}x^3 + \ldots\right]. \qquad \text{...(5)}$$

Putting $m = 0$, we get

$$y = a_0\left[1 - \frac{x^2}{4} - \frac{x^3}{12} + \frac{5x^4}{96} - \ldots\right] + a_1x\left[1 - \frac{1}{6}x^2 - \frac{1}{24}x^3 + \ldots\right] \qquad \text{...(6)}$$

which have two arbitrary constants, so it is the complete primitive of the given equation.

If we take $m = 1$ from (5) we obtain

$$y = a_0 x\left[1 - \frac{1}{6}x^2 - \frac{1}{24}x^3 + \frac{x^4}{24} + \ldots\right]$$

which is a constant multiple of the second series in the solution given by (6).

Example 6: Solve in series $2x(1-x)\frac{d^2y}{dx^2} + (1-x)\frac{dy}{dx} + 3y = 0.$ ...(1)

Solution: Since x=0 is a regular point of (1), let its series solution be

$$y = x^m\left(a_0 + a_1 x + a_2 x^2 + a_3 x^3 + \ldots\right) = \sum_{r=0}^{\infty} a_r x^{m+r}. \quad \text{...(2)}$$

Then $\frac{dy}{dx} = \sum_{r=0}^{\infty}(m+r)a_r x^{m+r-1}$

and $\frac{d^2y}{dx^2} = \sum_{r=0}^{\infty}(m+r)(m+r-1)a_r x^{m+r-2}.$

Substituting the values of y, $\frac{dy}{dx}$ and $\frac{d^2y}{dx^2}$, we get

$$\Rightarrow \sum_{r=0}^{\infty}\left[2(m+r)^2 - 2(m+r) + (m+r)\right]a_r x^{m+r-1}$$
$$- \sum_{r=0}^{\infty}\left[2(m+r)^2 - 2(m+r) + (m+r) - 3\right]a_r x^{m+r} = 0$$

$$\Rightarrow \sum_{r=0}^{\infty}\left[2(m+r)^2 - (m+r)\right]a_r x^{m+r-1} - \sum_{r=0}^{\infty}\left[2(m+r)^2 - (m+r) - 3\right]a_r x^{m+r} = 0$$

$$\Rightarrow \sum_{r=0}^{\infty}\left[(m+r)(2m+2r-1)\right]a_r x^{m+r-1} - \sum_{r=0}^{\infty}\left[2(m+r)^2 - (m+r) - 3\right]a_r x^{m+r} = 0$$

$$\Rightarrow \left[m(2m-1)a_0 x^{m-1} + (m+1)(2m+1)a_1 x^m + (m+2)(2m+3)a_2 x^{m+1} + (m+3)(2m+5)a_3 x^{m+2} + \ldots\right] - \left[(2m^2 - m - 3)a_0 x^m + \{2(m+1)^2 - (m+1) - 3\}a_1 x^{m+1} + \{2(m+2)^2 - (m+2) - 3\}a_2 x^{m+2} + \ldots\right] = 0.$$

Equate to zero the coefficient of lowest power of x, i.e. x^{m-1}, we get $m(2m-1)a_0 = 0 \Rightarrow m(2m-1) = 0$, which is called indicial equation

$$\Rightarrow \quad m = 0, \frac{1}{2}, \text{ as } a_0 \neq 0$$

Equate to zero the coefficient of x^m, we obtain

$$(m+1)(2m+1)a_1 - (2m^2 - m - 3)a_0 = 0$$

$$\Rightarrow \quad a_1 = \frac{2m^2 - m - 3}{(m+1)(2m+1)} a_0$$

Coefficient of $x^{m+1} = 0$ gives

$$(m+2)(2m+3)a_2 - \left[2(m+1)^2 - (m+1) - 3\right]a_1 = 0$$

$$\Rightarrow \quad (m+2)(2m+3)a_2 - (2m^2 + 3m - 2)a_1 = 0$$

$$\Rightarrow \quad a_2 = \frac{(2m^2 + 3m - 2)}{(m+2)(2m+3)} a_1 = \frac{(2m^2 - m - 3)(2m^2 + 3m - 2)}{(m+1)(m+2)(2m+1)(2m+3)} a_0$$

Coefficient of $x^{m+2} = 0$ gives

$$(m+3)(2m+5)a_3 - \left[2(m+2)^2 - (m+2) - 3\right]a_2 = 0$$

$$\Rightarrow \quad (m+2)(2m+5)a_3 - (2m^2 + 7m + 5)a_2 = 0$$

$$\Rightarrow \quad a_3 = \frac{2m^2 + 5m + 5}{(m+3)(2m+5)} a_2$$

$$\Rightarrow \quad a_3 = \frac{(2m^2 - m - 3)(2m^2 + 3m - 2)(2m^2 + 7m + 5)}{(m+1)(m+2)(m+3)(2m+1)(2m+3)(2m+5)} a_0, \text{ etc.}$$

Substitute these values in equation (2), we get

$$y = a_0 x^m \left[1 + \frac{(2m^2 - m - 3)}{(m+1)(2m+1)} x + \frac{(2m^2 - m - 3)(2m^2 + 3m - 2)}{(m+1)(m+2)(2m+1)(2m+3)} x^2 \right.$$

$$\left. + \frac{(2m^2 - m - 3)(2m^2 + 3m - 2)(2m^2 + 7m + 5)}{(m+1)(m+2)(m+3)(2m+1)(2m+3)(2m+5)} x^2 + \ldots \right] \quad \ldots(3)$$

Putting $m = 0$ in equation (3), we obtain first solution

$$y_1 = a_0 \left[1 - 3x + x^2 + \frac{1}{3} x^3 + \ldots\right]. \quad \ldots(4)$$

Putting $m = \frac{1}{2}$ in equation (3), we get second series solution

$$y_2 = a_0 x^{1/2} [1 - x]$$

The complete solution of the given differential equation is

$$y = c_1 y_1 + c_2 y_2 = A\left[1 - 3x + x^2 + \frac{x^3}{3} + \ldots\right] + Bx^{1/2}[1 - x + \ldots]$$

where $A = c_1 a_0, B = c_2 a_0$.

Example 7: Solve $x^2 \dfrac{d^2y}{dx^2} + (x+x^2)\dfrac{dy}{dx} + (x-9)y = 0$ near to $x = 0$.

Solution: We have $x^2 \dfrac{d^2y}{dx^2} + (x+x^2)\dfrac{dy}{dx} + (x-9)y = 0$ ...(1)

Here $x = 0$ is a singular point as $\dfrac{P_1(x)}{P_0(x)} = \dfrac{x+1}{x}, \dfrac{P_2(x)}{P_0(x)} = \dfrac{x-9}{x^2}$.

In fact $x = 0$ is a regular singular point. Let the series solution be

$y = \sum_{r=0}^{\infty} a_r x^{m+r}$, so that

$$\frac{dy}{dx} = \sum_{r=0}^{\infty} a_r (m+r) x^{m+r-1}, \frac{d^2y}{dx^2} = \sum_{r=0}^{\infty} a_r (m+r)(m+r-1) x^{m+r-2}.$$

Substituting these values in (1), we have

$$\sum_{r=0}^{\infty} a_r \Big[x^2 x^{m+r-2}(m+r)(m+r-1) + (x+x^2)(m+r) x^{m+r-1} + (x-9)x^{m+r} \Big] = 0$$

$$\Rightarrow \quad \sum_{r=0}^{\infty} a_r \Big[\{(m+r)^2 - 9\} x^{m+r} + (m+r+1) x^{m+r+1} \Big] = 0. \quad ...(2)$$

Equating to zero the coefficient of lowest power of x, i.e. x^m we have

$$(m^2 - 9)a_0 = 0 \Rightarrow m = 3, -3 \text{ as } a_0 \neq 0. \quad ...(3)$$

Equating to zero the coefficient of x^{m+r}, we have

$$\left[(m+r)^2 - 9\right] a_r + a_{r-1}(m+r) = 0$$

$$\Rightarrow a_r = -\frac{(m+r)}{(m+r+3)(m+r-3)} a_{r-1}.$$

Putting $r = 1, 2, 3, \ldots\ldots$, we get

$$a_1 = \frac{(m+1)}{(m+4)(m-2)} a_0, \; a_2 = \frac{(m+1)(m+2)}{(m+5)(m-1)(m+4)(m-2)} a_0,$$

$$a_3 = \frac{-(m+3)(m+2)(m+1)}{(m+6)(m+5)(m+4)m(m-1)(m-2)} a_0.$$

So that, we have

$$y = a_0 x^m \Big[1 - \frac{m+1}{(m-2)(m+4)} x - \frac{(m+1)(m+2)}{(m-1)(m-2)(m+4)(m+5)} x^2$$

$$-\frac{(m+1)(m+2)(m+3)}{m(m-1)(m-2)(m+4)(m+5)(m+6)}x^3+...].$$

Putting $m=3$, replacing a_0 by a, we get

$$y=ax^3\left[1-\frac{4}{7}x+\frac{4.5}{2.1.7.8}x^2-\frac{4.5.6}{3.2.1.7.8.9}x^3+...\right]=au.$$

Putting $m=-3$ and replacing a_0 by b, we obtain

$$y=bx^{-3}\left[1-\frac{2}{5}x+\frac{1}{20}x^2\right]=bv.$$

So the required solution is $y=au+bv$, where a and b are arbitrary constants.

Example 8: (Indicial roots are equal: $r_1=r_2=r$) Solve

$$x\frac{d^2y}{dx^2}+\frac{dy}{dx}-xy=0 \qquad ...(1)$$

Solution: $x=0$ is a regular singular point. Let the series solution be

$y=\sum_{r=0}^{\infty}a_r\,x^{m+r}$, so that

$$\frac{dy}{dx}=\sum_{r=0}^{\infty}a_r(m+r)x^{m+r-1},\ \frac{d^2y}{dx^2}=\sum_{r=0}^{\infty}a_r(m+r)(m+r-1)x^{m+r-2}.$$

Substituting these values in (1), we have

$$\sum_{r=0}^{\infty}a_r\left[x^{m+r-1}(m+r)(m+r-1)+(m+r)x^{m+r-1}-x^{m+r+1}\right]=0$$

$$\Rightarrow\quad \sum_{r=0}^{\infty}a_r\left[(m+r)^2x^{m+r-1}-x^{m+r+1}\right]=0 \qquad ...(2)$$

Equating to zero the coefficient of lowest power of x, i.e. x^{m-1} we have

$$m^2a_0=0\Rightarrow m=0,0 \text{ as } a_0\neq 0. \qquad ...(3)$$

The indicial root is a double root.

Equating to zero the coefficient of x^m we have

$$(m+1)^2a_1=0\Rightarrow a_1=0, \text{ as } m=0.$$

Equating to zero the coefficient to x^{m+r+1}, we have

$$(m+r+2)^2a_{r+2}-a_r=0\Rightarrow a_{r+2}=\frac{1}{(m+r+2)^2}a_r$$

Putting $r=1,3,.....$, we get $a_3=a_5=......=0$

Putting $r = 0, 2, 4, \ldots\ldots$, we get $a_2 = \frac{1}{(m+2)^2} a_0$,

$$a_4 = \frac{1}{(m+4)^2} a_2 = \frac{1}{(m+2)^2 (m+4)^2} a_0,$$

$$a_6 = \frac{1}{(m+6)^2} a_4 = \frac{1}{(m+2)^2 (m+4)^2 (m+6)^2} a_0$$

So that, we have

$$y = a_0 x^m \left[1 + \frac{1}{(m+2)^2} x^2 + \frac{1}{(m+2)^2 (m+4)^2} x^4 + \ldots\ldots \right]. \qquad \ldots(4)$$

For $m = 0$, we have $y_1(x) = a_0 \left[1 + \frac{1}{2^2} x^2 + \frac{1}{2^2 4^2} x^4 + \ldots\ldots \right] = a_0 u(x)$.

Let us now substitute $y(x)$ in (1), we get

$$x \frac{d^2y}{dx^2} + \frac{dy}{dx} - xy = a_0 m^2 x^{m-1} \qquad \ldots(5)$$

where right hand side is simply the indicial equation.

Consider m as a parameter and differentiating (5) w. r. t. 'm', we get

$$\frac{\partial}{\partial m}\left(x \frac{d^2y}{dx^2} + \frac{dy}{dx} - xy \right) = \frac{\partial}{\partial m}\left(a_0 m^2 x^{m-1} \right)$$

$$\Rightarrow \quad \left(x \frac{d^2}{dx^2} + \frac{d}{dx} - x \right) \frac{\partial y}{\partial m} = a_0 \left(2m x^{m-1} + m^2 x^{m-1} \log x \right).$$

At $m = 0$ right hand side vanishes, from which it follows that $\left(\frac{\partial y}{\partial m} \right)_{m=0}$ is also a solution of (1).

Now differentiating (5) w. r. t. 'm', and putting $m = 0$, we get

$$y_2(x) = \left(\frac{\partial y}{\partial m} \right)_{m=0} = (\log x) y_1(x) - a_0 \left[\frac{x^2}{4} + \frac{3}{128} x^4 + \ldots\ldots \right]$$

$$= (\log x) a_0 u(x) - a_0 \left[\frac{x^2}{4} + \frac{3}{128} x^4 + \ldots\ldots \right]$$

$$= a_0 \left[(\log x) u(x) - \left\{ \frac{x^2}{4} + \frac{3}{128} x^4 + \ldots\ldots \right\} \right] = a_0 v(x)$$

So the required solution is $y = a y_1(x) + b y_2(x) = A u(x) + B v(x)$, where $A = a a_0$ and $B = b a_0$ are arbitrary constants.

EXERCISE 4.2

Solve the following differential equations in series:

1. $2x(1-x)y''+(5-7x)y'-3y=0.$
2. $2x^2 y''+xy'-(x+1)y=0.$
3. $9x(1-x)y''-12y'+4y=0.$
4. $2x^2 y''-xy'+(1-x^2)y=0.$
5. $xy''+(x-1)y'-y=0.$
6. $x(1-x)y''+4y'+2y=0.$
7. $x(2+x^2)y''-y'-6xy=0.$
8. $y''-\frac{1}{4x}y'+\frac{1}{8x^2}y=0.$
9. $x(x-1)y''+(3x-1)y'+y=4x.$
10. $x\frac{d^2y}{dx^2}+3\frac{dy}{dx}+4x^3y=0.$
11. Find the power series in the powers of $(x-1)$ of the initial value problem $x\frac{d^2y}{dx^2}+\frac{dy}{dx}+2y=0, y(1)=2, y'(1)=2$.
12. Find the power series of the I. V. P. $(x^2-1)\frac{d^2y}{dx^2}+3x\frac{dy}{dx}+xy=0, y(2)=4, y'(2)=6.$
13. Solve (i) $(1-x^2)\frac{d^2y}{dx^2}-x\frac{dy}{dx}+4y=0.$

    (ii) $(1-x^2)\frac{d^2y}{dx^2}-2x\frac{dy}{dx}+n(n+1)y=0,$ where n is any constant.

    (iii) $(1-x^2)y''+2xy'+y=0$ about $x=0$.

ANSWERS 4.2

1. $y=a_0\left(1+\frac{3}{5}x+\frac{3}{7}x^2+\frac{3}{9}x^4 \ldots\right)+b_0\, x^{-3/2}.$
2. $y=a_0x\left(1+\frac{1}{5}x+\frac{1}{70}x^2+\ldots\right)+b_0\, x^{-1/2}\left(1-x-\frac{x^2}{2}+\ldots\right).$

3. $y = a_0\left(1 + \frac{x}{3} + \frac{1.4}{3.6}x^2 + \frac{1.4.7}{3.6.9}x^3 + ...\right)$

$+ b_0 x^{7/3}\left(1 + \frac{8}{10}x + \frac{8.11}{10.13}x^2 + ...\right).$

4. $y = a_0 x\left(1 + \frac{x^2}{10} + \frac{x^4}{360} + ...\right) + b_0 x^{1/2}\left(1 + \frac{x^2}{6} + \frac{x^4}{168} + ...\right).$

5. $y = a_0\left(1 - x + \frac{x^2}{2!} - \frac{x^3}{3!} + ...\right) + b_0\left(x^2 + \frac{2x^3}{3!} + \frac{2x^4}{4!} - ...\right).$

6. $y = a_0\left(1 - \frac{x}{2} + \frac{x^2}{10}\right) + a_0\, x^{-3}\left(1 - 5x + 10x^2\right).$

7. $y = a_0\left(1 + 3x^2 + \frac{3}{5}x^4 - \frac{1}{15}x^6 - ...\right) + b_0 x^{3/2}\left(1 + \frac{3}{8}x^2 - \frac{1.3}{8.16}x^4 + ...\right).$

8. $y = a_0\sqrt{x} + b_0\, x^{1/4}.$

9. $y = a_0\left(1 + x + x^2 + x^3 + ...\right) + b_0\left(1 + x + x^2 + x^3 + ...\right)\log x - \frac{x^2}{1-x}$

10. $y = a_0 x^{-2}\left[1 - \frac{x^4}{2!} + \frac{x^8}{4!} - ...\right] + b_0\, x^{-2}\left[x^2 - \frac{x^6}{3!} + \frac{x^{10}}{5!} ...\right].$

11. $y = 1 + 2(x-1) - 2(x-1)^2 + \left(\frac{2}{3}\right)(x-1)^3 - \frac{1}{6}(x-1)^4 + \frac{1}{15}(x-1)^5 - ....$

12. $y = 4 + 6(x-2) - \frac{22}{3}(x-2)^2 + \left(\frac{169}{27}\right)(x-2)^3 + \left(\frac{344}{81}\right)(x-2)^4 + ...$

13. (i) $y = a_0(1 - 2x) + a_1\left(x - \frac{x^3}{2} - \frac{x^5}{8} - \frac{x^7}{16} - ...\right)$

(ii) $y = a_0\left[1 - \frac{n(n+1)}{2!}x^2 + \frac{(n-2)n(n+1)(n+3)}{4!}x^4 - ...\right]$

$+ a_1\left[x - \frac{(n-1)(n+2)}{3!}x^3 + \frac{(n-3)(n-1)(n+2)(n+4)}{5!}x^5 - ...\right].$

(iii) $y = a_0\left[1 + \frac{x^2}{2} - \frac{x^4}{8} - ......\right] + a_1\left[x - \frac{x^3}{3} - \frac{x^5}{40} - ......\right]$

4.5 LEGENDRE'S EQUATION

The differential equation

$$\left(1-x^2\right)\frac{d^2y}{dx^2}-2x\frac{dy}{dx}+n(n+1)=0 \qquad ...(1)$$

is called Legendre's differential equation.

It can be solved in series of ascending or descending powers of x. The solution in descending powers of x is more important than the one in ascending powers. Any solution of (1) is called Legendre function. The study of these and other higher functions now occurring in calculus is called the theory of special functions. The Legendre differential equation appears in applied mathematics, particularly in boundary value problems of spheres.

Consider a series in descending powers of x as follows

$$y=\sum_{r=0}^{\infty} a_r\, x^{m-r}.$$

Therefore $$\frac{dy}{dx}=\sum_{r=0}^{\infty}(m-r)a_r\, x^{m-r-1}$$

and $$\frac{d^2y}{dx^2}=\sum_{r=0}^{\infty}(m-r)(m-r-1)a_r\, x^{m-r-2}.$$

Substituting the values of $y, \frac{dy}{dx}$ and $\frac{d^2y}{dx^2}$ in (1), we get

$$\left(1-x^2\right)\sum_{r=0}^{\infty}(m-r)(m-r-1)a_r x^{m-r-2}-2x\sum_{r=0}^{\infty}(m-r)a_r x^{m-r-1}$$
$$+n(n+1)\sum_{r=0}^{\infty}a_r x^{m-r}=0$$

$$\Rightarrow \quad \sum_{r=0}^{\infty}\left[(m-r)(m-r-1)\right]a_r x^{m-r-2}-\sum_{r=0}^{\infty}\left[(m-r)(m-r-1)+2\right.$$
$$\left.(m-r)-n(n+1)\right]a_r x^{m-r}=0$$

$$\Rightarrow \quad \sum_{r=0}^{\infty}\left[(m-r)(m-r-1)\right]a_r\, x^{m-r-2}-\sum_{r=0}^{\infty}\left[(m-r)^2-n^2\right.$$
$$\left.+(m-r-n)\right]a_r r^{m-r}=0$$

$$\Rightarrow \quad \sum_{r=0}^{\infty}\left[(m-r)(m-r-1)\right]a_r x^{m-r-2} - \sum_{r=0}^{\infty}\left[(m-r-n)\right.$$

$$\left.(m-r+n+1)\right]a_r x^{m-r} = 0. \quad ...(3)$$

Now (3) being an identity, we equate to zero the coefficient of various powers of x.

Equating to zero the coefficient of highest power of x i.e., x^m corresponding to $r=0$, we get

$$(m-n)(m+n+1)a_0 = 0 \Rightarrow (m-n)(m+n+1) = 0, \text{ since } a_0 \neq 0$$

$$\Rightarrow \quad m = n, -(n+1).$$

Equating to zero the coefficient of the next lower power of x i.e., x^{m-1}, we get

$$(m+n)(m-n-1)a_1 = 0 \Rightarrow a_1 = 0,$$

since neither $(m-n-1)$ nor $(m+n)$ is zero.

Equating to zero the coefficient of the general term x^{m-r}, we get

$$\left[(m-r+2)\right]\left[(m-r+1)\right]a_{r-2} - \left[(m-r-n)(m-r+n+1)\right]a_r = 0$$

$$\Rightarrow \quad a_r = \frac{(m-r+2)(m-r+1)}{(m-r-n)(m-r+n+1)}a_{r-2}$$

$$= -\frac{(m-r+2)(m-r+1)}{(n-m+r)(n+m-r+1)}a_{r-2}. \quad ...(4)$$

Putting $r = 3, 5, 7, \ldots$ in (4), we get $a_3 = a_5 = a_7 = .. = 0$.
Now two cases arise.

Case I: When $m = n$, the recurrence relation reduces to

$$a_r = -\frac{(n-r+2)(n-r+1)}{r(2n-r+1)}a_{r-2}.$$

Putting $r = 2, 4, 6, \ldots$, we get $a_2 = -\dfrac{n(n-1)}{2(2n-1)}a_0$

$$a_4 = -\frac{(n-2)(n-3)}{4(2n-3)}a_2$$

$$= \frac{n(n-1)(n-2)(n-3)}{2\,4(2n-1)(2n-3)}a_0 \text{ and so on.}$$

Therefore

$$y_1 = a_0\left[x^n - \frac{n(n-1)}{2(2n-1)}x^{n-2} + \frac{n(n-1)(n-2)(n-3)}{2.4(2n-1)(2n-3)}x^{n-4} - \ldots\right] \quad \ldots(5)$$

which is one solution of Legendre's equation.

Case II: When $m = -(n+1)$, the recurrence relation reduces to

$$a_r = \frac{(n+r-1)(n+r)}{r(2n+r+1)}a_{r-2}.$$

Putting $r = 2, 4, 6, \ldots$ etc. $a_2 = \frac{(n+1)(n+2)}{2(2n+3)}a_0$

$$a_4 = \frac{(n+3)(n+4)}{4(2n+5)}a_2$$

$$= \frac{(n+1)(n+2)(n+3)(n+4)}{2.4(2n+3)(2n+5)}a_0 \text{ and so on.}$$

Therefore $y_2 = a_0\left[x^{-n-1} + \frac{(n+1)(n+2)}{2.(2n+3)}x^{-n-2}\right.$

$$\left. + \frac{(n+1)(n+2)(n+3)(n+4)}{2.4(2n+3)(2n+5)}x^{-n-5} + \ldots\right]$$

which is other solution of Legendre's equation.

4.5.1 Legendre's Polynomials of First Kind $P_n(x)$

The solution of Legendre's differential equation is called Legendre's function.

Now if n is positive integer and $a_0 = \frac{13.5\ldots(2n-1)}{n!}$, the first solution given by equation (5) is called the Legendre's function of first kind and is denoted by $P_n(x)$.

Therefore

$$P_n(x) = \frac{1.3.5\ldots(2n-1)}{n!}\left[x^n - \frac{n(n-1)}{2(2n-1)}x^{n-2}\right.$$

$$\left. + \frac{n(n-1)(n-2)(n-3)}{2.4(2n-1)(2n-3)}x^{n-4} - \ldots\right].$$

$P_n(x)$ is a terminating series so it called Legendre's polynomial of degree n for different values of n. Total number of terms in $P_n(x)$ are given as:

When n is even, it contains $\left(\frac{n}{2}+1\right)$ terms and

The last term $=(-1)^{n/2}.\dfrac{n(n-1)(n-2)...2.1}{(2.4\ 6...n)[(2n-1)(2n-3)...(n+1)]}$

When n is odd, its contains $\left(\frac{n+1}{2}\right)$ terms and

The last term

$$=(-1)^{(n-1)/2}\frac{n(n-1)(n-2)..3.2}{[2.4...(n-1)][(2n-1)(2n-3)...(n+2)]}$$

Note: $P_n(x)$ is that solution of Legendre's differential equation which is equal to unity when $x=1$.

4.5.2 Legendre's Polynomials of Second Kind $Q_n(x)$

When n is a positive integer and $a_0=\dfrac{n!}{1.3.5...(2n+1)}$, then the solution is denoted by $Q_n(x)$ and is called the Legendre's function of the second kind of degree n.

$$\text{Therefore } Q_n(x)=\frac{n!}{1.3...(2n+1)}[x^{-n-1}+\frac{(n+1)(n+2)}{2(2n+3)}x^{-n-3}$$

$$+\frac{(n+1)(n+2)(n+3)(n+4)}{2.4(2n+3)(2n+5)}x^{-n-5}+...].$$

$Q_n(x)$ is an infinite series, so there is no last term. Since $P_n(x)$and $Q_n(x)$ are two independent solutions of the given Legendre's differential equation. The most general solution of the Legendre's equation is $y=AP_n(x)+BQ_n(x)$, where A and B are arbitrary constants.

4.5.3 Generating Function for $P_n(x)$

Prove that $P_n(x)$ is the coefficient of z^n in the expression of $\left(1-2xz+z^2\right)^{-1/2}$ in ascending powers of z, i.e.,

$$\left(1-2xz+z^2\right)^{-1/2}=\sum_{r=0}^{\infty}P_n(x).z^n.$$

Proof: We can write $\left(1-2xz+z^2\right)^{-1/2}=[1-z(2x-z)]^{-1/2}$.

Expanding R.H.S. by binomial theorem, we get

$$\left[1-z(2x-z)\right]^{-1/2}=1+\frac{z}{2}(2x-z)+\frac{1.3}{2.4}z^2(2x-z)^2+\frac{1.3.5}{2.4.6}$$

$$z^3(2x-z)^3+\ldots+\frac{1.3\ldots(2n-3)}{2.4.6\ldots(2n-2)}z^{n-1}(2x-z)^{n-1}+\frac{1.3\ldots(2n-1)}{2.4\ldots(2n)}$$

$$z^n(2x-z)^n+\ldots.$$

Now the coefficient of z^n in $\dfrac{1.3.5\ldots(2n-1)}{2.4\ldots(2n)}z^n(2x-z)^n$

$$=\frac{1.3.5\ldots(2n-1)}{2.4\ldots(2n)}\left[{}^nc_0(2x)^n(-z)^0\right]$$

$$=\frac{1.3.5\ldots(2n-1)}{2^n(n!)}2^nx^n,\ \left[\because {}^nc_0=1,(-z)^0=1\right]$$

$$=\frac{1.3.5\ldots(2n-1)}{(n!)}x^n.$$

The coefficient of z^n in $\left(\dfrac{1.3\ldots(2n-3)}{2.4\ldots(2n-2)}z^{n-1}(2x-z)^{n-1}\right)$

$$=-\frac{1.3\ldots(2n-3)}{2^{n-1}(n-1)!}\times(n-1)2^{n-2}\times x^{n-2}\times\frac{2n}{2n}$$

$$=-\frac{1.3..(2n-3)(2n-1)}{n!}\times\frac{n(n-1)}{2(2n-1)}x^{n-2}.$$

The coefficient of z^n in $\dfrac{1.3.5\ldots(2n-5)}{2.4\ldots(2n-4)}z^{n-2}(2x-z)^{n-2}$

$$=\frac{1.3.5\ldots(2n-1)}{n!}\ \frac{n(n-1)(n-2)(n-3)}{2.4(2n-1)(2n-3)}x^{n-4}.$$

Similarly, we can find other coefficient of z^n in

$$\frac{1.3.5\ldots(2n-2r+1)}{2.4\ldots(2n-2r)}z^{n-r}(2x-z)^{n-r}.$$

Therefore, the coefficient of z^n in the expansion of $\left(1-2xz+z^2\right)^{1/2}$ is

$$=\frac{1.3\ldots(2n-1)}{n!}\left[x^n-\frac{n(n-1)}{2(2n-1)}x^{n-2}+\frac{n(n-1)(n-2)(n-3)}{2.4(2n-1)(2n-3)}.x^{n-4}-\ldots\right]$$

$$=P_n(x).$$

Thus $\sum_{r=0}^{\infty} z^n P_n(x) = (1-2xz+z^2)^{-1/2}$.

The function $(1-2xz+z^2)^{-1/2}$ is called the generating function of the polynomials $P_n(x)$.

4.5.4 Orthogonal Properties of Legendre's Polynomials

(i) $\int_{-1}^{1} P_m(x) P_n(x) dx = 0 \text{ if } m \neq n$

(ii) $\int_{-1}^{1} [P_n(x)]^2 dx = \frac{2}{2n+1}$ *i.e.,* $\int_{-1}^{1} P_m(x) P_n(x) dx = \frac{2}{2n+1}$ if $m = n$.

Proof:

(i) Legendre's equation may be written as

$$\frac{d}{dx}\left\{(1-x^2)\frac{dy}{dx}\right\} + n(n+1)y = 0 \qquad \text{...(1)}$$

We know that the $P_m(x)$ and $P_n(x)$ are the solution of Legendre's equation, so we have

$$\frac{d}{dx}\left\{(1-x^2)\frac{dP_m}{dx}\right\} + m(m+1)P_m = 0 \qquad \text{...(2)}$$

and $$\frac{d}{dx}\left\{(1-x)^2\frac{dP_n}{dx}\right\} + n(n+1)P_n = 0. \qquad \text{...(3)}$$

Multiplying (2) by P_n and (3) by P_m and subtracting, we obtain

$$P_n \frac{d}{dx}\left\{(1-x^2)\frac{dP_m}{dx}\right\} - P_m \frac{d}{dx}\left\{(1-x^2)\frac{dP_n}{dx}\right\} + \{m(m+1) - n(n+1)\}P_m P_n = 0.$$

Integrating between the limits – 1 to 1, we have

$$\int_{-1}^{1} P_n \frac{d}{dx}\left\{(1-x^2)\frac{dP_m}{dx}\right\}dx - \int_{-1}^{1} P_m \frac{d}{dx}\left\{(1-x^2)\frac{dP_n}{dx}\right\}dx + \{m(m+1) - n(n+1)\}\int_{-1}^{1} P_m P_n dx = 0$$

$$\Rightarrow \left[P_n(1-x^2)\frac{dP_m}{dx}\right]_{-1}^{1} - \int_{-1}^{1} \frac{dP_n}{dx}\left\{(1-x^2)\frac{dP_m}{dx}\right\}dy - \left[P_m(1-x^2)\frac{dP_n}{dx}\right]_{-1}^{1}$$

$$+ \int_{-1}^{1} \frac{dP_m}{dx}\left((1-x^2)\frac{dP_n}{dx}\right)dx + \left[(m^2-n^2)+(m-n)\right]\int_{-1}^{1} P_m P_n dx = 0$$

$$\Rightarrow \quad \left[(m+n)(m-n)+(m-n)\right]\int_{-1}^{1} P_m P_n dx = 0$$

$$\Rightarrow \quad (m-n)(m+n+1)\int_{-1}^{1} P_m P_n dx = 0.$$

Hence, for $m \neq n$, we have $\int_{-1}^{1} P_m(x) P_n(x) dx = 0.$

(ii) We have $\left(1-2xz+z^2\right)^{-1} = \sum_{n=0}^{\infty} P_n(x) z^n.$

Squaring both sides, we get

$$\left(1-2xz+z^2\right)^{-1} = \sum_{n=0}^{\infty} P_n^2(x) z^{2n} + 2 \sum_{\substack{m=0\\n=0\\m\neq n}}^{\infty} P_n(x) P_m(x) z^{m+n}.$$

Integrating both sides w.r.t. 'x', between the limits –1 to 1, we have

$$\sum_{n=0}^{\infty} \int_{-1}^{1} \left[P_n(x)\right]^2 z^{2n}\, dx + 2 \sum_{\substack{m=0\\n=0\\m\neq n}}^{\infty} P_m(x) P_n(x) z^{m+n}\, dx = \int_{-1}^{1} \frac{dx}{\left(1-2xz+z^2\right)}$$

$$\Rightarrow \sum_{n=0}^{\infty} \int_{-1}^{1} \left[P_n(x)\right]^2 z^{2n} dx = \int_{-1}^{1} \frac{dx}{\left(1-2xz+z^2\right)}$$

$$= -\frac{1}{2z}\left[\log\left(1-2xz+z^2\right)\right]_{-1}^{1}$$

$$= -\frac{1}{2z}\left[\log(1-z)^2 - \log(1+z)^2\right]$$

$$= +\frac{1}{z}\left[\log\left(\frac{1+z}{1-z}\right)\right] = \frac{2}{z}\left[z + \frac{z^2}{3} + \frac{z^5}{5} + \ldots\right]$$

$$= 2\left[1 + \frac{z^2}{3} + \frac{z^4}{5} + \ldots + \frac{z^{2n}}{2n+1} + \ldots\right] = \sum_{n=0}^{\infty} \frac{2z^{2n}}{2n+1}$$

Equating the coefficient of z^{2n}, we obtain

$$\int_{-1}^{1} \left[P_n(x)\right]^2 dx = \frac{2}{2n+1}$$

4.5.5 Recurrence Formulae for Legendre's Polynomials

I. $(2n+1) x P_n(x) = (n+1) P_{n+1}(x) + n P_{n-1}(x).$

Proof: We know that $\left(1-2xz+z^2\right)^{-1/2} = \sum_{n=0}^{\infty} P_n(x) z^n.$

Differentiating both sides w.r.t 'z', we obtain

$$-\frac{1}{2}\left(1-2xz+z^2\right)^{-3/2}(-2x+2z) = \sum_{n=0}^{\infty} n z^{n-1} P_n(x)$$

$$\Rightarrow \left(1-2xz+z^2\right)^{-3/2}(x-z) = \sum_{n=0}^{\infty} n z^{n-1} P_n(x)$$

$$\Rightarrow \left(1-2xz+z^2\right)^{-1/2}(x-z)=\left(1-2xz+z^2\right)\sum_{n=0}^{\infty} nz^{n-1}P_n(x)$$

$$\Rightarrow [x-z]\sum_{n=0}^{\infty} z^n P_n(x)=\left(1-2xz+z^2\right)\sum_{n=0}^{\infty} nz^{n-1}P_n(x).$$

Equating the coefficient of z^n on both sides, we have

$$xP_n(x)-P_{n-1}(x)=(n+1)P_{n+1}(x)-2nxP_n(x)+(n-1)P_{n-1}(x)$$

$$\Rightarrow (2n+1)xP_n(x)=(n+1)P_{n+1}(x)+nP_{n-1}(x)$$

II. $nP_n(x)=xP_n'(x)-P'_{n-1}(x)$.

Proof: We have $\left(1-2xz+z^2\right)^{-1/2}=\sum_{n=0}^{\infty} z^n P_n(x)$. ...(1)

Differentiating w.r.t. z, we get

$$(x-z)\left(1-2xz+z^2\right)^{-3/2}=\sum_{n=0}^{\infty} nz^{n-1}P_n(x). \quad ...(2)$$

Differentiating w. r. t 'x' equation (1), we get

$$-\frac{1}{2}\left(1-2xz+z^2\right)^{-3/2}(-2z)=\sum_{n=0}^{\infty} z^n P_n'(x)$$

$$\Rightarrow z\left(1-2xz+z^2\right)^{-3/2}=\sum_{n=0}^{\infty} z^n P_n'(x)$$

$$\Rightarrow z(x-z)\left(1-2xz+z^2\right)^{-3/2}=(x-z)\sum_{n=0}^{\infty} z^n P_n'(x).$$

From (2) and (3), we have $z\sum_{n=0}^{\infty} nz^{n-1}P_n(x)=(x-z)\sum_{n=0}^{\infty} z^n P_n'(x)$.

Equating the coefficient of z^n on both sides, we get

$$nP_n(x)=xP_n'(x)-P'_{n-1}(x)$$

III. $(2n+1)P_n(x)=P'_{n+1}(x)-P'_{n-1}(x)$.

Proof: From recurrence relation I, we have

$$(2n+1)x\,P_n(x)=(n+1)P_{n+1}(x)+nP_{n-1}(x). \quad ...(1)$$

Differentiating w.r.t., 'x', we have

$$(2n+1)xP_n'(x)+(2n+1)P_n(x)=(n+1)P'_{n+1}(x)+nP_{n-1}(x). \quad ...(2)$$

From recurrence formula II, we have

$$xP_n'(x)=nP_n(x)+P'_{n-1}(x). \quad ...(3)$$

Eliminating $x\,P'_n(x)$ from (2) and (3), we have

$$(2n+1)\left[nP_n(x)+P'_{n-1}(x)\right]+(2n+1)P_n(x)=(n+1)P'_{n+1}(x)$$
$$+nP'_{n-1}(x)$$

$$\Rightarrow \quad (2n+1)(n+1)P_n(x)+(2n+1)P'_{n-1}(x)=(n+1)P'_{n+1}(x)$$
$$+nP'_{n-1}(x)$$

$$\Rightarrow \quad (2n+1)(n+1)P_n(x)=(n+1)P'_{n+1}(x)-(n+1)P'_{n-1}(x)$$

$$\Rightarrow \quad (2n+1)P_n(x)=P'_{n+1}(x)-P'_{n-1}(x).$$

IV. $(n+1)P_n(x)=P'_{n+1}(x)-xP'_n(x)$.

Proof: From recurrence formulae II and III, we have

$$nP_n(x)=xP'_n(x)-P'_{n-1}(x) \quad \text{...(1)}$$

and $(2n+1)P_n(x)=P'_{n+1}(x)-P'_{n-1}(x)$. ...(2)

Subtracting (1) from (2), we get

$$(n+1)P_n(x)=P'_{n+1}-xP'_n(x)$$

V. $(1-x^2)P_n'(x)=n\left[P_{n-1}(x)-xP_n(x)\right]$.

Proof: Replacing n by (n-1) in recurrence formula IV, we have

$$nP_{n-1}(x)=P_n'(x)-xP'_{n-1}(x) \quad \text{...(1)}$$

Rewriting recurrence formula II, we have

$$nP_n(x)=xP_n'(x)-P'_{n-1}(x). \quad \text{...(2)}$$

Multiplying (2) by x and then subtracting from (1), we obtain

$$n\left[P_{n-1}(x)-xP_n(x)\right]=(1-x^2)P_n'(x)$$

$$\Rightarrow \quad (1-x^2)P'_n(x)=n\left[P_{n-1}(x)-xP_n(x)\right]$$

VI. $(1-x^2)P_n'(x)=(n+1)\left[xP_n(x)-P_{n+1}(x)\right]$.

Proof: From recurrence formula I, we have

$$(2n+1)x\,P_n(x)=(n+1)P_{n+1}(x)+n\,P_{n-1}(x)$$

It may be written as

$$(n+1)xP_n(x)+nxP_n(x)=(n+1)P_{n+1}(x)+nP_{n-1}(x)$$

$$\Rightarrow \quad (n+1)\left[xP_n(x)-P_{n+1}(x)\right]=n\left[P_{n-1}(x)-xP_n(x)\right]. \quad \text{...(1)}$$

From recurrence formula V, we have

$$(1-x^2)P_n'(x)=n\left[P_{n-1}(x)-xP_n(x)\right]. \quad \text{...(2)}$$

From (1) and (2), we have

$$\left(1-x^2\right)P_n'(x)=(n+1)\left[xP_n(x)-P_{n+1}(x)\right].$$

4.5.6 Rodrigues' Formula for $P_n(x)$

To prove that $P_n(x)=\dfrac{1}{n!\,2^n}\dfrac{d^n}{dx^n}\left(x^2-1\right)^n.$

Proof: Let $y=\left(x^2-1\right)^n$.

On differentiating both sides w.r.t. 'x', we get

$$\frac{dy}{dx}=n\left(x^2-1\right)^{n-1}\times 2x.$$

Multiplying both sides by $\left(x^2-1\right)$, we get

$$\left(x^2-1\right)y_1=2nxy \Rightarrow \left(1-x^2\right)y_1+2nxy=0.$$

Differentiating $(n+1)$ times by Leibnitz theorem, we obtain

$$\left[\left(1-x^2\right)y_{n+2}+(n+1)(-2x)y_{n+1}+\frac{n(n+1)}{2!}(-2)y_n\right]+2n\left[xy_{n+1}+(n+1)y_n\right]=0$$

$$\Rightarrow \quad \left(1-x^2\right)y_{n+2}-2xy_{n+1}+n(n+1)y_n=0$$

$$\Rightarrow \quad \left(1-x^2\right)\frac{d^2y_n}{dx^2}-2x\frac{d(y_n)}{dx}+n(n+1)y_n=0$$

which is Legendre's equation.

Hence its solution is $y_n=cP_n(x)$, where c is a constant.

$$\text{Now} \quad cP_n(x)=\frac{d^n}{dx^n}\left(x^2-1\right)^n \qquad \text{...(1)}$$

$$=\frac{d^n}{dx^n}\left[(x+1)(x-1)\right]^n$$

Differentiating w. r. t 'x', n times by Leibnitz theorem, we get

$$cP_n(x)=(x-1)^n\cdot\frac{d^n}{dx^n}(x+1)^n+{c_1}^n\cdot n(x-1)^{n-1}\frac{d^{n-1}}{dx^{n-1}}(x+1)^n+\ldots+(x+1)^n\frac{d^n}{dx^n}(x-1)^n$$

$$\Rightarrow cP_n(x)=(x-1)^n\,n!+c_1^n\cdot n(x-1)^{n-1}\cdot\frac{d^{n-1}}{dx^{n-1}}(x+1)^n+\ldots+(x+1)^n\,n!$$

$=(x+1)^n\, n! +$ terms containing powers of $(x-1)$.

Putting $x=1$, we have $cP_n(1)=n!.2^n \Rightarrow c=n!.2^n$, since $P_n(1)=1$.

Substituting the value of c in (1), we get $P_n(x)=\frac{1}{2^n n!}\frac{d^n}{dx^n}(x^2-1)^n$

Legendre Polynomial for Different values of *n*: By Rodrigues' formula, we have

$$P_n(x)=\frac{1}{(2)^n\, n!}\frac{d^n}{dx^n}(x^2-1)^n.$$

For $n=0$, $P_0(x)=1$

For $n=1$,
$P_1(x)=x$

For $n=2$, $P_2(x)=\frac{1}{2^2 2!}\cdot\frac{d^2}{dx^2}(x^2-1)^2=\frac{1}{2}(3x^2-1)$.

For $n=3$, $P_3(x)=\frac{1}{2}(5x^3-3x)$

For $n=4$, $P_4(x)=\frac{1}{8}(35x^4-30x^2+3)$ and so on.

Example 1: Show that

$$P_0(x)=1, P_1(x)=x, P_2(x)=\frac{(3x^2-1)}{2}, P_3(x)=\frac{1}{2}(5x^3-5x).$$

Solution: We know that $\sum_{n=0}^{\infty} z^n P_n(x)=(1-2xz+z^2)^{-1/2}$

$$\Rightarrow \quad \sum_{n=0}^{\infty} z^n P_n(x)=[1-z(2x-z)]^{-1/2}=1+\frac{z}{2}(2x-z)+\frac{1.3}{2.4}z^2$$

$$(2x-z)^2+\frac{1.3.5}{2.4.6}z^2(2x-z)^2+\ldots$$

$$\Rightarrow \quad P_0(x)+zP_1(x)+z^2P_2(x)+z^2P_3(x)+\ldots$$

$$=1+xz+\frac{1}{2}(3x^2-1)z^2+\frac{1}{2}(5x^2-3x)z^3+\ldots.$$

Equating the coefficients of like powers of z, we have

$$\Rightarrow \quad P_0(x)=1, P_1(x)=x, P_2(x)=\frac{1}{2}(3x^2-1)P_3(x)=\frac{1}{2}(5x^3-3x)$$

Example 2: Show that

(i) $P_{2n}(0) = (-1)^n \dfrac{1.3.5\ldots(2n-1)}{2.4.6\ldots 2n}$ (ii) $P_{2n+1}(0) = 0$

Solution: We know that $\sum z^n P_n(x) = (1 - 2xz + z^2)^{-1/2}$.

$$\Rightarrow \quad \sum z^n P_n(0) = (1 + z^2)^{-1/2}$$

$$= 1 + \left(-\frac{1}{2}\right)z^2 + \frac{\left(-\frac{1}{2}\right)\left(-\frac{3}{2}\right)}{2!}(z^2)^2 + \frac{\left(-\frac{1}{2}\right)\left(-\frac{3}{2}\right)\left(-\frac{5}{2}\right)}{3!}(z^2)^3 + \ldots$$

$$+ \frac{\left(-\frac{1}{2}\right)\left(-\frac{3}{2}\right)\left(-\frac{5}{2}\right)\ldots\left(-\frac{1}{2} - n + 1\right)}{n!}(z^2)^n + \ldots$$

(i) Equating the coefficient of z^{2n} both sides, we have

$$P_{2n}(0) = \frac{\left(-\frac{1}{2}\right)\left(-\frac{3}{2}\right)\left(-\frac{5}{2}\right)\ldots\left(-\frac{1}{2} - n + 1\right)}{n!}$$

$$= \frac{(-1)^n.1.3\ldots(2n-1)}{2^n(n!)} = (-1)^n \frac{1.3\ldots(2n-1)}{2.4.6\ldots(2n)}.$$

(ii) Equating coefficients of z^{2n+1} both sides, we have

$$P_{2n+1}(0) = 0.$$

Example 3: Show that (i) $P_n(1) = 1$ (ii) $P_n(-x) = (-1)^n P_n(x)$. Hence deduce that $P_n(-1) = (-1)^n$.

Solution: (i) We know that $\sum_{n=0}^{\infty} z^n P_n(x) = (1 - 2xz + z^2)^{-1/2}$.

Putting $x = 1$, we get

$$\sum_{n=0}^{\infty} z^n P_n(1) = (1 - 2z + z^2)^{-1/2} = (1 - z)^{-1}$$

$$= 1 + z + z^2 + z^3 + \ldots z^n + \ldots = \sum_{n=0}^{\infty} z^n.$$

Equating the coefficient of z^n, we have $P_n(1) = 1$.

(ii) We have $(1 - 2xz + z^2)^{-1/2} = \sum_{n=0}^{\infty} z^n P_n(x)$.

Therefore $\left(1+2xz+z^2\right)^{-1/2} = \left[1-2x(-z)+(-z)^2\right]^{-1/2}$

$$= \sum_{n=0}^{\infty} (-z)^n P_n(x)$$

$$\neq \sum_{n=0}^{\infty} (-1)^n z^n \, P_n(x). \qquad ...(1)$$

Again $\quad \left(1+2xz+z^2\right)^{-1/2} = \left[1-2(-x)z+z^2\right]^{-1/2}$

$$= \sum_{n=0}^{\infty} z^n \, P_n(-x). \qquad ...(2)$$

From (1) and (2), we have $\sum_{n=0}^{\infty} z^n \, P_n(-x) = \sum_{n=0}^{\infty} (-1)^n z^n \, P_n(x).$

Equating the coefficient of z^n, we get

$$P_n(-x) = (-1)^n P_n(x).$$

Putting $x=1$, we have $P_n(-1) = (-1)^n \, P_n(1) = (-1)^n$ *as* $P_n(1) = 1.$

Example 4: Express $P(x) = x^4 + 2x^3 + 2x^2 - x - 3$ in terms of Legendre's polynomials.

Solution: By Rodrigues' formula, we have

$$P_n(x) = \frac{1}{(n!)2^n} \frac{d^n}{dx^n}\left(x^2-1\right)^n.$$

Putting $n = 0, 1, 2, 3, 4, \ldots$ in Rodrigues' formula

$$P_0(x) = 1, P_1(x) = x, \qquad ...(1)$$

$$P_2(x) = \frac{1}{2}\left(3x^2-1\right), \qquad ...(2)$$

$$P_3(x) = \frac{1}{2}\left(5x^3-3x\right) \qquad ...(3)$$

$$P_4(x) = \frac{1}{8}\left(35x^4-30x^2+3\right). \qquad ...(4)$$

From (4),(3), (2) and (1), we have

$$x^4 = \frac{8}{35} P_4(x) + \frac{6}{7} x^2 - \frac{3}{35},$$

$$x^3=\frac{2}{5}P_3(x)+\frac{3}{5}x,\ x^2=\frac{2}{3}P_2(x)+\frac{1}{3},\ x=P_1(x), 1=P_0(x).$$

Substituting these values in given polynomial. we get

$$P(x)=\frac{8}{35}P_4(x)+\frac{4}{5}P_3(x)+\frac{40}{21}P_2(x)+\frac{1}{5}P_1(x)-\frac{224}{105}P_0(x)$$

which is the required solution.

4.5.7 Beltrami's Result

To prove that $(2n+1)(x^2-1)P_n'(x)=n(n+1)(P_{n+1}(x)-P_{n-1}(x))$.

Proof: From recurrence formulae V and VI, we get

$$(1-x^2)P_n'(x)=n(P_{n-1}(x)-xP_n(x)) \qquad ...(1)$$

and $$(1-x^2)P_n'(x)=(n+1)(xP_n(x)-P_{n+1}(x)). \qquad ...(2)$$

Multiplying (1) by $(n+1)$ and (2) by n and adding, we get

$$[(n+1)+n](1-x^2)P_n'(x)=n(n+1)[P_{n-1}(x)-P_{n+1}(x)]$$

$$\Rightarrow \quad (2n+1)(x^2-1)P_n'(x)=n(n+1)[P_{n+1}(x)-P_{n-1}(x)].$$

4.5.8 Christoffel's Expansion Formula

To prove that

$$P_n'(x)=(2n-1)P_{n-1}(x)+(2n-5)P_{n-3}(x)+(2n-9)P_{n-5}(x)+...$$

with last term of the series being $3P_1$ or P_0 according as n is even or odd.

Proof: From recurrence formula (III), we have

$$P'_{n+1}(x)=(2n+1)P_n(x)+P'_{n-1}(x). \qquad ...(1)$$

Replacing n by $(n - 1)$, we have

$$P_n'(x)=(2n-1)P_{n-1}(x)+P'_{n-2}(x). \qquad ...(2)$$

Replacing n by $(n - 2)$ in the above equation, we get

$$P'_{n-2}(x)=(2n-5)P_{n-3}(x)+P'_{n-4}(x). \qquad ...(3)$$

Replacing n by $(n - 4)$ ..., 4, 2 in (2), we get

$$P'_{n-4}(x)=(2n-9)P_{n-5}(x)+P'_{n-6}(x) \qquad ...(4)$$

....

....

when n is even $P_2' = 3P_1 + P_0' = 3P_1$, since $P_0' = 0$ ...(5)

Adding (2), (3), (4), etc., we obtain

$$P_n'(x) = (2n-1)P_{n-1}(x) + (2n-5)P_{n-3} + (2n-9)P_{n-5} + \dots + 3P_1.$$

When n is odd, we have

$$P_3' = 5P_2 + P_1' = 5P_2 + P_0, \; [\because P_1' = 1 = P_0]$$

Adding as before, we get

$$P_n'(x) = (2n-1)P_{n-1}(x) + (2n-5)P_{n-3}(x) + \dots + 5P_2 + P_0.$$

Hence $P_n'(x) = (2n-1)P_{n-1}(x) + (2n-5)P_{n-3}(x) + (2n-9)P_{n-5}(x) + \dots$

with last term of the series being $3P_1$ *or* P_0 according as n is even or odd.

4.5.9 Expansion of a Function

Here we expand a function of x in a series of Legendre polynomials which is known as Fourier–Legendre series. It is possible due to the orthogonal property of Legendre polynomial in $x = -1$ *to* $x = 1$.

For this purpose, consider

$$f(x) = \sum_{n=0}^{\infty} a_n P_n(x) = a_0 P_0(x) + a_1 P_1(x) + a_2 P_2(x) + \dots \qquad \text{...(1)}$$

To determine the coefficient a_n, we multiply (1) by $P_n(x)$ and then integrate between the limits -1 to 1 with respect to x. Thus we get

$$\int_{-1}^{1} f(x) P_n(x) dx = a_n \int_{-1}^{1} [P_n(x)]^2 dx = a_n \frac{2}{2n+1}$$

$$\Rightarrow \quad \Rightarrow a_n = \frac{2n+1}{2} \int_{-1}^{1} f(x) P_n(x) dx. \qquad \text{...(2)}$$

Thus, the expansion of $f(x)$ is given by (1) & (2). This expansion (1) is called Fourier-Legendre series of $f(x)$.

4.5.10 Christoffel's Summation Formula

$$\sum_{r=0}^{\infty} (2r+1) P_r(x) \theta_r(y) = (n+1) \frac{P_{n+1}(x)\theta_n(y) - P_n(x)\theta_{n+1}(y)}{x-y}.$$

Proof: From recurrence formula (I) for P_n and θ_n, we have

$$(2r+1) x P_r(x) = (r+1) P_{r+1}(x) + r P_{r-1}(x) \qquad \text{...(1)}$$

and $(2r+1)y\theta_r(y)=(r+1)\theta_{r+1}(y)+r\theta_{r-1}(y).$ ...(2)

Multiplying (1) by $\theta_r(y)$ and (2) by $P_r(x)$ and then subtracting, we get

$$(2r+1)(x-y)P_r(x)\theta_r(y)=(r+1)\left[P_{r+1}(x)\theta_r(y)-P_r(x)\theta_{r+1}(y)\right]$$

$$-r\left[P_r(x)\theta_{r-1}(y)-P_{r-1}(x)\theta_r(y)\right] \quad \text{...(3)}$$

Putting $r = 0, 1, 2, 3, \ldots (n-1), n$ in (3), we obtain

$$(x-y)P_0(x)\theta_0(y)=P_1(x)\theta_0(y)-P_0(x)\,\theta_1(y)$$

$$3(x-y)P_1(x)\theta_1(y)=2\left[P_2(x)\theta_1(y)-P_1(x)\theta_2(y)\right]$$

$$-1\left[P_1(x)\theta_0(y)-P_0(x)\theta_1(y)\right]$$

$$5(x-y)P_2(x)\theta_2(y)=3\left[P_3(x)\theta_2(y)-P_2(x)\theta_3(y)\right]$$

$$-2\left[P_2(x)\theta_1(y)-P_2(x)\theta_2(y)\right]$$

....

....

$$(2n-1)(x-y)P_{n-1}(x)\theta_{n-1}(y)=n\left[P_n(x)\theta_{n-1}(y)-P_{n-1}(x)\theta_n(y)\right]$$

$$-(n-1)\left[P_{n-1}(x)\theta_{n-2}(y)-P_{n-2}(x)\theta_{n-1}(y)\right]$$

$$(2n+1)(x-y)P_n(x)\theta_n(y)=(n+1)\left[P_{n+1}(x)\theta_n(y)-P_n(x)\theta_{n+1}(y)\right]$$

$$-n\left[P_n(x)\theta_{n-1}(y)-P_{n-1}(x)\theta_n(y)\right]$$

Adding above equations, we get

$$\sum_{r=0}^{n}(2r+1)(x-y)P_r(x)\theta_r(y)=(n+1)\left[P_{n+1}(x)\theta_n(y)-P_n(x)\theta_{n+1}(y)\right].$$

$$\Rightarrow \quad \sum_{r=1}^{n}(2r+1)P_r(x)\theta_r(y)=(n+1)\frac{P_{n+1}(x)\theta_n(y)-P_n(x)\theta_{n+1}(y)}{x-y}.$$

4.5.11 Laplace's First Integral for $P_n(x)$

To prove that $P_n(x)=\dfrac{1}{\pi}\displaystyle\int_0^{\pi}\left[x\pm\sqrt{x^2-1}\cos\phi\right]^n d\phi.$

Proof: We know that $\int_0^{\pi} \frac{d\phi}{a \pm b\cos\phi} = \frac{\pi}{\sqrt{a^2 - b^2}}$, where $a > b$.

Putting $a = 1 - xz$ and $b = z\sqrt{(x^2 - 1)}$

so that $a^2 - b^2 = (1 - xz)^2 - z^2(x^2 - 1) = 1 - 2xz + z^2$.

We have

$$\pi(1 - 2xz + z^2)^{-1/2} = \int_0^{\pi} \left[1 - xz \pm z\sqrt{x^2 - 1}\cos\phi\right]^{-1} d\phi$$

$$= \int_0^{\pi} \left[1 - z\left\{x \pm \sqrt{x^2 - 1}\cos\phi\right\}\right]^{-1} d\phi$$

$$= \int_0^{\pi} [1 - zt]^{-1} d\phi \text{ where } t = x \pm \sqrt{x^2 - 1}\cos\phi$$

$$\Rightarrow \quad \pi \sum_{n=0}^{\infty} z^n P_n(x) = \int_0^{\pi} \left[1 + zt + z^2t^2 + z^3t^3 + \ldots + z^n t^n + \ldots\right] d\phi.$$

Equating the coefficient of z^n, we get

$$\pi P_n(x) = \int_0^{\pi} t^n d\phi$$

$$\Rightarrow \quad P_n(x) = \frac{1}{\pi} \int_0^{\pi} \left[x \pm \sqrt{x^2 - 1}\cos\phi\right]^n d\phi.$$

4.5.12 Laplace's Second Integral for $P_n(x)$

To prove that $P_n(x) = \frac{1}{\pi} \int_0^{\pi} \frac{d\phi}{\left[x \pm \sqrt{x^2 - 1}\cos\phi\right]^{n+1}}$.

Proof: We know that $\int_0^{\pi} \frac{d\phi}{a \pm b\cos\phi} = \frac{\pi}{\sqrt{a^2 - b^2}}$, where $a > b$. ...(1)

Putting $a = xz - 1$ and $b = z\sqrt{x^2 - 1}$, so that $a^2 - b^2 = 1 - 2xz + z^2$

Substituting the above values in equation (1), we get

$$\pi(1 - 2xz + z^2)^{-1/2} = \int_0^{\pi} \left[-1 + xz \pm z\sqrt{x^2 - 1}\cos\phi\right]^{-1} d\phi$$

$$\Rightarrow \quad \pi(1 - 2xz + z^2)^{-1/2} = \int_0^{\pi} \left[-1 + xz \pm z\sqrt{x^2 - 1}\cos\phi\right]^{-1} d\phi$$

$$\Rightarrow \quad \frac{\pi}{z} \sum_{n=0}^{\infty} \frac{1}{z^n} P_n(x) = \int_0^{\pi} (t - 1)^{-1} d\phi \text{ where } t = z\left[x \pm \sqrt{x^2 - 1}\cos\phi\right]$$

$$= \int_0^\pi \frac{1}{t}\left[1-\frac{1}{t}\right]^{-1} d\phi = \int_0^\pi \frac{1}{t}\left[1+\frac{1}{t}+\frac{1}{t^2}+\ldots+\frac{1}{t^n}+\ldots\right] d\phi$$

$$= \int_0^\pi \sum_{n=0}^{\infty} \frac{1}{t^{n+1}}\, d\phi = \int_0^\pi \sum_{n=0}^{\infty} \frac{d\phi}{z^{n+1}\left\{x \pm \sqrt{x^2-1}\cos\phi\right\}^{n+1}}$$

$$\Rightarrow \quad \pi \sum_{n=0}^{\infty} \frac{1}{z^{n+1}} P_n(x) = \int_0^\pi \sum_{n=0}^{\infty} \frac{d\phi}{z^{n+1}\left\{x \pm \sqrt{x^2-1}\cos\phi\right\}^{n+1}}.$$

Equating the coefficients of $\frac{1}{z^{n+1}}$ on both sides, we get

$$P_n(x) = \frac{1}{\pi}\int_0^\pi \frac{d\phi}{\left[x \pm \sqrt{x^2-1}\cos\phi\right]^{n+1}}.$$

Example 5: If $f(x) = \begin{cases} 0, -1 < x < 0 \\ x, 0 < x < 1 \end{cases}$, then show that

$$f(x) = \frac{1}{4}P_0(x) + \frac{1}{2}P_1(x) + \frac{5}{16}P_2(x) - \frac{3}{32}P_4(x) + \ldots.$$

Solution: We know that $f(x) = \sum_{n=0}^{\infty} a_n P_n(x)$,

$$\text{where } a_n = \left(n+\frac{1}{2}\right)\int_{-1}^{0} f(x)P_n(x)\,dx + \left(n+\frac{1}{2}\right)\int_0^1 f(x)P_n(x)\,dx$$

$$= \left(\frac{2n+1}{2}\right)\int_0^1 f(x)P_n(x)\,dz. \qquad \ldots(2)$$

We find the values of $a_0, a_1, a_2, a_3, \ldots$ by putting $n = 1, 2, 3, 4, \ldots$ in (2) as follows:

$$a_0 = \frac{1}{2}\int_0^1 xP_0(x)\,dx = \frac{1}{2}\left(\frac{x^2}{2}\right)_0^1 = \frac{1}{4}$$

$$a_1 = \frac{3}{2}\int_0^1 xP_1(x)\,dx = \frac{3}{2}\left(\frac{x^3}{3}\right)_0^1 = \frac{1}{2}$$

$$a_2 = \frac{5}{2}\int_0^1 xP_2(x)\,dx = \frac{5}{2}\int_0^1 x\left[\frac{3x^2-1}{2}\right]dx = \frac{5}{16}$$

$$a_3 = \frac{7}{2}\int_0^1 xP_3(x)\,dx = \frac{7}{2}\int_0^1 x\left[\frac{5x^3-3x}{2}\right]dx = 0.$$

$$\text{Similarly } a_4 = \frac{9}{2}\int_0^1 xP_4(x)\,dx = \frac{9}{2}\int_0^1 x\frac{(35x^4-30x^2+3)}{8}\,dx = -\frac{3}{32}.$$

Substituting these values in (1), we get the required result.

$$f(x)=\frac{1}{4}P_0(x)+\frac{1}{2}P_1(x)+\frac{5}{16}P_2(x)-\frac{3}{32}P_5(x)+\ldots.$$

Example 6: Prove that

(i) $\int_{-1}^{1} P_n(x)dx=0, n\neq 0$

(ii) $\int_{-1}^{1} P_0(x)dx=2.$

Solution: (i) From Rodrigues' formula, we have

$$P_n(x)=\frac{1}{n!2^n}\cdot\frac{d^n}{dx^n}(x^2-1)^n.$$

Therefore, $\int_{-1}^{1} P_n(x)dx=\frac{1}{2^n n!}\int_{-1}^{1}\frac{d^n}{dx^n}(x^2-1)^n\,dx$

$$=\frac{1}{2^n n!}\left[\frac{d^{n-1}}{dx^{n-1}}(x^2-1)^n\right]_{-1}^{1} \quad \ldots(1)$$

Now $\frac{d^{n-1}}{dx^{n-1}}(x^2-1)^n=\frac{d^{n-1}}{dx^{n-1}}\left[(x+1)(x-1)\right]^n$

$$=\frac{d^{n-1}}{dx^{n-1}}(x+1)^n(x-1)^n,$$

Expanding by Leibnitz's theorem

$$=(x+1)^n\frac{d^{n-1}}{dx^{n-1}}(x-1)^n+(n-1).n(x+1)^{n-1}\frac{d^{n-2}}{dx^{n-2}}(x-1)^n+\ldots$$

$$+(x-1)^n\frac{d^{n-1}}{dx^{n-1}}(x+1)^n$$

$$=(x)^n\frac{n!}{1!}(x-1)+n(n-1)(x+1)^{n-1}\frac{n!}{2!}(x-1)^2+\ldots+(x-1)^n n!(x+1)=0,$$

when $x=-1$ or 1

Therefore from (1), we have $\int_{-1}^{1} P_n(x)dx=0.$

(ii) We know that $P_0(x)=1.$

Therefore $\int_{-1}^{1} P_0(x)dx=\int_{-1}^{1}dx=[x]_{-1}^{1}=2.$

Example 7: Prove that $\int_{-1}^{1}(x^2-1)P_{n+1}(x)P_n'(x)dx=\frac{2n(n+1)}{(2n+1)(2n+3)}.$

Solution: From recurrence formula V, we get

$$(x^2-1)P_n'(x)=n\left[xP_n(x)-P_{n-1}(x)\right]$$

Therefore $\int_{-1}^{1}(x^2-1)P_{n+1}\,P_n'\,dx=\int_{-1}^{1}n\left[xP_n-P_{n-1}\right]P_{n+1}dx$

$=n\int_{-1}^{1}x\,P_nP_{n+1}dx$, because $\int_{-1}^{1}P_m\,P_ndx=0$, for $m\neq n$

$=n\int_{-1}^{1}\frac{(n+1)P_{n+1}+nP_{n-1}}{2n+1}P_{n+1}dx$, from recurrence formula I

$$=\frac{n(n+1)}{2n+1}\int_{-1}^{1}P_{n+1}^2\,dx+\frac{n^2}{2n+1}\int_{-1}^{1}P_{n-1}\,P_{n+1}dx$$

$$=\frac{n(n+1)}{(2n+1)}\cdot\frac{2}{\left[2(n+1)+1\right]}=\frac{2n(n+1)}{(2n+1)(2n+3)}.$$

Example 8: Prove that

$$\int_{-1}^{1}x^2\,P_{n+1}(x).P_{n-1}(x)dx=\frac{2n(n+1)}{(2n-1)(2n+1)(2n+3)}.$$

Solution: The recurrence formula I is

$$(2n+1)xP_n=(n+1)P_{n+1}+nP_{n-1}. \qquad ..(1)$$

Replacing n by $(n+1)$ and $(n-1)$ respectively, we get

$$(2n+3)xP_{n+1}=(n+2)P_{n+2}+(n+1)P_n \qquad ...(2)$$

$$(2n-1)xP_{n-1}=nP_n+(n-1)P_{n-2}. \qquad ...(3)$$

Multiplying (2) and (3) and then integrating between the limits -1 to +1 w. r. t. 'x', we have

$$(2n+3)(2n-1)\int_{-1}^{1}x^2\,P_{n+1}\,P_{n-1}\,dx=n(n+1)\int_{-1}^{1}P_n^2dx+n(n+2)$$

$$\int_{-1}^{1}P_nP_{n+2}dx\;+(n^2-1)\int_{-1}^{1}P_n\,P_{n-2}dx+(n-1)(n+2)\int_{-1}^{1}P_{n+2}\,P_{n-2}\,dx$$

$=n(n+1)\int_{-1}^{1}P_n^2dx$, as all other integrals being zero

$$=\frac{2n(n+1)}{(2n+1)}$$

Therefore $\int_{-1}^{1}x^2\,P_{n+1}\,P_{n-1}\,dx=\frac{2n(n+1)}{(2n-1)(2n+1)(2n+3)}$

Example 9: Prove that $\int_{-1}^{1}(1-x^2)\left[P_n'(x)\right]^2dx=\frac{2n(n+1)}{2n+1}$.

Solution: From Christoffel's expansion, we have

$$P_n'=(2n-1)P_{n-1}+(2n-5)P_{n-3}+(2n-9)P_{n-5}\cdots\cdots \qquad ...(1)$$

Also from Beltrami's result, we get

$$(1-x^2)P_n'(x)=\frac{n(n+1)}{(2n+1)}\left[P_{n-1}(x)-P_{n+1}(x)\right]. \quad ...(2)$$

Multiplying (1) and (2) and then integrating between the limits − 1 to +1, we have

$$\int_{-1}^{1}(1-x^2)(P_n')^2\,dx=\frac{n(n+1)}{2n+1}\int_{-1}^{1}[(2n-1)P_{n-1}^2-(2n-1)P_{n-1}P_{n+1}$$

$$-(2n-5)P_{n-3}P_{n+1}-...]dx$$

$$=\frac{n(n+1)(2n-1)}{2n+1}\int_{-1}^{1}P_{n-1}^2(x)dx,\ \text{since}\int_{-1}^{1}P_m P_n\,dx=0,\ m\neq n$$

$$=\frac{n(n+1)(2n-1)}{2n+1}\frac{2}{\left[2(n-1)+1\right]}=\frac{2n(n+1)}{2n+1},$$

Therefore $\int_{-1}^{1}(1-x^2)\left[P_n'(x)\right]^2dx=\frac{2n(n+1)}{2n+1}.$

Example 10: Prove that all the roots of $P_n(x)=0$ are real and lie between − 1 to +1.

Solution: By Rodrigues' formula, we have

$$P_n(x)=\frac{1}{2^n n!}\cdot\frac{d^n}{dx^n}(x^2-1)^n$$

Now $(x^2-1)^n=\left[(x+1)(x-1)\right]^n=(x+1)^n(x-1)^n.$

So $(x^2-1)^n$ vanishes n times at $x=\pm1$. Therefore, by the theory of equations, $\frac{d^n}{dx^n}(x^2-1)$ will have n roots all real lying between − 1 to +1.

Hence $P_n(x)=0$ have n real roots between − 1 to +1.

Example 11: Prove that all roots of $P_n(x)=0$ are distinct.

Solution: If possible, let all roots of $P_n(x)=0$ are not distinct and consider α be a repeated root, then we have

$$P_n(\alpha)=0 \text{ and } P_n'(\alpha)=0. \quad ...(1)$$

But $P_n(x)$ is the solution of Legendre's equation, so that

$$(1-x^2)\frac{d^2}{dx^2}P_n(x)-2x\frac{d}{dx}P_n(x)+n(n+1)P_n(x)=0.$$

On differentiating r times using Leibnitz's theorem, we have

$$(1-x^2)\frac{d^{r+2}}{dx^{n2}}P_n(x)-2x\,{}^rC_1\frac{d^{r+1}}{dx^{r+1}}P_n(x)-2\,{}^rC_2\frac{d^r}{dx^r}P_n(x)-2x\frac{d^{r+1}}{dx^{r+1}}P_n(x)$$

$$P_n(x)-2\,{}^rC_1\frac{d^r}{dx^r}P_n(x)+n(n+1)\frac{d^r}{dx^r}P_n(x)=0$$

$$\Rightarrow\ (1-x^2)\frac{d^{r+2}}{dx^{r+2}}P_n(x)-2x(r+1)\frac{d^{r+1}}{dx^{r+1}}P_n(x)-\left[r(r+1)-n(n+1)\right]$$

$$\frac{d^r}{dx^r}P_n(x)=0. \quad ...(2)$$

Putting $r=0$ and $x=\alpha$, we get

$(1-\alpha^2)\frac{d^2}{dx^2}P_n(\alpha)-2\alpha\frac{d}{dx}P_n(\alpha)+n(n+1)P_n''(\alpha)=0 \Rightarrow P_n''(\alpha)=0$, using (1)

Similarly, putting $r = 1, 2, 3, \ldots$ by (3) and simplifying, we get

$$P_n''(\alpha)=P_n'''(\alpha)=\ldots=P_n^{(n)}(\alpha)=0. \quad ...(3)$$

But $$P_n(x)=\frac{1.3.5\ldots(2n-1)}{n!}\left[x^n-\frac{n(n-1)x^{n-2}}{(2n-1)}+\frac{n(n-1)(n-2)(n-3)}{2.4(2n-1)(2n-3)}x^{n-4}+\ldots\right].$$

Therefore, $P_n^{(n)}(\alpha)=\frac{1.3.5\ldots(2n-1)}{n!}.n!\neq 0.$

This contradicts (3). Hence, our assumption that $P_n(x)=0$ has a repeated root is wrong and therefore, $P_n(x)=0$ has all roots distinct.

Example 12: Prove that

$$P'_{n+1}(x)+P_n'(x)=P_0+3P_1+5P_2+\ldots+(2n+1)P_n.$$

Solution: From recurrence formula III, we have

$$(2n+1)P_n(x)=P'_{n+1}(x)-P'_{n-1}(x). \quad ...(1)$$

Putting $n = 1, 2, 3, 4, \ldots, (n-2)$ in (1), we get

$$3P_1=P_2'-P_0'$$
$$5P_2=P_3'-P_1'$$
$$7P_3=P_4'-P_3'$$
$$\ldots\ldots\ldots\ldots\ldots\ldots$$
$$(2n-3)P_{n-2}=P'_{n-1}-P'_{n-3}$$
$$(2n-1)P_{n-1}=P'_n-P'_{n-2}$$

$$(2n+1)P_n = P'_{n+1} - P'_{n-1}.$$

Adding all, we get

$$3P_1 + 5P_2 + \ldots(2n+1)P_n = P'_n + P'_{n+1} - P'_0 - P'_1$$

$$= P'_n + P'_{n+1} - 0 - P_0, \text{ since } P'_0(x) = 0 \text{ and } P'_1 = 1 = P_0$$

Hence , we have $P_0 + 3P_1 + 5P_2 = \ldots + (2n+1)P_n = P'_{n+1} + P'_n.$

Example 13: Prove that $\int_{-1}^{1} x^m\, P_n(x)dx = 0,\ if\ m < n.$

Solution: Here $\int_{-1}^{1} x^m\, P_n(x)dx = \int_{-1}^{1} x^m \dfrac{1}{2^n\, n!}\dfrac{d^n}{dx^n}(x^2-1)^n\, dx$

(using Rodrigues' formula)

$$= \frac{1}{2^n n!}\int_{-1}^{1} x^m \frac{d^n}{dx^n}(x^2-1)^n\, dx$$

$$= \frac{1}{2^n\, n!}\left[\left\{x^m \frac{d^{n-1}}{dx^{n-1}}(x^2-1)^n\right\}_{-1}^{1} - \int_{-1}^{1} mx^{m-1}\frac{d^{n-1}}{dx^{n-1}}(x^2-1)^n\, dx\right]$$

Integrating by parts

$$= 0 - \frac{m}{2^n n!}\int_{-1}^{1} x^{m-1}\frac{d^{n-1}}{dx^{n-1}}(x^2-1)^n\, dx$$

Further integration yields

$$\int_{-1}^{1} x^m\, P_n(x)dx = (-1)^2\frac{m(m-1)}{2^n\, n!}\int_{-1}^{1} x^{m-2}\frac{d^{n-2}}{dx^{n-2}}(x^2-1)^n\, dx$$

Integrating $(m-2)$ times in all, we get

$$\int_{-1}^{1} x^m\, P_n(x)dx = (-1)^m\frac{m(m-1)\ldots 1}{2^n\, n!}\int_{-1}^{1}\frac{d^{n-m}}{dx^{n-m}}(x^2-1)^n\, dx$$

$$= \frac{(-1)^m\, m!}{2^n\, n!}\int_{-1}^{1}\frac{d^{n-m}}{dx^{n-m}}(x^2-1)^n\, dx$$

$$= \frac{(-1)^m\, m!}{2^n\, n!}\left[\frac{d^{n-m-1}}{dx^{n-m-1}}(x^2-1)^n\right]_{-1}^{1} = 0.$$

Example 14: Prove that $\dfrac{P_{n+1} - P_{n-1}}{2n+1} = \int P_n\, dx + c.$

Solution: From recurrence relation III, we have

$$P'_{n+1} - P'_{n-1} = (2n+1)P_n.$$

Integrating both sides w. r. t. 'x', we get $\dfrac{P_{n+1} - P_{n-1}}{2n+1} = \int P_n dx + c.$

Example 15: Prove that $P_n(x) = P_{-n-1}(x)$.

Solution: We know that $P_n(x) = \frac{1}{\pi}\int_0^\pi \left(x \pm \sqrt{x^2-1}\cos\phi\right)^n d\phi$

Replacing n by $-n-1$, we get

$$P_{-n-1}(x) = \frac{1}{\pi}\int_0^\pi \left(x \pm \sqrt{x^2-1}\cos\phi\right)^{-n-1} d\phi$$

$$= \frac{1}{\pi}\int_0^\pi \frac{d\phi}{\left(x \pm \sqrt{x^2-1}\cos\phi\right)^{n+1}} = P_n(x).$$

Example 16: Prove that

$$P'_{2n+1} = (4n+1)P_{2n} + (4n-3)P_{2n-2} + (4n-7)P_{2n-4} + \ldots + 5P_2 + P_0.$$

Solution: By Christoffel's expansion formula, we have

$$P'_n = (2n-1)P_{n-1} + (2n-5)P_{n-3} + (2n-9)P_{n-5} + \ldots\ldots + 5P_2 + P_0$$

(when n is odd)

Replacing n by $2n+1$, then we get

$$P'_{2n+1} = (4n+1)P_{2n} + (4n-3)P_{2n-2} + \ldots + 5P_2 + P_0.$$

Example 17: Prove that $xP'_n = nP_n + (2n-3)P_{n-2} + (2n-7)P_{n-4} + \ldots$

Solution: From recurrence relation (2), we have

$$xP'_n = n\,P_n + P'_{n-1} \quad \ldots(1)$$

From recurrence relation (3), we have

$$P'_{n+1} = (2n+1)P_n + P'_{n-1} \quad \ldots(2)$$

Replacing n by $n-2$ in (2), we get

$$P'_{n-1} = (2n-3)P_n + P'_{n-3} \quad \ldots(3)$$

Replacing n by $n-4$ in (2), we get

$$P'_{n-3} = (2n-7)P_{n-4} + P'_{n-5} \quad \ldots(4)$$

From (1), (3) and (4), we obtain

$$xP'_n = nP_n + (2n-3)P_n + (2n-7)P_{n-4} + \ldots$$

Example 18: Prove that $\int_{-1}^{1} xP_n\,P'_n\,dx = \frac{2n}{2n+1}$.

Solution: L.H.S. = $\int_{-1}^{1} P_n(xP'_n)dx$

$$= \int_{-1}^{1} P_n\left[nP_n + (2n-3)P_{n-2} + (2n-7)P_{n-4} + \ldots\right]dx$$

$$= \int_{-1}^{1} nP_n^2\, dx + (2n-3)\int_{-1}^{1} P_n\, P_{n-2}\, dx + (2n-7)\int_{-1}^{1} P_n\, P_{n-4}\, dx + \ldots$$

$$= n \cdot \frac{2}{2n+1} + 0 + 0 + \ldots \quad \text{(using orthogonality property)}$$

$$= \frac{2n}{2n+1} = R.H.S.$$

Example 19: Prove that $\int_{-1}^{1} (1-x^2) P_m'(x) P_n'(x)\, dx = 0$ if $m \neq n$ and m, n are positive integers.

Solution: Now integrating by parts, we get

$$I = \int_{-1}^{1} (1-x^2) P_m'(x) P_n'(x)\, dx = \left[(1-x^2) P_m'(x) P_n(x)\right]_{-1}^{1} - \int_{-1}^{1} P_n(x) \frac{d}{dx}\left[(1-x^2) P_m'(x)\right] dx$$

Since by Legendre's differential equation, we have

$$\frac{d}{dx}\left[(1-x^2) P'_m(x)\right] + m(m+1) P_m(x) = 0$$

Using it, we get

$$I = -\int_{-1}^{1} P_n(x)\{-m(m+1) P_m(x)\}\, dx$$

$$= m(m+1)\int_{-1}^{1} P_n(x) P_m(x)\, dx$$

$$= 0, \left[\text{since} \int_{-1}^{1} P_n(x) P_m(x)\, dx = 0 \text{ if } m \neq n\right].$$

Example 20: Prove that $\sum_{n=0}^{\infty} \frac{x^{n+1}}{n+1} P_n(x) = \frac{1}{2} \log\left(\frac{1+x}{1-x}\right)$.

Solution: By generating function relation, we know that

$$\sum_{n=0}^{\infty} z^n P_n(x) = (1 - 2xz + z^2)^{-1/2}$$

Integrating with respect to z from 0 to z, we get

$$\sum_{n=0}^{\infty} \frac{z^{n+1}}{n+1} P_n(x) = \int_0^z \frac{dz}{\sqrt{(1-2xz+z^2)}}$$

$$= \int_0^z \frac{dz}{\sqrt{(z-x)^2 + (1+x^2)}}, \text{ if } |x| < 1.$$

$$= \log \frac{(z-x) + \sqrt{z^2 - 2zx + 1}}{1-x}$$

Putting $z = x$, we obtain

$$\sum_{n=0}^{\infty} \frac{x^{n+1}}{n+1} P_n(x) = \log \frac{\sqrt{x^2 - 2x^2 + 1}}{1-x} = \log\left(\frac{\sqrt{1-x^2}}{1-x}\right) = \frac{1}{2}\log\left(\frac{1+x}{1-x}\right).$$

Example 21: If α is a root of $P_n(x)$, then $P_{n-1}(\alpha)$ and $P_{n+1}(\alpha)$ are of opposite signs.

Solution: From recurrence relation, we know that

$$(2n+1)x\,P_n(x) = (n+1)P_{n+1}(x) + n\,P_{n-1}(x)$$

Putting $x = \alpha$, we get

$$(2n+1)\alpha\,P_n(\alpha) = (n+1)P_{n+1}(\alpha) + n\,P_{n-1}(\alpha) \qquad \text{...(1)}$$

But α is a root of $P_n(x)$, so that $P_n(\alpha) = 0$. Hence (1) gives

$$(n+1)P_{n+1}(\alpha) = -\,nP_{n-1}(\alpha)$$

$$\Rightarrow \quad \frac{P_{n+1}(\alpha)}{P_{n-1}(\alpha)} = -\frac{n}{n+1},$$ a negative quantity.

Hence $P_{n+1}(\alpha)$ and $P_{n-1}(\alpha)$ are of opposite signs.

Example 22: Prove that

$$\int_{-1}^{1} x^2 P_n^2(x)\,dx = \frac{1}{8(2n-1)} + \frac{3}{4(2n+1)} + \frac{1}{8(2n+3)}$$

Solution: Form recurrence relation I, we know that

$$(2n+1)xP_n = (n+1)P_{n+1} + nP_{n-1}.$$

Squaring both sides, we obtain

$$(2n+1)^2 x^2 P_n^2 = (n+1)^2 P_{n+1}^2 + n^2 P_{n-1}^2 + 2(n+1)nP_{n+1}P_{n-1}$$

Integrating with respect to x between -1 to 1, we get

$$(2n+1)^2 \int_{-1}^{1} x^2 P_n^2\,dx = (n+1)^2 \int_{-1}^{1} P^2{}_{n+1}\,dx + n^2 \int_{-1}^{1} P^2{}_{n-1}\,dx$$
$$+ 2n(n+1)\int_{-1}^{1} P_{n-1}P_{n+1}\,dx$$
$$= (n+1)^2 \cdot \frac{2}{[2(n+1)+1]} + n^2 \cdot \frac{2}{[2(n-1)+1]} + 0$$

This gives $\displaystyle\int_{-1}^{1} x^2 P_n^2\,dx = \frac{2}{(2n+1)^2}\left[\frac{(n+1)^2}{2n+3} + \frac{n^2}{2n-1}\right]$

$$= \frac{1}{8(2n-1)} + \frac{3}{4(2n+1)} + \frac{1}{8(2n+3)},$$ resolving into partial fractions.

Example 23: Show that $\displaystyle\int_{-1}^{1} P_n(x)(1-2zx+z^2)^{-1/2}\,dx = \frac{2z^n}{2n+1}$, when $|z| < 1$ and $|x| \le 1$.

Solution: We know that, $(1-2zx+z^2)^{-1/2} = \sum_{0}^{\infty} z^n P_n(x)$ ...(1)

Multiplying both sides of (1) by $P_n(x)$, we get

$$P_n(x)(1-2zx+z^2)^{-1/2} = P_n(x)\left[P_0(x)+zP_1(x)+z^2 P_2(x)\right.$$
$$\left.+z^3 P_3(x)+\ldots+z^n P_n(x)+\ldots\right]$$

Integrating both sides w.r.t. x between limits -1 to 1, we get

$$\int_{-1}^{1} P_n(x)(1-2zx+z^2)^{-1/2} dx = \int_{-1}^{1} P_0(x)P_n(x)dx$$
$$+z\int_{-1}^{1} P_1(x)P_n(x)dx+\ldots\ldots+z^n\int_{-1}^{1}\left[P_n(x)\right]^2 dx$$
$$+z^{n+1}\int_{-1}^{1} P_{n+1}(x)P_n(x)dx+\ldots$$

$=\dfrac{2z^n}{2n+1}$, as all other integrals vanish.

Example 24: Prove that $\int_{-1}^{1}\left[P_n'(x)\right]^2 dx = n(n+1)$.

Solution: By Christoffel's expansion formula, we have

$$\int_{-1}^{1}\left[P_n'(x)\right]^2 dx$$
$$=\int_{-1}^{1}\left[(2n-1)P_{n-1}+(2n-5)P_{n-3}+(2n-9)P_{n-5}+\ldots\right]^2 dx$$
$$=\int_{-1}^{1}(2n-1)^2 P_{n-1}^2 dx+\int_{-1}^{1}(2n-5)^2 P^2{}_{n-3} dx+\int_{-1}^{1}(2n-9)^2 P_{n-5}^2 dx+\ldots$$
$$+2\int_{-1}^{1}(2n-1)(2n-5)P_{n-1}P_{n-3}dx$$
$$+2\int_{-1}^{1}+2\int_{-1}^{1}(2n-1)(2n-9)P_{n-1}P_{n-5}dx+\ldots$$
$$=(2n-1)^2\cdot\frac{2}{2(n-1)+1}+(2n-5)^2\cdot\frac{2}{2(n-3)+1}+(2n-9)^2\cdot$$
$$\frac{2}{2(n-5)+1}+\ldots\ldots+0+0+0+\ldots\ldots, \text{ by orthogonal properties}$$
$$=(2n-1)^2\cdot\frac{2}{2n-1}+(2n-5)^2\cdot\frac{2}{2n-5}+(2n-9)^2\cdot\frac{2}{2n-9}+\ldots$$
$$=2\left[(2n-1)+(2n-5)+(2n-9)+\ldots+1\right]$$

Here using the formula $l = a+(N-1)d$, we get

$$1=(2n-1)+(N-1)(-4) \Rightarrow N=\frac{n+1}{2}.$$

Therefore the number of terms in above series $=\dfrac{n+1}{2}$

Now $S_{\frac{n+1}{2}}=2.\dfrac{1}{2}\left(\dfrac{n+1}{2}\right)\left[(2n-1)+1\right]$, since $S_n=\dfrac{n}{2}(a+l)$

$$=\left(\frac{n+1}{2}\right).2n=n(n+1)$$

Hence we have $\int_{-1}^{1}\left[P_n'(x)\right]^2 dx=n(n+1)$.

Example 25: Prove that $(1-2xz+z^2)^{-1/2}$ is a solution of the equation

$$z\frac{\partial^2}{\partial z^2}(zv)+\frac{\partial}{\partial x}\left[(1-x^2)\frac{\partial v}{\partial x}\right]=0.$$

Solution: We know that $v=\left(1-2xz+z^2\right)^{-1/2}=\sum_{n=0}^{\infty} z^n P_n(x)$

$$\Rightarrow \quad zv=\sum_{n=0}^{\infty} z^{n+1} P_n(x)$$

Therefore $z\dfrac{\partial^2}{\partial z^2}(zv)=\sum_{n=0}^{\infty}(n+1)nz^n P_n(x)$ ...(1)

Also, $\dfrac{\partial v}{\partial x}=\sum_{n=0}^{\infty} z^n P_n'(x)$

Now $\dfrac{\partial}{\partial x}\left\{(1-x^2)\dfrac{\partial v}{\partial x}\right\}=\dfrac{\partial}{\partial x}\left[(1-x^2)\sum_{n=0}^{\infty} z^n P_n'(x)\right]$

$$=(1-x^2)\sum_{n=0}^{\infty} z^n P_n''(x)-2x\sum_{n=0}^{\infty} z^n P_n'(x) \quad ...(2)$$

Substituting in L.H.S. of the given equation, we have

$$\sum_{n=0}^{\infty}\left[(n+1)n\,z^n P_n(x)+(1-x^2)z^n P_n''(x)-2xz^n P_n'(x)\right]$$

$$=\sum_{n=0}^{\infty} z^n\left[(1-x^2)P_n''(x)-2x\,P_n'(x)+n(n+1)P_n(x)\right]$$

$=0$, since $P_n(x)$ is a solution of Legendre's equation.

EXERCISE 4.3

1. Prove that

(i) $x^4+x^3+x^2+x+1=\dfrac{8}{35}P_4(x)+\dfrac{2}{5}P_3(x)+\dfrac{26}{21}P_2(x)+\dfrac{3}{5}P_1(x)+\dfrac{161}{105}P_0(x)$

(ii) $x^3 = \frac{2}{5} P_3(x) + \frac{3}{5} P_1(x)$

(iii) $1 + x - x^2 = -\frac{2}{3} P_2(x) + P_1(x) + \frac{2}{3} P_0(x)$

(iv) $x^2 = \frac{1}{3} P_0(x) + \frac{2}{3} P_2(x)$

(v) $x^3 + 1 = \frac{2}{5} P_3(x) + \frac{3}{5} P_1(x) + P_0(x).$

2. Prove that $P_n'(x) - P'_{n-2}(x) = (2n-1) P_{n-1}$.

   [Hint. Replace n by $(n-1)$ is recurrence formula III]

3. Using Laplace's integral of first kind, prove that $P_2(x) = (3x^2 - 1)/2$.

4. Express x^7 as a series in Legendre's polynomials.

5. Prove that

(i) $P_1(x) = \frac{1}{\pi} \int_0^{\pi} \left[x + \sqrt{x^2 - 1} \cos \theta \right] d\theta.$

(ii) $\int_{-1}^{1} x P_n(x) . P_{n-1}(x) dx = \frac{2n}{4n^2 - 1}.$

(iii) $\int_x^1 P_n(x) dx = \frac{1}{2n+1} \left[P_{n-1}(x) - P_{n+1}(x) \right].$

(iv) $\int_{-1}^{1} P_n(x) dx = 0, (n \neq 0).$

(v) $\int_0^1 P_n(x) dx = \frac{(-1)^{(n-1)/2} (n-1)!}{2^n \{(n+1)/2\}! \{(n-1)/2\}!}$, if n is odd.

6. Use Rodrigues' formula $P_n(x) = \frac{1}{2^n n!} \cdot \frac{d^n}{dx^n} (x^2 - 1)^n$ for Legendre's polynomial $P_n(x)$, prove that

$$\int_{-1}^{1} f(x) P_n(x) dx = \frac{(-1)^n}{2^n n!} \int_{-1}^{1} f^n(x).(x^2 - 1)^n dx$$

in which $f(x)$ is a continuous function on $(-1, 1)$, and $f^n(x)$ denotes the nth derivative of $f(x)$.

Hence deduce that $\int_{-1}^{1} x^m P_n(x) dx = \begin{cases} 0, m < n \\ = \dfrac{2^{n+1} (n!)^2}{(2n+1)!}, m = n \end{cases}$

7. Prove that

(i) $\int_0^1 P_n^2(x)dx = \frac{1}{2n+1}$ (ii) $P'_n(-1) = (-1)^{n-1}\frac{n}{2}(n+1)$

(iii) $P'_n(1) = \frac{1}{2}n(n+1)$. (iv) $\int_0^1 P_n(x)dx = \frac{1}{n+1}P_{n-1}(0)$.

(v) $\int_0^1 P_{2n}(x)P_{2n+1}(x)dx = \int_0^1 P_{2n}(x)P_{2n-1}(x)dx$.

ANSWER 4.3

4. $\frac{16}{429}P_7(x) + \frac{8}{39}P_5(x) + \frac{14}{33}P_3(x) + \frac{1}{3}P_1(x)$.

4.6 BESSEL'S DIFFERENTIAL EQUATION

The differential equation $x^2\frac{d^2y}{dx^2} + x\frac{dy}{dx} + (x^2 - n^2)y = 0$ ...(1)

is called Bessel's differential equation and its particular solution is called Bessel's function of order n.

Since $x = 0$ is a regular singular point of (1), let its series solution be

$$y = x^m(a_0 + a_1x + a_2x^2 + \dots + a_rx^r + \dots) = \sum_{r=0}^{\infty} a_rx^{m+r} \quad \text{...(2)}$$

Then $\frac{dy}{dx} = \sum_{r=0}^{\infty}(m+r)a_rx^{m+r-1}$

and $\frac{d^2y}{dx^2} = \sum_{r=0}^{\infty}(m+r)(m+r-1)a_rx^{m+r-2}$.

Substituting the values of y, $\frac{dy}{dx}$ and $\frac{d^2y}{dx^2}$ in (1), we get

$$x^2\sum_{r=0}^{\infty}[(m+r)(m+r-1)]a_rm^{m+r-2} + x\sum_{r=0}^{\infty}(m+r)a_rx^{m+r-1}$$

$$+(x^2 - n^2)\sum_{r=0}^{\infty}a_rx^{m+r} = 0$$

$$\Rightarrow \sum_{r=0}^{\infty}[(m+r)(m+r-1) + (m+r) - n^2]a_rx^{m+r} + \sum_{r=0}^{\infty}a_rx^{m+r+2} = 0$$

$$\Rightarrow \sum_{r=0}^{\infty}\left[(m+r)^2 - n^2\right]a_rx^{m+r} + \sum_{r=0}^{\infty}a_rx^{m+r+2} = 0. \quad \text{...(3)}$$

Since relation (3) is an identity, the coefficients of various powers of x must be zero. Therefore, equating to zero the coefficient of lowest power of x, i.e. the coefficient of $x^m = 0$ corresponding to $r = 0$, we get $(m^2 - n^2)a_0 = 0 \Rightarrow m^2 - n^2 = 0$, which is an indicial equation.

$$\Rightarrow \quad m = \pm n$$

Again equating to zero the coefficient of x^{m+1}, we have

$$\left[(m+1)^2 - n^2\right]a_1 = 0.$$

But $(m+1)^2 - n^2 \neq 0$, as $m = \pm n$, so we must have $a_1 = 0$.

Again equating to zero the coefficient of general term, i.e. x^{m+r+2}, we get recurrence relation.

$$\left[(m+r+2)^2 - n^2\right]a_{r+2} + a_r = 0$$

$$\Rightarrow \quad a_{r+2} = -\frac{a_r}{\left[(m+r+n+2)\right]\left[m+r-n+2\right]}. \qquad \text{...(4)}$$

Putting $r = 1, 3, 5, ...$ in (4), we get $a_3 = a_5 = a_7 = ... = 0$.
Putting $r = 0, 2, 4, 6, ...$ in (4), we have

$$a_2 = -\frac{a_0}{(m+n+2)(m-n+2)}$$

$$a_4 = -\frac{a_2}{[m+n+4][m-n+4]}$$

$$= \frac{a_0}{(m+n+2)(m-n+2)(m+n+4)(m-n+4)} \text{ etc.}$$

Substituting these values in (2), we get

$$y = a_0 x^m \left[1 - \frac{x^2}{(m-n+2)(m+n+2)} + \frac{x^4}{(m-n+2)(m+n+2)(m-n+4)(m+n+4)} - ...\right] \qquad \text{...(5)}$$

Case I: When n is not integer or zero, we get two independent solutions for $m = n, -n$. For $m = n$, from (5), we get

$$y_1 = a_0 x^n \left[1 - \frac{x^2}{2(2n+2)} + \frac{x^4}{2.4(2n+2)(2n+4)} - ...\right]$$

$$= a_0 x^n \left[1 + (-1)^1 \frac{x^2}{1!.2^2(n+1)} + (-1)^2 \frac{x^4}{2!.2^4(n+1)(n+2)} + ...\right]$$

$$= a_0 x^n \sum_{r=0}^{\infty} (-1)^r \frac{x^{2r}}{r!(n+1)(n+2)...(n+r)\, 2^{2r}}$$

$$\Rightarrow \quad y_1 = a_0 x^n \sum_{r=0}^{\infty} (-1)^r \frac{\Gamma(n+1) x^{2r}}{r!.\Gamma(n+r+1)\, 2^{2r}}$$

If $a_0 = \dfrac{1}{2n\,\Gamma(n+1)}$, this solution is called $J_n(x)$, known as Bessel's function of the first kind of order n, i.e. $J_n = \sum_{r=0}^{\infty} \dfrac{(-1)^r .(x/2)^{n+2r}}{r!\Gamma(n+r+1)}$ ...(6)

Case II: For $m = -n$, the series can be obtained by replacing n by $-n$ in equation (6) as

$$J_{-n}(x) = \sum_{r=0}^{\infty} (-1)^r \frac{(x/2)^{-n+2r}}{r!.\Gamma(-n+r+1)}.$$

This is called Bessel's function of the second kind of order $-n$.

The most general solution of Bessel's equation is

$$y = AJ_n(x) + BJ_{-n}(x),$$

where A, B are arbitrary constants.

4.6.1 Integration of Bessel's Equation in Series for $n = 0$

The differential equation $x\dfrac{d^2y}{dx^2} + \dfrac{dy}{dx} + xy = 0$ ...(1)

is called Bessel's equation of order zero $(i.e., n = 0)$. Let its solution be

$$y = \sum_{r=0}^{\infty} a_r x^{m+r}. \quad ...(2)$$

Then $\dfrac{dy}{dx} = \sum_{r=0}^{\infty} (m+r) a_r x^{m+r-1}$

and $\dfrac{d^2y}{dx^2} = \sum_{r=0}^{\infty} (m+r)(m+r-1) a_r\, x^{m+r-2}$

Substituting these values of $y, \dfrac{dy}{dx}$ and $\dfrac{d^2y}{dx^2}$ in (1), we get

$$x \sum_{r=0}^{\infty} (m+r)(m+r-1) a_r\, x^{m+r-2} + \sum_{r=0}^{\infty} (m+r) a_r\, x^{m+r-1}$$

$$+ x . \sum_{r=0}^{\infty} a_r\, x^{m+r} = 0$$

$$\Rightarrow \quad \sum_{r=0}^{\infty}\left[(m+r)(m+r-1)+(m+r)\right]a_r\, x^{m+r-1}+\sum_{r=0}^{\infty} a_r\, x^{m+r+1}=0$$

$$\Rightarrow \quad \sum_{r=0}^{\infty}(m+r)^2\, a_r\, x^{m+r-1}+\sum_{r=0}^{\infty} a_r\, x^{m+r+1}=0. \qquad ...(3)$$

Equating to zero the coefficient of lowest power of x, i.e. x^{m-1} corresponding to $r=0$, we get

$$m^2\, a_0=0 \Rightarrow m=0,0\,;\text{ since } a_0\neq 0$$

Equating to zero the coefficient of x^m corresponding to $r=1$, we get $(m+1)^2\, a_1=0.$

But $(m+1)^2\neq 0$, as $m=0$, so we must have $a_1=0$.

Again equating to zero the coefficient of the general term, i.e. x^{m+r+1} in (3), we have

$$(m+r+2)^2\, a_{r+2}+a_r=0 \Rightarrow a_{r+2}=-\frac{a_r}{(m+r+2)^2}. \qquad ...(4)$$

Putting $r=1, 3, 5, ...,$ in (4), we get $a_3=a_5=a_7=...=0.$

Putting $r=0, 2, 4, \ldots$ in (4), we get $a_2=-\dfrac{a_0}{(m+2)^2}$

$$a_4=-\frac{a_2}{(m+4)^2}=\frac{a_0}{(m+2)^2(m+4)^2} \text{ and so on.}$$

Putting these values in (2), we get

$$y=a_0x^m\left[1-\frac{x^2}{(m+2)^2}+\frac{x^4}{(m+2)^2(m+4)^2}-\ldots\right] \qquad ...(5)$$

When $m=0$, we get

$$y_1=a_0\left[1-\frac{x^2}{2^2}+\frac{x^4}{2^2.4^2}-\frac{x^6}{2^2.4^2.6^2}+\ldots\right]$$

This is called Bessel's function of zeroth order and denoted by $J_0(x)$ for $a_0=1$.

$$\text{Thus} \quad J_0(x)=1-\frac{x^2}{2^2}+\frac{x^4}{2^2.4^2}-\frac{x^6}{2^2.4^2.6^2}+\ldots, \quad (\text{if } a_0=1)$$

Differentiating equation (5) w. r. t. 'm', (partially) and then taking limit $m\to 0$, we get

$$y=\left(\frac{\partial y}{\partial m}\right)_{m=0}=J_0(x)\log x+\left[\frac{x^2}{2^2}-\frac{1}{2^2.4^2}\left(1+\frac{1}{2}\right)x^4+\ldots\right]$$

$$=J_0(x)\log x+\left[\left(\frac{x}{2}\right)^2+\frac{(-1)}{(2!)^2}\left(1+\frac{1}{2}\right)\left(\frac{x}{2}\right)^4+\ldots\right]$$

$$=J_0(x)\log x+\sum_{r=1}^{\infty}\frac{(-1)^{r+1}}{(r!)^2}\left(1+\frac{1}{2}+\frac{1}{3}+\ldots+\frac{1}{r}\right)\left(\frac{x}{2}\right)^{2r}$$

This is called Bessel function of second kind of zeroth order and denoted by $Y_0(x)$, then

$$Y_0(x)=J_0(x)\log x+\sum_{r-1}^{\infty}\frac{(-1)^{r+1}}{(r!)^2}\left(1+\frac{1}{2}+\ldots+\frac{1}{r}\right)\left(\frac{x}{2}\right)^{2r}.$$

The complete solution of Bessel's differential equation of order zero is $y=AJ_0(x)+BY_0(x)$, where A, B are arbitrary constants.

4.6.2 Bessel Functions of the Second Kind of Order *n* (for *n* is an integer)

The Bessel's differential equation

$$x^2\frac{d^2y}{dx^2}+\frac{dy}{dx}+\left(x^2-n^2\right)y=0 \quad \ldots(1)$$

whose one solution is $J_n(x)$. Let $y=u(x)J_n(x)$ be a solution of (1), where n is integer.

Then $\frac{dy}{dx}=u'J_n+uJ_n'$

and $\frac{d^2y}{dx^2}=u''J_n+u'J_n'+u'J_n'+uJ_n''=u''J_n+2u'J_n'+uJ_n''.$

Substituting these values of y, $\frac{dy}{dx}$ *and* $\frac{d^2y}{dx^2}$ in (1), we have

$$x^2\left[u''J_n+2u'J_n'+uJ_n''\right]+x\left[u'J_n+uJ_n'\right]+\left(x^2-n^2\right)uJ_n=0$$

$$\Rightarrow \quad u\left[x^2J_n''+xJ_n'+\left(x^2-n^2\right)J_n\right]+x^2u''J_n+2x^2u'J_n'+xu'J_n=0$$

$$\Rightarrow \quad x^2u''J_n+2x^2u'J_n'+xu'J_n=0. \quad \ldots(2)$$

Since J_n is a solution of Bessel differential equation. Dividing both sides of (2) by $x^2u'J_n$, we get

$$\frac{u''}{u'}+2\frac{J_n'}{J_n}+\frac{1}{x}=0 \Rightarrow \frac{d}{dx}[\log u']+2\frac{d}{dx}[\log J_n]+\frac{d}{dx}[\log x]=0$$

$$\Rightarrow \quad \frac{d}{dx}[\log u' + 2\log J_n + \log x] = 0 \Rightarrow \frac{d}{dx}\left[\log\left(u' J_n^2 x\right)\right] = 0.$$

Integrating w.r.t. 'x', we get $\log\left(u' J_n^2 x\right) = \log B$

$$\Rightarrow \quad u' = \frac{B}{xJ_n^2} \Rightarrow \frac{du}{dx} = \frac{B}{xJ_n^2} \Rightarrow du = \frac{Bdx}{xJ_n^2}$$

Integrating again, we get $u = \int B\frac{dx}{xJ_n^2} + A$ (constant).

Putting the value of u in the assumed solution $y = u(x)J_n(x)$, we have

$$y = \left[B\int \frac{dx}{xJ_n^2} + A\right]J_n(x)$$

$$\Rightarrow \quad y = AJ_n(x) + BY_n(x), \text{ where } Y_n(x) = J_n(x)\int \frac{dx}{xJ_n^2}.$$

The function $Y_n(x)$ is known as Bessel's function of the second kind of order n.

4.6.3 Recurrence Formula for $J_n(x)$

I. $xJ_n'(x) = nJ_n(x) - x\,J_{n+1}(x)$

Proof: We know that

$$J_n(x) = \sum_{r=0}^{\infty} \frac{(-1)^r\left(\frac{x}{2}\right)^{n+2r}}{r!\,\Gamma(n+r+1)}$$

Differentiating w.r.t. 'x', we get

$$J_n'(x) = \sum_{r=0}^{\infty} \frac{(-1)^r(n+2r)\left(\frac{x}{2}\right)^{n+2r-1}\frac{1}{2}}{r!\,\Gamma(n+r+1)}$$

Multiplying both sides by x, we get

$$xJ_n'(x) = \sum_{r=0}^{\infty} \frac{(-1)^r(n+2r)\left(\frac{x}{2}\right)^{n+2r-1}\left(\frac{x}{2}\right)}{r!\,\Gamma(n+r+1)}$$

$$= n\sum_{r=0}^{\infty} \frac{(-1)^r\left(\frac{x}{2}\right)^{n+2r}}{r!\,\Gamma(n+r+1)} + \sum_{r=0}^{\infty} \frac{(-1)^r\,2r\left(\frac{x}{2}\right)^{n+2r-1}\left(\frac{x}{2}\right)}{r!\,\Gamma(n+r+1)}$$

$$= nJ_n(x) + x\sum_{r=1}^{\infty} \frac{(-1)^r \left(\frac{x}{2}\right)^{n+2r-1}}{(r-1)!\Gamma(n+r+1)}$$

$$= nJ_n(x) + x\sum_{s=0}^{\infty} \frac{(-1)^{s+1} \left(\frac{x}{2}\right)^{n+1+2s}}{s!\Gamma(n+1+s+1)}, \text{ (replacing } r-1 \text{ by } s)$$

$$= nJ_n(x) - x\sum_{s=0}^{\infty} \frac{(-1)^{s} \left(\frac{x}{2}\right)^{\overline{n+1}+2s}}{s!\Gamma(\overline{n+1}+s+1)} = nJ_n(x) - xJ_{n+1}(x)$$

Hence, we get $xJ_n'(x) = nJ_n(x) - xJ_{n+1}(x)$

II. $xJ_n'(x) = -nJ_n(x) + xJ_{n-1}(x)$.

Proof: We know that

$$J_n(x) = \sum_{r=0}^{\infty} \frac{(-1)^r \left(\frac{x}{2}\right)^{n+2r}}{r!\Gamma(n+r+1)}.$$

Differentiating w.r.t. 'x', we know

$$J_n'(x) = \sum_{r=0}^{\infty} \frac{(-1)^r (n+2r)\left(\frac{x}{2}\right)^{n+2r-1}\left(\frac{1}{2}\right)}{r!\Gamma(n+r+1)}$$

$$\Rightarrow \quad xJ_n'(x) = \sum_{r=0}^{\infty} \frac{(-1)^r (n+2r)\left(\frac{x}{2}\right)^{n+2r-1}\left(\frac{x}{2}\right)}{r!\Gamma(n+r+1)}$$

$$= \sum_{r=0}^{\infty} \frac{(-1)^r [-n+2n+2r]\left(\frac{x}{2}\right)^{n+2r-1}\left(\frac{x}{2}\right)}{r!\Gamma(n+r+1)}$$

$$= -n\sum_{r=0}^{\infty} \frac{(-1)^r \left(\frac{x}{2}\right)^{n+2r}}{r!\Gamma(n+r+1)} + x\sum_{r=0}^{\infty} \frac{(-1)^r (n+r)\left(\frac{x}{2}\right)^{n+2r-1}}{r!\Gamma(n+r+1)}$$

$$= -nJ_n(x) + x\sum_{r=0}^{\infty} \frac{(-1)^r \left(\frac{x}{2}\right)^{n+2r-1}}{r!\Gamma(n+r)}$$

$$= -nJ_n(x) + x\sum_{r=0}^{\infty} \frac{(-1)^r \left(\frac{x}{2}\right)^{\overline{n-1}+2r}}{r!\,\Gamma\left(\overline{n-1}+r+1\right)}$$

$$= -nJ_n(x) + xJ_{n-1}(x)$$

Hence, we obtain $xJ_n'(x) = -nJ_n(x) + xJ_{n-1}(x)$

III. $2\,J_n'(x) = J_{n-1}(x) - J_{n+1}(x)$.

Proof: We know that

$$J_n = \sum_{r=0}^{\infty} \frac{(-1)^2 \left(\frac{x}{2}\right)^{n+2r}}{r!\,\Gamma(n+r+1)}.$$

Differentiating w.r.t. 'x', we have

$$J_n'(x) = \sum_{r=0}^{\infty} \frac{(-1)^r (n+2r)\left(\frac{x}{2}\right)^{n+2r-1}\left(\frac{1}{2}\right)}{r!\,\Gamma(n+r+1)}$$

$$\Rightarrow \quad 2\,J_n'(x) = \sum_{r=0}^{\infty} \frac{(-1)^r [n+r+r]\left(\frac{x}{2}\right)^{n+2r-1}}{r!\,\Gamma(n+r+1)}$$

$$\Rightarrow \quad 2J_n'(x) = \sum_{r=0}^{\infty} \frac{(-1)^r (n+r)\left(\frac{x}{2}\right)^{n+2r-1}}{r!\,\Gamma(n+r+1)} + \sum_{r=0}^{\infty} \frac{(-1)^r r\left(\frac{x}{2}\right)^{n+2r-1}}{r!\,\Gamma(n+r+1)}$$

$$= \sum_{r=0}^{\infty} \frac{(-1)^r \left(\frac{x}{2}\right)^{\overline{n-1}+2r}}{r!\,\Gamma\left(\overline{n-1}+r+1\right)} + \sum_{r=1}^{\infty} \frac{(-1)^r \left(\frac{x}{2}\right)^{n+2r-1}}{(r-1)!\,\Gamma(n+r+1)}$$

$$= J_{n-1}(x) + \sum_{r=1}^{\infty} \frac{(-1)^r \left(\frac{x}{2}\right)^{n+2r-1}}{(r-1)!\,\Gamma(n+r+1)}$$

$$= J_{n-1}(x) + \sum_{r=0}^{\infty} \frac{(-1)^{n+1}\left(\frac{x}{2}\right)^{\overline{n+1}+2r}}{s!\,\Gamma\left(\overline{n+1}+s+1\right)} \quad \text{(on replacing } r-1 \text{ by } s\text{)}$$

$$= J_{n-1}(x) - J_{n+1}(x).$$

Hence, we have $2J_n'(x)=J_{n-1}(x)-J_{n+1}(x)$

Aliter: By recurrence formulae I and II , we have

$$xJ_n'(x)=nJ_n(x)-xJ_{n+1}(x)$$

and $xJ_n'(x)=-n\,J_n(x)+xJ_{n-1}(x)$.

Adding these, we get $2xJ_n'(x)=xJ_{n-1}(x)-xJ_{n+1}(x)$.

Hence, we have $2J_n'(x)=J_{n-1}(x)-J_{n+1}(x)$

IV. $2nJ_n(x)=x\left[J_{n-1}(x)+J_{n+1}(x)\right]$.

Proof: We know that $J_n(x)=\sum\limits_{r=0}^{\infty}\dfrac{(-1)^r\left(\dfrac{x}{2}\right)^{n+2r}}{r!\,\Gamma(n+r+1)}$

$$\Rightarrow \quad J_n(x)=\sum_{r=0}^{\infty}\frac{(-1)^r[2n+2r-2r]\left(\dfrac{x}{2}\right)^{n+2r}\left(\dfrac{1}{2n}\right)}{r!\,\Gamma(n+r+1)}$$

$$\Rightarrow \quad 2nJ_n(x)=\sum_{r=0}^{\infty}\frac{(-1)^r(2n+2r)\left(\dfrac{x}{2}\right)^{n+2r}}{r!\,\Gamma(n+r+1)}-\sum_{r=0}^{\infty}\frac{(-1)^r\,2r\left(\dfrac{x}{2}\right)^{n+2r}}{r!\,\Gamma(n+r+1)}$$

$$\Rightarrow \quad 2nJ_n(x)=x\sum_{r=0}^{\infty}\frac{(-1)^r\left(\dfrac{x}{2}\right)^{\overline{n-1}+2r}}{r!\,\Gamma(\overline{n-1}+r+1)}-x\sum_{r=0}^{\infty}\frac{(-1)^r\left(\dfrac{x}{2}\right)^{n+2r}}{(r-1)!\,\Gamma(n+r+1)}$$

$$\Rightarrow \quad 2nJ_n(x)=xJ_{n-1}\,x-x\sum_{r=0}^{\infty}\frac{(-1)^r\left(\dfrac{x}{2}\right)^{n+2r-1}}{(r-1)!\,\Gamma(n+r+1)}$$

$$\Rightarrow \quad 2nJ_n(x)=xJ_{n-1}(x)-x\sum_{r=0}^{\infty}\frac{(-1)^{s+1}\left(\dfrac{x}{2}\right)^{\overline{n+1}+2s}}{s!\,\Gamma(\overline{n+1}+s+1)},\ \text{using } r=s+1$$

$$\Rightarrow \quad 2nJ_n(x)=xJ_{n-1}(x)+xJ_{n-1}(x).$$

Hence, we have $2nJ_n(x)=x\left[J_{n-1}(x)+J_{n+1}(x)\right]$.

V. $\dfrac{d}{dx}\left[x^n J_n(x)\right]=x^n J_{n-1}(x)$.

Proof: Writing recurrence relation II, we have $xJ_n' = -nJ_n + xJ_{n-1}$.

Multiplying equation (1) by x^{n-1}, we get

$$x^n J_n' = -nx^{n-1} J_n + x^n J_{n-1}$$

$$\Rightarrow \quad x^n J_n' + nx^{n-1} J_n = x^n J_{n-1} \Rightarrow \frac{d}{dx}\left[x^n J_n(x)\right] = x^n J_{n-1}$$

Hence we have π.

Aliter: We have $\frac{d}{dx}\left[x^n J_n(x)\right] = nx^{n-1} J_n(x) + x^n J_n'(x)$

$$= x^{n-1}[nJ_n(x) + x J_n'(x)]$$

$$= x^{n-1}\left[nJ_n(x) - nJ_n(x) + xJ_{n-1}(x)\right]$$

(from recurrence formula II)

$$= x^{n-1}\left[xJ_{n-1}(x)\right] = x^n J_{n-1}(x)$$

Hence, we have $\frac{d}{dx}\left[x^n J_n(x)\right] = x^n J_{n-1}(x)$

VI. $\frac{d}{dx}\left[x^{-n} J_n(x)\right] = -x^{-n} J_{n+1}(x)$.

Proof: We have $\frac{d}{dx}\left[x^{-n} J_n(x)\right] = -nx^{-n-1} J_n(x) + x^{-n} J_n'(x)$

$$= x^{-n-1}\left[-nJ_n(x) + xJ_n'(x)\right]$$

$$= x^{-n-1}\left[-nJ_n(x) + nJ_n(x) - xJ_{n+1}(x)\right],$$

using recurrence formula I

$$= -x^{-n} J_{n+1}(x)$$

Hence, we have $\frac{d}{dx}\left[x^{-n} J_n(x)\right] = -x^{-n} J_{n+1}(x)$

4.6.4 Generating Function for $J_n(x)$

Prove that when n is a positive integer $J_n(x)$ is the coefficient of z^n in the expansion of $e^{\frac{x}{2}\left(z-\frac{1}{z}\right)}$ in ascending and descending powers of z. Also prove that $J_n(x)$ is the coefficient of z^{-n} multiplied by $(-1)^n$ in the expansion of above expression.

or Prove that $J_n(x)$ is the coefficient of z^n in the expansion of $e^{\frac{x}{2}\left(z-\frac{1}{z}\right)}$. Also prove that

(i) $e^{\frac{x}{2}\left(z-\frac{1}{z}\right)} = \sum_{n=0}^{\infty} z^n J_n(x)$

(ii) $e^{\frac{x}{2}\left(z-\frac{1}{z}\right)} = \sum_{n=\infty}^{\infty} (-1)^n z^{-n} J_n(x)$

That is, $(-1)^n J_n$ is the coefficient of z^{-n} in the above expansion.

Proof: We have $e^{\frac{x}{2}\left(z-\frac{1}{z}\right)} = e^{\frac{xz}{2}} . e^{-\frac{x}{2z}}$

$$= \left[1 + \left(\frac{xz}{2}\right) + \frac{1}{2!}\left(\frac{xz}{2}\right)^2 + \ldots + \frac{1}{n!}\left(\frac{xz}{2}\right)^n + \frac{1}{(n+1)!}\left(\frac{xz}{2}\right)^{n+1} + \frac{1}{(n+2)!}\left(\frac{xz}{2}\right)^{n+2} + \ldots\right] \times \left[1 - \left(\frac{x}{2z}\right) + \frac{1}{2}\left(\frac{x}{2z}\right)^2 - \ldots + \frac{(-1)^n}{n!}\left(\frac{x}{2z}\right)^n + \frac{(-1)^{n+1}}{(n+1)!}\left(\frac{x}{2z}\right)^2 \ldots\right] \quad \ldots(1)$$

Coefficient of z^n in this product

$$= \frac{1}{n!}\left(\frac{x}{2}\right)^n - \frac{1}{(n+1)!}\left(\frac{x}{2}\right)\left(\frac{x}{2}\right)^{n+1} + \frac{1}{(n+2)!} \cdot \frac{1}{2!}\left(\frac{x}{2}\right)^2\left(\frac{x}{2}\right)^{n+2} + \ldots$$

$$= \frac{(-1)^0}{n!}\left(\frac{x}{2}\right)^n + (-1)^1 \frac{1}{1!(n+1)!}\left(\frac{x}{2}\right)^{n+2} + \frac{(-1)^2}{2!(n+2)!}\left(\frac{x}{2}\right)^{n+4} + \ldots$$

$$= \sum_{r=0}^{\infty} (-1)^r \frac{1}{r!\,\Gamma(n+r+1)}\left(\frac{x}{2}\right)^{n+2r} = J_n(x).$$

Similarly, the coefficient of z^{-n} in the product (1)

$$= \frac{(-1)^n}{n!}\left(\frac{x}{2}\right)^n + \frac{(-1)^{n+1}}{(n+1)!}\left(\frac{x}{2}\right)\left(\frac{x}{2}\right)^{n+1} + \frac{(-1)^{n+2}}{n!.(n+2)!}\left(\frac{x}{2}\right)^2\left(\frac{x}{2}\right)^{n+2} + \ldots$$

$$= \left[\frac{(-1)^n}{0!\,\Gamma(n+1)}\left(\frac{x}{2}\right)^n + \frac{(-1)^{n+1}}{1!\,\Gamma(n+2)}\left(\frac{x}{2}\right)^{n+2} + \frac{(-1)^{n+2}}{2!.\,\Gamma(n+3)}\left(\frac{x}{3}\right)^{n+4} + \ldots\right]$$

$$= (-1)^n\left[\frac{(-1)^0}{0\,\Gamma(n+1)}\left(\frac{x}{2}\right)^n + \frac{(-1)^1}{1!.\,\Gamma(n+2)}\left(\frac{x}{2}\right)^{n+2} + \frac{(-1)^2}{2!\,\Gamma(n+3)}\left(\frac{x}{2}\right)^{n+4} + \ldots\right]$$

$$=(-1)^n \sum_{r=0}^{\infty} \cdot \frac{(-1)^r \left(\frac{x}{2}\right)^{n+2r}}{r!.\Gamma(n+r+1)} = (-1)^n J_n(x).$$

4.6.5 Orthogonality of Bessel's Functions

(i) $\int_0^p xJ_n(\alpha x)J_n(\beta x)dx = 0$, if $\alpha \neq \beta$

(ii) $\int_0^p xJ_n^2(\alpha x)dx = \frac{p^2}{2} J_{n+1}^2(\alpha p)$ *i.e.*, $\int_0^p xJ_n(\alpha x)J_n(\beta x)$

$$dx = \frac{p^2}{2} J_{n+1}^2(\alpha p) \text{ if } \alpha + \beta$$

where α, β are the roots of $J_n(xp) = 0$. This gives us the orthogonality of Bessel's functions over the interval (0, p).

Proof:

(i) We know that the solutions of the differential equation

$$x^2u'' + xu' + (\alpha^2 x^2 - n^2)u = 0 \quad \text{...(1)}$$

and $x^2v'' + xv' + (\beta^2 x^2 - n^2)v = 0$ ...(2)

are $u = J_n(\alpha x)$ and $v = J_n(\beta x)$, respectively.

Multiplying equation (1) by $\frac{v}{x}$ and (2) by $\frac{u}{x}$ and subtracting, we have

$$x(u''v - uv'') + (u'v - uv') + (\alpha^2 - \beta^2)xuv = 0$$

$$\Rightarrow \quad \frac{d}{dx}\left[x(u'v - uv')\right] + (\alpha^2 - \beta^2)xuv = 0$$

$$\Rightarrow \quad \frac{d}{dx}\left[x(u'v - uv')\right] = (\beta^2 - \alpha^2)x\,uv.$$

Integrating both sides w.r.t. 'x' over the limits 0 to p, we have

$$\left[x(u'v - uv')\right]_0^p = (\beta^2 - \alpha^2)\int_0^p x\,uv\,dx$$

$$\Rightarrow \quad (\beta^2 - \alpha^2)\int_0^p x\,uv\,dx = \left[x(u'v - uv')\right]_0^p = p\left[u'v - uv'\right]_0^p. \quad \text{...(3)}$$

Now $u = J_n(\alpha x)$, then $u' = \alpha J_n'(\alpha x)$.

Similarly, if $v = J_n(\beta x)$, then $v' = \beta J_n'(\beta x)$.

Putting the values of u, v, u' and v' in equation (3), we get

$$(\beta^2 - \alpha^2)\int_0^p xJ_n(\alpha x)J_n(\beta x)dx = p\left[\alpha J_n'(\alpha p)J_n(\beta p) - \beta J_n(\alpha p)J_n'(\beta p)\right] \text{...(4)}$$

Since α, β are the distinct roots of $J_n(xp)=0$, then $J_n(\alpha p)=0$ and $J_n(\beta p)=0$. Putting these values in equation (4), we get

$$(\beta^2-\alpha^2)\int_0^p xJ_n(\alpha x)J_n(\beta x)dx=0$$

$$\Rightarrow \quad \int_0^p x\,J_n(\alpha x)J_n(\beta x)dx=0, \text{ since } \alpha\neq\beta$$

(ii) From equation (4), we have

$$\int_0^p xJ_n(\alpha x)J_n(\beta x)dx=\frac{p\left[\alpha J_n'(\alpha p)J_n(\beta p)-\beta J_n(\alpha p)J_n'(\beta p)\right]}{\beta^2-\alpha^2} \quad ...(5)$$

Now if α be root of $J_n(\alpha x)$, then $J_n(\alpha p)=0$. Let β tends to α, then

$$\lim_{\beta\to\alpha}\int_0^p xJ_n(\alpha x)J_n(\beta x)dx=\lim_{\beta\to\alpha}\frac{p\alpha J_n'(\alpha p)J_n(\beta p)-0}{\beta^2-\alpha^2}, \left[\frac{0}{0}\, form\right]$$

Applying, L' Hospital's rule, we get

$$\int_0^p xJ_n^2(\alpha x)dx=\lim_{\beta\to\alpha}\frac{p^2\,\alpha J_n'(\alpha p)J_n'(\beta p)}{2\,\beta}=\frac{p^2}{2}\left[J_n'(\alpha p)\right]^2. \quad ...(6)$$

But we know that $xJ_n'(x)=xJ_n(x)-xJ_{n+1}(x)$

Thus $\alpha p\, J_n'(\alpha p)=0-\alpha pJ_{n+1}(\alpha p)$, since $J_n(\alpha p)=0$

$$\Rightarrow \quad J_n'(\alpha p)=J_{n+1}(\alpha p) \Rightarrow \left[J_n'(\alpha p)\right]^2=J_{n+1}^2(\alpha p).$$

Substituting this value into (6), we get

$$\int_0^p xJ_n^2(\alpha x)dx=\frac{p^2}{2}J_{n+1}^2(\alpha p).$$

Remark: For $p=1$, the orthogonality condition of Bessel's functions becomes

$$\int_0^1 xJ_n(\alpha x)J_n(\beta x)dx=0, \alpha\neq\beta$$

and $\int_0^1 x\,J_n^2(\alpha x)dx=\frac{1}{2}J_{n+1}^2(\alpha).$

4.6.6 Bessel's Integral

(i) $J_n(x)=\frac{1}{\pi}\int_0^\pi \cos(n\theta-x\sin\theta)d\theta$, where n is a positive integer.

(ii) $J_n(x)=\frac{1}{\pi}\int_0^\pi \cos(n\theta-x\sin\theta)d\theta$, where n is a any integer.

(iii) $J_0(x) = \frac{1}{\pi}\int_0^{\pi} \cos(x\sin\theta)d\theta = \frac{1}{\pi}\int_0^{\pi} \cos(x\cos\theta)d\theta$.

Proof:

(i) We shall use the following results

$$\int_0^{\pi} \cos m\theta \cos n\theta \, d\theta = \int_0^{\pi} \sin m\theta \sin n\theta \, d\theta = \begin{cases} \frac{\pi}{2}, \text{if } m = n \\ 0, \text{if } m \neq n \end{cases} \quad ...(1)$$

$$\cos(x\sin\theta) = J_0 + 2\cos 2\theta \, . \, J_2 + 2\cos 4\theta J_4 + ... \quad ...(2)$$

$$\sin(x\sin\theta) = 2\sin\theta J_1 + 2\sin 3\theta J_3 + ... \quad ...(3)$$

Multiplying (2) by $\cos n\theta$ and integrating w.r.t. 'θ', from 0 to π and using (1), we have

$$\int_0^{\pi} \cos(x\sin\theta)\cos n\theta \, . \, d\theta = \begin{cases} 0, \text{if } n \text{ is odd} & ...(4) \\ \pi J_n, \text{if } n \text{ is even} & ...(5) \end{cases}$$

Next, multiplying both sides of (3) by $\sin n\theta$ and then integrating w.r.t. 'θ', from 0 to π and using (1), we get

$$\int_0^{\pi} \sin(x\sin\theta)\sin n\theta \, d\theta = \begin{cases} \pi J_n, \text{if } n \text{ is odd} & (6) \\ 0, \text{if } n \text{ is even} & (7) \end{cases}$$

Let n be odd. Adding (4) and (6), we have

$$\int_0^{\pi} \left[\cos(x\sin\theta)\cos n\theta + \sin(x\sin\theta)\sin n\theta\right] d\theta = nJ_n$$

$$\Rightarrow \quad \int_0^{\pi} \cos(n\theta - x\sin\theta)d\theta = \pi J_n$$

Therefore $J_n(x) = \frac{1}{\pi}\int_0^{\pi} \cos(n\theta - x\sin\theta)d\theta$. ...(8)

Let n be even then adding (5) and (7), we get same result. Thus, the result holds for each positive integer.

(ii) Let n be a positive integer then from (8), we get

$$J_n(x) = \frac{1}{\pi}\int_0^{\pi} \cos(n\theta - x\sin\theta)d\theta \quad ...(9)$$

Let n be a negative integer so that $n = -m$, where m is a positive integer. We prove that

$$J_{-m}(x) = \frac{1}{\pi}\int_0^{\pi} \cos[-m\theta - x\sin\theta]d\theta. \quad ...(10)$$

Let $\theta = \pi - \phi$ and $d\theta = -d\phi$, we have

R.H.S. of (10) $= \frac{1}{\pi}\int_0^\pi \cos\left[-m(\pi-\phi) - x\sin(\pi-\phi)\right](-d\phi)$

$$= \frac{1}{\pi}\int_0^\pi \cos\left[(m\phi - x\sin\phi) - m\pi\right]d\phi$$

$$= \frac{1}{\pi}\int_0^\pi \left[\cos(m\phi - x\sin\theta)\cos m\pi + \sin(m\phi - x\sin\phi)\sin m\pi\right]d\phi$$

$$= \frac{1}{\pi}\int_0^\pi (-1)^m \cos(m\phi - x\sin\phi)d\phi \frac{1}{\pi}(-1)^m \int_0^\pi \cos(m\phi - x\sin\phi)d\phi$$

$$= (-1)^m J_m(x) = J_{-m}(x), \text{ since } J_{-m}(x) = (-1)^m J_m(x).$$

(iii) We know that $\int_0^\pi \cos p\theta \, d\theta = 0$, if p is an even integer. ...(11)

On integrating both sides of (2) w.r.t. 'θ', from 0 to π and using (11) we have

$$\int_0^\pi \cos(x\sin\theta)d\theta = J_0(x)\int_0^\pi d\theta + 0 + 0 \ldots = \pi J_0(r).$$

Therefore $J_0(x) = \frac{1}{\pi}\int_0^\pi \cos(x\sin\theta)d\theta.$ ...(12)

Replacing θ by $\frac{\pi}{2} - \theta$ in (2), we get

$$\cos(x\cos\theta) = J_0 - 2\cos 2\theta\, J_2 + 2\cos 4\theta\, J_4 - \ldots.$$

Integrating both sides w.r.t. 'θ' between the limits 0 to π and using (11), we get

$$\int_0^\pi \cos(x\cos\theta)d\theta = J_0(x).\pi - 0 + 0 \ldots = \pi J_0(x).$$

Therefore $\quad J_0(x) = \frac{1}{\pi}\int_0^\pi \cos(x\cos\theta)d\theta.$

4.6.7 Equations Reducible to Bessel's Equation

In many problems, the second order differential equations with variable coefficients can easily be reduced to Bessel's equation and therefore, can be solved by means of Bessel's function.

(i) Consider the differential equation

$$x^2\frac{d^2y}{dx^2} + x\frac{dy}{dx} + (k^2x^2 - n^2)y = 0. \quad \ldots(1)$$

Let $t = kx$, $\frac{dt}{dx} = k$, $\frac{dy}{dx} = \frac{dy}{dt}\frac{dt}{dx} = k\frac{dy}{dt}$ *and* $\frac{d^2y}{dx^2} = k^2\frac{d^2y}{dt^2}$

Putting these values in (1), we get

$$t^2\frac{d^2y}{dt^2} + t\frac{dy}{dt} + \left(t^2 - n^2\right)y = 0. \qquad ...(2)$$

The solution of (2) is $y = AJ_n(t) + BJ_{-n}(t)$.

Hence the solution of equation (1) is given by

$$y = AJ_n(kx) + BJ_{-n}(kx).$$

(ii) To reduce the differential equation

$$x\frac{d^2y}{dx^2} + a\frac{dy}{dx} + k^2 xy = 0. \qquad ...(3)$$

In Bessel's equation, we put $y = x^n z$, then

$$\frac{dy}{dx} = x^n\frac{dz}{dx} + nx^{n-1}z$$

and
$$\frac{d^2y}{dx^2} = x^n\frac{d^2z}{dx^2} + 2nx^{n-1}\frac{dz}{dx} + n(n-1)x^{n-2}z.$$

Substituting these values in (3), we get

$$x^{n+1}\frac{d^2z}{dx^2} + (2n+a)x^n\frac{dz}{dx} + \left[k^2x^2 + n^2 + (a-1)n\right]x^{n-1}z = 0.$$

Dividing both sides by x^{n-1}, we get

$$x^2\frac{d^2z}{dx^2} + (2n+a)x\frac{dz}{dx} + \left[k^2x^2 + n^2 + (a-1)n\right]z = 0.$$

Putting $2n + a = 1$, we get $x^2\frac{d^2z}{dx^2} + x\frac{dz}{dx} + \left(k^2x^2 - n^2\right)z = 0.$

Its solution be $z = AJ_n(kx) + BJ_{-n}(kx)$.

Hence the complete solution of (3) is

$$y = x^n\left[AJ_n(kx) + BJ_{-n}(kx)\right], \text{ where } n = (1-a)/2.$$

(iii) To reduce the differential equation

$$x\frac{d^2y}{dx^2} + c\frac{dy}{dx} + k^2x^r y = 0 \qquad ...(4)$$

in Bessel's equation, we put $x = t^m$, $t = x^{1/m}$ so that

$$\frac{dy}{dx} = \frac{dy}{dt}\cdot\frac{dt}{dx} = \frac{1}{m}t^{1-m}\frac{dy}{dt} \quad \text{and}$$

$$\frac{d^2y}{dx^2} = \frac{1}{m^2} t^{2-2m} \frac{d^2y}{dt^2} + \frac{1-m}{m^2} t^{1-2m} \frac{dy}{dt}.$$

Substituting the values of $x, \frac{dy}{dx}$ and $\frac{d^2y}{dx^2}$ in (4), we get

$$t^m \left[\frac{1}{m^2} t^{(2-2m)} \frac{d^2y}{dt^2} + \frac{1-m}{m^2} + t^{1-2m} \frac{dy}{dt} \right] + \frac{c}{m} t^{(1-m)} \frac{dy}{dt} + k^2 t^{mr} y = 0$$

$$\Rightarrow \quad \frac{1}{m^2} t^{2-m} \frac{d^2y}{dt^2} + \left(\frac{1-m+cm}{m^2} \right) t^{1-m} \frac{dy}{dt} + k^2 t^{mr} \, y = 0.$$

Multiplying both sides by $\left(\frac{m^2}{t^{1-m}} \right)$, we get

$$t \frac{d^2y}{dt^2} + (1-m+cm) \frac{dy}{dt} + (km)^2 \, t^{mr+m-1} \, y = 0. \qquad ...(5)$$

Putting $m = \frac{2}{r+1}$, we obtain $a = 1 - m + cm = \frac{r+2c-1}{r+1}$.

Equation (5) reduces to $t \frac{d^2y}{dt^2} + a \frac{dy}{dt} + (km)^2 \, ty = 0$.

Its solution is $y = t^n \left[C_1 J_n(kmt) + C_2 J_{-n}(kmt) \right]$.

The complete solution is

$$y = x^{n/m} \left[C_1 J_n\left(kmx^{1/m}\right) + C_2 J_{-n}\left(kmx^{1/m}\right) \right],$$

where $n = \frac{1-a}{2} = \frac{m-cm}{2} = \frac{m(1-c)}{2}$.

(iv) Consider the differential equation of the form

$$x^2 \frac{d^2y}{dx^2} + (1-2\alpha) x \frac{dy}{dx} + \left[\beta^2 \gamma^2 x^{2\gamma} + \left(\alpha^2 - n^2\gamma^2\right) \right] y = 0 \qquad ...(6)$$

where α, β, γ and n are constant.

Let $X = \beta x^{\gamma}$ and $Y = x^{-\alpha} y$, then equation (6) becomes

$$X^2 \frac{d^2Y}{dX^2} + X \frac{dY}{dX} + \left(X^2 - n^2\right) Y = 0 \qquad ...(7)$$

This is a Bessel's equation. Thus solution of (2) is given by

$$Y = \begin{cases} c_1 J_n(X) + c_2 J_{-n}(X) & \text{when } n \text{ is not an integer} \\ c_1 J_n(X) + c_2 Y_n(X) & \text{when } n \text{ is an integer} \end{cases}$$

$$\Rightarrow \quad y = \begin{cases} x^{\alpha} \left[c_1 J_n(\beta x^r) + c_2 J_{-n}(\beta x^r) \right] & \text{when } n \text{ is not an integer} \\ x^{\alpha} \left[c_1 J_n(\beta x^r) + c_2 Y_n(\beta x^r) \right] & \text{when } n \text{ is an integer} \end{cases}$$

4.6.8 Jacobi Series

Prove that the Jacobi Series

(i) $\cos(x \sin\theta) = J_0 + 2\cos 2\theta\, J_2 + 2\cos 4\theta\, J_4 + \ldots$

(ii) $\sin(x \sin\theta) = 2\sin\theta\, J_1 + 2\sin 3\theta\, J_3 + 2\sin 5\theta\, J_5 + \ldots$

Proof: From generating function of $J_n(x)$, we have

$$e^{x\left(z-\frac{1}{z}\right)/2} = J_0 + \left(z - \frac{1}{z}\right)J_1 + \left(z^2 + \frac{1}{z^2}\right)J_2 + \left(z^3 - \frac{1}{z^3}\right)J_3 + \ldots.$$

Putting $z = e^{i\theta}$, we get

$$e^{xi\sin\theta} = J_0 + \left(e^{i\theta} - e^{-i\theta}\right)J_1 + \left(e^{2i\theta} + e^{-2i\theta}\right)J_2 + \left(e^{3i\theta} - e^{-3i\theta}\right)J_3 + \ldots$$

$$\Rightarrow \quad \cos(x\sin\theta) + i\sin(x\sin\theta) = J_0 + 2i\sin\theta\, J_1 + 2\cos 2\theta J_2 + 2i\sin 3\theta J_3 + \ldots.$$

Equating real and imaginary parts, we obtain

$$\cos(x\sin\theta) = J_0 + 2\cos 2\theta\, J_2 + 2\cos 4\theta\, J_4 + \ldots$$

$$\sin(x\sin\theta) = 2\sin\theta\, J_1 + 2\sin 3\theta\, J_3 + 2\sin 5\theta\, J_5 + \ldots.$$

Remark: Putting $\theta = \frac{\pi}{2}$ in the Jacobi's series, we have

$$\cos x = J_0 - 2J_2 + 2J_4 - 2J_6 + \ldots$$

$$\sin x = 2J_1 - 2J_3 + 2J_5 - 2J_7 + \ldots.$$

Example 1: Prove that

(i) $J_{-n}(x) = (-1)^n J_n(x)$, where n is positive integer.

(ii) $J_n(-x) = (-1)^n J_n(x)$, for positive or negative integers.

Solution: (i) We know that $J_{-n}(x) = \sum_{r=0}^{\infty} \frac{(-1)^r \left(\frac{x}{2}\right)^{-n+2r}}{r!\,\Gamma(-n+r+1)}$.

Since Γ (a negative integer or zero) is infinity, each term in the summation is equal to zero till $-n+r+1 < 1$, i.e. $r < n$. Thus we have

$$J_{-n}(x) = 0 + \sum_{r=n}^{\infty} \frac{(-1)^r \left(\frac{x}{2}\right)^{-n+2r}}{r!\,\Gamma(-n+r+1)}$$

$$=\sum_{s=0}^{\infty}\frac{(-1)^{n+s}\left(\frac{x}{2}\right)^{n+2s}}{(n+s)!\Gamma(-n+n+s+1)}, \text{ putting } r=n+s,$$

$$=(-1)^n\sum_{s=0}^{\infty}\frac{(-1)^s\left(\frac{x}{2}\right)^{n+2s}}{s!\Gamma(n+s+1)}=(-1)^n J_n(x)$$

Hence, we have $J_{-n}(s)=(-1)^n J_n(x)$.

(ii) We know that $J_n(x)=\sum_{r=0}^{\infty}(-1)^r\frac{1}{r!\Gamma(n+r+1)}\left(\frac{x}{2}\right)^{n+2r}$

Let n be a positive integer. Replacing x to $-x$, we get

$$J_n(-x)=\sum_{r=0}^{\infty}\frac{(-1)^r\left(\frac{-x}{2}\right)^{n+2r}}{r!\Gamma(n+r+1)}=(-1)^{n+2r}\sum_{r=0}^{\infty}\frac{(-1)^r\left(\frac{x}{2}\right)^{n+2r}}{r!\Gamma(n+r+1)}$$

$$=(-1)^n\sum_{r=0}^{\infty}\frac{(-1)^r\left(\frac{x}{2}\right)^{n+2r}}{r!\Gamma(n+r+1)}, \quad \left[\because(-1)^{2r}=1\right]$$

$$=(-1)^n J_n(x).$$

Hence, we have $J_n(-x)=(-1)^n J_n(x)$.

Again let n be a negative integer, say $n=-m$, where m is a positive integer, we get

$$J_n(x)=J_{-m}(x)\ =(-1)^m J_m(x), \text{ [From part (i)]}$$

Replacing x by $-x$, we obtain

$$J_n(-x)=(-1)^m J_m(-x)\ =(-1)^m(-1)^m J_m(x)$$

$$=(-1)^{2m}(-1)^{-m}J_{-m}(x)$$

$$=(-1)^{-m}J_{-m}(x)\ =(-1)^n J_n(x).$$

Hence, we have $J_n(-x)=(-1)^n J_n(x)$ for positive or negative integers.

Example 2: Show that $\int xJ_0^2(x)dx=\frac{1}{2}x^2\left[J_0^2(x)+J_1^2(x)\right]$.

Solution: We have $\int xJ_0^2(x)dx=J_0^2(x)\frac{x^2}{2}-\int 2J_0(x)J_0'(x)\frac{x^2}{2}dx$

$$=\frac{x^2}{2}\cdot J_0^2(x)-\int x^2 J_0(x)J_0'(x)dx$$

$$=\frac{x^2}{2}J_0^2(x)-\int x^2 J_0(x)(-J_1(x))dx, \left[\because J_0'(x)=-J_1(x)\right]$$

$$=\frac{x^2}{2}J_0^2(x)+\int xJ_1(x).xJ_0(x)dx$$

$$=\frac{x^2}{2}J_0^2(x)+\int xJ_1(x).\frac{d}{dx}\left[xJ_1(x)\right]dx,$$

$$=\frac{x^2}{2}J_0^2(x)+\frac{\left[xJ_1(x)\right]^2}{2}=\frac{x^2}{2}\left[J_0^2(x)+J_1^2(x)\right].$$

Hence, we have $\int xJ_0^2(x)dx=\frac{x^2}{2}\left[J_0^2(x)+J_1^2(x)\right].$

Example 3: Prove that

(i) $J_{-1/2}(x)=\sqrt{\frac{2}{\pi x}}.\cos x$

(ii) $J_{1/2}(x)=\sqrt{\frac{2}{\pi x}}.\sin x$

(iii) $\left[J_{1/2}^2(x)+J_{-1/2}^2(x)\right]=\frac{2}{\pi x}.$

Solution: By definition of Bessel's function, we have

$$J_n(x)=\frac{x^n}{2^n\,\Gamma(n+1)}\left[1-\frac{x^2}{2(2n+2)}+\frac{x^4}{2.4.(2n+2)(2n+4)}-\ldots\right] \quad \ldots(1)$$

(i) Putting $n=-\frac{1}{2}$, we obtain

$$J_{-1/2}(x)=\frac{x^{-1/2}}{2^{-1/2}\,\Gamma(1/2)}\left[1-\frac{x^2}{2!}+\frac{x^4}{4!}+\right]$$

$$\Rightarrow \quad J_{-1/2}(x)=\sqrt{\frac{2}{\pi x}}\cos x. \quad \ldots(2)$$

(ii) Putting $n=\frac{1}{2}$ in (1), we have

$$J_{1/2}(x)=\frac{x^{1/2}}{2^{1/2}\,\Gamma(3/2)}\left[1-\frac{x^2}{2(1+2)}+\frac{x^4}{2.4(1+2)(1+4)}-\ldots\right]$$

$$=\sqrt{\frac{x}{2}}\cdot\frac{1}{\frac{1}{2}\sqrt{\pi}}\cdot\frac{1}{x}\left[x-\frac{x^3}{3!}+\frac{x^5}{5!}+\ldots\right]=\sqrt{\frac{2}{\pi x}}\sin x. \quad \ldots(3)$$

(iii) Square and add (2) and (3), we get

$$\left[J_{1/2}^2(x)+J_{-1/2}^2(x)\right]=\left[\frac{2}{\pi x}\left(\sin^2 x+\cos^2 x\right)\right]=\frac{2}{\pi x}.$$

Example 4: Prove that

(i) $J_0' = -J_1$ (ii) $J_2 = J_0'' - x^{-1} J_0'$ (iii) $J_2 - J_0 = 2J_0''$.

Solution: (i) From recurrence formula I, we have

$$xJ_n' = nJ_n - xJ_{n+1}. \qquad ...(1)$$

Putting $n = 0$, we get

$$xJ_0' = -xJ_1 \Rightarrow J_0' = J_1.$$

(ii) Putting $n = 1$ in (1), we have $xJ_1' = J_1 - xJ_2$. ...(2)

From (i), we have $J_0' = -J_1$. On differentiating w.r.t. 'x', we get

$$J_0'' = -J_1'.$$

Substituting these values in (2), we obtain

$$-xJ_0'' = -J_0' - xJ_2 \Rightarrow J_2 = J_0'' - x^{-1}J_0'$$

(iii) From recurrence formula III, we have

$$2J_n' = J_{n-1} - J_{n+1}. \qquad ...(3)$$

Differentiating both sides w.r.t. 'x' and multiplying by 2, we get

$$2^2 J_n'' = 2J_{n-1}' - 2J'_{n+1}. \qquad ...(4)$$

From (3), replacing n by $(n - 1)$ and $(n + 1)$, we have

$$2J'_{n-1} = J_{n-2} - J_n \quad \text{and} \quad 2J'_{n+1} = J_n - J_{n+2}$$

Substituting these values in (4), we get

$$2^2 J_n'' = (J_{n-2} - J_n) - (J_n - J_{n+2}) = J_{n-2} - 2J_n + J_{n+2}.$$

Putting $n = 0$, we get $2^2 J_0'' = J_{-2} - 2J_0 + J_2$

$$= (-1)^2 J_2 - 2J_0 + J_2 = 2J_2 - 2J_0.$$

Hence, we have $2J_0'' = J_2 - J_0$.

Example 5: Prove that $\dfrac{d}{dx}\left[xJ_nJ_{n+1}\right] = x\left[J_n^2 - J_{n+1}^2\right]$.

Solution: We have $\dfrac{d}{dx}\left[xJ_n J_{n+1}\right] = J_n J_{n+1} + x\left[J_n' J_{n+1} + J_n J'_{n+1}\right]$

$$= J_n J_{n+1} + (xJ_n')J_{n+1} + J_n\left(xJ'_{n+1}\right). \qquad ...(1)$$

From recurrence relations I and II, we have

$$xJ_n'(x) = nJ_n - xJ_{n+1} \qquad ...(2)$$

and $\quad xJ'_n(x) = -nJ_n + xJ_{n-1}.$ ...(3)

Replacing n by $(n+1)$ in (3), we get

$$xJ'_{n+1}(x) = -(n+1)J_{n+1} + xJ_n. \quad ...(4)$$

Substituting the values of xJ'_n and xJ'_{n+1} from (2) and (4) in (1), we get

$$\frac{d}{dx}\left[xJ_n J_{n+1}\right] = J_n J_{n+1} + \left(nJ_n - xJ_{n+1}\right)J_{n+1} + J_n\left[-(n+1)J_{n+1} + xJ_n\right]$$

Hence, we have $\frac{d}{dx}\left[xJ_n J_{n+1}\right] = x\left[J_n^2 - J_{n+1}^2\right].$

Example 6: Show that $J'_2(x) = \left(1 - \frac{4}{x^2}\right)J_1(x) + \frac{2}{x}J_0(x).$

Solution: From recurrence, formula II, we have

$$xJ'_n = -nJ_n + xJ_{n-1}. \quad ...(1)$$

Putting $n = 2$, in (1), we get

$$xJ'_2 = -2J_2 + xJ_1 \Rightarrow J'_2 = -\frac{2}{x}J_2 + J_1. \quad ...(2)$$

Now, from recurrence formula I, we have

$$xJ'_n = nJ_n - xJ_{n+1}. \quad ...(3)$$

From equations (1) and (3), we get

$$-nJ_n + xJ_{n-1} = nJ_n - xJ_{n+1}.$$

Putting $n = 1$, we get

$$-J_1 + xJ_0 = J_1 - xJ_2 \Rightarrow -\frac{1}{x}J_1 + J_0 = \frac{1}{x}J_1 - J_2$$

$$\Rightarrow \quad J_2 = \frac{2}{x}J_1 - J_0. \quad ...(4)$$

Substituting the value of J_2 in equation (2), we get

$$J'_2 = -\frac{2}{x}\left[\frac{2}{x}J_1 - J_0\right] + J_1 = -\frac{4}{x^2}J_1 + \frac{2}{x}J_0 + J_1.$$

Hence, we have $J'_2(x) = \left[1 - \frac{4}{x^2}\right]J_1(x) + \frac{2}{x}J_0(x).$

Example 7: Prove that $J_{3/2} = \sqrt{\frac{2}{\pi x}}\left[\frac{1}{x}\sin x - \cos x\right].$

Solution: Using $n = \frac{3}{2}$ in the expression of $J_n(x)$, we get

$$J_{3/2}(x)=\frac{x^{3/2}}{2^{3/2}\,\Gamma(5/2)}\left[1-\frac{x^2}{2.5}+\frac{x^4}{2.4.5.7}-\frac{x^6}{2.4.5.6.7.9}+\ldots\right]$$

$$=\frac{x\sqrt{x}}{2\sqrt{2}\times\frac{3}{2}\times\frac{1}{2}\times\sqrt{\pi}}\left[1-\frac{x^2}{2.5}+\frac{x^4}{2.4.5.7}-\frac{x^6}{2.4.5.6.7.9}+\ldots\right]$$

$$=\frac{\sqrt{2}\times x\sqrt{x}\times\sqrt{x}}{3\times\sqrt{\pi}\times\sqrt{x}}\left[1-\frac{x^2}{2.5}+\frac{x^4}{2.4.5.7}-\frac{x^6}{2.4.5.6.7.9}+\ldots\right]$$

$$=\sqrt{\frac{2}{\pi x}}\cdot\frac{1}{3}\left[x^2-\frac{x^4}{2.5}+\frac{x^6}{2.4.5.7}-\frac{x^8}{2.4.5.6.7.9}+\ldots\right]$$

$$=\sqrt{\frac{2}{\pi x}}\left[\frac{2x^2}{3!}-\frac{4x^4}{5!}+\frac{6x^6}{7!}-\frac{8x^8}{9!}+\ldots\right]$$

$$=\sqrt{\frac{2}{\pi x}}\left[\left(\frac{1}{2!}-\frac{1}{3!}\right)x^2-\left(\frac{1}{4!}-\frac{1}{5!}\right)x^4+\left(\frac{1}{6!}-\frac{1}{7!}\right)x^6-\left(\frac{1}{8!}-\frac{1}{9!}\right)x^8+\ldots\right]$$

$$=\sqrt{\frac{2}{\pi x}}\left[\left(\frac{x^2}{2!}-\frac{x^4}{4!}+\frac{x^6}{6!}-\ldots\right)+\left(-\frac{x^2}{3!}+\frac{x^4}{5!}-\frac{x^6}{7!}\ldots\right)\right]$$

$$=\sqrt{\frac{2}{\pi x}}\left[-\left(1-\frac{x^2}{2!}+\frac{x^4}{4!}+\frac{x^6}{6!}-..\right)+\frac{1}{x}\left(x-\frac{x^3}{3!}+\frac{x^5}{5!}-\ldots\right)\right].$$

Hence, we have $J_{3/2}(x)=\sqrt{\dfrac{2}{\pi x}}\left[\dfrac{1}{x}\sin x-\cos x\right].$

Example 8: Show that $\dfrac{d}{dx}\left[J_n^2+J_{n+1}^2\right]=\dfrac{2}{x}\left[n\,J_n^2-(n+1)J_{n+1}^2\right].$

Solution: We have $\dfrac{d}{dx}\left[J_n^2+J_{n+1}^2\right]=2J_nJ_n{}'+2J_{n+1}J'{}_{n+1}$. ...(1)

Writing recurrence formulae (I) and (II), we have

$$J_n'=\frac{n}{x}J_n-J_{n+1} \qquad ...(2)$$

and $$J_n'=-\frac{n}{x}J_n+J_{n-1}. \qquad ...(3)$$

Replacing n by $(n+1)$ in (3), we get

$$J_{n+1}'=-\frac{(n+1)}{x}J_{n+1}+J_n. \qquad ...(4)$$

Putting the values of J_n' and J_{n+1}' from (2) and (4) in (1), we get

$$\frac{d}{dx}\left[J_n^2+J_{n+1}^2\right]=2J_n\left[\frac{n}{x}J_n-J_{n+1}\right]+2J_{n+1}\left[-\frac{(n+1)}{x}J_{n+1}+J_n\right]$$

$$=2\left(\frac{n}{x}J_n^2-\frac{n+1}{x}J_{n+1}^2\right).$$

Hence, we have $\frac{d}{dx}\left[J_n^2+J_{n+1}^2\right]=\frac{2}{x}\left(nJ_n^2-(n+1)J_{n+1}^2\right).$

Example 9: Prove that $\int J_3(x)dx=-J_2(x)-\frac{2}{x}J_1(x).$

Solution: From the recurrence formula (VI), we have

$$\frac{d}{dt}\left[x^{-n}J_n(x)\right]=-x^{-n}J_{n+1}.$$

Therefore $\int x^{-n}J_{n+1}dx=-x^{-n}J_n.$ ...(1)

Now $\int J_3(x)dx=\int x^2\left(x^{-2}J_3\right)dx$

$$=x^2\left[-x^{-2}J_2\right]-\int 2x\left[-x^{-2}J_2\right]dx\text{ , [using (1) for } n=2]$$

$$=-J_2(x)+2\int x^{-1}J_2dx\quad=-J_2(x)+2\left[-x^{-1}J_1\right],$$

[using (1) for $n = 1$]

$$=-J_2(x)-\frac{2}{x}J_1(x).$$

Example 10: Prove that $4J_n''(x)=J_{n-2}(x)-2J_n(x)+J_{n+2}(x).$

Solution: From recurrence formula III, we have

$2J_n'=J_{n-1}-J_{n+1}.$ ...(1)

Differentiating (1), we get $2J_n''=J_{n-1}'-J_{n+1}'.$ ...(2)

Replacing n by $(n-1)$ and $(n+1)$ in (1), we have

$2J_{n-1}'=J_{n-2}-J_n$ ....(3)

and $2J_{n+1}'=J_n-J_{n+2}$ ....(4)

Subtracting (4) from (3), we get $2[J_{n-1}'-J_{n+1}']=J_{n-2}+J_{n+2}-2J_n.$

From (2), we get $4J_n''(x)=J_{n-2}(x)-2J_n(x)+J_{n+2}(x).$

Example 11: Use Jacobi series to show that

$$\left[J_0(x)\right]^2+2\left[J_1(x)\right]^2+2\left[J_2(x)\right]^2+\ldots=1.$$

Solution: We know that the Jacobi series are

$$J_0(x)+2J_2(x)\cos 2\theta+2J_4(x)\cos 4\theta+\ldots=\cos(x\sin\theta) \quad \ldots(1)$$

and $$2J_1(x)\sin\theta+2J_3(x)\sin 3\theta+\ldots=\sin(x\sin\theta). \quad \ldots(2)$$

Squaring (1) and (2) and integrating w.r.t. 'θ' between the limits 0 to π and using the integrals, if m, n are integers

$$\int_0^\pi \cos^2 n\theta\, d\theta=\int_0^\pi \sin^2 n\theta\, d\theta=\frac{\pi}{2}$$

and $$\int_0^\pi \cos m\theta \cos n\theta\, d\theta=\int_0^\pi \sin m\theta \sin n\theta\, d\theta=0,\ m\neq n.$$

We get $\pi[J_0(x)]^2+2\pi[J_2(x)]^2+2\pi[J_4(x)]^2+\ldots$

$$=\int_0^\pi \cos^2(x\sin\theta)d\theta \quad \ldots(3)$$

and $$2\pi[J_1(x)]^2+2\pi[J_3(x)]^2+\ldots=\int_0^\pi \sin^2(x\sin\theta)d\theta. \quad \ldots(4)$$

Adding (3) and (4), we have

$$\pi\left\{[J_0(x)]^2+2[J_1(x)]^2+2[J_2(x)]^2+\ldots\right\}=\int_0^\pi d\theta=\pi.$$

Hence, we have the required result.

Example 12: If $a>0$, prove that $\int_0^\infty e^{-ax} J_0(bx)dx=\dfrac{1}{\sqrt{a^2+b^2}}$.

Solution: We know that $J_0(x)=\dfrac{1}{\pi}\int_0^\pi \cos(x\sin\phi)d\phi$

$$\Rightarrow \quad J_0(bx)=\frac{1}{\pi}\int_0^\pi \cos(bx\sin\phi)d\phi.$$

Therefore $$\int_0^\infty e^{-ax} J_0(bx)dx=\int_0^\infty e^{-ax}\left\{\frac{1}{\pi}\int_0^\pi \cos(bx\sin\phi)d\phi\right\}dx$$

$$=\frac{1}{\pi}\int_0^\infty\left\{\int_0^\pi e^{-ax}\cos(bx\sin\phi)d\phi\right\}dx$$

$$=\frac{1}{\pi}\int_0^\pi\left\{\int_0^\infty e^{-ax}\cos(bx\sin\phi)dx\right\}d\phi$$

[on changing order of integration]

$$=\frac{1}{\pi}\int_0^\pi\left\{\int_0^\infty e^{-ax}\frac{e^{ibx\sin\phi}+e^{-ibx\sin\phi}}{2}dx\right\}d\phi$$

$$=\frac{1}{2\pi}\int_0^{\pi}\left\{\int_0^{\infty}\left[e^{-(a-ib\sin\phi)x}+e^{-(a+ib\sin\phi)x}\right]dx\right\}d\phi$$

$$=\frac{1}{2\pi}\int_0^{\pi}\left[\frac{e^{-(a-ib\sin\phi)x}}{-(a-ib\sin\phi)}+\frac{e^{-(a+ib\sin\phi)x}}{-(a+ib\sin\phi)}\right]_0^{\infty}d\phi$$

$$=\frac{1}{2\pi}\int_0^{\pi}\left[\frac{1}{(a-ib\sin\phi)}+\frac{1}{(a+ib\sin\phi)}\right]d\phi$$

$$=\frac{1}{2\pi}\int_0^{\pi}\frac{2a}{a^2+b^2\sin^2\phi}d\phi\ =\frac{a}{\pi}2\int_0^{\pi/2}\frac{\operatorname{cosec}^2\phi}{b^2+a^2\operatorname{cosec}^2\phi}d\phi$$

$$=\frac{2a}{\pi}\int_0^{\pi/2}\frac{\operatorname{cosec}^2\phi}{b^2+a^2\left(1+\cot^2\phi\right)}d\phi$$

$$=\frac{2a}{\pi}\int_0^{\pi/2}\frac{\operatorname{cosec}^2\phi}{\left(a^2+b^2\right)+a^2\cot^2\phi}d\phi$$

Let $a\cot\phi=t$, then $-\operatorname{cosec}^2\phi\, d\phi=\frac{dt}{a}$.

Limits: when $\phi=\frac{\pi}{2}$, $t=\cot\frac{\pi}{2}=0$ and $\phi=0$, $t=\cot 0=\infty$.

Hence, we have $\int_0^{\infty}e^{-ax}J_0(bx)dx=\frac{2a}{\pi}\int_{\infty}^{0}\frac{1}{\left(a^2+b^2\right)+t^2}\frac{-dt}{a}$

$$=\frac{2}{\pi}\int_0^{\infty}\frac{dt}{\left(\sqrt{a^2+b^2}\right)^2+t^2}dt\ =\frac{2}{\pi\sqrt{a^2+b^2}}\left[\tan^{-1}\frac{t}{\sqrt{a^2+b^2}}\right]_0^{\infty}$$

$$=\frac{2}{\pi\sqrt{a^2+b^2}}\left[\tan^{-1}\infty-\tan^{-1}0\right]\ =\frac{2}{\pi\sqrt{a^2+b^2}}\left(\frac{\pi}{2}-0\right)$$

$$=\frac{1}{\sqrt{a^2+b^2}}.$$

Example 13: Prove that $\int_0^{\pi/2}\sqrt{\pi x}\,J_{1/2}(2x)dx=1.$

Solution: Putting $2x=t,\ x=\frac{t}{2},\ dx=\frac{dt}{2}$, we have

$$\int_0^{\pi/2}\sqrt{\pi x}\,J_{1/2}(2x)dx=\int_0^{\pi}\sqrt{\pi}\left(\frac{t}{2}\right)^{1/2}J_{1/2}(t)\frac{dt}{2}$$

$$\int_0^{\pi}\sqrt{\pi}\,\frac{t^{1/2}}{2\sqrt{2}}J_{1/2}(t)dt\ =\frac{\sqrt{\pi}}{2\sqrt{2}}\int_0^{\pi}t^{1/2}J_{1/2}(t)dt \qquad \text{...(2)}$$

We know that $J_{1/2}(t)=\sqrt{\frac{2}{\pi t}}\sin t$. Hence from (2), we get

$$\int_0^\pi \sqrt{\pi x}\, J_{1/2}(2x)dx = \frac{\sqrt{\pi}}{2\sqrt{2}}\int_0^\pi t^{1/2}\sqrt{\frac{2}{\pi t}}\sin t\, dt$$

$$\frac{\sqrt{\pi}}{2\sqrt{2}}\cdot\frac{\sqrt{2}}{\sqrt{\pi}}\int_0^\pi \sin t\, dt$$

$$=\frac{1}{2}\left[-\cos t\right]_0^\pi = \frac{1}{2}\left[-\cos\pi+\cos 0\right]=\frac{1}{2}[1+1]=1.$$

Example 14: Prove that $J_4(x)=\left(\frac{48}{x^3}-\frac{8}{x}\right)J_1(x)+\left(1-\frac{24}{x^2}\right)J_0(x)$.

Solution: From recurrence relation IV, we have

$$2n\, J_n(x)=x[J_{n-1}(x)+J_{n+1}(x)]$$

This gives $J_{n+1}(x)=\frac{2n}{x}J_n(x)-J_{n-1}(x)$ ...(1)

Putting $n=1, 2, 3\ in (1)$, we get

$$J_2(x)=\frac{1}{x}\left[2J_1(x)-xJ_0(x)\right] \quad ...(2)$$

$$J_3(x)=\frac{1}{x}\left[4J_2(x)-xJ_1(x)\right] \quad ...(3)$$

$$J_4(x)=\frac{1}{x}\left[6J_3(x)-xJ_2(x)\right] \quad ...(4)$$

Putting value of $J_2(x)$ from (2) in (3), we get

$$J_3(x)=\frac{8}{x^2}J_1(x)-\frac{4}{x}J_0(x)-J_1(x)=\left(\frac{8-x^2}{x^2}\right)J_1(x)-\frac{4}{x}J_0(x)$$

Using values of $J_3(x)$and $J_2(x)$ in (4), we get

$$J_4(x)=\left(\frac{48-6x^2}{x^3}\right)J_1(x)-\frac{24}{x^2}J_0(x)-\frac{2}{x}J_1(x)+J_0(x)$$

$$=\left(\frac{48}{x^3}-\frac{8}{x}\right)J_1(x)+\left(1-\frac{24}{x^2}\right)J_0(x).$$

Example 15: Prove that $4J_0'''(x)+3J_0'(x)+J_3'(x)=0$

Solution: We know that $J_0'=-J_1$

Differentiating w.r.t. 'x', we get $J_0''(x)=-J_1'(x)$ ...(1)

From recurrence relation III, we have

$$2J_n'(x)=J_{n-1}(x)-J_{n+1}(x) \quad ...(2)$$

Putting $n = 1$, we get $2J_1'(x) = J_0(x) - J_2(x)$

Using it in (1), we get $J_0''(x) = -\frac{1}{2}\left[J_0(x) - J_2(x)\right]$

Again differentiating w.r.t. 'x', we get

$$J_0'''(x) = -\frac{1}{2}J_0'(x) + \frac{1}{2}J_2'(x) \quad ...(3)$$

Putting $n = 2$ in (2), we get $2J_2'(x) = J_1(x) - J_3(x)$

Thus (3) gives

$$J_0'''(x) = -\frac{1}{2}J_0'(x) + \frac{1}{4}\left[J_1(x) - J_3(x)\right]$$

$$= -\frac{1}{2}J_0'(x) - \frac{1}{4}J_0'(x) - \frac{1}{4}J_3(x) = -\frac{3}{4}J_0'(x) - \frac{1}{4}J_3(x)$$

$$\Rightarrow \quad 4J_0'''(x) + 3J_0'(x) + J_3(x) = 0.$$

Example 16: Prove that $\lim_{x \to 0} \frac{J_n(x)}{x^n} = \frac{1}{2^n \Gamma(n+1)}; n > -1.$

Solution: We know that

$$J_n(x) = \frac{x^n}{2^n \Gamma n+1}\left[1 - \frac{x^2}{2.(2n+2)} + \frac{x^4}{2.4.(2n+2)(2n+4)} - ....\right]$$

Therefore

$$\lim_{x \to 0} \frac{J_n(x)}{x^n} = \lim_{x \to 0} \frac{1}{2^n \Gamma n+1}\left[1 - \frac{x^2}{2.(2n+2)} + \frac{x^4}{2.4.(2n+2)(2n+4)} - ....\right]$$

$$= \frac{1}{2^n \Gamma n+1}.$$

Example 17: Prove that $J_{n+3} + J_{n+5} = \frac{2}{x}(n+4)J_{n+4}$.

Solution: From recurrence relation IV, we have

$$2nJ_n = x\left(J_{n-1} + J_{n+1}\right).$$

Replacing n by $n + 4$, we get

$$\frac{2}{x}(n+4)J_{n+4} = J_{n+3} + J_{n+5}.$$

Example 18: Prove that $J_2'(x) = \left(1 - \frac{4}{x^2}\right)J_1(x) + \frac{2}{x}J_0(x).$

Solution: From recurrence relation II, we have

$$xJ_n' = -nJ_n + xJ_{n-1} \quad ...(1)$$

Putting $n = 2$, we get $xJ_2' = -2J_2 + xJ_1$

$$\Rightarrow \quad J_2' = -\frac{2}{x} J_2 + J_1 \qquad \text{...(2)}$$

From recurrence relation (I), we have

$$xJ_n' = nJ_n - xJ_{n+1} \qquad \text{...(3)}$$

Equations (1), (2) and (3) give $-nJ_n + xJ_{n-1} = nJ_n - xJ_{n+1}$.

Putting $n = 1$, we get

$$-J_1 + xJ_0 = J_1 - xJ_2 \Rightarrow J_2 = \frac{2}{x} J_1 - J_0 \qquad \text{...(4)}$$

Using it in (2), we get

$$J_2' = -\frac{2}{x}\left(\frac{2}{x} J_1 - J_0\right) + J_1 = \left(1 - \frac{4}{x^2}\right) J_1 + \frac{2}{x} J_0 .$$

Example 19: Prove that $J_n(x) = 0$ has no repeated root expect at $x = 0$.

Solution: Suppose, if possible, α is a double root of $J_n(x) = 0$

Then, $J_n(\alpha) = 0 \; and \; J_n'(\alpha) = 0$ ...(1)

From recurrence relation I and II, we have

$$J_{n+1}(x) = \frac{n}{x} J_n(x) - J_n'(x)$$

and $$J_{n-1}(x) = \frac{n}{x} J_n(x) + J_n'(x)$$

Using (1), we get $J_{n+1}(\alpha) = 0$ and $J_{n-1}(\alpha) = 0$

which is inadmissible because two distinct power series cannot have the same sum function.

Hence $J_n(x)$ has no repeated root expect $x = 0$.

Example 20: Prove that

$$x^2 J_n''(x) = (n^2 - n - x^2) J_n(x) + xJ_{n+1}(x); n = 0, 1, 2.$$

Solution: From recurrence relation I, we have

$$xJ_n' = nJ_n - xJ_{n+1} \qquad \text{...(1)}$$

Differentiating w.r.t. 'x', we get

$$xJ_n'' + J_n' = nJ'_n - xJ_{n+1}' - J_{n+1}$$

$$\Rightarrow \quad x^2 J_n'' = (n-1)xJ_n' - x^2 J_{n+1}' - xJ_{n+1} \qquad \text{...(2)}$$

From recurrence relation II, we have

$$xJ_n' = -nJ_n + xJ_{n-1} \Rightarrow xJ_{n+1}' = -(n+1)J_{n+1} + xJ_n \qquad ...(3)$$

Using (1) and (3) it in (2), we get

$$x^2 J_n'' = (n-1)\left[nJ_n - xJ_{n+1}\right] - x\left[-(n+1)J_{n+1} + xJ_n\right] - xJ_{n+1}$$

$$\Rightarrow \quad x^2 J_n'' = \left(n^2 - n - x^2\right)J_n + xJ_{n+1}.$$

Example 21: If $n > -1$, show that

$$\int_0^x x^{-n} J_{n+1}(x)dx = \frac{1}{2^n \Gamma n+1} - x^{-n} J_n(x).$$

Solution: From recurrence relation VI, we have

$$\frac{d}{dx}\left[x^{-n} J_n(x)\right] = -x^{-n} J_{n+1}(x)$$

Integrating it between 0 to x, we get

$$\int_0^x x^{-n} J_{n+1}(x)dx = -\left[x^{-n} J_n(x)\right]_0^x$$

$$= -x^{-n} J_n(x) + \lim_{x\to 0}\left(\frac{J_n(x)}{x^n}\right)$$

$$= -x^{-n} J_n(x) + \frac{1}{2^n \Gamma n+1}, n > -1$$

Example 22: Prove that

$$\frac{x}{2} J_n = (n+1)J_{n+1} - (n+3)J_{n+3} + (n+5)J_{n+5} - ......$$

Solution: From recurrence relation IV, we have

$$2nJ_n = x\left(J_{n-1} + J_{n+1}\right)$$

Replacing n by $(n + 1)$, we get

$$2(n+1)J_{n+1} = x\left(J_n + J_{n+2}\right) \Rightarrow \frac{x}{2} J_n = (n+1)J_{n+1} - \frac{x}{2} J_{n+2} \qquad ...(1)$$

Replacing n by $(n + 2)$, we get

$$\frac{x}{2} J_{n+2} = (n+3)J_{n+3} - \frac{x}{2} J_{n+4} \qquad ...(2)$$

Using it in (1), we get

$$\frac{x}{2} J_n = (n+1)J_{n+1} - (n+3)J_{n+3} + \frac{x}{2} J_{n+4}$$

Continuing this process, we get

$$\frac{x}{2} J_n = (n+1)J_{n+1} - (n+3)J_{n+3} + (n+5)J_{n+5} - .....$$

Example 23: Solve the differential equation $xy'' - 3y' + xy = 0$ in terms of Bessel functions.

Solution: Multiplying the given equation by x becomes

$$x^2 y'' - 3xy' + x^2 y = 0$$

Comparing with the general form (6) of section 4.6.7, we get

$$1 - 2\alpha = -3, \beta^2 \gamma^2 = 1, 2\gamma = 2, \alpha^2 - n^2 \gamma^2 = 0$$

i.e., $\alpha = 2, \beta = \gamma = 1, n = 2$

Here n is integer, so the solution be $y = x^2 \left[c_1 J_2(x) + c_2 Y_2(x) \right]$.

Example 24: Obtain in terms of Bessel functions, the solution of differential equation

$$\frac{d^2y}{dx^2} + \left(9x - \frac{20}{x^2} \right) y = 0.$$

Solution: The given equation on multiplying by x^2 becomes

$$x^2 \frac{d^2y}{dx^2} + \left(9x^3 - 20 \right) y = 0$$

Comparing this with standard transformed equation

$$x^2 y'' + (1 - 2\alpha) xy' + \left\{ \gamma^2 \beta^2 x^{2\gamma} + \left(\alpha^2 - n^2 \gamma^2 \right) \right\} y = 0,$$

We get $1 - 2\alpha = 0, \beta^2 \gamma^2 = 9,$

$$2\gamma = 3, \text{ and } \alpha^2 - n^2 \gamma^2 = -20$$

This gives $\alpha = \frac{1}{2}, \gamma = \frac{3}{2}, \beta = 2, n = 3$

Here n is an integer, so the solution be

$$y = A \sqrt{x}\, J_3\left(2x^{3/2} \right) + B \sqrt{x}\, Y_3\left(2x^{3/2} \right).$$

Example 25: Solve the differential equation $y'' + \frac{y'}{x} + 4\left(x^2 - \frac{n^2}{x^2} \right) y = 0$ in terms of Bessel's functions.

Solution: The given equation can be written as

$$x^2 y'' + xy' + 4\left(x^4 - n^2 \right) y = 0 \qquad \text{...(1)}$$

Comparing with the general form, we get

$$1 - 2\alpha = 1, \ \beta^2 \gamma^2 = 4, \ 2\gamma = 4 \ \text{ and } \ \alpha^2 - m^2 \gamma^2 = -4n^2$$

On solving these equations, we get

$$\alpha = 0, \gamma = 2, \beta^2 = 1 \Rightarrow \beta = 1 \text{ and } 0 - m^2 (4) = -4n^2 \Rightarrow m = n$$

When n is not an integer, solution of (1) is

$$y = x^{\alpha} \left[c_1 J_m\left(\beta x^{\gamma} \right) + c_2 J_{-m}\left(\beta x^{\gamma} \right) \right]$$

$$= x^0\left[c_1\, J_n\left(x^2\right) + c_2\, J_{-n}\left(x^2\right)\right]$$

$$= c_1\, J_n\left(x^2\right) + c_2\, J_{-n}\left(x^2\right)$$

When n is an integer, the solution of (1) is

$$y = x^\alpha\left[c_1\, J_m\left(\beta x^\gamma\right) + c_2\, Y_m\left(\beta x^\gamma\right)\right]$$

$$= x^0\left[c_1\, J_n\left(x^2\right) + c_2\, Y_n\left(x^2\right)\right] = c_1\, J_n\left(x^2\right) + c_2\, Y_n\left(x^2\right).$$

Example 26: Solve the differential equation $4\,y'' + 9xy = 0$ in terms of Bessel's functions.

Solution: The given equation can be written as

$$x^2\, y'' + \frac{9}{4}\, x^3\, y = 0 \qquad ...(1)$$

Comparing with the general form (6) of section 4.6.7, we get

$$1 - 2\alpha = 0,\ \beta^2\, \gamma^2 = \frac{9}{4},\ 2\gamma = 3 \text{ and } \alpha^2 - n^2\, \gamma^2 = 0.$$

On solving these equations, we get

$$\alpha = \frac{1}{2}, \gamma = \frac{3}{2}, \beta^2 . \frac{9}{4} = \frac{9}{4} \Rightarrow \beta = 1$$

and $\left(\frac{1}{4}\right) - n^2\left(\frac{9}{4}\right) = 0 \Rightarrow n^2 = \frac{1}{9} \Rightarrow n = \frac{1}{3},$

Since n is not an integer, the solution of equation (1) is

$$y = x^\alpha\left[c_1 J_n\left(\beta x^\gamma\right) + c_2 J_{-n}\left(\beta x^\gamma\right)\right]$$

$$\Rightarrow \quad y = x^{1/2}\left[c_1\, J_{1/3}\left(x^{3/2}\right) + c_2\, J_{-1/3}\left(x^{3/2}\right)\right].$$

Example 27: Solve the differential equation $y'' + \frac{y'}{x} + \left(\frac{8}{x} - \frac{1}{x^2}\right)y = 0$ in terms of Bessel's functions.

Solution: The given equation is

$$x^2\, y'' + xy' + (8x - 1)y = 0 \qquad ...(1)$$

Comparing (1) with the general form

$$x^2\, y'' + (1 - 2\alpha)xy' + \left[\beta^2\, \gamma^2\, x^{2\gamma} + \left(\alpha^2 - n^2\, \gamma^2\right)\right]y = 0 \qquad ...(2)$$

where α, β, γ and n are constants, we get

$$1 - 2\alpha = 1,\ \beta^2\, \gamma^2 = 8,\ 2\gamma = 1 \text{ and } \alpha^2 - n^2\, \gamma^2 = -1$$

Solving these, we get

$$\alpha = 0,\ \gamma = \frac{1}{2}, \frac{\beta^2}{4} = 8 \Rightarrow \beta^2 = 32 \Rightarrow \beta = 4\sqrt{2},$$

and $\quad (0)^2 - n^2\left(\frac{1}{4}\right) = -1 \Rightarrow n^2 = 4 \Rightarrow n = 2.$

Here n is an integer so the solution be

$$y = x^{\alpha}\left[\, c_1 J_n\left(\beta x^{\gamma}\right) + c_2 Y_n\left(\beta x^{\gamma}\right)\right]$$

$$= x^{0}\left[\, c_1 J_2\left(4\sqrt{2}x^{1/2}\right) + c_2 Y_2\left(4\sqrt{2}\, x^{1/2}\right)\right]$$

$$\Rightarrow \quad y = c_1 J_2\left(4\sqrt{2x}\right) + c_2 Y_2\left(4\sqrt{2x}\right).$$

Example 28: Solve the differential equation $xy'' + y' + \frac{1}{4}y = 0$ in terms of Bessel's functions.

Solution: Multiplying by x, given equation becomes

$$x^2 y'' + xy' + \frac{1}{4}xy = 0 \qquad ...(1)$$

Comparing (1) with the general form (6) of section 4.6.7, we get

$$1 - 2\alpha = 1,\ \beta^2\gamma^2 = \frac{1}{4},\ 2\gamma = 1 \text{ and } \alpha^2 - n^2\gamma^2 = 0$$

On solving these, we get

$$\alpha = 0,\ \gamma = \frac{1}{2},\ \beta^2\left(\frac{1}{4}\right) = \frac{1}{4} \Rightarrow \beta^2 = 1 \Rightarrow \beta = 1,$$

and $\quad 0 - n^2\left(\frac{1}{4}\right) = 0 \Rightarrow n^2 = 0 \Rightarrow n = 0.$

Here n is an integer, Therefore the solution of (1) is

$$y = x^{\alpha}\left[\, c_1 J_n\left(\beta x^{\gamma}\right) + c_2 Y_n\left(\beta x^{\gamma}\right)\right] = x^{0}\left[\, c_1 J_0\left(x^{1/2}\right) + c_2 Y_0\left(x^{1/2}\right)\right]$$

$$= c_1 J_0\left(\sqrt{x}\right) + c_2 Y_0\left(\sqrt{x}\right).$$

EXERCISE 4.4

1. Prove that

(i) $J_{-3/2}(x) = -\sqrt{\frac{2}{\pi x}}\left[\frac{\cos x}{x} + \sin x\right]$

(ii) $J_{5/2}(x) = \sqrt{\frac{2}{\pi x}}\left[\left(\frac{3 - x^2}{x^2}\right)\sin x - \frac{3\cos x}{x}\right]$

(iii) $J_{-5/2}(x) = \sqrt{\frac{2}{\pi x}}\left[\left(\frac{3 - x^2}{x^2}\right)\cos x + \frac{3\sin x}{x}\right].$

2. Prove that

(i) $\int_0^x x^n J_{n-1}(x)dx = x^n J_n(x)$

(ii) $\int_0^x x^{n+1} J_n(x)dx = x^{n+1} J_{n+1}(x)$.

3. Prove that

(i) $J_n(x) = \frac{1}{2\pi}\int_0^{2\pi} \cos(x\sin\theta - n\theta)d\theta$

(ii) $\cos x = J_0 - 2J_2 + 2J_4 - ...$

(iii) $\sin x = 2J_1 - 2J_3 + 2J_5 - ....$

4. Prove that $J_0(x) = \frac{1}{\pi}\int_0^{\pi} \cos(x\sin\theta)d\theta = \frac{2}{\pi}\int_0^{\pi/2} \cos(x\cos\theta)d\theta$.

5. Prove that $\int J_0(x)J_1(x)dx = -\frac{1}{2}J_0^2(x)$.

6. Prove that

(i) $\frac{d}{dx}\left[x^n J_n(\alpha x)\right] = \alpha x^n J_{n-1}(\alpha x)$

(ii) $\frac{d}{dx}\left[J_n^2(x)\right] = \frac{x}{2n}\left[J_{n-1}^2(x) - J_{n-1}^2(x)\right]$.

7. Express $J_4(x)$ in terms of $J_0(x)$ and $J_1(x)$.

8. Solve the differential equation

$$\frac{d^2y}{dx^2} + \frac{1}{x}\cdot\frac{dy}{dx} + \left(8 - \frac{1}{x^2}\right)y = 0.$$

9. Solve the following differential equations in terms of Bessel's functions:

(i) $y'' - \frac{2}{x}y' + 9\left(1 - \frac{2}{x^2}\right)y = 0$ (ii) $xy'' - 3y' + xy = 0$

(iii) $y'' + \left(9 - \frac{20}{x^2}\right)y = 0$ (iv) $x^2y'' - xy' + 4x^2y = 0$.

10. Show that $x^n J_n$ is a solution of the equation $y'' + (1-2n)y' + xy = 0$.

ANSWERS 4.4

7. $J_4(x) = \left[\left(\frac{48}{x^3} - \frac{8}{x}\right)J_1(x) - \left(\frac{24}{x^2} - 1\right)J_0(x)\right]$.

8. $y = AJ_1(2\sqrt{2}\,x) + BY_1(2\sqrt{2}\,x)$

9. (i) $y = x^{3/2}\left[C_1 J_{9/2}(3x) + C_2 J_{-9/2}(3x)\right]$

(ii) $y = x^2 \left[C_1 J_2(x) + C_2 Y_2(x) \right]$

(iii) $y = A\sqrt{x}\, J_3\left(2x^{3/2}\right) + B\sqrt{x}\, Y_3\left(2x^{3/2}\right)$

(iv) $y = x\left[C_1 J_1(2x) + C_2 Y_1(2x) \right]$.

4.7 MODIFIED BESSEL'S EQUATION

The differential equation

$$x^2 \frac{d^2y}{dx^2} + x\frac{dy}{dx} - \left(x^2 + n^2\right)y = 0 \quad \text{...(1)}$$

is known as modified Bessel's equation of order n.

This equation can be re-written as

$$x^2 \frac{d^2y}{dx^2} + x\frac{dy}{dx} + \left(i^2 x^2 - n^2\right)y = 0$$

If n is not an integer, the solution is given by

$$y = c_1 J_n(ix) + c_2 J_{-n}(ix)$$

But $\quad J_n(ix) = \sum_{k=0}^{\infty} \frac{(-1)^k}{k!\,\Gamma(n+k+1)} \left(\frac{ix}{2}\right)^{n+2k}$

$$= i^n \sum_{r=0}^{\infty} \frac{1}{k!\,\Gamma(n+k+1)} \left(\frac{x}{2}\right)^{n+2k} = i^n I_n(x),$$

$$\left[\because (-1)^k (i)^{2k} = i^{2k}.\, i^{2k} = i^{4k} = 1 \right]$$

The series $\sum_{k=0}^{\infty} \frac{1}{k!\,\Gamma(n+k+1)} \left(\frac{x}{2}\right)^{n+2k}$ is a real function with all terms positive and is denoted by $I_n(x)$. $I_n(x)$ is called the modified Bessel function of the first kind of order n.

This gives $I_n(x) = i^{-n} J_n(ix)$

Clearly $I_n(x)$ is also a solution of (1), because i^{-n} is a constant.

If n is not an integer, a second independent solution of (1) is $I_{-n}(x)$, where

$$I_{-n}(x) = \sum_{r=0}^{\infty} \frac{1}{k!\,\Gamma(-n+k+1)} \left(\frac{x}{2}\right)^{-n+2k}$$

In this case, the complete solution of (1) is given by

$$y = c_1 I_n(x) + c_2 I_{-n}(x).$$

If n is a non-zero integer, a second independent solution of (1) is given by

$$K_n(x) = \frac{\pi/2}{\sin n\pi}\left[I_{-n}(x) - J_n(x) \right]$$

and is called modified function of second kind of order n.

In this case, the complete solution of (1) is given by

$$y = c_1 I_n(x) + c_2 K_n(x).$$

4.8 BER AND BEI FUNCTIONS

Let the differential equation is $x\frac{d^2y}{dx^2} + \frac{dy}{dx} - ixy = 0$ ...(1)

Comparing it with general form (6) of section 4.6.7, we get

$$\alpha = 0, n = 0, \gamma = 1 \text{ and } \beta^2 = -i \text{ or } \beta^2 = i^3 \Rightarrow \beta = i^{3/2}$$

Hence a solution of (1) is given by $J_0\left(i^{3/2} x\right)$. Replacing x by $i^{3/2}x$ in the series for $J_0(x)$, we have

$$J_0\left(i^{3/2} x\right) = 1 - \frac{i^3 x^2}{2^2} + \frac{i^6 x^4}{(2!)^2 . 2^4} - \frac{i^9 x^6}{(3!)^2 . 2^6} + \frac{i^{12} x^8}{(4!)^2 . 2^8} - \ldots\ldots$$

$$= \left[1 - \frac{x^4}{2^2 . 4^2} + \frac{x^8}{2^2 . 4^2 . 6^2 . 8^2} - \ldots\ldots\right]$$

$$+ i\left[\frac{x^2}{2^2} - \frac{x^6}{2^2 . 4^2 . 6^2} + \frac{x^{10}}{2^2 . 4^2 . 6^2 . 8^2 . 10^2} - \ldots\ldots\right]$$

The function $J_0\left(i^{2/3} x\right)$ is a complex function for real values of x. The real and the imaginary parts are denoted by ber (x) (Bessel-real) and bei (x) (Bessel-imaginary) functions respectively.

Thus $$ber(x) = 1 + \sum_{k=1}^{\infty} \frac{(-1)^k x^{4k}}{2^2 . 4^2 . 6^2 \ldots\ldots (4k)^2}$$

and $$bei(x) = 1 - \sum_{k=1}^{\infty} \frac{(-1)^k x^{4k-2}}{2^2 . 4^2 . 6^2 \ldots\ldots (4k-2)^2}$$

Hence a solution of (1) is $y = J_0\left(i^{3/2} x\right) = ber(x) + i\, bei(x)$.

4.9 FOURIER-BESSEL EXPANSION OF $f(x)$

Here we expand a function of x in a series of Bessel's function which is known as Fourier-Bessel series in the range 0 to p. It is possible due to the orthogonal property of Bessel's functions in the range 0 to p. For this purpose consider

$$f(x)=\sum c_i J_n(\lambda_i x)$$

$$= c_1 J_n(\lambda_1 x)+c_2 J_n(\lambda_2 x)+...+c_n J_n(\lambda_n x)+... \qquad ...(1)$$

where $\lambda_1, \lambda_2, ...$ are the roots of the equation $J_n(\lambda p)=0$.

To determine the coefficients c_i, we multiply (1) by $xJ_n(\lambda_i x)$ and then integrate 0 to p with respect to x. Thus, we get

$$\int_0^p xf(x)J_n(\lambda_i x)dx = c_i \int_0^p x J_n^2(\lambda_i x)dx = c_i \frac{p^2}{2} J_{n+1}^2(\lambda_i p)$$

$$\Rightarrow \qquad c_i = \frac{2}{p^2 J_{n+1}^2(\lambda_i p)} \int_0^p xf(x)J_n(\lambda_i x)dx \qquad ...(2)$$

Putting $i = 1, 2, 3, ...$ we can find $c_1, c_2, c_3, ...$ Thus the expansion of $f(x)$ is given by (1) and (2). This expansion is the Fourier-Bessel series of $f(x)$.

Example 1: If $\alpha_1, \alpha_2, ..., \alpha_n$ are the positive roots of $J_0(x)=0$, prove that $\frac{1}{2}=\sum_{n=1}^{\infty} \frac{J_0(\alpha_n x)}{\alpha_n J_1(\alpha_n)}$.

Solution: From Fourier-Bessel expansion of $f(x)$ we know that if

$$f(x)=\sum_{i=1}^{\infty} c_i J_n(\alpha_i x) \qquad ...(1)$$

then $c_i = \frac{2}{a^2 J_{n+1}^2(\alpha_i \alpha)} \int_0^a xf(x)J_n(\alpha_i x)dx$

Taking $f(x)=1, \alpha=1$ and $n=0$, we get

$$c_i = \frac{2}{J_1^2(\alpha_i)} \int_0^1 xJ_0(\alpha_1 x)dx = \frac{2}{J_1^2(\alpha_i)}\left[\frac{xJ_1(\alpha_i x)}{\alpha_i}\right]_0^1 = \frac{2}{\alpha_i J_1(\alpha_i)}$$

Therefore from (1), we obtain

$$1=\sum_{i=1}^{\infty} \frac{2}{\alpha_i J_1(\alpha_i)} J_0(\alpha_i x) \Rightarrow \frac{1}{2}=\sum_{n=1}^{\infty} \frac{J_0(\alpha_n x)}{\alpha_n J_1(\alpha_n)}.$$

Example 2: Show that

(a) $\frac{d}{dx}[x\, ber'(x)] = -x\, bei(x)$. (b) $\frac{d}{dx}[x\, bei'(x)] = x\, ber(x)$.

Solution: From definition of ber and bei functions, we know that

$$ber(x) = 1 - \frac{x^4}{2^2 . 4^2} + \frac{x^8}{2^2 . 4^2 . 6^2 . 8^2} - .... \quad ...(1)$$

$$bei(x) = \frac{x^2}{2^2} - \frac{x^6}{2^2 . 4^2 . 6^2} + \frac{x^{10}}{2^2 . 4^2 . 6^2 . 8^2 . 10^2} - .... \quad ...(2).$$

From (1), we have $ber'(x) = -\frac{x^3}{2^2 . 4} + \frac{x^7}{2^2 . 4^2 . 6^2 . 8} - ....$

$$\Rightarrow \quad x\, ber'(x) = -\frac{x^4}{2^2 . 4} + \frac{x^8}{2^2 . 4^2 . 6^2 . 8} - ....$$

$$\Rightarrow \quad \frac{d}{dx}\left[x\, ber'(x) \right] = -\frac{x^3}{2^2} + \frac{x^7}{2^2 . 4^2 . 6^2} - ....$$

$$= -x\left(\frac{x^2}{2^2} - \frac{x^6}{2^2 . 4^2 . 6^2} + ... \right) = -x\, bei(x).$$

Differentiating (2), we have

$$bei'(x) = \frac{x}{2} - \frac{x^5}{2^2 . 4^2 . 6^2} + \frac{x^9}{2^2 . 4^2 . 6^2 . 8^2 . 10} - ....$$

$$\Rightarrow \quad x\, bei'(x) = \frac{x^2}{2} - \frac{x^6}{2^2 . 4^2 . 6} + \frac{x^{10}}{2^2 . 4^2 . 6^2 . 8^2 . 10} - ....$$

$$\Rightarrow \quad \frac{d}{dx}\left[x\, bei'(x) \right] = x - \frac{x^5}{2^2 . 4^2} + \frac{x^9}{2^2 . 4^2 . 6^2 . 8^2} - ....$$

$$= x\left[1 - \frac{x^4}{2^2 . 4^2} + \frac{x^8}{2^2 . 4^2 . 6^2 . 8^2} - .... \right] = x\, ber(x).$$

Example 3: Show that the Fourier-Bessel series in $J_2(\lambda_n x)$ *for* $f(x) = x^2\,(0 < x < a)$, where λ_n are positive roots of $J_2(x) = 0$, is

$$x^2 = 2a^2 \sum_{n=1}^{\infty} \frac{J_2(\lambda_n x)}{a\lambda_n J_3(\lambda_n a)}.$$

Solution: Let the Fourier-Bessel series representing $f(x) = x^2$ be given by $x^2 = \sum_{n=1}^{\infty} c_n J_2(\lambda_n x)$.

Multiplying both sides by $x J_2(\lambda_n x)$ and integrating w. r. t. 'x', between the limits 0 to a, we get

$$\int_0^a x^3 J_2(\lambda_n x)\,dx = c_n \int_0^a x J_2^2(\lambda_n x)\,dx$$

$$\Rightarrow \left[\frac{x^3 J_3(\lambda_n x)}{\lambda_n}\right]_0^a = c_n \cdot \frac{a^2}{2} J_3^2(\lambda_n a) \Rightarrow \frac{a^3 J_3(\lambda_n a)}{\lambda_n} = c_n \cdot \frac{a^2}{2} J_3^2(\lambda_n a)$$

$$\Rightarrow \quad c_n = \frac{2a^2}{a\lambda_n} \cdot \frac{1}{J_3(\lambda_n a)}$$

Hence, we have $x^2 = 2a^2 \sum_{n=1}^{\infty} \frac{J_2(\lambda_n x)}{a \lambda_n J_3(\lambda_n a)}$.

EXERCISE 4.5

1. Show that under the transformation $y = \frac{u}{\sqrt{x}}$, Bessel's equation reduces to $u'' + \left[1 + \frac{1-4n^2}{4x^2}\right] u = 0$. Hence find the solution of the equation.
2. Expand $f(x) = 1$ over the interval $0 < x < 3$ in terms of the functions $J_0(\lambda_n x)$, where λ_n are determined by $J_0(3\lambda) = 0$.
3. Expand $f(x) = 4x - x^3$ over the interval (0, 2) in terms of Bessel functions of first kind of order one which satisfies the condition $[J_1(\lambda x)]_{x-2} = 0$.
4. By the use of the substitution $y = \frac{u}{\sqrt{x}}$, show that the solution of the equation $x^2 y'' + xy' + \left(x^2 - \frac{1}{4}\right) y = 0$ can be written in the form $y = c_1 \frac{\sin x}{\sqrt{x}} + c_2 \frac{\cos x}{\sqrt{x}}$.
5. If a is the root of the equation $J_0(x) = 0$, show that

   (i) $\int_0^1 J_1(ax)dx = \frac{1}{a}$ (ii) $\int_0^a J_1(x)dx = 1$.
6. If $\alpha_1, \alpha_2, \ldots\ldots, \alpha_n, \ldots\ldots\ldots$ are the positive roots of $J_1(x) = 0$, $(0 < x < 1)$ prove that

   (i) $x^2 = \frac{1}{2} + 4 \sum_{n=1}^{\infty} \frac{J_0(\alpha_n x)}{\alpha_n^2 J_0(\alpha_n)}$ (ii) $(1-x^2)^2 = \frac{1}{3} - 64 \sum_{n=1}^{\infty} \frac{J_0(\alpha_n x)}{\alpha_n^2 J_0(\alpha_n)}$.

ANSWERS 4.5

1. $y=\sqrt{x}\left[c_1 J_n(x)+c_2 J_{-n}(x)\right]$

2. $1=\frac{2}{3}\sum_{n=1}^{\infty}\frac{J_0(\lambda_n x)}{\lambda_n J_1(3\lambda_n)}$

3. $4x-x^3=8\sum_{n=1}^{\infty}\frac{J_3(2\lambda_n)}{\lambda_n^2 J_2^2(2\lambda_n)}\cdot J_1(\lambda_n x).$

OBJECTIVE QUESTIONS

1. A point $x=x_0$ is called an ordinary point of the equation $y''+P(x)y'+Q(x)y=0$ If
   (a) Both the functions $P(x)$ and $Q(x)$ are analytic at $x=x_0$.
   (b) Both the functions $P(x)$ and $Q(x)$ are not analytic at $x=x_0$.
   (c) The function $P(x)$ is analytic and $Q(x)$ is not analytic at $x=x_0$.
   (d) The function $P(x)$ is not analytic and $Q(x)$ analytic at $x=x_0$.

2. A point $x=x_0$ which is not ordinary point of the differential equation $y''+P(x)y'+Q(x)y=0$ is called
   (a) Singular point
   (b) Regular singular point
   (c) Irregular singular point
   (d) None of these.

3. A singular point $x=a$ of the differential equation $y''+P(x)y'+Q(x)y=0$ is called regular if
   (a) $(x-a)P(x)$ is analytic but $(x-a)^2 Q(x)$ is not analytic at $x=a$.
   (b) $(x-a)P(x)$ and $(x-a)^2 Q(x)$ are not analytic at $x=a$.
   (c) $(x-a)P(x)$ is analytic.
   (d) $(x-a)P(x) and (x-a)^2 Q(x)$ are analytic at $x=0$.

4. If $(x-a)P(x)$and$(x-a)^2 Q(x)$ are not analytic at $x=a$, then the point $x=a$ of the differential equation $y''+P(x)y'+Q(x)y=0$ is called
   (a) Regular singular point
   (b) Irregular singular point
   (c) Singular point
   (d) None of these

5. If all the singularities of a differential equation are regular, then this differential equation is called
   (a) Fuchsian
   (b) Bessel's equation
   (c) Frobenius
   (d) None of these.

6. In $P_n(x)$ is a solution of Legendre's equation then value of $P_0(x)$ will be:

(a) 1 (b) 0
(c) -1 (d) x.

7. If $P_n(x)$ is the Legendre's polynomial, then value of $P_1(x)$ is:

(a) 1 (b) 0
(c) -1 (d) x.

8. Legendre's equation is:

(a) $\frac{dy}{dx} = P + Qy + Ry^2$

(b) $\frac{d^2y}{dx^2} - 2x\frac{dy}{dx} + 2xy = 0$

(c) $(1-x^2)\frac{d^2y}{dx^2} - 2x\frac{dy}{dx} + n(n+1)y = 0$

(d) $(1-x^2)\frac{d^2y}{dx^2} - 2x\frac{dy}{dx} + n^2\,xy = 0$.

9. In the expansion of $(1-2xh+h^2)^{-1/2}$ in ascending powers of h, $P_n(x)$ is the coefficient of:

(a) h^n (b) x^n
(c) constant terms (d) h^{2n}

10. Value of $P_n(-1)$ is:

(a) 1
(b) $(-1)^n$
(c) -1
(d) 0

11. Value of $P_2(x)$ is:

(a) x (b) 1
(c) $\frac{1}{2}(3x^2-1)$ (d) $\left(\frac{5x^3-3x}{2}\right)$

12. Select the incorrect formula:

(a) $(2n+1)P_n(x) = P'_{n+1}(x) + P'_{n-1}(x)$
(b) $(n+1)P_n(x) = P'_{n+1}(x) - xP'(x)$
(c) $nP_n(x) = xP'_n(x) - P'_{n-1}(x)$
(d) $(1-x^2)P'_n(x) = n(P_{n-1}(x) - xP_n(x))$.

13. If $P_n(x)$ is a Legendre polynomial of degree n and α is such that $P_n(\alpha)=0$. Then $P_{n-1}(\alpha)$ and $P_{n+1}(\alpha)$ are:

(a) both of same sign (b) both zero

(c) both of opposite (d) none of these

14. Formula $P_n(x)=\frac{1}{n!2^n}\frac{d^n}{dx^n}(x^2-1)^n$ is called:

(a) Rodrigues' formula

(b) Beltrami's result

(c) Christophell's formula

(d) Recurrence formula

15. All the roots of $P_n(x)$ are real and lie between

(a) 0 and +1 (b) -1 and 0

(c) -1 and $+1$ (d) $-\frac{1}{2}$ and 1.

16. The value of $\int_{-1}^{1}\{P_n(x)\}^2 dx$ is:

(a) 1 (b) $\frac{2}{2n+1}$

(c) $2n^2$ (d) 0.

17. $(1-x^2)P_n - nP_{n-1} = \ldots.$

(a) P_n (b) $2x\,P_n$

(c) xP_n (d) xP_n'

18. The differential equation $(1-x^2)y''-2xy'+n(n+1)y=0$ is called

(a) Hermite's equation

(b) Bessel's equation

(c) Legendre's equation

(d) Laguerre's equation

19. All roots of $P_n(x)=0$ are:

(a) Real (b) some real

(c) 0 (d) complex

20. The value of $P_n(1)$ is:

(a) 0 (b) 1

(c) n (d) $\frac{1}{n}$

21. Value of $J_n(x)$ is:

(a) $\sum_{r=0}^{\infty}(-1)^r\left(\frac{x}{2}\right)^{n+2r}\frac{1}{r!\Gamma(n+r+1)}$

(b) $\sum_{r=0}^{\infty}(-1)^r\left(\frac{x}{2}\right)^{n+2r}\frac{1}{r!\Gamma(n+r)}$

(c) $\sum_{r=0}^{\infty}(-1)^r\left(\frac{x}{2}\right)^{n+r}\frac{1}{r!\Gamma(n+r+1)}$

(d) None of these

22. Value of $J_{1/2}(x)$ is:

(a) $\sqrt{\frac{2}{\pi x}}$ (b) $\sqrt{\frac{2}{\pi x}}\cos x$.

(c) $\sqrt{\frac{2}{\pi x}}\sin x$ (d) $\sqrt{\frac{\pi x}{2}}\cos x$.

23. Select whichever is correct:

(a) $\frac{d}{dx}\left[x^{-n}J_n(x)\right]=-x^{-n}J_{n+1}(x)$

(b) $2nJ_n'(x)=nJ_n(x)+x\,J_{n+1}(x)$

(c) $xJ_n'(x)=nJ_n(x)+xJ_{n+1}(x)$.

(d) $xJ_n'(x)=nJ_n(x)+xJ_{n-1}(x)$.

24. Choose the incorrect formula:

(a) $2\,J_n'(x)=J_{n-1}(x)-J_{n+1}(x)$

(b) $2nJ_n'(x)=x\left[J_{n+1}(x)+J_{n-1}(x)\right]$

(c) $\frac{d}{dx}\left[x^{-n}J_n(x)\right]=x^{-n}J_{n+1}(x)$

(d) $xJ_n'(x)=n\,J_n(x)+xJ_{n-1}(x)$.

25. When n is any integer, value of $J_n(-x)$ is:

(a) $J_n(x)$ (b) $-J_n(x)$

(c) $(-1)^n J_n(x)$ (d) None of these

26. Value of J_0' is:

(a) $-J_0$ (b) $-J_1$

(c) J_1 (d) J_0

27. $J_{n-1}(x) - J_{n+1}(x) = \ldots$

(a) $2\,J_n'(x)$ (b) $2\,J_n(x)$
(c) $n.\,J_n(x)$ (d) $n\,J_n'(x)$

28. If J_0 and J_1 are Bessel functions, then $J_1'(x)$ is given by

(a) $J_0(x) - \frac{1}{x}J_1(x)$ (b) $-J_0$
(c) $J_0(x) + \frac{1}{x}J_1(x)$ (d) $J_0(x) - \frac{1}{x^2}J_1(x)$.

29. The integral $\int_0^x x\,J_0(x)\,dx$ is equal to

(a) $xJ_1(x) - J_0(x)$ (b) $x\,J_1(x)$
(c) $J_1(x)$ (d) $x^2\,J_1(x)$

30. The value of $\int_{-1}^{1} P_n(x)\,dx$, when $n \neq 0$, is

(a) 0 (b) 2
(c) 1 (d) –1

31. The example of a differential equations whose all singularities are not regular are
(a) Legendre's equation (b) Hypergeometric equation
(c) Bessel's equations (d) All above.

ANSWERS

1. (a)	2. (a)	3. (d)	4. (b)
5. (a)	6. (a)	7. (d)	8. (c)
9. (a)	10. (b)	11. (c)	12. (a)
13. (c)	14. (a)	15. (c)	16. (b)
17. (c)	18. (c)	19. (a)	20. (b)
21. (a)	22. (b)	23. (a)	24. (d)
25. (c)	26. (b)	27. (a)	28. (a)
29. (b)	30. (a)	31. (c)	

5 Hypergeometric Functions

5.1 INTRODUCTION

The study of one-variable hypergeometric functions is more than 200 years old. Hypergeometric series were studied by Euler, but the first full systematic treatment was given by Gauss (1813). In the nineteenth century, Ernst Kummer and Bernhard Riemann studied hypergeometric function by means of the differential equation it satisfies. Integral representations of hypergeometric functions were studied by Barnes and Mellin, and their special properties by Schwarz and Goursat. The famous Gauss hypergeometric equation is ubiquitous in mathematical physics as many well-known partial differential equations may be reduced to Gauss' equation via separation of variables.

There are three possible ways in which one can characterize hypergeometric functions: as functions represented by series whose coefficients satisfy certain recursive properties; as solutions to a system of differential equations which is, in an appropriate sense, holonomic and has mild singularities; as functions defined by integrals such as the Mellin-Barnes integral.

The function ${}_2F_1(\alpha, \beta; \gamma; z)$ arises frequently in physical problems is known as hypergeometric function or, more explicitly, Gauss's hypergeometric function. The term "hypergeometric function" sometimes refers to the generalized hypergeometric function. In mathematics, the Gaussian or simply hypergeometric function ${}_2F_1(\alpha, \beta; \gamma; z)$ is a special function represented by the hypergeometric series, that includes many other special functions as specific or limiting cases. It is a solution of a second-order linear ordinary differential equation (ODE). Every second-order linear ODE with three regular singular points can be transformed into this equation.

Now we first define factorial function or Pochhammer symbol in order to describe generalized hypergeometric function, confluent hypergeometric function and their integral representations.

5.2 FACTORIAL FUNCTION

For the positive integer n, we use the notation

$$(\alpha)_n = \prod_{k=1}^{n} (\alpha + k - 1) = \alpha(\alpha+1)\ldots(\alpha+n-1), \text{ for } n \geq 1.$$

$$(\alpha)_0 = 1, \text{ for } \alpha \neq 0.$$

The function $(\alpha)_n$ is called the factorial function or Pochhammer symbol. This function is immediate generalization of the elementary factorial as $(1)_n = 1(1+1)(1+2)\cdots\cdots(1+n-1) = 1\,2\,3\,4\cdots\cdots n = n!$

Deductions: By definition, we have

(I) $(\alpha)_n = \alpha(\alpha+1)\ldots(\alpha+n-1)$

$$= \frac{1\,2\,3\ldots(\alpha-1)\alpha(\alpha+1)\ldots(\alpha+n-1)}{1\,2\,3\ldots(\alpha-1)} = \frac{\Gamma(\alpha+n)}{\Gamma\alpha}$$

where Γ represents gamma function and has the property $\Gamma n = (n-1)\Gamma(n-1)$.

Thus, we have $(\alpha)_n = \dfrac{\Gamma(\alpha+n)}{\Gamma\alpha}$.

(II) $(\alpha)_{n+1} = \alpha(\alpha+1)\ldots[\alpha+(n+1)-1]$

$$= \alpha[(\alpha+1)(\alpha+2)\ldots(\alpha+1+n-1)] = \alpha(\alpha+1)_n$$

Thus, we have

$(\alpha)_{n+1} = \alpha(\alpha+1)_n$.

(III) $(\alpha+n)(\alpha)_n = \alpha(\alpha+1)\ldots(\alpha+n-1)(\alpha+n)$

$$= \alpha(\alpha+1)\ldots(\alpha+n-1)(\alpha+n+1-1) = (\alpha)_{n+1}$$

Thus, we have $(\alpha+n)(\alpha)_n = (\alpha)_{n+1}$.

(IV) $(\alpha)_{2n} = \alpha(\alpha+1)\ldots(\alpha+2n-1)$

$$= 2^{2n}\,\frac{\alpha}{2}\,\frac{(\alpha+1)}{2}\cdots\cdots\frac{(\alpha+2n-2)}{2}\,\frac{(\alpha+2n-1)}{2}$$

$$= 2^{2n}\left\{\frac{\alpha}{2}\left(\frac{\alpha}{2}+1\right)\cdots\cdots\left(\frac{\alpha}{2}+n-1\right)\right\}\left\{\frac{\alpha+1}{2}\left(\frac{\alpha+1}{2}+1\right)\right.$$

$$\left.\cdots\cdots\left(\frac{\alpha+1}{2}+n-1\right)\right\}$$

$$= 2^{2n}\left(\frac{\alpha}{2}\right)_n\left(\frac{\alpha+1}{2}\right)_n$$

5.3 GENERAL HYPERGEOMETRIC FUNCTION

The general hypergeometric function is defined by

$${}_mF_n(\alpha_1, \alpha_2, ..., \alpha_m; \beta_1, \beta_2, ..., \beta_n; x) = \sum_{r=1}^{\infty} \frac{(\alpha_1)_r(\alpha_2)_r..(\alpha_m)_r}{(\beta_1)_r(\beta_2)_r...(\beta_n)_r}\frac{x^r}{r!}$$

The general hypergeometric function can also be denoted by

$${}_mF_n\begin{bmatrix}\alpha_1 & \alpha_2 & \cdots & \alpha_m; & \\ & & & & x \\ \beta_1 & \beta_2 & \cdots & \beta_n; & \end{bmatrix}$$

Remarks:

1. If $m = n = 1$, the function ${}_1F_1(\alpha; \beta; x)$ is called the confluent hypergeometric function or Kummer's function. Thus the confluent hypergeometric function ${}_1F_1(\alpha; \beta; x)$ is defined by

$${}_1F_1(\alpha; \beta; x) = \sum_{r=0}^{\infty} \frac{(\alpha)_r}{(\beta)_r}\frac{x^r}{r!}.$$

The confluent hypergeometric function ${}_1F_1(\alpha; \beta; x)$ can also be represented by $M(\alpha; \beta; x)$.

2. If $m = 2, n = 1$, the function ${}_2F_1(\alpha_1, \alpha_2; \beta; x)$ is called the hypergeometric function and is defined by

$${}_2F_1(\alpha_1, \alpha_2; \beta; x) = \sum_{r=0}^{\infty} \frac{(\alpha_1)_r(\alpha_2)_r}{(\beta)_r}\frac{x^r}{r!}.$$

5.4 GAUSS'S HYPERGEOMETRIC EQUATION

The differential equation

$$x(1-x)\frac{d^2y}{dx^2} + [\alpha - (\alpha + \beta + 1)x]\frac{dy}{dx} - \alpha\beta y = 0$$

is known as Gauss's hypergeometric equation or simple hypergeometric equation or Gauss's equation.

5.5 THE HYPERGEOMETRIC SERIES

The series

$$1 + \frac{\alpha\beta}{1!\gamma}x + \frac{\alpha(\alpha+1)\beta(\beta+1)}{2!\gamma(\gamma+1)}x^2 + \frac{\alpha(\alpha+1)(\alpha+2)\beta(\beta+1)(\beta+2)}{3!\gamma(\gamma+1)(\gamma+2)}x^3 + \cdots$$

is called the hypergeometric series and is denoted by ${}_2F_1(\alpha, \beta; \gamma; x)$. This series is also known as hypergeometric function. Hypergeometric series is frequently used in connection with the theory of spherical harmonics. We now study some of its properties after studying particular cases of hypergeometric functions.

5.6 PARTICULAR CASES OF HYPERGEOMETRIC SERIES

Here we are giving some well known series denoted by hypergeometric function.

1. $$(1+x)^n = 1 + nx + \frac{n(n-1)}{2!}x^2 + \frac{n(n-1)(n-2)}{3!}x^3 + \ldots$$

$$= 1 + \frac{(-n).1}{1!1}(-x) + \frac{(-n)(-n+1)(1)(1+1)}{2!.1(1+1)}(-x)^2$$

$$+ \frac{(-n)(-n+1)(-n+2)(1)(1+1)(1+2)}{3!.1(1+1)(1+2)}(-x)^3 + \cdots$$

$$= {}_2F_1(-n, 1; 1; -x).$$

2. $$\log(1+x) = x - x^2/2 + x^3/3 - x^4/4 \ldots$$

$$= x\left[1 - x/2 + x^2/3 - x^3/4 \ldots\right]$$

$$= x\left[1 + \frac{1\,1}{1!.\,2}(-x) + \frac{1(1+1)1(1+1)}{2!2(2+1)}(-x)^2\right.$$

$$\left. + \frac{1.(1+1)(1+2)1(1+1)(1+2)}{3!2(2+1)(2+2)}(-x)^3 + \ldots\right]$$

$$= x\ {}_2F_1(1, 1; 2; -x).$$

3. $$\tan^{-1} x = x - x^3/3 + x^5/5 - x^7/7 + \ldots = x\left[1 - x^2/3 + x^4/5 - x^6/7 + \ldots\right]$$

$$= x\left[1 + \frac{\frac{1}{2}1}{1!\frac{3}{2}}(-x^2) + \frac{\frac{1}{2}(\frac{1}{2}+1)1(1+1)}{2!\frac{3}{2}(\frac{3}{2}+1)}(-x^2)^2 + \ldots\right]$$

$$= x\ {}_2F_1\left(\tfrac{1}{2}, 1; \tfrac{3}{2}; -x^2\right).$$

4. $$\sin^{-1} x = x + 1^2\frac{x^3}{3!} + 1^2\,3^2\frac{x^5}{5!} + 1^2\,3^2\,5^2\frac{x^7}{7!} + \ldots$$

$$= x\left[1 + 1^2\frac{x^2}{3!} + 1^2\,3^3\frac{x^4}{4!} + 1^2\,3^2\,5^2\frac{x^6}{7!} + \ldots\right]$$

$$= x\left[1+\frac{\frac{1}{2}\frac{1}{2}}{1!\frac{3}{2}}x^2+\frac{\frac{1}{2}\frac{3}{2}\frac{1}{2}\frac{3}{2}}{2!\frac{3}{2}\left(\frac{3}{2}+1\right)}\left(x^2\right)^2+\ldots\right]$$

$$= x\,{}_2F_1\left(\tfrac{1}{2};\tfrac{1}{2};\tfrac{3}{2};x^2\right).$$

5.7 SOLUTION OF THE HYPERGEOMETRIC EQUATION

A differential equation of the type

$$x(1-x)\frac{d^2y}{dx^2}+\left[\gamma-(1+\alpha+\beta)x\right]\frac{dy}{dx}-\alpha\beta y=0 \qquad \ldots(1)$$

where α, β, γ are constants (it is assumed that γ is not an integer), is said to be hypergeometric differential equation. Dividing equation (1) by $x(1-x)$, we get

$$\frac{d^2y}{dx^2}+X_1\frac{dy}{dx}+X_2\,y=0,$$

where $X_1=\dfrac{\left[\gamma-(1+\alpha+\beta)x\right]}{x(1-x)}$, $X_2=-\dfrac{\alpha\beta}{x(1-x)}$.

Note that $X_1\to\infty$ when $x=0$ or 1 or ∞ and $X_2\to\infty$ when $x=0$ or 1 or ∞.

Hence $x=0, x=1$ and $x=\infty$ are the singular points. Since $x(1-x)X_1$ and $x(1-x)X_2$ remain finite for all finite values of x, so hypergeometric differential equation can be solved by series integration method.

We shall develop the series about $x=0, x=1$ and $x=\infty$.

(a) When $x = 0$: Let us consider the solution of (1) in ascending powers of x as

$$y=\sum_{r=0}^{\infty}a_r x^{k+r} \qquad \ldots(2)$$

so that $$\frac{dy}{dx}=\sum_{r=0}^{\infty}a_r(k+r)x^{k+r-1}$$

and $$\frac{d^2y}{dx^2}=\sum_{r=0}^{\infty}a_r(k+r)(k+r-1)x^{k+r-2} \qquad \ldots(3)$$

Substituting the values of y, $\dfrac{dy}{dx}$ and $\dfrac{d^2y}{dx^2}$ in (1), we obtain

$$\sum_{r=0}^{\infty} a_r \left[(x^2 - x)(k+r)(k+r-1)x^{k+r-2} \right.$$

$$\left. + \{(1+\alpha+\beta)x - \gamma\}(k+r)x^{k+r-1} + \alpha\beta x^{k+r} = 0 \right.$$

$$\Rightarrow \sum_{r=0}^{\infty} a_r \left[\{(k+r)^2 + (\alpha+\beta)(k+r) + \alpha\beta\} x^{k+r} \right.$$

$$\left. -(k+r)(k+r+\gamma-1)x^{k+r-1} \right] = 0 \qquad \ldots(4)$$

This equation must be an identity. So, equating to zero the coefficient of x^{k-1}, we obtain

$$a_0 k(k+\gamma-1) = 0$$

Now $a_0 \neq 0$, because it is the starting term of the series which can not be zero. Therefore we have

$$k = 0 \text{ or } \quad k = 1-\gamma \qquad \ldots(5)$$

Equating to zero the coefficient of the next higher power of x, i.e. x^k, we obtain

$$a_0 \left[k^2 + (\alpha+\beta)k + \alpha\beta \right] - (k+1)(k+\gamma)a_1 = 0.$$

$$\Rightarrow \quad a_1 = \frac{k^2 + (\alpha+\beta)k + \alpha\beta}{(k+1)(k+\gamma)} a_0 = \frac{(k+\alpha)(k+\beta)}{(k+1)(k+\gamma)} a_0.$$

Again equating to zero the coefficient of the general term x^{k+r}, we obtain

$$a_r \left[(k+r)^2 + (\alpha+\beta)(k+r) + \alpha\beta \right] - (k+r+1)(k+r+\gamma)a_{r+1} = 0.$$

$$\Rightarrow \quad a_{r+1} = \frac{(k+r+\alpha)(k+r+\beta)}{(k+r+1)(k+r+\gamma)} a_r. \qquad \ldots(6)$$

Case I: When $k = 0$, from (6), we have

$$a_{r+1} = \frac{(\alpha+r)(\beta+r)}{(r+1)(\gamma+r)} a_r.$$

Thus putting $r = 0, 1, 2, \ldots\ldots;$ we obtain

$$a_1 = \frac{(\alpha)(\beta)}{1(\gamma)} a_0 = \frac{(\alpha)_1 (\beta)_1}{1!(\gamma)_1} a_0$$

$$a_2 = \frac{(\alpha+1)(\beta+1)}{2(\gamma+1)} a_1 = \frac{(\alpha)_2 (\beta)_2}{2!(\gamma)_2} a_0$$

In general $a_n = \dfrac{(\alpha)_n (\beta)_n}{n!(\gamma)_n} a_0$, where $(\alpha)_n = \alpha(\alpha+1)...(\alpha+n-1)$

Hence from (2), we have

$$y = a_0 x^k + a_1 x^{k+1} + a_2 x^{k+2} + \cdots\cdots + a_n x^{k+n} + \cdots$$

$$= a_0 \left[1 + \frac{(\alpha)_1 . (\beta)_1}{1!(\gamma)_1} x + \frac{(\alpha)_2 . (\beta)_2}{2!(\gamma)_2} x^2 + \frac{(\alpha)_3 . (\beta)_3}{3!(\gamma)_3} x^3 + \cdots \right.$$

$$\left. + \frac{(\alpha)_n . (\beta)_n}{n!(\gamma)_n} x^n + \cdots \right], \text{ since k=0}$$

$$= a_0 \sum_{n=0}^{\infty} \frac{(\alpha)_n (\beta)_n}{n!(\gamma)_n} x^n = a_0 \, {}_2F_1(\alpha, \beta; \gamma; x). \qquad \ldots(7)$$

If $a_0 = 1$, this series is called hypergeometric series and its sum y denoted by ${}_2F_1(\alpha, \beta; \gamma; x)$ is called the hypergeometric function.

Case II: When $k = 1-\gamma$ from (6), we have

$$a_{r+1} = \frac{(1-\gamma+r+\alpha)(1-\gamma+r+\beta)}{(1-\gamma+r+1)(1-\gamma+r+\gamma)} a_r$$

$$\Rightarrow \quad a_{r+1} = \frac{(\alpha'+r)(\beta'+r)}{(\gamma'+r)(r+1)} a_r, \text{ where } \alpha' = 1-\gamma+\alpha,\ \beta' = 1-\gamma+\beta$$

and $\gamma' = 2-\gamma$.

Putting $r = 0, 1, 2, \ldots.$, we obtain

$$a_1 = \frac{\alpha' \beta'}{1.\gamma'} a_0 = \frac{(\alpha')_1 . (\beta')_1}{1!(\gamma')_1} a_0$$

$$\alpha_2 = \frac{(\alpha'+1)(\beta'+1)}{2.(\gamma'+1)} a_1 = \frac{(\alpha')_2 . (\beta')_2}{2!(\gamma')_2} a_0$$

In general, $a_n = \dfrac{(\alpha')_n . (\beta')_n}{n!(\gamma')_n} a_0$.

Hence from (2), we have

$$y = a_0 x^k + a_1 x^{k+1} + a_2 x^{k+2} + \cdots\cdots + a_r x^{k+r} + \cdots\cdots$$

$$= a_0 \, x^{1-\gamma} \left[1 + \frac{(\alpha')_1 . (\beta')_1}{1!(\gamma')_1} x + \frac{(\alpha')_2 . (\beta')_2}{2!(\gamma')_2} x^2 + \cdots\cdots \right.$$

$$\left. + \frac{(\alpha')_n . (\beta')_n}{n!(\gamma')_n} x^n + \ldots \right], \text{ since } k = 1-\gamma$$

$$= a_0\, x^{1-\gamma} \sum_{n=0}^{\infty} \frac{(\alpha')_n (\beta')_n}{n!(\gamma')_n} x^n = a_0\, x^{1-\gamma}\, {}_2F_1(\alpha'; \beta'; \gamma'; x)$$

$$= x^{1-\gamma}\, {}_2F_1(\alpha-\gamma+1, \beta-\gamma+1; 2-\gamma, x), \text{ if } a_0 = 1 \qquad \ldots(8)$$

Thus, we get two independent particular solutions (7) and (8) of the hypergeometric differential equation about $x = 0$.

Hence the general solution is given by

$$y = A\, {}_2F_1(\alpha, \beta; \gamma; x) + Bx^{1-\gamma}\, {}_2F_1(\alpha-\gamma+1, \beta-\gamma+1; 2-\gamma; x) \qquad \ldots(9)$$

Remarks:

1. In case $\alpha = 1$ and $\beta = \gamma$, the hypergeometric series becomes

   $y = 1 + x + x^2 + x^3 + \ldots$

   which is a geometric series. That is why it is known as hypergeometric series.

2. Radius of convergence of hypergeometric series.

   We know that the mth term of the series $y_m = \dfrac{(\alpha)_m \cdot (\beta)_m}{m!(\gamma)_m} x^m$

   and $(m+1)^{th}$ term of the series $y_{m+1} = \dfrac{(\alpha)_{m+1}(\beta)_{m+1}}{(m+1)!(\gamma)_{m+1}} x^{m+1}$

   $$\therefore \quad \frac{y_{m+1}}{y_m} = \frac{(\alpha)_{m+1}(\beta)_{m+1}}{(\alpha)_m (\beta)_m} \frac{m!(\gamma)_m}{(m+1)!(\gamma)_{m+1}} x = \frac{(\alpha+m)(\beta+m)}{(m+1)(\gamma+m)} x$$

   Therefore $$\lim_{m\to\infty} \left|\frac{y_{m+1}}{y_m}\right| = \lim_{m\to\infty} \left|\frac{(\alpha)_{m+1}(\beta)_{m+1}}{(\alpha)_m (\beta)_m} \frac{m!(\gamma)_m}{(m+1)!(\gamma)_{m+1}} x\right|$$

   $$= \lim_{m\to\infty} \left|\frac{(\alpha+m)(\beta+m)}{(m+1)(\gamma+m)} x\right| = \lim_{m\to\infty} \frac{\{1+(\alpha/m)\}\{1+(\beta/m)\}}{\{1+(1/m)\}\{1+(\gamma/m)\}} |x| = |x|$$

   But the series will be convergent if $|x| < 1$. Thus the radius of convergence of the series is unity.

3. γ cannot be unity because if $\gamma = 1$, the solution given by equation (8) becomes similar to (7).

4. γ cannot be negative integer because if $\gamma = -ve$ integer, $a_n = \infty$.

   So ${}_2F_1(\alpha; \beta; \gamma, x)$ cannot be obtained in this case.

(b) When $x = 1$: We use transformation $1 - x = z$. Thus we have $\dfrac{dz}{dx} = -1$.

Now $\dfrac{dy}{dx}=\dfrac{dy}{dz}\dfrac{dz}{dx}=-\dfrac{dy}{dz}$

and $\dfrac{d^2y}{dx^2}=\dfrac{d}{dx}\left(\dfrac{dy}{dx}\right)=\dfrac{d}{dx}\left(-\dfrac{dy}{dz}\right)=-\dfrac{d^2y}{dz^2}\dfrac{dz}{dx}=\dfrac{d^2y}{dz^2}$

Thus equation (1) reduces to

$$z(1-z)\frac{d^2y}{dz^2}+\left[\gamma-(\alpha+\beta+1)(-z+1)\right]\left(-\frac{dy}{dz}\right)-\alpha\beta y=0$$

$$\Rightarrow\quad z(1-z)\frac{d^2y}{dz^2}+\left[(\alpha+\beta-\gamma+1)-(\alpha+\beta+1)z\right]\frac{dy}{dz}-\alpha\beta y=0$$

which is identical to (1) with γ replaced by $\alpha+\beta-\gamma+1$ and x by $1-x$

Hence the required solution is given by

$$y=A\,{}_2F_1(\alpha,\beta;\alpha+\beta-\gamma+1;1-x)$$
$$+B(1-x)^{\gamma-\alpha-\beta}\,{}_2F_1(\gamma-\beta,\gamma-\alpha;\gamma-\alpha-\beta+1;1-x)$$
$$=A\,{}_2F_1(\alpha,\beta;\alpha+\beta-\gamma+1;1-x)$$
$$+B(1-x)^{\gamma-\beta-\alpha}\,{}_2F_1(\gamma-\alpha,\gamma-\beta;\gamma-\alpha-\beta+1;1-x)\quad ...(10)$$

(c) When $x=\infty$: In this case, we consider the solution of (1) as

$y=\sum_{r=0}^{\infty}a_r\,x^{-k-r}$ and on substituting it in (1), we have

$$\sum_{r=0}^{\infty}a_r\left[(-k-r)(-k-r-1+\gamma)x^{-k-r-1}-(-k-r-\alpha)\right.$$
$$\left.(-k-r+\beta)x^{-k-r}\right]=0\quad ...(11)$$

Equating to zero coefficient of x^{-k}, we obtain

$$a_0(-k+\alpha)(-k+\beta)=0$$

Now $\quad a_0\neq 0$, gives $k=\alpha, k=\beta$ ...(12)

Again equating to zero the coefficient of the general term x^{-k-r}, we obtain

$$a_{r-1}(-k-r+1)(-k-r+\gamma)-a_r(-k-r+\alpha)(-k-r+\beta)=0$$

$$\Rightarrow\quad a_r=\frac{(-k-r+1)(-k-r+\gamma)}{(-k-r+\alpha)(-k-r+\beta)}a_{r-1}$$

Case I: When $k=\alpha$, we have

$$a_r=\frac{(r+\alpha-1)(r+\alpha-\gamma)}{r(r+\alpha-\beta)}a_{r-1}$$

Putting $r = 1, 2, 3....n,$ we get

$$a_1 = \frac{\alpha.(1+\alpha-\gamma)}{1.(1+\alpha-\beta)} a_0 = \frac{(\alpha)_1 (1+\alpha-\gamma)_1}{1!(1+\alpha-\beta)_1} a_0$$

$$a_2 = \frac{(\alpha+1)(2+\alpha-\gamma)}{2.(2+\alpha-\beta)} a_1 = \frac{(\alpha)_2 (1+\alpha-\gamma)_2}{2!(1+\alpha-\beta)_2} a_0$$

$$a_3 = \frac{(\alpha+2)(3+\alpha-\gamma)}{3.(3+\alpha-\beta)} a_2$$

$$= \frac{\alpha(\alpha+1)(\alpha+2)(1+\alpha-\gamma)(2+\alpha-\gamma)(3+\alpha-\gamma)}{3!(1+\alpha-\beta)(2+\alpha-\beta)(3+\alpha-\beta)} a_0$$

$$= \frac{(\alpha)_3 (1+\alpha-\gamma)_3}{3!(1+\alpha-\beta)_3} a_0$$

In general, $$a_n = \frac{\begin{bmatrix} \alpha(\alpha+1)...(\alpha+r-1).(1+\alpha-\gamma)(1+\alpha-\gamma+1) \\ ...(1+\alpha-\gamma+r-1) \end{bmatrix}}{2!(1+\alpha-\beta)(1+\alpha-\beta+1)....(1+\alpha+\beta+r-1)} a_0$$

$$= \frac{(\alpha)_r (1+\alpha-\gamma)_r}{r!(\alpha-\beta+1)_r} a_0$$

Hence, we have

$$y = \sum_{r=0}^{\infty} \frac{(\alpha)_r (1+\alpha-\gamma)_r}{r!(\alpha-\beta+1)_r} x^{-\alpha-r}, \text{ when } a_0 = 1$$

$$= x^{-\alpha} \sum_{r=0}^{\infty} \frac{(\alpha)_r (\alpha-\gamma+1)_r}{r!(\alpha-\beta+1)_r} \left(\frac{1}{x}\right)^r = x^{-\alpha} {}_2F_1\left(\alpha, 1+\alpha-\gamma; 1+\alpha-\beta; \frac{1}{x}\right).$$

Case II: Similarly, when $k = \beta$, we have

$$y = x^{-\beta} {}_2F_1\left(\beta, 1+\beta-\gamma; 1+\beta-\alpha; \frac{1}{x}\right).$$

Hence the general solution is

$$y = A\, x^{-\alpha} {}_2F_1\left(\alpha, 1+\alpha-\gamma; 1+\alpha-\beta; \frac{1}{x}\right)$$

$$+ Bx^{-\beta} {}_2F_1\left(\beta, 1+\beta-\gamma; 1+\beta-\alpha; \frac{1}{x}\right) \quad ...(13)$$

5.8 CONVERGENCE OF HYPERGEOMETRIC SERIES

The nth and $(n+1)^{th}$ terms in the expansion of hypergeometric function ${}_2F_1(\alpha,\beta;\gamma;x)$ are, respectively, given by

$$u_n = \frac{\alpha(\alpha+1)(\alpha+2)...(\alpha+n-1)\beta(\beta+1)....(\beta+n-1)}{n!\gamma(\gamma+1)....(\gamma+n-1)}x^n$$

$$u_{n+1} = \frac{\alpha(\alpha+1)....(\alpha+n-1)(\alpha+n)\beta(\beta+1)....(\beta+n-1)(\beta+n)}{(n+1)\gamma(\gamma+1)....(\gamma+n-1)(\gamma+n)}x^{n+1}$$

$$= \frac{(\alpha+n)(\beta+n)}{(n+1)(\gamma+n)}xu_n$$

This implies that $\dfrac{u_n}{u_{n+1}} = \dfrac{(n+1)(\gamma+n)}{(\alpha+n)(\beta+n)x}$

Taking limit $n \to \infty$, we have $\lim\limits_{n\to\infty}\dfrac{u_n}{u_{n+1}} = \lim\limits_{n\to\infty}\left[\dfrac{(n+1)(\gamma+n)}{(\alpha+n)(\beta+n)x}\right] = \dfrac{1}{x}$.

Thus, by ratio test we conclude

(i) If $\dfrac{1}{x} > 1 \Rightarrow x < 1$, the series converges

(ii) If $\dfrac{1}{x} = 1 \Rightarrow x > 1$, the series diverges

(iii) If $\dfrac{1}{x} = 1 \Rightarrow x = 1$, the ratio test fails

Now for $x = 1$, we apply Rabbi's test

$$\frac{u_n}{u_{n+1}} - 1 = \frac{(n+1)(\gamma+n)}{(\alpha+n)(\beta+n)} - 1 = \frac{n(\gamma-\alpha-\beta+1)+(\gamma-\alpha\beta)}{(\alpha+n)(\beta+n)}.$$

Thus,

$$\lim_{n\to\infty}\left[n\left(\frac{u_n}{u_{n+1}} - 1\right)\right] = \lim_{n\to\infty}\left\{\frac{(1+\gamma-\alpha-\beta)+\dfrac{\gamma-\alpha\beta}{n}}{\left(1+\dfrac{\alpha}{\beta}\right)\left(1+\dfrac{\beta}{n}\right)}\right\} = 1+\gamma-\alpha-\beta.$$

Here we conclude

(i) If $1+\gamma-\alpha-\beta > 1$, *i.e.* $\gamma > \alpha+\beta$, the series converges

(ii) If $1+\gamma-\alpha-\beta < 1$, *i.e.* $\gamma < \alpha+\beta$, the series diverges

(iii) If $1+\gamma-\alpha-\beta = 1$, *i.e.* $\gamma = \alpha+\beta$, Rabbis' test fails

Now for $\gamma = \alpha + \beta$, we apply logarithmic test as follows:

When $\gamma = \alpha + \beta$, we have

$$n\left(\frac{u_n}{u_{n+1}} - 1\right) - 1 = \frac{n^2(1+\alpha+\beta-\alpha-\beta)+(\alpha+\beta-\alpha\beta)n}{(n+\beta)(n+\beta)} - 1$$

$$= \frac{n^2+(\alpha+\beta-\alpha\beta)n - n^2 - n\alpha - n\beta - \alpha\beta}{(n+\alpha)(n+\beta)} = \frac{-\alpha\beta n - \alpha\beta}{(n+\alpha)(n+\beta)}$$

Thus $$\lim_{n\to\infty}\left[\left\{n\left(\frac{u_n}{u_{n+1}} - 1\right) - 1\right\}\log n\right] = \lim_{n\to\infty}\left\{\frac{-\alpha\beta(n+1)}{(n+\alpha)(n+\beta)}\log n\right\}$$

$$= \lim_{n\to\infty}\left\{\frac{-\alpha\beta\left(1+\frac{1}{n}\right)}{\left(1+\frac{\alpha}{n}\right)\left(1+\frac{\beta}{n}\right)}\frac{\log n}{n}\right\} = 0, \text{ since } \lim_{n\to\infty}\frac{\log n}{n} = 0.$$

Hence the series diverges.

Thus, we have the following results.

(i) If $x < 1$, the series converges.

(ii) If $x > 1$, the series diverges

(iii) If $x = 1$, the series converges for $\gamma > \alpha + \beta$ and the series diverges for $\gamma \leq \alpha + \beta$.

5.9 SYMMETRIC PROPERTY OF HYPERGEOMETRIC FUNCTION

Hypergeometric function has symmetric properties with respect to parameters α and β, i.e. it does not change if the parameters α and β are interchanged, keeping γ fixed.

Proof: To show the symmetric property with respect to parameters α and β, we have to show that ${}_2F_1(\alpha, \beta; \gamma; x) = {}_2F_1(\beta, \alpha; \gamma; x)$.

By definition, we have

$${}_2F_1(\alpha, \beta; \gamma; x) = \sum_{r=0}^{\infty}\frac{(\alpha)_r(\beta)_r}{(\gamma)_r}\frac{x^r}{r!} = \sum_{r=0}^{\infty}\frac{(\beta)_r(\alpha)_r}{(\gamma)_r}\frac{x}{r!} = {}_2F_1(\beta, \alpha; \gamma; x)$$

Hence the result follows.

5.10 DIFFERENTIATION OF HYPERGEOMETRIC FUNCTION

We have ${}_2F_1(\alpha, \beta; \gamma; x) = \sum_{n=0}^{\infty}\frac{(\alpha)_n(\beta)_n}{n!(\gamma)_n}x^n$.

Differentiating both sides, w. r. t. 'x', we get

$$\frac{d}{dx}\left\{{}_2F_1(\alpha,\beta;\gamma;x)\right\}=\sum_{n=1}^{\infty}\frac{(\alpha)_n(\beta)_n}{n!(\gamma)_n}nx^{n-1}=\sum_{n=1}^{\infty}\frac{(\alpha)_n(\beta)_n}{(n-1)!(\gamma)_n}x^{n-1}$$

$$=\sum_{m=0}^{\infty}\frac{(\alpha)_{m+1}(\beta)_{m+1}}{m!(\gamma)_{m+1}}x^m \text{, Replacing } n-1 \text{ by } m$$

$$=\sum_{m=0}^{\infty}\frac{\alpha(\alpha+1)_m\beta(\beta+1)_m}{m!\gamma(\gamma+1)_m}x^m$$

[Since $(\alpha)_{m+1}=\alpha(\alpha+1)...(\alpha+m+1-1)=\{\alpha(\alpha+1)(\alpha+2)$

$...(\alpha+1+m-1)\}=\alpha(\alpha+1)_m$]

$$=\frac{\alpha\beta}{\gamma}\sum_{m=0}^{\infty}\frac{(\alpha+1)_m(\beta+1)_m}{m!(\gamma+1)_m}x^m.$$

Therefore $\frac{d}{dx}\left\{{}_2F_1(\alpha,\beta;\gamma;x)\right\}=\frac{\alpha\beta}{\gamma}\,{}_2F_1(\alpha+1,\beta+1;\gamma+1;x)$

Differentiating again, w. r. t. 'x', we get

$$\frac{d^2}{dx^2}\,{}_2F_1(\alpha,\beta;\gamma;x)=\frac{\alpha\beta}{\gamma}\frac{d}{dx}\,{}_2F_1(\alpha+1,\beta+1;\gamma+1;x)$$

$$=\frac{\alpha\beta}{\gamma}\frac{d}{dx}\sum_{m=0}^{\infty}\frac{(\alpha+1)_m(\beta+1)_m}{m!(\gamma+1)_m}x^m$$

$$=\frac{\alpha\beta}{\gamma}\sum_{m=1}^{\infty}\frac{(\alpha+1)_m(\beta+1)_m}{(m-1)!(\gamma+1)_m}x^{m-1}$$

$$=\frac{\alpha\beta}{\gamma}\sum_{m=1}^{\infty}\frac{(\alpha+1)(\alpha+2)_{m-1}(\beta+1)(\beta+2)_{m-1}}{(m-1)!(\gamma+1)(\gamma+2)_{m-1}}x^{m-1}$$

$$=\frac{\alpha(\alpha+1)\beta(\beta+1)}{\gamma(\gamma+1)}\sum_{n=0}^{\infty}\frac{(\alpha+2)_n(\beta+2)_n}{n!(\gamma+2)_n}$$

[replacing $m-1$ by n]

$$=\frac{\alpha(\alpha+1)\beta(\beta+1)}{\gamma(\gamma+1)}\,{}_2F_1(\alpha+2,\beta+2;\gamma+2;x)$$

$$=\frac{(\alpha)_2(\beta)_2}{(\gamma)_2}\,{}_2F_1(\alpha+2;\beta+2;\gamma+2;x).$$

In general, we have

$$\frac{d^r}{dx^r}\left\{{}_2F_1(\alpha,\beta;\gamma;x)\right\}=\frac{(\alpha)_r(\beta)_r}{(\gamma)_r}\ {}_2F_1(\alpha+r,\beta+r;\gamma+r;x)$$

Remarks: If $x=0$, we have

(i) $3\left[{}_2F_1(\alpha,\beta;\gamma;x)\right]_{x=0}=\left[1+\frac{\alpha\beta}{\gamma}x+\frac{\alpha(\alpha+1)\beta(\beta+1)}{2!\gamma(\gamma+1)}x^2+\ldots\right]_{x=0}=1.$

(ii) $\left[\frac{d}{dx}\ {}_2F_1(\alpha,\beta;\gamma;x)\right]_{x=0}=\frac{\alpha\beta}{\gamma}\left[{}_2F_1(\alpha+1,\beta+1;\gamma+1;x)\right]_{x=0}$

$$=\frac{\alpha\beta}{\gamma}\left[1+\frac{(\alpha+1)(\beta+1)}{(\gamma+1)}x+\ldots\right]_{x=0}=\frac{\alpha\beta}{\gamma}.$$

(iii) $\left[\frac{d^2}{dx^2}\ {}_2F_1(\alpha,\beta;\gamma;x)\right]_{x=0}=\frac{(\alpha)_2(\beta)_2}{(\gamma)_2}\left[{}_2F_1(\alpha+2,\beta+2;\gamma+2;x)\right]_{x=0}$

$$=\frac{(\alpha)_2(\beta)_2}{(\gamma)_2}$$

(iv) $\left[\frac{d^p}{dx^p}\ {}_2F_1(\alpha,\beta;\gamma;x)\right]_{x=0}=\frac{(\alpha)_p\cdot(\beta)_p}{(\gamma)_p}$

5.11 INTEGRAL REPRESENTATION FOR THE HYPERGEOMETRIC FUNCTION

If $|x|<1$, the integral representation of hypergeometric function is given by

$$F(\alpha,\beta;\gamma;x)=\frac{\Gamma(\gamma)}{\Gamma(\beta)\Gamma(\gamma-\beta)}\int_0^1 t^{\beta-1}(1-t)^{\gamma-\beta-1}(1-xt)^{-\alpha}\,dt,$$

if $\text{Re}\,\gamma>\text{Re}\,\beta>0$.

Proof: We know that

$${}_2F_1(\alpha,\beta;\gamma;x)=\sum_{n=0}^{\infty}\frac{(\alpha)_n(\beta)_n}{n!(\gamma)_n}x^n \qquad \ldots(1)$$

We can write $(\alpha)_n=\alpha(\alpha+1)\ldots(\alpha+n-1)$

$$=\frac{1\,2\ldots(\alpha-1)\alpha(\alpha+1)(\alpha+n-1)}{1\,2\ldots(\alpha-1)}=\frac{\Gamma(\alpha+n)}{\Gamma(\alpha)}$$

Therefore, we have

$$\frac{(\beta)_n}{(\gamma)_n}=\frac{\Gamma(\beta+n)}{\Gamma\beta}\cdot\frac{\Gamma\gamma}{\Gamma(\gamma+n)}=\frac{\Gamma(\gamma)}{\Gamma(\beta)\Gamma(\gamma-\beta)}\cdot\frac{\Gamma(\beta+n)\Gamma(\gamma-\beta)}{\Gamma(\beta+n+\gamma-\beta)}$$

$$= \frac{\Gamma(\gamma)}{\Gamma(\beta)\Gamma(\gamma-\beta)} \int_0^1 (1-t)^{\gamma-\beta-1} t^{\beta+n-1} dt$$

$$\left[\text{since } \frac{\Gamma(m)\Gamma(n)}{\Gamma(m+n)} = B(m,n) = \int_0^1 (1-t)^{n-1} t^{m-1}\, dt\right]$$

Hence from (1), we have

$$ {}_2F_1(\alpha,\beta;\gamma;x) = \sum_{n=0}^{\infty} \frac{\Gamma(\gamma)}{\Gamma(\beta)\Gamma(\gamma-\beta)} \int_0^1 t^{\beta+n-1}(1-t)^{\gamma-\beta-1}\left(\frac{(\alpha)_n x^n}{n!}\right) dt$$

$$= \sum_{n=0}^{\infty} \frac{\Gamma(\gamma)}{\Gamma(\beta)\Gamma(\gamma-\beta)} \int_0^1 t^{\beta-1}(1-t)^{\gamma-\beta-1} \frac{(\alpha_n)(xt)^n}{n!} dt$$

$$= \frac{\Gamma(\gamma)}{\Gamma(\beta)\Gamma(\gamma-\beta)} \int_0^1 t^{\beta-1}(1-t)^{\gamma-\beta-1} \times \left[\sum_{n=0}^{\infty} \frac{(\alpha)_n (xt)^n}{n!}\right] dt$$

$$= \frac{\Gamma(\gamma)}{\Gamma(\beta)\Gamma(\gamma-\beta)} \int_0^1 t^{\beta-1}(1-t)^{\gamma-\beta-1} . \left\{1+\alpha.\, xt + \frac{\alpha(\alpha+1)}{2!}(xt)^2 + \ldots\right\} dt$$

$$\Rightarrow \quad {}_2F_1(\alpha,\beta;\gamma;x) = \frac{\Gamma(\gamma)}{\Gamma(\beta)\Gamma(\gamma-\beta)} \int_0^1 t^{\beta-1}(1-t)^{\gamma-\beta-1}(1-xt)^{-\alpha}\, dt$$

This is known as the *integral formula* for hypergeometric function and is valid only if $|x| < 1$, $Re\,\gamma > Re\,\beta > 0$.

5.12 GAUSS'S THEOREM

Putting $x = 1$ in the integral formula, we obtain

$$ {}_2F_1(\alpha,\beta;\gamma;1) = \frac{\Gamma(\gamma)}{\Gamma(\beta)\Gamma(\gamma-\beta)} \int_0^1 t^{\beta-1}.(1-t)^{\gamma-\beta-1}.(1-t)^{-\alpha}\, dt$$

$$= \frac{\Gamma(\gamma)}{\Gamma(\beta)\Gamma(\gamma-\beta)} \int_0^1 t^{\beta-1}.(1-t)^{(\gamma-\beta-\alpha)-1}\, dt$$

$$= \frac{\Gamma(\gamma)}{\Gamma(\beta)\Gamma(\gamma-\beta)} B(\beta, \gamma-\beta-\alpha)$$

$$= \frac{\Gamma(\gamma)}{\Gamma(\beta)\Gamma(\gamma-\beta)} \frac{\Gamma(\beta)\Gamma(\gamma-\beta-\alpha)}{\Gamma(\gamma-\alpha)}, \text{ since } B(m,n) = \frac{\Gamma(m)\Gamma(n)}{\Gamma(m+n)}$$

$$= \frac{\Gamma(\gamma)\Gamma(\gamma-\beta-\alpha)}{\Gamma(\gamma-\alpha)\Gamma(\gamma-\beta)}.$$

Hence Gauss's formula is

$$ {}_2F_1(\alpha,\beta;\gamma;1)=\frac{\Gamma(\gamma)\Gamma(\gamma-\beta-\alpha)}{\Gamma(\gamma-\alpha)\Gamma(\gamma-\beta)}. $$

It is valid only when $\mathrm{Re}(\gamma-\alpha-\beta)>0$ and $\gamma\neq 0$ & γ is not a negative integer.

5.13 LINEAR RELATION BETWEEN THE SOLUTIONS OF THE HYPERGEOMETRIC EQUATION

The series in the solution (9) of hypergeometric differential equation is convergent if $|x|<1$, i.e. in interval (–1, 1) and the series in the solution (10) is convergent in interval (0, 2). Thus in interval (0, 1) all the four series (9) and (10) are convergent. Since only two solutions of the differential equation (1) may be linearly independent in interval (0, 1), so there must exist a linear relation between the solutions given by (9) and (10) in interval (0, 1).

Thus, we let ${}_2F_1(\alpha,\beta;\gamma;x)=A_2\,F_1(\alpha,\beta;\alpha+\beta-\gamma+1;1-x)$

$$ +B(1-x)^{\gamma-\alpha-\beta}\,{}_2F_1(\gamma-\alpha,\gamma-\beta;\gamma-\alpha-\beta+1;1-x). \quad \ldots(1) $$

To determine the constants A and B; we suppose that $\alpha+\beta<\gamma<1$, so that three series

$$ {}_2F_1(\alpha,\beta;\gamma;1),\ {}_2F_1(\alpha,\beta;\alpha+\beta-\gamma+1;1), $$

$$ {}_2F_1(\gamma-\alpha,\gamma-\beta;\gamma-\alpha-\beta+1;1) $$

are convergent.

Then putting $x=0$ and $x=1$ in (1), we obtain

$$ 1=A_2\,F_1(\alpha,\beta;\alpha+\beta-\gamma+1;1)+B\,{}_2F_1(\gamma-\alpha,\gamma-\beta;\gamma-\alpha-\beta+1;1) $$

and ${}_2F_1(\alpha,\beta;\gamma;1)=A \Rightarrow A=\dfrac{\Gamma(\gamma)\Gamma(\gamma-\alpha-\beta)}{\Gamma(\gamma-\alpha)\Gamma(\gamma-\beta)}$ $\quad \ldots(2)$

$$ \text{Now } 1=A\frac{\Gamma(\alpha+\beta-\gamma+1)\Gamma(1-\gamma)}{\Gamma(\beta-\gamma+1)\Gamma(\alpha-\gamma+1)}+B\frac{\Gamma(\gamma-\alpha-\beta+1)\Gamma(1-\gamma)}{\Gamma(1-\beta)\Gamma(1-\alpha)}, $$

(by using Gauss formula)

$$ \Rightarrow \quad 1=\frac{\Gamma(\gamma)\Gamma(\gamma-\alpha-\beta)}{\Gamma(\gamma-\alpha)\Gamma(\gamma-\beta)}\frac{\Gamma(\alpha+\beta-\gamma+1)\Gamma(1-\gamma)}{\Gamma(\beta-\gamma+1)\Gamma(\alpha-\gamma+1)}, $$

$$ +B\frac{\Gamma(\gamma-\alpha-\beta+1)\Gamma(1-\gamma)}{\Gamma(1-\beta)\Gamma(1-\alpha)} \quad \text{using (2)} $$

$$ \Rightarrow \quad 1=\frac{\{\Gamma(\gamma)\Gamma(1-\gamma)\}\{\Gamma(\gamma-\alpha-\beta)\Gamma(\alpha+\beta-\gamma+1)\}}{\{\Gamma(\gamma-\alpha).\Gamma(\alpha-\gamma+1)\}.\{\Gamma(\gamma-\beta)\Gamma(\beta-\gamma+1)\}} $$

$$+ B\frac{\Gamma(\gamma-\alpha-\beta+1)\Gamma(1-\gamma)}{\Gamma(1-\beta)\Gamma(1-\alpha)}$$

$$\Rightarrow \quad 1=\frac{(\pi\ \text{cosec}\ \pi\gamma)\pi\ \text{cosec}\ \pi(\gamma-\alpha-\beta)}{\pi\ \text{cosec}\ \pi(\gamma-\alpha)\pi\ \text{cosec}\ \pi(\gamma-\beta)}$$

$$+ B\frac{\Gamma(\gamma-\alpha-\beta+1)\Gamma(1-\gamma)}{\Gamma(1-\beta)\Gamma(1-\alpha)}$$

[since $\Gamma(\alpha)\Gamma(1-\alpha)=\pi\ \text{cosec}\ \pi\,\alpha$]

$$\Rightarrow \quad B\frac{\Gamma(\gamma-\alpha-\beta+1)\Gamma(1-\gamma)}{\Gamma(1-\beta)\Gamma(1-\alpha)}=1-\frac{\sin\pi(\gamma-\alpha)\sin\pi(\gamma-\beta)}{\sin\pi\gamma\sin\pi(\gamma-\alpha-\beta)}$$

$$=\frac{\sin\pi\gamma.\sin\pi(\gamma-\alpha-\beta)-\sin\pi(\gamma-\alpha).\sin\pi(\gamma-\beta)}{\sin\pi\gamma.\sin\pi\ (\gamma-\alpha-\beta)}$$

$$=\frac{\cos\pi(\alpha+\beta)-\cos\pi(2\gamma-\alpha-\beta)-\cos\pi(\alpha-\beta)+\cos\pi(2\gamma-\alpha-\beta)}{2\sin\pi\gamma.\sin\pi(\gamma-\alpha-\beta)}$$

$$=\frac{\cos\pi(\alpha+\beta)-\cos\pi(\alpha-\beta)}{2\sin\pi\gamma.\sin\pi(\gamma-\alpha-\beta)}=\frac{-2\sin\pi\beta\sin\pi\alpha}{2\sin\pi\gamma.\sin\pi(\gamma-\alpha-\beta)}$$

$$\Rightarrow B\frac{\Gamma(\alpha)\Gamma(\beta)}{\Gamma(\gamma)\Gamma(\alpha+\beta-\gamma)}\ \frac{\{\Gamma(\alpha+\beta-\gamma)\Gamma(\gamma-\alpha-\beta+1)\}\{\Gamma(\gamma)\Gamma(1-\gamma)\}}{\{\Gamma(\beta)\Gamma(1-\beta)\}\{\Gamma(\alpha).\Gamma(1-\alpha)\}}$$

$$=\frac{\sin\pi\alpha\sin\pi\beta}{\sin\pi\gamma,\sin(\alpha+\beta-\gamma)}$$

$$\Rightarrow \quad B\frac{\Gamma(\alpha)\Gamma(\beta)}{\Gamma(\gamma)\Gamma(\alpha+\beta-\gamma)}\ \frac{\{\pi\ \text{cosec}\ \pi(\alpha+\beta-\gamma)\}\{\pi\ \text{cosec}\ \pi\gamma\}}{\{\pi\ \text{cosec}\ \pi\beta\}\{\pi\ \text{cosec}\ \pi\alpha\}}$$

$$=\frac{\sin\pi\alpha\sin\pi\beta}{\sin\pi\gamma\sin\pi(\alpha+\beta-\gamma)}$$

$$\Rightarrow \quad B=\frac{\Gamma(\gamma)\Gamma(\alpha+\beta-\gamma)}{\Gamma(\alpha)\Gamma(\beta)}$$

Hence from (1), we have

$${}_2F_1(\alpha,\beta;\gamma;x)=\frac{\Gamma(\gamma)\Gamma(\gamma-\alpha-\beta)}{\Gamma(\gamma-\alpha)\Gamma(\gamma-\beta)}\,{}_2F_1(\alpha,\beta;\alpha+\beta-\gamma+1;1-x)$$

$$+\frac{\Gamma(\gamma)\Gamma(\alpha+\beta-\gamma)}{\Gamma(\alpha)\Gamma(\beta)}.(1-x)^{\gamma-\alpha-\beta}\,{}_2F_1(\gamma-\alpha,\gamma-\beta;\gamma-\alpha-\beta+1;1-x)$$

5.14 KUMMER'S THEOREM

If $\mathrm{Re}(\frac{\beta}{2}-\alpha+1)>0$, then

$$_2F_1(\alpha,\beta;\beta-\alpha+1;-1)=\frac{\Gamma(\beta-\alpha+1)\Gamma(\beta/2+1)}{\Gamma(\beta+1)\Gamma(\beta/2-\alpha+1)}$$

Proof: Putting $x=-1$ and $\gamma=\beta-\alpha+1$ in the integral formula for hypergeometric function, we obtain

$$_2F_1(\alpha,\beta;\beta-\alpha+1;-1)=\frac{\Gamma(\beta-\alpha+1)}{\Gamma(\beta)\Gamma(1-\alpha)}\int_0^1 t^{\beta-1}(1-t)^{-\alpha}(1+t)^{-\alpha}\,dt$$

$$=\frac{\Gamma(\beta-\alpha+1)}{\Gamma(\beta)\Gamma(1-\alpha)}\int_0^1 t^{\beta-1}\left(1-t^2\right)^{-\alpha}\,dt$$

$$=\frac{\Gamma(\beta-\alpha+1)}{2\,\Gamma(\beta)\Gamma(1-\alpha)}\int_0^1 z^{\frac{\beta}{2}-1}(1-z)^{-\alpha}\,dz\text{, putting } t^2=z$$

$$=\frac{\Gamma(\beta-\alpha+1)}{2\,\Gamma(\beta)\Gamma(1-\alpha)}B\left(\frac{\beta}{2},1-\alpha\right)$$

$$=\frac{\Gamma(\beta-\alpha+1)\Gamma(\beta/2)\Gamma(1-\alpha)}{2\,\Gamma(\beta)\Gamma(1-\alpha)\Gamma\left[(\beta/2)+1-\alpha\right]}=\frac{\Gamma(\beta-\alpha+1)\frac{\beta}{2}\Gamma(\beta/2)}{\beta\,\Gamma(\beta)\Gamma\left(\frac{\beta}{2}-\alpha+1\right)}$$

$$=\frac{\Gamma(\beta-\alpha+1)\Gamma\left(\frac{\beta}{2}+1\right)}{\Gamma(\beta+1)\Gamma\left(\frac{\beta}{2}-\alpha+1\right)}.$$

5.15 VANDERMONDE THEOREM

Putting $\alpha=-n$ in Gauss's formula, we obtain

$$_2F_1(-n,\beta;\gamma;1)=\frac{\Gamma(\gamma)\Gamma(\gamma-\beta+n)}{\Gamma(\gamma+n)\Gamma(\gamma-\beta)}$$

$$=\frac{\Gamma(\gamma)(\gamma-\beta+n-1)(\gamma-\beta+n-2)\ldots(\gamma-\beta)\Gamma(\gamma-\beta)}{(\gamma+n-1)(\gamma+n-2)\ldots\gamma\,\Gamma(\gamma)\Gamma(\gamma-\beta)}$$

$$=\frac{(\gamma-\beta)(\gamma-\beta+1)\ldots(\gamma-\beta+n-1)}{\gamma(\gamma+1)\ldots(\gamma+n-1)}=\frac{(\gamma-\beta)_n}{(\gamma)_n}$$

Hence, we have the result ${}_2F_1(-n, \beta; \gamma; 1) = \dfrac{(\gamma-\beta)_n}{(\gamma)_n}$.

This is known as *Vandermonde formula.*

5.16 THE CONFLUENT HYPERGEOMETRIC FUNCTIONS

We know that the hypergeometric differential equation is

$$x(1-x)\frac{d^2y}{dx^2} + \{\gamma - (1+\alpha+\beta)x\}\frac{dy}{dx} - \alpha\,\beta\,\gamma = 0 \qquad \ldots(1)$$

Replacing x by $\dfrac{x}{\beta}$, we obtain

$$\frac{x}{\beta}\left(1-\frac{x}{\beta}\right)\beta^2\,\frac{d^2y}{dx^2} + \left\{\gamma - (1+\alpha+\beta)\frac{x}{\beta}\right\}\beta\,\frac{dy}{dx} - \alpha\,\beta\,y = 0$$

or

$$x\left(1-\frac{x}{\beta}\right)\frac{d^2y}{dx^2} + \left\{\gamma - \left(\frac{1+\alpha}{\beta}+1\right)x\right\}\frac{dy}{dx} - \alpha y = 0 \qquad \ldots(2)$$

Since ${}_2F_1(\alpha, \beta; \gamma; x)$ is the solution of (1), therefore ${}_2F_1\left(\alpha, \beta; \gamma; \dfrac{x}{\beta}\right)$ be the solution of (2).

Now if $\beta \to \infty$, we have $\underset{\beta\to\infty}{Lim}\; {}_2F_1\left(\alpha, \beta; \gamma; \dfrac{x}{\beta}\right)$ is the solution of differential equation

$$x\,\frac{d^2x}{dx^2} + (\gamma - x)\frac{dy}{dx} - \alpha y = 0 \qquad \ldots(3)$$

But $\displaystyle\lim_{\beta\to\infty} {}_2F_1\left(\alpha, \beta, \gamma, \frac{x}{\beta}\right) = \lim_{\beta\to\infty}\sum_{r=0}^{\infty}\frac{(\alpha)_r.(\beta)_r}{(\gamma)_r\, r!}.\left(\frac{x}{\beta}\right)^r$

$$= \sum_{r=0}^{\infty}\frac{(\alpha)_r}{(\gamma_r)}\frac{x^r}{r!} = {}_1F_1(\alpha; \gamma; x), \text{ since } \lim_{\beta\to\infty}\frac{(\beta)_r}{\beta^r} = 1.$$

The function ${}_1F_1(\alpha; \gamma; x)$ is called confluent hypergeometric function and the equation (3) is called confluent hypergeometric equation or Kummer's equation.

5.17 WHITTAKER'S CONFLUENT HYPERGEOMETRIC FUNCTIONS

The confluent hypergeometric equation is

$$x\,\frac{d^2y}{dx^2} + (\gamma - x)\frac{dy}{dx} - \alpha y = 0 \qquad \ldots(1)$$

and its general solution is $y = Ay_1(x) + By_2(x)$ $\qquad \ldots(2)$

where A and B are arbitrary constants and

$$y_1(x) = {}_1F_1(\alpha; \gamma; x) \qquad \ldots(3)$$

and $$y_2(x) = x^{1-\alpha}\ {}_1F_1(\alpha - \gamma + 1, 2 - \gamma, x). \qquad \ldots(4)$$

Putting $y(x) = x^{-\gamma/2} e^{x/2} W(x)$ $\qquad \ldots(5)$

In equation (1), we obtain

$$x\left[x^{-\gamma/2} e^{x/2} \frac{d^2W}{dx^2} + 2\left(-\frac{\gamma}{2} x^{(-(\gamma/2)-1)} e^{x/2} + \frac{1}{2} x^{-\gamma/2} e^{x/2}\right)\frac{dW}{dx}\right.$$

$$\left. + W(x)\left\{\frac{\gamma}{2}\left(\frac{\gamma}{2}+1\right)x^{\left(-\frac{\gamma}{2}-2\right)} e^{x/2} - \frac{\gamma}{2} x^{(-\gamma/2)-1} . e^{x/2} + \frac{1}{4} x^{-\gamma/2} . e^{x/2}\right\}\right]$$

$$+(\gamma - x)\left[x^{-\gamma/2} e^{x/2} \frac{dW}{dx} + W(x)\left(-\frac{\gamma}{2} x^{-(\gamma/2)-1} e^{x/2} + \frac{1}{2} x^{-\frac{\gamma}{2}} . e^{x/2}\right)\right]$$

$$- \alpha x^{-\gamma/2} e^{x/2} W(x) = 0$$

$$\Rightarrow \quad \frac{d^2W}{dx^2} + \left[-\frac{1}{4} + \frac{\frac{\gamma}{2} - \alpha}{x} + \frac{\left(-\frac{\gamma^2}{2} + \frac{\gamma}{2} + \frac{\gamma^2}{4}\right)}{x^2}\right] W(x) = 0$$

$$\Rightarrow \quad \frac{d^2W}{dx^2} + \left[-\frac{1}{4} + \frac{k}{x} + \frac{\frac{1}{4} - m^2}{x^2}\right] W(x) = 0;$$

where $k = \frac{\gamma}{2} - \alpha$ and $m = \frac{1}{2} - \frac{\gamma}{2}$ $\qquad \ldots(6)$

The solutions $W(x)$ of the equation (6) are known as Whittaker's confluent hypergeometric functions.

If $2m$ is neither 1 nor an integer, the solutions of the equation (1) corresponding to equation (6) are given by (2) with

$$\gamma = 1 + 2m \text{ and } \alpha = \frac{\gamma}{2} - k = \frac{1}{2} + m - k$$

Thus the solutions of equation (6) are the Whittaker functions

$$M_{k.m}(x) = -x^{\gamma/2} e^{-x/2} y_1(x) = x^{(1/2)+m} e^{-x/2}\ {}_1F_1\left(\frac{1}{2} - k + m; 1 + 2m; x\right)$$

and $$M_{k,-m} = x^{(1/2)-m} e^{-x/2}\ {}_1F_1\left(\frac{1}{2} - k - m; 1 - 2m; x\right).$$

5.18 DIFFERENTIATION OF CONFLUENT HYPERGEOMETRIC FUNCTION

We have ${}_1F_1(\alpha;\gamma;x)=\sum_{n=0}^{\infty}\frac{(\alpha)_n}{(\gamma)_n}=\frac{x^n}{n!}$

Differentiating both sides w. r. t. 'x', we obtain

$$\frac{d}{dx}\{{}_1F_1(\alpha;\gamma;x)\}=\sum_{n=1}^{\infty}\frac{(\alpha)_n}{(\gamma)_n}\cdot\frac{nx^{n-1}}{n!}=\sum_{n=1}^{\infty}\frac{(\alpha)_n}{(\gamma)_n}\cdot\frac{x^{n-1}}{(n-1)!}$$

$$=\sum_{m=0}^{\infty}\frac{(\alpha)_{m+1}}{(\gamma)_{m+1}}\cdot\frac{x^m}{m!}, \text{ replacing } n-1=m$$

$$=\frac{\alpha}{\gamma}\cdot\sum_{m=0}^{\infty}\frac{(\alpha+1)_m}{(\gamma+1)_m}\cdot\frac{x^m}{m!}, \text{ since} (\alpha)_{m+1}=\alpha.(\alpha+1)_m]$$

Therefore $\frac{d}{dx}\{{}_1F_1(\alpha;\gamma;x)\}=\frac{\alpha}{\gamma}\,{}_1F_1(\alpha+1;\gamma+1;x)$

Differentiating again w. r .t. 'x', we obtain

$$\frac{d^2}{dx^2}\{{}_1F_1(\alpha;\gamma;x)\}=\frac{\alpha(\alpha+1)}{\gamma(\gamma+1)}\,{}_1F_1(\alpha+2;\gamma+2;x).$$

In general, we obtain

$$\frac{d^p}{dx^p}\{{}_1F_1(\alpha;\gamma;x)\}=\frac{(\alpha)_p}{(\gamma)_p}\,{}_1F_1(\alpha+p;\gamma+p;x).$$

5.19 INTEGRAL REPRESENTATION OF THE CONFLUENT HYPERGEOMETRIC FUNCTION ${}_1F_1(\alpha;\gamma;x)$

Prove that ${}_1F_1(\alpha;\gamma;x)=\frac{\Gamma(\gamma)}{\Gamma(\alpha)\Gamma(\gamma-\alpha)}\int_0^1(1-t)^{\gamma-n-1}t^{n-1}e^{xt}\,dt$

Proof: We have ${}_1F_1(\alpha;\gamma;x)=\sum_{n=0}^{\infty}\frac{(\alpha)_n}{(\gamma)_n}\frac{x^n}{n!}$ ...(1)

But we know that $(\alpha)_n=\frac{\Gamma(\alpha+n)}{\Gamma(\alpha)}$.

Now $\frac{(\alpha)_n}{(\gamma)_n}=\frac{\Gamma(\alpha+n)}{\Gamma(\alpha)}\frac{\Gamma(\gamma)}{\Gamma(\gamma+n)}=\frac{\Gamma(\gamma)}{\Gamma(\alpha)\Gamma(\gamma-\alpha)}\frac{\Gamma(\gamma-\alpha)\Gamma(\alpha+n)}{\Gamma(\alpha+n+\gamma-\alpha)}$

$$=\frac{\Gamma(\gamma)}{\Gamma(\alpha)\Gamma(\gamma-\alpha)}B(\alpha+n,\gamma-\alpha)$$

$$=\frac{\Gamma(\gamma)}{\Gamma(\alpha)\Gamma(\gamma-\alpha)}\int_0^1(1-t)^{\gamma-\alpha-1}t^{\alpha+n-1}dt$$

Therefore from (1), we have

$${}_1F_1(\alpha;\gamma;x)=\sum_{n=0}^{\infty}\frac{\Gamma(\gamma)}{\Gamma(\alpha)\Gamma(\gamma-\alpha)}\int_0^1(1-t)^{\gamma-\alpha-1}t^{\alpha+n-1}\frac{x^n}{n!}dt$$

$$=\frac{\Gamma(\gamma)}{\Gamma(\alpha)\Gamma(\gamma-\alpha)}\int_0^1(1-t)^{\gamma-\alpha-1}t^{\alpha-1}\left\{\sum_{n=0}^{\infty}\frac{(xt)^n}{n!}\right\}dt$$

$$=\frac{\Gamma(\gamma)}{\Gamma(\alpha)\Gamma(\gamma-\alpha)}\int_0^1(1-t)^{\gamma-\alpha-1}t^{\alpha-1}e^{xt}\,dt$$

$$\Rightarrow\quad {}_1F_1(\alpha;\gamma;x)=\frac{1}{B(\alpha,\gamma-\alpha)}\int_0^1(1-t)^{\gamma-\alpha-1}t^{\alpha-1}e^{xt}\,dt$$

5.20 KUMMER'S FIRST FORMULA

If β is neither zero nor negative integer, then

$${}_1F_1(\alpha;\beta;z)=e^z\,{}_1F_1(\beta-\alpha;\beta;-z)$$

Proof: Replacing α by $\beta-\alpha,\gamma$ by β and z by $-z$ in the integral representation formula for the confluent hypergeometric function, we obtain

$${}_1F_1(\beta-\alpha;\beta;-z)=\frac{\Gamma(\beta)}{\Gamma(\beta-\alpha)\Gamma(\alpha)}\int_0^1 t^{\beta-\alpha-1}(1-t)^{\alpha-1}e^{-zt}\,dt.$$

Putting $1-t=u$, we obtain

$${}_1F_1(\beta-\alpha;\beta;-z)=\frac{\Gamma(\beta)}{\Gamma(\beta-\alpha)\Gamma(\alpha)}\int_1^0(1-u)^{\beta-\alpha-1}u^{\alpha-1}e^{uz-z}(-du)$$

$$=\frac{\Gamma(\beta)e^{-z}}{\Gamma(\beta-\alpha)\Gamma(\alpha)}\int_0^1 u^{\alpha-1}(1-u)^{\beta-\alpha-1}e^{uz}\,du$$

$$=e^{-z}\,{}_1F_1(\alpha;\beta;z).$$

5.21 CONTIGUOUS HYPERGEOMETRIC FUNCTIONS

Gauss has defined the contiguous hypergeometric functions. According to him, the function $F(a',b';c';z)$ is said to be contiguous to $F(a,b;c;z)$ if it is obtained by increasing or decreasing one and only one of the parameters a,b,c by unity.

Thus there are six hypergeometric functions contiguous to $F(a,b;c;z)$ and are denoted as

$$F(a+1,b;c;z)=F_{\alpha+},F(a-1,b;c;z)=F_{a-}$$
$$F(a,b+1;c;z)=F_{b+},(a;b-1;c;z)=F_{b-}$$
$$F(a,b;c+1;z)=F_{c+},F(a,b;c-1;z)=F_{c-}$$

Theorem 1: There exists a linear relation between $F(a,b;c;z)$ and only two hypergeometric functions contiguous to it, i.e.

$$(a-b)F=aF_{a+}-bF_{b+}$$

Proof: We have $(a-b)\dfrac{\Gamma(a)\Gamma(b)}{\Gamma(c)}F(a,b;c;z)$

$$=(a-b)\frac{\Gamma(a)\Gamma(b)}{\Gamma(c)}\sum_{n=0}^{\infty}\frac{(a)_n(b)_n}{(c)_n}\frac{z^n}{n!}$$

$$=\sum_{n=0}^{\infty}\{(a+n)-(b+n)\}\frac{\Gamma(a+n)\Gamma(b+n)}{\Gamma(c+n)}\frac{z^n}{n!}$$

$$=\sum_{n=0}^{\infty}\frac{\Gamma(a+n+1)\Gamma(b+n)}{\Gamma(c+n)}\frac{z^n}{n!}-\sum_{n=0}^{\infty}\frac{\Gamma(a+n)\Gamma(b+n+1)}{\Gamma(c+n)}\frac{z^n}{n!}$$

$$=\frac{\Gamma(a+1)\Gamma(b)}{\Gamma(c)}\sum_{n=0}^{\infty}\frac{(a+1)_n(b)_n}{(c)_n}\frac{z^n}{n!}-\frac{\Gamma(a)\Gamma(b+1)}{\Gamma(c)}\sum_{n=0}^{\infty}\frac{(a)_n(b+1)_n}{(c)_n}\frac{z^n}{n!}$$

$$=\frac{\Gamma(a)\Gamma(b)}{\Gamma(c)}a\,F(a+1,b;c;z)-\frac{\Gamma(a)\Gamma(b)}{\Gamma(c)}b\,F(a,b+1;c;z)$$

$$=\frac{\Gamma(a)\Gamma(b)}{\Gamma(c)}\{aF_{a+}-bF_{B+}\}.$$

$\therefore\ (a-b)F=a\,F_{a+}-b\,F_{b+}$, which holds when $|z|<1$.

Hence, in general, the theorem holds.

5.22 DIXON'S THEOREM

Prove that

$${}_3F_2\left[\begin{matrix}\alpha,\beta,\gamma; \\ 1+\alpha-\beta,1+\alpha-\gamma;\end{matrix}1\right]$$

$$=\frac{\Gamma\left(1+\frac{\alpha}{2}\right)\Gamma\left(1+\frac{\alpha}{2}-\beta-\gamma\right)\Gamma(1+\alpha-\beta)\Gamma(1+\alpha-\gamma)}{\Gamma(1+\alpha)(1+\alpha-\beta-\gamma)\Gamma\left(1+\frac{\alpha}{2}-\beta\right)\Gamma\left(1+\frac{\alpha}{2}-\gamma\right)}$$

Proof: We have

$$\frac{\Gamma(\alpha)\Gamma(\beta)\Gamma(\gamma)}{\Gamma(1+\alpha-\beta)\Gamma(1+\alpha-\gamma)}\,{}_3F_2\left[\begin{matrix}\alpha,\beta,\gamma;\\1+\alpha-\beta,1+\alpha-\gamma;\end{matrix}\,1\right]$$

$$=\frac{\Gamma(\alpha)\Gamma(\beta)\Gamma(\gamma)}{\Gamma(1+\alpha-\beta)\Gamma(1+\alpha-\gamma)}\sum_{n=0}^{\infty}\frac{(\alpha)_n(\beta)_n(\gamma)_n}{(1+\alpha-\beta)_n(1+\alpha-\gamma)_n}\frac{1}{n!}$$

$$=\sum_{n=0}^{\infty}\frac{\Gamma(\alpha+n)\Gamma(\beta+n)\Gamma(\gamma+n)}{n!\,\Gamma(1+\alpha+2n)\Gamma(1+\alpha-\beta-\gamma)}$$

$$\times\left\{\frac{\Gamma(1+\alpha+2n)\Gamma(1+\alpha-\beta-\gamma)}{\Gamma(1+\alpha-\beta+n)\Gamma(1+\alpha-\gamma+n)}\right\}$$

$$=\sum_{n=0}^{\infty}\frac{\Gamma(\alpha+n)\Gamma(\beta+n)\Gamma(\gamma+n)}{n!\,\Gamma(1+\alpha+2n)\Gamma(1+\alpha-\beta-\gamma)}\times{}_2F_1(\beta+n,\gamma+n;1+\alpha+2n;1),$$

$$\left[\text{since from Gauss's theorem }{}_2F_1(\alpha,\beta;\gamma;1)=\frac{\Gamma(\gamma)\Gamma(\gamma-\alpha-\beta)}{\Gamma(\gamma-\alpha)\Gamma(\gamma-\beta)}\right]$$

$$=\sum_{n=0}^{\infty}\frac{\Gamma(\alpha+n)\Gamma(\beta+n)\Gamma(\gamma+n)}{n!\,\Gamma(1+\alpha+2n)\Gamma(1+\alpha-\beta-\gamma)}\left\{\sum_{m=0}^{\infty}\frac{(\beta+n)_m(\gamma+n)_m}{(1+\alpha+2n)_m}\frac{1}{m!}\right\}$$

$$=\sum_{n=0}^{\infty}\sum_{m=0}^{\infty}\frac{\Gamma(\alpha+n)\{\Gamma(\beta+n)(\beta+n)_m\}\{\Gamma(\gamma+n)(\gamma+n)_m\}}{\{\Gamma(1+\alpha+2n)(1+\alpha+2n)_m\}\Gamma(1+\alpha-\beta-\gamma)n!\,m!}$$

$$=\sum_{n=0}^{\infty}\sum_{m=0}^{\infty}\frac{\Gamma(\alpha+n)\Gamma(\beta+n+m)\Gamma(\gamma+n+m)}{\Gamma(1+\alpha+2n+m)\Gamma(1+\alpha-\beta-\gamma)n!\,m!}$$

$$=\sum_{m=0}^{\infty}\sum_{n=0}^{\infty}\frac{\Gamma(\beta+n+m)\Gamma(\gamma+n+m)\Gamma(\alpha+n)}{\Gamma(1+\alpha-\beta-\gamma)\Gamma(1+\alpha+2n+m)n!\,m!},$$

changing the order of summation

$$=\sum_{p=0}^{\infty}\frac{\Gamma(\beta+p)\Gamma(\gamma+p)}{\Gamma(1+\alpha-\beta-\gamma)}\sum_{n=0}^{\infty}\frac{\Gamma(\alpha+n)}{n!(p-n)!\Gamma(1+\alpha+n+p)}$$

putting $p=n+m$

$$=\sum_{p=0}^{\infty}\frac{\Gamma(\beta+p)\Gamma(\gamma+p)}{\Gamma(1+\alpha-\beta-\gamma)}\sum_{n=0}^{\infty}\frac{\Gamma(\alpha+n)}{n!\,\Gamma(1+\alpha+n+p)}\times(-1)^n\frac{(-p)_n}{p!}$$

$$\text{since }\frac{1}{(p-n)!}=(-1)^n\frac{(-p)_n}{p!}$$

$$=\sum_{p=0}^{\infty}\frac{\Gamma(\beta+p)\Gamma(\gamma+p)\Gamma(\alpha)}{p!\,\Gamma(1+\alpha-\beta-\gamma)\Gamma(1+\alpha+p)}\times\sum_{n=0}^{\infty}\frac{(\alpha)_n(-p)_n}{(1+\alpha+p)_n}\frac{(-1)^n}{n!}$$

$$=\sum_{p=0}^{\infty}\frac{\Gamma(\alpha)\Gamma(\beta+p)\Gamma(\gamma+p)}{p!\,\Gamma(1+\alpha-\beta-\gamma)\Gamma(1+\alpha+p)}\times{}_2F_1(\alpha,-p;1+\alpha+p;-1)$$

$$=\sum_{p=0}^{\infty}\frac{\Gamma(\alpha)\Gamma(\beta+p)\Gamma(\gamma+p)}{p!\,\Gamma(1+\alpha-\beta-\gamma)\Gamma(1+\alpha+p)}\times\frac{\Gamma(1+\alpha+p)\Gamma\left(\frac{\alpha}{2}+1\right)}{\Gamma(\alpha+1)\Gamma\left(\frac{\alpha}{2}+p+1\right)},$$

using Kummer's theorem

$$=\sum_{p=0}^{\infty}\frac{\Gamma(\alpha)\Gamma(\beta+p)\Gamma(\gamma+p)\Gamma\left(\frac{\alpha}{2}+1\right)}{p!\,\Gamma(1+\alpha-\beta-\gamma)\Gamma(\alpha+1)\Gamma\left(\frac{\alpha}{2}+p+1\right)}$$

$$=\frac{\Gamma(\alpha)\Gamma(\beta)\Gamma(\gamma)}{\Gamma(1+\alpha)\Gamma(1+\alpha-\beta-\gamma)}\sum_{p=0}^{\infty}\frac{(\beta)_p(\gamma)_p}{\left(\frac{\alpha}{2}+1\right)_p}\frac{1}{p!}$$

$$=\frac{\Gamma(\alpha)\Gamma(\beta)\Gamma(\gamma)}{\Gamma(1+\alpha)\Gamma(1+\alpha-\beta-\gamma)}\,{}_2F_1\left(\beta,\gamma;\frac{\alpha}{2}+1;1\right)$$

$$=\frac{\Gamma(\alpha)\Gamma(\beta)\Gamma(\gamma)}{\Gamma(1+\alpha)\Gamma(1+\alpha-\beta-\gamma)}\frac{\Gamma\left(\frac{\alpha}{2}+1\right)\Gamma\left(\frac{\alpha}{2}+1-\beta-\gamma\right)}{\Gamma\left(\frac{\alpha}{2}+1-\beta\right)\Gamma\left(\frac{\alpha}{2}+1-\gamma\right)},$$

using Gauss's theorem

Hence we have the required result

$${}_3F_2\left[\begin{matrix}\alpha,\beta,\gamma;\\1+\alpha-\beta,1+\alpha-\gamma;\end{matrix}1\right]$$

$$=\frac{\Gamma\left(1+\frac{\alpha}{2}\right)\Gamma\left(1+\frac{\alpha}{2}-\beta-\gamma\right)\Gamma(1+\alpha-\beta)\Gamma(1+\alpha-\gamma)}{\Gamma(1+\alpha)(1+\alpha-\beta-\gamma)\Gamma\left(1+\frac{\alpha}{2}-\beta\right)\Gamma\left(1+\frac{\alpha}{2}-\gamma\right)}.$$

5.23 TRANSFORMATIONS OF ${}_2F_1[\alpha,\beta;\gamma;z]$

There are two types of transformations: one is simple one and another is quadratic.

(a) Simple transformations

(i) ${}_2F_1\left[\alpha,\beta;\gamma;z\right] = (1-z)^{-\alpha}\ {}_2F_1\left[\alpha,\gamma\quad\beta;\gamma;\frac{-z}{(1-z)}\right]$

(ii) ${}_2F_1\left[\alpha,\beta;\gamma;z\right] = (1-z)^{\gamma-\alpha-\beta}\ {}_2F_1\left[\gamma-\alpha,\gamma-\beta;\gamma;z\right]$

Proof: We have

$$(1-z)^{-\alpha}\ {}_2F_1\left[\alpha,\gamma-\beta;\gamma;\frac{-z}{(1-z)}\right] = \sum_{n=0}^{\infty}\frac{(\alpha)_n(\gamma-\beta)_n}{(\gamma)_n\, n!}\frac{(-1)^n z^n}{(1-z)^{n+\alpha}}$$

$$= \sum_{n=0}^{\infty}\frac{(\alpha)_n(\gamma-\beta)_n(-1)^n z^n}{(\gamma)_n\, n!}\sum_{m=0}^{\infty}\frac{(n+\alpha)_m\, z^m}{m!},$$

$$\left[\text{since}\,(1-z)^{-(n+\alpha)} = \sum_{m=0}^{\infty}\frac{(n+\alpha)_m\, z^m}{m!}\right]$$

$$= \sum_{n=0}^{\infty}\sum_{m=0}^{\infty}\frac{(\alpha)_n(\gamma-\beta)_n(-1)^n(\alpha)_{n+m}\, z^{n+m}}{(\alpha)_n(\gamma)_n\, n!\, m!},$$

$$\left[\text{since}\,(n+\alpha)_m = \frac{(\alpha)_{n+m}}{(\alpha)_n}\right]$$

$$= \sum_{n=0}^{\infty}\sum_{m=0}^{\infty}\frac{(\gamma-\beta)_n(\alpha)_{n+m}(-1)^n z^{n+m}}{(\gamma)_n\, n!\, m!}$$

Applying $\sum_{m=0}^{\infty}\sum_{n=0}^{\infty}A(n,m) = \sum_{m=0}^{\infty}\sum_{n=0}^{m}A(n,m-n)$, we obtain

$$= \sum_{m=0}^{\infty}\sum_{n=0}^{m}\frac{(\gamma-\beta)_n(\alpha)_{n+m-n}(-1)^n z^{n+m-n}}{(\gamma)_n\, n!\, m-n!}$$

$$= \sum_{m=0}^{\infty}\sum_{n=0}^{m}\frac{(\gamma-\beta)_n(\alpha)_m(-1)^n z^m}{(\gamma)_n\, n!\, m-n!}$$

$$= \sum_{m=0}^{\infty}\sum_{n=0}^{m}\frac{(-m)_n(\gamma-\beta)_n(\alpha)_m z^m}{(\gamma)_n\, n!\, m!},$$

$$\left[\text{since}\ (m-n)! = \frac{(-1)^n m!}{(-m)_n}, 0 \le n \le m\right]$$

But $\sum_{n=0}^{m}\frac{(-m)_n(\gamma-\beta)_n}{(\gamma)_n\, n!}$ is a terminating hypergeometric series of

${}_2F_1(-m,\gamma-\beta;\gamma;1)$, so we have

$$(1-z)^{-\alpha}\ {}_2F_1\left[\alpha,\gamma-\beta;\gamma;\frac{-z}{(1-z)}\right]=\sum_{m=0}^{\infty}\ {}_2F_1(-m,\gamma-\beta;\gamma;1)\frac{(\alpha)_m z^m}{m!}$$

Using Vandermonde theorem, ${}_2F_1(-n,\beta;\gamma;1)=\dfrac{\Gamma\gamma\,\Gamma(\gamma-\beta+n)}{\Gamma(\gamma+n)\,\Gamma(\gamma-\beta)}$,

we get

$$(1-z)^{-\alpha}\ {}_2F_1\left[\alpha,\gamma-\beta;\gamma;\frac{-z}{(1-z)}\right]=\sum_{m=0}^{\infty}\frac{\Gamma\gamma\,\Gamma(\beta+m)}{\Gamma(\gamma+m)\,\Gamma\beta}\frac{(\alpha)_m z^m}{m!}$$

$$=\sum_{m=0}^{\infty}\frac{(\alpha)_m(\beta)_m}{(\gamma)_m}\frac{z^m}{m!}={}_2F_1(\alpha,\beta;\gamma;z)$$

which is valid under the conditions $|z|<1$ and $|z/(1-z)|<1$.

(ii) Applying the above transformation repeatedly for

${}_2F_1\left[\alpha,\gamma-\beta;\gamma;\dfrac{-z}{1-z}\right]$ we obtain

$${}_2F_1[\alpha,\beta;\gamma;z]=(1-z)^{-\alpha}(1-z)^{\gamma-\beta}\ {}_2F_1[\gamma-\alpha,\gamma-\beta;\gamma;z]$$

$$=(1-z)^{\gamma-\alpha-\beta}\ {}_2F_1[\gamma-\alpha,\gamma-\beta;\gamma;z]$$

which is valid for $|z|<1$.

(b) Quadratic transformations

(i) $(1+z)^{-2\alpha}\ F\left[\begin{matrix}\alpha,\beta; & \dfrac{4z}{(1+z)^2}\\ 2\beta; & \end{matrix}\right]=F\left[\begin{matrix}\alpha,\alpha-\beta+\frac{1}{2}; & z^2\\ \beta+\frac{1}{2}; & \end{matrix}\right]$

(ii) $(1-y)^{-\alpha}F\left[\begin{matrix}\frac{1}{2}\alpha,\frac{1}{2}+\frac{1}{2}\alpha & \dfrac{y^2}{(1-y)^2}\\ \beta+\dfrac{1}{2}; & \end{matrix}\right]=F\left[\begin{matrix}\alpha,\beta; & 2y\\ 2\beta; & \end{matrix}\right]$

Proof: On putting $\gamma=2\beta$ in the hypergeometric differential equation, we obtain

$$z(1-z)\frac{d^2y}{dz^2}+[2\beta-(1+\alpha+\beta)z]\frac{dy}{dz}-\alpha\beta y=0 \qquad ...(1)$$

One of its solution be ${}_2F_1(\alpha,\beta;2\beta;z)$. On using transformation

$z=\dfrac{4x}{(1+x)^2}$, we get

$$\frac{dy}{dz}=\frac{dy}{dx}\frac{dx}{dz}=\frac{(1+x)^3}{4(1-x)}\frac{dy}{dx}$$

and $$\frac{d^2y}{dz^2}=\frac{(1+x)^6}{16(1-x)^2}\frac{d^2y}{dx^2}+\frac{(1+x)^5(2-x)}{8(1-x)^3}\frac{dy}{dx}$$

Substituting these values of $\frac{dy}{dz}$ and $\frac{d^2y}{dz^2}$ in (1), we obtain

$$x(1-x)(1+x)^2\frac{d^2y}{dx^2}+2(1+x)\left[\beta-2\alpha x+\beta x^2-x^2\right]\frac{dy}{dx}$$
$$-4\alpha\beta(1-x)y=0 \quad ...(2)$$

Its one solution is $y={}_2F_1\left[\begin{matrix}\alpha,\beta; & \frac{4x}{(1+x)^2}\\ 2\beta; & \end{matrix}\right]$ ...(3)

Taking $y=(1+x)^{2\alpha}w$, we have

$$\frac{dy}{dx}=(1+x)^{2\alpha}\frac{dw}{dx}+2\alpha(1+x)^{2\alpha-1}w$$

and $$\frac{d^2y}{dx^2}=(1+x)^{2\alpha}\frac{d^2w}{dx^2}+4\alpha(1+x)^{2\alpha-1}\frac{dw}{dx}+2\alpha(2\alpha-1)(1+x)^{2\alpha-2}w$$

Putting values of y, $\frac{dy}{dx}$ and $\frac{d^2y}{dx^2}$ in equation (2), we get

$$x(1-x^2)\frac{d^2w}{dx^2}+2\left[\beta-(2\alpha-\beta+1)x^2\right]\frac{dw}{dx}-2\alpha x(1+2\alpha-2\beta)w=0 \quad ...(4)$$

Its one solution is $w=(1+x)^{-2\alpha}{}_2F_1\left[\begin{matrix}\alpha,\beta; & \frac{4x}{(1+x)^2}\\ 2\beta; & \end{matrix}\right]$ ...(5)

It should be noted that the equation (4) will remain unchanged if we replace x by $-x$. Therefore substituting $v=x^2$ in (4) we obtain

$$v(1-v)\frac{d^2w}{dv^2}+2\left[\beta+\tfrac{1}{2}-(2\alpha-\beta+\tfrac{3}{2})v\right]\frac{dw}{dv}-\alpha(\alpha-\beta+\tfrac{1}{2})w=0 \quad ...(6)$$

This is identical to equation (1) and has a convergent hypergeometric series as solution for $|v|<1$.

The general solution of (6) is given by

$$w=A\,{}_2F_1\left[\begin{matrix}\alpha,\alpha-\beta+\frac{1}{2}; & v\\ \beta+\frac{1}{2}; & \end{matrix}\right]+B\,v^{\frac{1}{2}-\beta}\,{}_2F_1\left[\begin{matrix}\alpha-\beta+\frac{1}{2},\alpha+1-2\beta; & v\\ \frac{3}{2}-\beta; & \end{matrix}\right] \quad ...(7)$$

From this discussion it is clear that for some constants A and B when $|x|<1$, $\left|\dfrac{4x}{(1+x)^2}\right|<1$, and 2β is neither zero nor a negative integer, then the following equality must hold

$$(1+x)^{-2\alpha}\,{}_2F_1\left[\begin{matrix}\alpha,\beta; & \dfrac{4x}{(1+x)^2}\\ 2\beta; & \end{matrix}\right]=A\,{}_2F_1\left[\begin{matrix}\alpha,\alpha-\beta+\frac{1}{2}; & x^2\\ \beta+\frac{1}{2}; & \end{matrix}\right]$$

$$+B\,x^{1-2\beta}\,{}_2F_1\left[\begin{matrix}\alpha-\beta+\frac{1}{2},\alpha+1-2\beta; & x^2\\ \frac{3}{2}-\beta; & \end{matrix}\right] \quad \text{...(8)}$$

For $x=0$, we see that the last term has singularity due to the presence of $x^{1-2\beta}$, so we must have $B=0$. Thus we get

$$(1+x)^{-2\alpha}\,{}_2F_1\left[\begin{matrix}\alpha,\beta; & \dfrac{4x}{(1+x)^2}\\ 2\beta; & \end{matrix}\right]=A\,{}_2F_1\left[\begin{matrix}\alpha,\alpha-\beta+\frac{1}{2}; & x^2\\ \beta+\frac{1}{2}; & \end{matrix}\right] \quad \text{...(9)}$$

Again putting $x=0$, we get $A=1$. Thus if 2β is neither zero nor a negative integer, and if $|x|<1$ and $\left|\dfrac{4x}{(1+x)^2}\right|<1$ hold, then

$$(1+z)^{-2\alpha}\,F\left[\begin{matrix}\alpha,\beta; & \dfrac{4z}{(1+z)^2}\\ 2\beta; & \end{matrix}\right]=F\left[\begin{matrix}\alpha,\alpha-\beta+\frac{1}{2}; & z^2\\ \beta+\frac{1}{2}; & \end{matrix}\right]$$

(ii) To prove this, we apply the method of series manipulation by considering the function

$$(1-y)^{-\alpha}\,F\left[\begin{matrix}\frac{1}{2}\alpha,\frac{1}{2}+\frac{1}{2}\alpha; & \dfrac{y^2}{(1-y)^2}\\ \beta+\frac{1}{2}; & \end{matrix}\right]=\sum_{m=0}^{\infty}\frac{\left(\frac{1}{2}\alpha\right)_m\left(\frac{1}{2}+\frac{1}{2}\alpha\right)_m}{\left(\frac{1}{2}+\beta\right)_m m!}\frac{y^{2m}}{(1-y)^{\alpha+2m}}$$

$$=\sum_{m=0}^{\infty}\frac{(\alpha)_{2m}}{2^{2m}\left(\beta+\frac{1}{2}\right)_m m!}\frac{y^{2m}}{(1-y)^{\alpha+2m}},$$

$$\left[\text{since } (\alpha)_{2m}=2^{2m}\left(\tfrac{1}{2}\alpha\right)_m\left(\tfrac{1}{2}+\tfrac{1}{2}\alpha\right)_m\right]$$

$$=\sum_{m=0}^{\infty}\sum_{n=0}^{\infty}\frac{(\alpha)_{2m}(\alpha+2m)_n\,y^{2m}y^n}{2^{2m}\left(\beta+\frac{1}{2}\right)_m\,m!\,n!},$$

$$\left[\text{since } (1-y)^{-(\alpha+2m)}=\sum_{n=0}^{\infty}\frac{(\alpha+2m)_n}{n!}y^n\right]$$

$$= \sum_{m=0}^{\infty} \sum_{n=0}^{\infty} \frac{(\alpha)_{n+2m} \, y^{n+2m}}{2^{2m} \left(\beta + \frac{1}{2}\right)_m m! n!}$$

$$= \sum_{n=0}^{\infty} \sum_{m=0}^{[n/2]} \frac{(\alpha)_n \, y^n}{2^{2m} \left(\beta + \frac{1}{2}\right)_m m!(n-2m)!},$$

$$\left[\text{since} \sum_{m=0}^{\infty} \sum_{n=0}^{\infty} A(m,n) = \sum_{n=0}^{\infty} \sum_{m=0}^{[n/2]} A(m, n-2m)\right]$$

Also, we know that

$$(n-2m)! = \frac{(-1)^{2m} \, n!}{(-n)_{2m}} = \frac{n!}{2^{2m} \left(-\frac{1}{2}n\right)_m \left(-\frac{1}{2}n + \frac{1}{2}\right)_m}$$

This gives

$$(1-y)^{-\alpha} F\left[\begin{matrix} \frac{1}{2}\alpha, \frac{1}{2} + \frac{1}{2}\alpha; & \dfrac{y^2}{(1-y)^2} \\ \beta + \frac{1}{2}; & \end{matrix}\right]$$

$$= \sum_{n=0}^{\infty} \sum_{m=0}^{[n/2]} \frac{(\alpha)_n y^n \, 2^{2m} \left(-\frac{1}{2}n\right)_m \left(-\frac{1}{2}n + \frac{1}{2}\right)_m}{2^{2m} \left(\beta + \frac{1}{2}\right)_m m! n!}$$

$$= \sum_{n=0}^{\infty} \sum_{m=0}^{[n/2]} \frac{(\alpha)_n y^n \left(-\frac{1}{2}n\right)_m \left(-\frac{1}{2}n + \frac{1}{2}\right)_m}{\left(\beta + \frac{1}{2}\right)_m m! \, n!}$$

$$= \sum_{n=0}^{\infty} F\left[\begin{matrix} -\frac{1}{2}n, \frac{1}{2} - \frac{1}{2}n; & 1 \\ \beta + \frac{1}{2}; & \end{matrix}\right] \frac{(\alpha)_n \, y^n}{n!} \qquad \ldots(10)$$

But by Gauss's theorem, we have

$$F\left[\begin{matrix} -\frac{1}{2}n, \frac{1}{2} - \frac{1}{2}n; & 1 \\ \beta + \frac{1}{2}; & \end{matrix}\right] = \frac{\Gamma\left(\beta + \frac{1}{2}\right) \Gamma(\beta + n)}{\Gamma\left(\beta + \frac{1}{2}n + \frac{1}{2}\right) \Gamma\left(\beta + \frac{1}{2}n\right)}$$

$$= \frac{2^{2\beta + n - 1} \Gamma\left(\beta + \frac{1}{2}\right) \Gamma(\beta + n)}{\Gamma\left(\frac{1}{2}\right) \Gamma(2\beta + n)}$$

$$= \frac{2^{2\beta + n - 1} \Gamma\left(\beta + \frac{1}{2}\right) \Gamma 2\beta \, \Gamma\beta \, \Gamma(\beta + n)}{\Gamma 2\beta \, \Gamma\beta \, \Gamma(2\beta + n) \Gamma\left(\frac{1}{2}\right)}$$

$$= \frac{2^n (\beta)_n 2^{n-1} \Gamma\left(\beta + \frac{1}{2}\right) \Gamma\beta}{(2\beta)_n \Gamma(2\beta) \Gamma\left(\frac{1}{2}\right)} = \frac{2^n (\beta)_n}{(2\beta)_n}$$

Thus equation (10) gives

$$(1-y)^{-\alpha} F\left[\begin{matrix}\frac{1}{2}\alpha, \frac{1}{2}+\frac{1}{2}\alpha; \\ \beta+\frac{1}{2};\end{matrix} \frac{y^2}{(1-y)^2}\right]$$

$$= \sum_{n=0}^{\infty} \frac{2^n (\beta)_n}{(2\beta)_n} \frac{(\alpha)_n y^n}{n!} = F\left[\begin{matrix}\alpha, \beta; \\ 2\beta;\end{matrix} 2y\right] \qquad \ldots(11)$$

Example 1: Prove the following

(i) $e^x = {}_1F_1(\alpha; \alpha; x)$

(ii) $(1-x)^{-\alpha} = {}_2F_1(\alpha, \beta; \beta; x)$

Solution: (i) By definition, we have

$$ {}_1F_1(\alpha; \beta; x) = \sum_{r=0}^{\infty} \frac{(\alpha)_r x^r}{(\beta)_r r!} = 1 + \frac{\alpha}{\beta}\frac{x}{1!} + \frac{\alpha(\alpha+1)}{\beta(\beta+1)}\frac{x^2}{2!} + \cdots\cdots + \cdots\cdots \infty. \qquad \ldots(1)$$

Replacing β by α in (1), we have

$$ {}_1F_1(\alpha; \alpha; x) = 1 + \frac{x}{1!} + \frac{x^2}{2!} + \cdots\cdots + \cdots \infty = e^x .$$

(ii) By definition, we have

$$ {}_2F_1(\alpha; \beta; \gamma; x) = 1 + \frac{\alpha\,\beta}{\gamma}\frac{x}{1!} + \frac{\alpha(\alpha+1)\beta(\beta+1)}{\gamma(\gamma+1)}\frac{x^2}{2!} + \cdots\cdots$$

Replacing γ by β, we obtain

$$ {}_2F_1(\alpha, \beta; \beta; x) = 1 + \frac{\alpha\,\beta}{\beta}\frac{x}{1!} + \frac{\alpha(\alpha+1)\beta(\beta+1)}{\beta(\beta+1)}\frac{x^2}{2!} + \cdots$$

$$= 1 + \frac{\alpha\,x}{1!} + \alpha(\alpha+1)\frac{x^2}{2!} + \alpha(\alpha+1)(\alpha+2)\frac{x^3}{3!} + \cdots$$

$$= 1 + (-\alpha)(-x) + \frac{(-\alpha)(-\alpha-1)}{2!}(-x)^2 + \frac{(-\alpha)(-\alpha-1)(-\alpha-2)}{3!}(-x)^3 + \cdots$$

$= (1-x)^{-\alpha}$, by the Binomial theorem.

Example 2: Prove that $F\left(\alpha, \beta; \gamma; \frac{1}{x}\right) = \frac{\Gamma(\gamma)\Gamma(\gamma-\alpha-\beta)}{\Gamma(\gamma-\alpha)\Gamma(\gamma-\beta)} x^{\alpha}$

$$ {}_2F_1(\alpha, \alpha-\gamma+1; \alpha+\beta-\gamma+1; 1-x) + \frac{\Gamma(\gamma)\Gamma(\alpha+\beta-\gamma)}{\Gamma(\alpha)\Gamma(\beta)}$$

$$\times x^{\beta}(x-1)^{\gamma-\alpha-\beta}\ {}_2F_1(\gamma-\alpha, 1-\alpha; \gamma-\alpha-\beta+1; 1-x)$$

where $1 < x < 2$ and $1 > \gamma > \alpha + \beta$.

Solution: From the linear relation between the solutions of the hypergeometric function, we have

$$ {}_2F_1(\alpha,\beta;\gamma;x)=\frac{\Gamma(\gamma)\Gamma(\gamma-\alpha-\beta)}{\Gamma(\gamma-\alpha)\Gamma(\gamma-\beta)}\,{}_2F_1(\alpha,\beta;\alpha+\beta-\gamma+1;1-x) $$

$$ +\frac{\Gamma(\gamma)\Gamma(\alpha+\beta-\gamma)}{\Gamma(\alpha)\Gamma(\beta)}(1-x)^{\gamma-\alpha-\beta}\,{}_2F_1(\gamma-\alpha,\gamma-\beta;\gamma-\alpha-\beta+1;1-x) $$

Replacing x by $\frac{1}{x}$, we obtain

$$ {}_2F_1\left(\alpha,\beta;\gamma;\frac{1}{x}\right)=\frac{\Gamma(\gamma)\Gamma(\gamma-\alpha-\beta)}{\Gamma(\gamma-\alpha)\Gamma(\gamma-\beta)}\,{}_2F_1\left(\alpha,\beta;\alpha+\beta-\gamma+1;1-\frac{1}{x}\right) $$

$$ +\frac{\Gamma(\gamma)\Gamma(\alpha+\beta-\gamma)}{\Gamma(\alpha)\Gamma(\beta)}\left(1-\frac{1}{x}\right)^{\gamma-\alpha-\beta}{}_2F_1\left(\gamma-\alpha,\gamma-\beta;\gamma-\alpha-\beta+1;1-\frac{1}{x}\right) \quad ...(1) $$

We know that

$$ {}_2F_1(\alpha,\beta;\gamma;x)=(1-x)^{-\alpha}\,{}_2F_1\left(\alpha,\gamma-\beta;\gamma;\frac{1}{x}\right) \quad ...(2) $$

Replacing γ by $\alpha+\beta-\gamma+1$ and x by $1-\frac{1}{x}$, we get

$$ {}_2F_1\left(\alpha,\beta;\alpha+\beta-\gamma+1;1-\frac{1}{x}\right)= $$

$$ x^{\alpha}{}_2F_1(\alpha,\alpha-\gamma+1;\alpha+\beta-\gamma+1;1-x). \quad ...(3) $$

Again replacing α by $\gamma-\alpha,\beta$ by $\gamma-\beta,\gamma$ by $\gamma-\alpha-\beta+1$ and x by $1-\frac{1}{x}$ in (2), we get

$$ {}_2F_1\left(\gamma-\alpha,\gamma-\beta;\gamma-\alpha-\beta+1;1+\frac{1}{x}\right) $$

$$ =x^{\gamma-\alpha}\,{}_2F_1(\gamma-\alpha;1-\alpha;\gamma-\alpha-\beta+1;1+x). \quad ...(4) $$

Hence from (1), (3) and (4), we obtain

$$ {}_2F_1\left(\alpha,\beta;\gamma;\frac{1}{x}\right)=\frac{\Gamma(\gamma)\Gamma(\gamma-\alpha-\beta)}{\Gamma(\gamma-\alpha)\Gamma(\gamma-\beta)}x^{\alpha} $$

$$ {}_2F_1(\alpha,\alpha-\gamma+1;\alpha+\beta-\gamma+1;1-x)+\frac{\Gamma(\gamma)\Gamma(\alpha+\beta-\gamma)}{\Gamma(\alpha)\Gamma(\beta)}. $$

$$ \left(1-\frac{1}{x}\right)^{\gamma-\alpha-\beta}x^{\gamma-\alpha}\times{}_2F_1(\alpha,\alpha-\gamma+1;\alpha+\beta-\gamma+1;1-x) $$

$$= \frac{\Gamma(\gamma)\Gamma(\gamma-\alpha-\beta)}{\Gamma(\gamma-\alpha)\Gamma(\gamma-\beta)} . x^{\alpha}\ {}_2F_1(\alpha, \alpha-\gamma+1; \alpha+\beta-\gamma+1; 1-x)$$

$$+ \frac{\Gamma(\gamma)\Gamma(\alpha+\beta-\gamma)\, x^{\beta}(x-1)^{\gamma-\alpha-\beta}}{\Gamma(\alpha)\Gamma(\beta)}\ {}_2F_1(\alpha, \alpha-\gamma+1; \alpha+\beta-\gamma+1; 1-x)$$

Example 3: Prove that ${}_1F_1(\alpha;\gamma;x) = e^x\ {}_1F_1(\gamma-\alpha;\gamma;-x)$

Solution: We know that

$${}_1F_1(\alpha;\gamma;x) = \frac{1}{B(\alpha,\gamma-\alpha)} \int_0^1 (1-t)^{\gamma-\alpha-1}\, t^{\alpha-1}\, e^{xt}\, dt. \qquad \ldots(1)$$

Replacing x by $-x$ and α by $\gamma-\alpha$, we have

$${}_1F_1(\gamma-\alpha;\gamma;-x) = \frac{1}{B(\gamma-\alpha,\alpha)} \int_0^1 (1-t)^{\alpha-1}\, t^{\gamma-\alpha-1}\, e^{-xt}\, dt$$

$$= \frac{e^{-x}}{B(\alpha,\gamma-\alpha)} \int_0^1 (1-t)^{\alpha-1}\, t^{\gamma-\alpha-1}\, e^{x(1-t)}\, dt$$

Putting $1-t=u$ and $dt=-du$, we obtain

$${}_1F_1(\gamma-\alpha;\gamma;-x) = \frac{e^{-x}}{B(\alpha,\gamma-\alpha)} \int_0^1 u^{\alpha-1}(1-u)^{\gamma-\alpha-1}\, e^{ux}\, du$$

$$= e^{-x}\ {}_1F_1(\alpha;\gamma;x).$$

This implies that ${}_1F_1(\alpha;\gamma;x). = e^x\ {}_1F_1(\gamma-\alpha;\gamma;-x)$.

Example 4: Show that $P_n(\cos\theta) = \cos^2\theta\ {}_2F_1\left(-\frac{n}{2} - \frac{n-1}{2}; 1; -\tan^2\theta\right)$.

Solution: From Laplace's first integral, we have

$$P_n(\cos\theta) = \frac{1}{\pi}\int_0^{\pi}\left[\cos\theta + \sqrt{(\cos^2\theta-1)}\cos\phi\right]^n d\phi$$

$$= \frac{1}{\pi}\int_0^{\pi}[\cos\theta + i\sin\theta\cos\phi]^n\, d\phi$$

$$= \frac{1}{\pi}\cos^n\theta \int_0^{\pi}\left[1 + i\tan\theta\cos\phi + \frac{n(n-1)}{2!} i^2\tan^2\theta\cos^2\phi + \frac{n(n-1)(n-2)}{3!} i^3 \tan^3\theta\cos^3\phi + \ldots\right] d\phi$$

$$= \frac{1}{\pi}\cos^n\theta\left[\int_0^{\pi} d\phi + 0 + 2.\frac{n(n-1)}{2!} \times i^2\tan^2\theta \int_0^{\pi/2}\cos^2\phi\, d\phi + 0 + \cdots\cdots\right]$$

[by using definite integral property]

$$= \frac{1}{\pi} \cos^n \theta \left[\pi + i^2 \frac{n(n-1)\pi}{1!4} \tan^2 \theta + i^4 \frac{n(n-1)(n-2)(n-3)}{4} \right.$$

$$\left. 2 \frac{1.3}{2.4} \cdot \frac{\pi}{4} \tan^4 \theta + \cdots \right]$$

$$= \cos^n \theta \left[1 + \frac{\left(-\frac{n}{2}\right)\left(-\frac{n-1}{2}\right)}{1!1} \left(-\tan^2 \theta\right) + \frac{\left(-\frac{n}{2}\right)\left(-\frac{n}{2}+1\right).\left(-\frac{n-1}{2}\right)\left(-\frac{n-1}{2}+1\right)}{2!\ 2.1} \right.$$

$$\left. \left(-\tan^2 \theta\right)^2 + \cdots \right]$$

$$= \cos^n \theta \, {}_2F_1 \left(-\frac{n}{2}, -\frac{n-1}{2}; 1; -\tan^2 \theta \right).$$

Example 5: Prove that the following

(i) $P_n(\cos\theta) = {}_2F_1\left(-n, n+1; 1; \sin^2 \frac{\theta}{2}\right)$

(ii) $P_n(\cos\theta) = (-1)^n \, {}_2F_1\left(n+1, -n; 1; \cos^2 \frac{\theta}{2}\right)$

Solution: (i) From Rodrigue's formula, we have

$$P_n(x) = \frac{1}{2^n n!} \frac{d^n}{dx^n} \left(x^2 - 1\right)^n = \frac{1}{n!} \frac{d^n}{dx^n} \left[(x-1)^n \left\{ \frac{1}{2}(x+1) \right\}^n \right]$$

$$= \frac{(-1)^n}{n!} \frac{d^n}{dx^n} \left[(1-x)^n \left\{ 1 - \frac{1}{2}(1-x) \right\}^n \right]$$

$$= (-1)^n \frac{d^n}{dx^n} \left[(1-x)^n - \frac{n}{2}(1-x)^{n+1} + \frac{n(n-1)}{2!2^2}(1-x)^{n+2} \right.$$

$$\left. - \frac{n(n-1)(n-2)}{3!2^3}(1-x)^{n+3} + \ldots \right]$$

$$= \frac{(-1)^n}{n!} \left[(-1)^n n! - \frac{n}{2}(-1)^n \frac{(n+1)!}{1!}(1-x) \right.$$

$$\left. + \frac{n(n-1)}{2!} \frac{(-1)^2}{2^2} \frac{(n+2)!}{2!}(1-x)^2 - \ldots \right]$$

$$= 1 + \frac{(-n)(n+1)}{1.1!} \left(\frac{1-x}{2} \right) + \frac{(-n)(-n+1)}{2.1} \frac{(n+1)(n+2)}{2!} \left(\frac{1-x}{2} \right)^2 + \ldots$$

$$= {}_2F_1\left(-n, n+1; 1; \frac{1-x}{2}\right). \qquad \ldots(1)$$

Putting $x = \cos\theta$ in (1), we have

$$P_n(\cos\theta) = {}_2F_1\left(-n, n+1; 1; \sin^2\frac{\theta}{2}\right).$$

(ii) Putting $x = -\cos\theta$ in (1), we have

$$P_n(-\cos\theta) = {}_2F_1\left(-n, n+1; 1; \cos^2\frac{\theta}{2}\right)$$

$$\Rightarrow \quad (-1)^n P_n(\cos\theta) = {}_2F_1\left(-n, n+1; 1; \cos^2\frac{\theta}{2}\right),$$

since $P_n(-x) = (-1)^n P_n(x)$

$$\Rightarrow \quad P_n(\cos\theta) = (-1)^n {}_2F_1\left(-n, n+1; 1; \cos^2\frac{\theta}{2}\right)$$

$$\Rightarrow \quad P_n(\cos\theta) = (-1)^n {}_2F_1\left(n+1, -n; 1; \cos^2\frac{\theta}{2}\right)$$

Since in hypergeometric series α and β can be interchanged.

Example 6: Prove that $(\alpha - \gamma + 1)F(\alpha, \beta; \gamma; z) = \alpha F_{\alpha+} - (\gamma - 1)F_{\gamma-}$

Solution: We can write

$$(\alpha - \gamma + 1)\frac{\Gamma(\alpha)\Gamma(\beta)}{\Gamma(\gamma)} F(\alpha, \beta; \gamma; z)$$

$$= (\alpha - \gamma + 1)\frac{\Gamma(\alpha)\Gamma(\beta)}{\Gamma(\gamma)} \sum_{n=0}^{\infty} \frac{(\alpha)_n (\beta)_n z^n}{(\gamma)_n n!}$$

$$= \sum_{n=0}^{\infty} \{(\alpha + n) - (\gamma - 1 + n)\} \frac{\Gamma(\alpha + n)\Gamma(\beta + n)}{\Gamma(\gamma + n)} \frac{z^n}{n!}$$

$$= \sum_{n=0}^{\infty} \frac{\Gamma(\alpha + n + 1)\Gamma(\beta + n)}{\Gamma(\gamma + n)} \frac{z^n}{n!} - \sum_{n=0}^{\infty} \frac{\Gamma(\alpha + n)\Gamma(\beta + n)}{\Gamma(\gamma + n - 1)} \frac{z^n}{n!}$$

$$= \frac{\Gamma(\alpha + 1)\Gamma(\beta)}{\Gamma(\gamma)} \sum_{n=0}^{\infty} \frac{(\alpha + 1)_n (\beta)_n}{(\gamma)_n} \frac{z^n}{n!} - \frac{\Gamma(\alpha)\Gamma(\beta)}{\Gamma(\gamma - 1)} \sum_{n=0}^{\infty} \frac{(\alpha)_n (\beta)_n}{(\gamma - 1)_n} \frac{z^n}{n!}$$

$$= \frac{\Gamma(\alpha)\Gamma(\beta)}{\Gamma(\gamma)} \alpha F(\alpha + 1, \beta; \gamma; z) - \frac{\Gamma(\alpha)\Gamma(\beta)}{\Gamma(\gamma)} (\gamma - 1) F(\alpha, \beta; \gamma - 1; z)$$

$$= \frac{\Gamma(\alpha)\Gamma(\beta)}{\Gamma(\gamma)} . \{\alpha F_{\alpha+} - (\gamma - 1)F_{\gamma-}\}.$$

This implies that $(\alpha - \gamma + 1)F = \alpha F_{\alpha+,} - (\gamma - 1)F_{\gamma-}$

EXERCISE 5.1

1. Show that ${}_1F_1(\alpha;\gamma;x)=\lim_{\beta\to\infty}\left(\alpha,\beta;\gamma;\frac{x}{\beta}\right)$.

2. Show that the function $\omega={}_0F_1(-;\rho;z)$ is the solution of the equation

   $$z\frac{d^2\omega}{dz^2}+\rho\frac{d\omega}{dz}-\omega=0.$$

   Hence show that ${}_1F_1(\alpha;2\alpha;2z)=e^z\,{}_0F_1\left(-;\alpha+\tfrac{1}{2};\tfrac{z^2}{2}\right)$, provided $2\alpha>0$.

3. Show that ${}_2F_1(\alpha,\beta;\beta-\alpha+1;-1)=\dfrac{\Gamma(1+\beta-\alpha)\Gamma(1+\frac{1}{2}\beta)}{\Gamma(1+\beta)\Gamma(1+\frac{1}{2}\beta-\alpha)}$

   and deduce that ${}_2F_1(\alpha,1-\alpha;\gamma;\tfrac{1}{2})=\dfrac{2^{1-\gamma}\Gamma(\frac{1}{2}\gamma)\Gamma(\frac{1}{2}\gamma+\frac{1}{2})}{\Gamma(\frac{1}{2}\alpha+\frac{1}{2}\gamma)\Gamma(\frac{1}{2}-\frac{1}{2}\alpha+\frac{1}{2}\gamma)}$

4. Evaluate the integral $\int_0^\infty e^{-sx}\,{}_1F_1(\alpha;\beta;x)dx$

5. Prove that $F(a,b;a+b+\tfrac{1}{2};z)^2={}_3F_2\left[\begin{matrix}2a,a+b,2b;\\ a+b+\frac{1}{2},a+2b;\end{matrix}\;z\right]$

6. Prove that $F(\alpha,\beta+1;\gamma+1;x)-F(\alpha,\beta;\gamma;x)=\dfrac{\alpha(\gamma-\beta)}{\gamma(\gamma-1)}$

   $$xF(\alpha+1,\beta+1;\gamma+2;x)$$

7. Prove that ${}_2F_1(\alpha,\beta;\gamma;\tfrac{1}{2})=2^{\alpha+\beta-\gamma}{}_2F_1(\gamma-\alpha,\gamma-\beta;\gamma;\tfrac{1}{2})$.

8. Show that $\dfrac{d}{dx}F(\alpha,\beta;\gamma;x)=\dfrac{\alpha\beta}{\gamma}F(\alpha+1,\beta+1;\gamma+1;x)$

9. Show that ${}_1F_1(\alpha;\beta;x)=e^x\,F(\beta-\alpha;\beta;-x)$

10. The incomplete gamma function is defined by the equation

    $\gamma(\alpha,x)=\int_0^x e^{-t}\,t^{\alpha-t}\,dt,\ \alpha>0.$

    Prove that $\gamma(\alpha,x)=\alpha^{-1}x^{\alpha}\,F(\alpha;\alpha+1;-x)$.

11. Show that

    $$\int_0^{\pi/2}(1-k^2\sin^2\phi)^{-1/2}\,d\phi=\frac{\pi}{2}F\left(\frac{1}{2},\frac{1}{2};1;k^2\right),\text{ where }|k|<1.$$

12. Show that $\log\dfrac{1+x}{1-x}=2xF\left(\dfrac{1}{2},1;\dfrac{3}{2};x^2\right)$.

13. Show that ${}_2F_1\left(\alpha, \beta; \gamma; \frac{1}{2}\right) = 2^{\alpha}\ {}_2F_1(\alpha, \gamma - \beta; \gamma; -1)$.

14. Prove that

(i) $P_n(x) = \dfrac{(x-1)^n}{2^n}\ {}_2F_1\left(-n, -n; 1; \dfrac{x+1}{x-1}\right)$

(ii) $P_n(x) = {}_2F_1\left(-n, n+1; 1; \dfrac{1-x}{2}\right)$

ANSWER 5.1

4. $\left(\dfrac{1}{s}\right){}_2F_1(\alpha, 1; \beta; s)$.

6

Orthogonality of Functions and Sturm-Liouville Problems

6.1 INTRODUCTION

In this chapter, orthogonality of functions and a special kind of boundary value problems known as Sturm-Liouville problems are discussed. Such problems arise in physics, engineering and other applied sciences and help solving boundary value problems of partial differential equations. We also discuss the general theory of eigen values and eigen functions that exist as non-trivial solutions of the Sturm-Liouville problem which is one of the deepest and richest parts of modern mathematics.

6.2 SOME BASIC DEFINITIONS

Consider two real functions of $f(x)$ and $g(x)$ defined on an interval $a \le x \le b$.

6.2.1 Inner Product

Further, let the integral of the product of $f(x)$ and $g(x)$ on the interval $a \le x \le b$ exists and let it be denoted by (f, g). Then (f, g) is called the inner product of $f(x)$, $g(x)$ and defined as

$$(f, g) = \int_a^b f(x) g(x) dx. \qquad \ldots(1)$$

In case of complex valued functions, the inner product is defined as $(f, g) = \int_a^b f(x) \bar{g}(x) dx$.

6.2.2 Orthogonality

Two functions $f(x)$ and $g(x)$ are said to be orthogonal if

$$(f, g) = 0 \text{ i.e., } \int_a^b f(x) g(x) dx = 0. \qquad \ldots(2)$$

The norm of the function $f(x)$ is defined as non-negative square root of (f, f) and is generally denoted by $\| f \|$. Thus,

$$\text{norm of } f(x) = \| f \| = \sqrt{(f, f)} = \sqrt{\int_a^b f^2(x) dx} \qquad \ldots(3)$$

The function $f(x)$ is said to be normalized when

$$\|f\| = 1 \text{ i.e., } \int_a^b f^2(x)dx = 1. \quad \text{...(4)}$$

These terms arise in a natural way from an analogy with vectors in a vector space.

6.3 ORTHOGONAL SET OF FUNCTIONS

The set of functions $\{f_1(x), f_2(x), f_3(x), \ldots, f_n(x)\}$ is said to be orthogonal on the interval $a \le x \le b$, if

$$\int_a^b f_m(x) f_n(x)dx = 0, \text{ when } m \ne n.$$

6.4 ORTHOGONAL WITH RESPECT TO A WEIGHT OR DENSITY FUNCTION

Let $p(x) \ge 0$, then two functions $f(x)$ and $g(x)$ are said to be orthogonal with respect to the weight function $p(x)$ if

$$\int_a^b p(x) f(x) g(x)dx = 0. \quad \text{...(1)}$$

Further the norm of the function $f(x)$ is defined as

$$\| f \| = \text{Norm of } f(x) = \sqrt{\left(\int_a^b p(x) f^2(x)dx\right)} \quad \text{...(2)}$$

Again $f(x)$ is said to be normalized when

$$\int_a^b p(x) f^2(x)dx = 1.$$

Note: The orthogonality with respect to a weight function $p(x)$ can be reduced to the ordinary inner product of $\sqrt{p(x)}g(x)$ and $\sqrt{p(x)}f(x)$ as two functions.

6.5 ORTHOGONAL SET OF FUNCTIONS WITH RESPECT TO A WEIGHT FUNCTION

Let $p(x) \ge 0$, then the set $\{f_n(x): n = 1, 2, 3, \ldots, n\}$ is said to be an orthogonal set of functions on the interval $a \le x \le b$ with respect to the weight function $p(x)$ if $\int_a^b p(x) f_m(x) f_n(x)dx = 0$ when $m \ne n$.

6.6 ORTHONORMAL SET OF FUNCTIONS

A set $\{\phi_n(x): n = 1, 2, 3, \ldots\}$ of real functions, is said to be an orthonormal set of functions on the interval $a \le x \le b$, if

$$\int_a^b \phi_m(x)\phi_n(x)dx = \delta_{m.n} = \begin{cases} 0 \text{ when } m \neq n \\ 1 \text{ when } m = n \end{cases}$$

where the Kronecker delta $\delta_{m.n} = \begin{cases} 0 \textit{ when } m \neq n \\ 1 \textit{ when } m = n \end{cases}$

6.7 ORTHONORMAL SET OF FUNCTIONS WITH RESPECT TO A WEIGHT FUNCTION

Let $p(x) \geq 0$, then a set $\{\phi_n(n): n = 1, 2, 3, \ldots\ldots, n\}$ of real functions defined on an interval $a \leq x \leq b$, is said to be orthonormal with respect to weight function $p(x)$ if

$$\int_a^b p(x)\phi_m(x)\phi_n(x)dx = \delta_{m.n} = \begin{cases} 0 \textit{ when } m = n \\ 1 \textit{ when } m = n \end{cases}$$

Example 1: Show that the functions $1, \cos x, \cos 2x, \cos 3x, \ldots$ are orthogonal on the interval $[0, 2\pi]$ and determine the corresponding orthonormal set.

Solution: The given functions can be denoted by

$g_m(x) = \cos mx, m = 0, 1, 2, 3, \ldots$. Now, we have

$$(g_m, g_n) = \int_0^{2\pi} g_m(x)g_n(x)dx = \int_0^{2\pi} \cos mx \cos nx, \textit{ when } m \neq n$$

$$= 2\int_0^{\pi} \cos mx \cos nx \, dx, \text{ by property of definite integral}$$

$$= \int_0^{\pi} \left[\cos(m+n)x - \cos(m-n)x\right]dx$$

$$= \left[\frac{\sin(m+n)x}{m+n} - \frac{\sin(m-n)x}{m-n}\right]_0^{\pi} = 0.$$

Hence the given functions form an orthogonal set on $[0, 2\pi]$. Now we find the norm of the function $g_m(x)$ as

$$\|g_m(x)\| = \|\cos mx\| = \left\{\int_0^{2\pi} \cos^2 mx \, dx\right\}^{1/2} = \left\{2\int_0^{\pi} \cos^2 mx \, dx\right\}^{1/2}$$

$$= \begin{cases} \sqrt{2\pi}, \text{ when } m = 0 \\ \sqrt{\pi}, \text{ when } m = 1, 2, 3, \ldots \end{cases}$$

Hence the corresponding orthonormal set is

$$\frac{1}{\sqrt{2\pi}}, \frac{\cos x}{\sqrt{\pi}}, \frac{\cos 2x}{\sqrt{\pi}}, \frac{\cos 3x}{\sqrt{\pi}}, \ldots\ldots$$

Example 2: Show that the functions $1, \cos x, \sin x, \cos 2x, \sin 2x, \ldots$ form an orthogonal set on the interval $-\pi \leq x \leq \pi$ and find the corresponding orthogonal set.

Solution: Let $g_m(x) = \cos mx, m = 0, 1, 2, 3, ...$ and

$$f_n(x) = \sin nx, n = 1, 2, 3, ...$$

Then $\quad u_i = \int_a^b r(x) y_i^2(x) dx.$

$$= \frac{1}{2}\int_{-\pi}^{\pi}\left[\sin(m+n)x \pm \sin(n-m)x\right]dx,$$

when $n > m$ or $n < m$ and for all m and n

$$= \frac{1}{2}\left[-\frac{\cos(m+n)x}{m+n} \pm \frac{\cos(n-m)x}{n-m}\right]_{-\pi}^{\pi} = 0.$$

Similarly, $(g_m, g_n) = \int_{-\pi}^{\pi} \cos mx \cos nx \, dx$

$$= 0, \text{ when } m \neq n; m = 0, 1, 2, 3, ... ; n = 1, 2, 3, ...$$

and $(f_m, f_n) = \int_{-\pi}^{\pi} \sin mx \sin nx \, dx = 0$, when $m \neq n$; $m, n = 1, 2, 3, ...$

Hence the given functions form an orthogonal set.

Now, $\| g_0 \| = \| 1 \| = \left\{ \int_{-\pi}^{\pi} 1^2 \, dx \right\}^{1/2} = \sqrt{2\pi}$

$$\| g_m \| = \| \cos mx \| = \left\{ \int_{-\pi}^{\pi} \cos^2 mx \, dx \right\}^{1/2} = \sqrt{\pi}, m = 1, 2, 3, ...$$

and $\| f_n \| = \| \sin nx \| = \left\{ \int_{-\pi}^{\pi} \sin^2 nx \, dx \right\}^{1/2} = \sqrt{\pi}, n = 1, 2, 3, ...$

Hence the corresponding orthonormal set is

$$\frac{1}{\sqrt{2\pi}}, \frac{\cos x}{\sqrt{\pi}}, \frac{\sin x}{\sqrt{\pi}}, \frac{\cos 2x}{\sqrt{\pi}}, \frac{\sin 2x}{\sqrt{\pi}}, ...$$

Example 3: Show that the set of functions $\left\{\sin \frac{n\pi x}{c} : n = 1, 2, 3, ...\right\}$ is an orthogonal set on $(0, c)$ and find the corresponding orthonormal set.

Solution: Let $g_n(x) = \sin \frac{n \pi x}{c}, n = 1, 2, 3, ...$ Then

$$\Rightarrow \int_{-1}^{1} \left(x + Ax^2 + Bx^3\right) dx = 0$$

$$= \frac{1}{2}\int_0^c \left[\cos \frac{(m-n)\pi x}{c} - \cos \frac{(m+n)\pi x}{c}\right] dx$$

$$= \frac{1}{2}\left[\frac{c}{\pi(m-n)} \sin \frac{(m-n)\pi x}{c} - \frac{c}{\pi(m+n)} \sin \frac{(m+n)\pi x}{c}\right]_0^c = 0,$$

when $m \neq n$; $m, n = 1, 2, 3, ...$

Hence the given set forms an orthogonal set.

Now, $\| g_n \| = \| \sin \dfrac{n\pi x}{c} \| = \left\{ \int_0^c \sin^2 \dfrac{n\pi x}{c}\, dx \right\}^{1/2}$

$$= \frac{1}{\sqrt{2}} \left\{ \int_0^c \left(1 - \cos \frac{2n\pi x}{c} \right) dx \right\}^{1/2}$$

$$= \frac{1}{\sqrt{2}} \left\{ \left[x - \frac{c}{2n\pi} \sin \frac{2n\pi x}{c} \right]_0^c \right\}^{1/2} = \sqrt{\frac{c}{2}}.$$

Hence the corresponding orthonormal set is

$$\left\{ \sqrt{\frac{2}{c}} \sin \frac{n\pi x}{c} : n = 1, 2, 3, \ldots \right\}.$$

Example 4: Show that the functions $f_1(x) = 1$, $f_2(x) = x$ are orthogonal on the interval $(-1, 1)$ and determine the value of the constants A and B such that the function $f_3(x) = 1 + Ax + Bx^2$ is orthogonal to both the functions f_1 and f_2 on that interval.

Solution: The given functions are

$$f_1(x) = 1,\ f_2(x) = x,\ f_3(x) = 1 + Ax + Bx^2 .$$

Now we have

$$\int_{-1}^{1} f_1(x) f_2(x)\, dx = \int_{-1}^{1} 1\, x\, dx = \left[\frac{x^2}{2} \right]_{-1}^{1} = 0.$$

Hence the functions $f_1(x)$ and $f_2(x)$ are orthogonal on the interval $(-1, 1)$.

Let the function $f_3(x) = 1 + Ax + Bx^2$ be orthogonal to both the functions $f_1(x)$ and $f_2(x)$ on the interval $(-1, 1)$. Then we must have

$$\int_{-1}^{1} f_1(x) f_3(x) = 0 \text{ and } \int_{-1}^{1} f_2(x) f_3(x)\, dx = 0$$

Now, $\int_{-1}^{1} f_1(x) f_3(x)\, dx = 0 \Rightarrow \int_{-1}^{1} (1 + Ax + Bx^2)\, dx = 0$

$$\Rightarrow \left[x + A\frac{x^2}{2} + B\frac{x^3}{3} \right]_{-1}^{1} = 0 \Rightarrow 2 + \frac{2}{3} B = 0 \Rightarrow B = -3.$$

Also, $\int_{-1}^{1} f_2(x) f_3(x)\, dx = 0 \Rightarrow \int_{-1}^{1} (x + Ax^2 + Bx^3)\, dx = 0$

$$\Rightarrow \left[\frac{x^2}{2} + A\frac{x^3}{3} + B\frac{x^4}{4} \right]_{-1}^{1} = 0 \Rightarrow \frac{2}{3} A = 0 \Rightarrow A = 0.$$

Hence the function $f_3(x)$ will be orthogonal to both the given functions $f_1(x)$ and $f_2(x)$ if $A = 0$ and $B = -3$.

EXERCISE 6.1

1. Prove that the set of given functions on the given interval are orthogonal and find also the corresponding orthonormal set of the following:

   (i) $1, \cos x, \cos 2x, \cos 3x, ... -\pi \le x \le \pi$.

   (ii) $\sin x, \sin 2x, \sin 3x, ..., 0 \le x \le \pi$.

   (iii) $1, \cos 2x, \cos 4x, \cos 6x, ... 0 \le x \le \pi$.

   (iv) $\sin \pi x, \sin 2\pi x, \sin 3\pi x, ..., -1 \le x \le 1$.

   (v) $1, \cos\frac{\pi x}{c}, \cos\frac{2\pi x}{c}, \cos\frac{3\pi x}{c} ....., 0 < x < c$.

   (vi) $1, \cos\frac{2\pi x}{T}, \cos\frac{4\pi x}{T}, \frac{\cos 6\pi x}{T} ....., 0 \le x \le T$.

2. Show that the set of functions $\left\{\sin\frac{n\pi x}{c}\right\}(n = 1, 2, 3, ...)$ is orthogonal on the interval $(-c, c)$ and also find the corresponding orthonormal set.

3. Show that the set of functions $\left\{1, \cos\frac{n\pi x}{c}\right\}, (n = 1, 2, 3, ...)$ is orthogonal on the interval $(-c, c)$ and also find the corresponding orthonormal set.

4. Show that the set of functions
$$\left\{1, \cos\frac{m\pi x}{c}, \sin\frac{n\pi x}{c}\right\}, (m, n = 1, 2, 3, ...)$$
is orthogonal on the interval $(-c, c)$ and also find the corresponding orthonormal set.

5. Show that the functions $y = c_1 \cos\sqrt{\lambda}\, x + c_2 \sin\sqrt{\lambda}\, x$ and $f_2(x) = x^3$ are orthogonal on the interval $(-2, 2)$ and find the value of constants A and B such that the functions $f_3(x) = 1 + Ax + Bx^2$ is orthogonal to both the functions $f_1(x)$ and $f_2(x)$.

6. Show that the functions $1, 1 - x, 1 - 2x + \frac{1}{2}x^2$ are orthogonal with respect to weighted function e^{-x} on the interval $0 \le x \le \infty$. Find the corresponding orthonormal set.

7. Show that the functions $g_m(x) = \sin mx,\ m = 1,2,3,....$ are orthogonal on the interval $-\pi \le x \le \pi$ and find the corresponding orthonormal set.

ANSWERS 6.1

1. (i) $\frac{1}{\sqrt{2\pi}}, \frac{\cos x}{\sqrt{\pi}}, \frac{\cos 2x}{\sqrt{\pi}}, \frac{\cos 3x}{\sqrt{\pi}}, \ldots$

   (ii) $\sqrt{\frac{2}{\pi}} \sin x, \sqrt{\frac{2}{\pi}} \sin 2x, \sqrt{\frac{2}{\pi}} \sin 3x \ldots.$

   (iii) $\sqrt{\frac{1}{\pi}}, \sqrt{\frac{2}{\pi}} \cos 2x, \sqrt{\frac{2}{\pi}} \cos 4x \ldots.$

   (iv) $\sin \pi x, \sin 2\pi x, \sin 3\pi x, \ldots.$

   (v) $\frac{1}{\sqrt{c}}, \sqrt{\frac{2}{c}} \cos \frac{\pi x}{c}, \sqrt{\frac{2}{c}} \cos \frac{2\pi x}{c} \ldots$

   (vi) $\frac{1}{\sqrt{T}}, \sqrt{\frac{2}{T}} \cos \frac{2\pi x}{T}, \sqrt{\frac{2}{T}} \cos \frac{4\pi x}{T} \ldots.$

2. $\left\{\frac{1}{\sqrt{c}} \sin \frac{n\pi x}{c}\right\}, n = 1, 2, 3, \ldots.$

3. $\left\{\frac{1}{\sqrt{2c}}, \frac{1}{\sqrt{c}} \cos \frac{n\pi x}{c}\right\}, n = 1, 2, 3.$

4. $\left\{\frac{1}{\sqrt{2c}}, \frac{1}{\sqrt{c}} \cos \frac{m\pi x}{c}, \frac{1}{\sqrt{c}} \sin \frac{n\pi x}{c}\right\}, m, n = 1, 2, 3, \ldots,$

5. $A = 0$ and $B = -\frac{3}{4}$

6. $1, 1 - x, 1 - 2x + \frac{1}{2}x^2$

7. $\frac{\sin x}{\sqrt{\pi}}, \frac{\sin 2x}{\sqrt{\pi}}, \frac{\sin 3x}{\sqrt{\pi}}, \ldots.$

6.8 BOUNDARY VALUE PROBLEM

A differential equation with a set of conditions imposed at more than one distinct points is called a boundary value problem (BVP) and conditions are called boundary conditions (BCs).

6.9 STURM-LIOUVILLE'S EQUATION

The differential equation

$$\frac{d}{dx}\left\{p(x).\frac{dy}{dx}\right\} + \left[\lambda q(x) + r(x)\right]y = 0. \quad \ldots(1)$$

with $p(x)$ and $q(x)$ are positive functions; $p(x), q(x), r(x)$ are continuous real functions on $a \le x \le b$ and λ is the parameter independent of x, is called *Sturm-Liouville equation.*

The boundary conditions at the end points are given by

$$\alpha_1\, y(a)+\alpha_2\, y'(a)=0 \qquad \text{...(2)}$$

$$\beta_1\, y(b)+\beta_2\, y'(b)=0. \qquad \text{...(3)}$$

where $\alpha_1, \alpha_2, \beta_1, \beta_2$ are constants and neither α_1, α_2 are both zero nor β_1, β_2 are both zero together. The problem of finding the solution of (1) subject to boundary conditions (2) and (3); is called *Sturm-Liouville boundary value problem.*

The trivial solution $y=0$, for every value of parameter λ, is of no practical use. The non trivial solutions of Sturm-Liouville boundary value problem are called eigen functions corresponding to the eigen values.

Particular case: Putting $p(x)=1, q(x)=1, r(x)=0$ in (1), we have

$$\frac{d^2y}{dx^2}+\lambda y=0.$$

Now taking conditions as $\alpha_1=\beta_1=1$ and $\alpha_2=\beta_2=0$, we get

$$y(a) = 0 \text{ and } y(b) = 0$$

Hence $\left.\begin{array}{c} y''+\lambda y=0 \\ y(a)=0,\, y(b)=0 \end{array}\right\}$ is the simplest form of Sturm-Liouville problem.

For example:

1. The boundary value problem (BVP) $\frac{d^2y}{dx^2}+\lambda y=0,\, y(0)=0,\, y(\pi)=0$ is a Sturm-Liouville problem, as the equation can be written as $\frac{d}{dx}\left[1\frac{dy}{dx}\right]+[0+\lambda\, 1]y=0$ with the boundary conditions (BCs):

$$1y(0)+0y'(0)=0\,;\;\; 1y(\pi)+0y'(\pi)=0.$$

Note that $\lambda=1^2, 2^2, ..., n^2, ...$ are the eigen values and the corresponding solutions sin x, sin $2x$, ..., sin nx... are the eigen functions.

2. The boundary value problem

$$x^2y''+xy'+\lambda y=0,\, y'(1)=0=y'(e^{2\pi}),\, \lambda\ge 0$$

is a Sturm-Liouville problem because the given equation can be written as $[xy']'+\frac{\lambda}{x}y=0$ with the boundary conditions (BCs):

$$0y(1)+1y'(1)=0\,;\;\; 0y(e^{2\pi})+1y'(e^{2\pi})=0.$$

6.10 ORTHOGONALITY OF EIGEN FUNCTIONS

If $p(x), q(x), r(x)$ and r' are the usual functions in Sturm-Liouville equation and $\lambda_m(x), \lambda_n(x)$ are eigen functions of Sturm-Liouville problem, then $(\lambda_m - \lambda_n)\int_a^b q(x)y_m\, y_n\, dx = \left[y_m(p(x)y_n') - y_n(p(x)y_m')\right]_a^b$

$$= p(b)\left[y_n'(b)y_m(b) - y'_m(b)y_n(b)\right]$$

$$- p(a)\left[y_n'(a)y_m(a) - y_n'(a)y_n(a)\right] = 0, \text{ if}$$

(i) $y(a) = y(b)$

(ii) $y'(a) = y'(b)$

(iii) $\alpha_1\, y(a) + \alpha_2\, y'(a) = 0$

(iv) $\beta_1\, y(b) + \beta_2\, y'(b) = 0.$

Equation (1) gives $\int_a^b q(x)y_m\, y_n\, dx = 0 (m \neq n)$. It means that eigen functions y_m, y_n are orthogonal with the weight $q(x)$.

Corollary 1: (Reality of eigen values) All eigen values of a Sturm-Liouville problem are real.

Proof: Consider the Sturm-Liouville problem as

$$(p(x)y')^1 + (q(x) + \lambda r(x))y = 0, p(x) > 0, r(x) > 0 \; on[a, b]$$

with boundary conditions: $y(a) = y(b) = 0$.

Then the orthogonality relation of eigen functions of the problem with weight function $r(x)$ is

$$\int_a^b r(x)y_m(x)y_n(x)dx = 0. \qquad \ldots(4)$$

If possible suppose the eigen values are complex namely $\alpha + i\beta$ and $\alpha - i\beta$. Also suppose the corresponding eigen functions are $y_m = u + iv$, $y_n = u - iv$ respectively. Then, equation (4) becomes

$$\int_a^b r(x)(u^2 + v^2)dx = 0.$$

As $r(x) > 0$, this is possible only if $u = 0 = v$. But this contradicts the fact that eigen functions are non-zero. Thus eigen values cannot be complex.

Hence, we conclude that all eigen values of a Sturm-Liouville problem are real.

Corollary 2: (Eigen function expansion) If $\{y_i(x)\}$ is the set of eigen functions of the preceeding Sturm-Liouville problem (1) and $f(x)$ is a function on $[a, b]$ such that $f(a) = f(b) = 0$, then the expansion

$$f(x) = \sum_{i=1}^{\infty} c_i \, y_i(x) \qquad \ldots(5)$$

where $c_i = \frac{1}{u_i} \int_a^b r(x) f(x) y_i(x) dx$, $u_i = \int_a^b r(x) y_i^2(x) dx.$,

is called eigen function expansion of $f(x)$.

6.11 THE ADJOINT OPERATOR

Consider a linear differential equation of nth order as follows:

$$L_n \, y = a_0 \, y^{(n)} + a_1 \, y^{(n-1)} + \ldots + a_n \, y = 0$$

Here L_n is an operator defined by

$$L_n \equiv a_0 \left(\frac{d^n}{dx^n} \right) + a_1 \left(\frac{d^{n-1}}{dx^{n-1}} \right) + \ldots . \, a_{n-1} \left(\frac{d}{dx} \right) + a_n$$

The adjoint operator to L_n is denoted by $\overline{L}_n$ and is defined as

$$\overline{L}_n \equiv (-1)^n \frac{d^n}{dx^n} a_{0*} + (-1)^{n-1} \frac{d^{n-1}}{dx^{n-1}} a_{1*} + \ldots + (-1) \frac{d}{dx} a_{n-1*} + a_{n*}$$

It should be noted that if we operate this operator on some independent variable y, then * will be replace by y that is, adjoint equation is

$$\overline{L}_n y = (-1)^n \frac{d^n}{dx^n} (a_0 y) + (-1)^{n-1} \frac{d^{n-1}}{dx^{n-1}} (a_1 y) + \ldots + (-1) \frac{d}{dx} (a_{n-1} y) + a_n y = 0$$

Hence the adjoint differential equation to $L_n \, y = 0$ is given by $\overline{L}_n \, y = 0$. If the operators $L_n \equiv \overline{L}_n$, then the operator L_n is called self-adjoint operator.

Example 1: Obtain the eigen values and eigen functions of the Sturm Liouville problem

$$\frac{d^2 y}{dx^2} + \lambda y = 0; \, y(0) = 0 = y(\pi) \qquad \ldots(1)$$

Solution: We shall consider separately three cases $\lambda = 0, \lambda < 0$ and $\lambda > 0$.

Case 1: When $\lambda = 0$. In this case, the equation (1) reduces to

$$\frac{d^2y}{dx^2} = 0.$$

Hence the general solution is $y = c_1 x + c_2$.

Applying the boundary conditions (BC): $y(0) = 0, y(\pi) = 0,$ we get

$$0 = c_1 0 + c_2 \text{ and } 0 = c_1 \pi + c_2 \Rightarrow c_1 = 0 = c_2.$$

Thus only trivial solution $y(x) = 0$ exists.

Case 2: When $\lambda < 0$. Putting $\lambda = -\alpha^2 ; \alpha \in R,$ the auxiliary equation is $m^2 - \alpha^2 = 0 \Rightarrow m = \pm \alpha$.

Hence, the general solution is $y = c_1 e^{\alpha x} + c_2 e^{-\alpha x}$

Now the boundary conditions give

$$0 = c_1 + c_2 \text{ and } 0 = c_1 e^{\alpha\pi} + c_2 e^{-\alpha\pi} \Rightarrow c_1 = 0 = c_2$$

which gives again the trivial solution $y(x) \equiv 0$ only.

A non-trivial solution will exist if

$$\begin{vmatrix} 1 & 1 \\ e^{\alpha\pi} & e^{-\alpha\pi} \end{vmatrix} = 0, \text{ i.e. if } e^{\alpha\pi} = e^{-\alpha\pi}, \text{ i.e. if } \alpha = 0, \text{ i.e. if } \lambda = 0.$$

But $\lambda < 0$. Hence there exists no non-trivial solution.

Case 3: When $\lambda > 0$. Putting $\lambda = \alpha^2 ; \alpha \in R$, the auxiliary equation is $m = \pm i\alpha$

Hence the general solution is

$$y = c_1 \sin \alpha x + c_2 \cos \alpha x.$$

Now the boundary conditions give

$$c_1 \sin 0 + c_2 \cos 0 = 0 \text{ and } c_1 \sin \alpha\pi + c_2 \cos \alpha\pi = 0$$

$$\Rightarrow \quad c_2 = 0, c_1 \sin \alpha\pi = 0 \Rightarrow c_2 = 0, \text{either } c_1 = 0 \text{ or } \sin \alpha\pi = 0.$$

But $c_2 = 0 = c_1$ gives again the trivial solution $y(x) \equiv 0$. So, we choose

$$\sin \alpha\pi = 0 = \sin n\pi \text{ for } n = 0, \pm 1, \pm 2, ...$$

$$\Rightarrow \quad \alpha\pi = n\pi, n = 0, \pm 1, \pm 2, ...$$

$$\Rightarrow \quad \lambda = \alpha^2 = n^2, n = 0, 1, 2, 3, ...$$

Since c_1 depends on the value of λ, we may write a_n for c_1 and thus get non-trivial solutions

$$y = a_n \sin \sqrt{\lambda_n}\, x, \lambda = 1, 4, 9, 16, ..., n^2, ...$$

Clearly, the eigen values of the problem are $1^2, 2^2, 3^3, ..., n^2, ...$ and the eigen functions are $y(x) = a_n \sin \lambda_n x$, where $\lambda_n = 1, 2, 3, ...$, and a_n is a non-zero arbitrary constant.

Example 2: Find all eigen values and eigen functions of the following Stúrm-Liouville problem

$$\frac{d^2y}{dx^2} + \lambda y = 0 \qquad ...(1)$$

$$y(0) = 0, y'\left(\frac{\pi}{2}\right) = 0. \qquad ...(2)$$

Solution: Case I: When $\lambda = 0$. In this case, the equation (1) reduces to

$$\frac{d^2y}{dx^2} = 0.$$

It's general solution is $y(x) = c_1 x + c_2$

Therefore $y'(x) = c_1$.

Applying boundary conditions (2), we get

$$0 = y(0) = c_1 0 + c_2 \Rightarrow c_2 = 0$$

And $0 = y'\left(\frac{\pi}{2}\right) = c_1 \Rightarrow c_1 = 0.$

Thus only trivial solution $y(x) = 0$ exists.

Case II: When $\lambda < 0$. Putting $\lambda = -\alpha^2 ; \alpha \in R$, the auxiliary equation is $m^2 - \alpha^2 = 0 \Rightarrow m = \pm \alpha$.

Hence, the general solution is

$$y = c_1 e^{\alpha x} + c_2 e^{-\alpha x}$$

and $y = \alpha c_1 e^{\alpha x} - \alpha c_2 e^{-\alpha x}$

Now the boundary conditions give

$$0 = y(0) = c_1 + c_2$$

and $0 = y'\left(\frac{\pi}{2}\right) = \alpha c_1 e^{\alpha\pi/2} - \alpha c_2 e^{-\alpha\pi/2}$

Solving these two equations, we get $c_1 = 0 = c_2$

which gives again the trivial solution $y(x) \equiv 0$ only.

Case III: When $\lambda > 0$. Putting $\lambda = \alpha^2 ; \alpha \in R$, the auxiliary equation is $m = \pm i\alpha$

Hence the general solution is $y = c_1 \sin \alpha x + c_2 \cos \alpha x$.

And $y' = c_1 \alpha \cos \alpha x - c_2 \alpha \sin \alpha x$

Applying boundary conditions (2), we get

$$0 = y(0) = c_1 \sin 0 + c_2 \cos 0 \Rightarrow c_2 = 0$$

and $$0 = y'\left(\frac{\pi}{2}\right) = c_1 \alpha \cos \alpha \frac{\pi}{2} - c_2 \alpha \sin \alpha \frac{\pi}{2}$$

$\Rightarrow$ $$0 = c_1 \alpha \cos \alpha \frac{\pi}{2}, \text{ since } c_2 = 0$$

$\Rightarrow$ either $c_1 = 0$ or $\cos \alpha \frac{\pi}{2} = 0$

But $c_2 = 0 = c_1$ gives again the trivial solution $y(x) \equiv 0$. So, we choose

$$\cos \alpha \frac{\pi}{2} = 0 = \cos(2n-1)\frac{\pi}{2}, n = 1, 2, 3, \ldots$$

$\Rightarrow$ $$\alpha = 2n - 1, \ n = 1, 2, 3, \ldots$$

$\Rightarrow$ $$\lambda = (2n-1)^2, \ n = 1, 2, 3, \ldots.$$

Since c_1 depends on the value of λ, we may write c_n for c_1 and thus get non-trivial solution

$$y = c_n \sin \sqrt{\lambda_n}\, x, \lambda_n = 1^2, 3^2, 5^2, \ldots, (2n-1)^2, \ldots.$$

Hence the eigen values of the problem are $\lambda_n = (2n-1)^2, n = 1, 2,$ 3..., and the corresponding eigen functions are $y_n = c_n \sin(2n-1)x$, $n = 1, 2, 3, \ldots$, where b_n is a non-zero arbitrary constant.

Hence, on taking $c_n = 1$, for all $n = 1, 2, 3, ..$ the required eigen functions are

$$y_n = \sin(2n-1)x, n = 1, 2, 3, \ldots$$

Example 3: Obtain the eigen values and eigen functions of the Sturm-Liouville problem

$$x^2 \frac{d^2y}{dx^2} + x\frac{dy}{dx} + \lambda y = 0, y'(1) = 0 = y'(e^{2\pi}), \lambda \geq 0, x > 0. \quad \ldots(1)$$

Solution: We shall consider separately the cases $\lambda = 0, \lambda > 0$.

Case I: When $\lambda = 0$. In this case the equation (1) reduces to

$$x^2 \frac{d^2y}{dx^2} + x\frac{dy}{dx} = 0 \text{ i.e., } \frac{d}{dx}\left\{x\frac{dy}{dx}\right\} = 0.$$

Hence the general solution is

$$y = c \log x + c', \text{ where } c, c' \text{ being arbitrary constants.}$$

Applying boundary conditions, we obtain

$$0 = c \text{ and } 0 = \frac{c}{e^{2\pi}} \Rightarrow c = 0 \text{ but } c' \text{ is arbitrary.}$$

Therefore solution is $y = c'$, which is non-trivial as c' is a non-zero arbitrary constant.

Case II: When $\lambda > 0$.

Assume that $x = e^t$. Then $\dfrac{dx}{dt} = e^t$, so that

$$\frac{dy}{dx} = \frac{dy}{dt}\frac{dt}{dx} = e^{-t}\frac{dy}{dt}$$

and $$\frac{d^2y}{dx^2} = \frac{d}{dx}\left(e^{-t}\frac{dy}{dt}\right) = \frac{d}{dt}\left(e^{-t}\frac{dy}{dt}\right)\frac{dt}{dx} = e^{-2t}\left(\frac{d^2y}{dt^2} - \frac{dy}{dt}\right).$$

Thus the equation (1) reduces to the form

$$e^{2t}\, e^{-2t}\left(\frac{d^2y}{dt^2} - \frac{dy}{dt}\right) + e^t\, e^{-t}\frac{dy}{dt} + \lambda y = 0 \Rightarrow \frac{d^2y}{dt^2} + \lambda y = 0.$$

Since $\lambda > 0$, the general solution is $y = c_1 \sin\sqrt{\lambda}\, t + c_2 \cos\sqrt{\lambda}\, t$

or $$y = c_1 \sin\left(\sqrt{\lambda}\log x\right) + c_2 \cos\left(\sqrt{\lambda}\log x\right)$$

$$\therefore \quad \frac{dy}{dx} = \frac{c_1\sqrt{\lambda}}{x}\cos\left(\sqrt{\lambda}\log x\right) - \frac{c_2\sqrt{\lambda}}{x}\sin\left(\sqrt{\lambda}\log x\right).$$

Now the boundary conditions $y'(1) = 0 = y'\left(e^{2\pi}\right)$ give

$$c_1\sqrt{\lambda}\cos\left(\sqrt{\lambda}\log 1\right) - c_2\sqrt{\lambda}\sin\left(\sqrt{\lambda}\log 1\right) = 0 \Rightarrow c_1\sqrt{\lambda} = 0 \Rightarrow c_1 = 0$$

and $$c_1\sqrt{\lambda}\, e^{-2\pi}\cos\left(\sqrt{\lambda}\log e^{2\pi}\right) - c_2\sqrt{\lambda}\, e^{-2\pi}\sin\left(\sqrt{\lambda}\log e^{2\pi}\right) = 0$$

$$\Rightarrow \quad c_2\sqrt{\lambda}\, e^{-2\pi}\sin\left(2\pi\sqrt{\lambda}\right) = 0, \text{ since } c_1 = 0$$

$$\Rightarrow \quad c_2 = 0 \text{ or } \sin\left(2\pi\sqrt{\lambda}\right) = 0.$$

But $c_1 = 0$, $c_2 = 0$ lead to the trivial solution $y(x) = 0$.
Thus, for non-trivial solution, we must have

$$\sin\left(2\pi\sqrt{\lambda}\right) = 0 = \sin n\pi,\ n = 0, \pm 1, \pm 2, \ldots$$

$$\Rightarrow \quad \lambda = n^2/4,\ n = 0, 1, 2, 3, \ldots$$

Hence the required eigen values are $\lambda = \dfrac{1^2}{4}, \dfrac{2^2}{4}, \dfrac{3^2}{4}, \ldots, \dfrac{n^2}{4}, \ldots$ and the corresponding eigen functions are

$$y = c_n \cos\left(\frac{n}{2}\log x\right), n = 1, 2, 3, \ldots$$

Example 4: Find all eigen values and eigen functions of the following Sturm-Liouville problem

$$\frac{d^2y}{dx^2} + \lambda y = 0 \qquad \ldots(1)$$

$$y'(0) = 0, y'(c) = 0 \qquad \ldots(2)$$

Solution: Case I: When $\lambda = 0$, the equation (1) becomes $\frac{d^2y}{dx^2} = 0$

Its general solution is $y(x) = c_1 x + c_2 \Rightarrow y'(x) = c_1$.

Applying boundary conditions (2), we get

$$0 = y'(0) = c_1 \text{ and } 0 = y'(c) = c_1 \Rightarrow c_1 = 0.$$

On taking $c_2 = 1$, we have $y(x) = 1$.

Thus $y_0 = 1$ is the eigen function corresponded to the eigen value $\lambda = 0$.

Case II: When $\lambda < 0$. Assume $\lambda = -\alpha^2$, then equation (1) becomes

$$\frac{d^2y}{dy^2} - \alpha^2 y = 0$$

General solution is

$$y(x) = c_1 e^{\alpha x} + c_2 e^{-\alpha x} \Rightarrow y'(x) = c_1 \alpha e^{\alpha x} - c_2 \alpha e^{-\alpha x}.$$

Applying boundary conditions (2), we get

$$0 = y'(0) = c_1 \alpha - c_2 \alpha \text{ and } 0 = y'(c) = c_1 \alpha e^{\alpha c} - c_2 \alpha e^{-\alpha c}$$

Solving the above two equations, we get

$$c_1 = 0, c_2 = 0.$$

Thus only trivial solution $y(x) = 0$ exists.

Case III: When $\lambda > 0$. Assume $\lambda = \alpha^2$, then equation (1) becomes

$$\frac{d^2y}{dy^2} + \alpha^2 y = 0$$

General solution is

$$y(x) = c_1 \cos \alpha x + c_2 \sin \alpha x \Rightarrow y'(x) = -c_1 \alpha \sin \alpha x + c_2 \alpha \cos \alpha x.$$

Applying boundary conditions (2), we get

$$0 = y'(0) = c_2\,\alpha; \Rightarrow c_2 = 0$$

and $\quad 0 = y'(c) = -c_1\,\alpha \sin \alpha c \Rightarrow$ either $c_1 = 0$ or $\sin \alpha c = 0$.

But $c_2 = 0 = c_1$ gives again the trivial solution $y(x) \equiv 0$ only. So, for non-trivial solutions, we must have

$$\sin \alpha c = 0 = \sin n\pi,\ n = 0, \pm 1, \pm 2, \pm 3, \ldots$$

$$\Rightarrow \quad \lambda = \alpha^2 = \frac{n^2\pi^2}{c^2},\ n = 0, 1, 2, 3, \ldots.$$

Thus the eigen values are $\lambda = \dfrac{n^2\pi^2}{c^2},\ n = 1, 2, 3, \ldots.$

On taking $c_1 = 1$, the corresponding eigen functions are

$$y_n = \cos\frac{n\pi x}{c},\ n = 1, 2, 3, \ldots.$$

Example 5: Find all eigen values and eigen functions of the following Sturm-Liouville problem:

$$y'' + \lambda y = 0 \qquad \ldots(1)$$

$$y(0) + y'(0) = 0,\ y(1) + y'(1) = 0. \qquad \ldots(2)$$

Solution: Case I: When $\lambda = 0$. In this case, equation (1) becomes $y'' = 0$

whose solution is $\quad y(x) = c_1\,x + c_2, \qquad \ldots(3)$

and $\quad y'(x) = c_1. \qquad \ldots(4)$

Adding equations (3) and (4), we get

$$y(x) + y'(x) = (1 + x)c_1 + c_2.$$

Applying boundary conditions (2) to the above equation, we get

$$0 = y(0) + y'(0) = c_1 + c_2 \qquad \ldots(5)$$

and $\quad 0 = y(1) + y'(1) = 2c_1 + c_2. \qquad \ldots(6)$

Solving equations (5) and (6), we get $c_1 = 0$ and $c_2 = 0$.

Thus the solution is $y(x) \equiv 0$, which is not an eigen function.

Case II: When $\lambda < 0$. Assume $\lambda = -\alpha^2$, then from equation (1), we have

$$y'' - \alpha^2\,y = 0$$

General solution is $y(x) = c_1\,e^{\alpha x} + c_2\,e^{-\alpha x}. \qquad \ldots(7)$

Differentiating both sides w.r.t. 'x', we get

$$y'(x) = c_1\ \alpha e^{\alpha x} - c_2\ \alpha e^{-\alpha x} \qquad \ldots(8)$$

Adding equations (7) and (8), we get

$$y(x) + y'(x) = c_1(1+\alpha)e^{\alpha x} + c_2(1-\alpha)e^{-\alpha x} \qquad \ldots(9)$$

Applying boundary conditions (2), we get

$$0 = y(0) + y'(0) = c_1(1+\alpha) + c_2(1-\alpha) \qquad \ldots(10)$$

and $\quad 0 = y(1) + y'(1) = c_1(1+\alpha)e^{\alpha} + c_2(1-\alpha)e^{-\alpha} \qquad \ldots(11)$

Solving equations (10) and (11), we get either $c_1 = 0, \alpha = 1$, therefore

$y(x) = c_2\ e^{-x}$ or $c_2 = 0, \alpha = -1$, and therefore $y(x) = c_1\ e^{-x}$

Hence $\lambda = -\alpha^2 = -1$ is an eigen value and the corresponding eigen function is $y = e^{-x}$.

Case III: When $\lambda > 0$. Assume $\lambda = \alpha^2$, then equation (1) becomes $y'' = \alpha^2 y = 0$.

Genral solution is $\quad y(x) = c_1 \cos ax + c_2 \sin \alpha x. \qquad \ldots(12)$

Differentiating (12) w. r. t. 'x', we get

$$y'(x) = -c_1\ \alpha \sin \alpha x + c_2\ \alpha \cos \alpha x. \qquad \ldots(13)$$

Adding equations (12) and (13), we get

$$y(x) + y'(x) = (c_1 + \alpha c_2)\cos \alpha x + (c_2 - \alpha c_1)\sin \alpha x.$$

Applying boundary conditions (2), we get

$$0 = y(0) + y'(0) = c_1 + \alpha c_2 \Rightarrow c_1 = -\alpha c_2$$

And $\quad 0 = y(1) = y'(1) = (c_1 + \alpha c_2)\cos \alpha + (c_2 - \alpha c_1)\sin \alpha.$

Solving the above two equations, we get

$$0 = (c_2 + \alpha^2 c_2)\sin \alpha \Rightarrow 0 = c_2(1+\alpha^2)\sin \alpha \Rightarrow c_2 = 0 \text{ or } \sin \alpha = 0.$$

If $c_2 = 0$, then $c_1 = 0$ and so $y(x) = 0$ which is not an eigen function. Thus, for non-trivial solution, we must have

$$\sin \alpha = 0 \Rightarrow \sin \alpha = \sin n\pi,\ n = 0, \pm 1, \pm 2, \pm 3, \ldots$$

$$\Rightarrow \alpha = n\pi,\ n = 0, \pm 1, \pm 2, \pm 3, \ldots$$

$$\Rightarrow \lambda = \alpha^2 = n^2\pi^2,\ n = 0, 1, 2, 3, \ldots.$$

Therefore $y(x) = c_1 \cos n\pi\, x + c_2 \sin n\pi\, x.$

$$\Rightarrow \qquad y(x) = c_2(\sin n\pi\, x - n\pi \cos n\pi\, x), \text{ since } c_1 = -\alpha c_2 = -n\pi\, c_2.$$

Here $\lambda_n = n^2 \pi^2, n = 1, 2, 3, \ldots$ are eigen values and on taking $c_2 = 1$, the corresponding eigen functions are

$$y_n = \sin n\pi x - n\pi \cos n\pi x, n = 1, 2, 3, \ldots.$$

EXERCISE 6.2

Find all eigen values and eigen functions of the following Sturm-Liouville problems:

1. $\dfrac{d^2y}{dx^2} + \lambda y = 0, y'(0) = 0, y(c) = 0.$
2. $\dfrac{d^2y}{dx^2} + \lambda y = 0; y(0) = 0, y'(l) = 0.$
3. $\dfrac{d^2y}{dx^2} + \lambda y = 0; y(0) = 0 = y'(\pi).$
4. $\dfrac{d^2y}{dx^2} + \lambda y = 0, y'(-\pi) = 0 = y'(\pi).$
5. $(xy') + \lambda x^{-1} y = 0; y(1) = 0, y(-1) = 0.$
6. $\dfrac{d^2y}{dx^2} + \lambda y = 0, y(0) = 0;\ hy(1) + y'(1) = 0$, where h is a positive constant.
7. For the boundary value problem
$$\frac{d^2y}{dx^2} + \lambda y = 0, y(0) - y'(0) = 0; y(1) + y'(1) = 0.$$
Obtain the form of the eigen functions and the equation satisfied by eigen values. Can you obtain approximate eigen value of smallest absolute value?
8. Find the eigen values and the corresponding eigen functions of the eigen value problem
$$\frac{d^2y}{dx^2} + \lambda y = 0, y(0) = 0, y(1) - y'(1) = 0.$$
9. Find the eigen values and eigen functions for the differential equation $\dfrac{d^2s}{dx^2} + k^2 s = 0$. Prove that the eigen functions are orthogonal.
10. Solve the Sturm-Liouville problem
$$(xy')' + \lambda\left(\frac{1}{x}\right)y = 0;\ y'(1) = 0, y(b) = 0, (b > 1).$$
Also normalize the eigen functions.

11. Solve the following Sturm-Liouville problem

$$\frac{d}{dx}\left\{(x^2+1)\frac{dy}{dx}\right\}+\frac{\lambda}{x^2+1}y=0,\ y(0)=0=y(1):$$

12. Find all the eigen values and eigen function of the following Sturm-Liouville problem $y''+\lambda y=0$ with boundary conditions $y(0)=0, y(l)=0.$

ANSWERS 6.2

1. $\lambda_n=\dfrac{(2n-1)^2\pi^2}{4c^2}; y_n=\cos\left(\dfrac{2n-1}{2c}\right)\pi x; n=1,2,3,....$

2. $\lambda_n=\left[\dfrac{(2n+1)\pi}{2l}\right]^2; y_n=\cos\left(\dfrac{2n+1}{2l}\right)\pi x; n=1,2,3,...$

3. $\lambda_n=\dfrac{1}{4}(2n-1)^2; y_n=\sin\dfrac{(2n-1)}{2}x; n=1,2,3,...$

4. $y_0=1; y_n=\cos\left(\dfrac{n(\pi+x)}{2}\right); n=1,2,3,...$

5. $\lambda_n=n^2\pi^2, y_n=\sin(n\pi\log|x|), n=1,2,3,....$

6. $\lambda_n=\alpha_n^2; y_0=x, y_n=\sin\alpha_n x, n=1,2,3,...$ where α_n are the consecutive positive roots of the equation $\tan\alpha=-\alpha/h$.

7. $y_n(x)=\sin\mu_n x+\mu_n\cos\mu_n x, \lambda_n=\mu_n^2, n=1,2,3,...$ where μ_n are positive roots of $(1-\mu^2)\tan\mu=2\mu$.

8. $y_n=\sin\mu_n x, \lambda_n=\mu_n^2, n=1,2,3,...$ where μ_n are the positive roots of $\tan\mu=\mu$. Again $y(x)=x$ is the eigen function corresponding to the eigen value $\lambda=0$.

10. $\lambda_n=\dfrac{(2n-1)\pi}{2\log b}, Y_n=\cos\left\{\dfrac{(2n-1)\pi\log x}{2\log b}\right\}, n=1,2,3,....$

11. $\lambda_n=16n^2, y_n=\sin(4n\tan^{-1}x), n=1,2,3,....$

12. $\lambda_n=\dfrac{n^2\pi^2}{l^2}, n=0,1,2,3,....; y_n(x)=\sin\dfrac{n\pi x}{l}, n=1,2,3,...$

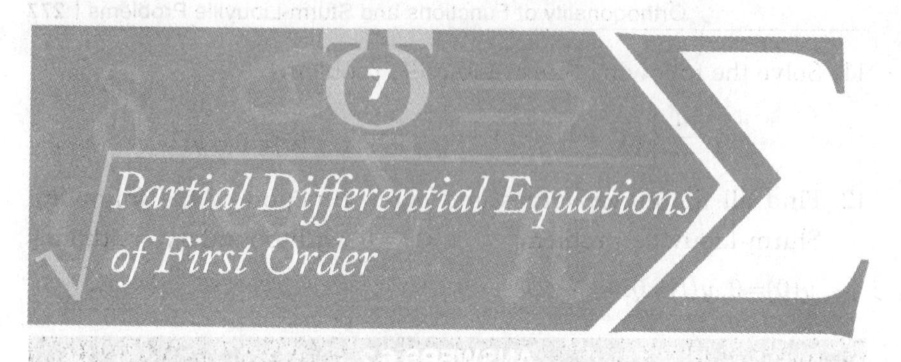

7.1 INTRODUCTION

So far the differential equations involving only one independent variable were discussed. Now the equations involving more than one independent variables and only one dependent variable denoted z will be considered. The independent variables, when the numbers of these variables are in number two, will be denoted by x and y and when these are more than two, these will be denoted by $x_1, x_2, \ldots\ldots, x_n$. In the former case the partial derivatives $\frac{\partial z}{\partial x}$ and $\frac{\partial z}{\partial y}$ will be denoted by p and q respectively and in the latter case the partial derivatives $\frac{\partial z}{\partial x_1}, \frac{\partial z}{\partial x_2}, \ldots\ldots$ etc., will be denoted by $p_1, p_2, \ldots\ldots$ etc.

A partial differential equation is a relation between the independent variables, the dependent variables and its partial derivatives with regard to them. For example

(i) $px - qy = (z - xy)^2$

(ii) $r + ka^2 t = 2as$, where $r = \frac{\partial^2 z}{\partial x^2}$, $t = \frac{\partial^2 z}{\partial y^2}$, $s = \frac{\partial^2 z}{\partial x \partial y}$.

The *order* of the partial differential equation is the order of the highest order partial derivative occurring in PDE. Thus the equation is of first order when all the partial derivatives occurring in it are of the order not higher than unity and is of second order when the partial derivatives are of order not higher than two and there is at least one partial derivative of order two and so on. In this chapter, we shall discuss the PDEs of first order only.

7.2 CLASSIFICATION OF FIRST ORDER PDEs

7.2.1 Linear Equation

A first order PDE is said to be a linear equation if it is linear in p, q and z, i.e. if it is of the form

$$P(x, y)\, p + Q(x, y)\, q + R(x, y)\, z = S(x, y),$$

where $p = \frac{\partial z}{\partial x}$, $q = \frac{\partial z}{\partial y}$

For example, $px - qy = x^2 z + xy^2$, $xp + yq = xy$ and $x^2 p + y^2 q = z$.

7.2.2 Semi-linear

A first order PDE is said to be a semi-linear equation if it is linear in p, q and the coefficients of p and q are functions of x and y only, i.e. if it is of the form

$P(x, y)\ p\ +\ Q(x, y)\ q\ =\ R(x,\ y,\ z)$, where $R(x, y, z)$ is not linear in z.

For example, $e^x\ p - xq = xz^3$.

7.2.3 Quasi-linear Equation

A first order PDE is said to be a quasi-linear equation if it is linear in p, q and is of the form

$$P(x, y, z)p + Q(x, y, z)q = R(x, y, z)$$

For example, $(x^2 + y^2)p - xyzq = xz^3 - x^2$.

7.2.4 Non-linear Equation

A PDE of the form $f(x, y, z, p, q) = 0$ which does not come under the given three types — linear, semi-linear and quasi-linear is known as non-linear PDE.

For example, $p^3 + q^3 = 27z$, $x^2 p^2 + y^2 q^2 = z^2$ and $p + \log q = 2z^3$.

7.2.5 Linear Homogeneous PDEs

A linear partial differential equation is said to be homogeneous if every term of the equation contains the dependent variable or its derivative. For example, $xp + yq = xyz$.

7.3 CLASSIFICATION OF INTEGRALS

By the integration of an equation means the derivation of all the values of z (as functions of x and y) which on substituting in the equation render an identity. Before dealing with the methods of integrations, we will classify the classes which include all possible integrals of the PDE. For this purpose only two independent variables will be taken into account while the results can be generalized to any number of independent variables.

7.3.1 Complete Integral

Consider a relation $f(x, y, z, a, b) = 0$ ...(1)

where 'a' and 'b' are arbitrary constants. Differentiating equation (1) partially w. r. t. 'x' and 'y', we get

$$\frac{\partial f}{\partial x}+\frac{\partial f}{\partial z}p=0 \text{ and } \frac{\partial f}{\partial y}+\frac{\partial f}{\partial z}q=0 \qquad ...(2)$$

From (1) and (2) the arbitrary constants 'a' and 'b' can be eliminated. The resulting equation (a relation between x, y, z, p and q) is a partial differential equation of first order. Let it be

$$F(x, y, z, p, q)=0 \qquad ...(3)$$

Conversely relation (1) is a solution of equation (3). It cannot have more than two arbitrary constants because if it has, these cannot be eliminated between equations (1) to (3) and to eliminate these, more equations are needed which are to be obtained by differentiating equations (2) and (3) and thus second and higher order derivatives will come into picture and the resulting equations will be of order higher than unity.

Thus equation (1) contains the maximum number of arbitrary constants that can be expected in a solution of equation (3). Relation (1) is known as complete integral of equation (3).

Thus the following definition of complete integral can be given for the general case.

Definition: Complete integral of an equation is a relation between the variables involving as many arbitrary constants as there are independent variables.

If particular values are given to the arbitrary constants, the solution obtained is known as a particular integral.

7.3.2 Singular Integral

In section (7.3.1) the suppositions was made that 'a' and 'b' are constants and the equation (3) was deduced from equations (1) and (2). But if 'a' and 'b' are assumed to be such functions of the independent variables that these do not alter the forms of p and q, the differential equation obtained by the elimination of the functions will be the same as in the case when 'a' and 'b' were arbitrary constants. Differentiating equation (1) partially w. r. t. 'x' and 'y' (regarding 'a' and 'b' as functions of 'x' and 'y'), we get

$$\frac{\partial f}{\partial x}+\frac{\partial f}{\partial z}p+\frac{\partial f}{\partial a}\frac{\partial a}{\partial x}+\frac{\partial f}{\partial b}\frac{\partial b}{\partial x}=0 \qquad ...(4)$$

and
$$\frac{\partial f}{\partial y}+\frac{\partial f}{\partial z}q+\frac{\partial f}{\partial a}\frac{\partial a}{\partial y}+\frac{\partial y}{\partial b}\frac{\partial b}{\partial y}=0 \qquad ...(5)$$

The forms of 'p' and 'q' will be the same as in (2) and (3) if

$$\frac{\partial f}{\partial a}\frac{\partial a}{\partial x}+\frac{\partial f}{\partial b}\frac{\partial b}{\partial x}=0 \qquad ...(6)$$

$$\frac{\partial f}{\partial a}\frac{\partial a}{\partial y}+\frac{\partial f}{\partial b}\frac{\partial b}{\partial y}=0 \qquad ...(7)$$

Therefore, if $J=\begin{vmatrix} \frac{\partial a}{\partial x} & \frac{\partial b}{\partial x} \\ \frac{\partial a}{\partial y} & \frac{\partial b}{\partial y} \end{vmatrix}$, then on solving equations (6) and (7), we obtain

$$J\frac{\partial f}{\partial a}=0 \text{ and } J\frac{\partial f}{\partial b}=0 \qquad ...(8)$$

If J is not zero, from equation (8), we have

$$\frac{\partial f}{\partial a}=0 \text{ and } \frac{\partial f}{\partial b}=0 \qquad ...(9)$$

Hence from the equations in (9), the values of 'a' and 'b' can be obtained in terms of the independent variables. The relation (1) is still a solution with the change in the quantities 'a' and 'b'; when these values thus obtained are substituted in (1) and a solution of equation (4) not involving any arbitrary constant, is obtained. This new solution is known as singular integral, which generally differs from any solution obtained from (1) by giving particular constant values to 'a' and 'b'.

Thus, we can sometimes obtain another solution by finding the envelope of the two-parameter family (1). This is obtained by eliminating a and b from the equations

$$f(x,y,z,a,b)=0,\ f_a=0,\ f_b=0 \qquad ...(10)$$

and is called the *singular integral* of the PDE (3).

Singular integral is a relation between the variables involving no arbitrary constant. Sometimes it occurs as a particular instance of a complete integral when special values are given to arbitrary constants but generally it is not the case and the singular integral (when it exists) is generally distinct from a complete integral.

7.3.3 General Integral

Equations (8) are also satisfied if $J=0$ *i.e.* 'a' and 'b' are functionally related. This functional relation may be arbitrary. Let the functional relation between 'a' and 'b' be

$$b=\phi(a) \qquad ...(11)$$

where ϕ is an arbitrary function. Multiplying equations in (8) by da and db respectively and adding, we obtain

$$\frac{\partial f}{\partial a}da+\frac{\partial f}{\partial b}db=0 \qquad ...(12)$$

But from equation (11), we have $db = \frac{\partial \phi}{\partial a} da$, so using it in (12), we get

$$\frac{\partial f}{\partial a} + \frac{\partial f}{\partial b} \frac{\partial \phi}{\partial a} = 0 \qquad \text{...(13)}$$

From equation (13) the value of 'a' involving the arbitrary functions ϕ may be obtained. Then 'b' is given by (11). When these values are substituted in (1) the solution takes a new form which is different from discussed in last two sections. This solution is called general integral.

Note.

(1) Evidently general integral can be obtained by eliminating 'a' and 'b' from equations (11) and (13) and singular integral by eliminating 'a' and 'b' from equations (1) and (9).

(2) Generally (but not universally) these three classes of integrals include all integrals. Exceptions may arise for partial differential equations of particular forms and when these do occur, the integrals of these PDEs are called special integrals.

7.4 GEOMETRICAL INTERPRETATION OF THREE TYPES OF INTEGRALS

7.4.1 Complete Integral

A complete integral, being a relation between x, y and z, is the equation of a surface. Since it contains two arbitrary parameters, it belongs to a double infinite t-system of surfaces or to a single infinite system of family of surfaces.

7.4.2 General Integral

Let a complete integral be $f(x, y, z, a, b) = 0$ ...(1)

A general integral is obtained by eliminating 'a' between (1) and the equations

$$b = \phi(a) \text{ and } \frac{\partial f}{\partial a} + \frac{\partial f}{\partial b} \phi'(a) = 0 \qquad \text{...(2)}$$

where ϕ is an arbitrary function.

The operation of elimination is performed by selecting representative surfaces from the system of families of surfaces and then finding surface of the family containing parameter 'a' by eliminating parameter 'b'. The equation obtained by eliminating 'a' between them is the envelope of the family. Hence the envelope touches the surface represented by (1) and $b = \phi(a)$ along the curve represented by equations (1) and (2). This curve is called the characteristic of the envelope and the general integral thus represents the envelope of a family of surfaces considered, as composed of its characteristics.

7.4.3 Singular Integral

The singular integral is obtained by eliminating 'a' and 'b' between equations $f(x, y, z, a, b) = 0$, $\frac{\partial f}{\partial a} = 0$ and $\frac{\partial f}{\partial b} = 0$. The operation of elimination is equivalent to find the envelope of all the surfaces included in the complete integral. The three equations give the point of contact of the particular surface represented by (1) with the general envelope. The Singular Integral thus represents the general envelope of all the surfaces included in the complete integral.

Note:

(1) It is necessary to ensure that the resulting equation obtained on eliminating 'a' and 'b' between equations $f(x, y, z, a, b) = 0$, $\frac{\partial f}{\partial a} = 0$ and $\frac{\partial f}{\partial b} = 0$ is that of the envelope and not that of any of the loci which are included in the same equations. The equations of such loci do not satisfy the differential equation. It is therefore desirable to substitute the result in the differential equation; it is to be retained only when it is a solution.

(2) It may happen that the entire system of surfaces does not admit a general envelope, in such a case the singular integral does not exist for the corresponding differential equation. Its non-existence will be indicated by the equations ordinarily used to obtain it.

7.5 SINGULAR INTEGRAL FROM PDEs DIRECTLY

Consider a partial differential equation $f(x, y, z, p, q) = 0$ ...(1)

Let its complete integral is $F(x, y, z, a, b) = 0$ (2)

where 'a' and 'b' are arbitrary constants. The singular integral is obtained by eliminating a and b from the equations (2), $\frac{\partial F}{\partial a} = 0$ and $\frac{\partial F}{\partial b} = 0$.

The values of z, p, q derived from (2) when substituted in (1) will render it as an identity and the substitution of the values of p and q (but not of z) will in general render (1) equivalent to the integral equation. Let this substitution be made so in (1) that p and q are replaced by functions of x, y, z, a and b. Then the singular integral is given by equation (1) and the equations obtained on differentiating (2) partially w. r. t. 'a' and 'b' *viz.*, the equations

$$\frac{\partial f}{\partial p}\frac{\partial p}{\partial a} + \frac{\partial f}{\partial q}\frac{\partial q}{\partial a} = 0 \quad ...(3)$$

$$\frac{\partial f}{\partial p}\frac{\partial p}{\partial b}+\frac{\partial f}{\partial q}\frac{\partial q}{\partial b}=0 \qquad \text{...(4)}$$

If $\frac{\partial f}{\partial p}\neq 0$ and $\frac{\partial f}{\partial q}\neq 0$, then equations (3) and (4) hold when $\frac{\partial p}{\partial a}\frac{\partial q}{\partial b}-\frac{\partial p}{\partial b}\frac{\partial q}{\partial a}=0$.

which implies that there exists a functional relation between p and q which does not contain a and b explicitly. Let this functional relation be

$$\phi(p,q)=0 \qquad \text{...(5)}$$

If both the constants a and b occur as function of p and q (which does not always happen), the equation (5) would imply that one of them is a function of the other and the equations using them give general integral which is now not concerned.

Equations (3) and (4) are also satisfied if $\frac{\partial f}{\partial p}=0$ and $\frac{\partial f}{\partial q}=0$.

The elimination of p and q between (1), $\frac{\partial f}{\partial p}=0$ and $\frac{\partial f}{\partial q}=0$ will furnish a relation between x, y, z independent of any arbitrary constant. If this relation satisfies the differential equation, it is the singular integral.

7.6 DERIVATION OF PDEs BY THE ELIMINATION OF ARBITRARY FUNCTIONS

Let u and v be two functions of x, y and z connected by the relation

$$f(u,v)=0 \qquad \text{...(1)}$$

where f is an arbitrary function. Differentiating (1) partially w. r. t. 'x' and 'y', respectively

$$\frac{\partial f}{\partial u}\left(\frac{\partial u}{\partial x}+p\frac{\partial u}{\partial z}\right)+\frac{\partial f}{\partial v}\left(\frac{\partial v}{\partial x}+p\frac{\partial v}{\partial z}\right)=0 \qquad \text{...(2)}$$

$$\frac{\partial f}{\partial u}\left(\frac{\partial u}{\partial y}+q\frac{\partial u}{\partial z}\right)+\frac{\partial f}{\partial v}\left(\frac{\partial v}{\partial y}+q\frac{\partial v}{\partial z}\right)=0 \qquad \text{...(3)}$$

Eliminating 'f' between equations (2) and (3), we get

$$\left(\frac{\partial u}{\partial x}+p\frac{\partial u}{\partial z}\right)\left(\frac{\partial v}{\partial y}+q\frac{\partial v}{\partial z}\right)-\left(\frac{\partial u}{\partial y}+q\frac{\partial u}{\partial z}\right)\left(\frac{\partial v}{\partial x}+p\frac{\partial v}{\partial z}\right)=0 \qquad \text{...(4)}$$

which can be written as $Pp+Qq=R$ ...(5)

where $P = \frac{\partial u}{\partial z}\frac{\partial v}{\partial y} - \frac{\partial u}{\partial y}\frac{\partial v}{\partial z} = \frac{\partial(u,v)}{\partial(z,y)}$, $Q = \frac{\partial u}{\partial x}\frac{\partial v}{\partial z} - \frac{\partial u}{\partial z}\frac{\partial v}{\partial x} = \frac{\partial(u,v)}{\partial(x,z)}$,

$$R = \frac{\partial u}{\partial y}\frac{\partial v}{\partial x} - \frac{\partial u}{\partial x}\frac{\partial v}{\partial y} = \frac{\partial(u,v)}{\partial(y,x)}.$$

Thus on eliminating one arbitrary function a partial differential equation of first degree in p and q is obtained. When the given relation between x, y and z contains more than one arbitrary function, the partial differential equation thus obtained will be in general of higher order.

7.7 SOLUTION OF PARTIAL DIFFERENTIAL EQUATIONS

A partial differential equation is said to be fully solved when its all three types of integrals (complete integral, singular integral and general integral) have been obtained otherwise it is not considered fully solved.

7.7.1 Equations Solvable by Direct Integration

The equations which contain only one partial derivative can be solved by direct integration. However, in place of the constants of integration, we must use arbitrary functions of the variable which kept constant. These arbitrary functions are obtained by using initial and boundary conditions.

Example 1. Form a partial differential equation by eliminating a, b from $z = (x+a)(y+b)$.

Solution. Here $\frac{\partial z}{\partial x} = p = (y+b)$ and $\frac{\partial z}{\partial y} = q = (x+a)$.

Substituting the values of $(x+a)$ and $(y+b)$ in (1), we obtain $z = pq$, which is the required differential equation.

Example 2. Form a partial differential equation by eliminating function f and F from $z = f(x+iy) + F(x-iy)$.

Solution. Given that $z = f(x+iy) + F(x-iy)$, so we have

$$\frac{\partial z}{\partial x} = f'(x+iy) + F'(x-iy), \quad \frac{\partial z}{\partial y} = i\,f'(x+iy) - i\,F'(x-iy)$$

$$\frac{\partial^2 z}{\partial x^2} = f''(x+iy) + F''(x-iy) \text{ and } \frac{\partial^2 z}{\partial x^2} = -f''(x+iy) - F''(x-iy)$$

Adding last two equations, we get

$$\frac{\partial^2 z}{\partial x^2} + \frac{\partial^2 z}{\partial y^2} = 0$$

Example 3. Solve $\frac{\partial^2 z}{\partial x^2} + z = 0$, when $\{z\}_{x=0} = e^y$ and $\left\{\frac{\partial z}{\partial x}\right\}_{x=0} = 1.$

Solution. If z is a function of x alone, then the solution of $\frac{\partial^2 z}{\partial x^2} + z = 0$ is given by $z = c_1 \cos x + c_2 \sin x$, where c_1 and c_2 are arbitrary constants. But z is a function of x and y, so we consider c_1 and c_2 as arbitrary functions of y. Therefore the solution of given PDE be $z = \phi(y)\cos x + \psi(y)\sin x$.

Using the condition $\{z\}_{x=0} = e^y$, we obtain $\phi(y) = e^y$.

Also using condition $\left\{\frac{\partial z}{\partial x}\right\}_{x=0} = 1$, we get $\psi(y) = 1$.

Hence the required solution is $z = e^y \cos x + \sin x$.

Example 4: Solve $\log s = x + y$.

Solution: Given equation can be written as $\frac{\partial^2 z}{\partial x\, \partial y} = e^{x+y}$.

Integrating w. r. t. 'x', we get $\frac{\partial z}{\partial y} = e^{x+y} + f(y)$, where the constant is taken as a function of y.

Again integrating w. r. t. 'y', we get

$$z = e^{x+y} + \int f(y)dy + \phi(x) = e^{x+y} + F(y) + \phi(x),$$

which is the required solution.

Example 5: Find the surface passing through the parabolas $z = 0, y^2 = 4ax$ and $z = 1, y^2 = -4ax$ and satisfying the differential equation $xr + 2p = 0$.

Solution: Given equation can be written as

$$x\frac{\partial^2 z}{\partial x^2} + 2p = 0 \Rightarrow x\frac{\partial p}{\partial x} + 2p = 0.$$

Integrating w. r. t. 'x', we get $x^2 p = f(y) \Rightarrow \frac{\partial z}{\partial x} = p = \frac{f(y)}{x^2}$.

Again integration w. r. t. 'x' yields, $z = -\frac{f(y)}{x} + g(y)$ ...(1)

Now we determine $f(y)$ and $g(y)$ by using given conditions.

Putting $z=0$ and $x=\dfrac{y^2}{4a}$ in (1), we get $0=-\dfrac{4a}{y^2}f(y)+g(y)$ ...(2)

Putting $z=1$ and $x=-\dfrac{y^2}{4a}$ in (1), we get $1=\dfrac{4a}{y^2}f(y)+g(y)$ ...(3)

Solving (2) and (3), we obtain $g(y)=\dfrac{1}{2}$ and $f(y)=\dfrac{y^2}{8a}$.

Substituting these values in (1), we obtain

$$z=-\frac{y^2}{8ax}+\frac{1}{2} \Rightarrow 8axz=4ax-y^2.$$

EXERCISE 7.1

1. Form the PDEs by eliminating the arbitrary constants and arbitrary functions.

   (i) $z=ax^2+bxy+cy^2$. (ii) $(x-h)^2+(y-k)^2+z^2=a^2$.

   (iii) $ax^2+by^2+cz^2=1$. (iv) $z=f(y/x)$.

   (v) $z=f(x)g(y)$. (vi) $f(x+y+z,x^2+y^2+z^2)=0$.

2. Solve the following PDEs.

   (i) $t=\sin(xy)$. (ii) $s=2x+2y$.

   (iii) $r=xy$. (iv) $\dfrac{\partial^2 z}{\partial x\,\partial t}=e^{-t}\cos x$.

   (v) $r=f(x,y)$. (vi) $s=4x\sin(3xy)$.

   (vii) $r=6x$. (viii) $\dfrac{\partial^3 z}{\partial x^2\,\partial y}=\cos(2x+3y)$

   (ix) $xr=(n-1)p$. (x) $s=(x+\beta y)/y$.

   (xi) $x\,y\,s=1$.

   (xii) $ys+p=\cos(x+y)-y\sin(x+y)$.

3. Find the surface satisfying $t=6x^2y$, containing the two lines $y=0=z,\ y=2=z$.

4. Solve $\dfrac{\partial^2 z}{\partial x\,\partial y}=\sin x\sin y$, when $\left\{\dfrac{\partial z}{\partial y}\right\}_{x=0}=-2\sin y$ and y is odd multiple of $\dfrac{\pi}{2}$ at $z=0$.

5. Solve $\dfrac{\partial^2 z}{\partial x^2}=a^2 z$ when $\dfrac{\partial z}{\partial x}=a\sin y$ and $\dfrac{\partial z}{\partial y}=0$ at $x=0$.

ANSWERS 7.1

1. (i) $x^2r + 2xys + y^2t = 2z$. (ii) $z^2(p^2 + q^2 + 1) = a^2$.
   (iii) $pq + zs = 0$. (iv) $px + qy = 0$.
   (v) $pq - zs = 0$. (vi) $(y - z)p + (z - x)q = x - y$.
2. (i) $z = -(1/x^2)\sin(xy) + y f(x) + \phi(x)$.
   (ii) $z = x^2y + xy^2 + f(y) + \phi(x)$.
   (iii) $z = \frac{1}{6}x^3 y + x f(y) + \phi(y)$.
   (iv) $z = -e^{-t}\sin x + f(x) + \phi(t)$.
   (v) $z = \int\left\{\int f(x, y)dx\right\}dx + x f(y) + \phi(y)$.
   (vi) $z = -\frac{4}{9y}\sin 3xy + f(x) + \phi(y)$.
   (vii) $z = x^3 + x f(y) + \phi(y)$.
   (viii) $z = -\frac{1}{12}\sin(2x + 3y) + f(x) + x\phi(y) + \psi(y)$.
   (ix) $z = x'' f(y) + \phi(y)$.
   (x) $z = \frac{1}{2}x^2 \log y + \beta xy + f(x) + \phi(y)$.
   (xi) $z = \log x \log y + f(x) + \phi(y)$.
   (xii) $yz = y\sin(x + y) + f(x) + \phi(y)$.
3. $z = x^2 y^3 + y(1 - 4x^2)$.
4. $z = (1 + \cos x)\cos y$.
5. $z = \sin x + e^y \cos x$.

7.8 LAGRANGE'S EQUATION

The partial differential equation of the form $Pp + Qq = R$ where P, Q, R are functions of x, y, z is the standard form of the linear partial differential equation of the order one and is called Lagrange's equation. Let u and v be two functions of x, y, z which are connected by the relation

$$f(u, v) = 0 \qquad \text{...(1)}$$

Differentiating, partially w. r. t. 'x' and 'y', we obtain

$$\frac{\partial f}{\partial u}\left(\frac{\partial u}{\partial x} + \frac{\partial u}{\partial z}p\right) + \frac{\partial f}{\partial v}\left(\frac{\partial v}{\partial x} + \frac{\partial v}{\partial z}p\right) = 0 \qquad \text{...(2)}$$

and $$\frac{\partial f}{\partial u}\left(\frac{\partial u}{\partial y}+\frac{\partial u}{\partial z}q\right)+\frac{\partial f}{\partial v}\left(\frac{\partial v}{\partial y}+\frac{\partial v}{\partial z}q\right)=0 \qquad ...(3)$$

Eliminating $\frac{\partial f}{\partial u}$ and $\frac{\partial f}{\partial v}$ from (2) and (3), we have

$$\left(\frac{\partial u}{\partial x}+\frac{\partial u}{\partial z}p\right)\left(\frac{\partial v}{\partial y}+\frac{\partial v}{\partial z}q\right)=\left(\frac{\partial u}{\partial y}+\frac{\partial u}{\partial z}q\right)\left(\frac{\partial v}{\partial x}+\frac{\partial v}{\partial z}p\right)$$

$$\Rightarrow \left(\frac{\partial u}{\partial y}\frac{\partial v}{\partial z}-\frac{\partial v}{\partial y}\frac{\partial u}{\partial z}\right)p+\left(\frac{\partial v}{\partial x}\frac{\partial u}{\partial z}-\frac{\partial u}{\partial x}\frac{\partial v}{\partial z}\right)q=\frac{\partial u}{\partial x}\frac{\partial v}{\partial y}-\frac{\partial u}{\partial y}\frac{\partial v}{\partial x}$$

$$\Rightarrow Pp+Qq=R \qquad ...(4)$$

where $P=\frac{\partial u}{\partial y}\frac{\partial v}{\partial z}-\frac{\partial v}{\partial y}\frac{\partial u}{\partial z}$, $Q=\frac{\partial v}{\partial x}\frac{\partial u}{\partial z}-\frac{\partial u}{\partial x}\frac{\partial v}{\partial z}$ and

$$R=\frac{\partial u}{\partial x}\frac{\partial v}{\partial y}-\frac{\partial u}{\partial y}\frac{\partial v}{\partial x}$$

Now, suppose $v=b$ and $u=a$, where a,b are constants, so that

$$dv=\frac{\partial v}{\partial x}dx+\frac{\partial v}{\partial y}dy+\frac{\partial v}{\partial z}dz=0$$

$$du=\frac{\partial u}{\partial x}dx+\frac{\partial u}{\partial y}dy+\frac{\partial u}{\partial z}dz=0$$

By cross-multiplication, we obtain

$$\frac{dx}{\frac{\partial u}{\partial y}\frac{\partial v}{\partial z}-\frac{\partial u}{\partial z}\frac{\partial v}{\partial y}}=\frac{dy}{\frac{\partial u}{\partial z}\frac{\partial v}{\partial x}-\frac{\partial u}{\partial x}\frac{\partial v}{\partial z}}=\frac{dz}{\frac{\partial u}{\partial x}\frac{\partial v}{\partial y}-\frac{\partial u}{\partial y}\frac{\partial v}{\partial x}}$$

$$\Rightarrow \frac{dx}{P}=\frac{dy}{Q}=\frac{dz}{R}$$

These equations are called Lagrange's auxiliary equations. Solutions of these auxiliary differential equations are $u=a$ and $v=b$. Hence $f(u,v)=0$ is the required solution of (4).

Method of multipliers: This method is used to solve the partial differential equation

$$Pp+Qq=R$$

We know that the auxiliary equations or subsidiary equations are

$$\frac{dx}{P}=\frac{dy}{Q}=\frac{dz}{R}.$$

We choose multiplier l,m,n (may be constants or functions of x,y,z) such that $lP+mQ+nR=0$.

Thus we have $\frac{dx}{P}=\frac{dy}{Q}=\frac{dz}{R}=\frac{ldx+mdy+ndz}{lP+mQ+nR}=\frac{ldx+mdy+ndz}{0}$

which implies $ldx+mdy+ndz=0$.

On integrating it, we get the solution (say) $u=C_1$.

Similarly, we choose another set of multipliers l_1, m_1, n_1 such that $l_1P+m_1Q+n_1R=0$

which implies $l_1dx+m_1\,dy+n_1\,dz=0$

Again on integrating, we get the solution (say) $v=C_2$

Then the required solution of the given differential equation be $f(u,v)=0$ or $u=f(v)$.

This procedure is known as method of multipliers.

Example 1: Solve $p+q=(z/a)$.

Solution: Comparing with Lagrange's equation $Pp+Qq=R$, we obtain $P=1, Q=1$ and $R=\frac{z}{a}$.

Therefore subsidiary equations are $\frac{dx}{P}=\frac{dy}{Q}=\frac{dz}{R} \Rightarrow \frac{dx}{1}=\frac{dy}{1}=\frac{dz}{z/a}$

Taking the first two terms, we get $dx=dy$.

Integration yields, $x=y+c_1 \Rightarrow x-y=c_1$.

Again taking the last two terms, we get $dy=\frac{dz}{z/a} \Rightarrow \frac{dy}{a}=\frac{dz}{z}$

On integrating we get $\frac{1}{a}y=\log z-\log c_2 \Rightarrow z=c_2\,e^{y/a}$.

Hence the general integral is $z=e^{y/a}\,f(x-y)$.

Example 2: Solve $pz-qz=z^2+(x+y)^2$.

Solution: The subsidiary equations are $\frac{dx}{z}=\frac{dy}{-z}=\frac{dz}{z^2+(x+y)^2}$.

Taking the first and second members, we have $dx+dy=0$.

Integration yields, $x+y=c_1$.

Taking the first and the second members, we obtain

$$\frac{z\,dz}{z^2+(x+y)^2}=dx \Rightarrow \frac{2z\,dz}{z^2+c_1^{\,2}}=2\,dx$$

On integrating, we get $\log\left(z^2+c_1^{\,2}\right)=2x+c_2$

$\Rightarrow \quad \log\left(z^2+x^2+y^2+2xy\right)-2x=c_2$.

Hence the general solution is

$$\log\left(z^2+x^2+y^2+2xy\right)-2x=f\left(x+y\right).$$

Example 3: Solve $yzp+zxq=xy.$

Solution: The subsidiary equations are $\dfrac{dx}{yz}=\dfrac{dy}{zx}=\dfrac{dz}{xy}$

Taking the first two members, we get $xdx-ydy=0.$

Integration yields $x^2-y^2=c_1.$

Similarly taking the first and the last members, we obtain $x^2-z^2=c_2.$

Hence the general integral is $f\left(x^2-y^2, x^2-z^2\right)=0.$

Example 4: Solve $\left(y+z\right)p+\left(z+x\right)q=x+y.$

Solution: The subsidiary equations are $\dfrac{dx}{y+z}=\dfrac{dy}{z+x}=\dfrac{dz}{x+y}$

$$\text{Each term}=\frac{dx+dy+dz}{2\left(x+y+z\right)}=\frac{dx-dy}{\left(-x+y\right)}=\frac{dy-dz}{-\left(y-z\right)}$$

Taking the first two members, we obtain $\dfrac{dx+dy+dz}{x+y+z}+2\dfrac{dx-dy}{x-y}=0$

Integration yields

$$\log\left(x+y+z\right)+2\log\left(x-y\right)=\log c_1 \Rightarrow \left(x+y+z\right)\left(x-y\right)^2=c_1$$

Again taking last two members, we have $\dfrac{dx-dy}{x-y}=\dfrac{dy-dz}{y-z}$

which gives $\log\left(x-y\right)=\log\left(y-z\right)+\log c_2 \Rightarrow \dfrac{x-y}{y-z}=c_2.$

Hence the general integral is $f\left[\dfrac{x-y}{y-z}, \left(x-y\right)^2\left(x+y+z\right)\right]=0.$

Example 5: Solve $\left(mz-ny\right)p+\left(nx-lz\right)q=ly-mx.$

Solution: The subsidiary equations are $\dfrac{dx}{mz-ny}=\dfrac{dy}{nx-lz}=\dfrac{dz}{ly-mx}.$

Using x, y, z as multipliers, we obtain *Each Fraction*

$$=\frac{x\,dx+y\,dy+z\,dz}{0}$$

$$\Rightarrow \qquad xdx+ydy+zdz=0.$$

On integrating, we get $x^2+y^2+z^2=c_1.$

Again using l, m, n as multipliers, we get

Each fraction $= \dfrac{l\,dx + m\,dy + n\,dz}{0}$

$\Rightarrow \quad l\,dx + m\,dy + n\,dz = 0$.

On integrating, we get $lx + my + nz = c_2$.

Hence the general integral is $f\left(lx + my + nz,\, x^2 + y^2 + z^2\right) = 0$.

Example 6: Solve $\left(y^2 + z^2 - x^2\right)p - 2xyq + 2zx = 0$.

Solution: The equation can be written as $\left(y^2 + z^2 - x^2\right)p - 2\,xyq = -\,2zx$

Here the subsidiary equations are $\dfrac{dx}{y^2 + z^2 - x^2} = \dfrac{dy}{-\,2xy} = \dfrac{dz}{-\,2zx}$.

Taking the last two members, we get $\dfrac{dy}{y} = \dfrac{dz}{z}$.

Integration yields $\log \dfrac{y}{z} = \log c_1 \Rightarrow \dfrac{y}{z} = c_1$

Again using x, y, z as multipliers, we get

Each fraction $= \dfrac{xdx + ydy + zdz}{-\,x\left(x^2 + y^2 + z^2\right)} = \dfrac{dz}{-\,2zx} \Rightarrow 2\dfrac{xdx + ydy + zdz}{x^2 + y^2 + z^2} = \dfrac{dz}{z}$

On integrating, we get

$$\log\left(x^2 + y^2 + z^2\right) = \log z + \log c_2 \Rightarrow x^2 + y^2 + z^2 = c_2\,z.$$

Hence the general solution is $x^2 + y^2 + z^2 = zf\left(\dfrac{y}{z}\right)$.

Example 7: Solve $\left(x^2 - yz\right)p - \left(y^2 - zx\right)q = z^2 - xy$.

Solution: The subsidiary equations are $\dfrac{dx}{x^2 - yz} = \dfrac{dy}{y^2 - zx} = \dfrac{dz}{z^2 - xy}$.

Each fraction

$$= \frac{dx - dy}{(x - y)(x + y + z)} = \frac{dy - dz}{(y - z)(x + y + z)} = \frac{dz - dx}{(z - x)(x + y + z)}.$$

Taking the first two members, we get $\dfrac{dx - dy}{x - y} = \dfrac{dy - dz}{y - z}$.

Integration yields $\log(x - y) = \log(y - z) + \log c_1 \Rightarrow \dfrac{x - y}{y - z} = c_1$

Similarly taking the last two members, we obtain $\dfrac{z - x}{y - z} = c_2$.

Hence the general solution is $f\left(\dfrac{x-y}{y-z}, \dfrac{z-x}{y-z}\right)=0.$

EXERCISE 7.2

Solve the following PDEs

1. $p-q=z/(x+y)$.
2. $xzp+yzq=xy$.
3. $x_2\, x_3\, zp_1+x_3\, x_1\, zp_2+x_1\, x_2\, zp_3=x_1\, x_2\, x_3$.
4. $p\tan x+q\tan y=\tan z$.
5. $p+q=1$.
6. $p_1+x_1\, p_2+x_1\, x_2\, p_3=x_1\, x_2\, x_3\sqrt{z}$.
7. $y^2p-xyq=x(z-2y)$.
8. $(y^2+z^2)p-xyq=-zx$.
9. $x^2\, p+y^2\, q=z^2$.
10. $\dfrac{y^2z}{x}p-xzq=y^2$.
11. $(x_1\, p_1+2\, x_2\, p_2+3\, x_3\, p_3+4\, x_4\, p_4)=0$.
12. $p_1+p_2+p_3=4z$
13. $(x+2z)p+(4\, zx-y)q=2x^2+y$.

ANSWERS 7.2

1. $(x+y)\log z=x+f(x+y)$
2. $xy-z^2=f\left(\dfrac{x}{y}\right)$
3. $f(x_2^{\ 2}-x_1^{\ 2}, x_3^{\ 2}-x_1^{\ 2}, z^2-x_1^{\ 2})=0$
4. $\dfrac{\sin z}{\sin y}=f\left(\dfrac{\sin x}{\sin y}\right)$
5. $f(x-z, y-z)=0$
6. $f(4\sqrt{z}-x_3^{\ 2}, 2x_3-x_2^{\ 2}, 2x_2-x_1^{\ 2})=0$
7. $f(x^2+y^2, zy-y^2)=0$
8. $\phi(y/z, x^2+y^2+z^2)=0$

9. $\frac{1}{x}-\frac{1}{z}=f\left(\frac{1}{y}-\frac{1}{z}\right)$

10. $f\left(x^2-z^2, x^3+y^3\right)=0$

11. $f\left(z, x_1^{2} x_2^{-1}, x_1^{3} x_3^{-1}, x_1^{4} x_4^{-1}\right)=0.$

12. $f\left(\frac{z}{e^{4x_1}}, \frac{z}{e^{4x_2}}, \frac{z}{e^{4x_3}}\right)=0.$

13. $f\left(xy-z^2, x^2-y-z\right)=0.$

7.9 SOLUTION OF NON-LINEAR PDEs OF ORDER ONE

The most general partial differential equation of order one may be denoted by

$$F(x, y, z, p, q)=0 \qquad ...(1)$$

To solve equations of this type, we use a general method due to Charpit and is well known as Charpit's method. The fundamental idea in this method is the introduction of the another partial differential equation of order one

$$f(x, y, z, p, q, a)=0 \qquad ...(2)$$

containing an arbitrary constant 'a' and satisfying the following conditions

(i) Equations (1) and (2) can be solved to give $p=p(x, y, z, a)$ and $q=q(x, y, z, a)$.

(ii) The equation

$$dz=p(x, y, z, a)dx+q(x, y, z, a)dy \qquad ...(3)$$

is integrable.

When a function 'f' satisfying the conditions (i) and (ii) has been found, the solution of equation in (3) containing two arbitrary constants (including 'a') will be a solution of equation (1). The condition (i) will be satisfied if

$$J=\frac{\partial(F, f)}{\partial(p, q)}=\begin{vmatrix} \frac{\partial F}{\partial p} & \frac{\partial f}{\partial p} \\ \frac{\partial F}{\partial q} & \frac{\partial f}{\partial q} \end{vmatrix} \neq 0 \qquad ...(4)$$

Assuming (4) is satisfied, equation (3) exists and the condition (ii) will be satisfied if

$$p\left(\frac{\partial q}{\partial z}\right)+q\left(-\frac{\partial p}{\partial z}\right)-\left(\frac{\partial p}{\partial y}-\frac{\partial q}{\partial x}\right)=0 \Rightarrow p\frac{\partial q}{\partial z}+\frac{\partial q}{\partial x}=q\frac{\partial p}{\partial z}+\frac{\partial p}{\partial y} \qquad ...(5)$$

Substituting the values of p and q as functions of x, y and z in equation (1) and (2) and differentiating with respect to x, we have

$$\frac{\partial F}{\partial x}+\frac{\partial F}{\partial p}\frac{\partial p}{\partial x}+\frac{\partial F}{\partial q}\frac{\partial q}{\partial x}=0$$

and $$\frac{\partial f}{\partial x}+\frac{\partial f}{\partial p}\frac{\partial p}{\partial x}+\frac{\partial f}{\partial q}\frac{\partial q}{\partial x}=0$$

Therefore

$$\left(\frac{\partial F}{\partial p}\frac{\partial f}{\partial q}-\frac{\partial f}{\partial p}\frac{\partial F}{\partial q}\right)\frac{\partial q}{\partial x}=\frac{\partial F}{\partial x}\frac{\partial f}{\partial p}-\frac{\partial F}{\partial p}\frac{\partial f}{\partial x}$$

$$\Rightarrow \quad \frac{\partial q}{\partial x}=\frac{1}{J}\left\{\frac{\partial F}{\partial x}\frac{\partial f}{\partial p}-\frac{\partial F}{\partial p}\frac{\partial f}{\partial x}\right\} \quad \text{...(6)}$$

Similarly $$\frac{\partial p}{\partial y}=\frac{1}{J}\left\{-\frac{\partial F}{\partial y}\frac{\partial f}{\partial q}+\frac{\partial F}{\partial q}\frac{\partial f}{\partial y}\right\}, \quad \text{...(7)}$$

$$\frac{\partial p}{\partial z}=\frac{1}{J}\left\{-\frac{\partial F}{\partial z}\frac{\partial f}{\partial q}+\frac{\partial F}{\partial q}\frac{\partial f}{\partial z}\right\} \quad \text{...(8)}$$

and $$\frac{\partial q}{\partial z}=\frac{1}{J}\left\{\frac{\partial F}{\partial z}\frac{\partial f}{\partial p}-\frac{\partial F}{\partial p}\frac{\partial f}{\partial z}\right\}. \quad \text{...(9)}$$

Substituting from (6) to (9) in equation (5), we get

$$\frac{1}{J}\left[p\left(\frac{\partial F}{\partial z}\frac{\partial f}{\partial p}-\frac{\partial F}{\partial p}\frac{\partial f}{\partial z}\right)+\left(\frac{\partial F}{\partial x}\frac{\partial f}{\partial p}-\frac{\partial F}{\partial p}\frac{\partial f}{\partial x}\right)\right]$$

$$=\frac{1}{J}\left[q\left(-\frac{\partial F}{\partial z}\frac{\partial f}{\partial q}+\frac{\partial F}{\partial q}\frac{\partial f}{\partial z}\right)+\left(-\frac{\partial F}{\partial y}\frac{\partial f}{\partial q}+\frac{\partial F}{\partial q}\frac{\partial f}{\partial y}\right)\right]$$

$$\Rightarrow \quad \left(-\frac{\partial F}{\partial p}\right)\frac{\partial f}{\partial x}+\left(-\frac{\partial F}{\partial q}\right)\frac{\partial f}{\partial y}+\left(-p\frac{\partial F}{\partial p}-q\frac{\partial F}{\partial q}\right)\frac{\partial f}{\partial z}$$

$$+\left(p\frac{\partial F}{\partial z}+\frac{\partial F}{\partial x}\right)\frac{\partial f}{\partial p}+\left(q\frac{\partial F}{\partial z}+\frac{\partial F}{\partial y}\right)\frac{\partial f}{\partial q}=0 \quad \text{...(10)}$$

The equation (10) being linear in variables x, y, z, p, q and so f has its subsidiary equations

$$\frac{dx}{-\frac{\partial F}{\partial p}}=\frac{dy}{-\frac{\partial F}{\partial q}}=\frac{dz}{-p\frac{\partial F}{\partial p}-q\frac{\partial F}{\partial q}}=\frac{dp}{\frac{\partial F}{\partial x}+p\frac{\partial F}{\partial z}}=\frac{dq}{\frac{\partial F}{\partial y}+q\frac{\partial F}{\partial z}} \quad \text{...(11)}$$

Any of the integrals of equation (11) satisfy equation (10). If such an integral involve p or q it can be taken for (1).

Once equation (2) has been found, the problem reduces to solving equations (2) and (3) for p and q and then integrate equation (3).

Example 1: Solve $px + qy = pq$.

Solution: Let $f = px + qy - pq = 0$ ...(1)

Hence the Charpit's auxiliary equations are

$$\frac{dp}{p} = \frac{dq}{q} = \frac{dz}{-p(x-q)-q(y-p)} = \frac{dx}{-(x-q)} = \frac{dy}{-(y-p)} = \frac{dF}{0}$$

Taking the first two members, we obtain $\frac{dp}{p} = \frac{dq}{q}$

On integrating, we get $\log p = \log q + \log a \Rightarrow p = aq$ ...(2)

Putting $p = aq$ in (1), we obtain $aqx + qy = aq^2 \Rightarrow q = (y + ax)/a$.

Therefore from (2), we get $p = aq = y + ax$.

Using p and q in $dz = p\,dx + q\,dy$, we obtain

$$dz = (y + ax)dx + \frac{(y+ax)}{a}dy$$

$$\Rightarrow \quad a\,dz = (y + ax)(dy + a\,dx).$$

On integrating, we get $az = \frac{1}{2}(y + ax)^2 + b$, which is the complete integral.

General integral: Let $b = \phi(a)$, we have $az = \frac{1}{2}(y + ax)^2 + \phi(a)$ ...(3)

Differentiating (3) w. r. t. 'a', we get $z = x(y + ax) + \phi'(a)$. ...(4)

General integral is obtained by eliminating a between (3) and (4).

Singular integral: Differentiating the complete integral partially w. r. t. 'a' and 'b' respectively, we obtain $z = x(y + ax)$ and $0 = 1$.

Hence there is no singular integral.

Example 2: Apply Charpit's method to find complete integral of $z^2(p^2 z^2 + q^2) = 1$.

Solution: Let $f \equiv p^2 z^4 + q^2 z^2 - 1 = 0$ ...(1)

Hence the Charpit's auxiliary equations are

$$\frac{dp}{0 + p.(4p^2 z^3 + 2q^2 z)} = \frac{dq}{q(4p^2 z^3 + 2q^2 z)} = \frac{dz}{-p(2pz^4) - q(2qz^2)}$$

$$= \frac{dx}{-2pz^4} = \frac{dy}{-2qz^2} = \frac{dF}{0}$$

Taking the first two members, we have $\dfrac{dp}{p}=\dfrac{dq}{q}$.

On integrating, we get $\log p=\log q+\log a \Rightarrow p=aq$

Putting $p=aq$ in (1), we obtain $z^2\left(a^2z^2q^2+q^2\right)=1$

$$\Rightarrow \quad q^2=\frac{1}{z^2\left(a^2z^2+1\right)} \Rightarrow q=\frac{1}{z\sqrt{\left(a^2z^2+1\right)}}$$

and $\quad p=\dfrac{a}{z\sqrt{\left(a^2z^2+1\right)}}$.

Substituting p and q in $dz=p\,dx+q\,dy$, we get

$$dz=\frac{a}{z\sqrt{\left(a^2z^2+1\right)}}dx+\frac{1}{z\sqrt{\left(a^2z^2+1\right)}}dy$$

$$\Rightarrow \quad z\sqrt{\left(a^2z^2+1\right)}dz=adx+dy \Rightarrow \frac{t^2}{a^2}dt=a\,dx+dy, \text{ using } a^2z^2+1=t^2.$$

On integrating, we get $\dfrac{1}{3a^2}t^3=ax+y+b$

$$\Rightarrow \quad \left(a^2z^2+1\right)^3=9a^4\left(ax+y+b\right)^2.$$

Example 3: Solve $\left(p^2+q^2\right)y=qz$.

Solution: Let $f\equiv\left(p^2+q^2\right)y-qz=0$ ...(1)

Now Charpit's auxiliary equations are

$$\frac{dp}{-pq}=\frac{dq}{\left(p^2+q^2\right)-q^2}=\frac{dz}{-2p^2y-2q^2y+qz}=\frac{dx}{-2py}=\frac{dy}{-2qy+z}=\frac{dF}{0}$$

Taking the first two members, we get $\dfrac{dp}{-pq}=\dfrac{dq}{p^2} \Rightarrow p\,dp+q\,dq=0$

Integration yields, $p^2+q^2=a^2$ (say) ...(2)

Using it in (1), we get $q=\dfrac{a^2y}{z}$

Hence from (2), we obtain

$$p=\sqrt{\left(a^2-q^2\right)}=\sqrt{\left\{a^2-\frac{a^4y^2}{z^2}\right\}}=\frac{a}{z}\sqrt{\left(z^2-a^2y^2\right)}$$

Thus from $dz=p\,dx+q\,dy$, we get

$$dz=\frac{a}{z}\sqrt{\left(z^2-a^2y^2\right)}dx+\frac{a^2y}{z}dy,$$

$$\Rightarrow \quad \frac{z\,dz - a^2\,y\,dy}{\sqrt{(z^2 - a^2\,y^2)}} = a\,dx \Rightarrow dt = a\,dx\text{, using } z^2 - a^2\,y^2 = t^2.$$

On integrating, we get $t = ax + b \Rightarrow z^2 = a^2\,y^2 + (ax+b)^2$ ...(3)
which is the complete solution.

General integral: Let $b = \phi(a)$, then the complete integral gives

$$z^2 = a^2\,y^2 + [ax + \phi(a)]^2. \quad \text{...(4)}$$

Differentiating (4) partially w. r. t. 'a', we get

$$0 = 2a\,y^2 + 2[ax + \phi(a)][x + \phi'(a)]. \quad \text{...(5)}$$

General integral is obtained by eliminating a between (4) and (5).

Singular integral: Differentiating (3) partially w. r. t. 'a' and 'b' respectively, we get

$$0 = 2\,ay^2 + 2(ax+b)x \quad \text{...(6)}$$

$$\text{and} \quad 0 = 2(ax+b). \quad \text{...(7)}$$

Eliminating a and b from (3), (6) and (7), we get $z = 0$ which clearly satisfies the given equation (1) and therefore is the singular integral.

7.10 STANDARD FORMS

7.10.1 Standard Form I

The equations of this form do not involve variables explicitly, *i.e.* these equations are of the form $F(p, q) = 0$ ...(1)

Hence Charpit's subsidiary equations for equation (1) are

$$\frac{dx}{F_p} = \frac{dy}{F_q} = \frac{dz}{p\,F_p + q\,F_q} = \frac{dp}{0} = \frac{dq}{0}$$

This gives $dp = 0 \Rightarrow p = \text{constant} = a$ (say) ...(2)

Substituting $p = a$ in equation (1), we get $F(a, q) = 0$.

This implies that q = constant (3)

If we set $q = b$, then it must satisfy $F(a, b) = 0$ (4)

Therefore $dz = pdx + qdy$ gives $dz = adx + bdy$

On integrating, we obtain $z = ax + by + c$ (5)

where c is an arbitrary constant. Equations (5) and (4) give complete integral of the equation (1).

General integral: Let equation (4) when solved for 'b' gives $b = \phi(a)$. Substituting the value of b in equation (5), we get $z = ax + \phi(a)y + c$.

To obtain general integral we let $c=\psi(a)$, where 'ψ' is an arbitrary function. Then general integral is obtained by eliminating 'a' between the equations

$$z = ax + \phi(a)y + \psi(a)$$

and $$0 = x + \phi'(a)y + \psi'(a).$$

Singular integral: Singular integral, if it exists, is determined by eliminating a and c between the equations

$$f(x, y, z, a, c) \equiv z - ax - \phi(a)y - c = 0$$

$$\frac{\partial f}{\partial a} \equiv 0 = x + \phi'(a)y$$

$$\frac{\partial f}{\partial c} \equiv 0 - 1$$

The last equation shows that singular integral does not exist.

7.10.2 Standard Form II

The equations of this form do not involve independent variables explicitly, *i.e.* these equations are of the form $F(z, p, q) = 0$ ...(1)

Hence Charpit's subsidiary equations for equation (1) are

$$\frac{dx}{F_p} = \frac{dy}{F_q} = \frac{dz}{pF_p + qF_q} = \frac{dp}{-pF_z} = \frac{dq}{-qF_z} \Rightarrow \frac{dp}{p} = \frac{dq}{q}$$

On integrating, we obtain $p = aq$, where 'a' is a constant.

Substituting it in equation (1), we get $F(z, aq, q) = 0 \Rightarrow q = f(a, z)$ ...(2)

Using these values of p and q in equation $dz = pdx + qdy$, we get

$$dz = a\, f(a, z)dx + f(a, z)dy \Rightarrow \frac{dz}{f(a, z)} = a\, dx + dy \quad ...(3)$$

Hence the complete integral becomes $\int \frac{dz}{f(a, z)} = a\,x + y + c$ ...(4)

Equation (4) gives the complete integral. General and singular integrals are obtained by usual methods.

7.10.3 Standard Form III (Separable type)

The equations of this form are $f(x, p) = h(y, q)$ ...(1)

Hence Charpit's subsidiary equations for equation (1) are

$$\frac{dx}{f_p} = \frac{dy}{-h_q} = \frac{dz}{pf_p - qh_q} = \frac{dp}{-f_x} = \frac{dq}{h_y}$$

Now $\frac{dx}{f_p} = \frac{dp}{-f_x} \Rightarrow \frac{\partial f}{\partial p}dp + \frac{\partial f}{\partial x}dx = 0 \Rightarrow df = 0$

On integrating, we get $f(x,p)=a$ (constant).
Therefore, equation (1) gives $f(x,p)=f(y,q)=a$ ...(2)
Let equation (2) on solving gives $p=F_1(a,x)$ and $q=F_2(a,y)$.
Therefore, equation $dz=p\,dx+q\,dy$ gives

$$dz=F_1(a,x)dx+F_2(a,y)dy \Rightarrow z=\int F_1(a,x)dx+\int F_2(a,y)dy+b, \qquad ...(3)$$

where 'b' is an arbitrary constant. Equation (3) is complete integral of equation (1). General and singular integrals are obtained by usual methods.

Steps of integrating equation (1)

(1) Replace in equation (1) 'q' by $\frac{'dz'}{dX}$ and 'p' by $a\frac{'dz'}{dX}$.

(2) Integrate the resulting equation to obtain z as a function of X.

(3) Replace X by $(y+ax)$.

The resulting equation will be complete integral.

7.10.4 Standard Form IV (Analogues to Clairut's form)

Here PDEs are of the form

$$z=px+qy+f(p,q). \qquad ...(1)$$

Hence Charpit's subsidiary equations for equation (1) are

$$\frac{dp}{-p+p}=\frac{dp}{-q+q} \Rightarrow \frac{dp}{0}=\frac{dq}{0} \Rightarrow dp=0=dq.$$

On integrating, we get $p=a$ and $q=b$, where a and b are arbitrary constants.
Substituting these values in (1), we get $z=ax+by+f(a,b)$ ...(2)
Also the equation $dz=pdx+qdy$ gives $z=ax+by+c$ ...(3)
which is similar to (2). Hence (2) is complete integral of the equation (1).
To obtain general integral, put $b=\phi(a)$, where ϕ is an arbitrary function.
Substituting it in (2), we get $z=ax+\phi(a)y+f[a,\phi(a)]$. ...(4)
Now the general integral is obtained by eliminating 'a' between equation (4) and

$$0=x+\phi'(a)\,y+f'(a,\phi(a)) \qquad ...(5)$$

Singular integral is obtained by eliminating 'a' and 'b' between equation (2) and

$$0=x+\frac{\partial f}{\partial a} \text{ and } 0=y+\frac{\partial f}{\partial b}.$$

Example 4: Solve $p^2 + q^2 = 1$.

Solution: The equation is of the form $f(p,q)=0$, so the solution is given by

$$z = ax + by + c \text{, where } a^2 + b^2 = 1 \text{ or } b = \sqrt{(1-a^2)}$$

Hence the complete integral is $z = ax + \sqrt{(1-a^2)}\,y + c$.

General integral: Let $c = \psi(a)$, then the general integral is obtained by eliminating a from

$$z = ax + \sqrt{(1-a^2)}\,y + \psi(a) \text{ and } 0 = x + \frac{-a}{\sqrt{(1-a^2)}}\,y + \psi'(a).$$

Example 5: Solve $q = 3p^2$.

Solution: The equation is of the form $f(p,q)=0$, so the complete integral is given by

$$z = ax + by + c \text{, where } b = 3a^2.$$

Hence the complete integral is $z = ax + 3a^2\,y + c$

Example 6: Solve $(x+y)(p+q)^2 + (x-y)(p-q)^2 = 1$.

Solution: Putting $x + y = X^2$, $x - y = Y^2$, we get

$$p = \frac{\partial z}{\partial x} = \frac{\partial z}{\partial X}\frac{\partial X}{\partial x} + \frac{\partial z}{\partial Y}\frac{\partial Y}{\partial x} = \frac{1}{2X}\frac{\partial z}{\partial X} + \frac{1}{2Y}\frac{\partial z}{\partial Y}$$

and $$q = \frac{\partial z}{\partial y} = \frac{\partial z}{\partial X}\frac{\partial X}{\partial y} + \frac{\partial z}{\partial Y}\frac{\partial Y}{\partial y} = \frac{1}{2X}\frac{\partial z}{\partial X} - \frac{1}{2Y}\frac{\partial z}{\partial Y}$$

Therefore $p + q = \frac{1}{X}\frac{\partial z}{\partial X}$ and $p - q = \frac{1}{Y}\frac{\partial z}{\partial Y}$.

Using these values in the given equation, we get

$$\left(\frac{\partial z}{\partial X}\right)^2 + \left(\frac{\partial z}{\partial Y}\right)^2 = 1,$$

which is of the form of standard I. Its solution is given by

$$z = aX + bY + c \text{, where } a^2 + b^2 = 1 \text{ or } b = \sqrt{(1-a^2)}.$$

Hence the complete integral is $z = a\sqrt{(x+y)} + \sqrt{(1-a^2)}\sqrt{(x-y)} + c$.

Example 7: Find the complete and singular integral of $p^3 + q^3 = 27z$. where $X = x + ay$.

Solution: Putting $z = f(x + ay) = f(X)$, we get $p = \frac{\partial z}{\partial x} = \frac{dz}{dX}$ and

$$q = \frac{\partial z}{\partial y} = a\frac{dz}{dX}$$

Hence the equation reduces to $(1 + a^3)\left(\frac{dz}{dX}\right)^3 = 27z$

$$\Rightarrow \quad (1 + a^3)^{1/3}\frac{dz}{dX} = 3z^{1/3} \Rightarrow (1 + a^3)^{1/3}\frac{2}{3}z^{-1/3}\,dz = 2\,dX$$

Integration yields, $z^{2/3}(1 + a^3)^{1/3} = 2X + c = 2(X + b)$

$$\Rightarrow \quad (1 + a^3)z^2 = 8(x + ay + b)^3 \qquad ...(1)$$

which is the complete integral.

To find the singular integral, we differentiate (1) partially w. r. t. 'a' and 'b', we get

$$3a^2 z^2 = 24\,y(x + ay + b)^2 \qquad ...(2)$$

and $\quad 0 = 24(x + ay + b)^2 \qquad ...(3)$

Eliminating a, b from (2) and (3), we obtain, $z = 0$, which is the required singular solution.

Example 8: Find the complete and singular integral of $z^2(p^2 z^2 + q^2) = 1$.

Solution: Let $z = f(x + ay) = f(X)$, where $X = x + ay$, then

$$p = \frac{\partial z}{\partial x} = \frac{dz}{dX} \text{ and } q = \frac{\partial z}{\partial y} = a\frac{dz}{dX}.$$

Thus given equation becomes $z^2\left[\left(\frac{dz}{dX}\right)^2 z^2 + a^2\left(\frac{dz}{dX}\right)^2\right] = 1$

$$\Rightarrow \quad z^2(z^2 + a^2)\left(\frac{dz}{dX}\right)^2 = 1 \Rightarrow z\sqrt{(z^2 + a^2)}\,dz = dX.$$

On integrating, we get

$$\frac{1}{3}(z^2 + a^2)^{3/2} = X + b \Rightarrow 9(x + ay + b)^2 = (z^2 + a^2)^3 \qquad ...(1)$$

which is the required complete integral.

Singular integral: Differentiating (1) partially w. r. t. 'a' and 'b', we obtain

$$18(x + ay + b)y = 6(z^2 + a^2)^2\,a \qquad ...(2)$$

and $\quad 18(x+ay+b)=0$ ...(3)

Solving (2) and (3), we get $a=0$ ...(4)

From (1), (3) and (4), we have $z=0$

which may be the singular integral.

But when $z=0$, we have $p=0, q=0$, which does not satisfy the given equation.

Hence $z=0$ is not singular integral, *i.e.* given equation does not have any singular integral.

Example 9: Solve $pz=1+q^2$.

Solution: Putting $z=f(x+ay)=f(X)$ where $X=x+ay$, we get

$$p=\frac{\partial z}{\partial x}=\frac{dz}{dX} \text{ and } q=\frac{\partial z}{\partial y}=a\frac{dz}{dX}$$

Hence the equation reduces to

$$z\frac{dz}{dX}=1+a^2\left(\frac{dz}{dX}\right)^2 \Rightarrow a^2\left(\frac{dz}{dX}\right)^2-z\frac{dz}{dX}+1=0$$

$$\Rightarrow \quad \frac{dz}{dX}=\frac{z\pm\sqrt{(z^2-4a^2)}}{2a^2} \Rightarrow \frac{dz}{z\pm\sqrt{(z^2-4a^2)}}=\frac{dX}{2a^2}$$

$$\Rightarrow \quad \frac{z\pm\sqrt{(z^2-4a^2)}}{4a^2}dz=\frac{dX}{2a^2} \Rightarrow \left\{z\pm\sqrt{(z^2-4a^2)}\right\}dz=2dX.$$

On integrating, we get

$$\frac{z^2}{2}\pm\left[\frac{z}{2}\sqrt{(z^2-4a^2)}-\frac{4a^2}{2}\log\left\{z+\sqrt{(z^2-4a^2)}\right\}\right]=2X+c$$

$$\Rightarrow \quad z^2\pm\left[z\sqrt{(z^2-4a^2)}-4a^2\log\left\{z+\sqrt{(z^2-4a^2)}\right\}\right]=4x+4ay+2c$$

which is the complete integral.

Example 10: Solve $9(p^2 z+q^2)=4$.

Solution: It is of the form $F(z,p,q)=0$, so putting $z=f(x+ay)=f(X)$

and $X=x+ay$ we get

$$p=\frac{\partial z}{\partial x}=\frac{dz}{dX} \text{ and } q=\frac{\partial z}{\partial y}=a\frac{dz}{dX}.$$

Therefore the equation reduces to $9\left[\left(\frac{dz}{dx}\right)^2 z+a^2\left(\frac{dz}{dX}\right)^2\right]=4$

$$\Rightarrow \quad 9\left(\frac{dz}{dX}\right)^2 (z+a^2) = 4 \Rightarrow \quad 3\sqrt{(z+a^2)}\,dz = 2dX$$

On integrating, we get $(z+a^2)^{3/2} = X + b \Rightarrow (z+a^2)^{3/2} = x + ay + b$

$\Rightarrow \quad (z+a^2)^3 = (x+ay+b)^2$, which is the required solution.

Example 11: Solve $p^2 = z^2(1-pq)$.

Solution: Let $z = f(x+ay) = f(X)$, where $X = x + ay$, then

$$p = \frac{\partial z}{\partial x} = \frac{dz}{dX} \text{ and } q = \frac{\partial z}{\partial y} = a\frac{dz}{dX}.$$

Hence the given equation reduces to $\left(\frac{dz}{dX}\right)^2 = z^2\left\{1 - \frac{dz}{dX}a\frac{dz}{dX}\right\}$

$$\Rightarrow \quad (1+az^2)\left(\frac{dz}{dX}\right)^2 = z^2 \Rightarrow \frac{\sqrt{(1+az^2)}}{z}dz = dX$$

$$\Rightarrow \quad \frac{1+az^2}{z\sqrt{(1+az^2)}}dz = dX \Rightarrow \left(\frac{1}{z\sqrt{(1+az^2)}} + \frac{az}{\sqrt{(1+az^2)}}\right)dz = dX$$

Integration yields $\frac{1}{\sqrt{a}}\log\left[z\sqrt{a} + \sqrt{a(1+az^2)}\right] + \sqrt{(1+az^2)} = X + c$

$$\Rightarrow \quad \frac{1}{\sqrt{a}}\log\left[z\sqrt{a} + \sqrt{a(1+az^2)}\right] + \sqrt{(1+az^2)} = x + ay$$

which is the complete integral.

Example 12: Solve $p(1+q^2) = q(z-a)$.

Solution: Let $z = f(x+ay) = f(X)$, where $X = x + ay$, then we get

$$p = \frac{\partial z}{\partial X} = \frac{dz}{dX} \text{ and } q = \frac{\partial z}{\partial y} = a\frac{dz}{dX}.$$

Hence the equation becomes $\frac{dz}{dX}\left[1 + a^2\left(\frac{dz}{dX}\right)^2\right] = a\frac{dz}{dX}(z-a)$

$$\Rightarrow \quad a^2\left(\frac{dz}{dX}\right)^2 = a(z-a) - 1 \Rightarrow \frac{adz}{\sqrt{[a(z-a)-1]}} = dX$$

On integrating, we have

$$2\sqrt{[a(z-a)-1]} = X + b \Rightarrow 4a(z-a) - 4 = (x + ay + b)^2$$

$$\Rightarrow \quad 4a(z-a) = 4 + (x + ay + b)^2$$

Example 13: Solve $\sqrt{p} + \sqrt{q} = 2x$.

Solution: The given equation can be written as $\sqrt{p} - 2x = -\sqrt{q}$.

Clearly it is of the form $f(p, x) = h(q, y)$, so we let $\sqrt{p} - 2x = a = -\sqrt{q}$

$$\Rightarrow \quad p = (a + 2x)^2 \text{ and } q = a^2.$$

Putting these values in $dz = pdx + qdy$, we obtain

$$dz = (a + 2x)^2\, dx + a^2\, dy.$$

On integrating, we get $z = \frac{1}{6}(a + 2x)^3 + a^2 y + b$, which is complete integral.

Example 14: Solve $p^2 + q^2 = x + y$.

Solution: Let $p^2 - x = y - q^2 = a$, then we have $p = \sqrt{(x + a)}$ and $q = \sqrt{(y - a)}$.

Putting these values in $dz = p\, dx + q\, dy$, we get

$$dz = \sqrt{(x + a)}\, dx + \sqrt{(y - a)}\, dy.$$

On integrating, we obtain $z = \frac{2}{3}(x + a)^{3/2} + \frac{2}{3}(y - a)^{3/2} + b$.

Example 15: Solve $pe^y = qe^x$.

Solution: The given equation can be written as $pe^{-x} = qe^{-y} = a$ (say).

Thus we get $p = ae^x$ and $q = ae^y$. Now from $dz = pdx + qdy$, we have $dz = ae^x\, dx + ae^y\, dy \Rightarrow z = ae^x + ae^y + b$.

Example 16: Solve $yp = 2\, yx + \log q$.

Solution: The given equation can be written as $p = 2x + \frac{1}{y}\log q$

$$\Rightarrow \quad p - 2x = \frac{1}{y}\log q = a.$$

Then we have $p = 2x + a$ and $\log q = ay \Rightarrow q = e^{ay}$

Using these in $dz = p\, dx + q\, dy$, we obtain $dz = (2x + a)dx + e^{ay}\, dy$.

On integrating, we get

$$z = x^2 + ax + \frac{1}{a} e^{ay} + b \Rightarrow az = ax^2 + a^2 x + e^{ay} + ab \, .$$

Example 17: *Solve* $z(p^2 - q^2) = x - y$.

Solution: The given equation can be written as

$$\left(\sqrt{z}\frac{\partial z}{\partial x}\right)^2 - \left(\sqrt{z}\frac{\partial z}{\partial y}\right)^2 = x - y$$

Putting $\sqrt{z}dz = dZ$ and $Z = \frac{2}{3} z^{3/2}$, the given equation becomes

$$\left(\frac{\partial Z}{\partial x}\right)^2 - \left(\frac{\partial Z}{\partial y}\right)^2 = x - y$$

$\Rightarrow \quad P^2 - Q^2 = x - y$ where $P = \frac{\partial Z}{\partial x}, Q = \frac{\partial Z}{\partial y}$

$\Rightarrow \quad P^2 - x = Q^2 - y = a$, which represents the standard form III.

Thus $P = \sqrt{(x+a)}$ and $Q = \sqrt{(y+a)}$. Using the values of P and Q in $dZ = Pdx + Qdy$, we obtain

$$dZ = \sqrt{(x+a)}\,dx + \sqrt{(y+a)}\,dy.$$

On integrating, we get $Z = \frac{2}{3}(x+a)^{3/2} + \frac{2}{3}(y+a)^{3/2} + b$

$\Rightarrow \quad z^{3/2} = (x+a)^{3/2} + (y+a)^{3/2} + b.$

Example 18: Solve $4z^2(p^2 + q^2) = x^2 + y^2$.

Solution: The given equation can be written as

$$\left(2z\frac{\partial z}{\partial x}\right)^2 + \left(2z\frac{\partial z}{\partial y}\right)^2 = x^2 + y^2.$$

Putting $2z\,dz = dZ$, we obtain $P^2 + Q^2 = x^2 + y^2$, where $P = \frac{dZ}{dx}$ and

$$q = \frac{dZ}{dy}$$

$\Rightarrow P^2 - x^2 = -Q^2 + y^2 = a(\text{say})$, which represents the standard form III.

Thus $P = \sqrt{(a + x^2)}$ and $Q = \sqrt{(y^2 - a)}$. Putting these values in $dZ = P\,dx + Q\,dy$, we get

$$dZ = \sqrt{(a + x^2)}\,dx + \sqrt{(y^2 - a)}\,dy.$$

On integrating, we get

$$Z=\frac{x}{2}\sqrt{(a+x^2)}+\frac{a}{2}\log\left\{x+\sqrt{(a+x^2)}\right\}$$

$$+\frac{y}{2}\sqrt{(y^2-a)}-\frac{a}{2}\log\left\{y+\sqrt{(y^2-a)}\right\}+b$$

$$\Rightarrow \quad 2z^2=x\sqrt{(a+x^2)}+a\log$$

$$\left\{x+\sqrt{(a+x^2)}\right\}+y\sqrt{(y^2-a)}-a\log\left\{y+(y^2-a)\right\}+c$$

Example 19: Find the singular integral of $z=px+qy+\log pq$.

Solution: This is of the type standard form IV, so the complete integral is $z=ax+by+\log ab$

Differentiating (1) partially w. r. t. *'a'* and *'b'*, we obtain $0=x+\frac{1}{a}$

and $\quad 0=y+\frac{1}{b}$

$$\Rightarrow \quad a=-\frac{1}{x} \text{ and } b=-\frac{1}{y}$$

Putting values of a and b in (1), the required singular integral be

$$z=-1-1+\log\frac{1}{xy} \Rightarrow z=-2-\log xy.$$

Example 20: Solve $z=px+qy+c\sqrt{(1+p^2+q^2)}$.

Solution: This is of the standard form IV, so the complete integral is

$$z=ax+by+c\sqrt{(1+a^2+b^2)}. \qquad ...(1)$$

Singular integral: Differentiating (1) partially w. r. t. *'a'* and *'b'*, we obtain

$$0=x+\frac{ac}{\sqrt{(1+a^2+b^2)}} \qquad ...(2)$$

and $$0=y+\frac{bc}{\sqrt{(1+a^2+b^2)}} \qquad ...(3)$$

From (2) and (3), we get

$$x^2+y^2=\frac{(a^2+b^2)c^2}{1+a^2+b^2} \Rightarrow c^2-x^2-y^2=\frac{c^2}{1+a^2+b^2}$$

$$\Rightarrow \quad 1+a^2+b^2=\frac{c^2}{c^2-x^2-y^2} \qquad ...(4)$$

Therefore from (2) and (4), we get

$$a = -\frac{x\sqrt{(1+a^2+b^2)}}{c} = \frac{-x}{\sqrt{(c^2-x^2-y^2)}}$$

Similarly from (3) and (4), we get

$$b = -\frac{y\sqrt{(1+a^2+b^2)}}{c} = \frac{-y}{\sqrt{(c^2-x^2-y^2)}}$$

Putting these values of a and b in (1), the singular integral is given by

$$z = \frac{-x^2}{\sqrt{(c^2-x^2-y^2)}} - \frac{y^2}{\sqrt{(c^2-x^2-y^2)}} + \frac{c^2}{\sqrt{(c^2-x^2-y^2)}}$$

$$= \frac{(c^2-x^2-y^2)}{\sqrt{(c^2-x^2-y^2)}} = \sqrt{(c^2-x^2-y^2)}. \Rightarrow x^2+y^2+z^2=c^2$$

Example 21: Solve $z = px + qy - 2\sqrt{(pq)}$.

Solution: This is of the standard form IV, so the complete integral is

$$z = ax + by - 2\sqrt{(ab)} \qquad ...(1)$$

Singular integral: Differentiating (1) partially w. r. t. 'a' and 'b', we obtain

$$0 = x - \frac{2}{2\sqrt{(ab)}}b \Rightarrow x = \sqrt{\left(\frac{b}{a}\right)} \qquad ...(2)$$

and $$0 = y - \frac{2}{2\sqrt{(ab)}}a \Rightarrow y = \sqrt{\left(\frac{a}{b}\right)} \qquad ...(3)$$

Eliminating a and b from (1) (2) and (3), the singular integral is $z = 0$.

EXERCISE 7.3

Solve the following PDEs.

1. $z = p^2x + q^2y$
2. $pxy + pq + qy = yz$
3. $p^2 - y^2q = y^2 - x^2$
4. $zpq = p + q$
5. $1 + p^2 = qz$
6. $2z + p^2 + qy + qy^2 = 0$
7. $q - xp = p^2$
8. $yzp^2 - q = 0$

ANSWERS 7.3

1. $z = \left\{\sqrt{ax} + \sqrt{y} + b\right\}^2 / (1+a)$

2. $\log(z - ax) = y - a\log(a + y) + b$

3. $z = \frac{x}{2}\sqrt{(a^2 - x^2)} + \frac{a^2}{2}\sin^{-1}\frac{x}{a} - \frac{a^2}{y} - y + b$

4. $4az = (1 - a)(x + ay + b)$

5. $\frac{z^2}{2} \pm \left\{\frac{z}{2}\sqrt{z^2 - 4a^2} - 2a^2\log\left(z + \sqrt{z^2 - 4a^2}\right)\right\} = 2ax + 2y + b$

6. $y^2\{(x - a)^2 + y^2 + 2z\} = b$

7. $z = -\frac{x^2}{4} \pm \frac{1}{2}\left[\frac{x}{2}\sqrt{(x^2 + 4a)} + 2a\log\left\{x + \sqrt{(x^2 + 4a)}\right\}\right] + ay + b$

8. $z^2(a - y^2) = (x + b)^2$

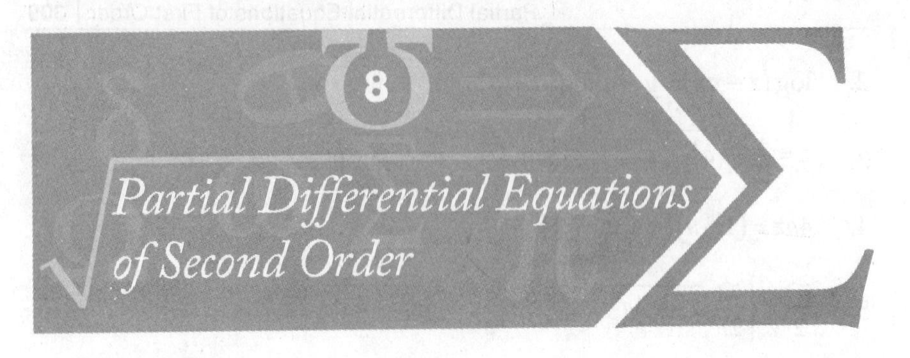

8.1 INTRODUCTION

In this chapter those partial differential equations of second order will be considered which are linear with respect to the dependent variable and there differential coefficients are constants only. It will be assumed that there are only two independent variables. The general form of an equation to be considered is

$$f\left(\frac{\partial}{\partial x}, \frac{\partial}{\partial y}\right)z = V(x, y) \qquad ...(1)$$

where f is a rational algebriacal function with constant coefficient, and 'V' may be any function of the independent variables x and y.

The complete integral of the equation is the most general relation (possible) between x, y and z such that when the value of z and the associated differential coefficient are substituted in the differential equation, the latter becomes an identity. No condition is annexed to the definition in regard to the form of the complete integral which may involve in its expression either arbitrary constants or arbitrary functions or both. As in the case of ordinary differential equations the complete integral of equation (1) consists of the sum of two parts:

(i) The most general integral of $f\left(\frac{\partial}{\partial x}, \frac{\partial}{\partial y}\right)z = 0$ is known as **complementary function.**

(ii) Any particular solution of equation (1), which is known as **Particular Integral.**

For convenience, we denote $\frac{\partial}{\partial x}$ and $\frac{\partial}{\partial y}$ by D and D' respectively and complementary function by 'C.F.' and particular integral by 'P.I.'.

8.2 CLASSIFICATION OF SECOND ORDER PDEs AND CANONICAL FORMS

8.2.1 Classification of Second Order PDEs

Linear: A PDE of second order of the form

$$F_1(x,y)z_{xx}+F_2(x,y)z_{yy}+F_3(x,y)z_{xy}+F_4(x,y)z_x+F_5(x,y)z_y+F_6(x,y)z=F_7(x,y)$$

where $F_1, F_2, F_3, F_5, \ldots, F_7$ are continuous functions of x, y is called a linear PDE.

Semi-linear: A PDE of second order of the form

$$R(x,y)z_{xx}+S(x,y)z_{xy}+T(x,y)z_{yy}+f(x,y,z,p,q)=0$$

where $R^2+S^2+T^2 \neq 0, R, S, T$ are continuous functions of x, y and $p=\frac{\partial z}{\partial x}, q=\frac{\partial z}{\partial y}$; is called a semi-linear PDE.

Quasi-linear: A PDE of second order of the form

$$R(x,y,z,p,q)z_{xx}+S(x,y,z,p,q)z_{xy}+T(x,y,z,p,q)z_{yy}+f(x,y,z,p,q)=0$$

is called a quasi-linear PDE.

8.2.2 Classification of Second Order Linear PDE in R^2

Consider a second order semi-linear PDE $Lu+g(x, y, z, z_x, z_y)=0$ ...(1)

where $Lu=R(x, y)z_{xx}+S(x, y)z_{xy}+T(x, y)z_{yy}$ ...(2)

and $R^2+S^2+T^2 \neq 0, R, S, T$ are continuous functions of x, y.

Here Lu is known as principal part of PDE (1). If we replace z_{xx}, z_{yy} by α^2, β^2 respectively, then from the principal part, we get a polynomial $P(\alpha, \beta)$ defined as

$$P(\alpha, \beta)=R\alpha^2+S\alpha\beta+T\beta^2.$$

Mathematically, the properties of the PDE (1) are determined by the algebraic property of $P(\alpha, \beta)$ and are classified as follows:

(i) If $S^2-4RT>0$, then (1) is called hyperbolic type or hyperbolic equation.

(ii) If $S^2-4RT=0$, then (1) is called parabolic type or parabolic equation.

(iii) If $S^2-4RT<0$, then (1) is called elliptic type or elliptic equation.

Example 1: Classify the PDE $u_{xx}+2yu_{xy}+xu_{yy}-u_x+4=0$.

Solution: Here the principal part is $Lu=u_{xx}+2yu_{xy}+xu_{yy}$.

Hence

$$P(\alpha,\beta)=\alpha^2+2y\alpha\beta+x\beta^2$$

And $\quad S^2-4RT=4y^2-4x=4(y^2-x).$

Thus, we conclude as follows

(i) If $y^2 - x > 0$, the equation is hyperbolic.

(ii) If $y^2 - x = 0$, the equation is parabolic.

(iii) If $y^2 - x < 0$, the equation is elliptic.

8.2.3 Reduction to Canonical Forms

Consider equation of the type

$$Ar + 2Bs + Ct + f(x, y, z, p, q) = 0 \qquad ...(1)$$

where $A = A(x, y), B = B(x, y), C = C(x, y)$ are continuous functions of x and y and possessing continuous partial derivatives of as high an order as necessary. By a suitable change of independent variables, we now show that (1) can be transformed into one of three canonical forms as discussed below, which are readily integrable.

Let $\quad u = u(x, y), v = v(x, y) \qquad ...(2)$

Then $p = \dfrac{\partial z}{\partial x} = \dfrac{\partial z}{\partial u}\dfrac{\partial u}{\partial x} + \dfrac{\partial z}{\partial v}\dfrac{\partial v}{\partial x}$ etc.

Substituting the values of p, q, r, s, t so obtained, (1) takes the form

$$A_1 \frac{\partial^2 z}{\partial u^2} + 2B_1 \frac{\partial^2 z}{\partial u\, \partial v} + C_1 \frac{\partial^2 z}{\partial v^2} = f_1\left(u, v, z, \frac{\partial z}{\partial u}, \frac{\partial z}{\partial v}\right) \qquad ...(3)$$

where $A_1 = A\left(\dfrac{\partial u}{\partial x}\right)^2 + 2B\dfrac{\partial u}{\partial x}\dfrac{\partial u}{\partial y} + C\left(\dfrac{\partial u}{\partial y}\right)^2 \qquad ...(4)$

$$B_1 = A\frac{\partial u}{\partial x}\frac{\partial v}{\partial x} + B\left(\frac{\partial u}{\partial x}\frac{\partial v}{\partial y} + \frac{\partial u}{\partial y}\frac{\partial v}{\partial x}\right) + C\frac{\partial u}{\partial y}\frac{\partial v}{\partial y} \qquad ...(5)$$

$$C_1 = A\left(\frac{\partial v}{\partial x}\right)^2 + 2B\frac{\partial v}{\partial x}\frac{\partial v}{\partial y} + C\left(\frac{\partial v}{\partial y}\right)^2 \qquad ...(6)$$

and $f_1\left(u, v, z, \dfrac{\partial z}{\partial u}, \dfrac{\partial z}{\partial v}\right)$ is the transformed form of $f(x, y, z, p, q)$.

Now we shall find out u and v so that (3) reduces to the simplest possible form. The determination of desired values of u and v becomes easy when the discriminate $B^2 - AC$ of the quadratic equation

$$A\lambda^2 + 2B\lambda + C = 0 \qquad ...(7)$$

is everywhere either positive, negative or zero. To affect this, we shall consider the following three cases.

Case I: $B^2 - AC > 0$. *In this case, the roots* λ_1, λ_2 *of (7) are real and distinct:*

Choose u and v such that $\dfrac{\partial u}{\partial x} = \lambda_1 \dfrac{\partial u}{\partial y} \qquad ...(8)$

$$\frac{\partial u}{\partial x} = \lambda_2 \frac{\partial u}{\partial y} \qquad ...(9)$$

Therefore, λ_1 is a root of (7) and $A\lambda_1^2 + 2B\lambda_1 + C = 0.$ ...(10)
From (4) by of using (8), we get

$$A_1 = \left(A\lambda_1^2 + 2B\,\lambda_1 + C\right)\left(\frac{\partial u}{\partial y}\right)^2 = 0, \text{[using (10)]}.$$

Similarly, we get $C_1 = 0$.
For (8), the Lagrange's auxiliary equations are

$$\frac{dx}{1} = \frac{dy}{-\lambda_1} = \frac{du}{0}. \qquad ...(11)$$

From third member of (11), we have $du = 0 \Rightarrow u = c_1$. ...(12)

From first and second members of (11), we have $\frac{dy}{dx} + \lambda_1 = 0.$ ...(13)

Let the solution of (13) be $f_1(x, y) = c_2$. ...(14)
Therefore, the general solution of (8) is $u = f_1(x, y)$.
Similarly, the general solution of (9) is $v = f_2(x, y)$.

We can easily show that $A_1C_1 - B_1^2 = \left(AC - B^2\right)\left(\frac{\partial u}{\partial x}\frac{\partial v}{\partial y} - \frac{\partial u}{\partial y}\frac{\partial v}{\partial x}\right)^2$

$$\Rightarrow \qquad B_1^2 = \left(B^2 - AC\right)\left(\frac{\partial u}{\partial x}\frac{\partial v}{\partial y} - \frac{\partial u}{\partial y}\frac{\partial v}{\partial x}\right)^2, \text{ when } A_1 = C_1 = 0. \qquad ...(15)$$

Since $B^2 - AC > 0$, it follows that $B_1^2 > 0$.

Hence we may divide both sides of (3) by B_1^2. Due to the above facts and $A_1 = 0, C_1 = 0$, the equation (3) becomes

$$\frac{\partial^2 z}{\partial u\,\partial v} = \phi\left(u, v, z, \frac{\partial z}{\partial u}, \frac{\partial z}{\partial v}\right) \qquad ...(16)$$

which is the desired canonical form of (1).

Case II: $B^2 - AC = 0$. *In this case, the roots λ_1, λ_2 of (7) are real and equal. Now set u exactly as in case I and take v to be any function of x, y which is independent of u. Then we have $A_1 = 0$ as in case I and so from (15) $B_1 = 0$, but $C_1 \neq 0$, otherwise v would be a function of x.*

Putting $A_1 = 0, B_1 = 0$ in (3) and dividing by C_1, (1) takes the form

$$\frac{\partial^2 z}{\partial v^2} = \phi\left(u, v, z, \frac{\partial z}{\partial u}, \frac{\partial z}{\partial v}\right) \qquad ...(17)$$

which is the desired canonical form of (1).

Case III: $B^2 - AC < 0$. *In this case, the roots of (7) are two conjugate complex roots* $\alpha + i\beta$ and $\alpha - i\beta$. *Now proceeding as in Case I we shall get*

$$\frac{\partial^2 z}{\partial u\,\partial v} = \phi\left(u, v, z, \frac{\partial z}{\partial u}, \frac{\partial z}{\partial v}\right).$$

To obtain a real canonical form we make further transformation

$$\alpha = \frac{1}{2}(u+v) \text{ and } \beta = \frac{1}{2}(v-u). \qquad ...(18)$$

Then we can show that $\dfrac{\partial^2 z}{\partial u\,\partial v} = \dfrac{1}{4}\left(\dfrac{\partial^2 z}{\partial \alpha^2} + \dfrac{\partial^2 z}{\partial \beta^2}\right)$. ...(19)

Now using (18) and (19) in (16), the desired canonical form is

$$\frac{\partial^2 z}{\partial \alpha^2} + \frac{\partial^2 z}{\partial \beta^2} = \psi\left(\alpha, \beta, z, \frac{\partial z}{\partial \alpha}, \frac{\partial z}{\partial \beta}\right). \qquad ...(20)$$

Example 2: Classify the following equation and solve it

$$y^2 r - 2xys + x^2 t - \frac{y^2}{x} p - \frac{x^2}{y} q = 0.$$

Solution: The given PDE is $y^2 r - 2xys + x^2 t - \dfrac{y^2}{x} p - \dfrac{x^2}{y} q = 0.$ ...(1)

Comparing (1) with $Ar + 2Bs + Ct + f(x, y, z, p, q) = 0$, we get

$$A = y^2, B = -xy, C = x^2.$$

Since $B^2 = (-xy)^2 = x^2 y^2, AC = y^2 x^2.$

Therefore $B^2 - AC = 0$ and so the given equation (1) is parabolic at all points.

The λ – quadratic $A\lambda^2 + 2B\lambda + C = 0$ gives

$$y^2\lambda^2 - 2xy\lambda + x^2 = 0 \text{ or } (y\lambda - x)^2 = 0.$$

Therefore $\lambda = \dfrac{x}{y}, \dfrac{x}{y}.$

Here $\dfrac{dy}{dx} + \lambda = 0$ becomes $\dfrac{dy}{dx} + \dfrac{x}{y} = 0$

$\Rightarrow 2y\,dy + 2x\,dx = 0 \Rightarrow d(x^2 + y^2) = 0.$

Integrating, we get $x^2 + y^2 = C_1$.

Therefore to reduce (1) to canonical form, we change x, y to u, v by taking $u = x^2 + y^2$.

Again we choose $v = x^2 - y^2$, which is independent of u. Thus,

$$u = x^2 + y^2 \text{ and } v = x^2 - y^2 \qquad ...(2)$$

Therefore $p=\frac{\partial z}{\partial x}=\frac{\partial z}{\partial u}\frac{\partial u}{\partial x}+\frac{\partial z}{\partial v}\frac{\partial v}{\partial x}=\frac{\partial z}{\partial u}2x+\frac{\partial z}{\partial v}2x=2x\left(\frac{\partial z}{\partial u}+\frac{\partial z}{\partial v}\right).$

Similarly, $q=\frac{\partial z}{\partial y}=2y\left(\frac{\partial z}{\partial u}-\frac{\partial z}{\partial v}\right).$

Also,

$$r=\frac{\partial}{\partial x}\left(\frac{\partial z}{\partial x}\right)=\frac{\partial}{\partial x}\left[2x\left(\frac{\partial z}{\partial u}+\frac{\partial z}{\partial v}\right)\right]=2\left(\frac{\partial z}{\partial u}+\frac{\partial z}{\partial v}\right)+2x\frac{\partial}{\partial x}\left(\frac{\partial z}{\partial u}+\frac{\partial z}{\partial v}\right)$$

$$=2\left(\frac{\partial z}{\partial u}+\frac{\partial z}{\partial v}\right)+2x\left[\frac{\partial}{\partial u}\left(\frac{\partial z}{\partial u}+\frac{\partial z}{\partial v}\right)\cdot\frac{\partial u}{\partial x}+\frac{\partial}{\partial v}\left(\frac{\partial z}{\partial u}+\frac{\partial z}{\partial v}\right)\cdot\frac{\partial v}{\partial x}\right]$$

$$=2\left(\frac{\partial z}{\partial u}+\frac{\partial z}{\partial v}\right)+4x^2\left(\frac{\partial^2 z}{\partial u^2}+2\frac{\partial^2 z}{\partial u\,\partial v}+\frac{\partial^2 z}{\partial v^2}\right)$$

$$t=\frac{\partial}{\partial y}\left(\frac{\partial z}{\partial y}\right)=\frac{\partial}{\partial y}\left[2y\left(\frac{\partial z}{\partial u}-\frac{\partial z}{\partial v}\right)\right]=2\left(\frac{\partial z}{\partial u}-\frac{\partial z}{\partial v}\right)+2y\frac{\partial}{\partial y}\left(\frac{\partial z}{\partial u}-\frac{\partial z}{\partial v}\right)$$

$$=2\left(\frac{\partial z}{\partial u}-\frac{\partial z}{\partial v}\right)+2y\left[\frac{\partial}{\partial u}\left(\frac{\partial z}{\partial u}-\frac{\partial z}{\partial v}\right)\cdot\frac{\partial u}{\partial y}+\frac{\partial}{\partial v}\left(\frac{\partial z}{\partial u}-\frac{\partial z}{\partial v}\right)\cdot\frac{\partial v}{\partial y}\right]$$

$$=2\left(\frac{\partial z}{\partial u}-\frac{\partial z}{\partial v}\right)+4y^2\left(\frac{\partial^2 z}{\partial u^2}-2\frac{\partial^2 z}{\partial u\,\partial v}+\frac{\partial^2 z}{\partial v^2}\right)$$

$$s=\frac{\partial}{\partial x}\left(\frac{\partial z}{\partial y}\right)=\frac{\partial}{\partial u}\left[2y\left(\frac{\partial z}{\partial u}-\frac{\partial z}{\partial v}\right)\right]=2y\frac{\partial}{\partial x}\left(\frac{\partial z}{\partial u}-\frac{\partial z}{\partial v}\right)$$

$$=2y\left[\frac{\partial}{\partial u}\left(\frac{\partial z}{\partial u}-\frac{\partial z}{\partial v}\right)\cdot\frac{\partial u}{\partial x}+\frac{\partial}{\partial v}\left(\frac{\partial z}{\partial u}-\frac{\partial z}{\partial v}\right)\frac{\partial v}{\partial x}\right]=4xy\left(\frac{\partial^2 z}{\partial u^2}-\frac{\partial^2 z}{\partial v^2}\right).$$

Substituting these values in (1) and simplifying, we get $\frac{\partial^2 z}{\partial v^2}=0$, which is the canonical form of (1).

Integrating (3) w. r. t. 'v', we get $z=v\phi_1(u)+\phi_2(u)$, where $\phi_2(u)$ is another arbitrary function of u.

Hence, by using (2), the required solution of (1) is

$$z=(x^2-y^2)\phi_1(x^2+y^2)+\phi_2(x^2+y^2).$$

Example 3: Classify and solve the equation $\frac{\partial^2 z}{\partial x^2}-\frac{\partial^2 z}{\partial y^2}=0.$

Solution: The given partial differential equation can be written as

$r-t=0.$ ...(1)

Comparing (1) with $Ar + 2Bs + Ct + f(x, y, z, p, q) = 0$, we get $A = 1, B = 0, C = -1$.

Since $B^2 = 0, AC = 1(-1) = -1$. Therefore $B^2 - AC > 0$.

Hence the given equation is hyperbolic everywhere.

The λ – quadratic $A\lambda^2 + 2B\lambda + C = 0$ gives

$\lambda^2 - 1 = 0$ so that $\lambda = \pm 1$.

Here $\lambda_1 = 1$ and $\lambda_2 = -1$ (real and distinct roots)

Hence $\frac{dy}{dx} + \lambda_1 = 0$ and $\frac{dy}{dx} + \lambda_2 = 0$ give $\frac{dy}{dx} + 1 = 0$ and $\frac{dy}{dx} - 1 = 0$

$\Rightarrow dx + dy = 0$ and $dx - dy = 0 \quad \Rightarrow d(x + y) = 0$ and $d(x - y) = 0$.

Integrating these, we get $x + y = c_1$ and $x - y = c_2$.

Therefore, to reduce (1) to canonical form we put

$$u = x + y \text{ and } v = x - y. \qquad \text{...(2)}$$

Hence

$$p = \frac{\partial z}{\partial x} = \frac{\partial z}{\partial u}\frac{\partial u}{\partial x} + \frac{\partial z}{\partial v}\frac{\partial v}{\partial x} = \frac{\partial z}{\partial u} + \frac{\partial z}{\partial v}$$

$$q = \frac{\partial z}{\partial y} = \frac{\partial z}{\partial u}\frac{\partial u}{\partial y} + \frac{\partial z}{\partial v}\frac{\partial v}{\partial y} = \frac{\partial z}{\partial u} - \frac{\partial z}{\partial v}$$

$$r = \frac{\partial^2 z}{\partial x^2} = \frac{\partial}{\partial x}\left(\frac{\partial z}{\partial x}\right) = \frac{\partial}{\partial x}\left(\frac{\partial z}{\partial u} + \frac{\partial z}{\partial v}\right)$$

$$= \frac{\partial}{\partial u}\left(\frac{\partial z}{\partial u} + \frac{\partial z}{\partial v}\right)\cdot\frac{\partial u}{\partial x} + \frac{\partial}{\partial v}\left(\frac{\partial z}{\partial u} + \frac{\partial z}{\partial v}\right)\cdot\frac{\partial v}{\partial x} = \frac{\partial^2 z}{\partial u^2} + 2\frac{\partial^2 z}{\partial u\,\partial v} + \frac{\partial^2 z}{\partial v^2}$$

$$t = \frac{\partial^2 z}{\partial y^2} = \frac{\partial}{\partial y}\left(\frac{\partial z}{\partial y}\right) = \frac{\partial}{\partial y}\left(\frac{\partial z}{\partial u} - \frac{\partial z}{\partial v}\right) = \frac{\partial}{\partial u}\left(\frac{\partial z}{\partial u} - \frac{\partial z}{\partial v}\right)\cdot\frac{\partial u}{\partial y} + \frac{\partial}{\partial v}\left(\frac{\partial z}{\partial u} - \frac{\partial z}{\partial v}\right)\cdot\frac{\partial v}{\partial y}$$

$$= \frac{\partial^2 z}{\partial u^2} - 2\frac{\partial^2 z}{\partial u\,\partial v} + \frac{\partial^2 z}{\partial v^2}.$$

Substituting these values of r and t in (1), we get

$$4\frac{\partial^2 z}{\partial u\,\partial v} = 0 \Rightarrow \frac{\partial^2 z}{\partial u\,\partial v} = 0 \Rightarrow \frac{\partial^2 z}{\partial v\,\partial u} = 0 \qquad \text{...(3)}$$

which is the canonical form of (1). Integrating w. r. t. 'v', we get

$$\frac{\partial z}{\partial u} = F(u) \qquad \text{...(4)}$$

where $F(u)$ is an arbitrary function of u. Integrating w. r. t. 'u', we get

$$z = \int F(u)du + \phi_2(v) \text{ or } z = \phi_1(u) + \phi_2(v).$$

where $\phi_1(u)=\int F(u)du$ and $\phi_2(v)$ are arbitrary functions.

Hence, the solution (1) is $z=\phi_1(x+y)+\phi_2(x-y)$, where ϕ_1 and ϕ_3 are arbitrary functions.

Example 4: Classify the following P.D.E. and reduce to canonical form

$$\frac{\partial^2 z}{\partial x^2}+x^2\frac{\partial^2 z}{\partial y^2}=0.$$

Solution: The given P.D.E. can be written as $r+x^2t=0$. ...(1)

Comparing (1) with $Ar+2Bs+Ct+f(x,y,z,u,p,q)=0$, we get

$$A=1, B=0, C=x^2.$$

Since $B^2=0, AC=1\,x^2=x^2$, therefore $B^2-AC<0$ at all points where $x\neq 0$.

Hence, the given equation is elliptic at all points except on the y-axis.

Also, the $\lambda-$quadratic $A\lambda^2+2B\lambda+C=0$ gives

$$\lambda^2+x^2=0 \text{ so that } \lambda=\pm\, ix \text{ (complex roots)}$$

Here $\lambda_1=ix$ and $\lambda_2=-\,ix$.

Hence $\frac{dy}{dx}+\lambda_1=0$ and $\frac{dy}{dx}+\lambda_2=0$ give $\frac{dy}{dx}+ix=0$ and $\frac{dy}{dx}-ix=0$

Integrating these, we get $y+\frac{1}{2}ix^2=c_1$ and $y-\frac{1}{2}ix^2=c_2$

Now to reduce (1) to canonical form, we change x, y to u, v and u, v to α, β by taking

$$u=y+\frac{1}{2}ix^2=\alpha+i\beta,\ v=y-\frac{1}{2}ix^2=\alpha-i\beta.$$

Solving for α,β, we get $\alpha=y$ and $\beta=\frac{1}{2}x^2$.

Therefore $p=\frac{\partial z}{\partial x}=\frac{\partial z}{\partial \alpha}\frac{\partial \alpha}{\partial x}+\frac{\partial z}{\partial \beta}\frac{\partial \beta}{\partial x}=x\frac{\partial z}{\partial \beta}$

$$q=\frac{\partial z}{\partial y}=\frac{\partial z}{\partial \alpha}\frac{\partial \alpha}{\partial y}+\frac{\partial z}{\partial \beta}\frac{\partial \beta}{\partial y}=\frac{\partial z}{\partial \alpha} \text{ so } \frac{\partial}{\partial y}=\frac{\partial}{\partial \alpha}$$

$$r=\frac{\partial}{\partial x}\left(\frac{\partial z}{\partial x}\right)=\frac{\partial}{\partial x}\left(x\frac{\partial z}{\partial \beta}\right)=1\frac{\partial z}{\partial \beta}+x\frac{\partial}{\partial x}\left(\frac{\partial z}{\partial \beta}\right)$$

$$=\frac{\partial z}{\partial \beta}+x\left[\frac{\partial}{\partial \alpha}\left(\frac{\partial z}{\partial \beta}\right)\frac{\partial \alpha}{\partial x}+\frac{\partial}{\partial \beta}\left(\frac{\partial z}{\partial \beta}\right)\frac{\partial \beta}{\partial x}\right]=\frac{\partial z}{\partial \beta}+x^2\frac{\partial^2 z}{\partial \beta^2}$$

$$t = \frac{\partial}{\partial y}\left(\frac{\partial z}{\partial y}\right) = \frac{\partial}{\partial \alpha}\left(\frac{\partial z}{\partial \alpha}\right) = \frac{\partial^2 z}{\partial \alpha^2}$$

Putting these values of r and t in (1), we get

$$\frac{\partial z}{\partial \beta} + x^2 \frac{\partial^2 z}{\partial \beta^2} + x^2 \frac{\partial^2 z}{\partial \alpha^2} = 0 \Rightarrow \frac{\partial^2 z}{\partial \alpha^2} + \frac{\partial^2 z}{\partial \beta^2} + \frac{1}{2\beta}\frac{\partial z}{\partial \beta} = 0$$

which is the required canonical form of the given equation.

Example 5: Classify and reduce the equation $\frac{\partial^2 z}{\partial x^2} = x^2 \frac{\partial^2 z}{\partial y^2}$ to canonical form.

Solution: The given partial differential equation can be written as $r - x^2 t = 0$. ...(1)

Comparing (1) with $Ar + 2Bs + Ct + f(x, y, z, p, q) = 0$, we get

$$A = 1, B = 0, C = -x^2$$

Since $B^2 = 0, AC = 1(-x^2) = -x^2$, so we have $B^2 - AC > 0$ except when $x = 0$.

Hence the given equation is hyperbolic except on y-axis.

Therefore the λ – quadratic $A\lambda^2 + 2B\lambda + C = 0$ gives $\lambda^2 - x^2 = 0$ so that $\lambda = \pm x$.

Here $\lambda_1 = x$ and $\lambda_2 = -x$ (real and distinct roots). Hence

$$\frac{dy}{dx} + \lambda_1 = 0 \text{ and } \frac{dy}{dx} + \lambda_2 = 0 \text{ become}$$

$$\frac{dy}{dx} + x = 0 \text{ and } \frac{dy}{dx} - x = 0.$$

Integrating these, we get $y + \frac{1}{2}x^2 = c_1$ and $y - \frac{1}{2}x^2 = c_2$.

Now to reduce (1) to canonical form, we change x, y to u, v by taking

$$u = y + \frac{1}{2}x^2 \text{ and } v = y - \frac{1}{2}x^2.$$

Therefore $p = \frac{\partial z}{\partial x} = \frac{\partial z}{\partial u}\frac{\partial u}{\partial x} + \frac{\partial z}{\partial v}\frac{\partial v}{\partial x} = x\frac{\partial z}{\partial u} - x\frac{\partial z}{\partial v}$

$$q = \frac{\partial z}{\partial y} = \frac{\partial z}{\partial u}\frac{\partial u}{\partial y} + \frac{\partial z}{\partial v}\frac{\partial v}{\partial y} = \frac{\partial z}{\partial u} + \frac{\partial z}{\partial v}$$

$$r = \frac{\partial^2 z}{\partial x^2} = \frac{\partial}{\partial x}\left(\frac{\partial z}{\partial x}\right) = \frac{\partial}{\partial x}\left[x\left(\frac{\partial z}{\partial u} - \frac{\partial z}{\partial v}\right)\right] = x\,\frac{\partial}{\partial x}\left(\frac{\partial z}{\partial u} - \frac{\partial z}{\partial v}\right) + 1\left(\frac{\partial z}{\partial u} - \frac{\partial z}{\partial v}\right)$$

$$= x\left[\frac{\partial}{\partial u}\left(\frac{\partial z}{\partial u} - \frac{\partial z}{\partial v}\right)\cdot\frac{\partial z}{\partial x} + \frac{\partial}{\partial v}\left(\frac{\partial z}{\partial u} - \frac{\partial z}{\partial v}\right)\cdot\frac{\partial v}{\partial x}\right] + \frac{\partial z}{\partial u} - \frac{\partial z}{\partial v}$$

$$= x^2\left(\frac{\partial^2 z}{\partial u^2} - 2\,\frac{\partial^2 z}{\partial u\,\partial v} + \frac{\partial^2 z}{\partial v^2}\right) + \frac{\partial z}{\partial u} - \frac{\partial z}{\partial v}$$

$$t = \frac{\partial^2 z}{\partial y^2} = \frac{\partial}{\partial y}\left(\frac{\partial z}{\partial y}\right) = \left(\frac{\partial}{\partial u} + \frac{\partial}{\partial v}\right)\left(\frac{\partial z}{\partial u} + \frac{\partial z}{\partial v}\right) = \frac{\partial^2 z}{\partial u^2} + 2\,\frac{\partial^2 z}{\partial u\,\partial v} + \frac{\partial^2 z}{\partial v^2}.$$

Substituting these values of r and t in (1), we get

$$x^2\left(\frac{\partial^2 z}{\partial u^2} - 2\,\frac{\partial^2 z}{\partial u\,\partial v} + \frac{\partial^2 z}{\partial v^2}\right) + \frac{\partial z}{\partial u} - \frac{\partial z}{\partial v} - x^2\left(\frac{\partial^2 z}{\partial u^2} + 2\,\frac{\partial^2 z}{\partial u\,\partial v} + \frac{\partial^2 z}{\partial v^2}\right) = 0$$

$$\Rightarrow \quad \frac{\partial^2 z}{\partial u\,\partial v} = \frac{1}{4x^2}\left(\frac{\partial z}{\partial u} - \frac{\partial z}{\partial v}\right) or\ \frac{\partial^2 z}{\partial u\,\partial v} = \frac{1}{4(u-v)}\left(\frac{\partial z}{\partial u} - \frac{\partial z}{\partial v}\right)$$

which is the required canonical form of the given equation.

Example 6: Classify and reduce to canonical form to the equation $\frac{\partial^2 z}{\partial x^2} + 2\,\frac{\partial^2 z}{\partial x\,\partial y} + \frac{\partial^2 z}{\partial y^2} = 0$ and hence solve it.

Solution: The given equation can be written as $r + 2s + t = 0$. ...(1)

Comparing (1) with $Ar + 2Bs + Ct + f(x, y, z, p, q) = 0,$ we get

$A = 1, B = 1, C = 1.$

Since $B^2 = 1, AC = 1.1 = 1,$ so we have $B^2 - AC = 0.$

Hence the given equation is parabolic at all points.

Therefore the $\lambda -$ quadratic $A\lambda^2 + 2B\lambda + C = 0$ gives

$\lambda^2 + 2\lambda + 1 = 0$ so that $\lambda = -1, -1$ (equal roots).

Therefore, $\frac{dy}{dx} + \lambda = 0$ becomes $\frac{dy}{dx} - 1 = 0$

Integrating it, we get $x - y = c_1.$

Now to reduce (1) to canonical form, we change x, y to u, v by taking $u = x - y.$

Again we choose $v = x + y,$ which is independent of $u.$ Thus $u = x - y$ and $v = x + y.$ ...(2)

Therefore $p = \frac{\partial z}{\partial x} = \frac{\partial z}{\partial u}\frac{\partial u}{\partial x} + \frac{\partial z}{\partial v}\frac{\partial v}{\partial x} = \frac{\partial z}{\partial u} + \frac{\partial z}{\partial v}$ so $\frac{\partial}{\partial x} = \frac{\partial}{\partial u} + \frac{\partial}{\partial v}$

$$r = \frac{\partial}{\partial x}\left(\frac{\partial z}{\partial x}\right) = \left(\frac{\partial}{\partial u} + \frac{\partial}{\partial v}\right)\left(\frac{\partial z}{\partial u} + \frac{\partial z}{\partial v}\right) = \frac{\partial^2 z}{\partial u^2} + 2\frac{\partial^2 z}{\partial u\,\partial v} + \frac{\partial^2 z}{\partial v^2}$$

$$t = \frac{\partial}{\partial y}\left(\frac{\partial z}{\partial y}\right) = \left(\frac{\partial}{\partial v} - \frac{\partial}{\partial u}\right)\left(\frac{\partial z}{\partial v} - \frac{\partial z}{\partial u}\right) = \frac{\partial^2 z}{\partial v^2} - 2\frac{\partial^2 z}{\partial u \partial v} + \frac{\partial^2 z}{\partial u^2}$$

$$s = \frac{\partial}{\partial x}\left(\frac{\partial z}{\partial y}\right) = \left(\frac{\partial}{\partial u} + \frac{\partial}{\partial v}\right)\left(-\frac{\partial z}{\partial u} + \frac{\partial z}{\partial v}\right) = -\frac{\partial^2 z}{\partial u^2} + \frac{\partial^2 z}{\partial v^2}.$$

Substituting these values of r, s, t in (1) and simplifying, we obtain the required canonical form as

$$\frac{\partial^2 z}{\partial v^2} = 0. \qquad ...(3)$$

Integrating (3) w. r. 'v', we get $\frac{\partial z}{\partial v} = \phi_1(u)$, where $\phi_1(u)$ is an arbitrary function of u. ...(4)

Integrating w. r. 'v', once again, we get $z = v\phi_1(u) + \phi_2(u)$, where $\phi_2(u)$ is another function of u.

Hence by using (2), the required solution of (1) is $z = (x+y)\phi_1(x-y) + \phi_2(x-y)$.

Example 7: Classify and reduce the equation $(n-1)^2 \frac{\partial^2 z}{\partial x^2} - y^{2n}\frac{\partial^2 z}{\partial y^2} = ny^{2n-1}\frac{\partial z}{\partial y}$ to canonical form, and find its general solution.

Solution: The given equation can be written as

$$(n-1)^2 r - y^{2n} t - ny^{2n-1} q = 0 \qquad ...(1)$$

Comparing (1) with $Ar + 2Bs + Ct + f(x, y, z, p, q) = 0$, we get

$$A = (n-1)^2, B = 0, C = -y^{2n}.$$

Since $B^2 = 0$, $AC = -(n-1)^2 y^{2n}$, so that $B^2 - AC > 0$ except when $y = 0$. Hence the given equation is hyperbolic at all points except on x-axis. Therefore the λ – quadratic $A\lambda^2 + 2B\lambda + C = 0$ gives

$$(n-1)^2 \lambda^2 - y^{2n} = 0 \Rightarrow \lambda = \pm(n-1)^{-1} y^n$$

Here $\lambda_1 = (n-1)^{-1} y^n$, $\lambda_2 = -(n-1)^{-1} y^n$. (Real and distinct roots).

Hence $\frac{dy}{dx} + \lambda_1 = 0$ and $\frac{dy}{dx} + \lambda_2 = 0$ become

$$\frac{dy}{dx}+(n-1)^{-1}y^{n}=0 \text{ and } \frac{dy}{dx}-(n-1)^{-1}y^{n}=0$$

$$\Rightarrow \quad dx+(n-1)y^{-n}\,dy=0 \text{ and } dx-(n-1)y^{-n}\,dy=0.$$

Integrating, we get $x-y^{-n+1}=c_1$ and $x+y^{-n+1}=c_2$.

Now to reduce (1) to canonical form we change x, y to u, v by taking $u=x-y^{-n+1}$ and $v=x+y^{-n+1}$. ...(2)

Therefore $p=\dfrac{\partial z}{\partial x}=\dfrac{\partial z}{\partial u}\dfrac{\partial u}{\partial x}+\dfrac{\partial z}{\partial v}\dfrac{\partial v}{\partial x}=\dfrac{\partial z}{\partial u}+\dfrac{\partial z}{\partial v}$ so $\dfrac{\partial}{\partial x}\equiv\dfrac{\partial}{\partial u}+\dfrac{\partial}{\partial v}$

$$q=\frac{\partial z}{\partial y}=\frac{\partial z}{\partial y}\frac{\partial u}{\partial y}+\frac{\partial z}{\partial v}\frac{\partial v}{\partial y}=(n-1)y^{-n}\left(\frac{\partial z}{\partial u}-\frac{\partial z}{\partial v}\right)$$

$$r=\frac{\partial}{\partial x}\left(\frac{\partial z}{\partial x}\right)=\left(\frac{\partial}{\partial u}+\frac{\partial}{\partial v}\right)\left(\frac{\partial z}{\partial u}+\frac{\partial z}{\partial v}\right)=\frac{\partial^2 z}{\partial u^2}+2\frac{\partial^2 z}{\partial u\,\partial v}+\frac{\partial^2 z}{\partial v^2}$$

$$t=\frac{\partial}{\partial y}\left(\frac{\partial z}{\partial y}\right)=\frac{\partial}{\partial y}\left[(n-1)y^{-n}\left(\frac{\partial z}{\partial u}-\frac{\partial z}{\partial v}\right)\right]$$

$$=-n(n-1)y^{-n-1}\left(\frac{\partial z}{\partial u}-\frac{\partial z}{\partial v}\right)+(n-1)y^{-n}\frac{\partial}{\partial u}\left(\frac{\partial z}{\partial u}-\frac{\partial z}{\partial v}\right)$$

$$=-n(n-1)y^{-n-1}\left(\frac{\partial z}{\partial u}-\frac{\partial z}{\partial v}\right)+(n-1)y^{-n}$$

$$\left[\frac{\partial}{\partial u}\left(\frac{\partial z}{\partial u}-\frac{\partial z}{\partial v}\right).\frac{\partial u}{\partial y}+\frac{\partial}{\partial v}\left(\frac{\partial z}{\partial u}-\frac{\partial z}{\partial v}\right).\frac{\partial v}{\partial y}\right]$$

$$=-n(n-1)^{-n-1}\left(\frac{\partial z}{\partial u}-\frac{\partial z}{\partial v}\right)+(n-1)^2y^{-2n}\left(\frac{\partial^2 z}{\partial u^2}-2\frac{\partial^2 z}{\partial u\,\partial v}+\frac{\partial^2 z}{\partial v^2}\right).$$

Substituting these values of r, t, q in (1), we obtain

$$\frac{\partial^2 z}{\partial u\,\partial v}=0 \text{ or } \frac{\partial^2 z}{\partial v\,\partial u}=0 \qquad \text{...(3)}$$

which is the required canonical form of the given equation (1). Integrating (3) w. r. t. 'v', we get

$\dfrac{\partial z}{\partial u}=F(u)$, where $F(u)$ is an arbitrary function of u. ...(4)

Again integrating, we obtain $z=\phi_1(u)+\phi_2(v)$, where $\phi_1(u)=\int F(u)\,du$ and $\phi_2(v)$ are arbitrary functions. Hence, the desired general solution of (1) is

$$z=\phi_1\left(x-y^{-n+1}\right)+\phi_2\left(x+y^{-n+1}\right).$$

EXERCISE 8.1

Classify the following PDEs reduce to canonical forms and solve:

1. $x^2r + xs - yt = 0.$
2. $x^2r - y^2t = xp - yp.$
3. $(r-s)x = (t-s)y.$
4. $(r-s)y + (s+t)x = q - p.$
5. $y^2r + x^2t = 0.$
6. $y^2r - x^2t = 0.$
7. $(1-x^2)r - t = 0.$
8. $r + \frac{2}{x}p - a^2t = 0.$
9. $y^2r + 2xys + x^2t = 0.$
10. $x(xy-1)r - (x^2y^2 - 1)s + y(xy-1)t + (x-1)p + (y-1)q = 0.$
11. $x^2(y-1)r - x(y^2-1)s + y(y-1)t + xyp - q = 0.$
12. $(y-1)r - (y^2-1)s + y(y-1)t + p - q = 2ye^{2x}(1-y)^3.$
13. $x^2r - 2xys + y^2t - xp + 3yq = 8y/x.$

ANSWERS 8.1

1. Elliptic for $y < -\frac{1}{4}$, hyperbolic for $y > -\frac{1}{4}$, parabolic for $x = 0$ *or for* $y = -\frac{1}{4}$.
2. Hyperbolic; $z = y^2\,\phi_1\left(\frac{y}{x}\right) + \phi_2(xy).$
3. Hyperbolic; $z = \phi_1(x+y) + \phi_2(x^2 - y^2).$
4. Hyperbolic; $z = \phi_1(x+y) + (x+y)\phi_2\left(\frac{y}{x}\right).$
5. Elliptic at all points except on x-axis and y-axis;
$$\frac{\partial^2 z}{\partial u^2} + \frac{\partial^2 z}{\partial v^2} + \frac{1}{2u}\frac{\partial z}{\partial u} + \frac{1}{2v}\frac{\partial z}{\partial v} = 0;\ u = y^2,\ v = x^2.$$
6. Hyperbolic at all points except on x-axis and y-axis;
$$2(u^2 - v^2)\frac{\partial^2 z}{\partial u\,\partial v} + v\frac{\partial z}{\partial u} + u\frac{\partial z}{\partial v} = 0.$$
7. Hyperbolic for $-1 < x < 1$, elliptic for $x < -1$ *or* $x > 1$, parabolic for $x = -1$ *and* $x = 1$; $2\frac{\partial^2 z}{\partial u\,\partial v} = \tan(u - v)\left\{\frac{\partial z}{\partial u} - \frac{\partial z}{\partial v}\right\}.$
8. Hyperbolic; $xz = \phi_1(y + ax) + \phi_2(y - ax).$

9. Parabolic $\left(v^2 - u^2\right)\dfrac{\partial^2 z}{\partial v^2} + u\dfrac{\partial z}{\partial u} = 0.$

10. Hyperbolic at all points except for

$$xy = 1; \frac{\partial^2 z}{\partial u\, \partial v} = 0;\, z = \phi_1\left(ye^x\right) + \phi_2\left(xe^y\right).$$

11. Hyperbolic at all points except for

$$x = 0 \text{ or } y = 1; \frac{\partial^2 z}{\partial u\, \partial v} = 0;\, z = \phi_1\left(xy\right) + \phi_2\left(xe^y\right).$$

12. Hyperbolic at all points except for y = 1;

$$\frac{\partial^2 z}{\partial u\, \partial v} = 2v;\; z = \left(x+y\right)y^2 e^{2x} + \phi_1\left(x+y\right) + \phi_2\left(ye^x\right)$$

13. $v\dfrac{\partial^2 z}{\partial v^2} + 2\dfrac{\partial z}{\partial v} = 2,\; z = \dfrac{y}{x} + x^2\phi_1\left(xy\right) + \phi_2\left(y/x\right).$

8.3 HOMOGENEOUS LINEAR EQUATIONS WITH CONSTANT COEFFICIENTS

Let $f\left(D, D'\right) = A_0\, D^n + A_1\, D^{n-1}D' + A_2\, D^{n-2}\, D'^2 + \ldots + A_n\, D'^n$ ...(1)

where $A_1, A_2, \ldots, A_n$ are constants.

Then equation (1) is known as homogeneous equation and takes the form

$$\left(A_0 D^n + A_1 D^{n-1}\, D' + A_2\, D^{n-2}\, D'^2 + \ldots + A_n D'^n\right)z = 0. \quad \text{...(2)}$$

8.3.1 To Find Complementary Function

Consider an equation

$$\left(A_0 D^n + A_1 D^{n-1}\, D' + A_2 D^{n-2}\, D'^2 + \ldots + A_n D'^n\right)z = 0 \quad \text{...(3)}$$

Let the solution of (3) be $z = \phi\left(y + mx\right)$ ...(4)

Then $\quad D^r z = m^r\, \phi^{(r)}\left(y + mx\right),\; D'^{(s)} z = \phi^{(s)}\left(y + mx\right)$ and

$$D^r\, D'^s\, z = m^r\, \phi^{(r+s)}\left(y + mx\right).$$

Therefore, substituting (4) in equation (3), we obtain

$$\left(A_0 m^n + A_1\, m^{n-1} + A_2 m^{n-2} + \ldots + A_n\right)\phi^{(n)}\left(y + mx\right) = 0$$

which will be satisfied if $A_0 m^n + A_1 m^{n-1} + A_2 m^{n-2} + \ldots + A_n = 0.$ ...(5)

Equation (5) is known as the **auxiliary equation.**

Let $m_1, m_2, \ldots, m_n$ be the roots of the equation (5). Two different cases arise:

Case I: ***When all the roots are distinct.*** *Let $m_1, m_2, \ldots, m_n$ be distinct.*

Part of C.F. corresponding to $m = m_1$ is $z = \phi_1\left(y + m_1 x\right)$, where ϕ_1 is an arbitrary function.

Part of C.F. corresponding to $m = m_2$ is $z = \phi_2(y + m_2 x)$, where 'ϕ_2' is an arbitrary function, etc.

Since equation (3) being linear, the sum of the solutions is also a solution.
Hence

$$C.F. = \phi_1(y + m_1 x) + \phi_2(y + m_2 x) + \ldots + \phi_n(y + m_n x). \qquad \ldots(6)$$

Case II: *When the roots of auxiliary equation are equal (repeated). Let* $m_1 = m_2 = \ldots = m_r = m$ *i.e.* m *be a r-times repeated root, then* $(D - mD')^r z = 0$

If $r = 2$, we have $(D - mD')^2 z = 0$

Putting $(D - mD')z = u$, we have $(D - mD')u = 0$.

Its solution is $u = f(y + mx)$.

Using it in $(D - mD')z = u$, we have $(D - mD')z = f(y + mx)$

$\Rightarrow p - mq = f(y + mx)$.

Here the subsidiary equations are $\dfrac{dx}{1} = \dfrac{dy}{-m} = \dfrac{dz}{f(y + mx)}$.

From first two terms, we have $dy + mdx = 0 \Rightarrow y + mx = a$.

Again taking first and last terms, we have $dz = f(a)dx$, [by using $y + mx = a$]

$\Rightarrow \quad z = f(a)x + b \Rightarrow z = xf(y + mx) + \phi(y + mx)$.

Hence in general, when the root repeats r-times, we have

$$z = f_1(y + mx) + xf_2(y + mx) + \ldots + x^{r-1} f_r(y + mx).$$

Note: (i) Equation (5) is obtained by putting $D = m$ and $D' = 1$ in the expression (3).

(ii) If $A_0 = 0$ *and* $A_n \neq 0$, it is advisable to obtain auxiliary equation by putting $D = 1$ and $D' = m$. In this case

$$C.F. = \phi_1(x + m_1 y) + \phi_2(x + m_2 y) + \ldots + \phi_n(x + m_n y). \qquad \ldots(9)$$

(iii) If $A_0 = 0 = A_n$, auxiliary equation will be of degree $(n - 1)$. In this case if auxiliary equation is obtained by (i), the nth term of C.F. of the form $\phi(x)$ and is auxiliary equation is obtained by (ii), the nth term is of the form $\phi(y)$.

8.4 PARTICULAR INTEGRAL

Consider a PDE $f(D, D')z = V(x, y)$, where $f(D, D')$ is given by (1). ...(10)

The operator $\dfrac{1}{f(D, D')}$ is defined by an identity

$$f(D,D')\frac{1}{f(D,D')}V(x,y)=V(x,y).$$

The P.I. is given by $\frac{1}{f(D,D')}V(x,y)$.

If $m_1, m_2, m_3, ...m_n$ be the roots of equation (5), expression (1) can be written as

$$f(D,D')=A_0(D-m_1)(D-m_2)...(D-m_n) \quad ...(11)$$

$$\text{Therefore } P.I.=\frac{1}{A_0(D-m_1)(D-m_2)...(D-m_n)}V(x,y). \quad ...(12)$$

8.4.1 P.I. when $V(x,y)$ is Rational Integral of Algebraic Function

If $V(x,y)$ is a rational integral of algebraic function, (12) can be evaluated by expanding the symbolic function $\frac{1}{f(D,D')}$ as an infinite series of ascending powers of D or D'.

8.4.2 P.I. in Case $V(x,y)=\phi(ax+by)$

If $V(x,y)$ is a function of $(ax+by)$, shorter methods may be used. To find particular integral, we use the method based on the following theorem.

Theorem 8.1: If $f(D, D')$ is of degree n, then

$$\frac{1}{f(D,D')}\phi^{(n)}(ax+by)=\frac{1}{f(a,b)}\phi(ax+by) \quad ...(1)$$

provided $f(a,b)\neq 0$.

Proof: By direct differentiation $D^r\,\phi(ax+by)=a^r\,\phi^{(r)}(ax+by)$

$$D'^s\,\phi(ax+by)=b^s\,\phi^{(s)}(ax+by)$$

and $D^r D'^s\,\phi(ax+by)=a^r b^s\,\phi^{(r+s)}(ax+by)$.

Therefore $f(D,D')\phi(ax+by)=f(a,b)\phi^{(n)}(ax+by)$.

Operating upon both sides by $\frac{1}{f(D,D')}$, we obtain

$$\phi(ax+by)=f(a,b)\frac{1}{f(D,D')}\phi^{(n)}(ax+by).$$

If $f(a,b)\neq 0$, then on dividing by $f(a,b)$, we have

$$\frac{1}{f(a,b)}\phi(ax+by)=\frac{1}{f(D,D')}\phi^{(n)}(ax+by)$$

$$\Rightarrow \frac{1}{f(D,D')}\phi^{(n)}(ax+by)=\frac{1}{f(a,b)}\phi(ax+by).$$

Corollary: If $f(a,b)=0$, then let $f(D,D')=(bD-aD')^r\, g(D,D')$ where $g(D,D')$ is of degree $n-r$ and $f(a,b)\neq 0$.

Then $\dfrac{1}{f(D,D')}\phi^{(n)}(ax+by)=\dfrac{1}{(bD-aD')^r}\dfrac{1}{g(D,D')}\phi^{(n)}(ax+by)$

$$=\frac{1}{(bD-aD')^r}\frac{1}{g(a,b)}\phi^{(r)}(ax+by)$$

$$=\frac{1}{g(a,b)}\frac{1}{(bD-aD')^r}\phi^{(r)}(ax+by). \qquad ...(2)$$

Let $\dfrac{1}{(bD-aD')}\phi^{(r)}(ax+by)=U$

So that $b\dfrac{\partial U}{\partial x}-a\dfrac{\partial U}{\partial y}=\phi^{(r)}(ax+by) \qquad ...(3)$

which has its subsidiary equations $\dfrac{dx}{b}=\dfrac{dy}{-a}=\dfrac{dU}{\phi^{(r)}(ax+by)}. \qquad ...(4)$

Two independent integrals of equation (4) are

$ax+by=\text{constant}$ and $U=\dfrac{x}{b}\phi^{(r)}(ax+by)$. Therefore

$$\frac{1}{(bD-aD')^2}\phi^{(r)}(ax+by)=\frac{1}{(bD-aD')}U=\frac{1}{bD-aD'}\frac{x}{b}\phi^{(r)}(ax+by)$$

$$=\frac{x^2}{2!\,b^2}\phi^{(r)}(ax+by).$$

In general,

$$\frac{1}{(bD-aD')^r}\phi^{(r)}(ax+by)=\frac{x^r}{r!}\frac{1}{b^r}\phi^{(r)}(ax+by).$$

Substituting in (2), we get

$$\frac{1}{f(D,D')}\phi^{n}(ax+by)=\frac{1}{g(a,b)}\frac{x^r}{r!}\frac{1}{b^r}\phi^{(r)}(ax+by).$$

8.4.3 General Method

Now we proceed to find the general method for getting a P.I.

Let us consider the equation

$(D-mD')z=\phi(x,y)$ or $p-mq=\phi(x,y)$.

The subsidiary equations are $\frac{dx}{1}=\frac{dy}{-m}=\frac{dz}{\phi(x,y)}$.

First two members give $y+mx=a$ (constant)

From the first and the last members, we have $dz=\phi(x,y)dx=\phi(x,a-mx)dx$.

Therefore $z=\int \phi(x,a-mx)dx$.

and $P.I.=\frac{1}{(D-mD')}\phi(x,y)=\int \phi(x,a-mx)dx$.

After integrating, a is replaced by $y+mx$.

Working rule: Take the P.I. corresponding to

$\frac{1}{(D-mD')}\phi(x,y)$ as $\int \phi(x,a-mx)dx$,

and replace a by $y+mx$ after integration.

8.5 NON-HOMOGENEOUS LINEAR EQUATIONS WITH CONSTANT COEFFICIENTS

Let us consider simplest case $(D-mD'-\alpha)z=0$ or $p-mq=\alpha z$.

The subsidiary equations are $\frac{dx}{1}=\frac{dy}{-m}=\frac{dz}{\alpha z}$.

The first two members give $y+mx=a$ (constant).

Again from the first and the third relations, we have $\frac{dz}{z}=\alpha\, dx$.

$\Rightarrow \quad \log z=\alpha x+\log b$ or $z=be^{\alpha x} \Rightarrow \quad z=e^{\alpha x} f(y+mx)$.

Similarly, it can easily be shown that the integral of

$$(D-m_1D'-\alpha_1)\ldots(D-m_2D'-\alpha_2)\ldots(D-m_nD'-\alpha_n)z=0$$

is $$z=e^{\alpha_1 x} f_1(y+m_1x)+e^{\alpha_2 x} f_2(y+m_2 x)+\ldots+e^{\alpha_n x} f_n(y+m_nx).$$

Also in case of repeated factors, e.g. $(D-mD'-\alpha)^r z=0$ the integral is

$$z=e^{\alpha x} f_1(y+mx)+xe^{\alpha x} f_2(y+mx)+\ldots+x^{r-1} e^{\alpha x} f_r(y+mx).$$

Note: If $f(D,D')$ cannot be resolved into factors linear in D and D', the equation cannot be integrated by the above methods. In such cases a trail method is used to find solutions.

8.6 PARTICULAR INTEGRALS

The methods of obtaining particular integrals of non-homogeneous partial differential equations are very similar to those used in solving linear equations with constant coefficients. Here we are considering few cases only:

Case I: *If the function* $V(x, y)$ *is of the form* e^{ax+by}, *then*

$$\frac{1}{f(D, D')}e^{ax+by} = \frac{1}{f(a, b)}e^{ax+by} \text{ provided } f(a, b) \neq 0, \text{ i.e. put}$$

$$D = a \text{ and } D' = b.$$

Case II: *If the function* $V(x, y)$ *is of the form* $x^m y^n$, *where* m *and* n *are positive integers, then*

$$\frac{1}{f(D, D')}x^m y^m = [f(D, D')]^{-1} x^m y^n.$$

Case III: *If the function* $V(x, y)$ *is of the form* $e^{ax+by} F$, *then*

$$\frac{1}{f(D, D')}\left(e^{ax+by}.F\right) = e^{ax+by}\frac{1}{f(D+a, D'+b)}F.$$

8.7 EQUATIONS REDUCIBLE TO HOMOGENEOUS LINEAR FORM

An equation, in which the coefficient of derivative of any order is a multiple of the variables of the same degree, may be transformed into the partial differential equations with constant coefficients.

For this we substitute $x = e^X, y = e^Y$ so that, X = log x and Y = log y,

$$\frac{\partial z}{\partial x} = \frac{\partial z}{\partial X}\frac{\partial X}{\partial x} = \frac{1}{x}\frac{\partial z}{\partial X}$$

Therefore $x\frac{\partial z}{\partial x} = \frac{\partial z}{\partial X} \Rightarrow x\frac{\partial}{\partial x} \equiv \frac{\partial}{\partial X} \equiv D$ (say).

Now $x\frac{\partial}{\partial x}\left(x^{n-1}\frac{\partial^{n-1} z}{\partial x^{n-1}}\right) = x^n\frac{\partial^n z}{\partial x^n} + (n-1)x^{n-1}\frac{\partial^{n-1} z}{\partial x^{n-1}}$

$$\Rightarrow \quad x^n\frac{\partial^n z}{\partial x^n} = \left(x\frac{\partial}{\partial x} - n + 1\right)x^{n-1}\frac{\partial^{n-1} z}{\partial x^{n-1}}.$$

Putting $n = 2, 3, \ldots$, we obtain

$$x^2\frac{\partial^2 z}{\partial x^2} = (D-1)x\frac{\partial z}{\partial x} = D(D-1)z,$$

$$x^3\frac{\partial^3 z}{\partial x^3} = (D-2)x^2\frac{\partial^2 z}{\partial x^2} = (D-2)(D-1)Dz, \text{ etc.}$$

Similarly $y\frac{\partial z}{\partial y} = \frac{\partial z}{\partial Y} = D'z$ and $xy\frac{\partial^2 z}{\partial x\,\partial y} = DD'z$ etc.

Substituting in the given equation, it reduces to the form $F(D, D')z = V$, which is an equation having constant coefficients and can easily be solved by the method discussed above.

Example 1: Solve $\frac{\partial^3 z}{\partial x^3} - 3\frac{\partial^3 z}{\partial x^2 \partial y} + 2\frac{\partial^3 z}{\partial x \partial y^2} = 0.$

Solution: A.E. is $m^3 - 3m^2 + 2m = 0 \Rightarrow m(m-1)(m-2) = 0 \Rightarrow m = 0, 1, 2.$
Hence the solution is $z = f_1(y) + f_2(y+x) + f_3(y+2x).$

Example 2: Solve $25r - 40s - 16t = 0.$

Solution: A.E. is $25m^2 - 40m + 16 = 0 \Rightarrow (5m-4)^2 = 0 \Rightarrow m = \frac{4}{5}, \frac{4}{5}.$

Hence the solution is $z = f_1\left(y + \frac{4}{5}x\right) + xf_2\left(y + \frac{4}{5}x\right)$

$$\Rightarrow \quad z = \phi_1(5y + 4x) + x\phi_2(5y + 4x).$$

Example 3: Solve $\frac{\partial^3 z}{\partial x^3} - 7\frac{\partial^3 z}{\partial x \partial y^2} + 6\frac{\partial^3 z}{\partial y^3} = 0.$

Solution: A.E. is $m^3 - 7m + 6 = 0 \Rightarrow (m-1)(m-2)(m+3) = 0$

$$\Rightarrow m = 1, 2, -3.$$

Hence the solution is $z = f_1(y+x) + f_2(y+2x) + f_3(y-3x).$

Example 4: Solve $\frac{\partial^4 z}{\partial x^4} - \frac{\partial^4 z}{\partial y^4} = 0.$

Solution: A.E. is $m^4 - 1 = 0 \Rightarrow (m-1)(m+1)(m^2+1) = 0$

$$\Rightarrow m = 1, -1, \pm i.$$

Hence the solution is $z = f_1(y+x) + f_2(y-x) + f_3(y+ix) + f_4(y-ix).$

Example 5: Solve $(D^3 - 6D^2D' + 11DD'^3 - 6D'^3)z = 0.$

Solution: Here A.E. is

$$m^3 - 6m^2 + 11m - 6 = 0 \Rightarrow (m-1)(m-2)(m-3) = 0 \Rightarrow m = 1, 2, 3.$$

Hence the solution is $z = f_1(y+x) + f_2(y+2x) + f_3(y+3x).$

Example 6: Solve $2r + 5s + 2t = 0.$

Solution: A.E. is $2m^2 + 5m + 2 = 0 \Rightarrow (2m+1)(m+2) = 0 \Rightarrow m = -\frac{1}{2}, -2.$

Therefore the solution is $z = f\left(y - \frac{x}{2}\right) + \psi(y - 2x)$

$$\Rightarrow \quad z = \phi(2y - x) + \psi(y - 2x).$$

Example 7: Solve $\left(2D^2D' - 3DD'^2 + D'^3\right)z = 0.$

Solution: A.E. (changing D' to m and D to unity) is

$$m^3 - 3m^2 + 2m = 0$$

$$\Rightarrow \quad m(m-1)(m-2) = 0 \Rightarrow m = 0, 1, 2.$$

Therefore $C.F. = \phi_1(x) + \phi_2(x+y) + \phi_3(x+2y).$

Example 8: Solve $\left(D^2 + 3DD' + 2D'^2\right)z = x + y.$

Solution: A.E. is $m^2 + 3m + 2 = 0 \Rightarrow (m+1)(m+2) = 0 \Rightarrow m = -1, -2$

Therefore C.F. $= f_1(y-x) + f_2(y-2x).$

$$\text{Now } P.I. = \frac{1}{D^2 + 3DD' + 2D'^2}(x+y) = \frac{1}{D^2}\left(1 + \frac{3D'}{D} + 2\frac{D'^2}{D^2}\right)^{-1}.(x+y)$$

$$= \frac{1}{D^2}\left(1 - \frac{3D'}{D} \ldots\right)\ldots(x+y) = \frac{1}{D^2}(x+y) - \frac{3}{D^3}D'(x+y)$$

$$= \frac{x^3}{6} + \frac{x^2}{2}y - \frac{3}{D^3}1 = \frac{x^3}{6} + \frac{1}{2}x^2y - 3\frac{x^3}{6}$$

$$= -\frac{1}{3}x^3 + \frac{1}{2}x^2y.$$

Therefore the general solution is

$$z = f_1(y-x) + f_2(y-2x) - \frac{1}{3}x^3 + \frac{1}{2}x^2y.$$

Example 9: Solve $r + (a+b)s + abt = xy.$

Solution: The given equation can be written as

$$\left\{D^2 + (a+b)DD' + abD'^2\right\}z = xy$$

A.E. is $m^2 + (a+b)m + ab = 0 \Rightarrow (m+a)(m+b) = 0, \Rightarrow m = -a, -b.$

Therefore $C.F. = f_1(y-ax) + f_2(y-bx)$

$$\text{Now} \quad P.I. = \frac{1}{D^2 + (a+b)DD' + abD'^2}xy$$

$$= \frac{1}{D^2}\left\{1 + (a+b)\frac{D'}{D} + ab\frac{D'^2}{D^2}\right\}^{-1}xy$$

$$= \frac{1}{D^2}\left\{1 - (a+b)\frac{D'}{D}\ldots\right\}.xy$$

$$= \frac{1}{D^2}xy - (a+b)\frac{1}{D^3}x = \frac{1}{6}x^3y - \frac{1}{24}(a+b)x^4.$$

Hence the solution is $z = f_1(y-ax) + f_2(y-bx) + \frac{1}{6}x^3 y - \frac{1}{24}(a+b)x^4$.

Example 10: Solve $(2D^2 - 5DD' + 2D'^2)z = 24(y-x)$.

Solution: Here A.E. is $2m^2 - 5m + 2 = 0$

$$\Rightarrow \quad (2m-1)(m-2)=0, \Rightarrow m = \frac{1}{2}, 2$$

Therefore $C.F. = f_1(2y+x) + f_2(y+2x)$.
Now

$$P.I. = \frac{1}{2D^2 - 5DD' + 2D'^2} 24(y-x)$$

$$= \frac{1}{2D^2}\left(1 + \frac{5D'}{2D} - \frac{D'^2}{D^2} \dots\right) 24(y-x)$$

$$= \frac{1}{2D^2}\left(1 + \frac{5D'}{2D}\right) 24(y-x) = \frac{1}{2D^2} 24(y-x) + \frac{5}{4D^3} 24$$

$$= 12\left(\frac{x^2}{2}y - \frac{x^3}{6}\right) + \frac{5}{4} 24 \frac{x^3}{6} = 6x^2y + 3x^3$$

Hence the solution is $z = f_1(2y+x) + f_2(y+2x) + 6x^2y + 3x^3$.

Example 11: Solve $(D^2 - DD' - 6D'^2)z = xy$.

Solution: A.E. is $m^2 - m - 6 = 0 \text{ or } (m-3)(m+2) = 0 \Rightarrow m = 3, -2$.

Therefore $C.F. = \phi_1(y+3x) + \phi_2(y-2x)$

Now $$P.I. = \frac{1}{D^2 - DD' - 6D'^2} xy = \frac{1}{D^2}\left(1 - \frac{D'}{D} - 6\frac{D'^2}{D^2}\right)^{-1} xy$$

$$= \frac{1}{D^2}\left(1 + \frac{D'}{D} + \dots\right) xy = \frac{1}{D^2}\left(xy + \frac{x^2}{2}\right) = \frac{1}{6}x^3y + \frac{x^4}{24}.$$

Hence the solution is $z = \phi_1(y+3x) + \phi_2(y-2x) + \frac{1}{6}x^3y + \frac{x^4}{24}$.

Example 12: Solve $(D^3 - D'^3)z = x^3 y^3$.

Solution: A.E. is $m^3 - 1 = 0 \Rightarrow m = 1, \omega, \omega^2$.

Therefore $C.F. = \phi_1(y+x) + \phi_2(y+x\omega) + \phi_3(y+x\omega^2)$

Now $P.I. = \frac{1}{D^3 - D'^3} x^3 y^2 = \frac{1}{D^3}\left(1 - \frac{D'^3}{D^3}\right)^{-1} x^3 y^3$

$$= \frac{1}{D^3}\left(1 + \frac{D'^3}{D^3} + \ldots\right)x^3\, y^3 = \frac{1}{D^3}\left(x^3\, y^3 + \frac{1}{20}\, x^6\right) = \frac{x^6 y^3}{120} + \frac{x^9}{10080}.$$

Hence the solution is

$$z = \phi_1 (y + x) + \phi_2 (y + \omega x) + \phi_3 (y + \omega^2 x) + \frac{x^6\, y^3}{120} + \frac{x^9}{10080}.$$

Example 13: Find a real function V of x and y, reducing to zero when $y = 0$ and satisfying $\frac{\partial^2 V}{\partial x^2} + \frac{\partial^2 V}{\partial y^2} = -4\pi (x^2 + y^2)$.

Solution: The real function V will be given by the P.I. of the given equation. Now

$$P.I. = \frac{1}{D^2 + D'^2}\{-4\pi(x^2 + y^2)\} = \frac{1}{D^2}\left[1 + \frac{D'^2}{D^2}\right]^{-1} .\{-4\pi(x^2 + y^2)\}$$

$$= \frac{1}{D^2}\left[1 - \frac{D'^2}{D^2} \ldots\right].\{-4\pi(x^2 + y^2)\} = \frac{1}{D^2}\{-4\pi(x^2 + y^2)\} - \frac{1}{D^4}.\{-4\pi\, 2\}$$

$$= -4\pi\left[\frac{x^4}{12} + \frac{x^2\, y^2}{2}\right] - \left[-8\pi\, \frac{x^4}{24}\right] = -\frac{\pi x^4}{3} - 2\pi\, x^2 y^2 + \frac{\pi x^4}{3} = -2\pi\, x^2\, y^2.$$

Therefore the required solution is $V = -2\pi\, x^2\, y^2$

Example 14: Solve $(D^2 - 6DD' + 9D'^2)z = 12x^2 + 36xy$.

Solution: Here A.E. is $m^2 - 6m + 9 = 0 \Rightarrow (m-3)^2 = 0 \Rightarrow m = 3, 3$.

Therefore $C.F. = f_1 (y + 3x) + x f_2 (y + 3x)$

Now

$$P.I. = \frac{1}{D^2 - 6DD' + 9D'^2} .(12x^2 + 36xy) = \frac{1}{D^2}\left(1 - \frac{3D'}{D}\right)^{-2} .(12\, x^2 + 36xy)$$

$$= \frac{1}{D^2}\left(1 + \frac{6D'}{D} + 27\, \frac{D'^2}{D^2} + \ldots\right)(12x^2 + 36xy)$$

$$= \frac{1}{D^2}(12x^2 + 36xy) + \frac{6}{D^3}\, D'(12\, x^2 + 36xy)$$

$$= x^4 + 6x^3 y + \frac{6}{D^3}\, 36\, x = x^4 + 6x^3 y + 9x^4 = 10x^4 + 6x^3 y.$$

Hence the general solution is $z = f_1 (y + 3x) + x f_2 (y + 3x) + 10x^4 + 6x^3 y$.

Example 15: Solve $(D^3 - 4D^2 D' + 4DD'^2)z = 4 \sin(2x + y)$.

Solution: Here A.E. is

$$m^3 - 4m^2 + 4m = 0 \Rightarrow m(m-2)^2 = 0 \Rightarrow m = 0, 2, 2.$$

Therefore $C.F. = f_1(y) + f_2(y+2x) + xf_3(y+2x)$

$$P.I. = \frac{1}{D^3 - 4D^2D' + 4DD'^2} 4\sin(2x+y) = \frac{1}{D(D-2D')^2} 4\sin(2x+y)$$

$$= \frac{1}{(D-2D')^2}\frac{1}{D} 4\sin(2x+y) = \frac{1}{(D-2D')^2}\{-2\cos(2x+y)\}$$

$$= \frac{x^2}{2!}\{-2\cos(2x+y)\} = -x^2\cos(2x+y).$$

Hence the solution is

$$z = f_1(y) + f_2(y+2x) + xf_3(y+2x) - x^2\cos(2x+y).$$

Example 16: Solve $(D^2 - 2DD' + D'^2)z = e^{x+2y} + x^3$.

Solution: Here A.E. is $m^2 - 2m + 1 = 0 \Rightarrow (m-1)^2 = 0 \Rightarrow m = 1, 1$

Therefore $C.F. = f_1(y+x) + xf_2(y+x)$

$$P.I. = \frac{1}{D^2 - 2DD' + D'^2} e^{x+2y} + \frac{1}{D^2 - 2DD' + D'^2} x^3$$

$$= \frac{1}{1^2 - 2\cdot1\cdot2 + 2^2} e^{x+2y} + \frac{1}{D^2}\left(1 - \frac{D'}{D}\right)^{-2} x^3$$

$$= e^{x+2y} + \frac{1}{D^2}\left(1 + \frac{2D'}{D}\ldots\right)x^3 = e^{x+2y} + \frac{1}{20}x^5.$$

Hence the solution is $z = f_1(y+x) + xf_2(y+x) + e^{x+2y} + \frac{1}{20}x^5$.

Example 17: Solve $\frac{\partial^2 z}{\partial x^2} + \frac{\partial^2 z}{\partial y^2} = \cos mx \cos ny$.

Solution: The given equation can be written as

$$(D^2 + D'^2)z = \frac{1}{2}[\cos(mx+ny) + \cos(mx-ny)].$$

A.E. is $m^2 + 1 = 0 \Rightarrow m = \pm i$

Therefore $C.F. = f_1(y+ix) + f_2(y-ix)$

$$P.I. = \frac{1}{2}\frac{1}{(D^2 + D'^2)}\cos(mx+ny) + \frac{1}{2}\frac{1}{(D^2 + D'^2)}\cos(mx-ny)$$

$$= \frac{1}{2}\frac{1}{(m^2+n^2)}\{-\cos(mx+ny)\} + \frac{1}{2}\frac{1}{(m^2+n^2)}\{-\cos(mx-ny)\}$$

Hence the solution is

$$z = f_1(y+ix) + f_2(y-ix) - \frac{1}{2(m^2+n^2)}\left[\cos(mx+ny) + \cos(mx-ny)\right].$$

Example 18: Solve $(D-D')(D+2D')z = (y+1)e^x$.

Solution: A.E. is $(m-1)(m+2) = 0 \Rightarrow m = 1, -2$.

Therefore $C.F. = \phi_1(y+x) + \phi_2(y-2x)$

Now $P.I. = \dfrac{1}{(D-D')(D+2D')}(y+1)e^x$

$$= \frac{1}{D-D'}\left(\int (a+2x+1)e^x\,dx\right), \text{ where } a = y-2x$$

$$= \frac{1}{D-D'}\left\{(a+2x+1)e^x - 2e^x\right\} = \frac{1}{D-D'}(y-1)e^x$$

$$= \int (a-x-1)e^x\,dx = \left\{(a-x-1)e^x + e^x\right\}, \text{ where } a = y+x$$

$$= ye^x.$$

Hence the solution is $z = \phi_1(y+x) + \phi_2(y-2x) + ye^x$

Example 19: Solve $(D^2 - 2DD' - 15D'^2)z = 12xy$.

Solution: A.E. is $m^2 - 2m - 15 = 0 \;\Rightarrow (m-5)(m+3) = 0 \;\Rightarrow m = 5, -3$

Therefore $C.F. = f_1(y+5x) + f_2(y-3x)$

Now $P.I. = \dfrac{1}{D^2 - 2DD' - 15DD'^2}12xy = \dfrac{1}{(D+3D')}\dfrac{1}{(D-5D')}12xy$

$$= \frac{12}{(D+3D')}\int x(a-5x)\,dx, \text{ where } y+5x = a$$

$$= \frac{12}{(D+3D')}\left(\frac{1}{2}ax^2 - \frac{5}{3}x^3\right) = \frac{12}{(D+3D')}\left\{\frac{1}{2}(y+5x)x^2 - \frac{5}{3}x^3\right\}$$

$$= \frac{2}{(D+3D')}.(3x^2y + 5x^3) = 2\int\left\{3x^2(3x+b) + 5x^3\right\}dx, \text{ where } y-3x = b$$

$$= 2\int (14x^3 + 3x^2b)\,dx = 2\left(\frac{14}{4}x^4 + x^3b\right)$$

$$= 7x^4 + 2x^3(y-3x) = x^4 + 2x^3y.$$

Hence the solution is $z = f_1(y+5x) + f_2(y-3x) + x^4 + 2x^3y$.

Aliter.

$$P.I. = \frac{1}{D^2 - 2DD' - 15D'^2} 12\,xy = \frac{1}{D^2}\left(1 - \frac{2D'}{D} - \frac{15D'^2}{D^2}\right)^{-1} .12xy$$

$$= \frac{1}{D^2}\left(1 + \frac{2D'}{D} + ...\right).12xy = \frac{1}{D^2}(12xy) + \frac{2}{D^3}12x = 2x^3y + x^4$$

Example 20: Solve

$$\left(D^2 - D\,D' - 2D'^2\right)z = \left(2x^3 + xy - y^2\right)\sin xy - \cos xy.$$

Solution: A.E. is $m^2 - m - 2 = 0 \Rightarrow (m-2)(m+1) = 0 \Rightarrow m = 2, -1.$

Therefore $C.F. = \phi_1(y + 2x) + \phi_2(y - x)$

Now $P.I. = \dfrac{1}{(D - 2D')(D + D')}\left\{\left(2x^3 + xy - y^3\right)\sin xy - \cos xy\right\}$

$$= \frac{1}{(D - 2D')(D + D')}\left\{(2x - y)(x + y)\sin xy - \cos xy\right\}$$

$$= \frac{1}{(D - 2D)}\left[\int_{a = y - x}\left\{(x - a)(2x + a)\sin x(a + x) - \cos x(a + x)\right\}dx\right]$$

$$= \frac{1}{(D - 2D')}\Big[-(x - a)\cos\left(ax + x^2\right)$$

$$+ \int \cos\left(ax + x^2\right)dx - \int \cos\left(ax + x^2\right)dx\ \Big]_{a = y - x}$$

$$= \frac{1}{(D - 2D')}(y - 2x)\cos xy = \int_{a = y + 2x}(a - 4x)\cos\left(ax - 2x^2\right)dx$$

$$= \left\{\sin\left(ax - 2x^2\right)\right\}_{a = y + 2x} = \sin xy$$

Hence the solution is $z = \phi_1(y + 2x) + \phi_2(y - x) + \sin xy.$

Example 21: Solve $\left(D^2 - 4D'^2\right)z = \dfrac{4x}{y^2} - \dfrac{y}{x^2}.$

Solution: A.E. is $m^2 - 4 = 0 \Rightarrow m = 2, -2$

Therefore $C.F. = \phi_1(y + 2x) + \phi_2(y - 2x)$

Now $P.I. = \dfrac{1}{(D + 2D')(D - 2D')}\left(\dfrac{4x}{y^2} - \dfrac{y}{x^2}\right)$

$$= \frac{1}{(D + 2D)^2}\int_{a = y + 2x}\left\{\frac{4x}{(a - 2x)^2} - \frac{(a - 2x)}{x^2}\right\}dx$$

$$\frac{1}{(D + 2D)'}\int_{a = y + 2x}\left\{\frac{-2}{(a - 2x)} + \frac{2a}{(a - 2x)^2} - \frac{a}{x^2} + \frac{2}{x}\right\}.dx$$

$$= \frac{1}{(D+2D)'}\left\{\log(a-2x) + \frac{a}{(a-2x)} + \frac{a}{x} + 2\log x\right\}_{a=y+2x}$$

$$= \frac{1}{(D+2D)'}\left[\log y + \frac{(y+2x)}{y} + \frac{(y+2x)}{x} + 2\log x\right]$$

$$= \int_{a=y-2x}\left\{\log(a+2x) + 1 + \frac{2x}{(a+2x)} + \frac{(a+2x)}{x} + 2 + 2\log x\right\}dx$$

$$= \left\{x\log(a+2x) + 5x + a\log x + 2x\log x - 2x\right\}_{a=y-2x}$$

$$= \left\{x\log y + 3x + y\log x\right\}.$$

Hence the solution is

$$z = \phi_1(y+2x) + \phi_2(y-2x) + x\log y + 3x + y\log x.$$

Example 22: Solve $(D^2 - D'^2 + D - D')z = 0$.

Solution: The given equation is $(D - D')(D + D' + 1)z = 0$.

Therefore the solution is $z = f_1(y+x) + e^{-x} f_2(y-x)$.

Example 23: Solve $(D^2 - D'^2 - 3D + 3D')z = 0$.

Solution: The equation can be written as $(D - D')(D + D' - 3)z = 0$.

Therefore $C.F. = \phi_1(y+x) + e^{3x}\phi_2(y-x) = \psi_1(y+x) + e^{3y}\psi_2(x-y)$.

Example 24: Solve $D\,D'(D - 2D' - 3)z = 0$.

Solution: Here $z = e^{0.x} f_1(y) + e^{0.y} f_2(0-x) + e^{3x} f_3(y+2x)$

$$= f_1(y) + \phi(x) + e^{3x} f_3(y+2x).$$

Example 25: Solve $r + 2s + t + 2p + 2q + z = 0$.

Solution: The given equation can be written as

$$(D^2 + 2\,D\,D' + D'^2 + 2\,D + 2\,D' + 1)z = 0$$

$$\Rightarrow \quad (D + D' + 1)^2 z = 0$$

Therefore solution is $z = e^{-x} f_1(y-x) + xe^{-x} f_2(y-x)$.

Example 26: Solve $(2\,D - D' + 4)(D + 2\,D' + 1)^2 z = 0$.

Solution: Part of C.F. corresponding to the factor

$(2D - D' + 4)$ is $e^{4y}\,\phi(x+2y)$.

Part of C.F. corresponding to the factor

$(D + 2D' + 1)^2$ is $e^{-x}\{x\,\phi_2(2x-y) + \phi_1(2x-y)\}$.

Hence, solution is $z = e^{4x}\,\phi(x+2y) + e^{-x}\{x\,\phi_2(2x-y) + \phi_1(2x-y)\}$.

Example 27: Solve $(2D^4 - 3D^2D' + D'^2)z = 0$.

Solution: The given equation can be written as

$$(2D^2 - D')(D^2 - D')z = 0.$$

Let $z = \sum Ae^{hx+ky}$ be the C.F. corresponding to $(D^2 - D')z$, than

$$(D^2 - D')z = \sum Ah^2e^{hx+ky} - \sum Ake^{hx+ky} = \sum A(h^2 - k)e^{hx+ky}$$

Hence the given equation (1) will be satisfied by this substitution if $h^2 - k = 0$.

Thus the C.F. corresponding to $(D^2 - D')z$ is $\sum Ae^{hx+h^2y}$.

Similarly C.F. corresponding to $(2D^2 - D')z$ is

$\sum Be^{h'x+k'y}$, where $2h'^2 - k' = 0$

$\Rightarrow \quad CF = \sum Be^{h'x+2h'^2y}$.

Hence the most general solution of the given equation is

$$z = \sum Ae^{hx+h^2y} + \sum Be^{h'x+2h'^2y}$$

Example 28: Solve $(D - 2D' - 1)(D - 2D'^2 - 1)z = 0$.

Solution: C.F. corresponding to the first factor is $z = e^x f_1(y+2x)$ and C.F. corresponding to 2nd factor is $\sum Ae^{hx+ky}$, where

$h - 2k^2 - 1 = 0$ or $h = 2k^2 + 1$.

Hence, $z = e^x f_1(y+2x) + \sum Ae^{ky+(2k^2+1)x}$.

Example 29: Solve $(D^2 - D'^2 - 3D + 3D')z = e^{x-2y}$.

Solution: Equation (1) is equivalent to $(D - D')(D + D' - 3)z = e^{x-2y}$

Therefore $C.F. = \phi_1(y+x) + e^{3x}\,\phi_2(y-x)$

Now $\quad P.I. = \dfrac{1}{(D-D')(D+D'-3)}e^{x-2y} = -\dfrac{1}{12}e^{x-2y}$.

Hence $z = \phi_1(y+x) + e^{3x}\,\phi_2(y-x) - \dfrac{1}{12}e^{x-2y}$.

Example 30: Solve $(D-D'-1)(D-D'-2)z=e^{2x-y}+x.$

Solution: Here C.F. $=e^{x}\,f_1(y+x)+e^{2x}\,f_2(y+2x)$

and P.I. corresponding to e^{2x-y}

$$=\frac{1}{(D-D'-1)(D-D'-2)}.e^{2x-y}$$

$$=\frac{1}{(2+1-1)(2+1-2)}.e^{2x-y}=\frac{1}{2}.e^{2x-y}.$$

P.I. corresponding to $x=\dfrac{1}{(D-D'-1)(D-D'-2)}.x$

$$=\frac{1}{2}(1-D+D')^{-1}\left(1-\frac{1}{2}D+\frac{1}{2}D'\right)^{-1}.x$$

$$=\frac{1}{2}(1+D-D'...)\left(1+\frac{D}{2}-\frac{D'}{2}...\right).x=\frac{1}{2}\left(1+D+\frac{D}{2}+...\right).x$$

$$=\frac{1}{2}x+\frac{1}{2}\frac{3}{2}Dx=\frac{1}{2}x+\frac{3}{4}.$$

Hence, the solution is

$$z=e^{x}\,f_1(y+x)+e^{2x}\,f_2(y+x)+\frac{x}{2}+\frac{3}{4}+\frac{1}{2}e^{2x-y}.$$

Example 31: Solve $(D^3-3\,D\,D'+D'+1)z=e^{2x+3y}$

Solution: Here $D^3-3\,D\,D'+D'+1$ cannot be resolved into factors linear in D and D'

Therefore $C.F.=\sum Ae^{hx+ky}$, where $h^3-3\,h\,k+k+1=0$.

$$P.I.=\frac{1}{D^3-3\,D\,D'+D'+1}.e^{2x+3y}=\frac{e^{2x+3y}}{2^3-3.2.3+3+1}=-\frac{1}{6}e^{2x+3y}.$$

Therefore the solution is $z=-\dfrac{1}{6}e^{2x+3y}+\sum Ae^{hx+ky}$, where

$h^3-3h\,k+k+1=0.$

Example 32: Solve $(D+D')(D+D'-2)z=\sin(x+2y).$

Solution: Here C.F. $=\phi_1(y-x)+e^{2x}\,\phi_2(y-x)$

And $P.I.=\dfrac{1}{(D+D')(D+D'-2)}\sin(x+2y)$

$$= \frac{1}{D^2 + 2\,D\,D' + D'^2 - 2\,D - 2\,D'} \sin(x + 2y)$$

$$= \frac{1}{(-1) + 2(-2) + (-4) - 2(D + D')} \sin(x + 2y)$$

$$= -\frac{1}{9 + 2\,D + 2\,D'} \sin(x + 2y)$$

$$= -(9 - 2\,D - 2\,D')\frac{1}{81 - 4(D^2 + D'^2 + 2\,D\,D')} \sin(x + 2y)$$

$$= (2\,D + 2\,D' - 9)\frac{1}{81 + 4(1 + 4 + 4)} \sin(x + 2y)$$

$$= \frac{1}{117}\{6\cos(x + 2y) - 9\sin(x + 2y)\}.$$

Hence $z = \phi_1(y - x) + e^{2x}\,\phi_2(y - x) + \frac{1}{117}\left[6\cos(x + 2y) - 9\sin(x + 2y)\right].$

Example 33: Solve $(D\,D' + D - D' - 1)z = xy$.

Solution: Equation is equivalent to $(D' + 1)(D - 1)z = xy$.

Therefore $C.F. = e^{-y}\,\phi_1(x) + e^{x}\,\phi_2(y)$

$$P.I. = \frac{1}{(D' + 1)(D - 1)} xy = -\{1 + D'\}^{-1}\{1 - D\}^{-1} xy$$

$$= -\{1 - D' + D'^2 - D'^3 + \ldots\}\{1 + D + D^2 + D^3 + \ldots\} xy$$

$$= -\{1 + D - D' - D\,D' \ldots\} xy = -(xy + y - x - 1)$$

Hence the solution is $z = e^{-y}\,\phi_1(x) + e^{x}\,\phi_2(y) - (xy + y - x - 1)$.

Example 34: Solve $D(D + D' - 1)(D + 3\,D' - 2)z = x^2 - 4\,x\,y + 2y^2$

Solution: Here C.F. $= \phi_1(y) + e^{x}\,\phi_2(y - x) + e^{2x}\,\phi_3(y - 3x)$

$$P.I. = \frac{1}{D(D + D' - 1)(D + 3\,D' - 2)}(x^2 - 4\,x\,y + 2y^2)$$

$$= \frac{1}{2\,D}\{1 - (D + D')\}^{-1}\left\{1 - \frac{D + 3\,D'}{2}\right\}^{-1}(x^2 - 4\,x\,y + 2y^2)$$

$$= \frac{1}{2\,D}\{1 + (D + D') + (D + D')^2 + \ldots\}\left\{1 + \frac{D + 3\,D'}{2} + \left(\frac{D + 3\,D'}{2}\right)^2 + \ldots\right\}$$

$$\left(x^2-4xy+2y^2\right)$$

$$=\frac{1}{2D}\left\{1+(D+D')+(D+D')^2+\frac{D+3D'}{2}+\frac{(D+D')(D+3D')}{2}+\frac{(D+3D')^2}{4}+\ldots\right\}$$

$$(x^2-4xy+2y^2)$$

$$=\frac{1}{2D}\left\{\left(x^2-4xy+2y^2\right)-2x-5x+4y-2-1+\frac{7}{2}\right\}$$

$$=\frac{1}{2D}\left\{x^2-4xy+2y^2-7x+4y+\frac{1}{2}\right\}$$

$$=\frac{1}{2}\left\{\frac{x^3}{3}-2x^2y+2y^2x-\frac{7}{2}x^2+4xy+\frac{1}{2}x\right\}$$

Hence $z=\phi_1(y)+e^{x}\phi_2(y-x)+e^{2x}\phi_3(y-3x)$

$$+\frac{1}{2}\left\{\frac{x^3}{3}-2x^2y+2y^2x-\frac{7}{2}x^2+4xy+\frac{1}{2}x\right\}$$

Example 35: Solve

$$\left(D^2-DD'-2D'^2+2D+2D'\right)z=e^{2x+3y}+\sin(2x+y)+xy.$$

Solution: Given equation is equivalent to

$$\left(D+D'\right)\left(D-2D'+2\right)z=e^{2x+3y}+\sin(2x+y)+xy$$

Here $C.F.=\phi_1(y-x)+e^{-2x}\phi_2(2x+y)$

and $P.I.=\dfrac{1}{(D+D')(D-2D'+2)}\left\{e^{2x+3y}+\sin(2x+y)+xy\right\}$

P.I. corresponding to $e^{2x+3y}=\dfrac{1}{(D+D')(D-2D'+2)}e^{2x+3y}$

$$=\frac{1}{(2+3)(2-6+2)}e^{2x+3y}=-\frac{1}{10}e^{2x+3y}.$$

P.I. corresponding to sin $(2x+y)$

$$=\frac{1}{D^2-DD'-2D'^2+2D+2D'}\sin(2x+y)$$

$$=\frac{1}{(-4)-(-2)-2(-1)+2D+2D'}\sin(2x+y)$$

$$=\frac{1}{2(D+D')}\sin(2x+y)=\frac{1}{2}\frac{(D-D')}{(D^2-D'^2)}\sin(2x+y)$$

$$=\frac{1}{2}(D-D').\frac{1}{(-4)-(-1)}\sin(2x+y)$$

$$= -\frac{1}{6}\{2\cos(2x+y) - \cos(2x+y)\} = -\frac{1}{6}\cos(2x+y).$$

P.I. corresponding to $xy = \dfrac{1}{(D+D')(D-2D'+2)}xy$

$$= \frac{1}{2D}\left\{1+\frac{D'}{D}\right\}^{-1}\left\{1+\frac{D-2D'}{2}\right\}^{-1} xy$$

$$= \frac{1}{2D}\left\{1-\frac{D'}{D}+\frac{D'^2}{D^2}-\frac{D'^3}{D^3}+\ldots\right\}\left\{1-\frac{D-2D'}{2}+\frac{(D-2D')^2}{4}\ldots\right\}xy$$

$$= \frac{1}{2D}\left\{1-\frac{D'}{D}+\frac{D'^2}{D^2}-\frac{D'^3}{D^2}+\ldots\right\}\left\{xy-\frac{y}{2}+x-1\right\}$$

$$= \frac{1}{2D}\left\{\left(xy-\frac{y}{2}+x-1\right)-\frac{x^2}{2}+\frac{x}{2}\right\} = \frac{1}{2D}\left\{xy-\frac{y}{2}+\frac{3}{2}x-\frac{x^2}{2}-1\right\}$$

$$= \frac{1}{2}\left\{\frac{x^2y}{2}-\frac{yx}{2}+\frac{3}{4}x^2-\frac{x^3}{6}-x\right\} = \frac{x}{24}\{6xy-6y+9x-2x^2-12\}.$$

Hence $z = \phi_1(y-x) + e^y\,\phi_2(2x+y) - \frac{1}{10}e^{2x+3y} - \frac{1}{6}\cos(2x+y)$

$$+\frac{x}{24}(6xy-6y+9x-2x^2-12).$$

Note: On proceeding as in last examples, reader may think

$$P.I. = \frac{1}{2D}\left(1+\frac{D'}{D}\right)^{-1}\left(1+\frac{D-2D'}{2}\right)^{-1} xy$$

$$= \frac{1}{2D}\left\{1-\frac{D'}{D}+\frac{D'^2}{D^2}-\ldots\right\}\left\{1-\frac{(D-2D')}{2}+\frac{(D-2D')^2}{4}-\frac{(D\ \ 2D')^3}{8}+\ldots\right\}xy$$

$$= \frac{1}{2D}\left\{1-\frac{D-2D'}{2}-DD'-\frac{D'}{D}+\frac{D'}{2}-\frac{1}{4}DD+\ldots\right\}xy$$

$$= \frac{1}{2D}\left\{1-\frac{D}{2}+\frac{3}{2}D'-\frac{5}{4}DD'-\frac{D'}{D}+\ldots\right\}xy = \frac{1}{2D}\left\{xy-\frac{y}{2}+\frac{3}{2}x-\frac{5}{4}-\frac{x^2}{2}\right\}$$

$$= \frac{1}{2}\left\{\frac{x^2y}{2}-\frac{1}{2}xy+\frac{3}{4}x^2-\frac{5}{4}x-\frac{x^3}{6}\right\} = \frac{1}{24}x\{6xy-6y+9x-2x^2-15\}$$

which is different form what has been obtained above. Moreover, this does not satisfy the equation. But in fact

$$P.I. = \frac{1}{2D}\left\{1-\frac{D'}{D}+\frac{D'^2}{D^2}\ldots\right\}$$

$$\left\{1-\left(\frac{D-2D'}{2}\right)+\left(\frac{D-2D'}{2}\right)^2-\left(\frac{D-2D'}{2}\right)^3+\ldots\right\}xy$$

$$=\frac{1}{2D}\left\{1-\frac{D-2D'}{2}+\frac{D^2}{4}-DD'-\frac{D'}{D}+\frac{D'}{2}-\frac{1}{4}DD'\ldots\right\}xy$$

$$=\frac{1}{2D}\left\{1-\frac{D}{2}+\frac{3D'}{2}+\frac{D^2}{4}+\frac{5}{4}DD'-\frac{D'}{D}+\ldots\right\}xy$$

$$=\frac{1}{2}\left\{\frac{1}{D}-\frac{1}{2}+\frac{3}{2}\frac{D'}{D}+\frac{1}{4}D-\frac{5}{4}D'-\frac{D'}{D^2}+\ldots\right\}xy$$

$$=\frac{1}{2}\left\{\frac{x^2y}{2}-\frac{1}{2}xy+\frac{3}{2}\frac{x^2}{2}+\frac{1}{4}y-\frac{5}{4}x-\frac{x^3}{6}\right\}$$

$$=\frac{1}{24}\left\{6x^2y-6xy+9x^2+3y-15x-2x^3\right\}$$

$$=\frac{1}{24}x\left\{6xy-6y+9x-2x^2-12\right\}+\frac{1}{24}3(y-x)$$

$=\frac{1}{24}x\left(6xy-6y+9x-2x^2-12\right)$, as the last term being contained in C.F. can be neglected.

Example 36: Solve $(D-3D'-2)^2 z=2e^{2x}\tan(y+3x)$.

Solution: Here $C.F.=e^{2x}\left\{\phi_1(3x+y)+x\,\phi_2(3x+y)\right\}$

and

$$P.I=2\frac{1}{(D-3D'-2)^2}e^{2x}\tan(y+3x)=2e^{2x}\frac{1}{(D-3D')^2}\tan(y+3x)$$

$$=2e^{2x}\frac{1}{D-3D}\int_{a=y+3x}\tan a\,dx=2e^{2x}\frac{1}{D-3D'}\left\{x\tan(y+3x)\right\}$$

$$=2e^{2x}\int_{a=y+3x}x\tan a\,dx=x^2e^{2x}\tan(y+3x).$$

Hence the solution is

$$z=e^{2x}\left\{\phi_1(y+3x)+x\,\phi_2(y+3x)\right\}+x^2e^{2x}\tan(y+3x).$$

Example 37: Solve $(3D^2-2D'^2+D-1)z=4e^{x+y}\cos(x+y)$.

Solution: Here $C.F.=\sum_{i=1}^{\infty}c_1 e^{a_i x+b_i y}$, where $3a_i^2-2b_i^2+a_i-1=0$

$$P.I. = 4\frac{1}{3D^2 - 2D'^2 + D - 1} e^{x+y} \cos(x+y)$$

$$= 4e^{x+y} \frac{1}{3(D+1)^2 - 2(D'+1)^2 + D + 1 - 1} \cos(x+y)$$

$$= 4e^{x+y} \frac{1}{3D^2 + 7D - 2D'^2 - 4D' + 1} \cos(x+y)$$

$$= 4e^{x+y} \frac{1}{(-3) + 7D - 2(-1) - 4D' + 1} \cos(x+y)$$

$$= 4e^{x+y} \frac{1}{7D - 4D'} \cos(x+y)$$

$$= 4e^{x+y} (7D + 4D') \frac{1}{49D^2 - 16D'^2} \cos(x+y)$$

$$= 4e^{x+y} (7D + 4D') \frac{1}{-49 + 16} \cos(x+y)$$

$$= \frac{4e^{x+y}}{-33}\left[-7\sin(x+y) - 4\sin(x+y)\right] = \frac{4}{3} e^{x+y} \sin(x+y).$$

Hence the solution is $z = C.F. + P.I.$

$$= \sum_{i=1}^{\infty} c_1 e^{a_i x + b_i y} + \frac{4}{3} e^{x+y} \sin(x+y), \text{ where } 3a_i^2 - 2b_i^2 + a_i - 1 = 0$$

Example 38: Solve $x^2 \frac{\partial^2 z}{\partial x^2} + 2xy \frac{\partial^2 z}{\partial x \partial y} + y^2 \frac{\partial^2 z}{\partial y^2} = 0.$

Solution: Substituting $x = e^X, y = e^Y$

So that $X = \log x, Y = \log y$

And denoting $\frac{\partial}{\partial X}$ and $\frac{\partial}{\partial Y}$ by D and D' respectively the given equation reduces to

$$[D(D-1) + 2DD' + D'(D'-1)]z = 0$$

$$\Rightarrow (D + D')(D + D' - 1)z = 0.$$

Hence the solution is

$$z = f_1(Y - X) + e^X f_2(Y - X) = f_1(\log y - \log x) + x f_2(\log y - \log x)$$

$$= f_1\left(\log\frac{y}{x}\right) + x f_2\left(\log\frac{y}{x}\right) = \phi_1\left(\frac{y}{x}\right) + x\phi_2\left(\frac{y}{x}\right).$$

Example 39: Solve $x^2 \dfrac{\partial^2 z}{\partial x^2} - y^2 \dfrac{\partial^2 z}{\partial y^2} - y\dfrac{\partial z}{\partial y} + x\dfrac{\partial z}{\partial x} = 0$

Solution: Substituting $x = e^X$, $y = e^Y$ so that $X = \log x$, $Y = \log y$,

and denoting $\dfrac{\partial}{\partial X}$ and $\dfrac{\partial}{\partial Y}$ by D and D' respectively the given equation reduces to

$$\left[D(D-1) - D'(D'-1) - D' + D\right]z = 0 \Rightarrow \left(D^2 - D'^2\right)z = 0.$$

Therefore $z = f_1(Y+X) + f_2(Y-X)$

$= f_1(\log y + \log x) + f_2(\log y - \log x)$

$= f_1(\log xy) + f_2\left\{\log \dfrac{y}{x}\right\}$

$= \phi_1(xy) + \phi_2(y/x).$

Example 40: Solve

$$x^2 \frac{\partial^2 z}{\partial x^2} - 4xy\frac{\partial^2 z}{\partial x \partial y} + 4y^2 \frac{\partial^2 z}{\partial y^2} + 6y\frac{\partial z}{\partial y} = x^3 y^4$$

Solution: Substituting $x = e^X$, $y = e^Y$, and denoting

$\dfrac{\partial}{\partial X}$ and $\dfrac{\partial}{\partial Y}$ by D and D' the equation reduces to

$$\left[D(D-1) - 4DD' + 4D'(D'-1) + 6D'\right]z = e^{3X+4Y}$$

$$\Rightarrow (D - 2D')(D - 2D' - 1)z = e^{3X+4Y}$$

Therefore $C.F. = f_1(Y + 2X) + e^X f_2(Y + 2X)$

$= f_1(\log y + 2\log x) + x\, f_2(\log y + 2\log x)$

$= f_1(\log y x^2) + x\, f_2(\log y x^2) = \phi_1(y x^2) + x\,\phi_2(y x^2).$

$$P.I. = \frac{1}{(D - 2D')(D - 2D' - 1)} e^{3X+4Y} = \frac{e^{3X+4Y}}{(3-8)(3-8-1)} = \frac{1}{30} x^3 y^4.$$

Hence the equation is $z = \phi_1(y x^2) + x\,\phi_2(y x^2) + \dfrac{1}{30} x^3 y^4$.

Example 41: Solve

$$x^2 \frac{\partial^2 z}{\partial x^2} + 2xy\frac{\partial^2 z}{\partial x \partial y} + y^2 \frac{\partial^2 z}{\partial y^2} - nx\frac{\partial z}{\partial x} - ny\frac{\partial z}{\partial y} + nz = x^2 + y^2$$

Solution: Substituting $x = e^{X}$, $y = e^{Y}$ If D and D' denote $\frac{\partial}{\partial X}$ and $\frac{\partial}{\partial Y}$, the given equation reduces to

$$\left[D(D-1) + 2DD' + D'(D'-1) - nD - nD' + n \right] z = e^{2X} + e^{2Y}$$

$$\Rightarrow \quad (D + D' - 1)(D + D' - n) z = e^{2X} + e^{2Y}$$

Therefore $C.F. = e^{X} f_1(Y - X) + e^{nX} f_2(Y - X)$

$$= x f_1(\log y - \log x) + x^n f_2(\log y - \log x)$$

$$= x f_1\left(\log \frac{y}{x}\right) + x^n f_2\left(\log \frac{y}{x}\right) = x\,\phi_1\left(\frac{y}{x}\right) + x^n \phi_2\left(\frac{y}{x}\right)$$

$$P.I. = \frac{1}{(D + D' - 1)(D + D' - n)} e^{2X} + \frac{1}{(D + D' - 1)(D + D' - n)} e^{2Y}$$

$$= \frac{e^{2X}}{(2 + 0 - 1)(2 + 0 - n)} + \frac{e^{2Y}}{(0 + 2 - 1)(0 + 2 - n)}$$

$$= \frac{e^{2X} + e^{2Y}}{2 - n} = \frac{x^2 + y^2}{2 - n}$$

Therefore the solution is $z = x\,\phi_1\left(\frac{y}{x}\right) + x^n \phi_2\left(\frac{y}{x}\right) + \frac{x^2 + y^2}{2 - n}$.

EXERCISE 8.2

Solve the following PDEs:

1. $(4D^2 + 12DD' + 9D'^2) z = 0$

2. $2\frac{\partial^2 z}{\partial x^2} - 3\frac{\partial^2 z}{\partial x\,\partial y} - 2\frac{\partial^2 z}{\partial y^2} = 0.$

3. $\frac{\partial^2 z}{\partial x^3} - 4\frac{\partial^3 z}{\partial x^2\,\partial y} + 4\frac{\partial^3 z}{\partial x\,\partial y^2} = 0.$

4. $r - 4s + 4t = 0$.

5. $\frac{\partial^3 z}{\partial x^3} - 7\frac{\partial^3 z}{\partial x\,\partial y^2} + 6\frac{\partial^3 z}{\partial y^3} = 0.$

6. $(D^4 - 2D^3 D' + 2DD'^3 - D'^4) z = 0.$

7. $\frac{\partial^2 z}{\partial x^2} - \frac{\partial^2 z}{\partial y^2} = x - y$

8. $\dfrac{\partial^2 z}{\partial x^2}+3\dfrac{\partial^2 z}{\partial x\,\partial y}+2\dfrac{\partial^2 z}{\partial y^2}=(2x+3y)$

9. $\dfrac{\partial^2 z}{\partial x^2}+3\dfrac{\partial^2 z}{\partial x\,\partial y}+2\dfrac{\partial^2 z}{\partial y^2}=6(x+y)$.

10. $(D^2+3DD'+2D'^2)z=12xy$.

11. $(D^2-a^2D'^2)z=x^2$.

12. $4r-4s+t=16\log(x+2y)$.

13. $(D^2-5DD'+4D'^2)z=\sin(4x+y)$

14. $\dfrac{\partial^3 z}{\partial x^3}-4\dfrac{\partial^3 z}{\partial x^2\,\partial y}+4\dfrac{\partial^3 z}{\partial x\,\partial y^2}=\sin(2x+y)$.

15. $(D^3-4D^2D'+4DD'^2)z=\cos(2x+y)$.

16. $\dfrac{\partial^2 z}{\partial x^2}+\dfrac{\partial^2 z}{\partial x\,\partial y}-6\dfrac{\partial^2 z}{\partial y^2}=y\sin x$.

17. $(D^2+2DD'+D'^2)z=2\cos y-x\sin y$.

18. $(D+D'-1)(D+2D'-2)z=0$.

19. $t+s+q=0$.

20. $(D^2+DD'+D'-1)z=\sin(x+2y)$.

21. $\dfrac{\partial^2 z}{\partial x^2}+4\dfrac{\partial^2 z}{\partial x\,\partial y}+4\dfrac{\partial^2 z}{\partial y^2}-\dfrac{\partial z}{\partial x}-2\dfrac{\partial z}{\partial y}=e^{x+y}$.

22. $(D^2-DD'-2D)z=\sin(3x+4y)+x^2y$.

23. $(D^2-D')z=2y-x^2$.

24. $x^2\dfrac{\partial^2 z}{\partial x^2}-3xy\dfrac{\partial^2 z}{\partial x\,\partial y}+2y^2\dfrac{\partial^2 z}{\partial y^2}+5y\dfrac{\partial z}{\partial y}-2z=0$.

25. $(x^2D^2+2xyDD'+y^2D'^2)z=x^my^n$.

ANSWERS 8.2

1. $z=f_1(2y-3x)+x\,f_2(2y-3x)$.
2. $z=f_1(2y-x)+f_2(y+2x)$.
3. $z=f_1(y)+f_2(y+2x)+x\,f_3(y+2x)$.
4. $z=f_1(y+2x)+x\,f_2(y+2x)$.

5. $z = f_1(y+x) + f_2(y+2x) + f_3(y-3x)$.

6. $z = f_1(y-x) + f_2(y+x) + x\,f_3(y+x) + x^2\,f_4(y+x)$.

7. $z = f_1(y+x) + f_2(y-x) + \frac{1}{6}x^3 - \frac{1}{2}x^2\,y$.

8. $z = f_1(y-x) + f_2(y-2x) - \frac{7}{6}x^3 + \frac{3}{2}x^2\,y$.

9. $z = f_1(y-x) + f_2(y-2x) - 2x^3 + 3x^2\,y$.

10. $z = f_1(y-x) + f_2(y-2x) + 2\,x^3\,y - \frac{3}{2}x^4$.

11. $z = f_1(y+ax) + f_2(y-ax) + \frac{1}{12}x^4$

12. $z = f_1(2y+x) + x\,f_2(2y+x) + 2x^2\log(x+2y)$.

13. $z = f_1(y+x) + f_2(y+4x) - \frac{1}{3}x\cos(4x+y)$.

14. $z = f_1(y) + f_2(y+2x) + x\,f_3(y+2x) - \frac{x^2}{4}\cos(2x+y)$.

15. $z = f_1(y) + f_2(y+2x) + x\,f_3(y+2x) + \frac{1}{4}x^2\sin(2x+y)$.

16. $z = f_1(y-3x) + f_2(y+2x) - (y\sin x + \cos x)$.

17. $z = f_1(y-x) + x\,f_2(y-x) + x\sin y$.

18. $z = e^x\,f_1(y-x) + e^{2x}\,f_2(y-2x)$.

19. $z = f_1(x) + e^{-x}\,f_2(y-x)$.

20. $z = e^{-x}\,f_1(y) + e^x\,f_2(y-x) - \frac{1}{10}\left[\cos(x+2y) + 2\sin(x+2y)\right]$.

21. $z = f_1(y-2x) + e^x\,f_2(y-2x) + \frac{1}{6}e^{x+y}$

22. $z = f_1(y) + e^{2x}\,f_2(y+x) + \frac{1}{15}\sin(3x+4y) + \frac{2}{15}\cos(3x+4y)$

$$-\frac{1}{6}x^3\,y - \frac{1}{4}x^2\,y + \frac{1}{12}x^3 - \frac{1}{4}xy + \frac{1}{4}x^2 + \frac{3}{8}x.$$

23. $z = \sum A e^{hx + h^2 y} - \left(y^2 + \frac{x^4}{12}\right)$.

24. $z = x^2\,f_1(xy) + x^{-1}\,f_2(x^2\,y)$.

25. $z = f_1\left(\frac{y}{x}\right) + x\, f_2\left(\frac{y}{x}\right) + \frac{x^m\, y^n}{(m+n)(m+n-1)}$.

8.8 MONGE'S METHODS

8.8.1 Monge's Method I

For integrating the differential equation of the form $Rr + Ss + Tt = V$,

where $r = \frac{\partial^2 z}{\partial x^2}, s = \frac{\partial^2 z}{\partial x\, \partial y}, t = \frac{\partial^2 z}{\partial y^2}$; and R, S, T, V are functions of x, y, z, p and q.

We have the equation as

$$Rr + Ss + Tt = V \qquad ...(1)$$

We know that $dp = \frac{\partial p}{\partial x} dx + \frac{\partial p}{\partial y} dy = r\, dx + s\, dy$

and $dq = \frac{\partial q}{\partial x} dx + \frac{\partial q}{\partial y} dy = s\, dx + t\, dy$

Therefore $r = \frac{dp - s\, dy}{dx}$ and $t = \frac{dq - s\, dx}{dy}$. Substituting these values in (1), we have

$$R\left(\frac{dp - s\, dy}{dx}\right) + Ss + T\left(\frac{dq - s\, dx}{dy}\right) = V$$

$$\Rightarrow \quad R\, dy\,(dp - s\, dy) + Ss\, dx\, dy + T\,(dq - s\, dx)\,dx = V\, dx\, dy$$

$$\Rightarrow \quad (R\, dp\, dy + T\, dq\, dx - V\, dx\, dy) - s\left(R\, dy^2 - S\, dx\, dy + T\, dx^2\right) = 0 \qquad ...(2)$$

Now any relation between x, y, z, p, q which satisfies (2) makes each of the bracketed expressions vanish. Hence we must have

$$R\, dy^2 - Sdx\, dy + T\, dx^2 = 0 \qquad ...(3)$$

$$R\, dp\, dy + T\, dq\, dx - V\, dx\, dy = 0 \qquad ...(4)$$

The equations (3) and (4) are called Monge's subsidiary equations.

Let (3) can be reduced into two linear equations in dx and dy as

$$dy - m_1\, dx = 0 \qquad ...(5)$$

and $\quad dy - m_2\, dx = 0 \qquad ...(6)$

Now from (4) and (5), combined if necessary with $dz = p\, dx + q\, dy$, we obtain two integrals

$$u_1 = a \text{ and } v_1 = b \text{ (say)}$$

Then the relation $u_1 = f_1(v_1) \qquad ...(7)$

is the solution and is called an intermediate integral. Similarly, we can find $u_2 = f_2(v_2)$ ...(8)

From (7) and (8), find the values of p and q in terms of x and y. Putting these values in $dz = p\,dx + q\,dy$ and on integrating, we get the complete solution of equation (1).

Example 1: Solve $r = a^2 t$.

Solution: We have $dp = r\,dx + s\,dy$ and $dq = s\,dx + t\,dy$

which gives $r = \dfrac{dp - s\,dy}{dx}$ and $t = \dfrac{dq - s\,dx}{dy}$.

Putting these values of r in $r = a^2 t$, we get

$$\frac{dp - s\,dy}{dx} = a^2\,\frac{dq - s\,dx}{dy}$$

$$\Rightarrow dp\,dy - a^2\,dx\,dq - s\left(dy^2 - a^2\,dx^2\right) = 0.$$

Thus the Monge's equations are $dp\,dy - a^2\,dx\,dq = 0$ ...(1)

$dy^2 - a^2\,dx^2 = 0$ ...(2)

Equation (2) can be resolved into factors $dy - a\,dx = 0$ ...(3)

and $dy + a\,dx = 0$ ...(4)

Combining equation (3) with (1), we get $dp\,(a\,dx) - a^2\,dx\,dq = 0$

$\Rightarrow \quad dp - a\,dq = 0$ ...(5)

Equations (3) and (5) on integrating give respectively

$$\left.\begin{aligned} y - ax &= A \\ p - aq &= B \end{aligned}\right\} \Rightarrow p - aq = f_1(y - ax) \qquad \text{...(6)}$$

Similarly, combining (4) and (1), we get $p + aq = f_2(y + ax)$ ...(7)

On adding and subtracting (6) and (7), we get

$$p = \frac{1}{2}\left[\, f_1(y - ax) + f_2(y + ax)\,\right]$$

and $q = \dfrac{1}{2a}\left[\, f_2(y + ax) - f_1(y - ax)\,\right]$.

Substituting these values in $dz = p\,dx + q\,dy$, we obtain

$$dz = \frac{1}{2}\left[\, f_1(y - ax) + f_2(y + ax)\,\right]dx + \frac{1}{2a}\left[\, f_2(y + ax) - f_1(y - ax)\,\right]dy$$

$$= \frac{1}{2a}(dy + a\,dx)\,f_2(y + ax) - \frac{1}{2a}(dy - a\,dx)\,f_1(y - ax).$$

Integrating, we get $z = \frac{1}{2a}\phi_1(y+ax) - \frac{1}{2a}\phi_2(y-ax)$

$\Rightarrow \quad z = F_1(y+ax) + F_2(y-ax)$.

Example 2: Solve $r + (a+b)s + abt = xy$.

Solution: We have $r + (a+b)s + abt = xy$ ...(1)

Putting $r = \frac{dp - s\,dy}{dx}$ and $t = \frac{dq - s\,dx}{dy}$ in equation (1), we get

$dp\,dy + ab\,dq\,dx - xy\,dx\,dy = s\left[dy^2 - (a+b)\,dx\,dy + ab\,dx^2\right]$

and subsidiary equations are $dy^2 - (a+b)\,dx\,dy + ab\,dx^2 = 0$ ...(2)

$dp\,dy + ab\,dq\,dx - xy\,dx\,dy = 0$ ...(3)

From (2), we have

$dy - a\,dx = 0$ ...(4)

$dy - b\,dx = 0$ ...(5)

On integrating (4) and (5), we get $y - ax = c_1$ and $y - bx = c_2$

$\Rightarrow \quad y = c_1 + ax$ and $y = c_2 + bx$

Combining these with (3), we get $a\,dp + ab\,dq - ax(c_1 + ax)\,dx = 0$

And $b\,dp + ab\,dq - bx(c_2 + bx)\,dx = 0$

$\Rightarrow \quad p + bq - c_1\frac{x^2}{2} - \frac{ax^3}{3} = A$ and $p + aq - c_2\frac{x^2}{2} - \frac{bx^3}{3} = B$

$\Rightarrow \quad p + bq - (y-ax)\frac{x^2}{2} - \frac{ax^3}{3} = \phi_1(c_1) = \phi_1(y-ax)$

and $p + aq - (y-bx)\frac{x^2}{2} - \frac{bx^3}{3} = \phi_2(c_2) = \phi_2(y-bx)$.

Solving these, we get

$$p = \frac{1}{(a-b)}\left[\frac{yx^2}{2}(a-b) - (a^2-b^2)\frac{x^3}{6} + a\,\phi_1(y-ax) - b\,\phi_2(y-bx)\right]$$

and $$q = \frac{1}{(b-a)}\left[-\frac{x^3}{6}(a-b) + \phi_1(y-ax) - \phi_2(y-bx)\right].$$

Putting these values in $dz = p\,dx + q\,dy$, we obtain

$$dz = \left[\frac{yx^2}{2} - (a+b)\frac{x^3}{6} + \frac{a\,\phi_1(y-ax)}{(a-b)} - \frac{b\,\phi_2(y-bx)}{(a-b)}\right]dx$$

$$+ \left[\frac{x^3}{6} - \frac{\phi_1(y-ax)}{(a-b)} + \frac{\phi_2(y-bx)}{(a-b)}\right]dy$$

$$= -\frac{(a+b)x^3}{6}dx + \frac{3x^2 y\,dx + x^3\,dy}{6} - \frac{1}{(a-b)}[\phi_1(y-ax)(dy - a\,dx)]$$

$$+\frac{1}{(a-b)}[\phi_2(y-bx)(dy-b\,dx)]$$

Therefore $$z = -\frac{(a+b)x^4}{24} + \frac{yx^3}{6} + \psi_1(y-ax) + \psi_2(y-bx).$$

Example 3: Obtain the integral of $q^2 r - 2pqs + p^2 t = 0$ in the form $y + x\,f(z) = F(z)$ and show that this represents a surface generated by straight lines that are parallel to fixed plane.

Solution: Putting $r = \frac{dp - s\,dy}{dx}$ and $t = \frac{dq - s\,dx}{dy}$ in the given equation, we have

$$q^2\frac{dp - s\,dy}{dx} - 2p\,qs + p^2\frac{dq - s\,dx}{dy} = 0$$

$$\Rightarrow \quad (q^2\,dp\,dy + p^2\,dq\,dx) - s(q^2\,dy^2 + 2\,pq\,dx\,dy + p^2\,dx^2) = 0.$$

Therefore the Monge's subsidiary equations are

$$q^2\,dy^2 + 2\,pq\,dx\,dy + p^2\,dx^2 = 0 \quad \text{...(1)}$$

and $$q^2\,dp\,dy + p^2\,dq\,dx = 0 \quad \text{...(2)}$$

Equation (1) gives $(q\,dy + p\,dx)^2 = 0$

$$\Rightarrow \quad p\,dx + q\,dy = 0 \quad \text{...(3)}$$

Hence $dz = p\,dx + q\,dy = 0 \quad \Rightarrow z = A.$ ...(4)

From (2) and (3), we have

$$q\,dp - p\,dq = 0 \text{ or } \frac{dp}{p} = \frac{dq}{q} \text{ or } \log p = \log q + \log B$$

Therefore $p = q\,B$ ...(5)

Hence the intermediate integral is $p - q\,f(z) = 0$.

Therefore Lagrange's subsidiary equations are $\frac{dx}{1} = \frac{dy}{-f(z)} = \frac{dz}{0}$.

The last member gives $dz = 0 \Rightarrow z = c$.

Again from the first two members, we have

$$dy + f(z)\,dx = 0 \text{ or } dy + f(c)\,dx = 0.$$

Integrating, we get $y + x\,f(c) = c$ or $y + x\,f(z) = F(z)$ ...(6)

which is the required integral.

The integral of the given differential equation is the surface (6) which is the locus of the straight lines given by the intersection of planes $y + x f(c) = F(c)$ and $z = c$. These lines are all parallel to the plane $z = 0$ as they lie on the plane $z = c$ for varying values of c.

Example 4: Solve $pt - qs = q^3$.

Solution: Putting $t = \dfrac{dq - s\,dx}{dy}$, the given equation becomes

$$\frac{p(dq - s\,dx)}{dy} - ds = q^3 \Rightarrow (p\,dq - q^3\,dy) - s(p\,dx + q\,dy) = 0.$$

The Monge's subsidiary equations are

$$p\,dq - q^3\,dy = 0 \qquad \text{...(1)}$$

and $\quad p\,dx + q\,dy = 0 \qquad$ ...(2)

From (2), we have $dz = 0$, $\Rightarrow z = A \qquad$ ...(3)

Combining (1) and (2), we have $dq + q^2\,dx = 0$

$$\Rightarrow \quad \frac{dq}{q^2} + dx = 0 \Rightarrow -\frac{1}{q} + x = B$$

$$\Rightarrow \quad -\frac{1}{q} + x = f(z) \Rightarrow \frac{\partial y}{\partial z} - x = -f(z).$$

Integrating, we get

$$y = xz - \int f(z)\,dz + c \Rightarrow y = zx + f_1(z) + c$$

$$\Rightarrow \quad y = xz + f_1(z) + f_2(x)$$

Because c is a function of x as it regarded constant at the time of integration.

Example 5: Solve $2x^2 r - 5xys + 2y^2 t + 2(px + qy) = 0$, and hence find the surface satisfying the above equation and touching the hyperbolic paraboloid $z = x^2 - y^2$ along its section by the plane $y = 1$.

Solution: Putting $r = \dfrac{dp - s\,dy}{dx}$ *and* $t = \dfrac{dq - s\,dx}{dy}$, the given equation reduces to $2x^2 \dfrac{dp - s\,dy}{dx} - 5xys + 2y^2 \dfrac{dq - s\,dx}{dy} + 2(px + qy) = 0$

$$\Rightarrow \quad \{2x^2\,dp\,dy + 2y^2\,dq\,dx + 2(px + qy)\,dx\,dy\}$$

$$- s(2x^2\,dy^2 + 5xy\,dx\,dy + 2y^2\,dx^2) = 0.$$

Hence the Monge's subsidiary equations are

$$2\,x^2\,dp\,dy + 2\,y^2\,dq\,dx + 2\,(px + qy)\,dx\,dy = 0 \qquad ...(1)$$

and $2\,x^2\,dy^2 + 5\,xy\,dx\,dy + 2\,y^2\,dx^2 = 0$ ...(2)

From (2), we have $(2\,x\,dy + y\,dx)(x\,dy + 2\,y\,dx) = 0$. This gives

$$2\,x\,dy + y\,dx = 0 \qquad ...(3)$$

and $x\,dy + 2\,y\,dx = 0$ ...(4)

From (4), we have $\dfrac{dy}{y} + 2\dfrac{dx}{x} = 0$.

Integrating $\log y + 2\log x = \log A \Rightarrow y\,x^2 = A$.

Also from (4) and (1), we have $2\,x\,dp - y\,dq + 2\,p\,dx - q\,dy = 0$

$\Rightarrow \quad (2\,x\,dp + 2\,p\,dx) - (y\,dq + q\,dy) = 0$

$\Rightarrow \quad 2\,px - qy = B$

Hence $2\,p\,x - qy = f_1\left(y\,x^2\right)$ ...(5)

is one intermediate integral.

Similarly from (3) and (1) another intermediate integral is

$$p\,x - 2\,q\,y = f_2\left(x\,y^2\right) \qquad ...(6)$$

Solving (5) and (6), we have

$$p = \frac{1}{3\,x}\left[2\,f_1\left(x^2\,y\right) - f_2\left(x^2\right)\right] \text{and } q = \frac{1}{3\,y}\left[\,f_1\left(x^2\,y\right) - 2\,f_2\left(x\,y^2\right)\right].$$

Substituting these values in $dz = p\,dx + q\,dy$, we have

$$d\,z = \frac{1}{3\,x}\left[2\,f_1\left(x^2\,y\right) - f_2\left(x\,y^2\right)\right]dx + \frac{1}{3\,y}\left[\,f_1\left(x^2\,y\right) - 2\,f_2\left(x\,y^2\right)\right]dy$$

$$= \frac{1}{3}\,f_1\left(x^2\,y\right)\left(\frac{2\,dx}{x} + \frac{dy}{y}\right) - \frac{1}{3}\,f_2\left(x y^2\right)\left(\frac{dx}{x} + \frac{2\,dy}{y}\right)$$

$$= \frac{1}{3}\,f_1\left(x^2\,y\right)d\left[\log\left(x^2\,y\right)\right] - \frac{1}{3}\,f_2\left(x\,y^2\right)d\left[\log\left(x\,y^2\right)\right],$$

Hence

$$z = \frac{1}{3}\int f_1\left(x^2\,y\right)d\left(\log x^2\,y\right) - \frac{1}{3}\int f_2\left(x\,y^2\right)d\left(\log x\,y^2\right)$$

$\Rightarrow \quad z = \phi_1\left(x^2\,y\right) + \phi_2\left(x\,y^2\right) + c$ ...(7)

Consider the surface $z = x^2 - y^2$ ...(8)

II Part: For the surface satisfying (7) and touching the hyperbolic paraboloid $z = x^2 - y^2$, along its section by the plane $y = 1$, the values of p and q for surfaces (7) and (8) must be equal for any point on $y = 1$.

Now for $y=1$, we have $p=2xy\phi'_1(x^2y)+y^2\phi'_2(xy^2)=2x$ ...(9)

And $q=x^2\phi'_1(x^2y)+2xy\phi'_2(xy^2)=-2y$ ...(10)

Solving (9) and (10), we get $\phi'_1(x^2y)=\frac{4}{3y}+\frac{2y}{3x^2}$ ...(11)

And $\phi'_2(xy^2)=-\frac{2x}{3y^2}-\frac{4}{3x}$ ...(12)

Putting $y=1$ in (11), we have

$$\phi'_1(x^2)=\frac{4}{3}+\frac{2}{3x^2}\ or\ 2x\phi'_1(x^2)=\frac{8}{3}x+\frac{2}{3}\frac{2x}{x^2}.$$

Integrating, we get

$$\phi_1(x^2)=\frac{4}{3}x^2+\frac{2}{3}\log x^2.$$

Substituting x^2y for x^2 $(\because y=1)$, we have

$$\phi_1(x^2y)=\frac{4}{3}x^2y+\frac{2}{3}\log(x^2y) \quad ...(13)$$

Also putting $y=1$ in (12), we have $\phi'_2(x)=-\frac{2}{3}x-\frac{4}{3x}$.

Integrating, we obtain

$$\phi_2(x)=-\frac{1}{3}x^2-\frac{4}{3}\log x.$$

Substituting xy^2 for x $(\because y=1)$, we have

$$\phi_2(xy^2)=-\frac{1}{3}(xy^2)^2-\frac{4}{3}\log(xy^2) \quad ...(14)$$

Using (13) and (14) in (7), we have

$$z=\frac{4}{3}x^2y+\frac{2}{3}\log(x^2y)-\frac{1}{3}x^2y^4-\frac{4}{3}\log(xy^2)+c$$

$$\Rightarrow \quad z=\frac{4}{3}x^2y-\frac{1}{3}x^2y^4+\frac{2}{3}\log\{(x^2y)/(xy^2)^2\}+c$$

$$\Rightarrow \quad z=\frac{4}{3}x^2y-\frac{1}{3}x^2y^4-2\log y+c \quad ...(15)$$

when $y=1$, the values of z from (8) and (15) should be the same. Hence

$$\frac{4}{3}x^2-\frac{1}{3}x^2-2\log 1+c=x^2-1 \Rightarrow c=-1.$$

Hence the required surface is given by

$$z=\frac{4}{3}x^2y-\frac{1}{3}x^2y^4-2\log y-1.$$

Example 6: Solve $t - r\sec^2 y = 2q\tan y$.

Solution: Putting $r = \dfrac{dp - s\,dy}{dx}$ *and* $t = \dfrac{dp - s\,dx}{dy}$, the given equation gives

$$\frac{dq - s\,dx}{dy} - \frac{dp - s\,dy}{dx}\sec^4 y = 2q\tan y$$

$$\Rightarrow \left(dq\,dx - dp\,dy\sec^4 y - 2q\tan y\,dx\,dy\right) - s\left(dx^2 - dy^2\sec^4 y\right) = 0.$$

The Monge's subsidiary equations are

$$dq\,dx - dp\,dy\sec^4 y - 2q\tan y\,dx\,dy = 0 \quad ...(1)$$

and $dx^2 - dy^2\sec^4 y = 0$ ...(2)

From (2), we have $dx - dy\sec^2 y = 0$ ...(3)

And $dx + dy\sec^2 y = 0$ ...(4)

From (1) and (3), we have $dq - dp\sec^2 y - 2q\tan y\,dy = 0$

$$\Rightarrow \quad dp - \left(\cos^2 y\,dq - 2q\sin y\cos y\,dy\right) = 0.$$

Integrating, we get

$$p - q\cos^2 y = A.$$

Also integrating (3), we have $x - \tan y = B$.

Hence $p - q\cos^2 y = f_1(x - \tan y)$ ...(5)

This is an intermediate integral. Similarly, from (1) and (4), the other intermediate integral is

$$p + q\cos^2 y = f_2(x + \tan y) \quad ...(6)$$

Solving (5) and (6), we have $p = \frac{1}{2}\left[f_1(x - \tan y) + f_2(x + \tan y)\right]$

and $q = \frac{1}{2}\sec^2 y\left[f_2(x + \tan y) - f_1(x - \tan y)\right]$.

Substituting these values in $dz = p\,dx + q\,dy$, we have

$$dz = \frac{1}{2}\left[f_1(x - \tan y) + f_2(x + \tan y)\right]dx$$

$$+ \frac{1}{2}\sec^2 y\left[f_2(x + \tan y) - f_1(x - \tan y)\right]dy$$

$$= \frac{1}{2}f_1(x - \tan y)\left(dx - \sec^2 y\,dy\right) + \frac{1}{2}f_2(x + \tan y)\left(dx + \sec^2 y\,dy\right)$$

Therefore $z = \phi_1(x - \tan y) + \phi_2(x + \tan y)$.

Example 7: Solve $r+(a+b)s+a\,bt=xy$.

Solution: Putting $r=\dfrac{dp-s\,dy}{dx}$ *and* $t=\dfrac{dq-s\,dx}{dy}$ in the given equation,

we have $\dfrac{dp-s\,dy}{dx}+(a+b)s+ab\dfrac{dq-s\,dx}{dy}=xy$

$\Rightarrow (dp\,dy+a\,b\,dq\,dx-xy\,dx\,dy)-s\{dy^2-(a+b)\,dx\,dy+a\,b\,dx^2\}=0$.

Hence Monge's subsidiary equations are

$$dp\,dy+a\,b\,dq\,dx-x\,y\,dx\,dy=0 \quad ...(1)$$

and $dy^2-(a+b)\,dx\,dy+a\,b\,dx^2=0$ ...(2)

From (2), we have $dy-b\,dx=0$ ...(3)

and $dy-a\,dx=0$ ...(4)

From (3), we have $y-b\,x=A$.

From (1) and (3), we have

$$dp+a\,dq-x\,y\,dx=0 \text{ or } dp+a\,dq-x\,(b\,x+A)\,dx=0.$$

Therefore $p+a\,q-b\dfrac{x^3}{3}-\dfrac{A\,x^2}{2}=B$.

Hence one intermediate integral is

$$p+a\,q-\frac{b}{3}x^3-\frac{1}{2}(y-b\,x)\,x^2=f_1\,(y-b\,x)$$

$$\Rightarrow \quad p+a\,q+\frac{1}{6}b\,x^3-\frac{1}{2}x^2\,y=f_1\,(y-b\,x) \quad ...(5)$$

Similarly from (1) and (4) other intermediate integral is

$$p+b\,q+\frac{1}{6}a\,x^3-\frac{1}{2}x^2\,y=f_2\,(y-a\,x) \quad ...(6)$$

Solving (5) and (6), we have

$$p=\frac{1}{2}x^2\,y-\frac{1}{6}(a+b)\,x^3+\frac{1}{(a-b)}\left[a\,f_2\,(y-a\,x)-b\,f_1\,(y-b\,x)\right]$$

and $q=\dfrac{1}{6}x^3-\dfrac{1}{(a-b)}\left[\,f_2\,(y-a\,x)-f_1\,(y-b\,x)\right]$.

Substituting these values in $dz=p\,dx+q\,dy$, we have

$$dz=\frac{1}{2}x^2\,y\,dx+\frac{1}{6}x^3\,dy-\frac{1}{6}(a+b)\,x^3\,dx$$

$$+\frac{1}{(a+b)}\left[a\,f_2(y-ax)-bf_1(y-bx)\right]dx-\frac{1}{(a-b)}\left[\,f_2\,(y-a\,x)-f_1\,(y-b\,x)\,\right]dy$$

$$= \frac{1}{6}\left(3x^2 y\,dx + x^3\,dy\right) - \frac{1}{6}(a+b)\,x^3\,dx$$

$$+ \frac{1}{(b-a)}\left[\, f_2(y - a\,x)(dy - a\,dx) + f_1(y - b\,x)(dy - b\,dx)\,\right].$$

Integrating, we have

$$z = \frac{1}{6}x^3 y - \frac{1}{24}(a+b)x^4 + \frac{1}{b-a}\left[\phi_2(y-ax) + \phi_1(y-bx)\right]$$

$$= \frac{1}{6}x^3\,y - \frac{1}{24}(a+b)\,x^4 + \psi_2(y-ax) + \psi_1(y-bx).$$

Example 8: Solve $(b+c\,q)^2\,r - 2(b+c\,q)(a+c\,p)\,s + (a+c\,p)^2\,t = 0$.

Solution: Putting $r = \dfrac{dp - s\,dy}{dx}$ *and* $t = \dfrac{dq - s\,dx}{dy}$, the given equation becomes

$$(b+c\,q)^2\left(\frac{dp - s\,dy}{dx}\right) - 2(b+c\,q)(a+c\,p)\,s + (a+c\,p)^2\left(\frac{dq - s\,dx}{dy}\right) = 0$$

$$\Rightarrow \left\{(b+c\,q)^2\,dp\,dy + (a+c\,p)^2\,dq\,dx\right\} - s$$
$$\left\{(b+c\,q)^2\,dy^2 + 2(b+c\,q)(a+c\,p)\,dx\,dy + (a+c\,p)^2\,dx^2\right\} = 0.$$

The Monge's subsidiary equations are

$$(b+c\,q)^2\,dp\,dy + (a+c\,p)^2\,dq\,dx = 0 \qquad \text{...(1)}$$

and $(b+c\,q)^2\,dy^2 + 2(b+c\,q)(a+c\,p)\,dx\,dy + (a+c\,p)^2\,dx^2 = 0$ ...(2)

Equation (2) gives $\left[(b+c\,q)\,dy + (a+c\,p)\,dx\right]^2 = 0$ ..(3)

$\Rightarrow \quad b\,dy + a\,dx + c\,(p\,dx + q\,dy) = 0$

$\Rightarrow \quad a\,dx + b\,dy + c\,dz = 0$.

Therefore $ax + by + cz = A$.

From (1) and (3), we have $(b+cq)dp - (a+cp)dq = 0$

$$\Rightarrow \quad \frac{dp}{a+cp} - \frac{dq}{b+cq} = 0$$

Integrating, we get $\dfrac{a+cp}{b+cq} = B \Rightarrow (a+cp) = (b+cq)B$

Hence the intermediate integral is $a + cp = (b+cq)\,f(ax+by+cz)$

$\Rightarrow \quad cp - cf(ax+by+cz)q = -a + bf(ax+by+cz)$.

Therefore Lagrange's subsidiary equations are

$$\frac{dx}{c}=\frac{dy}{-cf\left(ax+by+cz\right)}=\frac{dz}{-a+bf\left(ax+by+cz\right)}$$

using a,b,c as multipliers, we have

$$\text{Each fraction } = \frac{a\,dx+b\,dy+c\,dz}{0}$$

$\Rightarrow \quad a\,dx+b\,dy+c\,dz=0 \;\Rightarrow\; ax+by+cz=A.$

Again taking the first two members, we have

$$dx=\frac{dy}{-f(A)} \text{ or } dy+f(A)\,dx=0.$$

$\Rightarrow \quad y+f(A)\,x=A$ (constant)

$\Rightarrow \quad y+x\,f\left(a\,x+b\,y+c\,z\right)=F\left(a\,x+b\,y+c\,z\right),$ which is the required solution.

Example 9: Solve $y^2\,r+2\,x\,y\,s+x^2\,t+p\,x+q\,y=0$.

Solution: Putting $r=\dfrac{dp-s\,dy}{dx}$ *and* $t=\dfrac{dq-s\,dx}{dy}$ in the given equation, we have

$$y^2\,\frac{dp-s\,dy}{dx}+2\,x\,y\,s+x^2\,\frac{dq-s\,dx}{dy}+p\,x+q\,y=0$$

$$\Rightarrow \quad \left\{y^2\,dp\,dy+x^2\,dq\,dx+\left(p\,x+p\,y\right)dx\,dy\right\}$$
$$-s\left\{y^2\,dy^2-2\,x\,y\,dx\,dy+x^2\,dx^2\right\}=0.$$

Hence Monge's subsidiary equations are

$$y^2\,dp\,dy+x^2\,dq\,dx+\left(p\,x+q\,y\right)dx\,dy=0 \qquad \text{...(1)}$$

And $y^2\,dy^2-2\,x\,y\,dx\,dy+x^2\,dx^2=0$ ...(2)

Equation (2) gives $\left(y\,dy-x\,dx\right)^2=0$ *or* $y\,dy-x\,dx=0$.

$\Rightarrow \quad -y^2+x^2=A.$ ...(3)

From (1) and (3), we have $y\,dp+x\,dq+p\,dy+q\,dx=0$

$\Rightarrow \quad \left(y\,dp+p\,dy\right)+\left(x\,dq+q\,dx\right)=0.$

Therefore $y\,p+x\,q=B$.

Hence one intermediate integral is $y\,p+x\,q=f\left(x^2-y^2\right)$.

Hence Lagrange's subsidiary equations are $\dfrac{dx}{y}=\dfrac{dy}{x}=\dfrac{dz}{f\left(x^2-y^2\right)}$

From the first two members, we have

$$x\,dx-y\,dy=0 \Rightarrow x^2-y^2=A.$$

Again from the last two members, we have

$$\frac{dy}{x} = \frac{dz}{f(A)} \Rightarrow dz = f(A)\frac{dy}{\sqrt{(y^2 + A)}}.$$

Hence $z = f(A)\log\left[y + \sqrt{A + y^2}\right] + A'$

$$\Rightarrow \quad z = f(x^2 - y^2)\log(y + x) + F(x^2 - y^2).$$

Example 10: Solve $q(1+q)r - (p+q+2pq)s + p(1+p)t = 0$.

Solution: Putting $r = \frac{dp - s\,dy}{dx}$ and $t = \frac{dq - s\,dx}{dy}$ in the given equation, we have

$$q(1+q)\frac{dp - s\,dy}{dx} - (p+q+2qp)s + p(1+p)\frac{dq - s\,dx}{dy} = 0$$

$$\Rightarrow \{q(1+q)\,dp\,dy + p(1+p)\,dq\,dx\}$$
$$- s\{q(1+q)\,dy^2 + (p+q+2pq)\,dx\,dy + p(1+p)\,dx^2\} = 0.$$

Hence Monge's subsidiary equations are

$$q(1+q)\,dp\,dy + p(1+p)\,dq\,dx = 0 \qquad ...(1)$$

and $q(1+q)\,dy^2 + (p+q+2qp)\,dx\,dy + p(1+p)\,dx = 0$ ...(2)

Equation (2) gives $p\,dx + q\,dy = 0$ ...(3)

and $(1+p)\,dx + (1+q)\,dy = 0$ ...(4)

From (3), we have $dz = p\,dx + q\,dy = 0 \Rightarrow z = A$.
Also from (1) and (3), we have

$$p(1+q)\,dp - p(1+q)\,dq = 0 \Rightarrow \frac{dp}{1+p} - \frac{dq}{1+q} = 0.$$

Therefore $\log(1+p) - \log(1+q) = \log B \Rightarrow (1+p) = (1+q)\,f_1(z)$ ...(5)

which is one intermediate integral.
Now from (4), we have $dx + dy + (p\,dx + q\,dy) = 0$

$$\Rightarrow \quad dx + dy + dz = 0 \Rightarrow x + y + z = A.$$

From (1) and (4), we have $q\,dp - p\,dq = 0 \Rightarrow \frac{dp}{p} = \frac{dq}{q}$

$$\Rightarrow \quad \log p = \log q + \log B$$

Hence $p = q\,f_2(x + y + z)$. ...(6)

Solving (5) and (6), we have $p = \frac{(f_1 - 1)f_2}{(f_2 - f_1)}$ and $q = \frac{(f_1 - 1)}{(f_2 - f_1)}$.

Substituting these values in $dz = p\,dx + q\,dy$, we have

$$dz = \frac{(f_1 - 1)\, f_2}{(f_2 - f_1)}\, dx + \frac{(f_1 - 1)}{(f_2 - f_1)}\, dy$$

$$\Rightarrow \quad (f_2 - f_1)\, dz = (f_1 - 1)\, f_2\, dx + (f_1 - 1)\, dy$$

$$\Rightarrow (f_2 - f_1)\, dz = (f_1 - 1)\, f_2\, dx + (f_1 - 1)(dx + dy + dz) - (f_1 - 1)(dx + dz)$$

$$\Rightarrow \quad (f_2 - 1) dz = (f_1 - 1)(f_2 - 1)\, dx + (f_1 - 1)(d\,x + d\,y + d\,z)$$

$$\Rightarrow \quad \frac{dz}{(f_1 - 1)} = d\,x + \frac{dx + dy + dz}{(f_2 - 1)}$$

$$\Rightarrow \quad \frac{dz}{f_1(z) - 1} = d\,x + \frac{dx + dy + dz}{f_2(x + y + z) - 1}.$$

Integrating, we have $\phi_1(z) = x + \phi_2(x + y + z)$, which is the required solution.

Example 11: Solve $y^2 r - 2\,y\,s + t = p + 6\,y$.

Solution: Putting $r = \dfrac{dp - s\,dy}{dx}$ and $t = \dfrac{dq - s\,dx}{dy}$ in the given equation, we have

$$y^2\, \frac{dp - s\,dy}{dx} - 2\,y\,s + \frac{dq - s\,dx}{dy} = p + 6y$$

$$\Rightarrow \left[y^2\, dp\, dy - (p + 6\,y)\, dx\, dy + dq\, dx\right] - s\left[y^2\, dy^2 + 2\,y\, dx\, dy + dx^2\right] = 0.$$

Hence the Monge's subsidiary equations are

$$y^2\, dp\, d\,y - (p + 6\,y)\, dx\, dy + dq\, dx = 0 \qquad \text{...(1)}$$

and $\quad y^2\, dy^2 + 2\,y\, dx\, dy + dx^2 = 0 \qquad$ ...(2)

From (2), we have $(y\, dy + dx)^2 = 0 \qquad$ ...(3)

Integrating, we get $y^2 + 2\,x = A \qquad$ ...(4)

From (1) and (3), we have

$$y\, dp + (p + 6\,y)\, dy - dq = 0 \Rightarrow (y\, dp + p\, dy) + 6\,y\, dy - dq = 0$$

Therefore $\quad p\,y + 3\,y^2 - q = B \qquad$ ...(5)

Hence from (4) and (5) the intermediate integral is

$$p\,y + 3\,y^2 - q = f\left(y^2 + 2\,x\right) \Rightarrow p\,y - q = -3\,y^2 + f\left(y^2 + 2\,x\right).$$

Lagrange's subsidiary equations are

$$\frac{dx}{y} = \frac{dy}{-1} = \frac{dz}{-3\,y^2 + f\left(y^2 + 2\,x\right)}.$$

From the first two members, we have

$y\,dy + dx = 0 \Rightarrow y^2 + 2x = A$.

Again from the second and third members, we have

$$dz = \left[3y^2 - f\left(y^2 + 2x\right)\right]dy \Rightarrow dz = 3y^2\,dy - f(A)\,dy.$$

Therefore $z = y^3 - y\,f(A) + B$

$\Rightarrow \quad z = y^3 - y\,f\left(y^2 + 2x\right) + F\left(y^2 + 2x\right)$, which is the required solution.

Example 12: Solve $r - t\cos^2 x + p\tan x = 0$.

Solution: Putting $r = \dfrac{dp - s\,dy}{dx}$ and $t = \dfrac{dq - s\,dx}{dy}$ in the given equation, we have

$$\frac{dp - s\,dy}{dx} - \frac{dq - s\,dx}{dy}\cos^2 x + p\tan x = 0$$

$$\Rightarrow \left(dp\,dy - \cos^2 x\,dq\,dx + p\tan x\,dx\,dy\right) - s\left(dy^2 - \cos^2 x\,dx^2\right) = 0.$$

Hence the Monge's subsidiary equations are

$$dp\,dy - \cos^2 x\,dq\,dx + p\tan x\,dx\,dy = 0 \quad ...(1)$$

and $\quad dy^2 - \cos^2 x\,dx^2 = 0 \quad ...(2)$

Equation (2) gives $dy - \cos x\,dx = 0 \quad ...(3)$

and $\quad dy + \cos x\,dx = 0. \quad ...(4)$

From (3), we have $y - \sin x = A$.

From (1) and (4), we have $dp - \cos x\,dq + p\tan x\,dx = 0$

$\Rightarrow \quad \sec x\,dp + p\tan x\sec x\,dx - dq = 0$.

Integrating, we get $\;p\sec x - q = B$

Hence $\;p\sec x - q = f_1\left(y - \sin x\right)$ Which is one intermediate integral. ...(5)

Similarly, the other intermediate integral is $p\sec x + q = f_1\left(y - \sin x\right)$ ...(6)

Solving (4) and (6), we have

$$p = \frac{1}{2}\cos x\left[\,f_1\left(y - \sin x\right) + f_2\left(y + \sin x\right)\right]$$

And $\;q = \dfrac{1}{2}\left[-f_1\left(y - \sin x\right) + f_2\left(y + \sin x\right)\right]$.

Substituting these values in $dz = p\,dx + q\,dy$, we have

$$dz = -\frac{1}{2} f_1(y - \sin x)(dy - \cos x\,dx) + \frac{1}{2} f_2(y + \sin x)(dy + \cos x\,dx)$$

$$= -\frac{1}{2} f_1(y - \sin x)\,d(y - \sin x) + \frac{1}{2} f_2(y + \sin x)\,d(y + \sin x)$$

Hence $z = \phi_1(y - \sin x) + \phi_2(y + \sin x)$, this is the complete solution.

EXERCISE 8.3

Solve the following PDEs by Monge's method:

1. $r = t$.
2. $x^2 r - y^2 t = 0$.
3. $2yq + y^2 t = 1$.
4. $r - 2s + t = \sin(2x + 3y)$.
5. $pt - qs = q^3$.
6. $(q+1)s = (p+1)t$.
7. $z(qs - pt) = pq^2$.
8. $rx^2 - 3sxy + 3ty^2 + px + 2py = x + 2y$.
9. $y^2 r + 2xys + x^2 t + px + qy = 0$.
10. $r - 2as + a^2 t = 0$.

ANSWERS 8.3

1. $z = f_1(x + y) + f_2(y - x)$.
2. $z = x f(y/x) + F(xy)$.
3. $yz = y \log y - f(x) + y F(x)$.
4. $z = f(x + y) + x F(x + y) - \sin(2x + 3y)$.
5. $y = zx + f(z) + F(x)$.
6. $z = F(x + y + z) + \phi(x)$.
7. $y = z F(z) + zG(x)$
8. $z = x + y + f(xy) + F(x^2 y)$.
9. $z = f(x^2 - y^2)\log(y + x) + \phi(x^2 - y^2)$.
10. $z = x f(y + ax) + F(y + ax)$.

8.8.2 Monge's Method II

For integrating $R\,r+S\,s+T\,t+U\left(r\,t-s^2\right)=V$, where R,S,T,U,V are functions of x,y,z,p,q.

We have $d\,p=\frac{\partial p}{\partial x}dx+\frac{\partial p}{\partial y}dy=r\,dx+s\,dy$

and $d\,q=\frac{\partial q}{\partial x}dx+\frac{\partial q}{\partial y}dy=s\,dx+t\,dy$.

Therefore $r=\frac{dp-s\,dy}{dx}$ *and* $t=\frac{dq-s\,dx}{dy}$

Putting these values in the equation, we have

$$R\frac{dp-s\,dy}{dx}+S\,s+T\frac{dq-s\,dx}{dy}+U\left\{\frac{(dp-s\,dy)(dq-s\,dx)}{dx\,dy}-s^2\right\}=V$$

$$(R\,dp\,dy+T\,dq\,dx+U\,dp\,dq-V\,dx\,dy)$$
$$-s\left(R\,dy^2-S\,dx\,dy+T\,dx^2+U\,dp\,dx+U\,dq\,dy\right)=0.$$

Hence Monge's subsidiary equations are

$$L\equiv R\,dp\,dy+T\,dq\,dx+U\,dp\,dq-V\,dx\,dy=0 \quad ...(1)$$

and $M\equiv R\,dy^2-S\,dx\,dy+T\,dx^2+U\,dp\,dx+U\,dq\,dy=0$. ...(2)

Now (2) may not be factorized, on account of the presence of the term $U\,dp\,dx+U\,dq\,dy$.

Hence let us try to factorize $M+\lambda L=0$

$$\Rightarrow \quad \left(R\,dy^2-S\,dx\,dy+T\,dx^2+U\,dp\,dx+U\,dq\,dy\right)$$
$$+\lambda\left(R\,dp\,dy+T\,dq\,dx+U\,dp\,dq-V\,dx\,dy\right)=0 \quad ...(3)$$

where λ is some multiplier to be determined.
Let the factors of (3) be

$$(R\,dy+m\,T\,dx+K\,U\,dp)\left(dy+\frac{1}{m}dx+\frac{\lambda}{K}dq\right)=0. \quad ...(4)$$

Therefore comparing (3) and (4), we have $\frac{R}{m}+m\,T=-(S+\lambda V)$,

$$K=m \text{ and } \frac{R\,\lambda}{K}=U. \quad ...(5)$$

From the last two, we have $m=\frac{R\,\lambda}{U}$.

Hence from (5), we have $\frac{U}{\lambda}+\frac{R\,\lambda\,T}{U}=-(S+\lambda V)$

$$\Rightarrow \quad \lambda^2 (UV + RT) + \lambda US + U^2 = 0. \qquad ...(6)$$

Let λ_1 and λ_2 be the values of λ from (6), when

$$\lambda = \lambda_1 \text{ then } m = \frac{R\lambda_1}{U}.$$

Therefore from (4), we have

$$\left(R\,dy + \frac{R\lambda_1}{U} T\,dx + R\lambda_1\,dp\right)\left(dy + \frac{U}{R\lambda_1} dx + \frac{U}{R} dq\right) = 0$$

$$\Rightarrow \quad (U\,dy + \lambda_1 T\,dx + \lambda_1 U\,dp)(U\,dx + \lambda_1 R\,dy + \lambda_1 U\,dq) = 0. \qquad ...(7)$$

Similarly, when $\lambda = \lambda_2$ $m = \dfrac{R\lambda_2}{U}$, from (4), we have

$$(U\,dy + \lambda_2 T\,dx + \lambda_2 U\,dp)(U\,dx + \lambda_2 R\,dy + \lambda_2 U\,dq) = 0. \qquad ...(8)$$

Now one factor of (7) is combined with one factor of (8) to give an intermediate integral and similarly other pair gives another intermediate integral.

Now let us suppose that two integrals $u_1 = a_1, v_1 = b_1$ are obtained from the equations

And
$$\left.\begin{aligned} U\,dy + \lambda_1 T\,dx + \lambda_1 U\,dp = 0 \\ U\,dx + \lambda_2 R\,dy + \lambda_2 U\,dq = 0 \end{aligned}\right\} \qquad ...(9)$$

and similarly let $u_2 = a_2, v_2 = b_2$ be two integrals obtained from the equations
and

$$\left.\begin{aligned} U\,dy + \lambda_2 T\,dx + \lambda_2 U\,dp = 0 \\ U\,dx + \lambda_1 R\,dy + \lambda_1 U\,dq = 0 \end{aligned}\right\}. \qquad ...(10)$$

Thus two intermediate integrals are

$$u_1 = f_1(v_1) \text{ and } u_2 = f_2(v_2).$$

Solving them, we obtain the values of p and q and then substitute in $dz = p\,dx + q\,dy$.

The complete integral is obtained by solving the above relation.

Remarks:

1. If two values of λ are equal then we get only one intermediate integrals, $u_1 = f_1(v_1)(say)$. This integral $u_1 = f_1(v_1)$ together with one of the integrals $v_1 = a_1$ and $v_1 = b_1$ gives the values of p and q such that $dz = p\,dx + q\,dy$ is solvable.
2. If it is not possible to find the suitable values of p and q from the two intermediate integrals $u_1 = f_1(v_1)$ and $u_2 = f_2(v_2)$ for the integration in $dz = p\,dx + q\,dy$.

Then we take one of the intermediate integrals $u_2 = f_2(v_2)$ say and one of the integrals from $u_1 = a_1$ and $v_1 = b_1$. Now substituting the values of p and q in $dz = p\,dx + q\,dy$ and integrating we get the required solution.

Example 1: Solve $s^2 - rt = a^2$

Solution: We have $s^2 - rt - a^2 = 0$ ...(1)

On comparing with the Monge's equation

$$R = 0, S = 0, T = 0, U = 1, V = -a^2.$$

Hence the equation in λ is

$$\lambda^2(-a^2) + \lambda\, 0 + 1 = 0 \Rightarrow \lambda = \pm\frac{1}{a}.$$

The two intermediate integrals are given by

$$\left.\begin{aligned} -dy - \frac{1}{a}dp = 0 \\ -dx + \frac{1}{a}dq = 0 \end{aligned}\right\} \quad ...(2)$$

and

$$\left.\begin{aligned} -dy + \frac{1}{a}dp = 0 \\ -dx - \frac{1}{a}dq = 0 \end{aligned}\right\}. \quad ...(3)$$

From (2), we have $\left.\begin{aligned} p + ay = F(\alpha) \\ q - ax = \alpha \end{aligned}\right\}$. ...(4)

From (3), we have $\left.\begin{aligned} p - ay = F(\beta) \\ q + ax = \beta \end{aligned}\right\}$ ...(5)

i.e., the two intermediate integrals are $p + ay = f(q - ax)$ ...(6)

and $p - ay = F(q + ax)$. ...(7)

Now it is not possible to find the values of p and q from (6) and (7). We forward as follows. Suppose α, β are not constants but parameters.

$$x = \frac{(\beta - \alpha)}{2a}, q = \frac{(\alpha + \beta)}{2} \quad ...(8)$$

$$p = \frac{1}{2}[F(\alpha) + f(\beta)] \quad ...(9)$$

$$y = \frac{1}{2a}[F(\alpha) - f(\beta)]. \quad ...(10)$$

Substituting these values in $dz = p\,dx + q\,dy$, we get

$$dz = \frac{1}{4a}[F(\alpha) + f(\beta)](d\beta - d\alpha) + \frac{(\alpha + \beta)}{4a}[F'(\alpha)\,d\alpha - f'(\beta)\,d\beta]$$

$$= \frac{1}{4a}\Big[\{F(\alpha)\,d\beta + \beta\,F'(\alpha)\,d\alpha\} - \{f(\beta)\,d\alpha + \alpha\,f'(\beta)\,d\beta\} +$$

$$\{F(\alpha)d\alpha + \alpha F'(\alpha)d\alpha\} - \{f(\beta)\,d\beta + \beta\,f'(\beta)\,d\beta\}\Big]$$

$$+\frac{1}{4a}\left[2f(\beta)d\beta - 2F(\alpha)d\alpha\right]$$

Hence

$$z = \frac{1}{4a}\left[\beta F(\alpha) - \alpha f(\beta) - \beta f(\beta) + \alpha F(\alpha)\right] + \frac{2}{4a}\int f(\beta)\,d\beta - \frac{2}{4a}\int F(\alpha)\,d\alpha$$

$$= \frac{1}{4a}\left[F(\alpha)(\alpha+\beta) - f(\beta)(\alpha+\beta)\right] + \frac{2}{4a}G(\beta) - \frac{2}{4a}\phi(\alpha)$$

$$= \frac{\alpha+\beta}{2}\left[\frac{F(\alpha) - f(\beta)}{2a}\right] + \frac{1}{2a}G(\beta) - \frac{1}{2a}\phi(\alpha)$$

$\Rightarrow\ z - q\,y = \psi_1(q + a\,x) + \psi_2(q - a\,x)$ [from equations (7) and (8)]

where $\psi_1(t) = \int \frac{f(t)}{2a}\,dt$ and $\psi_2(t) = -\int \frac{F(t)}{2a}\,dt$.

Hence the primitive is $z - q\,y = \psi_1(q + a\,x) + \psi_2(q - a\,x)$

$\Rightarrow\quad -y = \psi'_1(q + a\,x) + \psi'_2(q - a\,x)$

Example 2: Solve $2\,s + (r\,t - s^2) = 1$.

Solution: Comparing the given equation with the equation

$$R\,r + S\,s + T\,t + U\left(r\,t - s^2\right) = V$$

We have $R = 0, S = 2, T = 0, U = 1, V = 1$.

Therefore from λ equation $\lambda^2(U\,V + R\,T) + \lambda\,U\,S + U^2 = 0$.

We have $\lambda^2 + 2\lambda + 1 = 0 \Rightarrow \lambda_1 = -1$ and $\lambda_2 = -1$.

Putting $\lambda_1 = -1 = \lambda_2$ in one of pair of equations

$$U\,dy + \lambda_1\,T\,dx + \lambda_1\,U\,dp = 0$$

and $\quad U\,dx + \lambda_2\,R\,d\lambda + \lambda_2\,U\,dq = 0$.

We have $dy - dp = 0$ and $dx - dq = 0$

$\Rightarrow\quad y - p = a$ and $x - q = b$.

Therefore the intermediate integral is

$$y - p = f(x - q).$$

Also $p = y - a$ and $q = x - b$. Substituting these values of p and q in $dz = p\,dx + q\,dy$, we have

$$dz = (y-a)\,dx + (x-b)\,dy = (y\,dx + x\,dy) - a\,dx - b\,dy.$$

$\Rightarrow \quad z = xy - ax - by + c$, which is the complete integral.

Example 3: Solve $r + 3s + t + (rt + s^2) = 1$.

Solution: Comparing it with the equation

$Rr + Ss + Tt + U(rt - s^2) = V$, we have $R = 1, S = 3, T = 1, U = 1, V = 1$.

From $\lambda^2 (UV + RT) + US + U^2 = 0$, we have $2\lambda^2 + 3\lambda + 1 = 0$

$\Rightarrow \quad (2\lambda + 1)(\lambda + 1) = 0 \Rightarrow \lambda_1 = -1, \lambda_2 = -\dfrac{1}{2}$.

Hence first intermediate integral is given by

$$\begin{cases} U\,dy + \lambda_1 T\,dx + \lambda_1 U\,dp = 0 \\ U\,dx + \lambda_2 R\,dy + \lambda_2 U\,dq = 0 \end{cases}$$

$$\Rightarrow \quad \begin{cases} dy - dx - dp = 0 \\ dx - \dfrac{1}{2}dy - \dfrac{1}{2}dq = 0 \end{cases}$$

$$\Rightarrow \quad \begin{cases} dy - dx - dp = 0 \\ -2\,dx + dy + dq = 0. \end{cases}$$

$\Rightarrow \quad y - x - p = A$ and $-2x + y + q = B$.

So one intermediate integral is

$$y - x - p = f_1(y - 2x + q) = f_1(\alpha), \text{where } \alpha = y - 2x + q \qquad ...(1)$$

The second intermediate integral is given by

$$\begin{cases} U\,dx + \lambda_1 R\,dy + \lambda_1 U\,dq = 0 \\ U\,dy + \lambda_2 T\,dx + \lambda_2 U\,dp = 0 \end{cases}$$

$$\Rightarrow \quad \begin{cases} -dy + dx - dq = 0 \\ 2\,dy - dx - dp = 0. \end{cases}$$

Hence we have $-y + x - q = A$ and $2y - x - p = B$.

So the second intermediate integral is

$$2y - x - p = f_2(-y + x - q) = f_2(\beta), \text{ whene } \beta = -y + x - q \qquad ...(2)$$

Now $\alpha + \beta = -x \Rightarrow \quad d\alpha + d\beta = -dx$.

From (2) and (1), we get $y = f_2(\beta) - f_1(\alpha)$

$\Rightarrow \quad dy = f'_2(\beta)\,d\beta - f'_1(\alpha)\,d\alpha$

And $p = y - x - f_1(\alpha), q = x - y - \beta$.

Substituting in $dz = p\,dx + q\,dy$, we have

$$dz = [y - x - f_1(\alpha)]\,dx + (x - y - \beta)\,dy$$

$$= (y-x)\,dx - f_1(\alpha)\,dx + (x-y)\,dy - \beta\,dy$$
$$= -(x-y)(dx-dy) - f_1(\alpha)(-d\alpha - d\beta) - \beta\left[f'_2(\beta)\,d\beta - f'_1(\alpha)\,d\alpha\right]$$
$$= -(x-y)(dx-dy) + f_1(\alpha)\,d\alpha - \beta f'_2(\beta)\,d\beta + \{f_1(\alpha)\,d\beta + \beta f'_1(\alpha)\,d\alpha\}$$

on integrating, we get

$$z = -\frac{1}{2}(x-y)^2 + \int f_1(\alpha)\,d\alpha - \int \beta f'_2(\beta)\,d\beta + \int\{f_1(\alpha)d\beta + \beta f'_1(\alpha)d\alpha\}$$
$$= -\frac{1}{2}(x-y)^2 + \phi_1(\alpha) - \beta f_2(\beta) + \int f_2(\beta)\,d\beta + \beta f_1(\alpha)$$
$$= -\frac{1}{2}(x-y)^2 + \phi_1(\alpha) - \beta f_2(\beta) + \phi_2(\beta) + \beta f_1(\alpha),$$ which is the required solution.

Example 4: Solve $q\,r + (p+x)\,s + y\,t + y\left(r\,t - s^2\right) + q = 0$.

Solution: Comparing it with the equation

$$R\,r + S\,s + T\,t + U\left(r\,t - s^2\right) = V$$

We have $R = q, S = p + x, T = y, U = y, V = -q$.

Therefore λ equation is

$$\lambda^2(U\,V + R\,T) + \lambda\,U\,S + U^2 = 0 \Rightarrow \lambda = -\frac{y}{(p+x)}.$$

Substituting this in equation $U\,d\,y + \lambda\,T\,d\,x + \lambda\,U\,d\,p = 0$, we have

$$y\,dy - \frac{y^2}{(p+x)}dx - \frac{y^2}{(p+x)}dp = 0$$

$$\Rightarrow \quad \frac{dy}{y} = \frac{dx+dp}{x+p}. \qquad \text{...(1)}$$

Therefore $\log y = \log(p+x) + \text{constant}$.

$$\Rightarrow \quad \frac{(p+x)}{y} = a\,(\text{constant}). \qquad \text{...(2)}$$

Also form $R\,dy^2 - S\,dx\,dy + T\,dx^2 + U\,dp\,dx + U\,dq\,dy = 0$

We have $q\,dy^2 - (p+x)\,dx\,dy + y\,dx^2 + y\,dp\,dx + y\,dq\,dy = 0$

$$\Rightarrow \quad dy\,(q\,dy + y\,dq) = dx\left[(p+x)\,dy - y\,(dx+dp)\right]$$

$$\Rightarrow \quad dy\,(q\,dy + y\,dq) = 0, \text{ [using (1)]}$$

$$\Rightarrow \quad q\,dy + y\,dq = 0 \Rightarrow qy = b.$$

Hence the intermediate integral is $q\,y = f\left(\frac{p+x}{y}\right)$.

Charpit's auxiliary equations are

$$\frac{dx}{\frac{1}{y} f\left(\frac{p+x}{y}\right)} = \frac{dy}{y} = \frac{dp}{\frac{1}{y} f\left(\frac{p+x}{y}\right)}.$$

Taking the first and the last members, we get

$$dx + dp = 0 \Rightarrow p + x = c.$$

From $dz = p\,dx + q\,dy$, we have $dz = (c - x)\,dx + (1/y)\,f\,(c/y)\,dy$.

$\Rightarrow \quad z = c\,x - \frac{1}{2}x^2 + \phi\,(c/y) + \psi\,(c)$, which is the required solution.

Example 5: Solve $2\,r + t\,e^{\,x} - (r\,t - s^2) = 2\,e^{\,x}$.

Solution: Here we proceed directly. Putting $r = \frac{dp - s\,dy}{dx}$ and $t = \frac{dq - s\,dx}{dy}$, in the given equation, we get

$$2\,\frac{dp - s\,dy}{dx} + \frac{dq - s\,dx}{dy}\,e^x - \left(\frac{dp - s\,dy}{dx}\,\frac{dq - s\,dx}{dy} - s^2\right) = 2\,e^x$$

$$\Rightarrow \quad 2\,dp\,dy + e^x\,dq\,dx - dp\,dq - 2\,e^{\,x}\,dx\,dy - s$$
$$(2\,dy^2 + e^x\,dx^2 - dp\,dx - dq\,dy) = 0$$

Hence the Monge's equations are

$$2\,dp\,dy + e^x\,dq\,dx - dp\,dq - 2\,e^x\,dx\,dy = 0 \qquad ...(1)$$

and $\quad 2\,dy^2 + e^x\,dx^2 - dp\,dx - dq\,dy = 0 \qquad ...(2)$

Equation (1) gives $(2\,dy - dq)(dp - e^x\,dx) = 0$

$\Rightarrow \quad 2\,dy - dq = 0$ *or* $dp - e^x\,dx = 0$.

Integrating, we get $2\,y - q = A$ *and* $p - e^x = B$

$\Rightarrow \quad q = 2\,y - A$ *and* $p = e^x + B$.

Substituting in $d\,z = p\,d\,x + q\,d\,y$, we have

$$d\,z = (e^x + B)\,d\,x + (2\,y - A)\,d\,y.$$

Integrating, we get $z = e^x + B\,x + y^2 - A\,y + C$.

Alter: Taking $p - e^x = (-2\,y + q)\,m + n$, where m, n are constants

$\Rightarrow \quad p - q\,m = e^x - 2\,y\,m + n$.

Therefore Lagrange's auxiliary equations are

$$\frac{dx}{1} = \frac{dy}{-m} = \frac{dz}{e^x - 2\,y\,m + n} = \frac{-e^x\,dx - 2\,y\,dy + dz}{n}.$$

From the first two members, we have $y + m\,x = a$.

Again from the first and the last members, we have

$$-e^x\,dx - 2\,y\,dy + dz = n\,dx$$

$$\Rightarrow \quad dz = e^x\,dx + 2\,y\,dy + n\,dx.$$

On integrating, we get $z = e^x + y^2 + n\,x + b$

$$\Rightarrow \quad z = e^x + y^2 + n\,x + \phi\,(y + m\,x).$$

Example 6: Solve

$$z\left(1+q^2\right)r - 2\,p\,q\,z\,s + z\left(1+p^2\right)t + z^2\left(r\,t - s^2\right) + \left(1+p^2+q^2\right) = 0.$$

Solution: Comparing it with the equation

$R\,r + S\,s + T\,t + U\left(r\,t - s^2\right) = V$, we have

$$R = z\left(1+q^2\right), S = -2\,p\,q\,z, T = z\left(1+p^2\right), U = z^2 \text{ and } V = -\left(1+p^2+q^2\right).$$

Therefore from λ equation $\lambda^2\,(U\,V + R\,T) + \lambda\,U\,S + U^2 = 0$, we have

$$\left\{-z^2\left(1+p^2+q^2\right) + z^2\left(1+p^2\right)\left(1+q^2\right)\right\}\lambda^2 - 2\,p\,q\,z^3\,\lambda + z^4 = 0$$

$$\Rightarrow p^2\,q^2\,\lambda^2 - 2\,p\,q\,z\,\lambda + z^2 = 0 \Rightarrow (p\,q\,\lambda - z)^2 = 0 \Rightarrow \lambda_1 = \lambda_2 = \frac{z}{p\,q}.$$

Hence the intermediate integral is given by

$$U\,dy + \lambda\,T\,dx + \lambda\,U\,dp = 0 \text{ and } U\,dx + \lambda\,R\,dy + \lambda\,U\,dq = 0$$

i.e., $z^2\,dy + \dfrac{z}{pq}\,z\left(1+p^2\right)dx + \dfrac{z}{pq}\,z^2\,dp = 0$

and $z^2dy + \dfrac{z}{pq}\,z\left(1+q^2\right)dx + \dfrac{z}{pq}\,z^2dp = 0$

$$\Rightarrow \quad \begin{cases} pq\,dy + \left(1+p^2\right)dx + z\,dp = 0 \\ pq\,dx + \left(1+q^2\right)dx + z\,dq = 0 \end{cases} \qquad \text{...(1)}$$

From (1), we have $p(p\,dx + q\,dy) + z\,dp + dx = 0$

$$p\,dz + z\,dp + dx = 0, \text{ since } dz = p\,dx + q\,dy. \qquad \text{...(2)}$$

On integrating, we get $p\,z + x = A$. ...(3)

From (2), we have $q\,(p\,dx + q\,dy) + z\,dq + dy = 0$

$$\Rightarrow \quad q\,dz + z\,dq + dy = 0.$$

On integrating, we get $q\,z + y = B$. ...(4)

Substituting the values of p and q from (3) and (4), in $dz = p\,dx + q\,dy$, we have

$$dz = \frac{A - x}{z}\,dx + \frac{B - y}{z}\,dy$$

$\Rightarrow \quad z\,dz = (A-x)\,dx + (B-y)\,dy$.

Integrating, we have $z^2 = -(A-x)^2 - (B-y)^2 + c$

$\Rightarrow \quad z^2 + (A-x)^2 + (B-y)^2 = c$, which is the general solution.

Example 7: Solve $(rt - s^2) - s(\sin x + \sin y) = \sin x \sin y$.

Solution: Comparing with the equation

$$Rr + Ss + Tt + U(rt - s^2) = V \text{, we have}$$

$R = 0, S = -(\sin x + \sin y), T = 0, U = 1, V = \sin x \sin y$.

From λ equation $\lambda^2(UV + RT) + \lambda US + U^2 = 0$, we have

$\sin x \sin y\, \lambda^2 - (\sin x + \sin y)\lambda + 1 = 0$

$\Rightarrow (\lambda \sin x - 1)(\lambda \sin y - 1) = 0$.

Therefore, we have $\lambda_1 = \operatorname{cosec} y$ *and* $\lambda_2 = \operatorname{cosec} x$.

Substituting these values in $\begin{cases} U\,dy + \lambda_1 T\,dx + \lambda_1 U\,dp = 0 \\ U\,dx + \lambda_2 R\,dy + \lambda_2 U\,dq = 0, \end{cases}$ we have

$$\begin{cases} \sin x\,dx + dq = 0 \\ \sin y\,dy + dp = 0 \end{cases}$$

Hence, we have $-\cos y + p = A$ and $-\cos x + q = B$.

Now let $p - \cos y = m(q - \cos x) + n$, where m and n are constants, then we have

$$p - mq = \cos y - m\cos x + n.$$

Hence Lagrange's auxiliary equations are

$$\frac{dx}{1} = \frac{dy}{-m} = \frac{dz}{\cos y - m\cos x + n}.$$

Taking the first two members, we have

$$dy + m\,dx = 0 \Rightarrow y + mx = a$$

and from the first and the last members, we have

$dz = (\cos y - m\cos x + n)\,dx = \{\cos(a - mx) - m\cos x + n\}\,dx$.

On integrating, we get $z = -\dfrac{1}{m}\sin(a - mx) - m\sin x + nx + b$

$\Rightarrow \quad mz + \sin y + m^2 \sin x - mnx = m\,f(y + mx)$.

Example 8: Solve

$$(1+q^2)r - 2pqs + (1+p^2)t + (1+p^2+q^2)^{-1/2}(rt - s^2) = -(1+p^2+q^2)^{3/2}.$$

Solution: Here $R = 1 + q^2, S = -2pq, T = 1 + p^2$,

$$U=\left(1+p^2+q^2\right)^{-1/2}, V=-\left(1+p^2+q^2\right)^{3/2}.$$

From λ equation $\lambda^2(RT+UV)+US\lambda+U^2=0$, we have

$$\left[\left(1+p^2\right)\left(1+q^2\right)-\left(1+p^2+q^2\right)\right]\lambda^2-\frac{2pq}{\sqrt{\left(1+p^2+q^2\right)}}\lambda+\frac{1}{\left(1+p^2+q^2\right)}=0$$

$$\Rightarrow \quad p^2q^2\left(1+p^2+q^2\right)\lambda^2-2pq\sqrt{\left(1+p^2+q^2\right)}\lambda+1=0$$

$$\Rightarrow \quad \left\{pq\sqrt{1+p^2+q^2}\,\lambda-1\right\}=0 \Rightarrow \lambda=\frac{1}{pq\sqrt{1+p^2+q^2}}.$$

Therefore the intermediate function is given by

$$\begin{cases} U\,dy+\lambda T\,dx+\lambda U\,dp=0 \\ U\,dx+\lambda R\,dy+\lambda U\,dq=0 \end{cases}$$

$$\Rightarrow \begin{cases} \left(1+p^2+q^2\right)^{-1/2}dy+\dfrac{\left(1+p^2\right)}{pq\sqrt{1+p^2+q^2}}dx+\dfrac{dp}{pq\left(1+p^2+q^2\right)}=0 \\ \left(1+p^2+q^2\right)^{-1/2}dx+\dfrac{\left(1+q^2\right)}{pq\sqrt{\left(1+p^2+q^2\right)}}dx+\dfrac{dq}{pq\left(1+p^2+q^2\right)}=0 \end{cases}$$

This gives $pq\,dy+\left(1+p^2\right)dx+\dfrac{dp}{\sqrt{1+p^2+q^2}}=0$ ...(1)

and $pq\,dx+\left(1+q^2\right)dy+\dfrac{dq}{\sqrt{1+p^2+q^2}}=0$. ...(2)

Eliminating dy, we have

$$\left[\left(1+p^2\right)\left(1+q^2\right)-p^2q^2\right]dx+\frac{\left(1+q^2\right)dp}{\sqrt{\left(1+p^2+q^2\right)}}-\frac{pq\,dq}{\sqrt{\left(1+p^2+q^2\right)}}=0$$

$$\Rightarrow \quad dx+\frac{\left(1+q^2\right)dp-pq\,dq}{\left(1+p^2+q^2\right)^{3/2}}=0$$

$$\Rightarrow \quad dx+\frac{\left(1+p^2+q^2\right)dp}{\left(1+p^2+q^2\right)^{3/2}}-\frac{p^2\,dp+pq\,dq}{\left(1+p^2+q^2\right)^{3/2}}=0$$

$$\Rightarrow \quad dx+\left(1+p^2+q^2\right)^{-1/2}dp+p\,d\left(1+p^2+q^2\right)^{-1/2}=0.$$

On integrating, we get $x+p\left(1+p^2+q^2\right)^{-1/2}=A$. ...(3)

Similarly eliminating dx from (1) and (2) and simplifying, we have

$$y+q\left(1+p^2+q^2\right)^{-1/2}=B. \qquad ...(4)$$

From (3) and (4), we have $\dfrac{x-A}{y-B}=\dfrac{p}{q} \Rightarrow p=\dfrac{x-A}{y-B}q$. ...(5)

Putting in (3), we have $x+\dfrac{x-A}{y-B}q\left\{1+\left(\dfrac{x-A}{y-B}\right)^2 q^2+q^2\right\}^{-1/2}=A$

$$\Rightarrow \quad (x-A)+\frac{x-A}{y-B}q\left[1+\frac{\{(x-A)^2+(y-B)^2\}}{(y-B)^2}q^2\right]^{-1/2}=0$$

$$\Rightarrow \quad \left[1+\frac{[(x-A)^2+(y-B)^2]}{(y-B)^2}q^2\right]=\frac{q^2}{(y-B)^2}$$

$$\Rightarrow \quad (y-B)^2=q^2\left[1-\{(x-A)^2+(y-B)^2\}\right]$$

Therefore $q=\dfrac{(y-B)}{\sqrt{\left[1-\{(x-A)^2+(y-B)^2\}\right]}}$ and

$$p=\frac{(x-A)}{\sqrt{\left[1-\{(x-A)^2+(y-B)^2\}\right]}}$$

Substituting these values in $dz=p\,dx+q\,dy$, we have

$$dz=\frac{(x-A)\,dx+(y-B)\,dy}{\sqrt{\left[1-\{(x-A)^2+(y-B)^2\}\right]}}$$

Integrating, we obtain $z=\left[1-\{(x-A)^2+(y-B)^2\}\right]^{1/2}+C$,

$$\Rightarrow \quad (z-C)^2=1-\{(x-A)^2+(y-B)^2\}$$

$$\Rightarrow \quad (x-A)^2+(y-B)^2+(z-C)^2=1.$$

Example 9: $5r+6s+3t+2(rt-s^2)+3=0$.

Solution: Comparing it with the equation

$Rr+Ss+Tt+U(rt-s^2)=V$, we have

$R=5, S=6, T=3, U=2, V=-3$.

From λ equation $\lambda^2(UV+RT)+\lambda US+U^2=0$, we have

$$9\lambda^2+12\lambda+4=0 \Rightarrow (3\lambda+2)^2=0 \Rightarrow \lambda_1=\lambda_2=-\frac{2}{3}.$$

Putting these values in $\begin{cases} U\,dy+\lambda T\,dx+\lambda U\,dp=0 \\ U\,dx+\lambda R\,dy+\lambda U\,dq=0 \end{cases}$

On simplifying, we get

$$3\,dy - 3\,dx - 2\,dp = 0 \text{ and } -5\,dy + 3\,dx - 2\,dq = 0.$$

This gives $3y - 3x - 2p = A$ *and* $-5y + 3x - 2q = B$.

$$\Rightarrow \quad p = \frac{1}{2}(3y - 3x - A) \text{ and } q = \frac{1}{2}(-5y + 3x - B).$$

Substituting these values in $dz = p\,dx + q\,dy$, we have

$$dz = \frac{1}{2}(3y - 3x - A)\,dx + \frac{1}{2}(-5y + 3x - B)\,dy$$

$$\Rightarrow \quad 2\,dz = 3(y\,dx + x\,dy) - 3x\,dx - 5y\,dy - A\,dx - B\,dy.$$

On integrating, we get $2z = 3xy - \frac{3}{2}x^2 - \frac{5}{2}y^2 - Ax - By + c$

$\Rightarrow 4z = 6xy - 3x^2 - 5y^2 - 2Ax - 2By + c$, which is the required general solution.

EXERCISE 8.4

Solve the following equations:

1. $r + 4s + t + (rt - s^2) = 2$.
2. $rt - s^2 + 1 = 0$.
3. $3r + 4s + t + (rt - s^2) = 1$.
4. $7r - 8s - 3t + (rt - s^2) = 36$.
5. $qxr + (x + y)s + pyt + xy(rt - s^2) = 1 - pq$.
6. $(q^2 - 1)rz - 2pqzs + (p^2 - 1)zt + z^2(rt - s^2) = p^2 + q^2 - 1$.

ANSWERS 8.4

1. $z = -\frac{x^2}{2} - \frac{y^2}{2} + m_2 xy + Ax + \psi\{(m_2 - m_1)y + A\} + B$.
2. $z = xy + \frac{1}{2}\{\phi(\alpha) - \psi(\beta)\} + \beta y$,

   where $x = \frac{1}{2}(\alpha - \beta),\ y = \frac{1}{2}\{\psi(\beta) - \phi'(\alpha)\}$
3. $z = ax + by - \frac{1}{2}x^2 + 2xy - \frac{3}{2}y^2 + C$.
4. $2x = \alpha - \beta,\ 2y = \psi'(\beta) - \phi'(\alpha)$;

   $2z = 3x^2 - 6xy - 7y^2 + \phi(\alpha) - \psi(\beta) + 2\beta y$.
5. $z + mx + \frac{y}{m} - n\log x = f(x^m y)$.
6. $z^2 = x^2 + y^2 + 2Ax + 2By + C$ or $z^2 = x^2 + y^2 + 2nx + f(y - mx)$.

OBJECTIVE QUESTIONS

1. In order to find the solution of $P\,p + Q\,q = R$, the auxiliary equations are:

   (a) $\dfrac{dx}{P} = \dfrac{dy}{Q} = \dfrac{dz}{R}$ (b) $\dfrac{dx}{Q+R} = \dfrac{dy}{R+P} = \dfrac{dz}{P+Q}$

   (c) $\dfrac{dx}{Q-R} = \dfrac{dy}{R-P} = \dfrac{dz}{P-Q}$ (d) None of these

2. Solution of the D.E. $\left(D^2 - 2\,D\,D' + D'^2\right)z = 0$ is:

   (a) $z = C_1\,e^x + C_2\,e^y$ (b) $z = C_1e^x + C_2y$

   (c) $z = \phi_1\,(x+y) + \phi_2\,(x+y)$ (d) $z = \phi_1\,(y+x) + x\,\phi_2\,(y+x)$.

3. Monge's method is used to solve P.D.E. of:

   (a) n^{th} order (b) second order

   (c) first order (d) linear equation

4. Monge's subsidiary equations for $s^2 = a^2 t$ are:

   (a) $(dy)^2 + a^2\,(dx)^2 = 0$ and $dp\,dy - a^2\,dx\,dq = 0$

   (b) $(dy)^2 - a^2\,(dx)^2 = 0$ and $dp\,dy - a^2\,dx\,dq = 0$

   (c) $(dy)^2 - a^2\,(dx)^2 = 0$ and $dp\,dy + a^2\,dx\,dq = 0$

   (d) $(dy)^2 + a^2\,(dx)^2 = 0$ and $dp\,dy + a^2\,dx\,dq = 0$.

5. For the equation $z = p\,q$, Charpit's auxiliary equations are:

   (a) $\dfrac{dp}{p} = \dfrac{dq}{q} = \dfrac{dz}{2\,p\,q} = \dfrac{dx}{q} = \dfrac{dy}{p}$ (b) $\dfrac{dp}{q} = \dfrac{dq}{p} = \dfrac{dz}{p\,q} = \dfrac{dx}{p} = \dfrac{dy}{q}$

   (c) $\dfrac{dp}{p} = \dfrac{dq}{q} = \dfrac{dz}{p\,q} = \dfrac{dx}{p} = \dfrac{dy}{q}$ (d) $\dfrac{dp}{p} = \dfrac{dq}{q} = \dfrac{dz}{2\,p\,q} = \dfrac{dx}{p} = \dfrac{dy}{q}$.

6. Equation $r^2 + 2\,s - t^2 = 0$ is of order:

   (a) one (b) two

   (c) three (d) none of these

7. The solution of a P.D.E. contains:

   (a) arbitrary functions (b) arbitrary constants

   (c) both (a) and (b) (d) non of these

8. The equation $P\,p + Q\,q = R$ is known as:

   (a) Charpit's equation (b) Lagrange's equation

   (c) Clairaut's equation (d) None of these

9. The equation of envelope of surface represented by complete integral of the given P.D.E. is called:
   (a) singular solution (b) particular integral
   (c) general integral (d) none of these
10. Out of the following four PDEs, which equation is linear:
    (a) $\dfrac{\partial^3 z}{\partial x^3} - 3\dfrac{\partial^2 z}{\partial x^2}\dfrac{\partial z}{\partial y} + 8\dfrac{\partial^2 z}{\partial y^2} = \sin x$
    (b) $\dfrac{\partial^2 z}{\partial x^2} + \left(\dfrac{\partial z}{\partial y}\right)^2 + 9z = 0$
    (c) $\dfrac{\partial^2 z}{\partial x^2} + 5\dfrac{\partial^2 z}{\partial y^2} = 0$
    (d) None of these
11. The partial differential equation $\dfrac{\partial^2 y}{\partial t^2} = c^2\dfrac{\partial^2 y}{\partial x^2}$ is:
    (a) Parabolic (b) Elliptic
    (c) Hyperbolic (d) None of these
12. The solution of P.D.E. $(y - z)p + (z - x)q = x - y$ is:
    (a) $f(x + y + z) = x\,y\,z$
    (b) $f(x^2 + y^2 + z^2) = x\,y\,z$
    (c) $f(x^2 + y^2 + z^2\; x^2\; y^2\; z^2) = 0$
    (d) $f(x + y + z) = x^2 + y^2 + z^2$.
13. Number of arbitrary constants in singular solution of an equation of degree n are:
    (a) n (b) $n - 1$
    (c) 0 (d) 1.
14. Singular solution of $p^2 + q^2 = q^2$ is:
    (a) $x^2 + y^2 = a^2$ (b) $x\,y = a^2$
    (c) $y^2 = a\,x$ (d) None of these.
15. Singular solution of $z = p\,x + q\,y - 2\sqrt{p\,q}$ is:
    (a) $x^2 + y^2 = 1$ (b) $y^2 = x$
    (c) $x\,y = 1$ (d) $x = y^2$
16. P.I. in the solution of differential equation $(D^2 - 2\,D\,D' + D'^2)z = e^{x+2y}$ is:

(a) $x+2y$ (b) e^{x+2y}

(c) e^{2y+x} (d) None of these.

17. In the solution of equation $(D^2-6DD'+9D'^2)z=6x+2y$ the C.F. is:

(a) $\phi_1(y+3x)+\phi(y-3x)$

(b) $\phi_1(3y+x)+\phi_2(3y-x)$

(c) $\phi_1(3y+x)+x\phi_2(3y+x)$

(d) $\phi_1(y+3x)+x\phi_2(y+3x)$

18. Solution of the equation $z=px+qy+f(p,q)$ is:

(a) $z=f(ax,by)$ (b) $z=f(ax/by)$

(c) $z=ax+by+f(a,b)$ (d) None of these

19. Lagrange's subsidiary equations for P.D.E. $y^2zp+zx^2q=xy^2$ are:

(a) $\frac{dx}{y^2z}=\frac{dy}{zx^2}=\frac{dz}{y^2}$ (b) $\frac{dx}{x^2}=\frac{dy}{y^2}=\frac{dz}{zx}$

(c) $\frac{dx}{1/x^2}=\frac{dy}{1/y^2}=\frac{dz}{1/zx}$ (d) None of these.

20. The complete integral of equations of the type $f(p,q)=0$ is $z=ax+by+c$, where a and b are connected by the relation:

(a) $f(a,b)=0$ (b) $f(ab)=0$

(c) $f(b/a)=0$ (d) $f(a/b)=0$

21. P.I. of the P.D.E. $(D^3-3DD'+D'+1)z=e^{4x+5y}$ is:

(a) $\frac{1}{4}e^{4x+5y}$ (b) $\frac{1}{5}e^{4x+5y}$

(c) $\frac{1}{10}e^{4x+5y}$ (d) $\frac{1}{20}e^{4x+5y}$

22. The P.D.E. of the form $f(x,y,z,p,q,r,s,t)=0$ can be integrated by the method:

(a) Charpit's method (b) Monge's method
(c) Lagrange's method (d) Clairaut's method

23. The complete integral of $f(p,q)=0$ is:

(a) $z=ax+b$ (b) $z=ax+F(a)y+b$

(c) $z=ax+by+c$ (d) None of these

24. The order of the equation $r+2s-3t=xy$ is

(a) 1 (b) 2

(c) 3 (d) 4

25. The equation $Rr+Ss+Tt+f(x,y,z,p,q)=0$ is called elliptic if

(a) $S^2-4RT<0$ (b) $S^2-4RT=0$

(c) $S^2-4RT>0$ (d) None of these

26. The order of the equation $r-t=x-y$ is

(a) 1 (b) 2

(c) 3 (d) 0

ANSWERS

1.	(a)	2.	(d)	3.	(b)	4.	(b)
5.	(a)	6.	(b)	7.	(c)	8.	(b)
9.	(a)	10.	(c)	11.	(c)	12.	(d)
13.	(c)	14.	(d)	15.	(c)	16.	(b)
17.	(d)	18.	(c)	19.	(c)	20.	(a)
21.	(c)	22.	(b)	23.	(b)	24.	(b)
25.	(a)	26.	(b)				

9.1 INTRODUCTION

Laplace transform or Laplace transformation is one of the remarkable methods which can be used for solving linear differential equation arising in physics and engineering. It reduces to problem of solving differential equation to an algebraic equation. Laplace transform directly gives the solution of differential equations with given initial/ boundary conditions without necessity of first finding the general solution and then evaluating the arbitrary constants.

Definition. Let $F(t)$ be function defined for all positive values "t" then

$$L\{F(t)\} = f(p) = \int_0^\infty e^{-pt} F(t)\,dt,$$

provided the integral exists, is called the Laplace transform of $F(t)$.

9.2 SOME IMPORTANT DEFINITIONS

9.2.1 Piecewise (or Sectionally) Continuous Function

A function is said to be piecewise continuous on a closed interval if the interval can be divided into a finite number of subintervals for such that in each subinterval

(i) $F(t)$ is continuous

(ii) $F(t)$ approaches a finite limit at t tends to either end point from the interior, that is

$$\lim_{\substack{t \to t_i + \varepsilon \\ \varepsilon \to 0}} F(t) \text{ and } \lim_{\substack{t \to t_{i+1} - \varepsilon \\ \varepsilon \to 0}} F(t) \text{ exist}.$$

For example, the function $F(t) = \begin{cases} t, \text{ for } 0 \le t < 1 \\ 2, \text{ for } 1 < t < 2 \\ 10 - t^2, \text{ for } 2 < t \le 3 \end{cases}$ is piecewise continuous in the interval $[0,3]$ as show the following figure

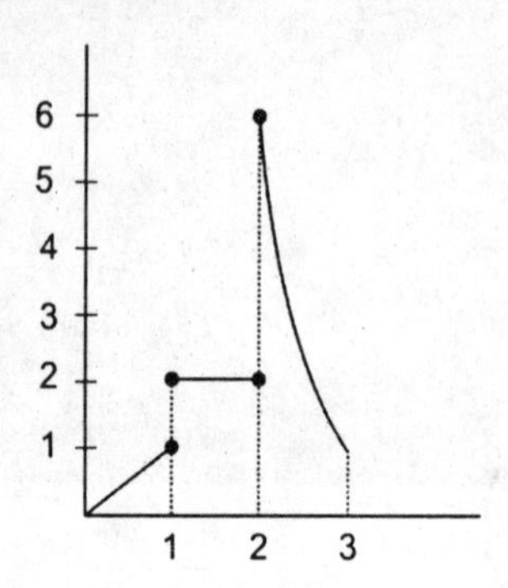

9.2.2 Existence Condition

The Laplace transform of a function $F(t), t \geq 0$ can be obtained if $\int_0^\infty e^{-pt} F(t)dt$ exists. This integral exists if $\lim_{\lambda \to \infty} \int_0^\lambda e^{-pt} F(t)dt$ exists. The following definitions will be helpful for the existence of the Laplace transform of $F(t)$.

9.2.3 Functions of Exponential Order

A function $F(t)$ is said to be of exponential order $'\alpha'$ as $t \to \infty$ if $\lim_{t\to\infty}\{e^{-\alpha t} F(t)\}$ exists and finite. That is, for a given $T, \exists$ a positive constant M such that

$$|F(t)| < M e^{\alpha t} \text{ or } |e^{-\alpha t} F(t)| < M, \text{for all } t \geq T.$$

If a function $F(t)$ is of exponential order $'\alpha'$ then it is also of exponential order β where $\beta > \alpha$.

9.2.4 A Function of Class "A"

A piecewise continuous function $F(t)$ on every finite interval in the range $t \geq 0$ is called a "function of class A" if $F(t)$ is of exponential order as $t \to \infty$, that is

$$\lim_{t\to\infty}\{e^{-\alpha t} F(t)\} \text{ exists and finite.}$$

9.2.5 Null Function

If $N(t)$ is a function of t such that $\int_0^t N(t)dt = 0$ for all positive values of t, then $N(t)$ is called a null function.

Theorem 9.1: If $F(t)$ is piecewise-continuous on every finite interval $t \geq 0$ and of exponential order $'\alpha'$ as $t \to \infty$, then the Laplace transform of $F(t)$ exists for all $p > \alpha$.

Proof. We know that $L\{F(t)\} = f(p) = \int_0^\infty e^{-pt} F(t)dt$

$$= \int_0^T e^{-pt} F(t)dt + \int_T^\infty e^{-pt} F(t)dt. \quad ...(1)$$

The first integral on right hand side exists because $F(t)$ is continuous on every finite interval in the range $t \geq 0$.

Now $\left| \int_T^\infty e^{-pt} F(t) dt \right| \leq \int_T^\infty \left| e^{-pt} F(t) \right| dt$

$< \int_T^\infty e^{-pt} e^{\alpha t} M \, dt$, since $F(t)$ is of exponential order α as $t \to \infty$

$$= M \int_T^\infty e^{-(p-\alpha)t} \, dt = \frac{M e^{-(p-\alpha)T}}{p-\alpha}, \text{ if } p > \alpha.$$

Since $\dfrac{M e^{-(p-\alpha)T}}{p-\alpha}$ can be made as small as we please by taking T large enough, the second integral on the right hand side of (1) also exists. Hence $\int_0^\infty e^{-pt} F(t) dt$ exists, for $p > \alpha$, *i.e.*, $f(p)$ exists for $p > \alpha$.

Remarks 1: This theorem can also be stated as

"If $F(t)$ is a function of class *A*, the Laplace transform of $F(t)$ exists for $p > \alpha$"

2. The condition of piecewise continuity in the above theorem is sufficient for the existence of the Laplace transform of $F(t)$, but not necessary. For example, the Laplace transform of $F(t) = \dfrac{1}{\sqrt{t}}$ exists even though $\dfrac{1}{\sqrt{t}}$ is not piecewise continuous on every finite interval in the range $t \geq 0$ since $f(t) \to \infty$ as $t \to 0$.

Example 1: Show that the Laplace transforms of the function $F(t) = t^n, -1 < n < 0$ exists, although it is not a function of class *A*.

Solution: Here $F(t) \to \infty$ as $t \to 0$ for $t \geq 0$, *i.e.*, the function is not piecewise continuous on every finite interval although in the range $t \geq 0$.

We have $\lim\limits_{t\to\infty} \left(e^{-at} F(t) \right) = \lim\limits_{t\to\infty} \left(\dfrac{t^n}{e^{at}} \right) = \lim\limits_{t\to\infty} \dfrac{n}{a\, e^{at}} t^{n-1}$, [by L'Hospital's rule]

$$= \lim_{t\to\infty} \frac{n!}{a^n e^{at}} = 0.$$

This shows that the function $F(t) = t^n$ is of exponential order. Since $F(t) = t^n$ is not sectionally continuous over finite interval in the range $t \geq 0$ hence it is not a function of class *A*.

But t^n is integrable from 0 to any positive number t_0.

Now $\quad L\{ F(t) \} = \int_0^\infty e^{-pt} F(t) dt$

$$= \int_0^\infty e^{-pt} t^n \, dt = \int_0^\infty e^{-x} \left(\frac{x}{p}\right)^n \frac{1}{p} dx,$$

(by using $pt = x \Rightarrow dt = \frac{dx}{p}$ for $p > 0$)

$$= \frac{1}{p^{n+1}} \int_0^\infty e^{-x} x^{(n+1)-1} dx$$

$$= \frac{\Gamma(n+1)}{p^{n+1}} \text{ if } p > 0 \text{ and } n+1 > 0 \text{ i.e., } n > -1.$$

Hence the Laplace transform of t^n, $0 > n > -1$ exists although it is not a function of class A.

9.3 LAPLACE TRANSFORMS OF SOME ELEMENTARY FUNCTIONS

1. $F(t) = 1$: $L\{F(t)\} = \int_0^\infty e^{-pt} 1 \, dt = \left[\frac{e^{-pt}}{-p}\right]_0^\infty = \frac{1}{p}$ if $p > 0$.

2. $F(t) = t^n$, if $n \geq 1$ and n is an integer:

$L\{t^n\} = \int_0^\infty e^{-pt} t^n \, dt$, let $pt = y$, then $t = \frac{y}{p}$ and $dt = \frac{dy}{p}$

$$= \int_0^\infty e^{-y} \left(\frac{y}{p}\right)^n \frac{dy}{p} = \frac{1}{p^{n+1}} \int_0^\infty e^{-y} y^{(n+1)-1} dy$$

$= \frac{\Gamma(n+1)}{p^{n+1}} = \frac{n!}{p^{n+1}}$, since if $n = 1$ is an integer, $\Gamma(n+1) = n!$.

Thus for $n = 1$, we get $L\{t\} = \frac{1}{p^2}$.

3. $F(t) = e^{at}$, if $p > a$

$$L\{e^{at}\} = \int_0^\infty e^{-pt} e^{at} \, dt = \int_0^\infty e^{-(p-a)t} \, dt = \left[\frac{e^{-(p-a)t}}{-(p-a)}\right]_0^\infty = \frac{1}{(p-a)}, p > a.$$

4. $F(t) = \sin at$ if $p > 0$: $L\{\sin at\} = \int_0^\infty e^{-pt} \sin at \, dt$

$$= \left[\frac{e^{-pt}(-p \sin at - a \cos at)}{(p^2 + a^2)}\right]_0^\infty,$$

since $\int e^{ax} \sin bx \, dx = \frac{e^{ax}}{(a^2 + b^2)}(a \sin bx - b \cos bx)$

$$= \frac{a}{(p^2 + a^2)}, p > 0.$$

5. $F(t) = \cos a\,t$ if $p > 0$:

$$L\{\cos a\,t\} = \int_0^\infty e^{-pt} \cos a\,t\,d\,t$$

$$= \left[\frac{e^{-pt}(-p\cos a\,t + a\sin a\,t)}{(p^2 + a^2)}\right]_0^\infty,$$

since $\int e^{ax}\cos b\,x\,d\,x = \frac{e^{ax}}{(a^2+b^2)}(a\cos b\,x + b\sin b\,x)$

$$= \frac{p}{(p^2 + a^2)}, p > 0$$

6. $F(t) = \sin h\,a\,t$ if $p > |a|$:

$$L\{\sin h\,a\,t\} = \int_0^\infty e^{-pt}\sin h\,a\,t\,d\,t = \int_0^\infty e^{-pt}\left(\frac{e^{at} - e^{-at}}{2}\right)d\,t$$

$$= \frac{1}{2}\left[\int_0^\infty e^{-(p-a)}\,d\,t - \int_0^\infty e^{-(p+a)t}\,d\,t\right]$$

$$= \frac{1}{2}\left[\frac{1}{-(p-a)}e^{-(p-a)t}\right]_0^\infty - \frac{1}{2}\left[\frac{e^{-(p+a)t}}{-(p+a)}\right]_0^\infty$$

$$= \frac{1}{2}\left[\frac{1}{(p-a)} - \frac{1}{(p+a)}\right] = \frac{a}{p^2 - a^2}, \text{ for } p > |a|.$$

7. $F(t) = \cos h\,a\,t$, *if* $p > |a|$:

$$L(\cos h\,a\,t) = \int_0^\infty e^{-pt}\cos h\,a\,t\,d\,t = \int_0^\infty e^{-pt}\frac{(e^{at} + e^{-at})}{2}d\,t$$

$$= \frac{1}{2}\int_0^\infty e^{-(p-a)t}\,d\,t + \frac{1}{2}\int_0^\infty e^{-(p+a)t}\,d\,t$$

$$= \frac{1}{2}\left[\frac{1}{-(p-a)}e^{-(p-a)t} + \frac{e^{-(p+a)t}}{-(p+a)}\right]_0^\infty$$

$$= \frac{1}{2}\left[\frac{1}{(p-a)} + \frac{1}{(p+a)}\right] = \frac{p}{(p^2 - a^2)}, \text{ for } p > |a|.$$

Laplace transforms of the above cases can be tabulated as given below:

$F(t)$	$L[F(t)] = f(p)$
1	$\frac{1}{p}, p>0$
t^n	$\frac{\lfloor n+1}{p^{n+1}}, p>0$
e^{at}	$\frac{1}{(p-a)}, p>a$
$\sin at$	$\frac{a}{(p^2+a^2)}, p>0$
$\cos at$	$\frac{p}{(p^2+a^2)}, p>0$
$\sinh at$	$\frac{a}{(p^2-a^2)}, p>\|a\|$
$\cosh at$	$\frac{p}{(p^2-a^2)}, p>\|a\|$

9.4 LINEARITY PROPERTY OF LAPLACE TRANSFORMATION

Theorem 9.2: The Laplace transformation is linear transformation, i.e.
$L\{C_1 F_1(t) \pm C_2 F_2(t)\} = C_1 L\{F(t)\} \pm C_2 L\{F_2(t)\} = C_1 f_1(p) \pm C_2 f_2(p)$
where $L\{F_1(t)\} = f_1(p), L\{F_2(t)\} = f_2(p)$ and C_1, C_2 are constants.

Proof: We know that $L\{F(t)\} = \int_0^\infty e^{-pt} F(t) dt$

Then $L\{C_1 F_1(t) \pm C_2 F_2(t)\} = \int_0^\infty e^{-pt} \{C_1 F_1(t) \pm C_2 F_2(t)\} dt$

$$= C_1 \int_0^\infty e^{-pt} F_1(t) dt \pm C_2 \int_0^\infty e^{-pt} F_2(t) dt$$

$$= C_1 L\{F_1(t)\} \pm C_2 L\{F_2(t)\} = C_1 f_1(p) \pm C_2 f_2(p).$$

Theorem 9.3: (First translation property or first shifting theorem).

If $L\{F(t)\} = f(p)$ then $L\{e^{at} F(t)\} = f(p-a)$.

Proof: We know that $L\{F(t)\} = \int_0^\infty e^{-pt} F(t) dt$.

Then $L\{e^{at} F(t)\} = \int_0^\infty e^{-pt} e^{at} F(t) dt = \int_0^\infty e^{-(p-a)t} F(t) dt$

$$= \int_0^\infty e^{-ut} F(t) dt, \text{(where } u \equiv p-a>0) = f(u) = f(p-a).$$

Similarly, we have $L\{e^{-at} F(t)\} = f(p+a)$.

Now applying this property, we get the following results:.

1. $L\{e^{at} t^n\} = \dfrac{n!}{(p-a)^{n+1}}$.
2. $L\{e^{at} \sin bt\} = \dfrac{b}{(p-a)^2 + b^2}$
3. $L\{e^{at} \cos bt\} = \dfrac{(p-a)}{(p-a)^2 + b^2}$
4. $L\{e^{at} \sinh bt\} = \dfrac{b}{(p-a)^2 - b^2}$
5. $L\{e^{at} \cosh bt\} = \dfrac{(p-a)}{(p-a)^2 - b^2}$

Theorem 9.4: (second translation property or Heaviside's shifting theorem).

If $L\{F(t)\} = f(p)$ and $G(t) = \begin{cases} F(t-a), & for\ t > a \\ 0, & for\ t < a \end{cases}$,

then $\quad L\{G(t)\} = e^{-ap} f(p)$.

Proof: We know that $L\{F(t)\} = \int_0^\infty e^{-pt} F(t)\,dt = f(p)$.

$$\begin{aligned} L\{G(t)\} &= \int_0^\infty e^{-pt} G(t)\,dt \\ &= \int_0^a e^{-pt} G(t)\,dt + \int_0^\infty e^{-pt} G(t)\,dt \\ &= \int_0^a e^{-pt}\, 0\, dt + \int_a^\infty e^{-pt} F(t-a)\,dt. \end{aligned}$$

Now putting $t - a = x$, we obtain

$$L\{G(t)\} = e^{-ap} \int_0^\infty e^{-px} F(x)\,dx = e^{-ap} L\{F(t)\} = e^{-ap} f(p).$$

Thus, $L\{G(t)\} = e^{-ap} f(p)$.

Theorem 9.5: (change of scale properties). If $L\{F(t)\} = f(p)$; then $L\{F(at)\} = \dfrac{1}{a} f\left(\dfrac{p}{a}\right)$.

Proof: We know by definition that

$$L\{F(t)\} = \int_0^\infty e^{-pt} F(t)\,dt.$$

Therefore $L\{F(at)\} = \int_0^\infty e^{-pt} F(at)\,dt$,

$$= \int_0^\infty e^{-p\left(\frac{x}{a}\right)} F(x) \frac{dx}{a}, \quad \left(\text{By using } at = x \text{ and } dt = \frac{dx}{a}\right)$$

$$=\frac{1}{a}\int_0^{\infty} e^{-(p/a)t} F(t)dt, \text{ since } \int_a^b F(x)dx = \int_a^b F(t)dt$$

$$=\frac{1}{a} f(p/a), \text{ since } f(p)=\int_0^{\infty} e^{-pt} F(t).$$

Example 2: Find (i) $L\{t^3 e^{-3t}\}$. (ii) $L\{e^{2t}(3\cos 6t - 5\sin 6t)\}$.

Solution: (i) we know that $L\{t^3\}=\frac{3!}{p^4}=\frac{6}{p^4}$.

Now from first shifting property, we get $L\{t^3 e^{-3t}\}=\frac{6}{(p+3)^4}$.

(ii) We have

$$L(3\cos 6t - 5\sin 6t)=3\frac{p}{(p^2+6^2)}-5\frac{6}{(p^2+6^2)}=\frac{(3p-30)}{(p^2+36)}=f(p).$$

Now from first shifting theorem, we have

$$L\{e^{-2t}(3\cos 6t - 5\sin 6t)=f(p+2)\}=\frac{3(p+2)-30}{(p+2)^2+36}=\frac{(3p-24)}{(p^2+4p+40)}.$$

Example 3: Find the Laplace transform of

$$F(t)=7e^{2t}+9e^{-2t}+5\cos t+7t^3+5\sin 3t+2.$$

Solution: We know that

$$L\{F(t)\}=L\{7e^{2t}+9e^{-2t}+5\cos t+7t^3+5\sin 3t+2\}$$

$$=7L\{e^{2t}\}+9L\{e^{-2t}\}+5L\{\cos t\}+7L\{t^3\}+5L\{\sin 3t\}+2L\{1\}$$

$$=7\frac{1}{(p-2)}+9\frac{1}{(p+2)}+5\frac{p}{(p^2+1)}+7\frac{3!}{p^4}+5\frac{3}{(p^2+9)}+2\frac{1}{p}$$

$$=\frac{7}{(p-2)}+\frac{9}{(p+2)}+\frac{5p}{(p^2+1)}+\frac{42}{p^4}+\frac{15}{(p^2+9)}+\frac{2}{p}.$$

Example 4: Find the Laplace transform of the following functions

(i) $1+\cos 2t$ (ii) $\sin 3t\cos 4t$

(iii) $\cos^2 2t$ (iv) $\cosh^3 2t$

(v) $(1+te^{-t})^3$.

Solution: (i) We have $L\{1+\cos 2t\}=L[1]+L\{\cos 2t\}=\frac{1}{p}+\frac{p}{(p^2+4)}$

(ii) Here $L\{\sin 3t\cos 4t\}=L\frac{1}{2}\{2\sin 3t\cos 4t\}$

$$= \frac{1}{2} L\{\sin 7t - \sin t\} = \frac{1}{2}[L\{\sin 7t\} - L\{\sin t\}]$$

$$= \frac{1}{2}\left(\frac{7}{p^2 + 7^2} - \frac{1}{p^2 + 1^2}\right) = \frac{1}{2}\left(\frac{7}{p^2 + 49} - \frac{1}{p^2 + 1}\right).$$

(iii) Now $L\{\cos^3 2t\} = L\{(\cos^2 2t)\cos 2t\} = L\left\{\frac{(\cos 4t + 1)}{2}\cos 2t\right\}$

$$= \frac{1}{2} L\{\cos 4t \cos 2t + \cos 2t\}$$

$$= \frac{1}{2} L\left\{\frac{1}{2}(\cos 6t + \cos 2t) + \cos 2t\right\}$$

$$= \frac{1}{4} L\{\cos 6t + 3\cos 2t\} = \frac{1}{4}[L\{\cos 6t\} + 3L\{\cos 2t\}]$$

$$= \frac{1}{4}\left(\frac{p}{p^2 + 36} + \frac{3p}{p^2 + 4}\right).$$

(iv) We know that $\cos h\, 6t = 4\cos h^3\, 2t - 3\cos h\, 2t$

$$\Rightarrow \cos h^3\, 2t = \frac{3}{4}\cos h\, 2t + \frac{1}{4}\cos h\, 6t.$$

Therefore $L\{\cos h^3\, 2t\} = L\left\{\frac{3}{4}\cos h\, 2t + \frac{1}{4}\cos h\, 6t\right\}$

$$= \frac{3}{4} L\{\cos h\, 2t\} + \frac{1}{4} L\{\cos h\, 6t\}$$

$$= \frac{3}{4}\left(\frac{p}{p^2 - 2^2}\right) + \frac{1}{4}\left(\frac{p}{p^2 - 6^2}\right) = \frac{3}{4}\left(\frac{p}{p^2 - 4}\right) + \frac{1}{4}\left(\frac{p}{p^2 - 36}\right).$$

(v) Here $L\{(1 + t e^{-t})^3\} = L\{1 + t^3 e^{-3t} + 3t e^{-t} + 3t^2 e^{-2t}\}$

$$= L\{1\} + L\{t^3 e^{-3t}\} + 3L\{t e^{-t}\} + 3L\{t^2 e^{-2t}\}$$

$$= \frac{1}{p} + \frac{3!}{(p+3)^4} + 3\frac{1}{(p+1)^2} + 3\frac{2!}{(p+2)^3}$$

Example 5: By using first shifting property find the Laplace transform of

(i) $t e^{-4t} \sin 3t$ (ii) $t^2 \sin 4t$ (iii) $t^2 e^{-t} \sin 4t$ (iv) $e^{-t} \cos t \cos 2t$

(v) $\cos h\, at \sin bt$.

Solution: (i). We know that $L(t) = \frac{1}{p^2}$

By first shifting property, we have

$$L\{t\,e^{3it}\}=\frac{1}{(p-3i)^2}=\frac{(p+3i)^2}{(p-3i)^2(p+3i)^2}$$

$$\Rightarrow L\{t\cos 3t+i\,t\sin 3t\}=\frac{(p^2-9)+6ip}{(p^2+9)^2}$$

On equating the imaginary parts, we get $L\{t\sin 3t\}=\dfrac{6p}{(p^2+9)^2}$.

Again applying first shifting property, we have

$$L\{e^{-4t}\,t\sin 3t\}=\frac{6(p+4)}{\{(p+4)^2+9\}^2}$$

(ii) We know that $L\{t^2\}=\dfrac{2}{p^3}$. By first shifting property, we have

$$L\{t^2\,e^{4it}\}=\frac{2}{(p-4i)^3}=\frac{2(p+4i)^3}{(p-4i)^3(p+4i)^3}=\frac{2\left[p^3-64i+12ip(p+4i)\right]}{(p^2+16)^3}$$

$$\Rightarrow \qquad L\{t^2\cos 4t+i\,t^2\sin 4t\}=\frac{2\left[(p^3-48p)+i(12p^2-64)\right]}{(p^2+16)^3}$$

On equating imaginary parts, we get $L\{t^2\sin 4t\}=\dfrac{8(3p^2-16)}{(p^2+16)}$ ...(1)

(iii) Applying first shifting property on (1), we get

$$L\{e^{-t}\,t^2\sin 4t\}=\frac{8\left[3(p+1)^2-16\right]}{\left[(p+1)^2+16\right]^3}$$

$$=\frac{8\left[3p^2+6p+3-16\right]}{(p^2+2p+1+16)^3}=\frac{8(3p^2+6p-13)}{(p^2+2p+17)^3}$$

(iv) Since $\cos t\cos 2t=\dfrac{1}{2}(\cos 3t+\cos t)$, we have

$$L\{e^{-t}\cos t\cos 2t\}=\frac{1}{2}\left[L\{e^{-t}\cos 3t\}+L\{e^{-t}\cos t\}\right]$$

$$=\frac{1}{2}\left[\frac{p+1}{(p+1)^2+9}+\frac{p+1}{(p+1)^2+1}\right]=\frac{p+1}{2}\left[\frac{1}{p^2+2p+10}+\frac{1}{p^2+2p+2}\right]$$

(v) Since $\cos h\,at\sin bt=\dfrac{1}{2}\left(\dfrac{e^{at}+e^{-at}}{2}\right)\sin bt$, we have

$$L\{\cos h\,a\,t\sin b\,t\}=\frac{1}{2}\left[L\{e^{at}\sin b\,t\}+L\{e^{-at}\sin b\,t\}\right]$$

$$= \frac{1}{2}\left[\frac{b}{(p-a)^2+b^2}+\frac{b}{(p-a)^2+b^2}\right]$$

Example 6: Given $L\{F(t)\}=\dfrac{(p^2-p+1)}{(2p+1)^2(p-1)}$ applying the change of scale property show that $L\{F(2t)\}=\dfrac{p^2-2p+4}{4(p+1)^2(p-2)}$.

Solution: We have $L\{F(t)\}=\dfrac{p^2-p+1}{(2p+1)^2(p-1)}=f(p)$.

Therefore

$$L\{F(2t)\}=\frac{1}{2}f(p/2)=\frac{1}{2}\frac{(p/2)^2-(p/2)+1}{\left[2(p/2)+1\right]^2(p/2-1)}=\frac{p^2-2p+4}{4(p+1)^2(p-2)}$$

Example 7: Find $L\{G(t)\}$ where $G(t)=\begin{cases} e^{t-a}, & t>a \\ 0, & t<a \end{cases}$

Solution: We know that if $L\{F(t)\}=f(p)$ and $G(t)=\begin{cases} e^{t-a}, & t>a \\ 0, & t<a \end{cases}$; then

$$L\{G(t)\}=e^{-ap}f(p)$$

For $F(t)=e^t$, we have $L\{F(t)\}=\dfrac{1}{(p-1)},(p>1)=f(p)$

and $G(t)=\begin{cases} F(t-a), & t>a \\ 0, & t<a \end{cases}=\begin{cases} e^{t-a}, & t>a \\ 0, & t<a \end{cases}$.

Therefore $L\{G(t)\}=e^{-ap}f(p)=\dfrac{e^{-ap}}{(p-1)}, p>1$.

Example 8: Find $L\{F(t)\}$, where $F(t)=\begin{cases} \cos\left(t-\frac{2}{3}\pi\right), & t>\frac{2\pi}{3} \\ 0, & t<\frac{2\pi}{3} \end{cases}$

Solution: Let $\phi(t)=\cos t$, then $F(t)=\begin{cases} \phi\left(t-\frac{2}{3}\pi\right), & t>\frac{2\pi}{3} \\ 0, & t<\frac{2\pi}{3} \end{cases}$

We have $L\{\phi(t)\}=L\{\cos t\}=\dfrac{p}{(p^2+1)}=f(p)$.

From second shifting theorem, we obtain

$$L\{F(t)\}=e^{\left(-\frac{2}{3}\pi\right)p} f(p)=e^{-\frac{2\pi p}{3}} \frac{p}{(p^2+1)}.$$

Aliter. We know that $L\{F(t)\}=\int_0^\infty e^{-pt} F(t)dt$.

Therefore $L\{G(t)\}=\int_0^{2\pi/3} e^{-pt} F(t)dt+\int_{2\pi/3}^\infty e^{-pt} F(t)dt$

$$=\int_0^{2\pi/3} e^{-pt}\, 0\, dt+\int_{2\pi/3}^\infty e^{-pt} \cos(t-2\pi/3)dt.$$

Now putting $(t-2\pi/3)=x$, we get

$$L\{G(t)\}=0+\int_0^\infty e^{-(x+2\pi/3)p} \cos x\, dx=e^{-\frac{2\pi p}{3}} \int_0^\infty e^{-px} \cos x\, dx$$

$$=e^{-\frac{2\pi p}{3}} L\{\cos x\}=e^{-\frac{2\pi p}{3}} \frac{p}{(p^2+1)}, p>0.$$

Example 9: Find the Laplace transform of $F(t)=\begin{cases} \frac{t}{T}, \text{ when } 0<t<T \\ 1, \text{ when } t>T \end{cases}$

Solution: We have

$$L\{F(t)\}=\int_0^\infty e^{-pt} F(t)dt=\int_0^T e^{-pt} \frac{t}{T} dt+\int_T^\infty e^{-pt} 1.dt.$$

$$=\frac{1}{T}\left[1.t\frac{e^{-pt}}{-p}\right]_0^T-\frac{1}{T}\int_0^T 1.\frac{e^{-pt}}{-p}dt+\left[\frac{e^{-pt}}{-p}\right]_T^\infty$$

$$=\frac{-e^{-pT}}{p}-\left(\frac{e^{-pT}-1}{Tp^2}\right)+\frac{e^{-pT}}{p}=\frac{1-e^{-pT}}{Tp^2}.$$

Example 10: If $L\{J_0\sqrt{(t)}\}=\frac{e^{-1/(4p)}}{p}$, *then find* $L\{J_0 2\sqrt{(t)}\}$.

Solution: By change of scale property

$$L\{J_0\sqrt{4t}\}=\frac{1}{4}\left\{\frac{e^{-1/(4p/4)}}{p/4}\right\} \Rightarrow L\{J_0(2\sqrt{t})\}=\frac{1}{p}e^{-1/p}$$

EXERCISE 9.1

Find the Laplace transform:

1. (a) $e^{-t}\cos 4t$ (b) $e^{3t}\sin 4t$. 2. $e^{2t}(3\sin 4t-4\cos 4t)$.

3. $e^{-t}(3\sin 2t-5\cosh 2t)$. 4. $e^{-4t}\cosh 2t$.

5. $e^{-3t}+5e^{-t}$.

6. $7e^{2t}+9e^{-2t}+5\cos t+7t^3+5\sin 3t+2$.

7. $-4e^{-3t}-2\sin 5t+3\cos 2t-2t^3+3t^4$.

8. (a) $2e^{2t}\sin 4t$ (b) $\dfrac{e^{-at}t^{n-1}}{(n-1)!}$

9. $(\sin t - \cos t)^2$.

10. $\sin h(t/2)\sin\left(\sqrt{3}\,t/2\right)$

11. Show that

(i) $L\{\sin kt + \sin h\,kt\} = \dfrac{(2p^2k)}{(p^4-k^4)}$ (ii) $L\{\sin^3 t\} = \dfrac{6}{(p^2+1)(p^2+9)}$

(iii) $L\{t\sin at\} = \dfrac{2ap}{(p^2+a^2)^2}$ (iv) $L\{t\cos at\}\dfrac{(p^2-a^2)}{(p^2+a^2)^2}$

12. Find the Laplace transform of $F(t)$ defined as

(i) $F(t)=\begin{cases} t, & 0<t<4 \\ 5, & t>4 \end{cases}$ (ii) $F(t)=\begin{cases} \sin t, & 0<t<\pi \\ 0, & t>\pi \end{cases}$

(iii) $F(t)=\begin{cases} \cos t, & 0<t<2\pi \\ 0, & t>2\pi \end{cases}$ (iv) $F(t)=\begin{cases} t^2, & 0<2 \\ (t-1), & 2<t<3 \\ 7, & t>3 \end{cases}$

(v) $F(t)=\begin{cases} e^t, & 0<t<5 \\ 3, & t>5 \end{cases}$ (vi) $F(t)=\begin{cases} \sin\left(t-\dfrac{2\pi}{3}\right), & t>2\pi/3 \\ 0, & t<2\pi/3 \end{cases}$

(vii) $F(t)=\begin{cases} \cos t, & 0<t<\pi \\ 0, & t>\pi \end{cases}$ (viii) $F(t)=\begin{cases} 1, & 0\le t<1 \\ t, & 1\le t<2 \\ t^2, & 2\le t<\infty \end{cases}$

13. If $L\{F(t)\} = \dfrac{1}{p}e^{-1/p}$ prove that $L\{e^{-t}F(3t)\} = \dfrac{e^{-3/(p+1)}}{(p+1)}$.

ANSWERS 9.1

1. (a) $\dfrac{4(p+1)}{\left[(p+1)^2+16\right]}$ (b). $\dfrac{4}{(p^2-6p+25)}$

2. $\dfrac{(20-4p)}{(p^2-2p+20)}$

3. $\dfrac{6}{(p^2+2p+5)} - \dfrac{(5p+5)}{(p^2+2p-3)}$.

4. $\dfrac{p+4}{p^2+8p+12}$

5. $\dfrac{2(3p+8)}{(p+1)(p+3)}, p>-1$.

6. $\left[\dfrac{16p-4}{p^2-4} + \dfrac{5p}{p^2+1} + \dfrac{42+2p^3}{p^4} + \dfrac{15}{p^2+9}\right]$.

7. $\left[-\frac{4}{p+3}-\frac{10}{p^2+25}+\frac{3p}{p^2+4}-\frac{12}{p^4}+\frac{72}{p^5}\right]$.

8. (a) $\frac{4}{p^2+4p+20}$ (b) $\frac{1}{(p+a)^n}$

9. $\left[\frac{1}{p}-\frac{2}{p^2+4}\right]$. 10. $\frac{\sqrt{3}\,p}{2(p^4+p^2+1)}$.

12. (i) $\left[\frac{1}{p^2}+e^{-4p}\left(\frac{1}{p}-\frac{1}{p^2}\right)\right]$ (ii) $\left[\frac{1+e^{-\pi/p}}{p^2+1}\right], p>0$

(iii) $\frac{p\left(1-e^{-2\pi p}\right)}{p^2+1}$ (iv) $\frac{2}{p^3}-\frac{e^{-2p}}{p^3}\left(2+3p+3p^2\right)+\frac{e^{-3p}}{p^2}(5p-1)$

(v) $\frac{1-e^{-5(p-1)}}{(p-1)}+\frac{3}{p}e^{-5p}$. (vi) $\frac{e^{-(2\pi/3)p}}{p^2+1}$.

(vii) $\frac{p(1+e^{-p\pi})}{p^2+1}$. (viii) $\frac{1}{p}+\frac{2e^{-2p}}{p}+\frac{3e^{-2p}}{p^2}+\frac{2e^{-2p}}{p^3}+\frac{e^{-p}}{p^2}$.

9.5 LAPLACE TRANSFORM OF THE DERIVATIVE OF $F(t)$

Theorem 9.6: Let $F(t)$ be continuous for all $t \geq 0$ and be of exponential order as $t \to \infty$ and if $F'(t)$ is of class A, then Laplace transform of the derivative $F'(t)$ exists when $p>a$ and

$$L\{F'(t)\}=pL\{F(t)\}-F(0).$$

Proof: If case $F'(t)$ is continuous for all $t \geq 0$, then

$$L\{F'(t)\}=\int_0^\infty e^{-pt}F'(t)dt=\left[e^{-pt}F(t)\right]_0^\infty+p\int_0^\infty e^{-pt}F(t)dt$$

$$=\lim_{t\to\infty} e^{-pt}F(t)-F(0)+pL\{F(t)\}. \qquad \text{...(1)}$$

Now $|F(t)| \leq Me^{at}$ for all $t \geq 0$ and for some constants a and M.

We have

$$\left|e^{-pt}F(t)\right|=e^{-pt}|F(t)| \leq e^{-pt}Me^{at}=Me^{-(p-a)t} \to 0 \text{ as } t\to\infty \text{ if } p>a.$$

Hence $\lim_{t\to\infty} e^{-pt}F(t)=0$, for $p>a$.

Therefore from (1), we conclude that $L\{F'(t)\}$ exists and

$$L\{F'(t)\}=pL\{F(t)\}-F(0).$$

Theorem 9.7: If $F(t)$ is continuous except for an ordinary discontinuity at $t=a(a>0)$ as given below, then

$$L\{F'(t)\}=pL\{F(t)\}-F(0)-e^{-ap}\left[F(a+0)-F(a-0)\right]$$

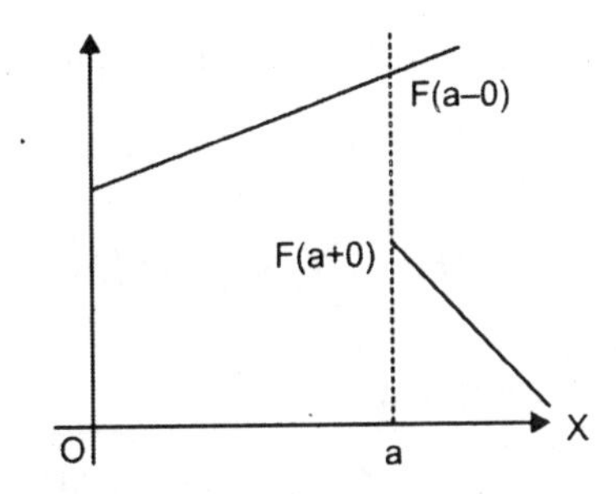

where $F(a+0)$ and $F(a-0)$ are the limits of F at $t=a$ as t approaches a from right and from left respectively, the quantity $F(a+0)-F(a-0)$ is called the jump of the discontinuity at $t=a$ and $e^{-pt}F(t)\to 0$ as $t\to\infty$.

Proof: Here

$$L\{F'(t)\}=\int_0^\infty e^{-pt}F'(t)dt=\int_0^a e^{-pt}F'(t)dt+\int_a^\infty e^{-pt}F'(t)dt$$

$$=\left[e^{-pt}F(t)\right]_0^a+p\int_0^a e^{-pt}F(t)dt+\left\{e^{-pt}F(t)\right\}_a^\infty+\int_a^\infty e^{-pt}F(t)dt$$

$$=e^{-ap}F(a-0)-F(0)+p\int_0^\infty e^{-pt}F(t)dt+\lim_{t\to\infty}\{e^{-pt}F(t)\}-e^{-pa}F(a+0)$$

which implies that

$$L\{F'(t)\}=pL\{F(t)\}-F(0)-e^{-ap}\left[F(a+0)-F(a-0)\right]$$

Theorem 9.8: [Laplace transform of the nth order derivative of $F(t)$]
Let $F(t)$ and its derivatives $F'(t), F''(t), \ldots\ldots, F^{n-1}(t)$ are continuous functions for all $t\geq 0$ and are of exponential orders as $t\to\infty$ and if $F^{n}(t)$ is of class A, then Laplace transform of $F^{n}(t)$ exists when $p>a$ and is given by

$$L\{F^{n}(t)\}=p^{n}L\{F(t)\}-p^{n-1}F(0)-p^{n-2}F'(0)\ldots.-F^{n-1}(0).$$

Proof: We have $L\{F'(t)\}=pL\{F(t)\}-F(0)$ ...(1)

Now applying the equation (1) for second order derivative $F''(t)$, we get

$$L\{F''(t)\}=pL\{F'(t)\}-F'(0)=p\left[pL\{F(t)\}-F(0)\right]-F'(0)$$

$$=p^{2}L\{F(t)\}-pF(0)-F'(0).\qquad\ldots(2)$$

Again applying equation (1) for third order derivative $F'''(t)$, we obtain

$$L\{F'''(t)\}=pL\{F''(t)\}-F''(0)=p\left[p^2 L\{F''(t)\}-pF(0)-F'(0)\right]-F''(0)$$
$$=p^3L\{F(t)\}-p^2F(0)-pF'(0)-F''(0).$$

Proceeding similarly for n^{th} order derivative, we get

$$L\{F^n(t)\}=p^n L\{F(t)\}-p^{n-1}F(0)-p^{n-2}F'(0).....F^{n-1}(0)$$

Or $$L\{F^n(t)\}=p^nL\{F(t)\}-\sum_{r=0}^{r=n-1}p^{n-1-r}F^r(0).$$

9.6 INITIAL VALUE THEOREM

If Laplace transform of $F(t)$ and $F'(t)$ exist and if $L\{F(t)\}=f(p)$, then

$$\lim_{p\to\infty}\left[pL\{F(t)\}\right]=\lim_{t\to 0}F(t).$$

Proof: We know that

$$L[F'(t)]=pf(p)-F(0), \text{ since } L\{F(t)\}=f(p).$$

then $$pf(p)=L[F'(t)]+F(0)=\int_0^\infty e^{-pt}F'(t)\,dt+F(0).$$

Taking limit $p\to\infty$ on both sides, we obtain

$$\lim_{p\to\infty}\{pf(p)\}=\lim_{p\to\infty}\int_0^\infty e^{-pt}F'(t)\,dt+F(0)=\int_0^\infty \lim_{p\to\infty}\{e^{-pt}F'(t)\}\,dt+F(0)$$

$$=0+F(0),\left(\text{as }\lim_{p\to\infty}\{e^{-pt}F'(t)\}=0\right)=\lim_{t\to 0}F(t).$$

Hence $$\lim_{p\to\infty}pf(p)=\lim_{t\to 0}F(t).$$

9.7 FINAL VALUE THEOREM

If $F(t)$ is continuous for all $t\geq 0$ and is of exponential order as $t\to\infty$ and if $F'(t)$ is of class *A*, then

$$\lim_{t\to\infty}F(t)=\lim_{p\to 0}p\,L\{F(t)\}.$$

Proof: We know that $L\{F'(t)\}=pf(p)-F(0)$, where $f(p)=L\{F(t)\}$

$\Rightarrow$ $$pf(p)=L\{F'(t)\}+F(0)=\int_0^\infty e^{-pt}F'(t)\,dt+F(0).$$

Taking limits as $p\to 0$ on both sides, we have

$$\lim_{p\to 0}\{pf(p)\}=\lim_{p\to 0}\int_0^\infty e^{-pt}F'(t)\,dt+F(0)=\int_0^\infty \lim_{p\to 0}\{e^{-pt}F'(t)\}\,dt+F(0)$$

$$= \int_0^\infty F'(t)\,dt + F(0) = \lim_{t\to\infty} \int_0^t F'(t)\,dt + F(0)$$

$$= \lim_{t\to\infty} \left[F(t) \right]_0^t + F(0) = \lim_{t\to\infty} \{F(t) - F(0)\} + F(0)$$

$$= \lim_{t\to\infty} F(t) - F(0) + F(0) = \lim_{t\to\infty} F(t).$$

Hence $\lim_{p\to 0} \{p\, f(p)\} = \lim_{t\to\infty} F(t).$

9.8 LEIBNITZ RULE FOR DIFFERENTIATION

To develop the theory of Laplace transform further, we state the following results for differentiation under the integral sign.

Let , where v_1 and v_2 may depend on the parameter α .

Then $\dfrac{d\phi}{d\alpha} = \displaystyle\int_{v_1}^{v_2} \frac{\partial f}{\partial \alpha}\, dx + f(v_2, \alpha)\frac{d v_2}{d\alpha} - f(v_1, \alpha)\frac{d v_1}{d\alpha}$ ...(1)

for $a \le \alpha \le b$ if $f(x, \alpha)$ and $\dfrac{\partial f}{\partial \alpha}$ are continuous in both x and α in some region of $x\alpha$ plane including $v_1 \le x \le v_2$, $a \le \alpha \le b$ and v_1 , v_2 are continuous and have continuous derivatives in interval (a, b). This expression (1) is known as Leibnitz rule for differentiation under the sign of integral.

Note. If v_1 , v_2 are constants, the last two terms in (1) are zero and so

$$\frac{d\phi}{d\alpha} = \int_{v_1}^{v_2} \frac{\partial f}{\partial \alpha}\, dx.$$

9.9 LAPLACE TRANSFORM OF INTEGRALS

If $L\{ F(t)\} = f(p)$, then $L\left\{ \int_0^t F(t)\,dt \right\} = \dfrac{1}{p} f(p).$

Proof: Let. $L\left\{\dfrac{\sin t}{t}\right\} = \cot^{-1} p$, then find $L\left\{ \displaystyle\int_0^t \left(\frac{\sin t}{t}\right) dt \right\}.$

Taking Laplace transform, we get

$$L\{G'(t)\} = p\, L\{G(t)\} - G(0) = p\, L\{G(t)\}$$

$$\Rightarrow \quad L\{G(t)\} = \frac{1}{p} L\{G'(t)\} = \frac{1}{p} L\{ F(t)\} = \frac{1}{p} f(p)$$

$$\Rightarrow \quad L\left[\int_0^t F(t)\,dt \right] = \frac{1}{p} f(p).$$

9.10 MULTIPLICATION BY POWERS OF t

Theorem 9.9: [Multiplication by t]. If $F(t)$ is a function of class A and if $L\{F(t)\}=f(p)$, then $L\{t\,F(t)\}=-f'(p)$.

Proof: We know that $L\{t\,F(t)\}=-f'(p)$.

Therefore $f'(p)=\dfrac{d}{dp}\int_0^\infty e^{-pt}F(t)\,dt=\int_0^\infty \dfrac{\partial}{\partial p}\{e^{-pt}F(t)\}\,dt$,

by Leibnitz rule for differentiation under the sign of integral.

$$=\int_0^\infty -t\,e^{-pt}F(t)\,dt=-\int_0^\infty e^{-pt}\{t\,F(t)\}\,dt=-L\{t\,F(t)\}.$$

Thus, $\quad L\{t\,F(t)\}=-f'(p)$

Theorem 9.10: [Multiplication by t^n]. If $F(t)$ is a function of class A and if $L\{F(t)\}=f(p)$.

Then $L\{t^n F(t)\}=(-1)^n \dfrac{d^n}{dp^n} f(p)$, where $n=1,2,3,\ldots$.

Proof: We shall prove this theorem by mathematical induction. We know that

$$L\{t\,F(t)\}=(-1)\frac{d}{dp}f(p).$$

Thus the theorem is true for $n=1$.

Now assume that the theorem is true for a particular value of n say s, then

$$L\{t^s F(t)\}=(-1)^s\frac{d^s}{dp^s}f(p)\Rightarrow\int_0^\infty e^{-pt}t^s F(t)\,dt=(-1)^s\frac{d^s}{dp^s}f(p).$$

Now differentiating both sides w. r. t. 'p', we obtain

$$\frac{d}{dp}\int_0^\infty e^{-pt}t^s F(t)\,dt=(-1)^s\frac{d^{s+1}}{dp^{s+1}}f(p)$$

$\Rightarrow \quad \displaystyle\int_0^\infty \frac{\partial}{\partial p}\{e^{-pt}t^s F(t)\}\,dt=(-1)^s\frac{d^{s+1}}{dp^{s+1}}f(p)$, by Leibnitz's rule for differentiation

$$\Rightarrow \quad -\int_0^\infty e^{-pt}t^{s+1}F(t)\,dt=(-1)^s\frac{d^{s+1}}{dp^{s+1}}f(p)$$

$$\Rightarrow \quad L\{t^{s+1}F(t)\}=(-1)^{s+1}\frac{d^{s+1}}{dp^{s+1}}f(p)$$

which shows that if the theorem is true for any particular value of n, it is true for next value of n. Therefore the theorem is true for every positive integral value of n.

Theorem 9.11: [Division by t]. If $L\{F(t)\}=f(p)$, then

$L\left\{\frac{1}{t}F(t)\right\}=\int_p^\infty f(p)dp$ provided the integral exists.

Proof: We know that $f(p)=\int_0^\infty e^{-pt}F(t)dt$.

Integrating both w. r. t. 'p' from p to ∞, we obtain

$\int_p^\infty f(p)dp=\int_p^\infty\left[\int_0^\infty e^{-pt}F(t)dt\right]dp.$

Since p and t are independent, changing the order of integration on the right hand side, we get

$$\int_p^\infty f(p)dp=\int_0^\infty\left[\int_p^\infty e^{-pt}dp\right]F(t)dt$$

$$=\int_0^\infty\left[\frac{e^{-pt}}{-t}\right]_p^\infty F(t)dt=\int_0^\infty e^{-pt}\frac{F(t)}{t}dt=L\left\{\frac{1}{t}F(t)\right\}.$$

Example 1: Find (i) $L\{t\cos at\}$ (ii) $L\{t\sin at\}$.

Solution: (i) We know that $L\{\cos at\}=\frac{p}{(p^2+a^2)}, p>0$. So we have

$$L\{t\cos t\}=-\frac{d}{dp}\left\{\frac{p}{p^2+a^2}\right\}=-\left[\frac{1}{(p^2+a^2)}-\frac{2p.p}{(p^2+a^2)^2}\right]=\frac{(p^2-a^2)}{(p^2+a^2)^2}.$$

(ii) We know that $L\{\sin at\}=\frac{a}{(p^2+a^2)}, p>0$.

Therefore we have

$$L\{t\sin at\}=-\frac{d}{dp}\left\{\frac{a}{(p^2+a^2)}\right\}=-\left[\frac{-2ap}{(p^2+a^2)^2}\right]=\frac{2ap}{(p^2+a^2)^2}.$$

Example 2: If $L\left\{\frac{\sin t}{t}\right\}=\cot^{-1}p$, then find $L\left\{\int_0^t\left(\frac{\sin t}{t}\right)dt\right\}$.

Solution: Using the Laplace transform of integral of the function $F(t)$ i.e., if $L\{F(t)\}=f(p)$, then $L\left[\int_0^t F(t)dt\right]=\frac{1}{p}f(p)$.

Therefore, we have $L\left[\int_0^t\frac{\sin t}{t}dt\right]=\frac{1}{p}\cot^{-1}p$.

Example 3: Show that $L\left\{\frac{\cos\sqrt{t}}{\sqrt{t}}\right\}=\sqrt{\left(\frac{\pi}{p}\right)}e^{-1/4p}$.

Solution: Let $F(t)=\sin\left(\sqrt{t}\right)$, then $F'(t)=\dfrac{\cos\left(\sqrt{t}\right)}{2\sqrt{t}}$ and $F(0)=0$.

Therefore, from $L\{F'(t)\}=p\,L\{F(t)\}-F(0)$, we get

$$L\left\{\frac{\cos\sqrt{t}}{2\sqrt{t}}\right\}=p\,L\{\sin\sqrt{t}\}=p\frac{\sqrt{\pi}}{2\,p^{3/2}}e^{-1/4p}=\frac{1}{2}\sqrt{\left(\frac{\pi}{p}\right)}e^{-1/4p}$$

Hence $$L\left\{\frac{\cos\sqrt{t}}{\sqrt{(t)}}\right\}=\sqrt{\left(\frac{\pi}{p}\right)}e^{-1/4p}$$

Example 4: Prove that $L\left(\dfrac{\sin t}{t}\right)=\tan^{-1}\dfrac{1}{p}$ and hence find $L\left(\dfrac{\sin a\,t}{t}\right)$.

Does the Laplace transform of $\dfrac{\cos a\,t}{t}$ exist?

Solution: Let $F(t)=\sin t$, then $L\{\sin t\}=\dfrac{1}{(p^2+1)}=f(p)(say)$.

Now $\lim\limits_{t\to 0}\dfrac{F(t)}{t}=\lim\limits_{t\to 0}\dfrac{\sin t}{t}=1$

Therefore $L\left(\dfrac{\sin t}{t}\right)=\int_p^\infty f(x)\,d\,x=\int_p^\infty\dfrac{d\,x}{(x^2+1)}=\left(\tan^{-1}x\right)_p^\infty$

$$=\frac{\pi}{2}-\tan^{-1}p=\cot^{-1}p=\tan^{-1}\frac{1}{p}.$$

Now, $$L\left\{\frac{\sin a\,t}{t}\right\}=a\,L\left\{\frac{\sin a\,t}{a\,t}\right\}$$

$$=a\frac{1}{a}\tan^{-1}\left(\frac{1}{p/a}\right),\ \left[\text{since if } L\{F(t)\}=f(p),\text{ then } L\{F(at)\}=\frac{1}{a}f\left(\frac{p}{a}\right)\right]$$

$$=\tan^{-1}\left(\frac{a}{p}\right).$$

Again since $L\{\cos a\,t\}=\dfrac{p}{(p^2+a^2)}=f(p)(say)$, so that

$$L\left\{\frac{\cos a\,t}{t}\right\}=\int_p^\infty\frac{x}{(x^2+a^2)}d\,x=\left[\frac{1}{2}\log\left(x^2+a^2\right)\right]_p^\infty$$

$$=\frac{1}{2}\lim_{z\to\infty}\log\left(x^2+a^2\right)-\frac{1}{2}\log\left(p^2+a^2\right)$$

which does not exist since $\lim_{x\to\infty}\log\left(x^2+a^2\right)$ is infinite. Hence $L\left\{\frac{\cos at}{t}\right\}$ does not exist.

Example 5: Show that $\int_0^\infty \frac{\sin t}{t}\,dt=\frac{\pi}{2}$.

Solution: Let $F(t)=\sin t$, then

$$f(p)=L\{F(t)\}=L\{\sin t\}=\frac{1}{\left(p^2+1\right)},\, p>0.$$

Therefore $\int_0^\infty e^{-pt}\frac{\sin t}{t}\,dt=L\left\{\frac{1}{t}\sin t\right\}=\int_p^\infty f(x)\,dx=\int_p^\infty \frac{1}{\left(x^2+1\right)}\,dx$

$$=\left[\tan^{-1}x\right]_p^\infty=\frac{\pi}{2}-\tan^{-1}p$$

Taking limits as $p\to 0$, we obtain

$$\int_0^\infty \frac{\sin t}{t}\,dt=\frac{\pi}{2}.$$

Example 6: Evaluate $\int_0^\infty\left(\frac{e^{-at}-e^{-bt}}{t}\right)dt.$

Solution: Let $F(t)=e^{-at}-e^{-bt}$, then

$$f(p)=L\{F(t)\}=L\{e^{-at}-e^{-bt}\}=\frac{1}{(p+a)}-\frac{1}{(p+b)}.$$

Now division by t, we have $L\left\{\frac{F(t)}{t}\right\}=\int_p^\infty f(x)\,dx$.

This gives $\int_0^\infty e^{-pt}\left(\frac{e^{-at}-e^{-bt}}{t}\right)dt=\int_p^\infty\left(\frac{1}{x+a}-\frac{1}{x+b}\right)dx$

$$=-\log\left(\frac{p+a}{p+b}\right)=\log\left(\frac{p+b}{p+a}\right).$$

Hence taking limit as $p\to 0$, we obtain $\int_0^\infty \frac{e^{-at}-e^{-bt}}{t}\,dt=\log(b/a)$.

Example 7: Evaluate (i) $\int_0^\infty e^{-pt}t^3\cos t\,dt$

(ii) $\int_0^\infty e^{-3t}\,t\sin t\,dt$.

Solution: (i) We know that $L\{\cos t\}=\frac{p}{(p^2+1)}$, so we have

$$L\{t^3\cos t\}=(-1)^3\frac{d^3}{dp^3}\left(\frac{p}{p^2+1}\right)=-\frac{d^2}{dp^2}\left\{\frac{(1-p^2)}{(p^2+1)^2}\right\}$$

$$=-\frac{d}{dp}\left\{\frac{-2p}{(p^2+1)^2}-\frac{(1-p^2)4p}{(p^2+1)^3}\right\}$$

$$=-\frac{d}{dp}\left[\frac{(2p^3-6p)}{(1+p^2)^3}\right]=-\left[\frac{(6p^2-6)}{(1+p^2)^3}-\frac{(2p^3-6p)6p}{(1+p^2)^4}\right]$$

$$=-\left[\frac{-6p^4+36p^2-6}{(1+p^2)^4}\right]=\frac{6(p^4-6p^2+1)}{(1+p^2)^4}$$

Hence, we obtain $\int_0^\infty e^{-pt}\,t^3\cos t\,dt=\frac{6(p^4-6p^2+1)}{(1+p^2)^4}$.

(ii) We know that $L\{\sin t\}=\frac{1}{(p^2+1)}, p>0$.

Therefore $L\{t\sin t\}=-\frac{d}{dp}\left[\frac{1}{p^2+1}\right]=\frac{2p}{(p^2+1)^2}$

$$\Rightarrow \quad \int_0^\infty e^{-pt}\,t\sin t\,dt=\frac{2p}{(p^2+1)^2}$$

Putting $p=3$, we get $\int_0^\infty e^{-3t}\,t\sin t\,dt=\frac{6}{100}=\frac{3}{50}$.

EXERCISE 9.2

1. Find the Laplace transform of

   (i) $t\sinh at$ (ii) $t\,e^{at}\sin at$

   (iii) $t\,e^{-t}\sin 2t$.

   (iv) $t^n e^{at}, p>a$ and n is a positive integer.

   (v) $(t^2-3t+2)\sin 3t$. (vi) $L\{t(3\sin 2t-2\cos 2t)\}$

   (vii) $L\{\sin at-at\cos at\}$.

2. If $L\left(\frac{1-\cos at}{a^2}\right)=\frac{1}{p(p^2+a^2)}$, then show that

$$L\left\{\frac{t(1-\cos at)}{a^2}\right\}=\frac{(3p^2+a^2)}{p^2(p^2+a^2)^2}.$$

3. Find the value of

(i) $\int_0^\infty \left(\frac{e^{-t}-e^{-3t}}{t}\right) dt$ (ii) $\int_0^\infty \frac{\cos 6t - \cos 4t}{t} dt$

(iii) $\int_0^\infty e^{-t} \frac{\sin t}{t} dt$ (iv) $\int_0^\infty \frac{1-\cos 2t}{t} e^{-pt} dt$

4. If $L\{F(t)\} = f(p)$, then prove that $L\{t^2 F(t)\} = \frac{d^2}{dp^2}\{f(p)\}$. Use this result to obtain $L(t^2 \cos at)$.

5. Prove that $\int_0^\infty \frac{e^{-t}\sin^2 t}{t} dt = \frac{1}{4}\log 5$.

6. Verify the initial value theorem for the functions
(i) eq. $f(t) = 1 + e^{-t}(\sin t + \cos t)$ eq. (ii) $f(t) = t^2 e^{-3t}$.

ANSWERS 9.2

1. (i) $\frac{2ap}{(p^2-a^2)^2}$ (ii) $\frac{2a(p-a)}{(p^2-2ap+2a^2)^2}$

(iii) $\frac{(4p+4)}{(p^2+2p+5)^2}$ (iv) $\frac{n!}{(p-a)^{n+1}}$

(v) $\frac{6p^4-18p^3+126p^2-162p+432}{(p^2+9)^3}$

(vi) $\frac{8+12p-2p^2}{(p^2+4)^2}$. (vii) $\frac{2a^3}{(p^2+a^2)^2}$.

3. (i) $\log 3$ (ii) $\log\frac{2}{3}$

(iii) $\frac{\pi}{4}$ (iv). $\frac{1}{2}\log\left(1+\frac{4}{p^2}\right)$.

4. $\frac{2p(p^2-3a^2)}{(p^2+a^2)^3}, p > 0.$

9.11 THE INVERSE LAPLACE TRANSFORM

Definition: If $L\{F(t)\} = f(p)$, then $F(t)$ is called the inverse Laplace transform of $f(p)$ and is denoted by

$$L^{-1}\{f(p)\} = F(t).$$

Here L^{-1} is called the inverse Laplace transformation operator.

Thus $\quad L^{-1}\left\{\frac{1}{p^2}\right\}=t$ and $L^{-1}\left\{\frac{a}{p^2+a^2}\right\}=\sin a\,t$.

The inverse Laplace transform given below follow at once from the results of Laplace transforms given earlier.

1. $L^{-1}\left\{\frac{1}{p}\right\}=1$

2. $L^{-1}\left\{\frac{1}{p-a}\right\}=e^{at}$

3. $L^{-1}\left\{\frac{1}{(p-a)^n}\right\}=\frac{e^{at}\,t^{n-1}}{(n-1)!}$

4. $L^{-1}\left\{\frac{1}{p^n}\right\}=\begin{cases}\frac{t^{n-1}}{(n-1)!}, & n \text{ is a positive integer}\\ \frac{t^{n-1}}{\Gamma n}, & n \text{ is not an integer}\end{cases}$

5. $L^{-1}\left\{\frac{1}{p^2+a^2}\right\}=\frac{1}{a}\sin a\,t$

6. $L^{-1}\left\{\frac{p}{p^2+a^2}\right\}=\cos a\,t$

7. $L^{-1}\left\{\frac{1}{p^2-a^2}\right\}=\frac{1}{a}\sinh a\,t$

8. $L^{-1}\left\{\frac{p}{p^2-a^2}\right\}=\cosh a\,t$

9.11.1 Methods of Finding Inverse Laplace Transform (Method of partial fractions)

We know that $L\{F(t)\}$ *i.e.*, $f(p)$ in each case, is a rational algebraic function. Hence to find the inverse Laplace transform, we first express the given function of 'p' into partial fractions, which will, then recognizable as one of the above standard forms. The function

$$\frac{N(p)}{D(p)}=\frac{a^0\,p^m+a^1\,p^{m-1}+a^2\,p^{m-2}+\ldots+a_m}{b_0\,p^n+b_1\,p^{n-1}+b_2\,p^{n-2}+\ldots+b_n}$$

in which m and n are positive integers, is called a rational algebraic function. When the numerator $N(p)$ is of the lower degree than the denominator $D(p)$, then it is called proper fraction.

To resolve a given function into partial fractions, we first factorize the denominator $D(p)$ into its simplest real factors. These will be either linear or quadratic or some of the factors may be repeated. We know from the algebra that the proper fraction $N(p)/D(p)$ can be resolved in one and only one way into a sum of partial fractions, which are of the following types:

(i) To every non-repeated linear factor $(p-a)$ in the denominator, corresponds a partial fraction of the form $A/(p-a)$.

(ii) To every repeated linear factor $(p-a)^r$ in which the denominator, corresponds a sum of 'r' partial fractions of the form

$$\frac{A_1}{(p-a)^2}+\frac{A_2}{(p-a)^2}+\frac{A_3}{(p-a)^3}+\ldots+\frac{A_r}{(p-a)^r}$$

(iii) To every non-repeated quadratic factor $(p^2 + lp + m)$ in the denominator, corresponds a partial fraction of the form $\frac{Ap+B}{p^2+lp+m}$.

(iv) To every repeated quadratic factor $(p^2 + lp + m)^r$ in the denominator, corresponds the sum of 'r' partial fractions of the form $\frac{A_1 p + B_1}{(p^2+lp+m)} + \frac{A_2 p + B_2}{(p^2+lp+m)^2} + \ldots + \frac{A_r p + B_r}{(p^2+lp+m)^r}$.

To obtain the partial fractions corresponding to the non-repeated linear factor $(p-a)$, put $p = a$ in the numerator and denominator everywhere in the given function except the factor $(p-a)$. Similarly, to obtain the partial fraction $\frac{A_r}{(p-a)^r}$ in the denominator, put $p = a$ everywhere in the given fraction, except in the factor $(p-a)^r$. It should be noted that rule only gives $\frac{A_r}{(p-a)^r}$. The partial fractions corresponding to $(p-a)^r$ should be found by other methods.

9.11.2 Inverse Laplace Transforms of Derivatives

If $L^{-1}\{f(p)\} = F(t)$, then

$$L^{-1}\{f^{(n)}(p)\} = L^{-1}\left[\frac{d^n}{dp^n}\{f(p)\}\right] = (-1)^n t^n F(t).$$

Proof: We know that $L\{t^n F(t)\} = (-1)^n \left\{\frac{d^n}{dp^n} f(p)\right\} = (-1)^n f^{(n)}(p)$.

Therefore $L^{-1}\{f^{(n)}(p)\} = (-1)^n t^n F(t)$.

9.11.3 Multiplication by p

If $L^{-1}\{f(p)\} = F(t)$ and $F(0) = 0$, then $L^{-1}\{p f(p)\} = F'(t)$.

Proof: We know that $L\{F'(t)\} = p f(p) - F(0) = p f\{p\}$, because $F(0) = 0$.

Thais implies that $L^{-1}\{p f(p)\} = F'(t)$.

Note 1: If $F(0) \neq 0$, then $L^{-1}\{p f(p)\} - F(0) = F'(t)$

$\Rightarrow \quad L^{-1}\{p\, f(p)\} = F'(t) + F(0)\delta t$, where $\delta(t)$ is the unit impulse function.

Note 2: In general $L^{-1}\{p^n\, f(p)\} = \dfrac{d^n}{d\,t^n}\{F(t)\}$ provided

$$F(0) = F(0)\ldots = F^{\,n-1}(0) = 0.$$

9.11.4 Division by p

If $L^{-1}\{f(p)\} = F(t)$, then

$$L^{-1}\left\{\frac{f(p)}{p}\right\} = \int_0^t f(u)\,d\,u$$

$$L^{-1}\left\{\frac{f(p)}{p^2}\right\} = \int_0^t \int_0^t f(u)\,d\,u\,d\,u$$

$$L^{-1}\left\{\frac{f(p)}{p^3}\right\} = \int_0^t \int_0^t \int_0^t f(u)\,d\,u\,d\,u\,d\,u.$$

Therefore in general

$$L^{-1}\left\{\frac{f(p)}{p^n}\right\} = \int_0^t \int_0^t \int_0^t \int_0^t \cdots \int_0^t f(u)\,du\,du\,du\,du\ldots n \text{ times}.$$

9.11.5 Linearity Property

If c_1 and c_2 are constants and $L\{F_1(t)\} = f_1(p)$ and $L\{F_2(t)\} = f_2(p)$,

then $\quad L^{-1}\{c_1\, f_1(p) \pm c_2\, f_2(p)\} = c_1\, L^{-1}\{f_1(p)\} \pm c_2\, L^{-1}\{f_2(p)\}.$

Proof: By definition

$$L\{c_1\, F_1(t) \pm c_2\, F_2(t)\} = \int_0^\infty e^{-pt}\{c_1\, F_1(t) \pm c_2\, F_2(t)\}d\,t$$

$$= c_1 \int_0^\infty e^{-pt}\, F_1(t)d\,t \pm c_2 \int_0^\infty e^{-pt}\, F_2(t)d\,t = c_1\, f_1(p) \pm c_2\, f_2(p)$$

$$\Rightarrow L^{-1}\{c_1\, f_1(p) \pm c_2\, f_2(p)\} = c_1\, F_1(t) \pm c_2\, F_2(t)$$

$$= c_1\, L^{-1}\{f_1(p)\} \pm c_2\, L^{-1}\{f_2(p)\}.$$

9.11.6 First Shifting Property

If $L^{-1}\{f(p)\} = F(t)$, then $\{f(p-a)\} = e^{at}\, F(t)$.

Proof: We know that $f(p) = \int_0^\infty e^{-pt}\, F(t)d\,t$ [by definition]

$$\therefore\; f(p-a) = \int_0^\infty e^{-(p-a)t}\, F(t)d\,t = \int_0^\infty e^{-pt}\, e^{at}\, F(t)d\,t = L\{e^{at}\, F(t)\}.$$

Hence, $L^{-1}\{f(p-a)\} = e^{at}\, F(t)$.

9.11.7 Second Shifting Property

If $L^{-1}\{f(p)\}=F(t)$ and $G(t)=\begin{cases}F(t-a), t>a\\ 0, t<a\end{cases}$, then

$$L^{-1}\{e^{-ap}f(p)\}=G(t)$$

Proof: We know that $L\{F(t)\}=f(p)=\int_0^\infty e^{-pt}F(t)dt$.

Therefore $e^{-ap}f(p)=\int_0^\infty e^{-ap}e^{-pt}F(t)dt$

$$=\int_0^\infty e^{-p(t+a)}F(t)dt,\ [\text{putting } t+a=u, t=u-a, dt=du]$$

$$=\int_a^\infty e^{-pu}F(u-a)du=\int_a^\infty e^{-pt}F(t-a)dt$$

$$=\int_0^a e^{-pt}\,0\,dt+\int_a^\infty e^{-pt}F(t-a)dt$$

$$=\int_0^\infty e^{-pt}G(t)dt=L\{G(t)\}, \text{ if } G(t)=\begin{cases}F(t-a), t>a\\ 0, t<a\end{cases}$$

Hence $L^{-1}\{e^{-ap}f(p)\}=G(t)$, where $G(t)=\begin{cases}F(t-a), t>a\\ 0, t<a\end{cases}$

Note: This theorem can also be defined as "If $L^{-1}\{f(p)\}=F(t)$, then

$$L^{-1}\{e^{-ap}f(p)\}=F(t-a)U(t-a),$$

where $U(t-a)$ is a Heaviside's unit step function defined as $U(t-a)=\begin{cases}1, t>a\\ 0, t<a\end{cases}$.

9.11.8 Change of Scale Property

If $L^{-1}\{f(p)\}=F(t)$, then $L^{-1}\{f(ap)\}=\frac{1}{a}F(t/a)$.

Proof: We know that $L\{f(t)\}=F(p)=\int_0^\infty e^{-pt}F(t)dt$

Now $f(ap)=\int_0^\infty e^{-apt}F(t)dt=\int_0^\infty e^{-pu}F(u/a)\frac{du}{a}$,

$$\left[\text{set } at=u, \text{ then } t=\frac{u}{a} \text{ and } dt=\frac{du}{a}\right]$$

$$=\frac{1}{a}\int_0^\infty e^{-pu}F(u/a)du=\frac{1}{a}\int_0^\infty e^{-pt}F(t/a)dt=L\left\{\frac{1}{a}F(t/a)\right\}.$$

Therefore, $L^{-1}\{f(ap)\}=\frac{1}{a}F(t/a)$.

Example 1: Find the inverse Laplace transform of following functions:

(i) $\left\{\dfrac{p}{p^2+2}+\dfrac{6p}{p^2-16}+\dfrac{3}{p-3}\right\}$ (ii) $\dfrac{2p^2-6p+5}{p^3-6p^2+11p-6}$.

(iii) $\dfrac{p-4}{p^2-4p+13}$ (iv) $\dfrac{2p+9}{p^2+9}+\dfrac{12}{4-3p}+\dfrac{1}{\sqrt{p}}$

(v) $\left\{\dfrac{3}{p^2-3}+\dfrac{3p+2}{p^3}-\dfrac{3p-27}{p^2+9}+\dfrac{6-30\sqrt{p}}{p^4}\right\}$

Solution: (i) $L^{-1}\left\{\dfrac{p}{p^2+2}+\dfrac{6p}{p^2-16}+\dfrac{3}{p-3}\right\}$

$$=L^{-1}\left\{\frac{p}{p^2+2}\right\}+6L^{-1}\left\{\frac{p}{p^2-4^2}\right\}+3L^{-1}\left\{\frac{1}{p-3}\right\}$$

$$=L^{-1}\left\{\frac{p}{p^2+\left(\sqrt{2}\right)^2}\right\}+6L^{-1}\left\{\frac{p}{p^2-4^2}\right\}+3e^{3t}L^{-1}\left\{\frac{1}{p}\right\}$$

$$=\cos\sqrt{2}\,t+6\cosh 4t+3e^{3t}.$$

(ii) $L^{-1}\left\{\dfrac{2p^2-6p+5}{p^3-6p^2+11p-6}\right\}=L^{-1}\left\{\dfrac{2p^2-6p+5}{(p-1)(p-2)(p-3)}\right\}$.

Now find the partial fraction as

$$\frac{2p^2-6p+5}{p^3-6p^2+11p-6}=\frac{A}{p-1}+\frac{B}{p-2}+\frac{C}{p-3}.$$

We get the values of A, B and C respectively on putting $p=1, 2$ and 3

$$A=\frac{2(1)^2-6(1)+5}{(1-2)(1-3)}=\frac{1}{2},\quad B=\frac{2(2)^2-6(2)+5}{(2-1)(2-3)}=-1,$$

$$C=\frac{2(3)^2-6(3)+5}{(4-1)(3-2)}=\frac{5}{2}.$$

Hence

$$L^{-1}\left\{\frac{2p^2-6p+5}{p^3-6p^2+11p-6}\right\}=L^{-1}\left\{\frac{1}{2}\left(\frac{1}{p-1}\right)-\left(\frac{1}{p-2}\right)+\frac{5}{2}\left(\frac{1}{p-3}\right)\right\}$$

$$=\frac{1}{2}e^{t}-e^{2t}+\frac{5}{2}e^{3t}.$$

(iii) $L^{-1}\left\{\dfrac{p-4}{p^2-4p+13}\right\}=L^{-1}\left\{\dfrac{p-4}{(p-2)^2+9}\right\}=L^{-1}\left\{\dfrac{p-2-2}{(p-2)^2+3^2}\right\}$

$$=L^{-1}\left\{\frac{(p-2)}{(p-2)^2+3^2}\right\}-2L^{-1}\left\{\frac{1}{(p-2)^2+3^2}\right\}$$

$$=e^{2t}\cos 3t-\frac{2}{3}e^{2t}\sin 3t.$$

(iv) $L^{-1}\left\{\dfrac{2p+9}{p^2+9}+\dfrac{12}{4-3p}+\dfrac{1}{\sqrt{p}}\right\}$

$$=L^{-1}\left\{\frac{2p+9}{p^2+9}\right\}+L^{-1}\left\{\frac{12}{4-3p}\right\}+L^{-1}\left\{\frac{1}{\sqrt{p}}\right\}$$

$$=2L^{-1}\left\{\frac{p}{p^2+3^2}\right\}+9L^{-1}\left\{\frac{1}{p^2+3^2}\right\}-L^{-1}\left\{\frac{4}{p-4/3}\right\}+L^{-1}\left\{\frac{1}{\sqrt{p}}\right\}$$

$$=2\cos 3t+3\sin 3t-4e^{4t/3}+\frac{1}{\sqrt{\pi t}}.$$

(v) $L^{-1}\left\{\dfrac{3}{p^2-3}+\dfrac{3p+2}{p^3}-\dfrac{3p-27}{p^2+9}+\dfrac{6-30\sqrt{p}}{p^4}\right\}$

$$=L^{-1}\left\{\frac{3}{p^2-3}\right\}+L^{-1}\left\{\frac{3p+2}{p^3}\right\}-L^{-1}\left\{\frac{3p-27}{p^2+9}\right\}+L^{-1}\left\{\frac{6-30\sqrt{p}}{p^4}\right\}$$

$$=\frac{3}{\sqrt{3}}L^{-1}\left\{\frac{\sqrt{3}}{p^2-(\sqrt{3})^2}\right\}+3L^{-1}\left\{\frac{1}{p^2}\right\}+2L^{-1}\left\{\frac{1}{p^3}\right\}-3L^{-1}\left\{\frac{p}{p^2+3^2}\right\}$$

$$+9L^{-1}\left\{\frac{3}{p^2+3^2}\right\}+6L^{-1}\left\{\frac{1}{p^4}\right\}-30L^{-1}\left\{\frac{1}{p^{7/2}}\right\}$$

$$=\frac{3}{\sqrt{3}}\sin h\sqrt{3}\,t+3t+2\frac{t^2}{2}-3\cos 3t+9\sin 3t+6\frac{t^3}{6}-30\frac{t^{7/2-1}}{\Gamma(7/2)}$$

$$=\frac{3}{\sqrt{3}}\sin h\sqrt{3}\,t+3t+t^2+t^3-16t^3\sqrt{t/\pi}-3\cos 3t+9\sin 3t.$$

Example 2: Prove that

$$L^{-1}\left\{\frac{5}{p^2}+\left(\frac{\sqrt{p}-1}{p}\right)^2-\frac{7}{3p+2}\right\}=1+6t-4\sqrt{\left(\frac{t}{\pi}\right)}-\frac{7}{3}e^{-2t/3}$$

Solution:

$$L^{-1}\left\{\frac{5}{p^2}+\left(\frac{\sqrt{p}-1}{p}\right)^2-\frac{7}{3}\frac{1}{p+2/3}\right\}=L^{-1}\left\{\frac{5}{p^2}+\frac{(p-2\sqrt{p}+1)}{p^2}-\frac{7/3}{p+2/3}\right\}$$

$$=6L^{-1}\left\{\frac{1}{p^2}\right\}+L^{-1}\left\{\frac{1}{p}\right\}-2L^{-1}\left\{\frac{1}{p^{3/2}}\right\}-\frac{7}{3}L^{-1}\left\{\frac{1}{p+2/3}\right\}$$

$$=6t+1-2\frac{t^{3/2-1}}{\Gamma(3/2)}-\frac{7}{3}e^{-2t/3}=6t+1-4\sqrt{t/\pi}-\frac{7}{3}e^{-2t/3}$$

Example 3: Find the inverse Laplace transform of

(i) $\dfrac{p+8}{p^2+4p+5}$ (ii) $\dfrac{1}{\sqrt{2p+3}}$

(iii) $\left\{\dfrac{p-2}{(p-2)^2+5^2}+\dfrac{p+4}{(p+4)^2+9^2}+\dfrac{1}{(p+2)^2+3^2}\right\}.$

Solution: (i) $L^{-1}\left\{\dfrac{p+8}{p^2+4p+5}\right\}=L^{-1}\left\{\dfrac{(p+2)+6}{(p+2)^2+1^2}\right\}$

$$=L^{-1}\left\{\frac{p+2}{(p+2)^2+1^2}\right\}+L^{-1}\left\{\frac{6}{(p+2)^2+1^2}\right\}$$

$$=e^{-2t}L^{-1}\left\{\frac{p}{p^2+1^2}\right\}+6e^{-2t}L^{-1}\left\{\frac{1}{p^2+1^2}\right\}=e^{-2t}(\cos t+6\sin t).$$

(ii) $L^{-1}\left\{\dfrac{1}{\sqrt{2p+3}}\right\}=L^{-1}\left\{\dfrac{1}{\sqrt{2}\sqrt{p+3/2}}\right\}=\dfrac{1}{\sqrt{2}}L^{-1}\left\{\dfrac{1}{\sqrt{p+3/2}}\right\}$

$$=\frac{1}{\sqrt{2}}e^{-3t/2}L^{-1}\left\{\frac{1}{\sqrt{p}}\right\}=\frac{1}{\sqrt{2}}e^{-3t/2}\frac{t^{1/2-1}}{\Gamma(1/2)}=\frac{1}{\sqrt{2\pi t}}e^{-3t/2}$$

(iii) $L^{-1}\left\{\dfrac{p-2}{(p-2)^2+5^2}+\dfrac{p+4}{(p+4)^2+9^2}+\dfrac{1}{(p+2)^2+3^2}\right\}$

$$=L^{-1}\left\{\frac{p-2}{(p-2)^2+5^2}\right\}+L^{-1}\left\{\frac{p+4}{(p+4)^2+9^2}\right\}+L^{-1}\left\{\frac{1}{(p+2)^2+3^2}\right\}$$

$$=e^{2t}L^{-1}\left\{\frac{p}{p^2+5^2}\right\}+e^{-4t}L^{-1}\left\{\frac{p}{p^2+9^2}\right\}+e^{-2t}L^{-1}\left\{\frac{1}{p^2+3^2}\right\}$$

$$=\frac{e^{2t}}{5}\cos 5t+e^{-4t}\cos 9t+\frac{1}{3}e^{-2t}\sin 3t.$$

9.12 HEAVISIDE EXPANSION FORMULA FOR INVERSE LAPLACE TRANSFORMS

If $f(p)$ and $g(p)$ are two polynomials in p and the degree of $f(p)$ is less than the degree of $g(p)$ and if $g(p)=(p-\alpha_1)(p-\alpha_2)....(p-\alpha_n)$, where $\alpha_1,\alpha_2,\alpha_3,...\alpha_n$ are distinct constants, real or complex, then

$$L^{-1}\left\{\frac{f(p)}{g(p)}\right\}=\sum_{r=1}^{n}\frac{f(\alpha_r)}{g'(\alpha_r)}e^{\alpha_r t}$$

Proof. By the method of partial fractions let, we get

$$\frac{f(p)}{g(p)}=\frac{A_1}{(p-\alpha_1)}+\frac{A_2}{(p-\alpha_2)}+...\frac{A_r}{(p-\alpha_r)}+...\frac{A_n}{(p-\alpha_n)}.$$

Multiplying both sides by $(p-\alpha_r)$ and taking $p \to \alpha_r$, we obtain

$$A_r = \lim_{p\to\alpha_r} \frac{f(p)(p-\alpha_r)}{g(p)}$$

$$A_r = \lim_{p\to\alpha_r} f(p) \lim_{p\to\alpha_r} \frac{(p-\alpha_r)}{g(p)} = \lim_{p\to\alpha_r} f(p) . \lim_{p\to\alpha_r} \frac{1}{g'(p)} = \frac{f(\alpha_r)}{g'(\alpha_r)}.$$

Therefore

$$\frac{f(p)}{g(p)} = \frac{f(\alpha_1)}{g(\alpha_1)} \frac{1}{(p-\alpha_1)} + \ldots + \frac{f(\alpha_r)}{g'(\alpha_r)} \frac{1}{(p-\alpha_r)} + \ldots + \frac{f(\alpha_n)}{g'(\alpha_n)} \frac{1}{(p-\alpha_n)}$$

Hence

$$L^{-1}\left\{\frac{f(p)}{g(p)}\right\} = \frac{f(\alpha_1)}{g'(\alpha_1)} e^{\alpha_1 t} + \frac{f(\alpha_2)}{g'(\alpha_2)} e^{\alpha_2 t} + \ldots + \frac{f(\alpha_r)}{g'(\alpha_r)} e^{\alpha_r t} + \ldots + \frac{f(\alpha_n)}{g'(\alpha_n)} e^{\alpha_n t}$$

$$= \sum_{r=1}^{n} \frac{f(\alpha_r)}{g'(\alpha_r)} e^{\alpha_r t}.$$

9.13 CONVOLUTION THEOREM

If $L^{-1}\{f(p)\} = F(t)$ and $L^{-1}\{g(p)\} = G(t)$, then

$$L^{-1}\{f(p)g(p)\} = \int_0^t F(u)G(t-u)\,du = F * G.$$

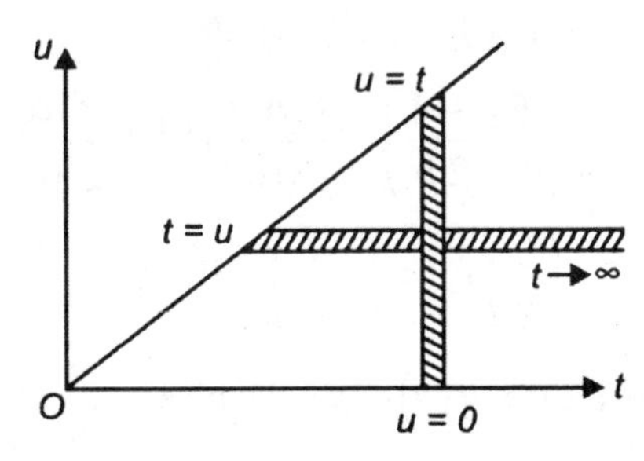

Proof: Let $\phi(t) = \int_0^t F(u)G(t-u)\,du$

Then $L\{\phi(t)\} = \int_0^\infty e^{-pt}\left[\int_0^t F(u)G(t-u)\,du\right]dt$

$$= \int_0^\infty \int_0^t e^{-pt} F(u)G(t-u)\,du\,dt.$$

On changing the order of integration, we get

$$L\{\phi(t)\} = \int_0^\infty \int_u^\infty e^{-pt} F(u)G(t-u)\,dt\,du$$

$$= \int_0^\infty e^{-pu} F(u)\left[\int_0^\infty e^{-pv} G(v)\,dv\right] du \text{, on putting } t-u=v$$

$$= \int_0^\infty e^{-pu} F(u) g(p)\,du = g(p)\int_0^\infty e^{-pu} F(u)\,du$$

$$= g(p) f(p) = f(p) g(p)$$

$$\Rightarrow \quad F * G = L^{-1}\{f(p)g(p)\} = \phi(t) = \int_0^t F(u)G(t-u)\,du.$$

We call $F * G$ the convolution of F and G. This theorem is known as convolution property.

Example 4: Apply Heaviside expansion theorem to obtain the value of

(i) $L^{-1}\left(\dfrac{2p^2+5p-4}{p^3+p^2-2p}\right)$ (ii) $L^{-1}\left(\dfrac{3p+1}{(p-1)(p^2+1)}\right)$.

Solution: (i) Let $f(p) = 2p^2+5p-4$ and

$$g(p) = p^3+p^2-2p = p(p-1)(p+2)$$

Now $g(p)=0$ gives $p=0,1,-2$

Let $\alpha_1 = 0, \alpha_2 = 1, \alpha_3 = -2$, then $f(\alpha_1) = -4, f(\alpha_2) = 3, f(\alpha_3) = -6$

Also $g'(p) = 3p^2+2p-2$ implies that

$$g'(\alpha_1) = -2, g'(\alpha_2) = 3, g'(\alpha_3) = 6.$$

Therefore $L^{-1}\left(\dfrac{2p^2+5p-4}{p^3+p^2-2p}\right) = \dfrac{f(\alpha_1)}{g'(\alpha_1)} e^{\alpha_1 t} + \dfrac{f(\alpha_2)}{g'(\alpha_2)} e^{\alpha_2 t} + \dfrac{f(\alpha_3)}{g'(\alpha_3)} e^{\alpha_3 t}$

$$= \left(\frac{-4}{-2}\right)e^{0.t} + \left(\frac{3}{3}\right)e^{1.t} + \left(\frac{-6}{6}\right)e^{-2t} = 2 + e^t - e^{-2t}.$$

(ii) Let $f(p) = 3p+1$ and $g(p) = (p-1)(p^2+1)$.

Now $g(p)=0$ gives $p = 1, i, -i$.

Let $\alpha_1 = 1, \alpha_2 = i, \alpha_3 = -i$, then

$f(\alpha_1) = 4, f(\alpha_2) = 3i+1, f(\alpha_3) = 1-3i$

Also $g'(p) = (p-1)(2p)+p^2+1$ implies that

$$g'(\alpha_1) = 2, g'(\alpha_2) = -2-2i, g'(\alpha_3) = 2i-2$$

Therefore $L^{-1}\left\{\dfrac{3p+1}{(p-1)(p^2+1)}\right\} = \left(\dfrac{4}{2}\right)e^t + \left(\dfrac{3i+1}{-2-2i}\right)e^{it} + \left(\dfrac{1-3i}{2i-2}\right)e^{-it}$

$$= 2e^t - \left(\frac{i}{2}+1\right)e^{it} + \left(\frac{i}{2}-1\right)e^{-it} = 2e^t - \frac{i}{2}\left(e^{it} - e^{-it}\right) - \left(e^{it} + e^{-it}\right)$$

$$= 2e^t - \frac{i}{2}2i\sin t - 2\cos t = 2e^t + \sin t - 2\cos t.$$

Example 5: Using Heaviside expansion formula, find the value of

(i) $L^{-1}\left\{\dfrac{p^2-6}{p^3+4p^2+3p}\right\}$ (ii) $L^{-1}\left\{\dfrac{2p^2-6p+5}{p^3-6p^2+11p-6}\right\}$.

Solution: (i) Let $f(p) = p^2 - 6$ and $g(p) = p^3 + 4p^2 + 3p$. Then

$$g(p) = p(p^2+4p+3) = p(p+3)(p+1)$$

Thus $g(p) = 0$ gives $p = 0, -1, -3$.

Let $\alpha_1 = 0, \alpha_2 = -1, \alpha_3 = -3$, then

$$f(\alpha_1) = -6, f(\alpha_2) = -5, f(\alpha_3) = 3.$$

Also $g'(p) = 3p^2 + 8p + 3$ implies that

$$g'(\alpha_1) = 3, g'(\alpha_2) = -2, g'(\alpha_3) = 6.$$

Now by Heaviside expansion formula, we have

$$L^{-1}\left\{\frac{f(p)}{g(p)}\right\} = \sum_{i=1}^{3}\frac{f(\alpha_1)}{g'(\alpha_i)}e^{\alpha_i t}.$$

$$\Rightarrow \quad L^{-1}\left\{\frac{(p^2-6)}{(p^3+4p^2+3p)}\right\} = \frac{f(\alpha_1)}{g'(\alpha_1)}e^{\alpha_1 t} + \frac{f(\alpha_2)}{g'(\alpha_2)}e^{\alpha_2 t} + \frac{f(\alpha_3)}{g(\alpha_3)}e^{\alpha_3 t}$$

$$= \left(-\frac{6}{3}\right)e^{0.t} + \left(\frac{-5}{-2}\right)e^{-1.t} + \left(\frac{3}{6}\right)e^{-3t} = -2 + \frac{5}{2}e^{-t} + \frac{1}{2}e^{-3t}$$

(ii) Let $f(p) = 2p^2 - 6p + 5$ and $g(p) = p^3 - 6p^2 + 11p - 6$. Then

$$g(p) = (p-1)(p-2)(p-3) \text{ and } g'(p) = 3p^2 - 12p + 11.$$

Also $g(p) = 0$ gives $p = 1, 2, 3$. Let $\alpha_1 = 1, \alpha_2 = 2, \alpha_3 = 3$.

By Heaviside expansion formula, we have

$$L^{-1}\left\{\frac{f(p)}{g(p)}\right\} = \sum_{i=1}^{3}\frac{f(\alpha_i)}{g'(\alpha_i)}e^{\alpha_i t}$$

$$\Rightarrow L^{-1}\left\{\frac{2p-6p+5}{p^3-6p^2+11p-6}\right\} = \frac{f(\alpha_1)}{g'(\alpha_1)}e^{\alpha_1 t} + \frac{f(\alpha_2)}{g'(\alpha_2)}e^{\alpha_2 t} + \frac{f(\alpha_3)}{g'(\alpha_3)}e^{\alpha_3 t}$$

$$= \frac{f(1)}{g'(1)}e^{1t} + \frac{f(2)}{g'(2)}e^{2t} + \frac{f(3)}{g'(3)}e^{3t} = \frac{1}{2}e^t - e^{2t} + \frac{5}{2}e^{3t}.$$

Example 6: Use convolution theorem to find

(i) $L^{-1}\left\{\dfrac{p}{(p^2+a^2)^2}\right\}$ (ii) $L^{-1}\left\{\dfrac{1}{p(p^2+4)}\right\}$ (iii) $L^{-1}\left\{\dfrac{p}{(p^2+4)^3}\right\}$.

Solution: (i) We know that $L^{-1}\left\{\dfrac{p}{(p^2+a^2)}\right\}=\cos a\,t$

and $$L^{-1}\left\{\frac{1}{p^2+a^2}\right\}=\frac{1}{a}\sin a\,t.$$

Now by convolution theorem

$$L^{-1}\left\{\frac{p}{(p^2+a^2)}\frac{1}{(p^2+a^2)}\right\}=\int_0^t \cos a\,u\,\frac{1}{a}\sin a(t-u)\,d\,u$$

$$=\frac{1}{a}\int_0^t(\cos a\,u\sin a\,t\cos a\,u-\cos a\,u\cos a\,t\sin a\,u)\,d\,u$$

$$=\frac{1}{a}\int_0^t\left[\sin a\,t.\left(\frac{1+\cos 2\,a\,u}{2}\right)-\frac{1}{2}\cos a\,t\sin 2\,a\,u\right]d\,u$$

$$=\frac{1}{a}\left[\frac{1}{2}\sin a\,t\left\{u+\frac{1}{2\,a}\sin 2\,a\,u\right\}_0^t+\frac{1}{4\,a}\cos a\,t\{\cos 2\,a\,u\}_0^t\right]$$

$$=\frac{1}{a}\left[\frac{1}{2}\sin a\,t\left\{t+\frac{1}{2\,a}\sin 2\,a\,t\right\}+\frac{1}{4\,a}\cos a\,t(\cos 2\,a\,t-1)\right]$$

$$=\frac{1}{a}\left[\frac{1}{2}t\sin a\,t+\frac{1}{4\,a}\cos(2\,a\,t-a\,t)-\frac{1}{4\,a}\cos a\,t\right]=\frac{1}{2\,a}t\sin a\,t.$$

(ii) $L^{-1}\left\{\dfrac{1}{p(p^2+4)^2}\right\}=L^{-1}\left\{\dfrac{1}{p^2}\dfrac{p}{(p^2+4)^2}\right\}$

Let $f(p)=\dfrac{1}{p^2}$, $g(t)=\dfrac{p}{(p^2+4)^2}$, then $L^{-1}\{f(p)\}=L^{-1}\left\{\dfrac{1}{p^2}\right\}=t$

$L^{-1}\{g(p)\}=L^{-1}\left\{\dfrac{p}{(p^2+4)^2}\right\}=\dfrac{1}{2.2}t\sin 2\,t$. [by result of (i)]

Therefore $L^{-1}\left\{\dfrac{1}{p(p^2+4)^2}\right\}=\displaystyle\int_0^t\left\{\frac{1}{4}u\sin 2\,u\right\}(t-u)\,d\,u$

$$=\frac{t}{4}\int_0^t u\sin 2\,u\,d\,u-\frac{1}{4}\int_0^t u^2\sin 2\,u\,d\,u$$

$$=\frac{1}{4}\left[t\left\{-\frac{u}{2}\cos 2u+\frac{1}{4}\sin 2u\right\}_0^t-\left\{\left(-\frac{u^2}{2}\cos 2u\right)_0^t+\int_0^t u\cos 2u\,du\right\}\right]$$

$$=\frac{1}{4}\left[\frac{1}{4}t\sin 2t-\frac{t}{2}\sin 2t-\frac{1}{4}(\cos 2t-1)\right]=\frac{1}{16}[1-\cos 2t-t\sin 2t].$$

(iii) We can write $L^{-1}\left\{\frac{p}{(p^2+4)^3}\right\}=L^{-1}\left\{\frac{p}{(p^2+4)^2}\frac{1}{(p^2+4)}\right\}$.

Let $f(p)=\frac{p}{(p^2+4)^2}$ and $g(p)=\frac{1}{(p^2+4)}$, then

$$L^{-1}\{f(p)\}=L^{-1}\left\{\frac{p}{(p^2+4)^2}\right\}=\frac{1}{2.2}t\sin 2t=\frac{t}{4}\sin 2t$$

[by the result of (i)]

$$L^{-1}\{g(p)\}=L^{-1}\left\{\frac{1}{(p^2+4)}\right\}=\frac{1}{2}\sin 2t.$$

By convolution theorem, we get

$$L^{-1}\{f(p).g(p)\}=\int_0^t\frac{1}{4}u\sin 2u\,.\frac{1}{2}\sin 2(t-u)\,du$$

$$=\frac{1}{16}\int_0^t u[\cos 2(t-2u)-\cos 2t]\,du$$

$$=\frac{1}{16}\left[-\frac{t^2}{2}\cos 2t+\left\{-\frac{u}{4}\sin 2(t-2u)+\frac{1}{16}\cos 2(t-2u)\right\}_0^t\right]$$

$$=\frac{1}{16}\left[-\frac{t^2}{2}\cos 2t+\frac{t}{4}\sin 2t+\frac{1}{16}(\cos 2t-\cos 2t)\right]$$

$$=\frac{t}{64}(\sin 2t-2t^2\cos 2t).$$

9.14. LAPLACE TRANSFORM OF PERIODIC FUNCTION

Let $F(t)$ be a periodic function with period T that is $F(t+T)=F(t)$, then by definition

$$L\{F(t)\}=\int_0^\infty e^{-pt}F(t)\,dt$$

$$=\int_0^T e^{-pt}F(t)\,dt+\int_T^{2T}e^{-pt}F(t)\,dt+\int_{2T}^{3T}e^{-pt}F(t)\,dt+\ldots$$

$$=I_1+I_2+I_3+I_4\ldots.$$

In I_2 putting $t=T+u$ so that $dt=du$, we get

$$I_2 = \int_T^{2T} e^{-pt} F(t)dt = \int_0^T e^{-p(T+u)} F(T+u)du$$

$$= e^{-pT} \int_0^T e^{-pu} F(u)du \qquad \text{[since } F(T+u) = F(u)\text{]}$$

$$= e^{-pT} I_1, \text{ where } I_1 = \int_0^T e^{-pt} f(t)dt.$$

Similarly in I_3 putting $t = 2T + u$ so that $dt = du$, we get

$$I_3 = \int_{2T}^{3T} e^{-pt} F(t)dt = \int_0^T e^{-p(2T+u)} F(2T+u)du$$

$$= e^{-2pT} \int_0^T e^{-pu} F(u)du, \text{ since } F(2T+u) = F(T+u) = F(u)$$

$$= e^{-2pT} I_1.$$

Proceeding in the same way, we get $I_4 = e^{-3pT} I_1, I_5 = e^{-4pT} I_1$ and so on.

Therefore $L\{F(t)\} = I_1 + e^{-pT} I_1 + e^{-2pT} I_1 + e^{-2pT} I_1 + \ldots$

$$= I_1\left(1 + e^{-pT} + e^{-2pT} + e^{-3pT} + \ldots\right)$$

$$= \frac{I_1}{1 - e^{-pT}}, \text{ [as the sum of an infinite G.P. since]}$$

Hence $\quad L\{F(t)\} = \dfrac{1}{1 - e^{-pT}} \displaystyle\int_0^T e^{-pT} F(t)dt.$

Example 7: Find the Laplace transform of a periodic function

$$f(t) = \begin{cases} A, 0 < t < L \\ -A, L < t < 2L \end{cases}.$$

Solution: The function $f(t)$ is $2L$ periodic therefore

$$L\{f(t)\} = \frac{1}{1 - e^{-2pL}} \int_0^{2L} e^{-pt} f(t)dt = \frac{1}{1 - e^{-2pL}}\left[\int_0^L Ae^{-pt}dt - \int_L^{2L} Ae^{-pt}dt\right.$$

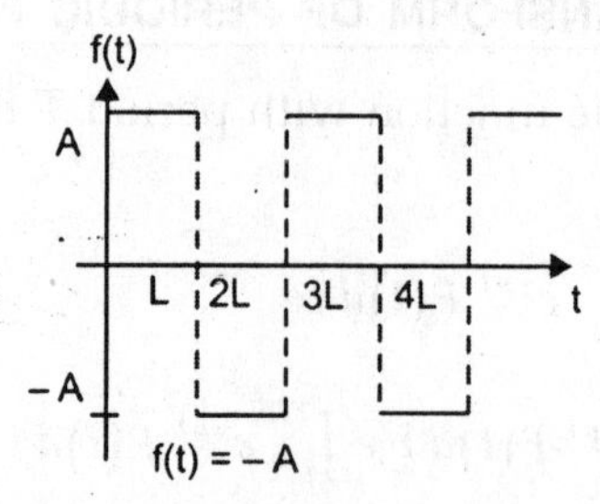

$$= \frac{A}{1 - e^{-2pL}}\left\{\left[\frac{e^{-pt}}{-p}\right]_0^L - \left[\frac{e^{-pt}}{-p}\right]_L^{2L}\right\}$$

$$=\frac{A}{p\left(1-e^{-2pL}\right)}\left[1-e^{-pL}-e^{-pL}+e^{-2pL}\right]$$

$$=\frac{A}{p\left[1-e^{-2pL}\right]}\left[1-2\,e^{-pL}+e^{-2pL}\right]$$

$$=\frac{A\left(1-e^{-pL}\right)^2}{p\left(1-e^{-pL}\right)\left(1+e^{-pL}\right)}=\frac{A\left(1-e^{-pL}\right)}{p\left(1+e^{-pL}\right)}=\frac{A}{p}\left(\frac{e^{pL/2}-e^{-pL/2}}{e^{pL/2}+e^{-pL/2}}\right)$$

$$=\frac{A}{p}\tan\text{h}\left(\frac{pL}{2}\right).$$

Example 8: Plot the 2π - periodic function $f(t)$ given by $f(t)$ = $\begin{cases} t\,,0<t<\pi \\ \pi-t\,,\pi<t<2\pi \end{cases}$ and find its Laplace transform.

Solution: Graph of the function $f(t)$ against t is as shown in the adjoining figure.

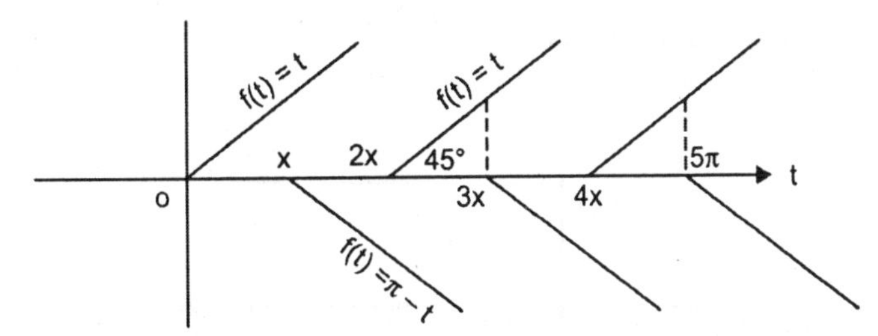

Since $f(t)$ is periodic function with period 2π, we have

$$L\{f(t)\}=\frac{1}{1-e^{-2p\pi}}\int_0^{2\pi}e^{-pt}f(t)\,dt$$

$$=\frac{1}{\left(1-e^{-2p\pi}\right)}\left[\int_0^{\pi}e^{-pt}\,t\,dt+\int_{\pi}^{2\pi}e^{-pt}(\pi-t)\,dt\right]$$

$$=\frac{1}{\left(1-e^{-2p\pi}\right)}\left\{\left[t\frac{e^{-pt}}{-p}\right]_0^{\pi}-\int_0^{\pi}\frac{e^{-pt}}{-p}dt+\left[(\pi-t)\frac{e^{-pt}}{-p}\right]_{\pi}^{2\pi}-\int_{\pi}^{2\pi}\frac{e^{-pt}(-1)}{-p}dt\right\}$$

$$=\frac{1}{\left(1-e^{-2p\pi}\right)}\left\{-\frac{\pi}{p}e^{-p\pi}+\left[\frac{e^{-pt}}{-p^2}\right]_0^{\pi}+\frac{\pi}{p}e^{-2p\pi}+\left[\frac{e^{-pt}}{p^2}\right]_{\pi}^{2\pi}\right\}$$

$$=\frac{1}{\left(1-e^{-2p\pi}\right)}\left\{\frac{\pi}{p}\left(e^{-2p\pi}-e^{-p\pi}\right)-\frac{1}{p^2}\left(1-2\,e^{-p\pi}+e^{-2p\pi}\right)\right\}$$

$$=\frac{1}{\left(1-e^{-p\pi}\right)\left(1+e^{-p\pi}\right)}\left[-\frac{\pi}{p}e^{-p\pi}\left(1-e^{-\pi p}\right)+\frac{1}{p^2}\left(1-e^{-p\pi}\right)^2\right]$$

$$=\frac{1}{\left(1+e^{-p\pi}\right)}\left[-\frac{\pi}{p}e^{-p\pi}+\frac{1}{p^2}\left(1-e^{-p\pi}\right)\right]=\frac{1}{p^2\left(1+e^{-p\pi}\right)}\left[1-(1+\pi p)e^{-\pi p}\right].$$

9.15. LAPLACE TRANSFORM OF SOME SPECIAL FUNCTIONS

Sometimes we find the solution of a differential equation of a physical system which is acted upon by

(i) A periodic force or periodic voltage.

(ii) An impulsive force or voltage acting instantaneously at a certain time or concentrated load acting at that time.

(iii) A force acting on a part of the system or voltage acting for a finite interval of time.

Let us study Laplace transform of such types of functions.

9.16. HEAVISIDE UNIT STEP FUNCTION

It is an extremely useful and simplest discontinuous function, denoted by $U(t-a)$ or $H(t-a)$ and is defined as

$$U(t-a)\left[\text{or } H(t-a)\right]=\begin{cases}0, & \text{for } t<a\\ 1, & \text{for } t\geq a\end{cases}$$

H(t-a)

f(t)

1

O

t

a

In particular if $a=0$, we have the unit step function defined as

$$U(t)\left[\text{or } H(t)\right]=\begin{cases}0, & \text{for } t<0\\ 1, & \text{for } t\geq 0\end{cases}$$

Now let us find Laplace transform of $U(t-a)$ as

$$L\{U(t-a)\}=\int_0^\infty e^{-pt}\,U(t-a)\,dt$$

$$=\int_0^a e^{-pt}\,0\,dt+\int_a^\infty e^{-pt}\,1\,dt=0+\left[\frac{e^{-pt}}{-p}\right]_a^\infty=\frac{1}{p}e^{-ap}.$$

Thus $L\{U(t-a)\}=\dfrac{e^{-ap}}{p}$.

In particular when $a=0$, we have $L\{U(t)\}=\dfrac{1}{p}$.

9.17. UNIT IMPULSE FUNCTION OR (DIRAC-DELTA FUNCTION)

The concept of very large force acting for a very short time comes from the mechanics. The unit impulse function is the limiting form of the function

$$\delta(t-a)=\begin{cases}\dfrac{1}{\varepsilon}, a\le t\le a+\varepsilon, \text{where } \varepsilon\to 0\\ 0, \text{otherwise}\end{cases}$$

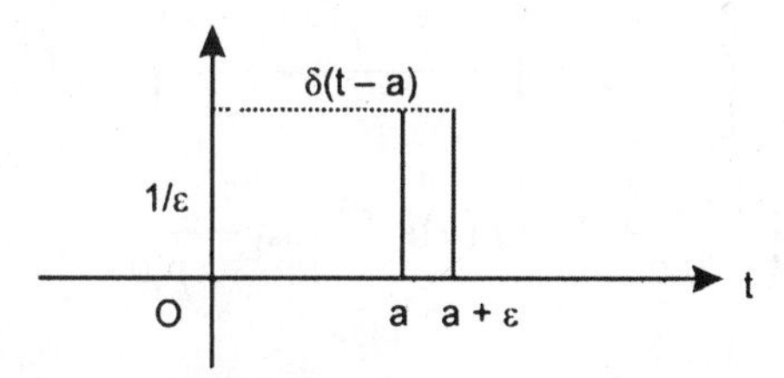

From the adjoining figure it is clear that smaller the ε more the height of strip because the area of the strip is always unity. The above fact can be stated as

$$\delta(t-a)=\begin{cases}\infty, \text{for } t=a\\ 0, \text{for } t\ne a\end{cases}$$

Such that $\int_0^\infty \delta(t-a)\,dt=1$ *for* $a\ge 0$.

9.17.1 Laplace Transform of the Dirac-Delta Function

$$L\{\delta(t-a)\}=\int_0^\infty e^{-pt}\,\delta(t-a)\,dt$$

$$=\int_0^a e^{-pt}\,0\,dt+\int_a^{a+\varepsilon} e^{-pt}\frac{1}{\varepsilon}\,dt+\int_{a+\varepsilon}^\infty e^{-pt}\,0\,dt$$

$$=\frac{1}{\varepsilon}\int_0^{a+\varepsilon} e^{-pt}\,dt=\frac{1}{\varepsilon}\left[\frac{e^{-pt}}{-p}\right]_a^{a+\varepsilon}=\frac{1}{p\,\varepsilon}\left[-e^{-p(a+\varepsilon)}+e^{-ap}\right]$$

$$=\frac{e^{-ap}}{p\,\varepsilon}\left(1-e^{-p\varepsilon}\right)=\frac{e^{-ap}}{p}\frac{\left(1-e^{-p\varepsilon}\right)}{\varepsilon}.$$

Taking limit as $\varepsilon\to 0$, we obtain

$$L\{\delta(t-a)\}=\frac{e^{-ap}}{p}\lim_{\varepsilon\to 0}\frac{1-e^{-\varepsilon p}}{\varepsilon}=\frac{e^{-ap}}{p}\lim_{\varepsilon\to 0}\frac{1}{\varepsilon}\left[1-\left(1-p\,\varepsilon+\frac{p^2\,\varepsilon^2}{2!}+\ldots\right)\right]$$

$$=\frac{e^{-ap}}{p}\lim_{\varepsilon\to 0}\left[p-\frac{p^2\,\varepsilon}{2!}+\frac{p^3\,\varepsilon^2}{3!}-\ldots\right]=\frac{e^{-ap}}{p}[p-0+0-\ldots]=e^{-ap}.$$

Thus $\quad L\{\delta(t-a)\}=e^{-ap}$.

In particular when $a=0$, we have $L\{\delta(t)\}=1$.

9.17.2 Laplace Transform of $f(t)\delta(t-a)$

$$L\{f(t)\delta(t-a)\}=\int_0^\infty e^{-pt}f(t)\delta(t-a)\,dt$$

$$=\int_0^a e^{-pt}f(t).0\,dt+\int_a^{a+\varepsilon}e^{-pt}f(t)\frac{1}{\varepsilon}\,dt+\int_{a+\varepsilon}^\infty e^{-pt}f(t).0\,dt$$

$$=\frac{1}{\varepsilon}\int_a^{a+\varepsilon}e^{-pt}f(t)\,dt=\frac{f(a)}{\varepsilon}\int_a^{a+\varepsilon}e^{-pt}\,dt,\text{ assuming } f(a+\varepsilon)=f(a)$$

$$=\frac{f(a)}{\varepsilon}\left[\frac{e^{-pt}}{-p}\right]_a^{a+\varepsilon}=f(a)\left(\frac{e^{-ap}-e^{-ap}e^{-p\varepsilon}}{p\varepsilon}\right)$$

$$=f(a)e^{-ap}\left(\frac{1-e^{-p\varepsilon}}{p\varepsilon}\right)=f(a)e^{-ap}\lim_{\varepsilon\to 0}\frac{1-e^{-p\varepsilon}}{p\varepsilon},\text{ taking limit } \varepsilon\to 0$$

$$=f(a)e^{-ap}.1$$

Thus $L\{f(t)\delta(t-a)\}=e^{-ap}f(a)$.

Note: One can easily prove that

$$L\{F(t-a)U(t-a)\}=e^{-pa}L\{F(t)\}.$$

Example 9: Find the Laplace transform of the function

$$f(t)=\begin{cases}t-1, 1<t<2\\ 3-t, 2<t<3\end{cases}$$

Solution: We can write

$$f(t)=(t-1)[U(t-1)-U(t-2)]+(3-t)[U(t-2)-U(t-3)]$$
$$=(t-1)U(t-1)-2(t-2)U(t-2)+(t-3)U(t-3).$$

Hence

$$L\{f(t)\}=L\{(t-1)U(t-1)\}-2L\{(t-2)U(t-2)\}+L\{(t-3)U(t-3)\}$$
$$=e^{-p}L\{t\}-2e^{-2p}L\{t\}+e^{-3p}L\{t\}$$
$$=\frac{e^{-p}}{p}\left(1-2e^{-p}+e^{-2p}\right)=\frac{e^{-p}}{p^2}\left(1-e^{-p}\right)^2.$$

Example 10: Find $L^{-1}\left\{\frac{e^{-3p}}{p^2+8p+25}\right\}$.

Solution: Let us write $L^{-1}\left\{\frac{e^{-3p}}{(p^2+8p+25)}\right\}=L^{-1}\{e^{-3p}f(p)\}$,

where $f(p)=\frac{1}{p^2+8p+25}=\frac{1}{(p+4)^2+3^2}$

Now $L^{-1}\{f(p)\}=L^{-1}\left\{\frac{1}{(p+4)^2+3^2}\right\}=e^{-4t}\frac{1}{3}\sin 3t=F\{t\}$

Therefore $F(t)=\frac{e^{-4t}}{3}\sin 3t$.

Now $F(t-3)=\frac{1}{3}e^{-4(t-3)}\sin 3(t-3)$.

We know that $L\{f(t-3)U(t-3)\}=e^{-3p}f(p)$, so that

$$L^{-1}\left\{\frac{e^{-3p}}{(p^2+8p+25)}\right\}=\frac{1}{3}e^{-4(t-3)}\sin(3t-9)U(t-3).$$

Example 11: Evaluate the following (i) $L\{t\,U(t-4)-t^3\,\delta(t-2)\}$ (ii) $L(\sin t\,U(t-\pi))$.

Solution: (i) For $t\,U(t-4)$, we write $t=(t-4)+4=F(t-4)$. So that $F(t)=t+4$.

Therefore $L\{F(t)\}=f(p)=\frac{1}{p^2}+4/p$.

Hence $L\{t\,U(t-4)\}=L\{F(t-4)U(t-4)\}$

$$=e^{-4p}\,L\{F(t)\}=e^{-4p}\left\{\frac{1}{p^2}+\frac{4}{p}\right\}.$$

Next for $t^3\,\delta(t-2)$, we consider $F(t)=t^3$, therefore $F(2)=8$.

Now $L\{t^3\,\delta(t-2)\}=F(2)e^{-2p}=8\,e^{-2p}$.

Therefore $L\{t(U-4)-t^3(t-2)\}=e^{-4p}\left(\frac{1}{p^2}+\frac{4}{p}\right)-8\,e^{-2p}$.

(ii) For $L\{\sin t\,U(t-\pi)\}$, we can write

$$\sin t=\sin(t-\pi+\pi)=-\sin(t-\pi).$$

Therefore $L[\sin t\,U(t-\pi)]=-L[\sin(t-\pi)U(t-\pi)]$

$$=-\left[e^{-ap}\,L(\sin t)\right]_{a=\pi}=-\frac{e^{-p\pi}}{p^2+1}.$$

EXERCISE 9.3

Find the Laplace inverse transform of fallowing functions:

1. $\frac{4}{(p-2)}$
2. $\frac{1}{p^4}$
3. $\frac{1}{p^{7/2}}$
4. $\frac{3P+7}{p^2-2p-3}$.
5. $\frac{p^2+p-2}{p(p+3)(p-2)}$.
6. $\frac{(p+3)}{(p^2+6p+13)^2}$.

7. $\dfrac{(p^2+2)}{p(p^2+4)}$.

8. $\dfrac{1}{p^3(p^2+1)}$.

9. $\dfrac{(5p+3)}{(p-1)(p^2+2p+5)}$.

10. $\dfrac{(p-4)}{4(p-3)^2+16}$.

11. $\dfrac{1}{(p-2)(p^2+1)}$

12. $\dfrac{(11p^2-2p+5)}{(2p^3-3p^2-3p+2)}$

13. $\dfrac{1}{10}\log\left(\dfrac{p^2+b^2}{p^2+a^2}\right)$.

14. $\dfrac{(p-1)}{(p^2-6p+25)}$.

15. $\dfrac{1}{(p+1)(p^2+1)}$.

16. $\dfrac{2p+1}{p(p+1)}$

17. Find the Laplace transform of the function (half wave rectifier)

$$f(t)=\begin{cases}\sin\omega t, \text{ for } 0<t<(\pi/\omega)\\ 0, \text{ for } (\pi/\omega)<t<(2\pi/\omega)\end{cases}.$$

18. Find the inverse Laplace transform of following functions using the convolution theorem

(i) $\dfrac{p}{(p^2+1)(p^2+4)}$

(ii) $\dfrac{1}{(p^2+1)^3}$

(iii) $\dfrac{a(p^2-2a^2)}{p^4+4a^4}$

(iv) $\dfrac{1}{p^2(p^2+a^2)}$.

(v) $\dfrac{1}{p^2(p^2-a^2)}$

19. Using Heaviside expansion formula find the inverse Laplace transform of the following:

(i) $\left\{\dfrac{3p+1}{(p-1)(p^2+1)}\right\}$

(ii) $\dfrac{2p^2-6p+5}{p^3-6p^2+11p-6}$

(iii) $\dfrac{19p+37}{(p+1)(p-2)(p+3)}$

(iv) $\dfrac{3p-2}{p^2-4p+20}$.

ANSWERS 9.3

1. $4e^{2t}$.

2. $\dfrac{t^3}{6}$.

3. $\dfrac{8}{15}t^2\sqrt{\dfrac{t}{\pi}}$.

4. $4e^{3t}-e^{-t}$

5. $\dfrac{1}{3}+\dfrac{4}{15}e^{3t}+\dfrac{2}{5}e^{-2t}$

6. $\dfrac{t}{4}e^{-3t}\sin 2t$

7. $\cos^2 t$

8. $\frac{t^2}{2} + \cos t - 1$

9. $e^t + \frac{3}{2} e^{-t} \sin 2t - e^{-t} \cos 2t$

10. $\frac{1}{4} e^{3t} \cos 2t - \frac{1}{8} e^{3t} \sin 2t$

11. $\frac{1}{5} e^{2t} - \frac{1}{5} \cos t - \frac{2}{5} \sin t$

12. $2e^{-t} + 5e^{2t} - \frac{3}{2} e^{t/2}$

13. $\frac{1}{t}(\cos at - \cos bt)$

14. $e^{3t} \cos 4t + \frac{1}{2} e^{3t} \sin 4t$

15. $\frac{1}{2}(\cos t - \sin t - e^{-t})$

16. $e^{-t} + 1$

17. $\frac{\omega}{(p^2 + \omega^2)\left(1 - e^{-\frac{\pi p}{\omega}}\right)}$.

18. (i) $\frac{1}{3}(\cos t - \cos 2t)$ (ii) $\frac{1}{8}\{(3 - t^2)\sin t - 3t \cos t\}$

(iii) $\cos at \sinh at$ (iv) $\frac{1}{a^3}(at - \sin at)$

(v) $\frac{1}{a^3}[-at + \sinh at]$

19. (i) $2e^t + \sin t - 2\cos t$ (ii) $\frac{1}{2} e^t - e^{2t} + \frac{5}{2} e^{3t}$

(iii) $5e^{2t} - 3e^{-t} - 2e^{-3t}$ (iv) $3e^{2t} \cos 4t + e^{2t} \sin 4t$.

OBJECTIVE QUESTIONS

1. The Laplace transform of $F(t)$ exists for all $p > a$ if
   (a) $F(t)$ is continuous (b) $F(t)$ is differentiable
   (c) $F(t)$ is a function of class A (d) None of these

2. Laplace transform of $\cos at$ is
   (a) $\frac{a}{(p^2 + a^2)}$. (b) $\frac{p}{(p^2 + a^2)}$.
   (c) $\frac{a}{(p^2 - a^2)}$. (d) $\frac{p}{(p^2 - a^2)}$

3. If $L\{F(t)\} = f(p)$, then $L\{t\,F(t)\}$ is
   (a) $f'(p)$. (b) $-f'(p)$.
   (c) $\int_p^\infty f(x)\,dx$. (d) $\frac{1}{p} f(p)$.

4. If $L^{-1}\{f(p)\} = F(t)$, then $L^{-1}\{f(p-a)\}$ is

(a) $e^{at} F(t)$. (b) $\frac{1}{a} F\left(\frac{t}{a}\right)$.

(c) $(-1)^n t^n F(t)$. (d) $\int_0^t F(u)du$.

5. If $L^{-1}\{f(p)\} = F(t)$, then $L^{-1}\left\{\frac{f(p)}{p}\right\}$ is

(a) $(-1)^n t^n F(t)$. (b) $\int_0^t F(u)du$.

(c) $F'(t)$. (d) $\frac{F(t)}{t}$.

6. If $L\{F(t)\} = f(p)$ and $G(t) = \begin{cases} F(t-a), t > a \\ 0, t < a \end{cases}$, then $L\{G(t)\}$ is

(a) $f(p-a)$. (b) $e^{-pa} f(p)$.

(c) $\frac{1}{a} f(p/a)$. (d) $f(ap)$.

7. If $L\{F(t)\} = f(p)$ and $F(t)$ is of class A, hence $L\{F'(t)\}$ is

(a) $p f(p) - F(0)$. (b) $p^2 f(p)$.

(c) $f(p) - F(0)$. (d) None of these.

8. The Laplace transform of cos *h at* is

(a) $\frac{p}{(p^2 - a^2)}$. (b) $\frac{p}{(p^2 + a^2)}$.

(c) $\frac{1}{(p^2 - a^2)}$. (d) $\frac{1}{(p^2 + a^2)}$.

9. If $L^{-1}\{f(p)\} = F(t)$, then $L^{-1}\{f(ap)\}$ is

(a) $F(at)$. (b) $1/a\, F(t/a)$.

(c) $F(t/a)$. (d) $a F(ap)$.

10. $L^{-1}\left\{\frac{1}{p^2 + a^2}\right\}$ is

(a) $\sin at$. (b) $\frac{1}{a} \sin at$.

(c) $\cos at$. (d) $\sin h\, at$

11. Laplace transform of e^t is

(a) $\frac{1}{(p-2)}$. (b) $\frac{1}{(p-2)^2}$.

(c) $\frac{1}{p^2}$. (d) None of these.

12. Value of $L^{-1}\left\{\dfrac{1}{(p-1)^3}\right\}$ is

(a) $t^2 e^{2t}$. (b) $\dfrac{t^2}{2} e^t$.

(c) $\dfrac{t^2}{2} e^{-t}$. (d) None of these.

13. $L^{-1}\left\{\dfrac{1}{p}\right\}$ is

(a) 1. (b) 0.

(c) t. (d) p.

14. The inverse Laplace transform of $f(p)=\dfrac{1}{p}+\dfrac{1}{(p+1)}$ is

(a) $1+e^{-t}$. (b) $1+e^t$.

(c) e^t. (d) e^{-t}.

15. The Laplace transform of $t^n e^{-at}$ is

(a) $\dfrac{\Gamma n}{(p+a)^n}$. (b) $\dfrac{\Gamma(n+1)}{(p+a)^{n+1}}$.

(c) $\dfrac{n!}{(p+a)^n}$. (d) $\dfrac{(n+1)!}{(p+a)^{n+1}}$.

16. A function which is sectionally continuous over every finite interval in the range $x \geq 0$ and is of exponential order as $x \to \infty$ is called

(a) Function of class A. (b) Transformation function.

(c) Laplace function. (d) Fourier function.

17. Laplace transform of $\sin at$ is

(a) $\dfrac{p}{p^2+a^2}$. (b) $\dfrac{a}{p^2+a^2}$.

(c) $\dfrac{p}{p^2-a^2}$. (d) $\dfrac{a}{p^2-a^2}$.

18. If $L\{F(t)\}=f(p)$, then $\{e^{at} F(t)\}$ is

(a) $f(p/a)$. (b) $f(p-a)$.

(c) $f'(p)$. (d) $f(p+a)$.

19. If $L\left\{\dfrac{\sin t}{t}\right\}=\tan^{-1}\left(\dfrac{1}{p}\right)$, then $L\left\{\dfrac{\sin 3t}{t}\right\}$ is

(a) $\dfrac{1}{3}\tan^{-1}\left(\dfrac{3}{p}\right)$. (b) $3\tan^{-1}(3p)$.

(c) $\tan^{-1}\left(\frac{3}{p}\right)$. (d) $\tan^{-1}(3p)$.

20. If $L\{F(t)\}=f(p)$, then $L\{F(at)\}$ is

(a) $f(p/a)$. (b) $\frac{1}{a}f(p/a)$.

(c) $f(ap)$. (d) $af(ap)$.

21. Laplace transform of $F(t)=1$ is

(a) $\frac{1}{p}$. (b) $\frac{1}{p+1}$.

(c) p. (d) $\frac{1}{p-1}$.

22. The inverse Laplace transform of $f(p)=\frac{3}{(p-2)^2+9}$ is

(a) $e^{2t}\sin 3t$. (b) $e^{3t}\sin 2t$.

(c) $e^{2t}\cos 3t$. (d) $e^{3t}\cos 2t$.

23. The Laplace transform of e^{at} is

(a) $\frac{1}{a-p}$. (b) $\frac{1}{p+a}$.

(c) $\frac{1}{p-a}$. (d) $\frac{1}{ap}$.

24. The inverse Laplace transform of $f(p)=\frac{1}{(p+a)^3}$ is

(a) $\frac{1}{2}t^2e^{-at}$. (b) 0.

(c) te^{at}. (d) $\frac{1}{2}te^{-at}$.

ANSWERS

1.	(c)	2	(b)	3	(b)
4	(a)	5.	(b)	6.	(b)
7.	(a)	8.	(a)	9.	(b)
10.	(b)	11.	(d)	12.	(b)
13.	(a)	14.	(a)	15.	(b)
16.	(a)	17.	(b)	18.	(b)
19.	(c)	20.	(b)	21.	(a)
22.	(a)	23.	(c)	24.	(a)

10

Applications of Laplace's Transformation

10.1 INTRODUCTION

Laplace transform is an important tool to solve ordinary and partial differential equations arising in various branches of Mathematics, Physics, Engineering and Technology namely Mechanics, Electricity and Magnetism, Electronics, Plasma Dynamics, Magnetohydrodynamics, Electrical and Mechanical Engineering, etc.

10.2 SOLUTION OF DIFFERENTIAL EQUATIONS BY LAPLACE TRANSFORM

In Mathematics, Laplace transform has been widely used to solve the ordinary and partial differential equations. So we shall discuss this method to find the solution of ordinary and simultaneous linear differential equations with constant coefficients. The advantage of this method is that it yields the particular solution directly without the necessity of first find the general solution and then evaluating the arbitrary constants.

We take the Laplace transform of both sides of the given differential equation and obtain an algebraic equation in $\tilde{y}$ and p, where $\tilde{y}$ is the Laplace transform of y, after using initial conditions,. We solve this algebraic equation for $\tilde{y}$ as a known function of 'p'. We then resolve the function of p into partial fractions and take the inverse Laplace transform of both sides of equation to yield y as a function of independent variable 't'. The function of 't' thus obtained is the desired solution of the given differential equation satisfying the given conditions.

The procedure to solve the simultaneous differential equations is similar to that discussed for the case of ordinary differential equations with constant coefficients.

If y is a function of 't', we denote Laplace transform of y as $L(y)$ or $\tilde{y}$.

Example 1: Using Laplace transform, find the solution of the differential equation

$$\frac{d^2 y}{d t^2} - 3\frac{d y}{d t} + 2y = 4t + 3e^t \text{ when } y(0) = 1 \text{ and } y(0) = -1.$$

Solution: We have $\dfrac{d^2 y}{d t^2} - 3\dfrac{d y}{d t} + 2y = 4t + 3e^t$. ...(1)

Taking the Laplace transform of both sides of equation (1), we get

$$\mathcal{L}\left\{\frac{d^2y}{dt^2}\right\} - 3\,\mathcal{L}\left\{\frac{dy}{dt}\right\} + 2\,\mathcal{L}\{y\} = 4\,\mathcal{L}\{t\} + 3\,\mathcal{L}\{e^t\}$$

$$\Rightarrow \quad \left[p^2\,\tilde{y} - p\,y(0) - y'(0) - 3\left[p\,\tilde{y} - y(0)\right]\right] + 2\,\tilde{y} = \frac{4}{p^2} + \frac{1}{p-3}$$

Now using the given conditions $y(0) = 1$ and $y'(0) = -1$, we have

$$\left(p^2 - 3p + 2\right)\tilde{y} - p + 1 + 3 = \frac{4}{p^2} + \frac{1}{p-3}$$

$$\Rightarrow \quad \left(p^2 - 3p + 2\right)\tilde{y} = \frac{4}{p^2} + \frac{1}{p-3} + p - 4$$

$$\Rightarrow \quad (p-1)(p-2)\,\tilde{y} = \frac{p^4 - 7p^3 + 13p^2 + 4p - 12}{p^2(p-3)}$$

$$\Rightarrow \quad \tilde{y} = \frac{p^4 - 7p^3 + 13p^2 + 4p - 12}{p^2(p-3)(p-1)(p-2)}$$

$$\Rightarrow \quad \tilde{y} = \frac{3}{p} + \frac{2}{p^2} - \frac{1}{2}\left(\frac{1}{p-1}\right) - 2\left(\frac{1}{p-2}\right) + \frac{1}{2}\left(\frac{1}{p-3}\right).$$

Taking Laplace inverse transform, we get

$$y = 3\,L^{-1}\left(\frac{1}{p}\right) + 2\,L^{-1}\left(\frac{1}{p^2}\right) - \frac{1}{2}\,L^{-1}\left(\frac{1}{p-1}\right) - 2\,L^{-1}\left(\frac{1}{p-2}\right) + \frac{1}{2}\,L^{-1}\left(\frac{1}{p-3}\right)$$

$$= 3 + 2t - \frac{1}{2}\left(e^t - e^{3t}\right) - 2e^{2t}.$$

Example 2: Using Laplace transform solve the following differential equation

$$\frac{d^2 x}{d t^2} + 9x = \cos 2t \text{ if } x(0) = 1 \text{ and } x\left(\frac{\pi}{2}\right) = -1.$$

Solution: We have

$$\frac{d^2 x}{d t^2} + 9x = \cos 2t. \qquad ...(1)$$

Taking Laplace transform of both the sides of equation (1), we get

$$L\left[\frac{d^2 x}{d t^2}\right] + 9\,L\{x\} = L\{\cos 2t\}$$

$$\Rightarrow \quad p^2\,\tilde{x} - p\,x(0) - x'(0) + 9\,\tilde{x} = \frac{p}{p^2 + 4}. \qquad ...(2)$$

On applying the given conditions $x(0)=1$ in (2), we get

$$p^2\,\bar{x} - p + 9\,\bar{x} - x'(0) = \frac{p}{p^2+4}$$

$$\Rightarrow \quad (p^2+9)\tilde{x} = p + \frac{p}{p^2+4} + x'(0) = \frac{p^3+5p}{p^2+4} + x'(0)$$

$$\Rightarrow \quad \tilde{x} = \frac{(p^3+5p)}{(p^2+4)(p^2+9)} + \frac{x'(0)}{p^2+9}$$

$$= \frac{1}{5}\frac{p}{p^2+4} + \frac{4}{5}\frac{p}{p^2+9} + \frac{x'(0)}{p^2+9}.$$

Taking the inverse Laplace transform, we get

$$x(t) = \frac{1}{5}L^{-1}\left\{\frac{p}{p^2+4}\right\} + \frac{4}{5}L^{-1}\left\{\frac{p}{p^2+9}\right\} + L^{-1}\left\{\frac{x'(0)}{p^2+9}\right\}$$

$$\Rightarrow \quad x(t) = \frac{1}{5}\cos 2t + \frac{4}{5}\cos 3t + \frac{x'(0)}{3}\sin 3t. \qquad \text{...(3)}$$

On putting $x(\pi/2) = -1$ in (3), we get

$$-1 = -\frac{1}{5} + 0 - \frac{x'(0)}{3} \Rightarrow x'(0) = \frac{12}{5}.$$

Hence putting $x'(0) = \frac{12}{5}$ in (3), we get

$$x = \frac{1}{5}\cos 2t + \frac{4}{5}\cos 3t + \frac{12}{5\times 3}\sin 3t$$

$$= \frac{1}{5}\cos 2t + \frac{4}{5}\cos 3t + \frac{4}{5}\sin 3t.$$

Example 3: Using Laplace transform find the solution of the initial value problem

$$y'' + 9y = 6\cos 3t \text{ if } y(0) = 2 \text{ and } y'(0) = 0.$$

Solution: We have $y'' + 9y = 6\cos 3t$. ...(1)

Taking Laplace transform of both sides of equation (1), we get

$$L(y'') + 9L\{y\} = 6L\{\cos 3t\}$$

$$\Rightarrow \quad p^2\,\tilde{y} - p\,y(0) - y'(0) + 9p\,\tilde{y} = 6\frac{p}{p^2+9}. \qquad \text{...(2)}$$

Putting the value $y(0)=2$ and $y'(0)=0$ in (2), we get

$$p^2\,\tilde{y} - 2p + 9\,\tilde{y} = \frac{6p}{p^2+9} \Rightarrow \tilde{y} = \frac{2p}{p^2+9} + \frac{6p}{(p^2+9)^2}.$$

Therefore, we have

$$y = L^{-1}\left\{\frac{2p}{p^2+9}\right\} + L^{-1}\left\{\frac{6p}{(p^2+9)^2}\right\} = 2\cos 3t - L^{-1}\left\{\frac{d}{dp}\left(\frac{3}{p^2+9}\right)\right\}$$

$$= 2\cos 3t + 3\frac{t\sin 3t}{3} = 2\cos 3t + t\sin 3t.$$

Example 4: Using Laplace transform find the solution of the initial value problem $y'' - 4y' + 4y = 64\sin 2t$ given that $y(0)=0$ and $y'(0)=1$.

Solution: We have $y'' - 4y' + 4y = 64\sin 2t$.

Taking Laplace transform of both sides of (1), we get

$$p^2\tilde{y} - p\,y(0) - y'(0) - 4\left(p\,\tilde{y} - y(0)\right) + 4\tilde{y} = 64\frac{2}{(p^2+4)}. \quad ...(1)$$

By using the value of $y(0)$ and $y'(0)$ in (1), we get

$$\tilde{y}(p^2 - 4p + 4) - 1 = \frac{128}{p^2+4} \Rightarrow \tilde{y} = \frac{1}{(p-2)^2} + \frac{128}{(p^2+4)(p-2)^2}$$

$$\Rightarrow \quad \tilde{y} = \frac{1}{(p-2)^2} - \frac{8}{(p-2)} + \frac{16}{(p-2)^2} + \frac{8p}{p^2+4}$$

Therefore, we have

$$y = L^{-1}\left[-\frac{8}{(p-2)} + \frac{17}{(p-2)^2} + \frac{8p}{p^2+4}\right] = -8e^{2t} + 17te^{2t} + 8\cos 2t.$$

Example 5: Solve the equation $\frac{d^2x}{dt^2} + 2\frac{dx}{dt} + 5x = e^{-t}\sin t$ with initial conditions $x(0)=0$, $x'(0)=1$.

Solution: We have $\frac{d^2x}{dt^2} + 2\frac{dx}{dt} + 5x = e^{-t}\sin t$.

The given equation can be written as $x'' + 2x' + 5x = e^{-t}\sin t$. Now taking Laplace transform of both sides, we get

$$\left[p^2\tilde{x} - p\,x(0) - x'(0)\right] + 2\left[p\,\tilde{x} - x(0)\right] + 5\tilde{x} = \frac{1}{(p+1)^2+1}. \quad ...(1)$$

Using the given conditions in equation (2), we get

$$(p^2 + 2p + 5)\tilde{x} - 1 = \frac{1}{p^2 + 2p + 2}$$

$$\Rightarrow \quad \tilde{x} = \frac{1}{(p^2+2p+2)(p^2+2p+5)} + \frac{1}{(p^2+2p+5)}$$

$$=\frac{1}{3}\left[\frac{1}{(p^2+2p+2)}+\frac{2}{p^2+2p+5}\right]$$

$$=\frac{1}{3}\left[\frac{1}{(p+1)^2+1}+\frac{2}{(p+1)^2+2^2}\right]$$

Taking Laplace inverse transform, we get

$$x=\frac{1}{3}L^{-1}\left[\frac{1}{(p+1)^2+1}+\frac{2}{(p+1)^2+2^2}\right]$$

$$=\frac{1}{3}e^{-t}\left[L^{-1}\left\{\frac{1}{p^2+1}\right\}+L^{-1}\left\{\frac{2}{p^2+2^2}\right\}\right]$$

$$=\frac{1}{3}e^{-t}\left[\sin t+\sin 2t\right]$$

Example 6: Solve the following equation by Laplace transform

$$\left(D^2+n^2\right)x=a\sin(nt+\alpha),\ x(0)=x'(0)=0.$$

Solution: We have $\left(D^2+n^2\right)x=a\sin(nt+\alpha)$. ...(1)

On taking Laplace transform of both sides, we get

$$L\{x''\}+n^2L\{x\}=L\{a\sin(nt+\alpha)\}$$

$$\Rightarrow \left[p^2\tilde{x}-p\,x(0)-x'(0)\right]+n^2\tilde{x}=a\cos\alpha\frac{n}{p^2+n^2}+a\sin\alpha\frac{p}{p^2+n^2}$$

$$\Rightarrow \quad \tilde{x}=\frac{a\,n\cos\alpha}{\left(p^2+n^2\right)^2}+\frac{a\,p\sin\alpha}{\left(p^2+n^2\right)^2} \qquad ...(2)$$

Taking the inverse Laplace transform of both sides, we get

$$x=(a\cos\alpha)L^{-1}\left\{\frac{n}{\left(p^2+n^2\right)^2}\right\}+(a\sin\alpha)L^{-1}\left\{\frac{p}{\left(p^2+n^2\right)^2}\right\} \qquad ...(3)$$

We know that $L^{-1}\left\{\frac{1}{\left(p^2+n^2\right)}\right\}=\frac{1}{n}\sin nt$ ...(4)

$$\Rightarrow \quad L^{-1}\frac{d}{dn}\left\{\frac{1}{\left(p^2+n^2\right)}\right\}=\frac{nt\cos nt-\sin nt}{n^2}$$

$$\Rightarrow \quad L^{-1}\left\{\frac{-2n}{\left(p^2+n^2\right)^2}\right\}=\frac{nt\cos nt-\sin nt}{n^2}$$

$$\Rightarrow \quad L^{-1}\left\{\frac{n}{\left(p^2+n^2\right)^2}\right\}=\frac{1}{2n^2}(\sin nt-nt\cos nt).$$

Again from (4), we have $L^{-1}\dfrac{d}{dp}\left\{\dfrac{1}{(p^2+n^2)}\right\}=-t\dfrac{1}{n}\sin nt$

$$\Rightarrow \quad L^{-1}\left\{-\frac{2p}{(p^2+n^2)^2}\right\}=-\frac{t}{n}\sin nt$$

$$\Rightarrow \quad L^{-1}\left\{\frac{p}{(p^2+n^2)^2}\right\}=\frac{t}{2n}\sin nt.$$

Now from (3), we have

$$x=(a\cos\alpha)\frac{1}{2n^2}(\sin nt-nt\cos nt)+(a\sin\alpha)\frac{t}{2n}\sin nt$$

$$=\frac{a}{2n^2}(\cos\alpha\sin t-nt\cos(\alpha+nt)).$$

Example 7: Voltage Ee^{-at} is applied at $t=0$ to a circuit of inductance L and resistance R. Show that the current at time t is

$$\frac{E}{R-aL}\left(e^{-at}-e^{-\frac{Rt}{L}}\right).$$

Solution: Let I be the current in the circuit at any time t. Then by Kirchhoff's law, we have

$$L\frac{dI}{dt}+RI=Ee^{-at}, \text{ where } I(0)=0. \qquad ...(1)$$

Taking Laplace transform of both sides of equation (1), we have

$$Lp\bar{I}-I(0)+R\bar{I}=\frac{E}{p+a}. \qquad ...(2)$$

Now using the given condition, equation (2) becomes

$$\bar{I}(Lp+R)=\frac{E}{p+a}$$

$$\Rightarrow \quad \bar{I}=\frac{E}{(p+a)(Lp+R)}=\frac{E}{R-aL}\left[\frac{1}{p+a}-\frac{L}{Lp+R}\right]$$

$$=\frac{E}{R-aL}\left(\frac{1}{p+a}-\frac{1}{p+R/L}\right).$$

Taking the inverse Laplace transform of both sides, we get

$$I=\frac{E}{R-aL}L^{-1}\left\{\frac{1}{p+a}-\frac{1}{p+R/L}\right\}=\frac{E}{R-aL}\left[e^{-at}-e^{-\frac{Rt}{L}}\right].$$

Example 8: The current $i(t)$ and the charge $q(t)$ on the capacitor in an electric circuit satisfy the equations $t\dfrac{di}{dt}+ri+\dfrac{q}{c}=e_0$ *and* $q=\int_0^t i(t)\,dt$

where l, r, c, e_0 are constants and 't' is the time after closing the switch in the circuit. If q and i are initially zero, then $i=\left[\dfrac{e_0}{w\,l}\right]e^{-bt}\sin wt$

where $b=\dfrac{r}{2l}$ and $w^2=\dfrac{1}{lc}-\dfrac{r^2}{4l^2}$.

Solution: Taking the Laplace transform of both sides of

$l\dfrac{di}{dt}+ri+\dfrac{q}{c}=e_0$, we get

$$l\left(p\bar{i}-i(0)\right)+r\bar{i}+\frac{1}{c}L\left\{\int_0^t i(t)\,dt\right\}=\frac{e^0}{p} \quad ...(1)$$

Since $q=\int_0^t i(t)\,dt$, where $\bar{i}=L\{i(t)\}$ so that $i=L^{-1}[\bar{i}]$.

So (1) becomes $l\left(p\bar{i}-i(0)\right)+r\bar{i}+\dfrac{1}{c}\left[\dfrac{1}{p}\bar{i}\right]=\dfrac{e^0}{p}$,

by using the result of Laplace transform of integrals.

Using the condition $i(0)=0$, we have

$$\left(lp+r+\frac{1}{cp}\right)\bar{i}=\frac{e_0}{p} \Rightarrow \left[\frac{lcp^2+rcp+1}{cp}\right]\bar{i}=\frac{e_0}{p}$$

Hence $$\bar{i}=\frac{c\,e_0}{lcp^2+rcp+1}=\frac{c\,e_0}{lc\left(p^2+\frac{r}{l}p+\frac{1}{lc}\right)}$$

$$=\frac{e_0}{l}\left[\frac{1}{\left(p+\frac{r}{2l}\right)^2+\left(\frac{1}{lc}-\frac{r^2}{4l^2}\right)}\right]$$

$$=\frac{e_0}{l}\left[\frac{1}{(p+b)^2+w^2}\right], \text{ where } b=\frac{r}{2l} \text{ and } w^2=\frac{1}{lc}-\frac{r^2}{4l^2}>0.$$

Taking the inverse Laplace transform of both sides, we get

$$i=\frac{e_0}{l}\frac{e^{-bt}}{w}\sin wt=\frac{e_0}{wl}e^{-bt}\sin wt.$$

Example 9: Using Laplace transformation, solve

$(D-2)x-(D+1)y=6e^{3t}$ and $(2D-3)x+(D-3)y=6e^{3t}$ given $x=3$,. $y=0$ when $t=0$.

Solution: We have the following simultaneous equation:

$$(D-2)x-(D+1)y=6e^{3t} \quad ...(1)$$

$$(2D-3)x+(D-3)y=6e^{3t}. \quad ...(2)$$

Taking the Laplace transform of both sides of (1) and (2), we get

$$[p\tilde{x}-x(0)-2\tilde{x}]-[p\tilde{y}-y(0)+\tilde{y}]=\frac{6}{p-3} \quad ...(3)$$

and $$[2\{p\tilde{x}-x(0)\}-3\tilde{x}]+[p\tilde{y}-y(0)-3\tilde{y}]=\frac{6}{p-3} \quad ...(4)$$

Simplifying (3) and (4) by using the conditions $x(0)=3, y(0)=0$, we have

$$(p-2)\tilde{x}-(p+1)\tilde{y}=\frac{3p-3}{p-3}$$

and $$(2p-3)\tilde{x}+(p-3)\tilde{y}=\frac{6p-12}{p-3}.$$

Solving these, we get

$$\tilde{x}=\frac{1}{(p-1)}+\frac{2}{(p-1)^2}+\frac{2}{(p-3)} \quad ...(5)$$

and $$\tilde{y}=\frac{1}{p-1}-\frac{1}{(p-1)^2}-\frac{1}{(p-3)} \quad ...(6)$$

Taking the inverse Laplace transforms of both sides of (5) and (6), we get

$$x=e^t+2te^t+2e^{3t}$$

and $$y=e^t-te^t-e^{3t}$$

Example 10: Solve the simultaneous equations $Dx-y=e^t$ ...(1)

$$Dy+x=\sin t \quad ...(2)$$

given that $x(0)=1, y(0)=0$.

Solution: Taking the Laplace transform of both sides (1) and (2), we get

$$p\tilde{x}-x(0)-\tilde{y}=\frac{1}{p-1} \Rightarrow p\tilde{x}-\tilde{y}=\frac{p}{p-1} \quad ...(3)$$

and $$p\tilde{y}-y(0)+\tilde{x}=\frac{1}{p^2+1} \Rightarrow \tilde{x}+p\tilde{y}=\frac{1}{p^2+1} \quad ...(4)$$

Solving (3) and (4), we get $\tilde{x} = \dfrac{p^2}{(p-1)(p^2+1)} + \dfrac{1}{(p^2+1)^2}$

$$= \frac{1}{2}\left[\frac{1}{p-1} + \frac{p}{p^2+1} + \frac{1}{p^2+1}\right] + \frac{1}{(p^2+1)^2}$$

$$\tilde{y} = \frac{p}{(p^2+1)^2} - \frac{p}{(p-1)(p^2+1)}$$

$$= \frac{p}{(p^2+1)^2} - \frac{1}{2}\left[\frac{1}{p-1} - \frac{p}{p^2+1} + \frac{1}{p^2+1}\right].$$

Taking the inverse Laplace transforms, we obtain

$$x = \frac{1}{2}\left(e^t + \cos t + \sin t\right) + \frac{1}{2}\left(\sin t - t\cos t\right),$$

$$\left[\text{by using } L^{-1}\left\{\frac{1}{(p^2+a^2)^2}\right\} = \frac{1}{2a^2}(\sin at - at\cos at)\right]$$

and $y = \dfrac{t}{2}\sin t - \dfrac{1}{2}\left(e^t - \cos t + \sin t\right)$

$$\left[\text{by using } L^{-1}\left\{\frac{p}{(p^2+a^2)^2}\right\} = \frac{t}{2a}\sin at\right]$$

Hence, the required solutions are

$$x = \frac{1}{2}\left[e^t + (1-t)\cos t + 2\sin t\right]$$

and $y = \dfrac{1}{2}\left[\cos t - (1-t)\sin t - e^t\right].$

Example 11: Solve the simultaneous equations.

$$\frac{d^2x}{dt^2} + 5\frac{dy}{dt} - x = t \text{ and } 2\frac{dx}{dt} - \frac{d^2y}{dt^2} + 4y = 2,$$

given that when $t = 0, x = 0, y = 0, \dfrac{dx}{dt} = 0, \dfrac{dy}{dt} = 0$.

Solution: Let $L\{x(t)\} = \tilde{x}(p)$ and $L\{y(t)\} = \tilde{y}(p)$

Then, taking Laplace transforms of given equations, we get

$$\left\{p^2\tilde{x} - x(0) - x'(0)\right\} + 5\left\{p\tilde{y} - y(0)\right\} - \tilde{x} = \frac{1}{p^2}$$

and $\quad 2\{p\,\tilde{x}-x(0)\}-\{p^2\,\tilde{y}-y(0)-y'(0)\}+4\,\tilde{y}=\dfrac{2}{p}$.

Using the given initial conditions, these equations reduce to

$$(p^2-1)\tilde{x}+5\,p\,\tilde{y}=\frac{1}{p^2} \qquad \text{...(1)}$$

and $$2\,p\,\tilde{x}-(p^2-4)\tilde{y}=\frac{2}{p} \qquad \text{...(2)}$$

Eliminating $\tilde{y}$ between (1) and (2), we find

$$\{(p^2-1)(p^2-4)+10\,p^2\}\tilde{x}=\frac{p^2-4}{p^2}+10$$

$$\Rightarrow \quad \tilde{x}=\frac{11\,p^2-4}{p^2(p^2+1)(p^2+4)}$$

$$=-\frac{1}{p^2}+\frac{5}{p^2+1}-\frac{4}{p^2+4}, \text{ on doing partial fractions}$$

Taking inverse Laplace transform, we get

$$x=-t+5\sin t-2\sin 2\,t.$$

Again eliminating $\tilde{x}$ between (1) and (2), we have

$$\{10\,p^2+(p^2-1)(p^2-4)\}\tilde{y}=\frac{2}{p}-\frac{2(p^2-1)}{p}$$

$$\Rightarrow \quad \tilde{y}=\frac{4-2\,p^2}{p(p^2+1)(p^2+4)}=\frac{1}{p}-\frac{2\,p}{p^2+1}+\frac{p}{p^2+4}.$$

Taking inverse Laplace transforms, we get $y=1-2\cos t+\cos 2\,t$. ...(4)

Thus (3) and (4) together constitute the desired solution.

Example 12: The co-ordinates (x, y) of a particle moving along a plane curve at any time t are given by

$$\frac{d\,y}{d\,t}+2\,x=\sin 2\,t \text{ and } \frac{d\,x}{d\,t}-2\,y=\cos 2\,t\,(t>0)$$

It is given that at $t=0$, $x=1$ and $y=0$. By using Laplace transform, show that the particle moves along the curve $4\,x^2+4\,x\,y+5\,y^2=4$.

Solution: The given equations are $\dfrac{d\,y}{d\,t}+2\,x=\sin 2\,t$, ...(1)

$$\frac{dx}{dt}-2y=\cos 2t \qquad \text{...(2)}$$

Taking Laplace transform of equation (1), we get

$$2\tilde{x} - p\tilde{y} - y(0) = \frac{2}{p^2+4}, \text{ where } \tilde{x} = L(x) \text{ and } \tilde{y} = L(y)$$

$$\Rightarrow \quad 2\tilde{x} + p\tilde{y} = \frac{2}{p^2+4}, \text{ since } y(0) = 0. \qquad ...(3)$$

Again, taking Laplace transform of both sides of equation (2), we get

$$p\tilde{x} - x(0) - 2\tilde{y} = \frac{p}{p^2+4}.$$

$$\Rightarrow \quad p\tilde{x} - 2\tilde{y} = \frac{p}{p^2+4} + 1, \text{ since } x(0) = 1, \qquad ...(4)$$

Multiplying equation (3) by 2 and equation (4) by p and then adding, we get

$$4\tilde{x} + p^2\tilde{x} = \frac{4}{p^2+4} + \frac{p^2}{p^2+4} + p$$

$$\Rightarrow \quad \tilde{x} = \frac{1+p}{4+p^2} = \frac{1}{p^2+4} + \frac{p}{p^2+4}.$$

Taking inverse Laplace transform, we get $x = \frac{1}{2}\sin 2t + \cos 2t$. ...(5)

Again multiplying equation (3) by p and equation (4) by 2 then subtracting equation (4) from (3), we get

$$p^2\tilde{y} + 4\tilde{y} = \frac{2p}{p^2+4} - \frac{2p}{p^2+4} - 2 \Rightarrow \tilde{y} = \frac{-2}{p^2+4}.$$

Taking inverse Laplace transform, we get $y = -\sin 2t$. ...(6)

Now, $$4x^2 = 4\left[\frac{1}{4}\sin^2 2t + \cos^2 2t + \sin 2t \cos 2t\right]$$

and $$5y^2 = 5\sin^2 2t.$$

Therefore $$4xy = 4\left[\left(\frac{1}{2}\sin 2t + \cos 2t\right)(-\sin 2t)\right]$$

$$= -\left(2\sin^2 2t + 4\sin 2t \cos 2t\right).$$

$$\Rightarrow \quad 4x^2 + 5y^2 + 4xy = 4\sin^2 2t + 4\cos^2 2t = 4.$$

EXERCISE 10.1

Solve the following differential equations using Laplace transform:

1. $(D^2 - 2D + 2)x = 0$, if $x = D(x) = 1$ at $t = 0$.
2. $x'' - 3x' + 2x = 1 - e^{2t}$, *if* $x(0) = 1, x'(0) = 0$.

3. $\dfrac{dx}{dt}+x=\sin\omega t$, if $x(0)=2$.

4. $(D^2-1)x=a\cosh t$, if $x(0)=x'(0)=0$.

5. $\dfrac{d^2y}{dt^2}+y=6\cos 2t$, where $y=3$, $\dfrac{dy}{dt}=1$ when $t=0$.

6. $(D^3+1)y=1, t>0$; if $y=Dy=D^2y=0$ when $t=0$.

7. $(D^4-1)y=1$, where $y(0)=y'(0)=y''(0)=y'''(0)=0$.

8. $(D^3-D^2+4D-4)x=68e^t\sin 2t$, with the given conditions $x(0)=1, x'(0)=-19, x''(0)=-37$.

9. $(D^2+9)y=18t$, if $y(0)=0$ and $y\left(\dfrac{\pi}{2}\right)=0$.

10. $(D^2+m^2)x=9\cos nt, t>0$, if $x=x_0$ and $Dx=x_1$ when $t=0$ and $n\neq m$.

11. $\dfrac{d^2y}{dt^2}+2\dfrac{dy}{dt}+2y=\dfrac{17}{2}\sin 5t+51\cos 5t$, if $y(0)=2, y'(0)=-2$.

12. $\dfrac{d^2y}{dt^2}+\dfrac{dy}{dt}-2y=3\cos 3t-11\sin 3t$, when $y(0)=0$ and $y'(0)=6$.

13. $(D^3-3D^2+3D-1)y=t^2e^{2t}$, if $y(0)=1, y'(0)=0, y''(0)=-2$.

14. $\dfrac{d^2y}{dt^2}+4y=0$, $y(0)=0$ and $y'(0)=1$.

15. $\dfrac{d^2y}{dt^2}+y=U(t-\pi)-U(t-2\pi)$ if $y(0)=y'(0)=0$.

Solve the following simultaneous differential equations using Laplace transform:

16. $(D+1)x+(D-1)y=e^{-t}$,

$(D+2)x+(D+1)y=e^t$, where $D=\dfrac{d}{dt}$ and $x(0)=1, y(0)=0$.

17. $\dfrac{dx}{dt}+y=\sin t$, $\dfrac{dy}{dt}+x=\cos t$. Given that $x=2$ and $y=0$ when $t=0$.

18. $3\frac{dx}{dt}+3\frac{dy}{dt}+5x=25\cos t,$

$2\frac{dx}{dt}-3\frac{dy}{dt}=5\sin t,$ with $x(0)=2, y(0)=3$.

19. Using Laplace transform solve the equations

$\frac{dx}{dt}+\frac{dy}{dt}=t, \frac{d^2}{dt^2}-y=e^{-t}$. Given that $x(0)=0\ y(0)=0\ x'(0)=0$.

20. Using Laplace transform method solve $(D^2-2)x-3y=e^{2t}$ and $(D^2-2)x-3y=e^{2t}$. If $x(0)=1, y(0)=1, x'(0)=0=y'(0)$.

21. A mechanical system with two degree of freedom satisfies the equations

$$2\frac{d^2x}{dt^2}+3\frac{dy}{dt}=4 \text{ and } 2\frac{d^2y}{dt^2}-3\frac{dx}{dt}=0.$$

Using Laplace transform, determine x and y at any instant given that $x, y, \frac{dx}{dt}, \frac{dy}{dt}$ all vanish at $t=0$.

22. Using the Laplace transform, solve the initial value problem

$$y''_1=y_1+3y_2 \text{ and } y''_2=4y_1-4e^t$$

where $y_1(0)=2, y'_1(0)=3, y_2(0)=1, y'_2(0)=2$.

23. A particle moves in a line so that its displacement x from a fixed point O at any time 't' is given by $\frac{d^2x}{dt^2}+4\frac{dx}{dt}+5x=80\sin 5t$. If initially particle is at rest, find its displacement at any time t

ANSWERS 10.1

1. $x=e^t\cos t$.

2. $x=\frac{1}{2}+\left(\frac{1}{2}-t\right)e^{2t}$.

3. $x=\left(\frac{2+\omega+2w^2}{\omega^2+1}\right)e^{-t}+\frac{(\sin\omega t-\omega\cos\omega t)}{\omega^2+1}$.

4. $x=\frac{1}{2}a\,t\sin\text{h}\,t$.

5. $y=5\cos t-2\cos 2t+\sin t$.

6. $y = 1 - \frac{1}{3}e^{-t} - \frac{2}{3}e^{t/2}\cos\left(\frac{t\sqrt{3}}{2}\right)$

7. $y = -1 + \frac{1}{2}(\cos h\, t + \cos t)$.

8. $x = \frac{1}{5}\left(e^{t} + 14\cos 2t - 3\sin 2t\right) - 2e^{t}\left(\cos 2t + 4\sin 2t\right)$.

9. $y = 2t + \pi\sin 3t$.

10. $x = x_0\cos mt + \frac{x_1}{m}\sin mt + \frac{9}{m^2 - n^2}(\cos nt - \cos mt)$.

11. $y = e^{-t}\left(4\cos t - \frac{1}{2}\sin t\right) - 2\cos 5t + \frac{1}{2}\sin 5t$.

12. $y = e^{t} - e^{-2t} + \sin 3t$.

13. $y = \left(t^2 - 6t + 12\right)e^{2t} - \left(\frac{3}{2}t^2 + 7t + 11\right)e^{t}$.

14. $y = \frac{1}{2}\sin 2t$ *for* $t > 0$.

15. $y = (1 + \cos t)U(t - \pi) - (1 - \cos t)U(t - 2\pi)$.

16. $x = e^{-3t}; y = \sin h\, t + e^{-t}\sin h\, 2t$.

17. $x = e^{t} + e^{-t}; y = e^{-t} - e^{t} + \sin t$.

18. $x = 2\cos t + 3\sin t,\ y = 3\cos t + 2\sin t$.

19. $x = \frac{1}{2}\left(e^{-t} - \cos t + \sin t\right) + \frac{t^2}{2} + \cos t - 1,$

$y = 1 - \frac{1}{2}\left(e^{-t} + \cos t + \sin t\right)$.

20. $x = \frac{3}{4}e^{t} + \frac{7}{4}e^{-t} + \frac{2}{5}e^{2t} + \frac{1}{10}(2\sin t - 19\cos t),$

$y = \frac{19}{10}\cos t - \frac{1}{5}\sin t - \frac{1}{15}e^{2t} - \frac{1}{4}e^{t} - \frac{7}{12}e^{-t}$.

21. $x = \frac{8}{9}\left[1 - \cos\left(\frac{3t}{2}\right)\right],\ y = \frac{8}{9}\left[\left(\frac{3t}{2}\right) - \sin\left(\frac{3t}{2}\right)\right]$.

22. $y_1 = e^{t} + e^{2t},\ y_2 = e^{2t}$.

23. $x = -2(\cos 5t + \sin 5t) + 2e^{-2t}(\cos t + 7\sin t)$.

10.3 PROBLEMS RELATED TO SOLUTIONS OF PARTIAL DIFFERENTIAL EQUATIONS

Theorem 10.1: If $y = y(x,t)$ then

(i) $L\left\{\dfrac{\partial y}{\partial t}\right\} = p\,\tilde{y}(x,p) - y(x,0)$

(ii) $L\left\{\dfrac{\partial^2 y}{\partial t^2}\right\} = p^2\,\tilde{y}(x,p) - p\,y(x,0) - y_t(x,0)$

(iii) $L\left\{\dfrac{\partial y}{\partial x}\right\} = \dfrac{d\,\tilde{y}}{d\,x}(x,p)$

(iv) $L\left\{\dfrac{\partial^2 y}{\partial x^2}\right\} = \dfrac{d^2\,\tilde{y}}{d\,x^2}(x,p)$

where $L\{y(x,t)\} = \tilde{y}(x,p),\, y_t(x,0) = \left(\dfrac{\partial y}{\partial t}\right)_{t=0}$

Proof: (i) $L\left\{\dfrac{\partial y}{\partial t}\right\} = \int_0^\infty e^{-pt}\dfrac{\partial y}{\partial t}\,d\,t = \lim_{a\to\infty}\int_0^a e^{-pt}\dfrac{\partial y}{\partial t}\,d\,t$

$$= \lim_{a\to\infty}\left[\left(e^{-pt}\,y(x,t)\right)_{t=0}^{a} + p\int_0^a e^{-pt}\,y(x,t)dt\right]$$

$$= -y(x,0) + p\int_0^\infty e^{-pt}\,y\,d\,t = p\,y(x,p) - y(x,0).$$

(ii) Let $v = \dfrac{\partial y}{\partial t} = y_t$, then $L\left\{\dfrac{\partial^2 y}{\partial t^2}\right\} = L\left\{\dfrac{\partial v}{\partial t}\right\} = p\,\tilde{v}(x,p) - v(x,0)$

$$= p\,L\{v\} - y_t(x,0) = p\,L\left\{\dfrac{\partial y}{\partial t}\right\} - y_t(x,0)$$

$$= p\left[p\,\tilde{y}(x,p) - y(x,0)\right] - y_t(x,0) \text{ [by (i)]}$$

$$= p^2\,\tilde{y}(x,p) - p\,y(x,0) - y_t(x,0).$$

(iii) Now $L\left\{\dfrac{\partial y}{\partial x}\right\} = \int_0^\infty e^{-pt}\dfrac{\partial y}{d\,x}d\,t = \dfrac{d}{d\,x}\int_0^\infty e^{-pt}\,y\,dt = \dfrac{d\,\tilde{y}}{d\,x}$

(iv) Again $L\left\{\dfrac{\partial^2 y}{\partial x^2}\right\} = L\left\{\dfrac{\partial u}{\partial x}\right\}$ where $u = \dfrac{\partial y}{\partial x}$, [by (iii)]

$$= \dfrac{d}{d\,x}\tilde{u} = \dfrac{d}{d\,x}L\left\{\dfrac{\partial y}{\partial x}\right\} = \dfrac{d}{d\,x}\left\{\dfrac{d\,\tilde{y}}{d\,x}\right\}, \text{ [by (iii)]}$$

$$= \dfrac{d^2\,\tilde{y}}{d\,x^2}.$$

Example 1: Using Laplace transform solve the partial differential equation $\frac{\partial y}{\partial t} = 2\frac{\partial^2 y}{\partial x^2}$, if $y(0,t) = 0 = y(5,t)$, $y(x,0) = 10 \sin 4\pi x$.

Solution: We have $\left\{\frac{\partial y}{\partial t}\right\} = 2\left\{\frac{\partial^2 y}{\partial x^2}\right\}$ ...(1)

Taking Laplace transform of both the sides of equation (1), we get

$$L\left\{\frac{\partial y}{\partial t}\right\} = 2L\left\{\frac{\partial^2 y}{\partial x^2}\right\} \Rightarrow p\tilde{y} - y(x,0) = 2\frac{d^2 \tilde{y}}{dx^2}.$$

Putting the value of $y(x, 0)$, we get

$$p\tilde{y} - 10 \sin 4\pi x = 2\frac{d^2\tilde{y}}{dx^2} \Rightarrow \frac{d^2\tilde{y}}{dx^2} - \frac{p}{2}\tilde{y} = -5 \sin 4\pi x.$$

General solution of this differential equation is

$$\tilde{y} = A e^{x\sqrt{(p/2)}} + B e^{-x(\sqrt{p/2})} + \frac{-5 \sin 4\pi x}{-(4\pi)^2 - p/2}$$

$$= A e^{x\sqrt{p/2}} + B e^{-x\sqrt{p/2}} + \frac{10 \sin 4\pi x}{32\pi^2 + p}. \quad ...(2)$$

Now applying the conditions $L\{y(0,t)\} = 0, L\{y(5,t)\} = 0$, we get

$$A + B = 0 \quad ...(3)$$

and $$0 = A e^{5\sqrt{p/2}} + B e^{-5\sqrt{p/2}} + \frac{10 \sin 20\pi}{32\pi^2 + p}.$$

$$\Rightarrow A e^{10\sqrt{p/2}} + B = 0 \; as \; \sin 20\pi = 0. \quad ...(4)$$

On solving (3) and (4), we obtain $A = 0, B = 0$.

Now (2) becomes $\tilde{y} = \frac{10 \sin 4\pi x}{32\pi^2 + p}$.

Taking inverse Laplace transform, we get $y = 10 e^{-32\pi^2 t} \sin 4\pi x$.

Example 2: Find the solution of

$$\frac{\partial^2 u}{\partial x^2} - \frac{\partial^2 u}{\partial t^2} = xt, \; if \; u = \frac{\partial u}{\partial t} = 0 \; when \; t = 0.$$

Solution: We have

$$\frac{\partial^2 u}{\partial x^2} - \frac{\partial^2 u}{\partial t^2} = xt. \quad ...(1)$$

Taking the Laplace transform of the given equation (1), we get

$$\frac{d^2\tilde{u}}{dx^2}-\left[p^2\tilde{u}-p\,u(x,0)-u_t(x,0)\right]=\frac{x}{p^2}\Rightarrow\frac{d^2\tilde{u}}{dx^2}-p^2\tilde{u}=\frac{x}{p^2}.$$

As $u(x,0)=0=u_t(x,0)=\left(\frac{\partial u}{\partial t}\right)_{t=0}$, we get $\left(D^2-p^2\right)\tilde{u}=\frac{x}{p^2}$. ...(2)

The general solution of this differential equation is

$$\tilde{u}=A\,e^{-px}+B\,e^{px}+\frac{1}{D^2-p^2}\frac{x}{p^2}.$$

But $\frac{1}{D^2-p^2}\frac{x}{p^2}=\frac{1}{p^2}\left(-\frac{1}{p^2}\right)\left(1-\frac{D^2}{p^2}\right)^{-1}x=-\frac{1}{p^4}\left(1+\frac{D^2}{p^2}+\ldots\right)x=-\frac{x}{p^4}.$

Hence from equation (2), we obtain $\tilde{u}=A\,e^{-px}+B\,e^{px}-\frac{x}{p^4}$

Since $u=0$ for $t=0\Rightarrow\tilde{u}=0\;\forall\,x\Rightarrow\tilde{u}=0$ as $x\to\infty$

$\Rightarrow B=0$ according to (1)

So (3) gives $\tilde{u}=A\,e^{-px}-\frac{x}{p^4}$.

Again applying $L\{u\}=0$ for $x=0$, we get $A=0$. Therefore, $\tilde{u}=-\frac{x}{p^4}$.

Consequently

$$u=-x\,L^{-1}\left\{\frac{1}{p^4}\right\}=-\frac{x\,t^3}{3!}=-\frac{x\,t^3}{6}.$$

Hence the required solution is $u=-\frac{x\,t^3}{6}$.

Example 3: Solve the partial differential equation

$$\frac{\partial y}{\partial t}=3\frac{\partial^2 y}{\partial t^2},\text{ where } y\left(\frac{\pi}{2},t\right)=0,\left(\frac{\partial y}{\partial x}\right)_{x=0}=0$$

using Laplace transform, where $y(x,0)=30\cos 5x$.

Solution: We have $\frac{\partial y}{\partial t}=3\frac{\partial^2 y}{\partial x^2}$. ...(1)

Taking Laplace transform of both sides, we get

$$p\,\tilde{y}-y(x,0)=3\frac{d^2\tilde{y}}{dx^2}\text{ where } L\{y\}=\tilde{y}$$

$$\Rightarrow\quad 3\frac{d^2\tilde{y}}{dx^2}-p\,\tilde{y}=-30\cos 5x\Rightarrow\left(D^2-\frac{p}{3}\right)\tilde{y}=-10\cos 5x.\quad ...(2)$$

A.E. is $m^2 - \frac{p}{3} = 0 \Rightarrow m = \pm\sqrt{\frac{p}{3}}$.

Hence $C.F. = A\,e^{x\sqrt{p/3}} + B\,e^{-x\sqrt{p/3}}$.

Now $P.I. = \frac{1}{\left(D^2 - \frac{p}{3}\right)}(-10\cos 5x) = \frac{1}{\left(-25 - \frac{P}{3}\right)}(-10\cos 5x)$

$$= \frac{3}{75+p}\,10\cos 5x = \frac{30}{75+p}\cos 5x.$$

Hence the general solution for the equation (2) is

$$\tilde{y} = A\,e^{x\sqrt{p/3}} + B\,e^{-x\sqrt{p/3}} + \frac{30\cos 5x}{75+p} \qquad ...(3)$$

Now

$$\frac{\partial y}{\partial x} = 0 \text{ when } x = 0 \Rightarrow L\left(\frac{\partial y}{\partial x}\right) = 0\; a\,t\; x = 0 \Rightarrow \frac{d\tilde{y}}{dx} = 0 \; at\; x = 0.$$

Again $y\left(\frac{\pi}{2}, t\right) = 0 \Rightarrow L\left\{y\left(\frac{\pi}{2}, t\right)\right\} = 0 \Rightarrow \tilde{y}\left(\frac{\pi}{2}, p\right) = 0$

Hence $\frac{d\tilde{y}}{dx} = \sqrt{\frac{p}{3}}\left[A\,e^{x\sqrt{p/3}} - B\,e^{-x\sqrt{p/3}} - \frac{150}{75+p}\sin 5x\right].$

Using $\frac{d\tilde{y}}{dx} = 0$ at $x = 0$ in (3), we have $0 = \sqrt{\frac{p}{3}}[A - B] \Rightarrow A = B$

Now equation (3) becomes $L\,y = A\left[e^{x\sqrt{p/3}} + e^{-x\sqrt{p/3}}\right] + \frac{30\cos 5x}{75+p}$. ...(4)

Using condition $\tilde{y}\left(\frac{\pi}{2}, p\right) = 0$ in (4), we have

$$0 = A\left[e^{(\pi/2)\sqrt{p/3}} + e^{-(\pi/3)\sqrt{p/3}}\right] + \frac{30\cos\left(\frac{5\pi}{2}\right)}{75+p}$$

$$\Rightarrow \qquad 2A\cosh\left[\frac{\pi}{2}\sqrt{\frac{p}{3}}\right] = 0 \Rightarrow A = 0$$

Hence equation (4) becomes $\tilde{y} = \frac{30\cos 5x}{75+p}$.

Taking inverse Laplace transform, we get

$$y = L^{-1}\left\{\frac{30\cos 5x}{75+p}\right\} \Rightarrow y = 30\,e^{-75t}\cos 5x.$$

Example 4: Find the solution of $\frac{\partial y}{\partial x}=2\frac{dy}{dt}+y, if\left(y,0\right)=6\,e^{-3x}$

Solution: We have $\frac{\partial y}{\partial x}=2\frac{dy}{dt}+y$. ...(1)

Taking Laplace transform of the given equation (1), we have

$$\frac{d\tilde{y}}{dx}=2\left[p\,\tilde{y}(x,p)-y(x,0)\right]+\tilde{y}, \text{ where } L(y)=\tilde{y}(x,p)$$

$$\Rightarrow \quad \frac{d\tilde{y}}{dx}=2\,p\,\tilde{y}-2\times 6\,e^{-3x}+\tilde{y} \Rightarrow \frac{d\tilde{y}}{dx}-(1+2\,p)\tilde{y}=-12\,e^{-3x}$$

which is a linear differential equation. Hence

$$I.F.=e^{\int pd\,x}=e^{-\int (1+2p)dx}=e^{-x(1+2p)}$$

The solution of equation (2) is given by

$$\tilde{y}\,e^{-(1+2p)x}=c+\int -12\,e^{-3x}\,e^{-(1+2p)x}\,dx$$

$$\Rightarrow \quad \tilde{y}\,e^{-(1+2p)x}=c-12\int e^{-(2p+4)x}\,dx=c+\frac{6\,e^{-(2p+4)}}{p+2}$$

$$\Rightarrow \quad \tilde{y}=c\,e^{(1+2p)x}+\frac{6\,e^{-3x}}{p+2}.$$

Since $y(x,t)$ is bounded $\forall\, x$ and so $\tilde{y}(x,p)$ is bounded and finite $\forall\, x$ and $\forall\, p>0$. This implies that $c = 0$. Thus we have

$$\tilde{y}=\frac{6\,e^{-3x}}{p+2} \Rightarrow y=L^{-1}\left\{\frac{6\,e^{-3x}}{p+2}\right\}$$

$$\Rightarrow \quad y=6\,e^{-3x}\,L^{-1}\left\{\frac{1}{p+2}\right\}=6\,e^{-3x}\,e^{-2t}=6\,e^{-(3x+2t)}.$$

EXERCISE 10.2

Solve the following partial differential equations by using Laplace transform method:

1. $\frac{\partial y}{\partial t}=\frac{\partial^2 y}{\partial x^2}-4y$, if $y(0,t)=0=y(\pi,t), y(x,o)=6\sin x-4\sin 2x$
2. $\frac{\partial y}{\partial t}=\frac{\partial^2 y}{\partial x^2}$ if $y(x,0)=\sin\pi x,\ y(0,t)=0=y(1,t), 0<x<1, t>0$.
3. $\frac{\partial^2 y}{\partial t^2}=16\frac{\partial^2 y}{\partial x^2}, y\left(\frac{1}{2},t\right)=0=y\left(\frac{3}{2},t\right),$

$y'(x,o) = 48\pi \cos \pi x + 192\pi \cos 3\pi x - 160\pi \cos 5\pi x,\ y(x,0) = 0.$

4. $\dfrac{\partial^2 u}{\partial t^2} = \dfrac{\partial^2 u}{\partial x^2}$, if $u(x,0) = \dfrac{1}{2}x(1-x),\ u'(x,0) = 0,\ u(0,t) = 0 = u(1,t)$

5. $\dfrac{\partial y}{\partial t} = \dfrac{\partial^2 y}{\partial x^2}$, if $y\left(\dfrac{\pi}{2}, t\right) = 0 = \left(\dfrac{\partial y}{\partial x}\right)_{x=0}$, $y(x,0) = \cos 5x.$

ANSWERS 10.2

1. $y = 6e^{-5t} \sin x - 4e^{-8t} \sin 2x$. 2. $y = e^{-\pi^2 t} \sin \pi x$.
3. $y = 12 \cos \pi x \sin 4\pi t + 16 \cos 3\pi x \sin 12\pi t - 8 \cos 5\pi x \sin 20\pi t$.
4. $u = \dfrac{1}{2}x(1-x) + \dfrac{t^2}{2}$. 5. $y = e^{-25t} \cos 5x$.

10.4 APPLICATION OF LAPLACE TRANSFORM TO SOLVE INTEGRAL EQUATIONS

In this section, we shall discuss solutions of integral equations using Laplace's transformation. First of all we define some standard integral equations as follows.

10.4.1 Integral Equation

An equation of the form

$$u(t) = f(t) + \int_{g_1(t)}^{g_2(t)} k(x,t)\,u(x)\,dx$$

in which the unknown function $u(x)$ occurs inside the integral sign, is called an integral equation. Here $f(t)$ and $k(x,t)$ are known and $g_1(t), g_2(t)$ are either constants or functions of t.

Hence the function $u(t)$ is to determined by convolution theorem (convolution property) we know that if $F(t)$ and $G(t)$ are two functions of class A and $L^{-1}\{f(p)\} = F(t), L^{-1}\{g(p)\} = G(t)$, then

$$F * G = \int_0^t F(x)G(t-x)\,dx = \int_0^t F(t-x)G(x)\,dx.$$

So the integral equations of the form

$$u(t) = f(t) + \int_0^t k(t-x)\,u(x)\,dx$$

can be written as $u(t) = f(t) + k(t) * u(t)$

This equation is called integral equation of convolution type.

10.4.2 Abel's Integral Equation

An equation of the form $G(t)=\int_0^t \frac{u(x)}{(t-x)^n}\,dx$ is known as Abel's integral equation where $u(x)$ is unknown and $G(t)$ is known and n is a constant such that $0<n<1$.

10.4.3 Volterra Integral Equation

An equation of the form

$\alpha(t)u(t)=f(t)+\lambda\int_0^t k(t,x)u(x)\,dx$ is called a Volterra integral equation.

10.4.4 Fredholm Integral Equation

An equation of the form

$\alpha(t)u(t)=f(t)+\lambda\int_a^b k(t,x)u(x)\,dx$ is called a Fredholm integral equation.

10.4.5 Integro-Differential Equation

An equation in which various derivatives of the unknown function $u(t)$ are present is called an integro-differential equation. For example

(i) $u''(t)=f(t)+u(t)+\lambda\int_0^x k(t,x)u(x)\,dx$

(ii) $u'(t)=f(t)+u(t)+\lambda\int_a^b k(t,x)u(x)\,dx$.

Working rule for solving integral equations using Laplace Transform

Step 1: Change the given integral equation into an integral equation of convolution type.

Step 2: Take Laplace transform of the integral equation of convolution type which was found in step1.

Step 3: Separate $L\{u(t)\}$ or Laplace transform of unknown as function of p.

Step 4: Finally take inverse Laplace transform to get $u(t)$ or unknown.

This procedure can be understood from following examples.

Example 1: Solve the integral equation $F(t)=t+2\int_0^t \cos(t-u)F(u)\,du$.

Solution: The integral equation in convolution type be

$$F(t)=t+2\cos t * F(t).$$

Taking Laplace transform of both sides of (1), we get

$$L\{F(t)\}=L\{t\}+2\,L\{\cos t * F(t)\}=L\{t\}+2\,L\{\cos t\}\;L\{F(t)\}$$

$$=\frac{1}{p^2}+\frac{2p}{p^2+1}L\{F(t)\}$$

$$\Rightarrow \quad f(p)=\frac{1}{p^2}+\frac{2p}{p^2+1}f(p), as\, L\{F(t)\}=f(p)$$

$$\Rightarrow \quad f(p).\left(1-\frac{2p}{p^2+1}\right)=\frac{1}{p^2}$$

$$\Rightarrow \quad f(p)=\frac{p^2+1}{p^2(p-1)^2}=\frac{1}{(p-1)^2}+\frac{1}{p^2(p-1)^2}$$

Now taking inverse Laplace transform, we get

$$F(t)=L^{-1}\left\{\frac{1}{(p-1)^2}\right\}+L^{-1}\left\{\frac{1}{p^2(p-1)^2}\right\},\ \left[\text{since } L^{-1}\{f(p)\}=F(t)\right] \quad (2)$$

$$\text{Now } L^{-1}\left\{\frac{1}{(p-1)^2}\right\}=e^t\,L^{-1}\left\{\frac{1}{p^2}\right\}=t\,e^t \quad ...(3)$$

$$\text{Therefore } L^{-1}\left\{\frac{1}{p(p-1)^2}\right\}=\int_0^t u\,e^u\,du=\left[u\,e^u-\int e^u\,du\right]_0^t$$

$$=\left[u\,e^u-e^u\right]_0^t=\left[(u-1)e^u\right]_0^t=e^t(t-1)+1$$

$$\Rightarrow \quad L^{-1}\left\{\frac{1}{p^2(p-1)^2}\right\}=\int_0^t\left[e^u(u-1)+1\right]du$$

$$=\int_0^t u\,e^u\,du-\int_0^t e^u\,du+\int_0^t du$$

$$=e^t(t-1)+1-(e^t-1)+t=e^t(t-2)+t+2. \quad ...(4)$$

Writing (2) with the help of (3) and (4), we get the required solution of the given integral equation as

$$F(t)=t\,e^t+e^t(t-2)+t+2=2\,e^t(t-1)+2+t.$$

Example 2: Solve the integral equation

$$y(t)=\frac{1}{2}t^2-\int_0^t(t-u)y(u)\,du$$ and verify your solution.

Solution: The given integral equation can be written as

$$y(t)=\frac{1}{2}t^2-t*y(t). \quad ...(1)$$

Let $L\{y(t)\}=Y(p)$. Then taking Laplace transformation on both sides of (1) and using convolution theorem, we get

$$L\{y(t)\}=\frac{1}{2}L\{t^2\}-L\{t*y(t)\} \Rightarrow Y(p)=\frac{1}{2}\frac{2!}{p^3}-L(t).L\{y(t)\}$$

$$\Rightarrow \quad Y(p)=\frac{1}{p^3}-\frac{1}{p^2}Y(p) \Rightarrow Y(p)\left(1+\frac{1}{p^2}\right)=\frac{1}{p^3}$$

$$\Rightarrow \quad Y(p)=\frac{1}{p(p^2+1)}=\frac{(p^2+1)-p^2}{p(p^2+1)} \Rightarrow Y(p)=\frac{1}{p}-\frac{p}{p^2+1}. \quad ...(2)$$

Taking inverse Laplace transformation of (2), we get

$$L^{-1}\{Y(p)\}=L^{-1}\left\{\frac{1}{p}\right\}-L^{-1}\left\{\frac{p}{p^2+1^2}\right\} or\ y(t)=1-\cos t.$$

Verification: Substitute the above value in the right-hand side of the given integral equation, we have

$$R.H.S.=\frac{1}{2}t^2-\int_0^t (t-u)(1-\cos u)\,du$$

$$=\frac{1}{2}t^2-\left[(t-u)(u-\sin u)\right]_0^t-\int_0^t (u-\sin u)\,du$$

$$=\frac{1}{2}t^2-0-\left[\frac{1}{2}u^2+\cos u\right]_0^t=\frac{1}{2}t^2-\left[\frac{1}{2}t^2+\cos t-1\right]$$

$$=1-\cos t=y(t)=L.H.S.$$

Example 3: Solve the integral equation

$$F(t)=a\sin t-2\int_0^t F(u)\cos(t-u)\,du.$$

Solution: The given integral equation may be expressed as

$$F(t)=a\sin t-2\,F(t)*\cos t.$$

Taking the Laplace transform of both sides, we have

$$L\{F(t)\}=a\,L\{\sin t\}-2\,L\{F(t)*\cos t\}=a\,L\{\sin t\}-2\,L\{F(t)\}L\{\cos t\}$$

$$=\frac{a}{p^2+1}-2\,L\{F(t)\}.\frac{p}{p^2+1}$$

$$\Rightarrow \quad \left(1+\frac{2p}{p^2+1}\right)L\{F(t)\}=\frac{a}{p^2+1}\ or\ L\{F(t)\}=\frac{a}{(p+1)^2}$$

Now taking inverse Laplace transform of both sides, we get

$$F(t)=a\,L^{-1}\left\{\frac{1}{(p+1)^2}\right\}=a\,e^{-t}\,L^{-1}\left\{\frac{1}{p^2}\right\}=a\,t\,e^{-t}$$

Example 4: Solve the integral equation

$F(t)=1+\int_0^t F(u)\sin(t-u)\,du$ and verify your solution.

Solution: The given integral equation is expressible as

$F(t)=1+F(t)*\sin t$.

Taking the Laplace transform, we have

$$f(p)=L\{1\}+L\{F(t)*\sin t\}=\frac{1}{p}+L\{F(t)\}L\{\sin t\}$$

$$=\frac{1}{p}+f(p)\frac{1}{p^2+1} \text{ where } L\{F(t)\}=f(p)$$

$$\Rightarrow \quad \left(1-\frac{1}{p^2+1}\right)f(p)=\frac{1}{p}\Rightarrow f(p)=\frac{p^2+1}{p^2\,p}=\frac{1}{p}+\frac{1}{p^3}.$$

Now taking inverse Laplace transform of both sides, we get

$$F(t)=1+\frac{t^2}{\Gamma(3)}=1+\frac{t^2}{2}.$$

Verification: We have $F(t)=1+\frac{t^2}{2}$ putting in the R.H.S. of the given equation, we have

$$R.H.S.=1+\int_0^t\left(1+\frac{u^2}{2}\right)\sin(t-u)\,du$$

$$=1+\left[\left(1+\frac{u^2}{2}\right)\cos(t-u)\right]_0^t-\int_0^t u\cos(t-u)\,du$$

$$=1+1+\frac{t^2}{2}-\cos t-\left[-u\sin(t-u)\right]_0^t-\int_0^t\sin(t-u)\,du$$

$$=2+\frac{t^2}{2}-\cos t-\left[\cos(t-u)\right]_0^t=2+\frac{t^2}{2}-\cos t-(1-\cos t)$$

$$=1+\frac{t^2}{2}=F(t)=L.H.S.$$

Example 5: Solve $\int_0^t \frac{F(u)\,du}{\sqrt{t-u}}=1+t+t^2$.

Solution: The given equation can be written as $F(t)*t^{-1/2}=1+t+t^2$.

Taking Laplace transform of both sides, we get

$$L\{F(t)*t^{-1/2}\}=L\{1\}+L\{t\}+L\{t^2\}$$

$$\Rightarrow \quad L\{F(t)\}L\{t^{-1/2}\}=\frac{1}{p}+\frac{\Gamma(2)}{p}+\frac{\Gamma(3)}{p^3}$$

$$\Rightarrow \quad f(p)\frac{\Gamma(1/2)}{p^{1/2}}=\frac{1}{p}+\frac{1}{p^2}+\frac{2}{p^3} \text{ or } \sqrt{\pi}\, f(p)=\frac{1}{p^{1/2}}+\frac{1}{p^{3/2}}+\frac{2}{p^{5/2}}.$$

Now taking inverse Laplace transform, we get

$$\sqrt{\pi}\, L^{-1}\{f(p)\}=L^{-1}\left\{\frac{1}{p^{1/2}}\right\}+L^{-1}\left\{\frac{1}{p^{3/2}}\right\}+2\,L^{-1}\left\{\frac{1}{p^{5/2}}\right\}$$

$$\Rightarrow \quad \sqrt{\pi}\, F(t)=\frac{t^{-1/2}}{\Gamma\left(\frac{1}{2}\right)}+\frac{t^{1/2}}{\Gamma\left(\frac{3}{2}\right)}+\frac{2\,t^{3/2}}{\Gamma\left(\frac{5}{2}\right)}$$

$$\Rightarrow \quad F(t)=\frac{1}{\pi}\left[t^{-1/2}+2\,t^{1/2}+2\frac{2}{3}\frac{2}{1}t^{3/2}\right]=\frac{1}{\pi}\left[t^{-1/2}+2\,t^{1/2}+\frac{8}{3}t^{3/2}\right].$$

Example 6: Solve the integral equation $\int_0^t \frac{F(u)\,du}{(t-u)^{-1/3}}=t(1+t)$.

Solution: The given integral equation is expressible as

$$F(t)*t^{-1/3}=t(1+t)=t+t^2$$

Taking Laplace transform of both sides, we have

$$L\{F(t)*t^{-1/3}\}=L\{t\}+L\{t^2\}$$

$$\Rightarrow \quad L\{F(t)\}L\{t^{-1/3}\}=\frac{\Gamma(2)}{p^2}+\frac{\Gamma(3)}{p^3}$$

$$\Rightarrow \quad f(p)\frac{\Gamma\left(-\frac{1}{3}+1\right)}{p^{-1/3+1}}=\frac{1}{p^2}+\frac{2}{p^3}$$

$$\Rightarrow \quad f(p)\Gamma\left(\frac{2}{3}\right)=p^{2/3}\left(\frac{1}{p^2}+\frac{2}{p^3}\right)=\frac{1}{p^{4/3}}+\frac{2}{p^{7/3}}=\frac{1}{p^{1/3+1}}+\frac{2}{p^{4/3+1}}.$$

Now taking inverse Laplace transform, we obtain

$$\Gamma\left(\frac{2}{3}\right)L^{-1}\{f(p)\}=L^{-1}\left\{\frac{1}{p^{1/3+1}}\right\}+2\,L^{-1}\left\{\frac{1}{p^{4/3+1}}\right\}$$

$$\Rightarrow \quad \Gamma\left(\frac{2}{3}\right)F(t)=\frac{t^{1/3}}{\Gamma\left(\frac{1}{3}+1\right)}+2\frac{t^{4/3}}{\Gamma\left(\frac{4}{3}+1\right)}=\frac{t^{1/3}}{\frac{1}{3}\Gamma\left(\frac{1}{3}\right)}+2\frac{t^{4/3}}{\frac{4}{3}\cdot\frac{1}{3}\Gamma\left(\frac{1}{3}\right)}$$

$$\Rightarrow \quad F(t)\Gamma\left(\frac{1}{3}\right)\Gamma\left(1-\frac{1}{3}\right)=3\,t^{1/3}+\frac{9}{2}t^{4/3}$$

$$\Rightarrow \quad F(t)\frac{\pi}{\sin(\pi/3)}=\frac{3\,t^{1/3}}{2}(2+3\,t)$$

$$\Rightarrow \quad F(t)=\frac{3\sqrt{3}}{4\,\pi}\,t^{1/3}\,(2+3\,t).$$

Example 7: Solve the integral equation $\int_0^t F(u)F(t-u)\,d\,u=16\sin 4\,t$.

Solution: The given integral equation may be expressed as

$$F(t)^*\,F(t)=16\sin 4\,t.$$

Taking the Laplace transform of both sides, we have

$$L\{F(t)^*\,F(t)\}=16\,L\{\sin 4\,t\}\Rightarrow L\{F(t)\}L\{F(t)\}=16\,\frac{4}{p^2+a^4}$$

$$\Rightarrow \quad L\{F(t)\}=\pm\frac{8}{\sqrt{p^2+4^2}}$$

$$\Rightarrow \quad F(t)=\pm 8\,L^{-1}\left\{\frac{1}{\sqrt{p^2+4\,a^2}}\right\}=\pm 8\,J_0(4\,t).$$

Example 8: Solve the following integral-differential equations using Laplace transform

(i) $u'(x)=\sin x+\int_0^x u(x-t)\cos t\,d\,t,\,u(0)=0$

(ii) $u''(x)+\int_0^x e^{2(x-t)}\,u'(t)d\,t=e^{2x},\,u(0)=0=u'(0).$

Solution: (i) Here $u'(x)=\sin x+u(x)^*\cos x$.

Taking Laplace transform of both sides, we have

$$L[u'(x)]=L[\sin x]+L[u(x)]L[\cos(x)]$$

$$\Rightarrow \quad p\,\bar{u}(p)-u(0)=\frac{1}{1+p^2}+\bar{u}(p)\frac{p}{1+p^2}$$

$$\Rightarrow \quad \left[p-\frac{p}{1+p^2}\right]\bar{u}(p)=\frac{1}{1+p^2}\quad\Rightarrow \bar{u}(p)=\frac{1}{p^3}$$

$$\Rightarrow \quad u(x)=L^{-1}\left[\frac{1}{p^3}\right]=\frac{x^2}{2!}\quad\Rightarrow u(x)=\frac{x^2}{2}.$$

(ii) Here $u''(x)+e^{2x}*u'(x)=e^{2x}$.

Taking Laplace transform of both sides, we have

$$L[u''(x)]+L[e^{2x}]L[u'(x)]=L[e^{2x}]$$

$$\Rightarrow \quad p^2\,\bar{u}(p)-p\,u'(0)-u(0)+\frac{1}{(p-2)}[p\,\bar{u}(p)-u(0)]=\frac{1}{p-2}$$

$$\Rightarrow \quad \left[p^2 + \frac{p}{p-2}\right]\bar{u}(p) = \frac{1}{p-2} \qquad [\because u(0) = 0 = u'(0)]$$

$$\Rightarrow \quad \bar{u}(p) = \frac{1}{p(p-1)^2} = \frac{1}{p} - \frac{1}{(p-1)} + \frac{1}{(p-1)^2}$$

$$\Rightarrow \quad u(x) = L^{-1}\left[\frac{1}{p} - \frac{1}{p-1} + \frac{1}{(p-1)^2}\right] = 1 - e^x + x\,e^x$$

Hence the solution is $u(x) = 1 - e^x + x\,e^x$.

Example 9: Solve $u'(x) = x + \int_0^x u(x-t)\cos t\,dt,\ u(0) = 4$ using Laplace transform.

Solution: Here $u'(x) = x + u(x) * \cos x$

$$\Rightarrow \quad L[u'(x)] = L[x] + L[u(x)]L[\cos x]$$

$$\Rightarrow \quad p\,L[u(x)] - u(0) = \frac{1}{p^2} + L[u(x)]\frac{p}{p^2+1}$$

$$\Rightarrow \quad p\left[1 - \frac{1}{p^2+1}\right]L[u(x)] = \frac{1}{p^2} + 4$$

$$\Rightarrow \quad L[u(x)] = \frac{p^2+1}{p^3}\left(4 + \frac{1}{p^2}\right) = \frac{4}{p} + \frac{5}{p^3} + \frac{1}{p^5}$$

$$\Rightarrow \quad u(x) = 4\,L^{-1}\left[\frac{1}{p}\right] + 5\,L^{-1}\left[\frac{1}{p^3}\right] + L^{-1}\left[\frac{1}{p^5}\right] = 4 + 5\frac{x^2}{2!} + \frac{x^4}{4!}$$

Hence the required solution is

$$u(x) = 4 + \frac{5}{2}x^2 + \frac{1}{24}x^4.$$

Example 10: Solve the following Abel's integral equations using Laplace transform method $\int_0^x \frac{u(t)\,dt}{\sqrt{(x-t)}} = \sin x$.

Solution: $\int_0^x (x-1)^{-1/2}\,u(t)\,dt = \sin x \Rightarrow x^{-1/2} * u(x) = \sin x$

Taking the Laplace transform of both sides, we have

$$L[x^{-1/2}]L[u(x)] = L[\sin x] \quad \Rightarrow \frac{\Gamma\left(-\frac{1}{2}+1\right)}{p^{1/2}}\bar{u}(p) = \frac{1}{1+p^2}$$

$$\Rightarrow \quad \bar{u}(p)=\frac{1}{\sqrt{\pi}}\frac{p^{1/2}}{1+p^2} \quad \Rightarrow u(x)=\frac{1}{\sqrt{\pi}}L^{-1}\left[\frac{p^{1/2}}{1+p^2}\right].$$

Now

$$L^{-1}\left[\frac{p^{1/2}}{(1+p^2)}\right]=L^{-1}\left[\frac{p}{p^{1/2}(1+p^2)}\right]$$

$$=L^{-1}\left[\frac{1}{p^{1/2}}\right]*L^{-1}\left[\frac{p}{1+p^2}\right]=\frac{x^{-1/2}}{\Gamma(1/2)}*\cos x$$

$$=\frac{1}{\Gamma 1/2}\int_0^x \frac{\cos t}{\sqrt{(x-t)}}dt=\frac{1}{\sqrt{\pi}}\int_0^x \frac{\cos t}{\sqrt{x-t}}dt$$

Hence $u(x)=\frac{1}{\pi}\int_0^x \frac{\cos t}{\sqrt{x-t}}dt$.

Example 11: Solve the Abel integral equation using Laplace transform method $F(x)=\int_0^x \frac{u(x)}{(x-t)^{\alpha}}dx, 0<\alpha<1$.

Solution: Here $F(x)=x^{-\alpha}*u(x)$

$$\Rightarrow \quad L[F(x)]=L[x^{-\alpha}]L[u(x)] \Rightarrow f(p)=\frac{\Gamma(1-\alpha)}{p^{1-\alpha}}L[u(x)]$$

$$\Rightarrow \quad L[u(x)]=\frac{p^{1-\alpha}f(p)}{\Gamma(1-\alpha)}=\frac{p\,\Gamma\alpha}{\Gamma\alpha\,\Gamma(1-\alpha)}p^{-\alpha}f(p)$$

$$=\frac{p\sin\pi\alpha}{\pi}\left[\Gamma\alpha\,p^{-\alpha}f(p)\right]=\frac{p\sin\pi\alpha}{\pi}L\left[x^{\alpha-1}*F(x)\right]$$

$$=\frac{\sin\pi\alpha}{\pi}pL\left[\int_0^x (x-t)F(t)dt\right] \qquad ...(1)$$

Let $G(x)=\int_0^x (x-t)^{\alpha-1}F(t)dt$, therefore $G(0)=0$

Hence $L[G'(x)]=pL[G(x)]-G(0)=pL[G(x)]$

$$\Rightarrow \quad pL\left\{\int_0^t (t-x)^{\alpha-1}F(x)dx\right\}=L\left\{\frac{d}{dx}G(t)\right\} \qquad ...(2)$$

From (1) and (2), we have

$$L\{u(x)\}=\frac{\sin\pi\alpha}{\pi}L\left\{\frac{d}{dx}G(x)\right\}=L\left\{\frac{\sin\pi\alpha}{\pi}\frac{d}{dx}G(x)\right\}$$

$$\Rightarrow \quad u(x)=\frac{\sin\pi\alpha}{\pi}\frac{d}{dx}\left\{\int_0^x (x-t)^{\alpha-1}F(t)dt\right\}.$$

Example 12: Solve the following integral equation using Laplace transform method $\int_0^x u(t)u(x-t)\,dt = \frac{x^3}{6}$.

Solution: Here $u(x)^* u(x)= \frac{x^3}{6}$.

Taking Laplace transform of both sides, we have

$$L[u(x)^* u(x)] = L\left[\frac{x^3}{6}\right] \Rightarrow [\bar{u}(p)]^2 = \frac{1}{6}\frac{3!}{p^{3+1}} = \frac{1}{p^4}$$

$$\Rightarrow \quad \bar{u}(p) = \pm\frac{1}{p^2} \Rightarrow u(x) = \pm L^{-1}\left[\frac{1}{p^2}\right] \Rightarrow u(x) = \pm x.$$

The functions $u_1(x) = x$, $u_2(x) = -x$ are the solutions of the integral equation.

Note: The solution of the integral equation is not unique.

Example 13: Solve the system of the integral equations using Laplace transform method

$$u_1(x) = 1 - 2\int_0^x e^{2(x-t)}\,u_1(t)\,dt + \int_0^x u_2(t)\,dt$$

$$u_2(x) = 4x - \int_0^x u_1(t)\,dt + 4\int_0^x (x-t)u_2(t)\,dt.$$

Solution: The system can be written as

$$u_1(x) = 1 - 2e^{2x} * u_1(x) + 1 * u_2(x) \qquad ...(1)$$

$$u_2(x) = 4x - 1 * u_1(x) + 4x * u_2(x) \qquad ...(2)$$

Taking Laplace transform of both sides of the system of equations, we have

$$\bar{u}_1(p) = \frac{1}{p} - 2\frac{1}{p-2}\bar{u}_1(p) + \frac{1}{p}\bar{u}_2(p) \qquad ...(3)$$

$$\bar{u}_2(p) = \frac{4}{p^2} - \frac{1}{p}\bar{u}_1(p) + 4\frac{1}{p^2}\bar{u}_2(p) \qquad ...(4)$$

Equation (3) and (4) can be written as

$$\frac{p}{p-2}\bar{u}_1(p) - \frac{1}{p}\bar{u}_2(p) = \frac{1}{p} \qquad ...(5)$$

$$\frac{1}{p}\bar{u}_1(p) + \left(1 - \frac{4}{p^2}\right)\bar{u}_2(p) = \frac{4}{p^2} \qquad ...(6)$$

Solving (5) and (6), we have

$$\bar{u}_1(p) = \frac{p}{(p+1)^2} = \frac{1}{(p+1)} - \frac{1}{(p+1)^2} \qquad ...(7)$$

$$\overline{u_2}(p)=\frac{3p+2}{(p-2)(p+1)^2}=\frac{8}{9}\frac{1}{(p-2)}+\frac{1}{3}\frac{1}{(p+1)^2}-\frac{8}{9}\frac{1}{(p+1)} \quad ...(8)$$

Taking inverse Laplace of (7) and (8), we have

$$u_1(x)=e^{-x}-x\,e^{-x} \quad ...(9)$$

$$u_2(x)=\frac{8}{9}e^{2x}+\frac{1}{3}x\,e^{-x}-\frac{8}{9}e^{-x} \quad ...(10)$$

which are solutions of the system of integral equations

Example 14: Solve the following integral equation for $y\ (t)$

$$y'(t)+5\int_0^t \cos 2(t-u)y(u)\,du=10,\ y(0)=2.$$

Solution: The given integral equation can be written as

$$y'(t)+5\cos 2t * y(t)=10,\ y(0)=2.$$

Taking Laplace transformation of both sides and using convolution theorem, we get

$$p\,Y(p)-y(0)+5\frac{p}{p^2+4}Y(p)=\frac{10}{p} \Rightarrow \left(p+\frac{5p}{p^2+4}\right)Y(p)=\frac{10}{p}+2$$

$$\Rightarrow Y(p)=\frac{10(p^2+4)}{p^2(p^2+9)}+\frac{2(p^2+4)}{p(p^2+9)}=\frac{40}{9p^2}+\frac{50}{9}\frac{1}{p^2+9}+\frac{8}{9}\frac{1}{p}+\frac{10}{9}\frac{p}{p^2+9}$$

Now taking inverse Laplace transformation, we get

$$L^{-1}\{Y(p)\}=\frac{40}{9}L^{-1}\left\{\frac{1}{p^2}\right\}+\frac{50}{9}L^{-1}\left\{\frac{1}{p^2+3^2}\right\}$$

$$+\frac{8}{9}L^{-1}\left\{\frac{1}{p}\right\}+\frac{10}{9}L^{-1}\left\{\frac{p}{p^2+3^2}\right\}$$

$$\Rightarrow \quad y(t)=\frac{40}{9}t+\frac{50}{9}\frac{\sin 3t}{3}+\frac{8}{9}1+\frac{10}{9}\cos 3t$$

$$\Rightarrow \quad y(t)=\frac{1}{9}\left(8+40t+\frac{50}{3}\sin 3t+10\cos 3t\right).$$

EXERCISE 10.3

Solve the following integral equations:

1. $F(t)=1+2\int_0^t F(t-u)e^{-2u}\,du.$

2. $F(t)=t+2\int_0^t \cos(t-u)F(u)\,du.$

3. $F(t)=1+2\int_0^t F(t-u)\cos u\,du.$

4. $2F(t)=2-t+\int_0^t F(t-u)F(u)\,du.$

5. $F'(t) = t + \int_0^t F(y-u)\cos u\, d\, u,\ F(0) = 4$.

6. $y(t) = t + \frac{1}{6}\int_0^t (t-u)^3\, y(u)\, d\, u$.

7. $y(t) = t - 1 + \int_0^t y(u)\sin(t-u)\, d\, u$.

8. $y(t) = a\, t + \int_0^t y(u)\sin(t-u)\, d\, u$.

9. $y(t) = a \sin b\, t + c\int_0^t y(u)\sin b(t-u)\, d\, u$, when $b > c > 0$.

10. Show that the solution of the integral equation

$F(t) = 4\, t - 3\int_0^t F(u)\sin(t-u)\, d\, u$ is $F(t) = t + \frac{3}{2}\sin 2\, t$.

11. Solve the following integral equations using Laplace transform method

(i) $\int_0^x e^{x-t}\, u(t)\, d\, t = x$ (ii) $\int_0^x J_0(x-t)\, u(t)\, d\, t = \sin x$.

12. Solve the integral equation $\int_0^1 \frac{y(u)}{\sqrt{t-u}}\, d\, u = \sqrt{t}$ with the help of Laplace transform.

13. Solve the integral equation $F(t) = e^{-t} - 2\int_0^t \cos(t-u) F(u)\, d\, u$.

14. Solve the integral equation $y(t) = t^2 + \int_0^t y(u)\sin(t-u)\, d\, u$.

ANSWERS 10.3

1. $F(t) = 1 + 2\, t$.
2. $F(t) = 2 + t + 2\, e^t(t-1)$.
3. $F(t) = 1 + 2\, t\, e^t$.
4. $F(t) = 1, -1$.
5. $F(t) = 4 + \frac{5\, t^2}{2} + \frac{t^4}{24}$.
6. $y(t) = \frac{1}{2}(\sinh t + \sin t)$.
7. $y(t) = -1 + t - \frac{1}{2}t^2 + \frac{1}{6}t^3$.
8. $y(t) = a\left(t + \frac{1}{6}t^3\right)$.
9. $y(t) = a\, b(b^2 - b\, c)^{-1/2}\sin\left(t\sqrt{b^2 - b\, c}\right)$.

11. (i) $u(x) = 1 - x$ (ii) $u(x) = J_0(x)$.

12. $y(t) = \frac{1}{2}$

13. $F(t) = e^{-t}(1 - 2\, t + t^2) = e^{-t}(1-t)^2$.

14. $y(t) = t^2 + \frac{t^2}{12}$

PART B
Calculus of Variations

11

Variational Problems with Fixed Boundaries

11.1 INTRODUCTION

Calculus of variation is one of the most important branch of theoretical and applied mathematics. Applications of calculus of variations are concerned mainly with the determination of maxima and minima of certain expressions involving unknown functions.

A problem to investigate extrema (maxima and minima) of a functional is called a *variational problem*. Functionals are called variable quantities and play an important role in many problems arising in analysis, mechanics, geometry, etc. Euler (1707–1783) was the first who gave important results in this area. Till now the "Calculus of functional" does not have methods comparable to the methods of classical analysis of calculus of functions.

The most developed branch of the "Calculus of functional" concerned with the variational problems of functional is called calculus of variations. Historically the calculus of variations has its origin in the generalization of the elementary theory of maxima and minima. The basic problem in the calculus of variations is to determine a function such that a certain definite integral involving that function and certain of its derivatives take on a maximum or minimum value.

11.2 VARIATIONAL PRINCIPLES

The laws which govern functional to attain extrema are called variational principles. Several physical laws can be deduced from concise mathematical principles to the effect that a certain functional in a given process attains a maximum or minimum.

Examples of variational principles are as follows:

(i) The principle of least square in Mechanics.

(ii) The principle of least action in Mechanics.

(iii) The principle of conservation of linear momentum in Mechanics.

(iv) The principle of conservation of angular momentum in Mechanics.

(v) The principle of Castigliano in the theory of electricity.

(vi) Fermat's principle in optics.

There are some examples of variational problems which are given below:

1. The brachistochrone (shortest time) problem which is posed by J. Bernoulli in 1696 A.D. According to which it is required to find the curve joining two points A and B those do not lie on a vertical line and possesses the property that a moving bead slides down on this curve from A to B in the shortest time. It turns out that the curve of quickest descent is not the straight line though it is the shortest distance between A and B but it is a cycloid.
2. A problem of finding a closed curve of given length l bounding a maximum area S is one of the problem which was solved in ancient Greece and is called isoperimetric. The problem essentially consists of the maximization of the area A bounded by a closed curve $r = r(\theta)$ of the given length l. The functional A is given by

   $$A = \frac{1}{2}\int_0^{2\pi} r^2 d\theta \text{ is maximum subject to } l = \int_0^{2\pi}\left[r^2 + \left(\frac{dr}{d\theta}\right)^2\right]^{1/2} d\theta.$$
3. A problem in differential geometry is to determine the curve of shortest length joining two given points (x_0, y_0, z_0) and (x_1, y_1, z_1) on a surface S given by $\phi(x,y,z) = 0$. Such shortest curves are called geodesics. This is a typical variational problem with a constraint since here we are required to minimize the arc length l joining the two points on S given by the functional

   $$l = \int_{x_0}^{x_1}\left[1 + \left(\frac{dy}{dx}\right)^2 + \left(\frac{dz}{dx}\right)^2\right]^{1/2} dx$$

   subject to the constraints $\phi(x,y,z) = 0$. In 1698, this problem was first solved by J. Bernoulli, but a general method of solving such problems was given by Euler.

Though calculus of variations began to develop in 1696 A.D. with brachistochrone problem, but it matured into an independent mathematical discipline after 1744 A.D. when Euler discovered the basic differential equations known as Euler's equations for a minimizing curve.

11.3 FUNCTIONALS

By a functional we mean a correspondence which assigns a definite real number to each function/curve belonging to some class, that is, a quantity whose values are determined by one or several functions is called a functional. Thus if M be a class of functions $y(x)$ and I be a definite number (by some law) associated to each function $y(x) \in M$, then

$$I = I[y(x)]$$

is called a functional with argument function $y(x)$.

The class M is called the set of admissible functions of the domain of definition of the functional. Thus the domain of a functional is set of admissible functions rather than a region of a co-ordinate space.

Most of the problems in physics and differential geometry can be formulated in variational forms in which a curve $x = x(t)$, $y = y(t)$, $a \le t \le b$, is found form a class of curves so that the integral

$$I = \int_a^b F[x(t), y(t), \dot{x}(t), \dot{y}(t)]\, dt$$

is maximum or minimum. The integral I instead of depending on independent variables varying over a given range, depends on the entire class of functions $x(t)$, $y(t)$ out of a certain class M.

Examples of Functionals

1. A simple example of functional is the shortest length of a curve through two points $P(x_1, y_1)$ and $Q(x_2, y_2)$, that is, the determination of the curve $y = y(x)$ on which the length l between two points (x_1, y_1) and (x_2, y_2) is minimum. Hence the integral

   $$l = \int_{x_1}^{x_2} \left[1 + \left(\frac{dy}{dx}\right)^2\right]^{1/2} dx$$ is known as functional.

   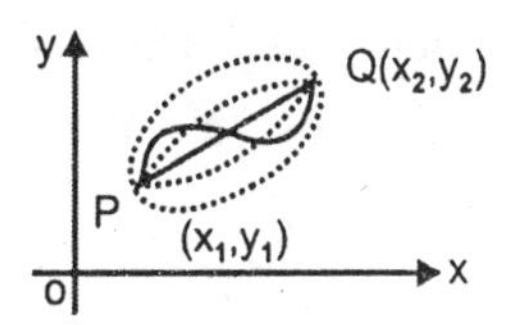

   In general, it is required to find the curve $y = y(x)$, where $y(x_1) = y_1$ and $y(x_2) = y_2$ such that for a given function $f\left(x, y, \frac{dy}{dx}\right)$,

   $$I[y(x)] = \int_{x_1}^{x_2} f\left[x, y, \frac{dy}{dx}\right] dx$$

   is maximum or minimum. This integral is known as functional.

2. The area S of a surface $z = z(x, y)$ bounded by a given curve C is an example of a functional. This area is determined by choice of the surface S, $z = z(x, y)$ as

   $$S[z(x,y)] = \iint_D \left[1 + \left(\frac{\partial z}{\partial x}\right)^2 + \left(\frac{\partial z}{\partial y}\right)^2\right]^{1/2} dx\, dy$$

where D is the projection of the area bounded by the curve C in the xy-plane.

3. Let $M=C[0,1]$ be the set of all continuous functions defined on the closed interval $[0,1]$, then for $y(x)\in M$,

   $I[y(x)]=\int_0^1 y(x)dx$ is a functional.

   In particular, $I[\cos \pi x]=0$ and $I[\sin \pi x]=\frac{2}{\pi}$

11.4 BASIC DEFINITIONS

Here we define some basic definitions related to spaces of functions. To study the functions of n variables, we consider a set of n numbers $(x_1, x_2, \ldots, x_n)$ as a point in the n-dimensional space.

11.4.1 Linear Space

A non-empty set V considering of elements x, y, z is called a linear space over a real number set R considering of real numbers α, β, γ, if the following axioms are satisfied.

(i) $x+y\in V, \forall\, x, y \in V$

(ii) $x+y=y+x$

(iii) $(x+y)+z=x+(y+z)$

(iv) $\exists$ an element $0\in V$ such that $x+0=x=0+x, \forall x\in V$

(v) For each $x\in V, \exists$ an element $-x\in V$ such that

$x+(-x)=0=(-x)+x$

(vi) $\alpha x\in V\ \forall\, x\in V,\ \alpha\in R$

(vii) $\alpha(\beta x)=(\alpha\beta)x$

(viii) $(\alpha+\beta)x=\alpha x+\beta x$

(ix) $\alpha(x+y)=\alpha x+\alpha y$

(x) $\exists\ 1\in R$ such that $1.x=x$

A linear space is also known as vector space.

11.4.2 Normed Linear Space

A linear space N is called a normed space if for each $x\in N$, we can assign a number $\|x\|$, called the norm of x, such that

(i) $\|x\|\geq 0$

(ii) $\|x\|=0$ iff $x=0$

(iii) $\| \alpha x \| = |\alpha| \| x \|$

(iv) $\| x+y \| \leq \| x \| + \| y \|$

11.4.3 Function Space

A linear space consisting of functions as its elements is called function space.

11.4.4 Linear Functional

Let N be a normed linear space of functions. The functional $I[y(x)]$ defined in the class N is called linear if

(i) $I[\alpha y(x)] = \alpha I[y(x)]$, α is a constant.

(ii) $I[y_1(x)+y_2(x)] = I[y_1(x)] + I[y_2(x)]$, where $y_1(x), y_2(x) \in N$.

(iii) $I[y(x)]$ is continuous for all $y(x) \in N$.

Examples

1. The functional $I[y(x)] = \int_a^b y(x)dx$ is a linear functional on the normed linear space $C[a,b]$.
2. The functional $I[y(x)] = \int_a^b [y(x)+y'(x)]dx$ is a linear functional on the normed linear space $C^1[a,b]$.
3. The functional $I[y(x)] = \int_a^b [p(x)y^2(x)+q(x)y'^2(x)]\,dx$ is a non-linear functional on the normed linear space $C^1[a,b]$.

11.5 THE CONCEPT OF A VARIATION/CONTINUITY OF A FUNCTIONAL

11.5.1 Variation of the Argument of a Functional

The variation or increment δy of the argument $y(x)$ of the functional $I[y(x)]$ is the difference between two functions $y(x)$ and $y_1(x)$ of the same class. Symbolically

$$\delta y = y(x) - y_1(x).$$

11.5.2 Continuity of a Functional

A functional $I[y(x)]$ is said to be continuous if a small change in $y(x)$ results a small change in $I[y(x)]$. In other words if the curve $y(x)$ and $y_1(x)$ are closed to each other, then the numbers $I[y(x)]$ and

$I[y_1(x)]$ should be closed to each other for all x for which $y(x)$ and $y_1(x)$ are defined. The closeness of $y(x)$ and $y_1(x)$ we mean $\| y(x)-y_1(x) \|$ small for all x. In this case, $y(x)$ is said to be close to $y_1(x)$ in the sense of zero-order proximity.

11.5.3 Proximity of Curves

Due to the above definition, the functional

$$I[y(x)] = \int_a^b F(x, y(x), y'(x))\, dx$$

which occurs in many applications is rarely continuous due to the presence of the argument $y'(x)$. This compels the notion of closeness of curves $y(x)$ and $y_1(x)$ such that both $\| y(x)-y_1(x) \|$ and $\| y'(x)-y_1'(x) \|$ are small for all x for which $y(x)$ and $y_1(x)$ are defined. In this case $y(x)$ is said to be close to $y_1(x)$ in the sense of first-order proximity.

In general the curve $y(x)$ and $y_1(x)$ are said to be close in the sense of nth order proximity if

$$\| y(x)-y_1(x) \|,\ \| y'(x)-y_1'(x) \|, \ldots, \| y^{(n)}(x)-y_1^{(n)}(x) \|$$

are small for all x for which $y\ (x)$ and $y_1(x)$ are defined.

It is clear that if two curves are close in the sense of nth order proximity then they must be close in the sense of any lower order proximity.

Out of following two figures, the figure (a) shows two curves which are close in the sense of zero-order proximity but not in the sense of first order proximity while the figure (b) shows two curves which are close in the sense of first order proximity.

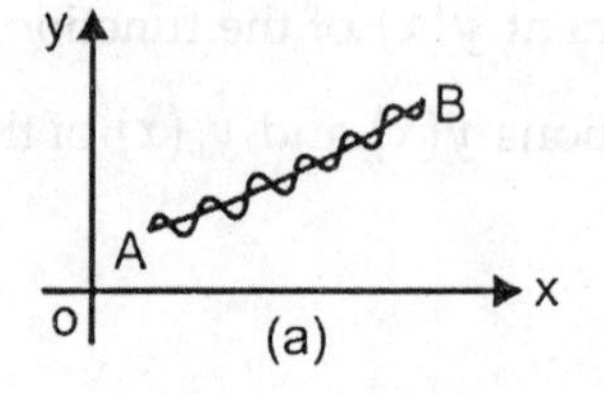

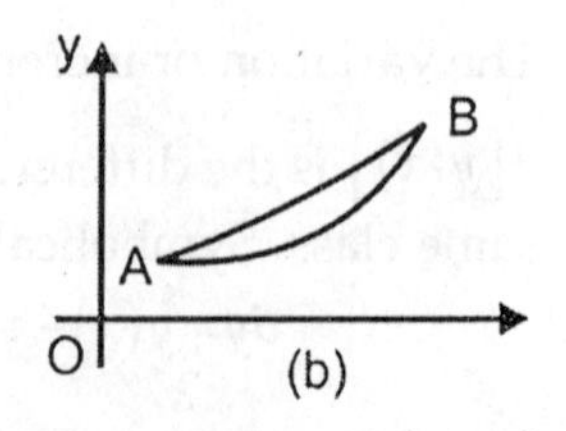

Definition (continuity of a functional): The functional $I[y(x)]$ is said to be continuous at $y_0(x)$ in the sense of nth order proximity, if for any given $\varepsilon > 0$, there exists a $\delta > 0$, such that

$$| I[y(x)] - I[y_0(x)] | < \varepsilon$$

for

$$\| y(x) - y_0(x) \| < \delta, \ \| y'(x) - y_0'(x) \| < \delta, \dots, \ \| y^{(n)}(x) - y_0^{(n)}(x) \| < \delta.$$

11.5.4 Increment of a Functional

Let $I[y(x)]$ be a functional on M. The expression

$$\Delta I = I[y(x) + \delta y] - I[y(x)]$$

is called the increment of the functional $I[y(x)]$ corresponding to the increment $\delta y = y(x) - y_1(x)$ in the argument function.

Theorem 1: If $\phi(x)$ is continuous on $[a,b]$ and $\int_a^b \phi(x) y(x)\, dx = 0$, for all $y(x) \in C[a, b]$ such that $y(a) = y(b) = 0$, then

$$\phi(x) = 0, \text{ for all } x \in [a,b].$$

Proof: Let the function ϕ be non-zero at some point say $p \in [a, b]$ that is $\phi(p) \neq 0$. Then there exists some interval $[x_1, x_2]$ such that $p \in [x_1, x_2] \subset [a, b]$ and $\phi(x)$ have the same sign in the interval $[x_1, x_2]$.
Without loss of generality we may assume that

$$\phi(x) > 0 \text{ in } [x_1, x_2]. \qquad \dots(1)$$

Set

$$y(x) = \begin{cases} (x - x_1)^2 (x_2 - x)^2, \text{ for } x \in [x_1, x_2] \\ 0, \text{ otherwise} \end{cases} \qquad \dots(2)$$

By the setting of $y(x)$ it is clear that $y(x)$ is continuous in $[x_1, x_2]$ and $y(x_1) = y(x_2) = 0$, so $y(x) \in C[a,b]$.
On the other hand

$$\int_a^b \phi(x) y(x) dx = \int_{x_1}^{x_2} \phi(x)(x - x_1)^2 (x_2 - x)^2 dx > 0.$$

Since the integrand is positive in the open interval (x_1, x_2). This contradicts the hypothesis of the theorem, therefore our assumption that $\phi(x) \neq 0$ at some point p is wrong.

Hence,

$$\phi(x) = 0, \text{ for all } x \in [a,b].$$

Theorem 2: If $\phi(x)$ is continuous in $[a, b]$ and $\int_a^b \phi(x) y'(x) dx = 0$, for all $y(x) \in C^1[a,b]$ such that $y(a) = y(b) = 0$, then $\phi(x) = c$, for all $x \in [a, b]$, where c is a constant.

Proof: Consider the constant c such that

$$\int_a^b [\phi(x) - c]\, dx = 0. \quad ...(1)$$

Set $\quad y(x) = \int_0^x [\phi(\xi) - c]\, d\xi \quad ...(2)$

Then $y(x)$ is differentiable and by the fundamental theorem of calculus, we have

$$y'(x) = \phi(x) - c, \text{ in } [a, b]. \quad ...(3)$$

This implies that

$$y(x) \in C^1[a, b].$$

Also from equations (1) and (2), we have

$$y(a) = y(b) = 0.$$

Thus the setting of $y(x)$ satisfies all the conditions of the theorem.

Now $\quad \int_a^b [\phi(x) - c]^2\, dx = \int_a^b [\phi(x) - c] y'(x)\, dx$

$$= \int_a^b \phi(x) y'(x) dx - c \int_a^b y'(x) dx$$

$$= 0 - c[y(b) - y(a)] = 0,$$

$$\left[\text{since by hypothesis} \int_a^b \phi(x) y'(x) dx = 0\right]$$

which implies that $\phi(x) - c = 0$, for all $x \in [a, b]$

$\Rightarrow \quad \phi(x) = c$, for all $x \in [a, b]$.

Theorem 3: If $\phi(x), \psi(x) \in C[a, b]$ and

$$\int_a^b [\phi(x) y(x) + \psi(x) y'(x)]\, dx = 0 \quad ...(1)$$

for each function $y(x) \in C^1(a, b)$ such that $y(a) = y(b) = 0$

then $\psi(x)$ is differentiable and $\psi'(x) = \phi(x)$, for all $x \in [a, b]$.

Proof: Let $G(x) = \int_a^x \phi(\xi)\, d\xi$, for all $x \in [a, b]$.

Then $G(a) = 0$ and $G'(x) = \phi(x)$, for all $x \in [a, b]$.

Now

$$\int_a^b \phi(x) y(x) dx = \left[y(x)\left(\int_a^x \phi(\xi)\, d\xi\right)\right]_a^b - \int_a^b \left[y'(x)\left\{\int_a^x \phi(\xi)\, d\xi\right\}\right] dx$$

$$= -\int_a^b G(x) y'(x) dx, \text{ since } y(a) = y(b) = 0 \quad ...(2)$$

From (1) and (2), we have

$$\int_a^b [-G(x)+\psi(x)].\, y'(x)dx = 0 \qquad \ldots(3)$$

for every function $y(x) \in C^1[a,b]$ such that $y(a) = y(b) = 0$. Applying Theorem 2 in relation (3), we have

$$-G(x) + \psi(x) = \text{constant}, \text{ for all } x \in [a,b].$$

Since $G(x)$ is differentiable, so $\psi(x)$ is differentiable and

$$\psi'(x) = G'(x) \Rightarrow \psi'(x) = \phi(x), \text{ for all } x \in [a,b].$$

This completes the proof.

11.6 VARIATION DIFFERENTIAL OF A FUNCTIONAL

Let us represent the increment ΔI of the functional $I[y(x)]$ as

$$\Delta I = L[y(x), \delta y] + \beta[y(x), \delta y] \max|\delta y|$$

where $L[y(x), \delta y]$ is a linear functional with respect to the argument δy and $\beta[y(x), \delta y] \to 0$ as the maximum value of $|\delta y| \to 0$, then $L[y(x), \delta y]$ is called the variation of the functional $I[y(x)]$ and is denoted by δI. Here δI is also called first variation of I.

In this case the functional $I[y(x)]$ is called differentiable at the point $y(x)$.

Note that the variation δI plays the same role for a functional I as δy does for a function $y(x)$.

A linear functional I is differentiable because

$$\Delta I = I[y(x) + \delta y(x)] - I[y(x)] = I[\delta y(x)]$$

$$= I[\delta y] + 0.\, \max|\delta y|.$$

Here $L = I$ and $\beta[y(x), \delta y] = 0.$

Lemma 1: The variation of a differentiable functional $I[y(x)]$ is given by the formula

$$\delta I = \left\{\frac{\partial I[y(x) + \alpha\,\delta y]}{\partial \alpha}\right\}_{\alpha=0}$$

Proof: Since the function $I[y(x)]$ is differentiable, we have

$$\Delta I = L[y(x), \delta y] + \beta[y(x), \delta y] \max|\delta y|$$

where $\beta \to 0$ as $|\delta y| \to 0$.

If we replace δy by $\alpha\,\delta y$ then the increment ΔI can be written as

$$\Delta I = I[y+\alpha\,\delta y]-I[y]=L[y(x),\alpha\,\delta y]+\beta[y(x),\alpha\,\delta y]|\alpha|\max|\delta y|$$

$$=\alpha L[y(x),\delta y]+\beta[y(x),\alpha\,\delta y]|\alpha|\max|\delta y|,\text{ since } L \text{ is linear.}$$

Now, the derivative of $I[y(x)+\alpha\,\delta y]$ with respect to α at $\alpha=0$ is given by

$$\left\{\frac{\partial I[y(x)+\alpha\,\delta y]}{\partial\alpha}\right\}_{\alpha=0} = \lim_{\Delta\alpha\to 0}\frac{\Delta I}{\Delta\alpha}=\lim_{\alpha\to 0}\frac{\Delta I}{\alpha}$$

$$=\lim_{\alpha\to 0}\frac{\alpha L[y(x),\delta y]+\beta[y(x),\alpha\,\delta y]|\alpha|\max|\delta y|}{\alpha}$$

$$=\lim_{\alpha\to 0}\frac{\alpha L[y(x),\delta y]}{\alpha}+\lim_{\alpha\to 0}\frac{\beta[y(x),\alpha\,\delta y]|\alpha|\max|\delta y|}{\alpha}$$

$$=L[y(x),\delta y]=\delta I\text{, since } \beta\to 0 \text{ as } \alpha\to 0$$

Hence variation $L[y,\delta y]$ of a functional I is $\left\{\dfrac{\partial I[y(x)+\alpha\,\delta y]}{\partial\alpha}\right\}_{\alpha=0}$

Theorem 4: Suppose the differential $L[y,\delta y]$ of a functional $I[y(x)]$ is a linear functional and $\dfrac{L[y(x),\delta y]}{|y(x)|}\to 0$ as $|y(x)|\to 0$.

then $\quad I[y(x)]=0,$ for all $y(x)$.

Proof: If possible, assume that $I[y_0]\neq 0,$ for all $y_0\neq 0$. then $\dfrac{I[y_0]}{|y_0|}\neq 0.$ ...(1)

Now, we define $y_n=\dfrac{y_0}{n}$, then $|y_n|\to 0$ as $n\to\infty$.

But $\quad \displaystyle\lim_{n\to\infty}\frac{I[y_n]}{|y_n|}=\lim_{n\to\infty}\frac{I[y_0/n]}{|y_0/n|}=\lim_{n\to\infty}\frac{\frac{1}{n}I[y_0]}{\frac{1}{n}[y_0]}$, since I is linear.

$$=\frac{I[y_0]}{|y_0|}\neq 0,\text{ from (1).}$$

This contradicts the hypothesis of the theorem, therefore $I[y]$ must be zero for all $y(x)$.

Theorem 5: A differential functional has the unique differential.

Proof: If possible assume that the differential of the functional $I[y]$ is not unique, that is

$$\Delta I[y] = L_1[y, \delta y] + \beta_1[y, \delta y]\max|\delta y|$$

and $\Delta I[y] = L_2[y, \delta y] + \beta_2[y, \delta y]\max|\delta y|$

where $L_1[y,\delta y]$ and $L_2[y, \delta y]$ are linear functional and $\beta_1, \beta_2 \to 0$ as $|\delta y| \to 0$. This implies that

$$L_1[y,\delta y] - L_2[y,\delta y] = \{\beta_2[y,\delta y] - \beta_1[y,\delta y]\}\max|\delta y|$$

$$\Rightarrow \quad \frac{L_1[y, \delta y] - L_2[y, \delta y]}{|\delta y|} = \beta_2[y,\delta y] - \beta_1[y,\delta y]$$

$\to 0$, as $|\delta y| \to 0$.

Thus from the Theorem 4, the functional

$$L_1[y,\delta y] - L_2[y,\delta y] = 0, \text{ identically}$$

$$\Rightarrow \quad L_1[y,\delta y] = L_2[y,\delta y], \text{ for all } y.$$

That is, the differential of the differentiable functional $I[y]$ is unique.

11.6.1 Extremals of the Functional

The functional $I[y(x)]$ is said to have an extremum on a curve $y = y_0(x)$ if the difference $I[y] - I[y_0]$ does not change its sign in some neighbourhood of the curve $y = y_0(x)$.

The functional $I[y(x)]$ attains a maximum on a curve $y = y_0(x)$ if the difference

$$\Delta I = I[y(x)] - I[y_0(x)] \leq 0$$

on each curve $y = y(x)$ close to $y = y_0(x)$.

Further if $\Delta I \leq 0$ and $\Delta I = 0$ only on $y = y_0(x)$, we say that strict maximum is attained on $y = y_0(x)$.

If $\Delta I \geq 0$ for all curves closed to $y = y_0(x)$, we say that the function I attains a maximum on the curve $y = y_0(x)$.

If $\Delta I \geq 0$ and $\Delta I = 0$ only on $y = y_0(x)$, we say that a strict minimum is attained on $y = y_0(x)$.

Theorem 6: Suppose a functional $I[y(x)]$ attains an extremum for $y = y_0(x)$, then $\delta I = 0$ at $y = y_0(x)$.

Proof: For fixed $y_0(x)$ and δy, the functional $I[y_0(x)+\alpha\delta y]$ is a function of α, let us denote it by $\psi(\alpha)$, that is

$$I[y_0(x)+\alpha\,\delta y]=\psi(\alpha)$$

This reaches a extremum i.e., maximum or minimum for $\alpha=0$.

Thus $\psi'(0)=0$

which implies $\left\{\frac{\partial}{\partial\alpha}I[y_0(x)+\alpha\,\delta y]\right\}_{\alpha=0}=0 \Rightarrow \delta I=0.$

This is the necessary condition for extremum of the functional

$$I[y(x)]=\int_a^b F(x,y,y')dx.$$

Example 1: If a functional $I[y(x)]=\int_0^1 y(x)dx$ is defined on the class $C[0,1]$, then prove that

$$I[1]=1 \text{ and } I[\sin\pi x]=\frac{2}{\pi}.$$

Solution: The given functional is $I[y(x)]=\int_0^1 y(x)dx.$

Thus, $I[1]=\int_0^1 1dx=[x]_0^1=1$

and

$$I[\sin\pi x]=\int_0^1 \sin\pi x\,dx=\left[-\frac{\cos\pi x}{\pi}\right]_0^1=\frac{-1}{\pi}[\cos\pi-\cos 0]=\frac{-1}{\pi}[-1-1]=\frac{2}{\pi}.$$

Example 2: Let a functional $I[y(x)]$ defined on the class $C^1[0,1]$ be given by

$$I[y(x)]=\int_0^1\sqrt{1+[y'(x)]^2}\,dx.$$

Then find $I[1]$, $I[x]$ and $I[x^2]$.

Solution: The given functional is $I[y]=\int_0^1\sqrt{1+\left(\frac{dy}{dx}\right)^2}\,dx.$

Therefore $I[1]=\int_0^1\sqrt{1+(0)^2}\,dx=\int_0^1 1dx=[x]_0^1=1;$

$$I[x]=\int_0^1\sqrt{1+\left(\frac{dy}{dx}\right)^2}\,dx=\int_0^1\sqrt{1+1}\,dx=\int_0^1\sqrt{2}\,dx=\sqrt{2}[x]_0^1=\sqrt{2};$$

and $I[x^2]=\int_0^1\sqrt{1+\left[\frac{d}{dx}(x^2)\right]^2}\,dx=\int_0^1\sqrt{1+4x^2}\,dx=2\int_0^1\sqrt{x^2+\left(\frac{1}{2}\right)^2}\,dx$

$$= 2\left[\frac{1}{2}x\sqrt{x^2+\left(\frac{1}{2}\right)^2}+\frac{1}{2}.\frac{1}{4}\sin h^{-1}\frac{x}{1/2}\right]_0^1$$

$$= 2\left[\frac{1}{2}\sqrt{1+\frac{1}{4}}+\frac{1}{8}\sin h^{-1}2-0\right]=\frac{\sqrt{5}}{2}+\frac{1}{4}\sin h^{-1}2.$$

Example 3: Show that the functional $I_1[y(x)] = \int_a^b [y'(x)+y(x)]dx$ is linear in the class $C^1[a,b]$, but the functional

$$I_2[y(x)] = \int_a^b \left[p(x)[y'(x)]^2 + q(x)y^2(x)\right]dx \text{ is non-linear.}$$

Solution: Let α be any constant and $y_1(x), y_2(x) \in C^1[a,b]$, then

$$I_1[\alpha y(x)] = \int_a^b \left[\frac{d}{dx}(\alpha y(x)) + \alpha y(x)\right] dx$$

$$= \alpha \int_a^b [y'(x)+y(x)]\, dx = \alpha I_1[y(x)]$$

and $$I_1[y_1(x)+y_2(x)] = \int_a^b [\{y_1(x)+y_2(x)\}' + y_1(x)+y_2(x)]\, dx$$

$$= \int_a^b [y'_1(x)+y_1(x)]dx + \int_a^b [y'_2(x)+y_2(x)]dx$$

$$= I_1[y_1(x)] + I_1[y_2(x)].$$

Thus, the functional $I_1[y(x)]$ is linear in $C^1[a, b]$.

Now $I_2[\alpha y(x)] = \int_a^b \left[p(x)[\alpha y'(x)]^2 + q(x)[\alpha\, y(x)]^2\right]dx$

$$= \alpha^2 \int_a^b \left[p(x)\{y'(x)\}^2 + q(x)y^2(x)\right] dx = \alpha^2 I_2\{y(x)\}.$$

Hence $I_2[y(x)]$ is not linear.

Example 4: Investigate the closeness of the following curves:

(a) $y(x) = \dfrac{\sin n^2 x}{n}$, $y_1(x) \equiv 0$ on $[0,\pi]$

(b) $y(x) = \dfrac{\sin nx}{n^2}$, $y_1(x) = 0$ on $[0,\pi]$.

Solution: (a) Here we have

$$|y(x)-y_1(x)| = \left|\frac{\sin n^2 x}{n} - 0\right| = \frac{|\sin n^2 x|}{n} \le \frac{1}{n} \to 0, \text{ as } n\to\infty$$

But $$|y'(x) - y_1'(x)| = n|\cos n^2\alpha| = n \text{ at } x = \frac{2\pi}{n^2}$$

$\to \infty$, as $n \to \infty$.

Hence the curves are close in the 0th order proximity but not in the first order proximity for large n.

(b) We have

$$|y(x)-y_1(x)| = \left|\frac{\sin nx}{n^2}\right| = \frac{|\sin nx|}{n^2} \le \frac{1}{n^2} \to 0, \text{ as } n \to \infty$$

$$|y'(x)-y_1'(x)| = \left|\frac{\cos nx}{n}\right| = \frac{|\cos nx|}{n} \le \frac{1}{n} \to 0, \text{ as } n \to 0.$$

Hence the curves are close in the first order proximity. However, it can easily be seen that they are not close in higher order proximity for large n.

Example 5: Find the distance between the curves $y = x$ and $y = x^2$ on the interval $[0, 1]$.

Solution: The given curves $y = x$ and $y = x^2$ intersect at (0, 0) and (1, 1).

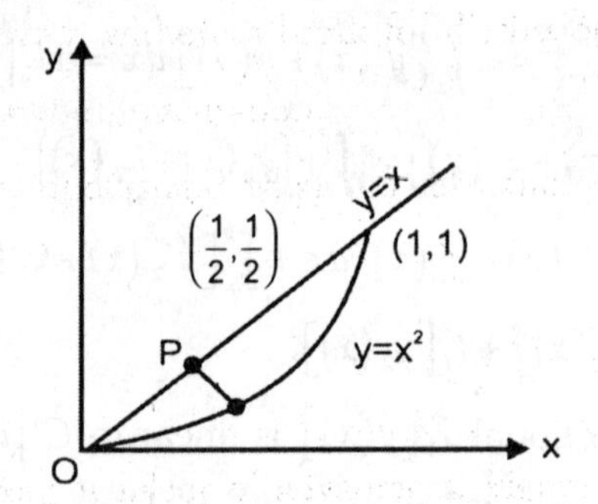

The required distance is $\rho = \max(x - x^2),\ 0 \le x \le 1$

Let $y(x) = x - x^2$, then $y'(x) = 0$ at $x = 1/2$

and $y''(1/2) = -2 < 0$.

This implies that $y(x)$ is maximum at $x = 1/2$. Hence, the maximum distance is $\rho = \frac{1}{2} - \frac{1}{4} = \frac{1}{4}$

Example 6: Find the first order distance between the curves $y = \log x$ and $y = x$ on the interval $[e^{-1}, e]$.

Solution: Let $Y(x) = x - \log x$, then the required first order distance is given by $\rho = \max\limits_{e^{-1} \le x \le e} \{|Y(x)|, |Y'(x)|\} = \max\limits_{e^{-1} \le x \le e} \{|x - \log x|, |1 - 1/x|\}$

$= \max\{1, e-1\} = e - 1.$

Example 7: Show that the functional $I[y(x)] = \int_0^1 x^3 [1+y^2(x)]^{1/2}\, dx$ defined on the set of functions $y(x) \in C[0, 1]$ is continuous on the function $y_0(x) = x^2$ in the sense of zero-order proximity.

Solution: Put $y(x) = x^2 + \alpha\eta(x)$, where $\eta(x) \in C[0,1]$ and α is arbitrary small.

Then $\quad I[y(x)] = I[x^2 + \alpha\eta(x)] = \int_0^1 x^3 \left[1 + (x^2 + \alpha\eta(x))^2\right]^{1/2} dx.$

Taking the limit $\alpha \to 0$, we obtain

$$\lim_{\alpha \to 0} I[y(x)] = \int_0^1 x^3 (1+x^4)^{1/2}\, dx = I[x^2]$$

and this establishes the continuity of the functional on $y_0(x) = x^2$ in the sense of zero-order proximity.

Example 8: Show that the functional $I[y(x)] = \int_0^1 [y(x) + 2y'(x)]\, dx$, $y(x) \in C^1[0,1]$ is continuous on the curve $y_0(x) = x$ in the sense of first order proximity.

Solution: We have

$$|I[y(x)] - I[y_0(x)]| = \left| \int_0^1 [y(x) + 2y'(x)] dx - \int_0^1 (x+2) dx \right|$$

$$= \left| \int_0^1 ([y(x) - x] + 2[y'(x) - 1]) dx \right|$$

$$\leq \int_0^1 [|y(x) - x| + 2|y'(x) - 1|]\, dx$$

Since $y(x)$ and $y'(x)$ are continuous on $[0, 1]$, so for a given $\varepsilon > 0$ we can choose $\delta > 0$ s. t.

$$|y(x) - x| < \delta \text{ and } |y'(x) - 1| < \delta,\ x \in [0,1].$$

Then $|I[y(x)] - I[y_0(x)]| < \int_0^1 [\delta + 2\delta]\, dx = 3\delta \int_0^1 dx = 3\delta = \varepsilon$, if $\delta = \dfrac{\varepsilon}{3}$.

Thus, for given $\varepsilon > 0$, $\exists\ \delta\ (= \varepsilon/3) > 0$ such that

$$|I[y(x)] - I[y_0(x)]| < \varepsilon,\ x \in [0,1].$$

Hence the functional $I[y(x)]$ is continuous on the function $y_0(x) = x$ in the sense of first order proximity.

Example 9: Find the variation of the functional

$$I[y(x)] = \int_a^b \int_a^b k(x,t)\, y(x) y(t)\, dx\, dt$$

where $k(x,t) \in L^2$ on $a \leq x, t \leq b; k(x,t) = k(t,x)$ and $y \in C[a,b]$.

Solution: Let $\psi(\alpha) = I[y(x) + \alpha\,\delta y]$. Then we have

$$\psi(\alpha) = \int_a^b \int_a^b k(x,t)\,\{y(x) + \alpha\,\delta y(x)\}\{y(t) + \alpha\,\delta\,y(t)\}\,dx\,dt.$$

Now $$\frac{\partial \psi}{\partial \alpha} = \int_a^b \int_a^b k(x,t)\{y(x) + \alpha\,\delta y(x)\}\delta y(t)\,dx\,dt$$
$$+ \int_a^b \int_a^b k(x,t)\{y(t) + \alpha\,\delta y(t)\}\delta y(x)\,dx\,dt.$$

Then

$$\delta I = \left(\frac{\partial \psi}{\partial \alpha}\right)_{\alpha=0} = \int_a^b \int_a^b k(x,t)y(x)\delta y(t)\,dx\,dt + \int_a^b \int_a^b k(x,t)y(t)\delta y(x)\,dx\,dt$$
$$= 2\int_a^b \int_a^b k(x,t)y(t)\delta y(x)\,dx\,dt,$$

because of the symmetry of $k(x,t)$, the first integral will be equal to the second integral.

Hence the variation of the given function is

$$\delta I = 2\int_a^b \int_a^b k(x,t)y(t)\delta\,y(x)\,dx\,dt.$$

EXERCISE 11.1

1. If a functional $I[y(x)] = \int_0^1 y(x)dx$ is defined on the class $C[0, 1]$, then prove that $I[x] = \frac{1}{2}$ and $I\left[\cos\frac{\pi}{2}x\right] = \frac{1}{\pi}.$

2. If a functional $I[y(x)] = \int_0^1 \sqrt{1+[y'(x)]^2}\,dx$ is defined on the class $C^1[a,b]$, then prove that

   $I\left[\frac{1}{2}\right] = 1$, $I[2x] = \sqrt{5}$ and $I[2x^2] = \frac{\sqrt{17}}{2} + \frac{1}{8}\sinh^{-1}4.$

3. If $\phi(x,y)$ is a continuous function on the domain $-1 \leq x \leq 1$, $-\infty < y < \infty$, then $\phi(x,y) \in C[-1,1]$ and we define the functional $I[y(x)]$ as $I[y(x)] = \int_{-1}^1 \phi(x, y(x))dx.$

   If $\phi(x,y) = \frac{x^2}{1+y^2}$, then show that $I[1] = \frac{1}{3}$ and $I[x] = 2 - \frac{\pi}{2}.$

4. If $k(x,t) \in L^2[a,b]$, then show that $I[y(x)] = \int_a^b k(x,t)y(x)dx$ is a linear functional on $C[a,b]$, where t is a parameter on $a \leq t \leq b$.

5. Prove that a functional $I[y(x)]$ is linear if it is continuous and satisfies $I[\alpha y(x)] = \alpha I[y(x)]$, where α is a parameter.
6. If f is differentiable with respect to its independent variable, then prove that the variation on the functional $I[y(x)] = \int_a^b f(x,y,y')dx$ is given by

$$\delta I = \int_a^b \left(\frac{\partial f}{\partial y}\delta y + \frac{\partial f}{\partial y'}\delta y' \right) dx.$$

11.7 EULER'S EQUATION

We first discuss the simplest form of variational problem containing first order derivation and one independent variable. In this case the functionals are on plane curve.

Theorem 7: Suppose $y(x) \in C^1[a,b]$ and $I[y] = \int_a^b F\left(x, y, \frac{dy}{dx}\right) dx$ ...(1)

is a functional defined on the set $C^1[a,b]$. Also assume that each $y(x) \in C^1[a,b]$ satisfies the boundary conditions

$$y(a) = A,\ y(b) = B. \qquad \text{...(2)}$$

Then a necessary condition for $I[y]$ in (1) to be maximum or minimum is that

$$\frac{\partial F}{\partial y} - \frac{d}{dx}\left(\frac{\partial F}{\partial y'}\right) = 0. \qquad \text{...(3)}$$

This equation (3) is known as Euler's equation.

Proof: Let $y = y(x)$ be the curve AB which makes the given functional $I[y]$ an extremum.

Now, if we slightly disturb the curve $y = y(x)$, then the functional $I[y]$ will certainly increase. Consider a family of disturbed curves (neighbouring curves) as

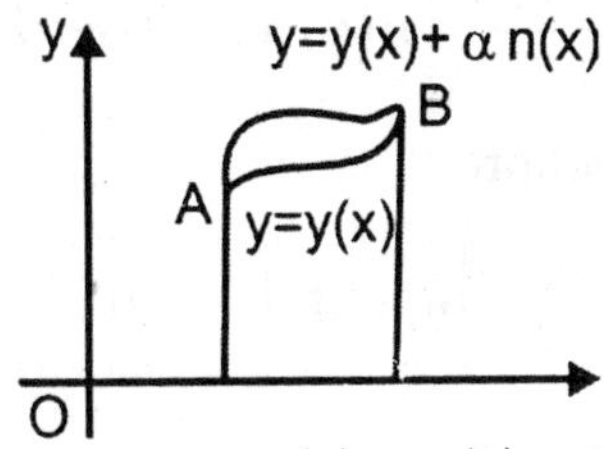

where α is a parameter and $\eta(a) = \eta(b) = 0$.

When $\alpha = 0$, the neighbouring curves become $Y = y(x)$ which is extremal. The family of neighbouring curves is known as the family of comparison functions.

On replacing y by $Y = y(x) + \alpha\eta(x)$, the definite integral $\int_a^b F(x, y, y')dx$ becomes $\int_a^b F[x, Y, Y']dx$ where stands for first order derivatives with respect to x, and is a function of α. So we denote it by $I(\alpha)$ i.e.,

$$I(\alpha) = \int_a^b F(x, Y, Y')dx. \qquad ...(4)$$

This takes on its extremum when $\alpha = 0$, which is possible only if

$$\left(\frac{\partial I}{\partial \alpha}\right)_{\alpha=0} = 0. \qquad ...(5)$$

Differentiating I under the integral sign by Leibnitz's rule, we have

$$\frac{\partial I}{\partial \alpha} = \int_a^b \frac{\partial F}{\partial \alpha} dx = \int_a^b \left(\frac{\partial F}{\partial x}\frac{\partial x}{\partial \alpha} + \frac{\partial F}{\partial Y}\frac{\partial Y}{\partial \alpha} + \frac{\partial F}{\partial Y'}\frac{\partial Y'}{\partial \alpha}\right) dx$$

$$= \int_a^b \left(\frac{\partial F}{\partial Y}\frac{\partial Y}{\partial \alpha} + \frac{\partial F}{\partial Y'}\frac{\partial Y'}{\partial \alpha}\right) dx \text{, since } x \text{ is independent of } \alpha \Rightarrow \frac{\partial x}{\partial \alpha} = 0 \qquad ...(6)$$

Now $Y = y(x) + \alpha\eta(x)$ implies that

$$Y' = y'(x) + \alpha\eta'(x), \ \frac{\partial Y}{\partial \alpha} = \eta(x), \ \frac{\partial Y'}{\partial \alpha} = \eta'(x).$$

Hence equation (6) becomes

$$\frac{\partial I}{\partial \alpha} = \int_a^b \left[\frac{\partial F}{\partial Y}\eta(x) + \frac{\partial F}{\partial Y'}\eta'(x)\right] dx$$

$$= \int_a^b \frac{\partial F}{\partial Y}\eta(x)dx + \left[\frac{\partial F}{\partial Y'}\eta(x)\right]_{x=a}^{b} - \int_a^b \frac{d}{dx}\left[\frac{\partial F}{\partial Y'}\right]\eta(x)\,dx,$$

using integration by parts

$$= \int_a^b \left[\frac{\partial F}{\partial Y} - \frac{d}{dx}\left(\frac{\partial F}{\partial Y'}\right)\right]\eta(x)\,dx \text{, since } \eta(a) = \eta(b) = 0$$

Using it in (5), we obtain

$$\int_a^b \left[\frac{\partial F}{\partial Y} - \frac{d}{dx}\left(\frac{\partial F}{\partial Y'}\right)\right]\eta(x)dx\bigg|_{\alpha=0} = 0$$

$$\Rightarrow \quad \int_a^b \left[\frac{\partial F}{\partial y} - \frac{d}{dx}\left(\frac{\partial F}{\partial y'}\right)\right]\eta(x)\,dx = 0, \qquad ...(7)$$

since, when $\alpha = 0, \partial Y = \partial y \text{ and } \partial Y' = \partial y'$

As $\eta(x)$ is an arbitrary continuous function, we have

$$\frac{\partial F}{\partial y} - \frac{d}{dx}\left(\frac{\partial F}{\partial y'}\right) = 0 \Rightarrow F_y - \frac{dF_{y'}}{dx} = 0 \qquad \text{...(8)}$$

which is the required Euler's equation. The integral curves that is, the solutions of this equation are known as extremals. Extremal is obtained by solving Euler's equation.

11.7.1 Other Forms

One can derive some other forms of Euler's equation as follows:

1. As F is a function of x, y, y', so by total differentiation

$$\frac{dF(x,y,y')}{dx} = \frac{\partial F}{\partial x}\frac{dx}{dx} + \frac{\partial F}{\partial y}\frac{dy}{dx} + \frac{\partial F}{\partial y'}\frac{dy'}{dx}$$

That is, $$\frac{dF}{dx} = \frac{\partial F}{\partial x} + \frac{\partial F}{\partial y}y' + \frac{\partial F}{\partial y'}y'' \qquad \text{...(9)}$$

On the other hand $$\frac{d}{dx}\left[y'\frac{\partial F}{\partial y'}\right] = y'\frac{d}{dx}\left(\frac{\partial F}{\partial y'}\right) + \frac{\partial F}{\partial y'}y''. \qquad \text{...(10)}$$

On subtracting (10) from (9), we obtain

$$\frac{dF}{dx} - \frac{d}{dx}\left(y'\frac{\partial F}{\partial y'}\right) = \frac{\partial F}{\partial x} + \frac{\partial F}{\partial y}y' - y'\frac{d}{dx}\left(\frac{\partial F}{\partial y'}\right)$$

$$\Rightarrow \frac{d}{dx}\left[F - y'\frac{\partial F}{\partial y'}\right] - \frac{\partial F}{\partial x} = y'\left[\frac{\partial F}{\partial y} - \frac{d}{dx}\left(\frac{\partial F}{\partial y'}\right)\right]$$

Using Euler's equation, we get

$$\frac{d}{dx}\left[F - y'\frac{\partial F}{\partial y'}\right] - \frac{\partial F}{\partial x} = 0. \qquad \text{...(11)}$$

This is the second form of Euler's equation.

2. Also note that $\frac{\partial F}{\partial y'}$ is a function of x, y, y'. Let us denote it by $\phi(x, y, y')$, so

$$\frac{d}{dx}\left(\frac{\partial F}{\partial y'}\right) = \frac{\partial \phi}{\partial x}\frac{dx}{dx} + \frac{\partial \phi}{\partial y}\frac{\partial y}{\partial x} + \frac{\partial \phi}{\partial y'}\frac{dy'}{dx}$$

$$= \frac{\partial \phi}{\partial x} + \frac{\partial \phi}{\partial y}y' + \frac{\partial \phi}{\partial y'}y'' = \frac{\partial^2 F}{\partial x \partial y'} + y'\frac{\partial^2 F}{\partial y\, \partial y'} + y''\frac{\partial^2 F}{\partial y'^2}$$

Using this value of $\frac{d}{dx}\left(\frac{\partial F}{\partial y'}\right)$ in Euler's equation, we obtain

$$\frac{\partial F}{\partial y}-\frac{\partial^2 F}{\partial x\,\partial y'}-y'\frac{\partial^2 F}{\partial y\,\partial y'}-y''\frac{\partial^2 F}{\partial y'^2}=0$$

or $\quad F_y - F_{xy'} - y'F_{yy'} - y''F_{y'y'} = 0.$ ...(12)

This is the third form of Euler's equation and is a second-order differential equation in $y(x)$. So its solution depends, in general, on two arbitrary constants, which are determined from the boundary conditions $y(a)=A,\ y(b)=B$.

11.7.2 Special Cases

It should also be noted that Euler's equation is not always readily integrable. Now we shall discuss some cases, where Euler's equation admits integrals.

Case I: *If F is independent of x, i.e. F is dependent on y and y′ only. In this case* $\frac{\partial F}{\partial x}=0$, *so by (11), we have*

$$\frac{d}{dx}\left[F-y'\frac{\partial F}{\partial y'}\right]=0 \qquad ...(13)$$

This implies that

$$F-y'\frac{\partial F}{\partial y'}=\text{constant.}$$

This equation may be integrated further after solving for y'

Case II: *If F is independent of y, i.e. F is dependent on x and y′ only. In this case* $\frac{\partial F}{\partial y}=0$, *so Euler's equation (8) becomes*

$$\frac{d}{dx}\left(\frac{\partial F}{\partial y'}\right)=0 \qquad ...(14)$$

This gives $\frac{\partial F}{\partial y'}=$ constant

$\Rightarrow F_{y'}=C(say).$ ...(15)

Since this relation does not involve y, it can be solved for y' as a function of x. Integration leads to a solution involving two arbitrary constants which can be found from the boundary conditions.

Case III: *If F is independent of y′, i.e. F is dependent on x and y only. In this case* $\frac{\partial F}{\partial y'}=0$, *so Euler's equation (8) becomes*

$$\frac{\partial F}{\partial y}=0. \qquad ...(16)$$

This is not a differential equation but an algebraic equation in x and y. The solution of this equation consists of one or more curves $y=y(x)$.

Case IV: *If F is a function of y′ only i.e., independent of x and y, then we have*

$$\frac{\partial F}{\partial x}=0 \text{ and } \frac{\partial F}{\partial y}=0 \Rightarrow \frac{\partial^2 F}{\partial x\,\partial y'}=0 \text{ and } \frac{\partial^2 F}{\partial y\,\partial y'}=0.$$

Thus Euler's equation (12) becomes

$$y''F_{y'y'}=0. \qquad \ldots(17)$$

If $\frac{\partial^2 F}{\partial y^2}\neq 0$, then $y''=0$. Solution of this differential equation is $y=mx+c$, so extremals are all straight lines.

Case V: *If* $F(x,y,y')$ *is of the form*

$$f(x,y)\sqrt{1+(y')^2} \text{ i.e., } F(x,y,y')=f(x,y)\sqrt{1+(y')^2}$$

Then we have

$$\frac{\partial F}{\partial y}-\frac{d}{dx}\left(\frac{\partial F}{\partial y'}\right)=\frac{\partial f(x,y)}{\partial y}\sqrt{1+(y')^2}-\frac{d}{dx}\left[f(x,y)\frac{y'}{\sqrt{1+(y')^2}}\right]$$

$$=\frac{\partial f}{\partial y}\sqrt{1+(y')^2}-\frac{\partial f}{\partial x}\frac{y'}{\sqrt{1+(y')^2}}-\frac{\partial f}{\partial y}\frac{(y')^2}{\sqrt{1+(y')^2}}-\frac{\partial}{\partial y'}\left(f\frac{y'}{\sqrt{1+(y')^2}}\right)\frac{dy'}{dx}$$

$$=f_y\sqrt{1+(y')^2}-f_x\frac{y'}{\sqrt{1+(y')^2}}-f_y\frac{(y')^2}{\sqrt{1+(y')^2}}-y''f\frac{\left[\sqrt{1+(y')^2}-\frac{(y')^2}{\sqrt{1+(y')^2}}\right]}{1+(y')^2}$$

$$=f_y\sqrt{1+(y')^2}-\frac{f_x y'}{\sqrt{1+(y')^2}}-\frac{f_y(y')^2}{\sqrt{1+(y')^2}}-\frac{y''f}{\left[1+(y')^2\right]^{3/2}}$$

$$=\frac{1}{\sqrt{1+(y')^2}}\left[f_x-y'f_x-y''f\right].$$

Therefore, in this case Euler's equation becomes

$$f_y-y'f_x-y''f=0. \qquad \ldots(18)$$

This is a differential equation of second order.

Case VI: *If* $F(x,y,y')=M(x,y)+N(x,y)y'$, *then*

$$I[y(x)]=\int_a^b F(x,y,y')dx=\int_a^b\left[M(x,y)+n(x,y)y'\right]dx. \qquad \ldots(19)$$

In this case

$$\frac{\partial F}{\partial y}=\frac{\partial M}{\partial y}+\frac{\partial N}{\partial y}.y' \text{ and } \frac{\partial F}{\partial y'}=N(x,y)$$

This implies that $\frac{d}{dx}\left(\frac{\partial F}{\partial y'}\right)=\frac{d}{dx}N(x,y)=\frac{\partial N}{\partial x}+\frac{\partial N}{\partial y}y'$.

Therefore Euler's equation becomes

$$\frac{\partial M}{\partial y}+\frac{\partial N}{\partial y}y'-\frac{\partial N}{\partial x}-\frac{\partial N}{\partial y}y'=0 \Rightarrow \frac{\partial M}{\partial y}-\frac{\partial N}{\partial x}=0.$$

This leads to an equation which does not satisfy the boundary conditions. Such type of variational problems does not have any solution.

On the other hand if $\frac{\partial M}{\partial y}-\frac{\partial N}{\partial x}\equiv 0$.

Then $M\,dx+N\,dy=0$ is an exact differential equation and so the functional

$$I=\int_a^b\left[M(x,y)+N(x,y)y'\right]dx$$

is independent of the path of integration. This implies that I has a constant value. So in this case variational problems are meaningless.

Example 1: Find the extremals of the functional

$$I[y(x)]=\int x\left(dx^2+dy^2\right)^{1/2}.$$

Solution: We have $I[y(x)]=\int x\left(dx^2+dy^2\right)^{1/2}=\int x\sqrt{1+y'^2}\,dx.$

Here $\quad F(x,y,y')=x\sqrt{1+y'^2}$, so

$$F_y=F_{yy'}=0,\ F_{xy'}=\frac{y'}{\sqrt{1+y'^2}} \text{ and } F_{y'y'}=\frac{x}{\left(1+y'^2\right)^{3/2}}.$$

Hence the Euler's equation $F_y-F_{xy'}-F_{yy'}y'-F_{y'y'}y''=0$ reduces to

$$\frac{y'}{\sqrt{1+y'^2}}+\frac{x}{\left(1+y'^2\right)^{3/2}}y''=0\Rightarrow xy''+y'\left(1+y'^2\right)=0.$$

Putting $y'=Y$, we obtain $xY'+Y\left(1+Y^2\right)=0 \Rightarrow \frac{dY}{Y\left(1+Y^2\right)}+\frac{dx}{x}=0$

Hence on integration, we have

$\int\frac{dY}{Y\left(1+Y^2\right)}+\int\frac{dx}{x}=\log a$, where $\log a$ is the constant of integration

or $\quad \int\left[\frac{1}{Y}-\frac{1}{2}\cdot\frac{2Y}{1+Y^2}\right]dY+\log x=\log a$

or $\quad 2\log Y-\log\left(1+Y^2\right)+2\log x=2\log a$

or $\quad x^2Y^2=a^2\left(1+Y^2\right)\Rightarrow\left(x^2-a^2\right)Y^2=a^2$

or $\quad y' = Y = \dfrac{a}{\sqrt{x^2 - a^2}} \Rightarrow dy = \dfrac{a}{\sqrt{x^2 - a^2}}\,dx.$

Hence on integration, we get

$$y = a\cos\text{h}^{-1}\frac{x}{a} + b \Rightarrow x = a\cos\text{h}\frac{y-b}{a},$$ which is a centenary.

Example 2: Test for an extremum of the functional

$$I[y(x)] = \int_0^1 \left(xy + y^2 - 2y^2y'\right)dx, y(0) = 1, y(1) = 2.$$

Solution: Here $F(x, y, y') = xy + y^2 - 2y^2y'$, therefore

$$F_y = x + 2y - 4yy' \text{ and } F_{y'} = -2y^2.$$

Hence Euler's equation $F_y - \dfrac{d}{dx}F_{y'} = 0$ gives

$$x + 2y - 4yy' - \frac{d}{dx}\left(-2y^2\right) = 0 \Rightarrow x + 2y - 4yy' + 4yy' = 0$$

$$\Rightarrow y = -\frac{x}{2}.$$

Clearly, this extremal cannot satisfy the boundary conditions $y(0) = 1, y(1) = 2$. Thus an extremum cannot be achieved for this functional.

Example 3: Test for extremum of the functional

$$I[y(x)] = \int_0^{\pi/2} \left(y'^2 - y^2\right)dx, y(0) = 0, y(\pi/2) = 1.$$

Solution: Here $F(x, y, y') = y'^2 - y^2$, therefore $F_y = -2y$ and $F_{y'} = 2y'$.

Hence Euler's equation $F_y - \dfrac{d}{dx}F_{y'} = 0$ gives

$$-2y - \frac{d}{dx}(2y') = 0 \Rightarrow y'' + y = 0 \qquad \ldots(1)$$

which is a linear differential equation of second order with constant coefficients. Hence its auxiliary equation is $m^2 + 1 = 0;$ *i.e.* $m = \pm i$.

Therefore, the general solution of (1) is $y = C_1\cos x + C_2\sin x$

Using the boundary conditions, $y(0) = 0, y(\pi/2) = 1,$ we find that $C_1 = 0, C_2 = 1$. Thus the extremum can be achieved only on the curve $y = \sin x$.

Example 4: Find the shortest curve joining two points (x_1, y_1) and (x_2, y_2).

Solution: We have to minimize the arc length integral

$$l[y(x)] = \int_{x_1}^{x_2} \sqrt{1+y'^2}\, dx.$$

Subject to the boundary conditions $y(x_1) = y_1$, and $y(x_2) = y_2$.

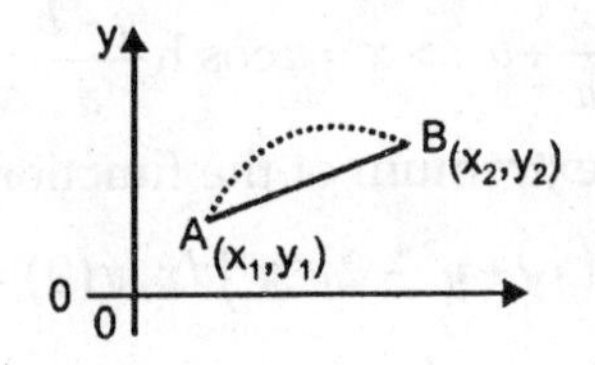

Here , $F(x,y,y') = \sqrt{1+y'^2}$ so $F_y = 0, F_{xy'} = 0$ and $F_{yy'} = 0.$

Hence Euler's equation reduces to the form $F_{y'y'}\, y'' = 0.$ ...(1)

Since $F_{y'y'} = \dfrac{\partial^2 F}{\partial y'^2} = \dfrac{1}{\left[1+y'^2\right]^{1/2}} \neq 0,$ so equation (1) gives $y'' = 0$ ...(2)

This on integration twice yields $y = C_1 x + C_2$. ...(3)

Now applying the boundary conditions, we get

$$y_1 = y(x_1) = C_1 x_1 + C_2 \quad ...(4)$$

and $$y_2 = y(x_2) = C_1 x_2 + C_2. \quad ...(5)$$

Solving (4) and (5), we obtain $C_1 = \dfrac{y_2 - y_1}{x_2 - x_1}, C_2 = y_1 - \dfrac{y_2 - y_1}{x_2 - x_1} x_1.$

Thus the extremum can be achieved on the curve

$$y - y_1 = \frac{y_2 - y_1}{x_2 - x_1}(x - x_1) \quad ...(6)$$

and this is of course the straight line joining the two given points (x_1, y_1) and (x_2, y_2). This analysis shows that l has extremum (a stationary value) and the corresponding stationary curve must be the straight line (6).

Example 5: Test the extremum of the functional

$$I[y(x)] = \int_0^4 \left[xy' - y'^2\right] dx, y(0) = 0, y(4) = 3.$$

Solution: Here $F(x,y,y') = xy' - y'^2$, therefore

$F_y = 0$ and $F_{y'} = x - 2y'$

Hence Euler's equation $F_y - \dfrac{d}{dx} F_{y'} = 0$ gives $\dfrac{d}{dx}(x - 2y') = 0$

which on integration yields $x - 2y' = C_1 \Rightarrow y' = \dfrac{1}{2}(x - C_1).$

Again integrating, we get $y = \dfrac{1}{4}x^2 - \dfrac{1}{2}C_1 x + C_2.$

Using the boundary conditions, we find that

$$0=y(0)=0-\frac{1}{2}C_1(0)+C_2 \;\Rightarrow C_2=0$$

and $3=y(4)=4-\frac{1}{2}C_1 4+C_2 \Rightarrow 2C_1-C_2=4-3 \Rightarrow C_1=\frac{1}{2}$.

Thus, the extremum can be achieved on the curve $y=\frac{1}{4}(x^2-x)$.

Example 6: Find the extremum of the functional

$$I[y(x)]=\int_{x_2}^{x_1}\tfrac{1}{x}\sqrt{(1+y'^2)}\,dx.$$

Solution: Here the functional I may be recognized as the time spent on translation along the curve $y=y(x)$ from one point to another, if the rate of motion $v=\frac{ds}{dt}$ is equal to x. In fact, we know from mechanics that $ds=\left(1+y'^2\right)^{1/2}dx$.

Since the functional is independent of y, Euler's equation $\frac{d}{dx}F_{y'}(x,y')=0$ leads to

$$\frac{d}{dx}\left\{\frac{1}{2x}\left(1+y^2\right)^{-1/2}.2y'\right\}=0 \;\Rightarrow \frac{d}{dx}\left\{\frac{y'}{x\sqrt{1+y'^2}}\right\}=0$$

which on integration yields

$$y'=C_1\,x\left(1+y'^2\right)^{1/2}. \qquad \ldots(1)$$

Putting $y'=\tan t$ in (1), we obtain

$$x=(1/C_1)\sin t=C'\sin t. \qquad \ldots(2)$$

Thus $dy=\tan t\,dx=\tan t\,d(C'\sin t)=\tan t.C'\cos t\,dt=C'\sin t\,dt$

This on integration leads to $y=-\overline{C_1}\cos t+C_2 \qquad \ldots(3)$

Elimination of t from (2) and (3), we obtain

$$x^2+(y-C_2)^2=\overline{C_1}^2,$$ which is a family of circles.

Example 7: Find the value of the curve $y=f(x)$ corresponding to which the extremum of the functional $I[y(x)]=\int_{x_1}^{x_2}x^n\left(\frac{dy}{dx}\right)^2dx$ exists.

Solution: Here $F=x^n\left(\frac{dy}{dx}\right)^2=x^n y'^2$ which is independent of y. Hence the reduced form of Euler's equation $\frac{d}{dx}\left(\frac{\partial F}{\partial y'}\right)=0$ gives

$$\frac{d}{dx}\left[\frac{\partial}{\partial y'}\left(x^n y'^2\right)\right]=0 \Rightarrow \frac{d}{dx}\left[2x^n y'\right]=0 \Rightarrow \frac{d}{dx}\left[x^n y'\right]=0$$

$\Rightarrow \quad x^n y' = a$, constant

$\Rightarrow \quad \dfrac{dy}{dx} = ax^{-n}$

Again integrating, we get

$$y = a\int x^{-n}dx + b, \; b \text{ being constant of integration}$$

or $\quad y = \dfrac{a\,x^{1-n}}{1-n} + b \Rightarrow \dfrac{a}{n-1}\cdot\dfrac{1}{x^{n-1}} + y = b.$

The above curve gives extremum of the functional $I[y(x)]$, where the constants a, b are obtained by using boundary conditions if they are given.

Example 8: If $f(y)$ is a given function and $u = \int_{x_0}^{x_1} f(y)\sqrt{1+y'^2}\,dx$

where the values of y at $x = x_0$ and $x = x_1$ are fixed, then show that the extremal value of u is given by the function $y = \phi(x)$ which is defined by

$$\frac{dy}{\sqrt{a\{f(y)\}^2 - 1}} = x - b$$

where a and b are constants, which are determined by the boundary conditions.

Solution: We have $u = I[y(x)] = \int_{x_0}^{x_1} f(y)\sqrt{1+y'^2}\,dx$ ...(1)

From (1) it is clear that $F = f(y)\sqrt{1+y'^2}$ depends on y and y' only. Hence Euler's equation becomes

$$F - y'F_{y'} = C, \; C \text{ being constant}$$

or $\quad f(y)\sqrt{1+y'^2} - y'\dfrac{\partial}{\partial y'}f(y)\sqrt{1+y'^2} = C$

or $\quad f(y)\sqrt{1+y^2} - y'f(y).\dfrac{1}{2\sqrt{1+y'^2}}.2y' = C$

or $\quad f(y)(1+y'^2) - f(y)y'^2 = C\sqrt{1+y'^2}$

or $\quad f(y) = C\sqrt{1+y'^2}$ or $y'^2 = \dfrac{1}{C^2}[f(y)]^2 - 1$

or $\quad \dfrac{dy}{dx} = \sqrt{a\{f(y)\}^2 - 1}$, where $a = \dfrac{1}{C^2}$

Integrating the above equation, we get

$\displaystyle\int \frac{dy}{\sqrt{a\{f(y)\}^2 - 1}} = \int dx - b = x - b$, b being constant of integration.

Hence the extremal value of u is given by the function $y=\phi(x)$ which is defined as

$$\int \frac{dy}{\sqrt{a\{f(y)\}^2-1}}=x-b.$$

Example 9: Find the curve with fixed boundary points (x_1, y_1) and (x_2, y_2) such that its rotation about x-axis gives rise to a surface of revolution of minimum surface area.

Solution: The area of the surface of revolution is

$$S[y(x)]=2\pi \int_{x_1}^{x_2} y\sqrt{1+y'^2}\,dx$$

where the end points A and B of the curve $y=y(x)$ have coordinates (x_1, y_1) and (x_2, y_2), respectively.

Here we have to minimize the functional S i.e., surface of revolution with minimum surface area. Since the integral is a function of y and y' only, a first integral of Euler's equation is

$$F-y'F_{y'}=C, \text{ where } F=2\pi y\sqrt{1+y'^2}$$

or $$2\pi y\sqrt{1+y'^2}-y'\frac{2\pi y\, 2y'}{2\sqrt{1+y'^2}}=C$$

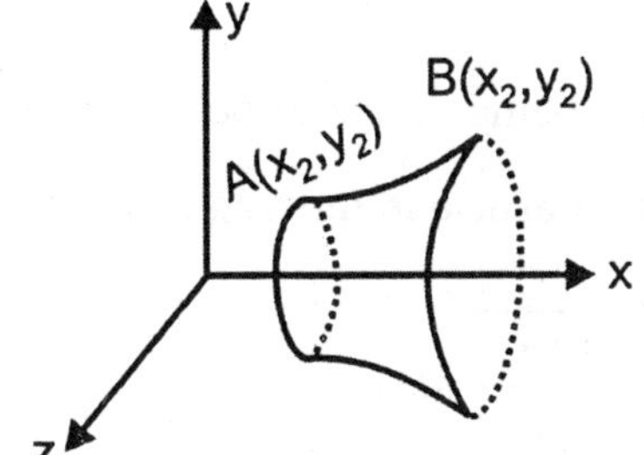

or $$y\sqrt{1+y'^2}-\frac{yy'^2}{\sqrt{1+y'^2}}=C_1, \text{ where } C_1=\frac{C}{2\pi}$$

or $$\frac{y}{\sqrt{1+y'^2}}=C_1.$$

To integrate this equation, we put $y'=\sin ht$. Then it is clear that

$$y=C_1\cos ht,\ y'=\frac{dy}{dx}=\sin ht$$

or $$y=C_1\cos ht,\ \frac{dx}{dt}=\frac{1}{\sin ht}\frac{dy}{dt}$$

or $$y=C_1\cos ht,\ \frac{dx}{dt}=\frac{1}{\sin ht}.C_1\sin ht$$

or $\quad y = C_1 \cosh t, \; dx = C_1\, dt$ ...(1)

On integrating, the second equation of (1), we have

$$y = C_1 \cosh t, \;\; x = C_1 t + C_2 \quad ...(2)$$

The elimination of t from (2) gives the required extremals

$$y = C_1 \cosh \frac{x - C_2}{C_1}$$

which constitutes a two parameter family of catenaries. The constants C_1 and C_2 are determined from the given boundary conditions that the given curve passes through the points A and B.

Example 10: Find the curve of quickest descent which is traversed by a particle moving under gravity from A to B. This is known as Brachistochrone problem.

Solution: Fix the origin at A with x-axis horizontally and y-axis vertically downward.

The speed of descent of the particle $\frac{ds}{dt}$ is given by

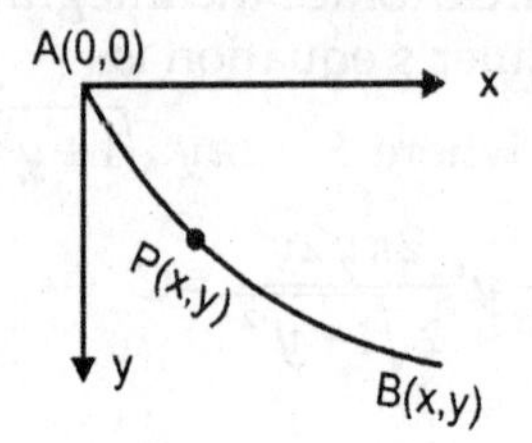

$$\frac{ds}{dt} = \sqrt{2gy}\text{, where } g \text{ being acceleration due to gravity.}$$

Thus the time taken by the particle in moving from A(0, 0) to B(x_1, y_1) is is

$$t[y(x)] = \frac{1}{\sqrt{2g}} \int_0^{x_1} \frac{\sqrt{1 + y'^2}}{\sqrt{y}}\, dx$$

This is an improper integral but convergent. Here the boundary conditions are

$$y(0) = 0, \; y(x_1) = y_1.$$

Since the integrand is independent of x, a first integral of Euler's equation is given by

$$F - y' F_{y'} = C \Rightarrow \frac{1}{\sqrt{2g}} \frac{\sqrt{1 + y'^2}}{\sqrt{y}} - y' \cdot \frac{1}{\sqrt{2g}} \cdot \frac{2y'}{2\sqrt{y}\sqrt{1 + y'^2}} = C$$

$$\Rightarrow \quad \frac{1}{\sqrt{y}\sqrt{1 + y'^2}} = \sqrt{2g}\, C \Rightarrow y(1 + y'^2) = C_1\text{, where } C_1 = \frac{1}{2gC^2}$$

Putting $y = \cot\theta$, θ being a parameter; we get

$$y = C_1 \sin^2\theta \Rightarrow y = \frac{C_1}{2}(1-\cos 2\theta).$$

Now, $dx = \dfrac{dy}{y'} = \dfrac{2C_1 \sin\theta \cos\theta \, d\theta}{\cot\theta} = C_1(1-\cos 2\theta)d\theta$

which on integration yields $x - C_2 = \dfrac{C_1}{2}(2\theta - \sin 2\theta)$.

Putting $2\theta = \theta_1$ and remembering that $y = 0$ at $x = 0$, we find that $C_2 = 0$. Thus the desired extremals are given by

$$x = \frac{C_1}{2}(\theta_1 - \sin\theta_1),\ y = \frac{C_1}{2}(1-\cos\theta_1)$$

which is a family of cycloids with $C_1/2$ as the radius of the rolling circle where C_1 is determined by the fact that the cycloid passes through $B(x_1, y_1)$.

Example11: Determine the shape of a solid of revolution moving in a flow of gas with least resistance.

Solution: Assume that the gas density is sufficiently small such that the gas molecules are mirror reflected from the surface of the solid. The component of gas pressure normal to the surface is

$$p = 2\rho v^2 \sin^2\theta \qquad \ldots(1)$$

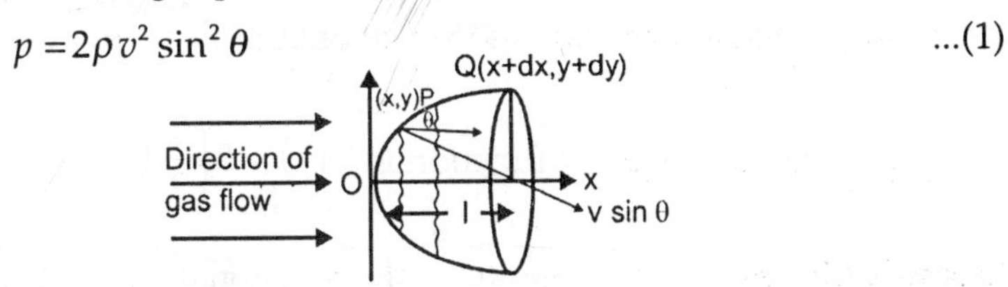

where $\rho =$ density of the gas

$v =$ velocity of gas relative to the solid

$\theta =$ the angle between the tangent at any point of the surface with the direction of flow.

Solid of revolution experiences the least pressure, when the pressure given by (1) is normal to the surface. Let dF be the force component along the x-axis acting on a ring PQ of width $\delta s \left(=\sqrt{1+y'^2}\,\delta x\right)$ and radius $y(x)$. Then

$$dF = 2\rho v^2 \sin^2\theta . \left[2\pi y\sqrt{1+y'^2}\right]\sin\theta \, dx. \qquad \ldots(2)$$

Hence the total force along the x direction is

$$F = \lim_{dx\to 0}\sum dF = \int_0^l 4\pi P v^2 y \sin^3\theta\sqrt{1+y'^2}\,dx \qquad \ldots(3)$$

Assume that $\sin\theta = \dfrac{y'}{(1+y'^2)^{1/2}} = y'$, where the slope y' is taken to be very small.

Thus from (3) the total resistance experienced by the body is

$$F = 4\pi\rho v^2 \int_0^l y'^3\, y\, dx. \qquad \ldots(4)$$

Now our variational problem is to find $y = y(x)$ for which F is minimum. Here the boundary conditions are

$$y(0) = 0,\ y(l) = R. \qquad \ldots(5)$$

Since the integral in (4) depends on y and y' only, a first integral of Euler's equation is $F - y'F_{y'} = C$, where $F = 4\pi\rho v^2 y'^3 y$ and C is being constant. Thus, we have

$$y'^3 y - y'.\, 3y'^2 y = \frac{C}{4\pi\rho v^2}$$

or $\quad y'^3 y = C_1^3$, *where* $C_1^3 = -\dfrac{C}{8\pi\rho v^2} \Rightarrow y^{1/3}\, dy = C_1\, dx$

On integrating, we obtain $y = (C_1 x + C_2)^{3/4}$.

Using boundary conditions (5), we obtain $C_1 = \dfrac{R^{4/3}}{l}$, $C_2 = 0$.

Therefore, the required function is $y(x) = R\left(\dfrac{x}{l}\right)^{3/4}$.

Example 12: Test for an extremum of the functional $I[y] = \int_0^2 (e^{y'} + 3)dx$, $y(0) = 0,\ y(2) = 1$.

Solution: In given problem $F(x, x, y') = e^{y'} + 3$, this is independent of y. This implies the Euler's equation is

$$\frac{d}{dx}\frac{\partial F}{\partial y'} = 0 \Rightarrow \frac{\partial F}{\partial y'} = \text{constant}$$

$$\Rightarrow \quad e^{y'} = c(say) \Rightarrow y' = \log c.$$

On integrating, we have $y = x\log c + c'$.

Applying given boundary conditions, we obtain

$$y(0) = 0 \Rightarrow c' = 0 \text{ and } y(2) = 1 \Rightarrow 1 = 2\log c \Rightarrow \log c = \frac{1}{2}.$$

Hence, the required extremal is $y = \dfrac{1}{2}x$.

Example 13: Find the extremals of $I[y(x)] = \int_1^2 \frac{\sqrt{1+x^2y'}}{x} dx$, subject to the boundary conditions $y(1)=0$ and $y(2)=1$.

Solution: Here $F=\frac{\sqrt{1+x^2y'}}{x}$, this is independent of y, so Euler's equation is

$$\frac{d}{dx}\left(\frac{\partial F}{\partial y'}\right)=0 \Rightarrow \frac{\partial F}{\partial y'}=\text{constant}$$

$$\Rightarrow \quad \frac{x^2}{2x\sqrt{1+x^2y'}}=c\ (say) \Rightarrow x=2c\sqrt{x^2y'+1}$$

$$\Rightarrow \quad \frac{x^2}{4c^2}=x^2y'+1$$

$$\Rightarrow \quad y'=\frac{x^2-4c^2}{4c^2x^2}.$$

On integrating, we obtain $y=\int \frac{1}{4c^2}dx-\int \frac{1}{x^2}dx+c'=\frac{x}{4c^2}+\frac{1}{x}+c'$
Using boundary conditions, we obtain

$$y(1)=0 \Rightarrow 0=\frac{1}{4c^2}+1+c' \text{ and } y(2)=1 \Rightarrow 1=\frac{2}{4c^2}+\frac{1}{2}+c'.$$

On solving these, we get $c=\frac{1}{\sqrt{6}}, c'=-\frac{5}{2}$.

Hence the required extremal is $y=\frac{3}{2}x+\frac{1}{x}-\frac{5}{2}$.

EXERCISE 11.2

1. Solve the Euler's equation for the functional
$$I[y(x)] = \int_a^b (x+y')y'dx.$$
2. Find the extremals of the functional $I[y(x)] = \int_a^b \frac{1+y^2}{y'^2} dx$.
3. Solve the Euler's equation for the functional
$$I[y(x)] = \int_a^b \left(y'+x^2y'^2\right)dx.$$
4. Find the extremum value of the following functional

   i. $I[y(x)] = \int_{1/2}^1 x^2y'\, dx$ under the boundary conditions
$$y(1/2)=1,\ y(1)=2.$$

ii. $I[y(x)] = \int_0^{\pi/2} (y'^2 - y^2)dx$ under the boundary conditions

$$y(0) = 0, y(\pi/2) = 1.$$

iii. $I[y(x)] = \int_1^2 \frac{\sqrt{1+y'^2}}{x} dx$ subject to conditions $y(1) = 0, y(2) = 1$.

iv. $I[y(x)] = \int_0^2 (x - y')^2 dx$ with boundary conditions

$$y(0) = 2, y(2) = 4.$$

v. $I[y(x)] = \int_0^1 (y'^2 + 12xy) dx$ under the boundary conditions

$$y(0) = 0, y(1) = 1.$$

vi. $I[y(x)] = \int_0^1 y y'^3 dx$ under the boundary conditions

$$y(0) = 0, y(1) = 1.$$

5. Minimize the functional $I[y] = \int_a^b (x - y)^2 dx$.

6. Suppose light travels in a medium from one point to another such that the time of travel is given by $\int \frac{ds}{v(x,y)}$ is maximum, where s is the arc length and $v(x,y)$ is the velocity of light in the medium. Show that the path of travel is the solution of the following differential equation

$$vy'' + \left[1 + (y')^2\right]\frac{\partial v}{\partial y} - y'\left[1 + (y')^2\right]\frac{\partial v}{\partial x} = 0.$$

7. In the above problem if $y = v$, then find the nature of the curve.

8. Prove that the extremals of the functional $I(x) = \int_a^b \sqrt{1+y'^2}\, dx$ are straight lines.

9. Prove that the extremals of the functional $I[y] = \int_a^b x\sqrt{1+y'^2}\, dx$ are parabolas.

10. Prove that the extremals of the functional $I[y] = \int_a^b y\sqrt{1+y'^2}\, dx$ are catenaries.

11. Find the extremals of the functional

$$I[y(x)] = \int_1^2 (y'^2 - 2xy)dx, y(1) = 0, y(2) = -1.$$

ANSWERS 11.2

1. $y = -\frac{x^2}{4} + c_1 x + c_2$. 2. $y = \sinh(c_1 x + c_2)$.

3. $y = c_1 x^{-1} + c_2$

4. (i) $y = -\frac{1}{x} + 3,\ 1$ (ii) $y = \sin x,\ 0$

(iii) $x^2 + y^2 - 4y - 1 = 0$, $\sec h^{-1}\frac{1}{\sqrt{5}} - \sec h^{-1}\frac{2}{\sqrt{5}}$

(iv) $y = \frac{x^2}{2} + x,\ 2$ (v). $y = x^3,\ \frac{21}{5}$

(vi) $y = x^{3/4},\ \frac{27}{64}$

5. $x - y = 0$, a straight line.

7. Required path is the arc of the circle $(x+b)^2 + y^2 = a^2$.

11. $y = \frac{1}{12}\left(-2x^3 + x + 2\right)$.

11.8 VARIATIONAL PROBLEMS FOR FUNCTIONALS OF SEVERAL VARIABLES

This is the generalizations of the simplest variational problems. Consider the functional I of n dependent variables as

$$I = \int_a^b F\left(x, y_1(x), y_2(x), \dots, y_n(x), y_1'(x), y_2'(x), \dots, y_n'(x)\right) dx \qquad \dots(1)$$

where the function F is differentiable three times with respect to all its arguments.

To determine the necessary conditions for the extremum of the above functional I, we consider the boundary conditions for $y_1(x), y_2(x), \dots y_n(x)$ as below.

$$\left.\begin{aligned} y_1(a) = a_1,\ y_2(a) = a_2, \dots, y_n(a) = a_n \\ y_1(b) = b_1,\ y_2(b) = b_2, \dots, y_n(b) = b_n \end{aligned}\right\}$$

where $a_1, a_2, \dots, a_n, b_1, b_2, \dots, b_n$ are constants.

Now, we are looking for an extremum of the functional such that (1) is defined on the set of smooth curves joining two fixed points in $(n+1)$ dimensional Euclidean space R^{n+1}. Some examples of this type problem are

(i) The problem of finding geodesics (shortest curve joining two points of some manifold).

(ii) In geometric optics the problem of finding the paths along which light rays propagate in an inhomogeneous media. The Fermat's principle states that the light goes from a point P_0 to a point P_1 along the path for which the transit time is smallest.

Theorem 8: Suppose $I = \int_a^b F\left(x, y_1, y_2, y_3, ..., y_n, y_1', y_2', ..., y_n'\right) dx$ ...(1)

is a functional defined on n continuously differentiable functions $y_i = y_i(x)$ for $i = 1, 2, ..., n$, with boundary conditions

$$y_i(a) = a_i,\ y_i(b) = b_i,\quad i = 1, 2, ..., n. \qquad ...(2)$$

Then a necessary condition for I to be extremum is that

$$\frac{\partial F}{\partial y_i} - \frac{d}{dx}\left(\frac{\partial F}{\partial y_i'}\right) = 0, \text{ for all } i = 1, 2, ..., n. \qquad ...(3)$$

Proof: In order to find the required condition, we calculate the variation δI. We replace each $y_i(x)$ by $y_i(x) + \varepsilon\eta_i(x)$, $\forall i = 1, 2, ..., n$, where

$$\eta_i(a) = 0 = \eta_i(b) \qquad ...(4)$$

By definition the variations δI of the functional I is linear in η_i and η_i' for $i = 1, 2, ..., n$, and which differs from the increment

$$\Delta I = I\left[y_1 + \varepsilon\eta_1, y_2 + \varepsilon\eta_2, ..., y_n + \varepsilon\eta_n\right] - I\left[y_1, y_2, ..., y_n\right] \qquad ...(5)$$

By a quantity of order higher than 1 relative to η_i and η_i', $i = 1, 2, ..., n$. Now

$$\Delta I = \int_a^b \left[F\left(x, y_1 + \varepsilon\eta_1 . y_2 + \varepsilon\eta_2, ..., y_n + \varepsilon\eta_n, y_1' + \varepsilon\eta_1', y_2' + \varepsilon\eta_2', ..., y_n' + \varepsilon\eta_n'\right)\right.$$

$$\left. - F\left(x, y_1, y_2, ..., y_n, y_1', y_2', ..., y_n'\right)\right] dx$$

By using Taylor's theorem, we obtain

$$\Delta I = \int_a^b \left\{\sum_{i=1}^n \left(\eta_i \frac{\partial F}{\partial y_i}\right) + \eta_i' \frac{\partial F}{\partial y_1'}\right\} dx + \text{terms containing higher powers of } \varepsilon \qquad ...(6)$$

The integral in the right hand side of (6) represents the principal linear part of the increment ΔI. Therefore by definition the variation

$$\delta I = \int_a^b \left\{\sum_{i=1}^n \eta_i \frac{\partial F}{\partial y_i} + \eta_i' \frac{\partial F}{\partial y_i'}\right\} dx. \qquad ...(7)$$

As the increments $\eta_i(x)$ are independent of each other, we can choose one of these arbitrarily as long as the boundary conditions are satisfied, setting all others equal to zero. Therefore, the necessary condition $\delta I = 0$ for an extremum gives

$$I\left[y(x)\right] = \int_a^b F\left(x, y, y', y'', y''', ..., y^{(n)}\right) dx \qquad ...(8)$$

which on using Theorem 3, implies

$$\frac{\partial F}{\partial y_i} = \frac{d}{dx}\left(\frac{\partial F}{\partial y_i'}\right) \Rightarrow \frac{\partial F}{\partial y_i} - \frac{d}{dx}\left(\frac{\partial F}{\partial y_i'}\right) = 0;\ i = 1,2,3,\ldots,n. \quad \ldots(9)$$

This equation represents a system of n second order differential equations and these equations are called Euler's equations. In general, the solution of (10) contains $2n$ arbitrary constants which are determined from the boundary conditions in (2). The solutions of n equations in (10) provide the family of extremals for the given variational problem.

11.8.1 Euler's Equations Related to Geodesics

Suppose S is a surface specified by a vector equation

$$\bar{r} = \bar{r}\,(u,v) \quad \ldots(1)$$

The shortest curve, that is, the curve of minimum length connecting any two points A and B of the surfaces S and lying on the surface, is called the geodesic on a given surface. For example, a geodesic on a plane is a straight line.

The differential equations for the geodesics of the surface S are the Euler's equations of the corresponding variational problem. "The problem of finding the minimum distance measured along the surface S between any two points of the surface S."

Suppose the parametric equations of a curve on the surface (1) are

$$u=u(t),\ v=v(t),\ \text{where } t \text{ being a parameter} \quad \ldots(2)$$

Let $\quad E = \bar{r}_u \cdot \bar{r}_u,\ F = \bar{r}_u \cdot \bar{r}_v,\ G = \bar{r}_v \cdot \bar{r}_v \quad \ldots(3)$

These quantities are called the coefficients of the first fundamental form of the surface (1). The arc length between the points $A(t_1)$ and $B(t_2)$ corresponding to the parameter t, is given by

$$I[u,v] = \int_{t_1}^{t_2} \sqrt{Ev'^2 + 2Fu'v' + Gv'^2}\, dt \quad \ldots(4)$$

[for details see differential geometry by C.E. Weatherburn] Therefore, Euler's equations for the functional (4) are

$$\frac{\partial}{\partial u}\left[\sqrt{Eu'^2 + 2Fu'v' + Gv'^2}\right] - \frac{d}{dt}\left[\frac{\partial}{\partial u'}\sqrt{Eu'^2 + 2Fu'v' + Gv'^2}\right] = 0$$

and

$$\frac{\partial}{\partial v}\left[\sqrt{Eu'^2 + 2Fu'v' + Gv'^2}\right] - \frac{d}{dt}\left[\frac{\partial}{\partial v'}\sqrt{Eu'^2 + 2Fu'v' + Gv'^2}\right] = 0.$$

These equations are equivalent to

$$\frac{E_u u'^2+2F_u u'v'+G_u v'^2}{\sqrt{Eu'^2+2Fu'v'+Gv'^2}}-\frac{d}{dt}\left[\frac{2(Eu'+Fv')}{\sqrt{Eu'^2+2Fu'v'+Gv'^2}}\right]=0$$

and $$\frac{E_v u'^2+2F_v u'v'+G_v v'^2}{\sqrt{Eu'^2+2Fu'v'+Gv'^2}}-\frac{d}{dt}\left[\frac{2(Fu'+Gv')}{\sqrt{Eu'+2Fu'v'+Gv'^2}}\right]=0.$$

Definition: Two functional are said to be equivalent if they have the same extremals.

Example 1: Derive the differential equations of the lines of propagation of light in an optically non-homogeneous medium with the speed of light $c=c(x,y,z)$.

Solution: According to Fermat's law, light propagation from one point to another point along a curve for which, the time T of passage of light will be minimum. Suppose the equation of the desired path of the light ray be $y=y(x)$ and $z=z(x)$, then it is clear that

$$T=\int_{x_1}^{x_2}\frac{ds}{c}=\int_{x_1}^{x_2}\frac{\sqrt{1+y'^2+z'^2}}{c(x,y,z)}dx$$

where ds is a line element on the path. Then the system of Euler's equation is

$$\frac{\sqrt{1+y'^2+z'^2}}{c^2}\frac{\partial c}{\partial y}+\frac{d}{dx}\left[\frac{y'}{c\sqrt{1+y'^2+z'^2}}\right]=0,$$

$$\frac{\sqrt{1+y'^2+z'^2}}{c^2}\frac{\partial c}{\partial z}+\frac{d}{dx}\left[\frac{z'}{c\sqrt{1+y'^2+z'^2}}\right]=0$$

which determines the path of the light propagation.

Example 2: Find the extremals of the functional

$$I=\int_0^{\pi/2}\left(y'^2+z'^2+2yz\right)dz \qquad \text{...(1)}$$

subject to the boundary conditions

$$y(0)=0,\ y(\pi/2)=1,\ z(0)=0,\ z(\pi/2)=-1 \qquad \text{...(2)}$$

Solution: Let $y_1(x)=y(x)$, $y_2(x)=z(x)$, then

$$F(x,y_1,y_2,y_1',y_2')=(y_1')^2+(y_2')^2+2y_1y_2.$$

Now the Euler's equations $\frac{\partial F}{\partial y_i}-\frac{d}{dx}\left(\frac{\partial F}{\partial y_i'}\right)=0,\ i=1,2$ be

$$2z-\frac{d}{dx}(2y')=0 \Rightarrow z-y''=0 \qquad \text{...(3)}$$

$$2y-\frac{d}{dx}(2z')=0 \Rightarrow y-z''=0. \qquad \text{...(4)}$$

From (3) and (4), we have $\frac{d^4z}{dx^4} - z = 0.$ ...(5)

Here, the A.E. is $m^4 - 1 = 0 \Rightarrow m = \pm 1, \pm i.$

Hence the general solution of (5) is

$z = c_1 e^x + c_2 e^{-x} + c_3 \cos x + c_4 \sin x$, where c_1, c_2, c_3 and c_4 are constants. ...(6)

Applying (6) in (4), we get

$$y = c_1 e^x + c_2 e^{-x} - c_3 \cos x - c_4 \sin x. \quad ...(7)$$

Using given boundary conditions (2), we have

$$c_1 + c_2 - c_3 = 0,\ c_1 e^{\pi/2} + c_2 e^{-\pi/2} - c_4 = 1$$

$$c_1 + c_2 + c_3 = 0,\ c_1 e^{\pi/2} + c_2 e^{-\pi/2} + c_4 = -1$$

which implies that $c_1 = c_2 = c_3 = 0, c_4 = -1.$

Hence the extremals of the given functional are $y(x) = \sin x$, $z(x) = -\sin x$, and the extremum value is

$$\int_0^{\pi/2} \left[\cos^2 x + \cos^2 x - 2\sin^2 x\right] dx = 2\int_0^{\pi/2} \cos 2x\, dx = 2\left[\frac{\sin 2x}{2}\right]_0^{\pi/2} = 0.$$

Example 3: Obtain the extremals of the functional

$$I[y,z] = \int_0^{\pi} \left(y'^2 - 2y^2 + 2yz - z'^2\right) dx$$

under the boundary conditions $y(0) = 0 = z(0),\ y(\pi) = 1 = z(\pi).$

Solution: Here the integrand $F = y'^2 - 2y^2 + 2yz - z'^2 \equiv F(x, y, z, y', z')$ has two dependent variables y, z; so Euler's equations

$$\frac{\partial F}{\partial y} - \frac{d}{dx}\left(\frac{\partial F}{\partial y'}\right) = 0 \text{ and } \frac{\partial F}{\partial z} - \frac{d}{dx}\left(\frac{\partial F}{\partial z'}\right) = 0$$

give $-4y + 2z - \frac{d}{dx}(2y') = 0 \Rightarrow \frac{d^2y}{dx^2} + 2y = z$ ...(1)

and $2y - \frac{d}{dx}(-2z') = 0 \Rightarrow \frac{d^2z}{dx^2} + y = 0.$ ...(2)

Using (1) and (2), we obtain

$$\frac{d^4y}{dx^4} + 2\frac{d^2y}{dx^2} + y = 0 \quad ...(3)$$

A.E. is $m^4 + 2m^2 + 1 = 0 \Rightarrow (m^2 + 1)^2 = 0 \Rightarrow m = \pm i, \pm i.$

Hence the general solution of (3) is

$$y = (c_1 + c_2 x)\cos x + (c_3 + c_4 x)\sin x. \quad ...(4)$$

Using (4) in (1), we get

$$z=\frac{d^2}{dx^2}\left[(c_1+c_2x)\cos x+(c_3+c_4x)\sin x\right]+2(c_1+c_2x)\cos x+2(c_3+c_4x)\sin x$$

$$=-(c_1+c_2x)\cos x-2c_2\sin x-(c_3+c_4x)\sin x+2c_4\cos x$$
$$+2(c_1+c_2x)\cos x+2(c_3+c_4x)\sin x$$

$$=2c_4\cos x-2c_2\sin x+(c_1+c_2x)\cos x+(c_3+c_4x)\sin x. \qquad ...(5)$$

Applying boundary conditions, we obtain

$$y(0)=0\Rightarrow c_1=0;$$
$$z(0)=0\Rightarrow c_1+2c_4=0\Rightarrow c_4=0;$$
$$y(\pi)=1\Rightarrow (c_1+c_2\pi)=1\Rightarrow c_2=-\frac{1}{\pi};$$
$$z(\pi)=1\Rightarrow -(c_1+c_2\pi)=1\Rightarrow c_2=-\frac{1}{\pi}.$$

Clearly value of c_3 cannot be found from given boundary conditions, so c_3 is arbitrary. By using values of c_1, c_2, c_4 in (4) and (5), the required extremals are given by

$$y=c_3\sin x-\frac{x}{\pi}\cos x$$
$$z=c_3\sin x-\frac{1}{\pi}(2\sin x-x\cos x).$$

Example 4: Find the geodesics of the sphere.

Solution: The parametric curve on the surface of the sphere is given by

$$\bar{r}=(a\sin\theta\cos\phi,\ a\sin\theta\sin\phi,\ a\cos\theta).$$

Consider θ as u, ϕ as v, then

$$\bar{r}_\theta=(a\cos\theta\cos\phi,\ a\cos\theta\sin\phi,\ -a\sin\theta)$$
$$\bar{r}_\phi=(-a\sin\theta\sin\phi,\ a\sin\theta\cos\phi,\ 0)$$

This implies that

$$E=\bar{r}_\theta\,.\,\bar{r}_\theta$$
$$=(a\cos\theta\cos\phi, a\cos\theta\sin\phi, -a\sin\theta).(a\cos\theta\cos\phi, a\cos\theta\sin\phi, -a\sin\theta)$$
$$=a^2$$
$$F=\bar{r}_\theta\,.\,\bar{r}_\phi=0$$
$$G=\bar{r}_\phi\,.\,\bar{r}_\phi=a^2\sin^2\theta.$$

Therefore, the functional $I=\int\sqrt{Eu'^2+2Fu'v'+Gv'^2}\,dt$ becomes

$$I=\int\left(\sqrt{a^2\left(\frac{d\theta}{dt}\right)^2+2.0.\frac{d\theta}{dt}\frac{d\phi}{dt}+a^2\sin^2\theta\left(\frac{d\phi}{dt}\right)^2}\right)dt$$

$$= \int_{\phi_1}^{\phi_2} a\left\{\sqrt{\left(\frac{d\theta}{d\phi}\right)^2 + \sin^2\theta}\right\}\frac{d\phi}{dt}.dt$$

$$= \int_{\phi_1}^{\phi_2} a\sqrt{\theta' + \sin^2\theta}\; d\phi, \quad \text{where } \theta' = \frac{d\theta}{d\phi}$$

Here the integral $F = \sqrt{\theta'^2 + \sin^2\theta}$ is independent of ϕ.

Therefore using case I of article 11.7.2, Euler's equation $F - \theta'\dfrac{\partial F}{\partial \theta'} = c$ gives

$$a\sqrt{\theta'^2 + \sin^2\theta} - \frac{a\theta'^2}{\sqrt{\theta'^2 + \sin^2\theta}} = c\ (say)$$

$$\Rightarrow \quad \frac{a\sin^2\theta}{\sqrt{\theta'^2 + \sin^2\theta}} = c$$

$$\Rightarrow \quad c^2\theta'^2 + c^2\sin^2\theta = a^2\sin^4\theta$$

$$\Rightarrow \quad c^2\theta'^2 = a^2\sin^4\theta - c^2\sin^2\theta \quad \Rightarrow \frac{d\theta}{d\phi} = \sqrt{\frac{(a^2 - c^2\operatorname{cosec}^2\theta)\sin^4\theta}{c^2}}$$

$$\Rightarrow \quad d\phi = -\frac{c\,\operatorname{cosec}^2\theta}{\sqrt{(a^2 - c^2) - c^2\cot^2\theta}}\,d\theta.$$

On integrating, we have

$$\phi = \cos^{-1}\left(\frac{c\cot\theta}{\sqrt{a^2 - c^2}}\right) + c_1, \text{ where } c_1 \text{ is a constant.}$$

$$\Rightarrow \quad \cos(\phi - c_1) = \frac{c\cot\theta}{\sqrt{a^2 - c^2}}$$

$$\Rightarrow \quad \cos\phi\cos c_1 + \sin\phi\sin c_1 = c_2\cot\theta, \text{ where } c_2 = \frac{c}{\sqrt{a^2 - c^2}}$$

$$\Rightarrow \quad \cos\phi\cos c_1 + \sin\phi\sin c_1 = c_2\cot\theta, \text{ where } c_2 = \frac{c}{\sqrt{a^2 - c^2}}$$

$$\Rightarrow \quad c_2\cos\theta = \cos c_1\sin\theta\cos\phi + \sin c_1\sin\theta\sin\phi$$

$$\Rightarrow \quad a\cos\theta = \frac{\cos c_1}{c_2}a\sin\theta\cos\phi + \frac{\sin c_1}{c_2}a\sin\theta\sin\phi$$

$$\Rightarrow \quad z = Ax + By$$

which is the equation of the plane passing through the centre (0, 0, 0) of the sphere. Thus it is intersecting the sphere along a great circle, i.e. geodesic on a sphere is an arc of a great circle.

Example 5: Find the geodesic on right circular cone.

Solution: Here we give an alternating method. Making use of spherical polar co-ordinates, the arc length ds between any two points on a right circular cone with semi-vertical angle α and radius r, is given by

$$ds = \sqrt{dr^2 + r^2 \sin^2 \alpha \, d\phi^2}.$$

Therefore, arc length

$$s = \int_{\phi_1}^{\phi_2} \sqrt{r'^2 + r^2 \sin^2 \alpha} \, d\phi \text{, where } r' = \frac{dr}{d\phi}.$$

Here the integrand $F = \sqrt{r'^2 + r^2 \sin^2 \alpha}$ is independent of ϕ, so Euler's equation $F - r'\frac{\partial F}{\partial r'} = c$ gives

$$\sqrt{r'^2 + r^2 \sin^2 \alpha} - \frac{r'^2}{\sqrt{r'^2 + r^2 \sin^2 \alpha}} = c$$

$$\Rightarrow \quad r^2 \sin^2 \alpha = c\sqrt{r'^2 + r^2 \sin^2 \alpha} \qquad \Rightarrow r'^2 = \frac{r^4}{c^2} \sin^4 \alpha - r^2 \sin^2 \alpha$$

$$\Rightarrow \quad \frac{dr}{d\phi} = \frac{r}{c} \sin\alpha \sqrt{r^2 \sin^2 \alpha - c^2}, \text{ since } \frac{dr}{d\phi} = r'$$

$$\Rightarrow \quad \frac{dr}{r \sin\alpha \sqrt{r^2 \sin^2 \alpha - c^2}} = \frac{1}{c} d\phi.$$

$$\Rightarrow \quad \frac{1}{\sin\alpha} \frac{dt}{t\sqrt{t^2 - c^2}} = \frac{1}{c} d\phi \text{, by using } r\sin\alpha = t$$

On integrating, we obtain

$$\frac{1}{\sin\alpha} \frac{1}{c} \sec^{-1} \frac{t}{c} = \frac{1}{c} \phi + \text{ constant}$$

$$\Rightarrow \quad \sec^{-1}\left(\frac{r \sin\alpha}{c}\right) = \phi \sin\alpha + k \text{, where } k \text{ is constant}$$

$$\Rightarrow \quad r \sin\alpha = c\sec(\phi \sin\alpha + k).$$

This is the required geodesic on right circular cone.

Example 6: Prove that geodesics on the plane are straight lines.

Solution: The arc length ds between any two points (x_1, y_1) and (x_2, y_2) on a plane is given by

$$ds = \sqrt{(dx)^2 + (dy)^2}.$$

Therefore the geodesic be the extremals of the functional

$$I[y] = \int_{x_1}^{x_2} \sqrt{1 + \left(\frac{dy}{dx}\right)^2} \, dx = \int_{x_1}^{x_2} \sqrt{1 + y'^2} \, dx.$$

Here the integrand $F=\sqrt{1+y'^2}$ is independent of y. Here, the corresponding Euler's equation $F_{y'}=c$ gives

$$\frac{1}{2}\frac{2y'}{\sqrt{1+y'^2}}=c \Rightarrow y'^2=c^2+c^2y'^2$$

$$\Rightarrow y'^2=\frac{c^2}{1-c^2}$$

$$\Rightarrow \quad y'=A, \text{ where } A=\frac{c}{\sqrt{1-c^2}}$$

$$\Rightarrow \quad y=Ax+B$$

which is a straight line in the plane. Hence, geodesics in a plane are straight lines.

11.9 VARIATIONAL PROBLEMS FOR FUNCTIONALS DEPENDING ON HIGHER ORDER DERIVATIVES

To find the extremals of the functional

$$I[y(x)]=\int_a^b F\left(x,y,y',y'',y''',\ldots,y^{(n)}\right)dx \qquad \ldots(1)$$

where F is differentiable $n+2$ times with respect to all its arguments and $y(x)\in C^n[a,b]$ with boundary conditions

$$y^{(i)}(a)=a_i,\, y^{(i)}(b)=b_i, \text{ for } i=0,1,2,3,\ldots,n. \qquad \ldots(2)$$

Solution: The general result that a necessary condition for a functional $I[y]$ to have an extremum is that the variation of $I[y]$ must vanish, i.e.

$$\delta I=0. \qquad \ldots(3)$$

Suppose ε is a parameter and $\eta(x)\in C^n[a,b]$ such that the boundary conditions (2) are satisfied, i.e.

$$\eta^{(i)}(a)=0=\eta^{(i)}(b); \quad \textit{for } i=0,1,2,\ldots,n. \qquad \ldots(4)$$

We replace $y(x)$ by the varied function $y(x)+\varepsilon\eta(x)$ in the functional (1), we obtain

$$I=\int_a^b F\left(x,y+\varepsilon\eta(x),\, y'+\varepsilon\eta'(x),\ldots,\, y^{(n)}+\varepsilon\eta^{(n)}(x)\right)dx.$$

Differentiating with respect to ε by Leibnitz rule of differentiation under the integral sign, we obtain

$$\frac{dI}{d\varepsilon}=\int_a^b\frac{\partial F}{\partial\varepsilon}dx=\int_a^b\left[\frac{\partial F}{\partial y}\frac{\partial y}{\partial\varepsilon}+\frac{\partial F}{\partial y'}\frac{\partial y'}{\partial\varepsilon}+\frac{\partial F}{\partial y''}\frac{\partial y''}{\partial\varepsilon}+\ldots.+\frac{\partial F}{\partial y^{(n)}}\frac{\partial y^{(n)}}{\partial\varepsilon}\right]dx$$

$$= \int_a^b \left[\eta \frac{\partial F}{\partial y} + \eta' \frac{\partial F}{\partial y'} + \eta'' \frac{\partial F}{\partial y''} + \ldots + \eta^{(n)} \frac{\partial F}{\partial y^{(n)}} \right] dx$$

For extremum $\left[\dfrac{dI}{d\varepsilon}\right]_{\varepsilon=0} = 0$ implies that

$$\int_a^b \left[\eta \frac{\partial F}{\partial y} + \eta' \frac{\partial F}{\partial y'} + \eta'' \frac{\partial F}{\partial y''} + \ldots. + \eta^{(n)} \frac{\partial F}{\partial y^{(n)}} \right] dx = 0$$

$$\Rightarrow \quad \int_a^b \left(\sum_{i=0}^{n} \eta^{(i)} \frac{\partial F}{\partial y^{(i)}} \right) dx = 0$$

$$\Rightarrow \quad \int_a^b \eta(x) \frac{\partial F}{\partial y} dx + \int_a^b \eta'(x) \frac{\partial F}{\partial y'} dx + \int_a^b \eta''(x) \frac{\partial F}{\partial y''} dx + \ldots. + \int_a^b \eta^{(n)}(x) \frac{\partial F}{\partial y^{(n)}} dx = 0.$$

Integrating by parts with respect to x and using $\eta^{(i)}(a) = 0 = \eta^{(i)}(b)$, $i = 0, 1, 2, \ldots, \eta$, we follow

$$\int_a^b \eta(x) \frac{\partial F}{\partial y} dx - \int_a^b \eta(x) \frac{d}{dx}\left(\frac{\partial F}{\partial y'}\right) dx + \int_a^b \eta(x) \frac{d^2}{dx^2}\left(\frac{\partial F}{\partial y''}\right) dx - \ldots. + (-1)^n \int_a^b \eta(x) \frac{d^n}{dx^n}\left(\frac{\partial F}{\partial y^{(n)}}\right) dx = 0 \quad \ldots(5)$$

because $\displaystyle\int_a^b \eta'(x) \frac{\partial F}{\partial y'} dx = \left[\eta(x) \frac{\partial F}{\partial y'}\right]_a^b - \int_a^b \eta(x) \frac{d}{dx}\left(\frac{\partial F}{\partial y'}\right) dx$

$$= -\int_a^b \eta(x) \frac{d}{dx}\left(\frac{\partial F}{\partial y'}\right) dx;$$

$$\int_a^b \eta''(x) \frac{\partial F}{\partial y''} dx = \left[\eta'(x) \frac{\partial F}{\partial y''}\right]_a^b - \int_a^b \eta'(x) \frac{d}{dx}\left(\frac{\partial F}{\partial y''}\right) dx$$

$$= -\left[\left\{ \eta(x) \frac{d}{dx}\left(\frac{\partial F}{\partial y''}\right) \right\}_a^b - \int_a^b \eta(x) \frac{d^2}{dx^2}\left(\frac{\partial F}{\partial y''}\right) dx \right]$$

$$= (-1)^2 \int_a^b \eta(x) \frac{d^2}{dx^2}\left(\frac{\partial F}{\partial y''}\right) dx.$$

Similarly, $\displaystyle\int_a^b \eta^{(i)}(x) \frac{\partial F}{\partial y^{(i)}} dx = (-1)^i \int_a^b \eta(x) \frac{d^i}{dx^i}\left(\frac{\partial F}{\partial y^{(i)}}\right) dx.$

Equation (5) implies that

$$\int_a^b \eta(x) \left[\frac{\partial F}{\partial y} - \frac{d}{dx}\left(\frac{\partial F}{\partial y'}\right) + \frac{d^2}{dx^2}\left(\frac{\partial F}{\partial y''}\right) - \ldots + (-1)^n \frac{d^n}{dx^n}\left(\frac{\partial F}{\partial y^{(n)}}\right) \right] dx = 0.$$

Since $\eta(x) \in C_{[a,b]}$ is an arbitrary function such that $\eta(a) = 0 = \eta(b)$, so by Theorem 1, we have

$$\frac{\partial F}{\partial y} - \frac{d}{dx}\left(\frac{\partial F}{\partial y'}\right) + \frac{d^2}{dx^2}\left(\frac{\partial F}{\partial y''}\right) - \frac{d^3}{dx^3}\left(\frac{\partial F}{\partial y'''}\right) + \ldots. + (-1)^n \frac{d^n}{dx^n}\left(\frac{\partial F}{\partial y^{(n)}}\right) = 0. \quad \ldots(6)$$

This is a required Euler's equation. Equation (6) is an ODE of order $2n$, so its general solution contains $2n$ constants, which can be determined from the $2n$ boundary conditions given by (4).

11.10 FUNCTIONAL DEPENDENT ON FUNCTIONS OF MORE THAN ONE INDEPENDENT VARIABLE

We now extend Euler's equation to the problem of determining the extrema of functional involving multiple integrals leading to one or more partial differential equations.

Let us consider the problem of finding an extremum of the functional

$$I[u(x,y)] = \iint_D F(x,y,u,u_x,u_y)\,dx\,dy \quad \ldots(1)$$

over a region of integration D by determining u which is continuous and has continuous partial derivatives with respect to independent variables x and y upto the second order, and takes on prescribed values on the boundary of D. We further assume that F is thrice differentiable. This problem is particularly useful in geometry.

Let the extremizing surface be $u = u(x,y)$, so that an admissible one-parameter surface can be taken as

$$u(x,y,\alpha) = u(x,y) + \alpha\eta(x,y)$$

where $\eta(x,y) = 0$ on Γ (the boundary of D). Then the necessary condition for an extremum is that the first variation must vanish, i.e.

$$\delta I = \left(\frac{\partial}{\partial\alpha} I[n + \alpha\eta]\right)_{\alpha=0} = 0. \quad \ldots(2)$$

Using (2) in (1), we obtain $\iint_D (F_u\eta + F_u\,\eta_x + F_u\,\eta_y)\,dx\,dy = 0$ ...(3)

which may be again transformed using integration by parts.

We now assume that the boundary Γ of D admits of a tangent, which turns piecewise continuously. Then using the Green's theorem, we have

$$\iint_D (\eta_x F_u + \eta_y F_u)\,dx\,dy = \int_\Gamma \eta\left(\frac{\partial}{\partial x}F_{u_x} + \frac{\partial}{\partial y}F_{u_y}\right)dx\,dy. \quad \ldots(4)$$

Thus from (3), in view of (4), we have

$$\iint_D \left[F_u - \frac{\partial}{\partial x} F_{u_x} - \frac{\partial}{\partial y} F_{u_y} \right] \eta \, dx\,dy + \int_\Gamma \eta \left(F_{u_x} \, dy - F_{u_y} \, dx \right) = 0. \quad ...(5)$$

Since $\eta = 0$ on Γ and (5) holds for any arbitrary continuously differentiable function η, it follows from (5) that by using the generalization of the fundamental lemma of section 9.3 that

$$F_u - \frac{\partial}{\partial x} F_{u_x} - \frac{\partial}{\partial y} F_{u_y} = 0 \quad ...(6)$$

The extremizing function $u(x,y)$ is determined from the solution of the second-order partial differential equation (6) which is known as Euler-Ostrogradsky equation.

In case, when the integrand of a functional I contains derivatives of order higher than two, then by a straight forward extension, we may derive a modified Euler-Ostrogradsky equation for determining extremals. For example, if we have the functional

$$I[u(x,y)] = \iint_D F(x, y, u, u_x, u_y, u_{xy}, u_{yy}) \, dx\,dy$$

Then the extremals are given by the equation

$$F_u - \frac{\partial}{\partial x} F_{u_x} - \frac{\partial}{\partial y} F_{u_y} + \frac{\partial^2}{\partial x^2} F_{u_{xx}} + \frac{\partial^2}{\partial x \partial y} F_{u_{xy}} + \frac{\partial^2}{\partial y^2} F_{u_{yy}} = 0.$$

Example 7: Determine the extremals of the functional

$$I[y(x)] = \int_{-l}^{l} \left(\frac{1}{2} \mu y''^2 + \rho y \right) dx$$

subject to $y(-l) = 0$, $y'(-l) = 0$, $y(l) = 0$, $y'(l) = 0$.

Solution: We observe that this variational problem arises in finding the axis of a flexibly bent cylindrical beam clamped at the ends. If the beam is homogeneous, ρ and μ are constants, then Euler-Poisson equation $F_y - \frac{d}{dx} F_{y'} + \frac{d^2}{dx^2} F_{y''} = 0$ gives

$$\rho - \frac{d}{dx}(0) + \frac{d^2}{dx^2}(\mu y'') = 0 \;\; or \;\; \frac{d^2}{dx^2}(\mu y'') = -\rho$$

This on integration twice with respect to x gives

$$\mu y'' = -\rho \frac{x^2}{2} + C_1 x + C_2.$$

Again integrating with respect to x, we get

$$\mu y' = -\frac{\rho x^3}{6} + C_1 \frac{x^2}{2} + C_2 x + C_3 \quad ...(1)$$

This on integration again gives

$$\mu y = -\frac{\rho x^4}{24} + C_1 \frac{x^2}{6} + C_2 \frac{x^2}{2} + C_3 x + C_4. \quad \ldots(2)$$

Using the boundary conditions $y'(-l) = 0,\ y'(l) = 0$, we get

$$0 = \mu y'(-l) = \frac{\rho l^3}{6} + C_1 \frac{l^2}{2} - C_2 l + C_3 \Rightarrow 3l^2 C_1 - 6C_2 l + 6C_3 + \rho l^3 = 0 \ldots(3)$$

and $$0 = \mu y'(l) = -\frac{\rho l^3}{6} + C_1 \frac{l^2}{2} + C_2 l + C_3$$

$$\Rightarrow 3l^3 C_1 + 6lC_2 + 6C_3 - \rho l^3 = 0 \quad \ldots(4)$$

Now boundary conditions $y(-l) = 0,\ \dot{y}(l) = 0$ and (2) give

$$0 = \mu y(-l) = -\frac{\rho l^4}{24} - C_1 \frac{l^3}{6} + C_2 \frac{l^2}{2} - C_3 l + C_4$$

$$\Rightarrow 4l^3 C_1 - 12l^2 C_2 + 24lC_3 - 24C_4 + \rho l^4 = 0 \quad \ldots(5)$$

and $$0 = \mu y(l) = -\frac{\rho l^4}{24} + C_1 \frac{l^3}{6} + C_2 \frac{l^2}{2} + C_3 l + C_4$$

$$\Rightarrow 4l^3 C_1 + 12l^2 C_2 + 24lC_3 = 24C_4 - \rho l^4 = 0 \quad \ldots(6)$$

Solving (3), (4), (5) and (6), we get $C_1 = 0,\ C_2 = \frac{\rho l^2}{12},\ C_3 = 0,\ C_4 = -\frac{\rho l^4}{24}$

Substituting these values in (2), the required extremals are given by

$$y = -\frac{\rho}{24\mu}\left(x^4 - 2l^2 x^2 + l^4\right).$$

Example 8: Find the extremals of the functional

$I[y] = \int_0^\pi \left(16y^2 - y''^2 + x^2\right)dx$ out of the functions of class C^2 satisfying the following boundary conditions

$$y(0) = 0 = y(\pi),\ y'(0) = 1 = y'(\pi).$$

Solution: The given integral $F = 16y^2 - y''^2 + x^2$ ...(1)

is a functional dependent on second order derivatives. So the Euler's

equation $\frac{\partial F}{\partial y} - \frac{d}{dx}\left(\frac{\partial F}{\partial y'}\right) + \frac{d^2}{dx^2}\left(\frac{\partial F}{\partial y''}\right) = 0$ gives

$$32y + \frac{d^2}{dx^2}(-2y'') = 0 \Rightarrow \frac{d^4 y}{dx^4} - 16y = 0 \quad \ldots(2)$$

A.E. is $m^4 - 16 + 0 \Rightarrow m = \pm 2, \pm 2i$.

Hence, the general solution is

$$y = c_1 e^{2x} + c_2 e^{-2x} + c_3 \cos 2x + c_4 \sin 2x \quad \ldots(3)$$

and $\quad y' = 2c_1 e^{2x} - 2c_2 e^{-2x} - 2c_3 \sin 2x + 2c_4 \cos 2x.$ ...(4)

Applying boundary conditions, we obtain

$$y(0) = 0 \Rightarrow c_1 + c_2 + c_3 = 0 \quad ...(5)$$

$$y(\pi) = 0 \Rightarrow c_1 e^{2\pi} + c_2 e^{-2\pi} + c_3 = 0 \quad ...(6)$$

$$y'(0) = 1 \Rightarrow 2c_1 - 2c_2 + 2c_4 = 1 \quad ...(7)$$

$$y'(\pi) = 1 \Rightarrow 2c_1 e^{2\pi} - 2c_2 e^{-2\pi} + 2c_4 = 1. \quad ...(8)$$

On solving (5), (6), (7) and (8), we get

$$c_1 = c_2 = c_3 = 0,\ c_4 = \frac{1}{2}.$$

Hence the required extremals are $y = \frac{1}{2}\sin 2x.$

Example 9: Find the surface of a minimum area, stretch over a given closed space curve C, enclosing the domain D in the xy-plane.

Solution: Here the problem is to determine the extremals of the functional

$$S[z(x,y)] = \iint_D \left(1 + z_x^2 + z_y^2\right)^{1/2} dx\, dy.$$

In this case, Euler-Ostrogradsky equation $F_z - \frac{\partial}{\partial x} F_{z_x} - \frac{\partial}{\partial y} F_{z_y} = 0$ gives

$$0 - \frac{\partial}{\partial x}\left[\frac{1}{2}\left(1 + z_x^2 + z_y^2\right)^{-1/2} 2z_x\right] - \frac{\partial}{\partial y}\left[\frac{1}{2}\left(1 + z_x^2 + z_y^2\right)^{-1/2} 2z_y\right] = 0$$

or $$\frac{\partial}{\partial x}\left[z_x\left(1 + z_x^2 + z_y^2\right)^{-1/2}\right] + \frac{\partial}{\partial y}\left[z_y\left(1 + z_x^2 + z_y^2\right)^{-1/2}\right] = 0$$

or $$z_x\left\{-\frac{1}{2}\left(1 + z_x^2 + z_y^2\right)^{-3/2} 2\left(z_x z_{xx} + z_y z_{xy}\right)\right\} + \left(1 + z_x^2 + z_y^2\right)^{-1/2} z_{xx}$$

$$+ z_y\left\{-\frac{1}{2}\left(1 + x_x^2 + z_y^2\right)^{-3/2} 2\left(z_x z_{yx} + z_y z_{yy}\right)\right\} + \left(1 + z_x^2 + z_y^2\right)^{-1/2} z_{yy} = 0$$

or $$-z_x^2 z_{xx} - z_x z_y z_{xy} + \left(1 + z_x^2 + z_y^2\right) z_{xx} - z_x z_y z_{yx} - z_y^2 z_{yy} + \left(1 + z_x^2 + z_y^2\right) z_{yy} = 0$$

or $$z_{xx}\left(1 + z_y^2\right) - 2z_x z_y z_{xy} + z_{yy}\left(1 + z_x^2\right) = 0$$

which is the required equation of minimal surfaces.

Example 10: Find the Euler-Ostrogradsky equation for the functional

$$I[u(x,y)] = \iint_D \left[\left(\frac{\partial u}{\partial x}\right)^2 + \left(\frac{\partial u}{\partial y}\right)^2\right] dx\, dy$$

where the values of u are prescribed on the boundary Γ of the domain D.

Solution: The Euler-Ostrogradsky equation $F_u - \frac{\partial}{\partial x} F_{u_x} - \frac{\partial}{\partial y} F_{u_y} = 0$ gives

$$0 - \frac{\partial}{\partial x}\left(2\frac{\partial u}{\partial x}\right) - \frac{\partial}{\partial y}\left(2\frac{\partial u}{\partial y}\right) = 0 \text{ i.e., } \nabla^2 u \equiv \frac{\partial^2 u}{\partial x^2} + \frac{\partial^2 u}{\partial y^2} = 0 \quad \ldots(1)$$

which is well known Laplace's equation. Thus we see that the variational problem is analogous to the solution of (1) subject to given value of u on the boundary Γ of the domain D. This problem is known as Dirichlet problem in mathematical physics.

Example 11: Find the extremizing function for the functional

$$J[z(x,y)] = \iint_D \left[\left(\frac{\partial^2 z}{\partial x^2}\right)^2 + \left(\frac{\partial^2 z}{\partial y^2}\right)^2 + 2\left(\frac{\partial^2 z}{\partial x \partial y}\right)^2 - 2z f(x,y)\right] dx\, dy$$

where $f(x,y)$ is a known function.

Solution: In this case, the extremizing function $z(x,y)$ satisfies the modified Euler-Ostrogradsky equation

$$F_z - \frac{\partial}{\partial x} F_{z_x} - \frac{\partial}{\partial y} F_{z_y} + \frac{\partial^2}{\partial x^2} F_{z_{xy}} + \frac{\partial^2}{\partial x \partial y} F_{z_{xy}} + \frac{\partial^2}{\partial y^2} F_{z_{yy}} = 0$$

or

$$-2f(x,y) - \frac{\partial}{\partial x}(0) - \frac{\partial}{\partial y}(0) + \frac{\partial^2}{\partial x^2}\left(2\frac{\partial^2 z}{\partial x^2}\right) + \frac{\partial^2}{\partial x \partial y}\left(4\frac{\partial^2 z}{\partial x \partial y}\right) + \frac{\partial^2}{\partial y^2}\left(2\frac{\partial^2 z}{\partial y^2}\right) = 0$$

or $$\frac{\partial^2 z}{\partial x^4} + 2\frac{\partial^4 z}{\partial x^2 \partial y^2} + \frac{\partial^4 z}{\partial y^4} = f(x,y),$$

which arises in the problem of deflection of a clamped plate in the theory of elasticity.

Example 12: Find the Euler-Ostrogradsky equation for the functional

$$I[u(x,y,z)] = \iiint_D \left[\left(\frac{\partial u}{\partial x}\right)^2 + \left(\frac{\partial u}{\partial y}\right)^2 + \left(\frac{\partial u}{\partial z}\right)^2 + 2uf\right] dx\, dy\, dz.$$

Solution: We know that for the extremals of

$$I[u(x,y,z)] = \iiint_D F(x,y,z,u,u_x,u_y,u_z)\, dx\, dy\, dz.$$

The extremizing function $u(x,y,z)$ satisfies the Euler-Ostrogradsky equation

$$F_u - \frac{\partial}{\partial x} F_{u_x} - \frac{\partial}{\partial y} F_{u_y} - \frac{\partial F}{\partial z} F_{u_z} = 0. \quad \ldots(1)$$

In the given problem

$$F=\left(\frac{\partial u}{\partial x}\right)^2+\left(\frac{\partial u}{\partial y}\right)^2+\left(\frac{\partial u}{\partial z}\right)^2+2uf=(u_x)^2+(u_y)^2+(u_z)^2+2uf.$$

Therefore, $F_u=2f,\ F_{u_x}=2u_x,\ F_{u_y}=2u_y,\ F_{u_z}=2u_z$.

Using these in Euler-Ostrogradsky equation (1), we obtain

$$2f-\frac{\partial}{\partial x}(2u_x)-\frac{\partial}{\partial y}(2u_y)-\frac{\partial}{\partial z}(2u_z)=0$$

$$\Rightarrow\quad u_{xx}+u_{yy}+u_{zz}=f$$

which is the required Euler-Ostrogradsky equation for the given functional.

EXERCISE 11.3

1. Find the geodesics of the right circular cylinder.
   [Hint. Here the parametric curve on the surface of the cylinder is $\bar{r}=(a\cos\theta, a\sin\theta, z)$, so on calculation, we get $I=\int_{\theta_1}^{\theta_2}\sqrt{a^2+z'}\,d\theta$, where $z'=\frac{dz}{d\theta}$]
2. Find the extremals of the following functionals.

i. $I=\int_{1/2}^{1}(y'^2-2xyz')dx$ with boundary conditions

$y\left(\frac{1}{2}\right)=2,\ z\left(\frac{1}{2}\right)=15,\ y(1)=1=z(1)$.

ii. $I=\int_1^2(z'^2-2xy'z)dx$, with boundary conditions

$y(1)=1=z(1),\ y(2)=-\frac{1}{6},\ z(2)=\frac{1}{2}$.

iii. $I=\int_0^1(y''^2+2y'^2+y^2)\,dx$, under the boundary conditions

$y(0)=0=y(1),\ y'(0)=1,\ y'(1)=-\sin 1$.

iv. $I=\frac{1}{2}\int_a^b y''^2\,dx$ under the boundary conditions

$y(a)=y(b)=0,\ y'(a)=y'(b)=0$.

3. Among the functions of class $C^3[a,b]$, find the one which can provide an extremum for the functional

$$I[y(x)]=\int_a^b(2xy+y'''^2)dx.$$

4. Find the Euler-Ostrogradsky equation for the functional

$$I=\iint_D\left[\left(\frac{\partial z}{\partial x}\right)^2-\left(\frac{\partial z}{\partial y}\right)^2\right]dx\,dy.$$

5. Prove that the Euler-Ostrogradsky equation of the functional

$$I[z(x,y)] = \iint_D [p^2 + q^2 + 2z f(x,y)]\, dx\, dy$$

is the Poisson equation $\nabla^2 z = f(x,y)$, where $\nabla^2 \equiv \frac{\partial^2}{\partial x^2} + \frac{\partial^2}{\partial y^2}$.

6. Find geodesic on the right circular cone.

7. Find the extremals of the following functional.

i. $I = \int_{-1}^{1} \left[\frac{1}{2} y''' + y\right] dx,\ y(\pm 1) = 0,\ y'(\pm 1) = 0.$

ii. $I = \int_0^{\pi/2} \left[y''^2 - y^2 + x^2\right] dx$ with $y(0) = 1 = -y\left(\frac{\pi}{2}\right), y'(0) = 0 = y'\left(\frac{\pi}{2}\right)$

iii. $I = \int_a^b \left[\frac{1}{2} myy'' + \frac{1}{2} ky^2\right] dx.$ Also interpret it.

ANSWERS 11.3

1. $z = c_1\theta + c_2$ a two parameter family of helical lines.

2. (i) $y = \frac{1}{x},\ z = \frac{2}{x^3} - 1.$ (ii) $y = \frac{4}{3x^3} - \frac{1}{3},\ z = \frac{1}{x}.$

(iii) $y = (1-x)\sinh x.$ (iv) $y(x) = 0.$

3. $y = \frac{x^7}{7!} + c_1 x^5 + c_2 x^4 + c_3 x^3 + c_4 x^2 + c_5 x + c_6.$

4. $\frac{\partial^2 z}{\partial x^2} - \frac{\partial^2 z}{\partial y^2} = 0.$

6. $r \sin\alpha = c \sec(\theta \sin\alpha + \beta)$, where α, β are constants.

7. (i) $y = -\frac{1}{24}(x^4 - 2x^2 + 1)$ (ii) $y = \cos x$

(iii) $y = c_1 \cos\omega x + c_2 \sin\omega x$, where $\omega = \sqrt{\frac{k}{m}}$.

11.11 VARIATIONAL PROBLEMS IN PARAMETRIC FORM

In many variational problems, it is more convenient and sometimes it is essential to make use of a parametric representation of a curve in the following form

$$x = \phi(t),\ y = \psi(t) \text{ for } t_0 \le t \le t_1 \qquad \ldots(1)$$

Let us consider the functional

$$I[x(t), y(t)] = \int_{t_0}^{t_1} F(t, x, y, \dot{x}, \dot{y})\, dt \qquad \ldots(2)$$

where $x(t), y(t) \in C^1[t_0, t_1]$, F is continuously differentiable function of the arguments, the integration is along the line (1) and a dot denotes derivative with respect to t.

In order that the values of the functional (2) depend only on the line, and not on the parameterization, it is both necessary and sufficient that the integrand in (2) does not contain t explicitly and that it is homogeneous of the first degree in x and y.

Thus $F(x, y, k\dot{x}, k\dot{y}) = kF(x, y, \dot{x}, \dot{y}),\ k > 0$ ...(3)

For example, take $I[x(t), y(t)] = \int_{t_0}^{t_1} \phi(x(t), y(t), \dot{x}(t), \dot{y}(t))\, dt$

where ϕ satisfies the homogeneity condition (3).

Now suppose that $\Phi(\alpha, \beta) = I[x(t) + \alpha\delta x, y(t) + \beta\delta y]$

then for an extremum, we have

$$\left[\frac{\partial\Phi}{\partial\alpha}\right]_{\alpha=\beta=0} = 0 \text{ and } \left[\frac{\partial\Phi}{\partial\beta}\right]_{\alpha=\beta=0} = 0 \qquad \text{...(4)}$$

These equations lead to the Euler-Lagrange equations

$$\Phi_x - \frac{d}{dt}\Phi_{\dot{x}} = 0,\ \Phi_y - \frac{d}{dt}\Phi_{\dot{y}} = 0. \qquad \text{...(5)}$$

Thus to find extremals for I, one has to solve Euler-Lagrange equations (5). However, these equations are not independent, because these must be satisfied by a certain solution $x = x(t),\ y = y(t)$, and for any other pairs of functions with a different parametric representation of the same curve, which, in the case of Euler-Lagrange equations being independent, would conflict with the theorem of existence and uniqueness of a solution of a system of differential equations. Thus in (5), any one equation is a consequence of the other and to find the extremals, one has to solve any one of the equations (5) along with the equation $\dot{x}^2 + \dot{y}^2 = 1$, which shows that the arc length of the curve is taken as the parameter.

The Weierstrassian form of Euler-Lagrange equations (5) is

$$\frac{1}{r} = \frac{\Phi_{x\dot{y}} - \Phi_{y\dot{x}}}{\Phi_1\left(\dot{x}^2 + \dot{y}^2\right)^{3/2}} \qquad \text{...(6)}$$

where r is the radius of curvature of the extremals, and Φ_1 is the common value of the ratios

$$\Phi_1 = \frac{\Phi_{\dot{x}\dot{x}}}{\dot{y}^2} = \frac{\Phi_{\dot{y}\dot{y}}}{\dot{x}^2} = -\frac{\Phi_{\dot{x}\dot{y}}}{\dot{x}\dot{y}}$$

Example 1: Find the extremals of the functional

$$I[x(t), y(t)] = \int_{t_0}^{t_1}\left[\left(x^2 + y^2\right)^{1/2} + a^2(xy - yx)\right] dt.$$

Solution: Here the integral $\Phi_{xy} = a^2$, $\Phi_{yx} = -a^2, \Phi_1 = \dfrac{1}{\left(x^2+y^2\right)^{3/2}}$

Hence $\dfrac{1}{r} = \dfrac{\Phi_{xy} - \Phi_{yx}}{\Phi_1\left(x^2+y^2\right)^{1/2}}$ *gives* $\dfrac{1}{r} = 2a^2$

which shows that the extremals are circles.

11.12 INVARIANCE OF EULER'S EQUATION UNDER CO-ORDINATE TRANSFORMATION

Suppose that a functional

$$I[y(x)] = \int_a^b F(x,y,y')dx$$

is transformed by replacing the independent variable or by a simultaneous replacement of the desired function and the independent variable. Then the extremals of the functional I are found from Euler's equation for the transformed integrand. This principle is known as the principle of invariance of Euler's equation.

Let the independent variable x and the desired function y be replaced by new variables u and v under the transformation

$$x = x(u, v),\ y = y(u, v).$$

Assume further that the Jacobian of the transformation $J = \dfrac{\partial(x,y)}{\partial(u,v)} = \begin{vmatrix} x_u & x_v \\ y_u & y_v \end{vmatrix} \neq 0$, then u and v can be solved in terms of x and y. Further, we see that

$$y' = \frac{dy}{dx} = \frac{dy/du}{dx/du} = \frac{y_u + y_v v'_u}{x_u + x_v v'_u}$$

and $\dfrac{dx}{du} = x_u + x_v v'_u \Rightarrow dx = (x_u + x_v v'_u)du$

Then

$$\int F(x,y,y')dx = \int F\left[\frac{x(u,v),\ y(u,v),\ y_u + y_v v'_u}{x_u + x_v v'_u}\right](x_u + x_v v'_u)du$$

$$= \int \Phi(u,v,\ v'_u)du.$$

Now the extremals of the original functional are determined from Euler's equation for $\int \Phi(u,v,v'_u)du$ that is, $\dfrac{\partial\Phi}{\partial v} - \dfrac{d}{du}\left(\dfrac{\partial\Phi}{\partial v'}\right) = 0$

Example 2: Verify invariance of Euler's equation under co-ordinate transformation in the problem of finding the extremals of the functional

$$I[y(x)] = \int_0^{\log 2}\left(e^{-x}y'^2 - e^x y^2\right)dx.$$

Solution: Euler's equation of the functional

$$I[y(x)] = \int_0^{\log 2} \left(e^{-x} y'^2 - e^x y^2\right) dx \text{ is given by}$$

$$F_y - \frac{d}{dx} F'_y = 0, \text{ where } F = e^{-x} y'^2 - e^x y^2$$

$$\Rightarrow \quad -2e^x y - \frac{d}{dx}\left(2e^{-x} y'\right) = 0 \Rightarrow -2e^x y + 2e^{-x} y' - 2e^{-x} y'' = 0$$

$$\Rightarrow \quad y'' - y' + e^{2x} y = 0$$

Using transformation $x = \log u$, $y = v$ the functional $I[y(x)]$ reduces to

$$I[v(u)] = \int_1^2 \left(v'^2 - v^2\right) du.$$

Its Euler's equation is given by

$$\frac{\partial \Phi}{\partial v} - \frac{d}{du}\left(\frac{\partial \Phi}{\partial v'}\right) = 0, \text{ where } \Phi(u, v, v') = v'^2 - v^2$$

$$\Rightarrow \quad -2v - \frac{d}{du}\left(\frac{\partial \Phi}{\partial v'}\right) = 0 \Rightarrow -2v - 2v'' = 0 \Rightarrow v'' + v = 0$$

Its solution is

$$v = C_1 \cos u + C_2 \sin u. \qquad \text{...(1)}$$

Further we see that in the original co-ordinates the required solution for extremals is

$$y = C_1 \cos\left(e^x\right) + C_2 \sin\left(e^x\right). \qquad \text{...(2)}$$

Since (1) and (2) are identical it verifies invariance of Euler's equation under co-coordinate transformation.

11.13 VARIATIONAL PROBLEMS WITH CONSTRAINT CONDITIONS

To find the necessary condition for extremum of the functional

$$I = \int_{x_1}^{x_2} F(x, y_1, y_2, y'_1, y'_2) dx$$

subject to $G(y_1, y_2) = 0$.

Solution: We know that the condition for extremum is $\delta I = 0$. This implies that

$$\delta \int_{x_1}^{x_2} F\, dx = 0 \Rightarrow \int_{x_1}^{x_2} \left[\left(\frac{\partial F}{\partial y_1}\delta y_1 + \frac{\partial F}{\partial y'_1}\delta y'_1\right) + \left(\frac{\partial F}{\partial y_2}\delta y_2 + \frac{\partial F}{\partial y'_2}\delta y'_2\right)\right] dx = 0$$

$$\int_{x_1}^{x_2} \left[\left\{\frac{d}{dx}\left(\frac{\partial F}{\partial y'_1}\right) - \frac{\partial F}{\partial y_1}\right\}\delta y_1\right] dx + \int_{x_1}^{x_2} \left[\left\{\frac{d}{dx}\left(\frac{\partial F}{\partial y'_2}\right) - \frac{\partial F}{\partial y_2}\right\}\delta y_2\right] dx = 0. \quad \text{...(1)}$$

From the given constraint $G(y_1, y_2) = 0$, we have

$$\delta G = 0 \Rightarrow \frac{\partial G}{\partial y_1}\delta y_1 + \frac{\partial G}{\partial y_2}\delta y_2 = 0 \qquad \ldots(2)$$

If a Lagrange's multiplier λ be a function of x, then by multiplying (2) with λ and integrating with respect to x, we get

$$-\int_{x_1}^{x_2} \lambda\left[\frac{\partial G}{\partial y_1}\delta y_1 + \frac{\partial G}{\partial y_2}\delta y_2\right] dx = 0 \qquad \ldots(3)$$

Adding (3) and (1), we obtain

$$\int_{x_1}^{x_2}\left[\left\{\frac{d}{dt}\left(\frac{\partial F}{\partial y_1'}\right) - \frac{\partial F}{\partial y_1} - \lambda\frac{\partial G}{\partial y_1}\right\}\delta y_1\right] dx + \int_{x_1}^{x_2}\left[\left\{\frac{d}{dt}\left(\frac{\partial F}{\partial y_1}\right) - \frac{\partial F}{\partial y_2} - \lambda\frac{\partial G}{\partial y_2}\right\}\delta y_2\right] dx = 0$$

which implies that the required conditions are

$$\frac{d}{dt}\left(\frac{\partial F}{\partial y_1'}\right) - \frac{\partial}{\partial y_1}(F + \lambda G) = 0$$

$$\frac{d}{dt}\left(\frac{\partial F}{\partial y_2'}\right) - \frac{\partial (F + \lambda G)}{\partial y_2} = 0.$$

Theorem 9: (Isoperimetric problems) Suppose $I[y] = \int_a^b F(x, y, y')dx$ $\ldots(1)$

is a functional with conditions $y(a) = A,\ y(b) = B$ $\ldots(2)$

and $K[y] = \int_a^b G(x, y, y')dx = 1$ $\ldots(3)$

where $K(y)$ is another functional and $y = y(x)$ is an extremals of $I[y]$. If $y = y(x)$ is not extremals of $K[y]$, then $\exists$ a constant λ such that $y = y(x)$ be an extremals of the functional

$$\int_a^b (F + \lambda G)dx. \qquad \ldots(4)$$

Proof: Choose $x_1,\ x_2 \in [a, b]$ where x_1 is arbitrary and x_2 may satisfy a condition to be stated later on. Now give an increment $\delta_1 y(x) + \delta_2 y(x)$ in $y(x)$, where the curve $\delta_1 y(x)$ is non zero only in the neighbourhood of x_1 and the curve $\delta_2 y(x)$ is non zero only in neighbourhood of x_2. Let

$$y^*(x) = y(x) + \delta_1 y(x) + \delta_2 y(x). \qquad \ldots(5)$$

Now we find the varied curve $y = y^*(x)$ satisfying the condition

$$K[y^*] = K[y] = l. \qquad \ldots(6)$$

Using variational derivatives, the increment

$$\Delta I = I\left[y^*\right] - I\left[y\right] \qquad \ldots(7)$$

of the functional I can be written as

$$\Delta I = \left\{ \frac{\delta F}{\delta y}\bigg|_{x=x_1} + \varepsilon_1 \right\} \int_a^b \left[\delta_1 y(x)\right] dx + \left\{ \frac{\delta F}{\delta y}\bigg|_{x=x_2} + \varepsilon_2 \right\} \int_a^b \left[\delta_2 y(x)\right] dx \qquad \ldots(8)$$

where $\varepsilon_1, \varepsilon_2 \to 0$ as $\int_a^b \left[\delta_1 y(x)\right] dx,\ \int_a^b \left[\delta_2 y(x)\right] dx \to 0.$ ...(9)

Similarly, we can obtain

$$\Delta K = K\left[y^*\right] - K\left[y\right]$$

$$= \left\{ \frac{\delta G}{\delta y}\bigg|_{x=x_1} + \varepsilon_1' \right\} \int_a^b \left[\delta_1 y(x)\right] dx + \left\{ \frac{\delta G}{\delta y}\bigg|_{x=x_1} + \varepsilon_2' \right\} \int_a^b \left[\delta_2 y(x)\right] dx \qquad \ldots(10)$$

where $\varepsilon_1', \varepsilon_2' \to 0$ as $\int_a^b \left[\delta_1 y(x)\right] dx,\ \int_a^b \left[\delta_2 y(x)\right] dx \to 0.$ ...(11)

Now, we choose x_2 such that

$$\frac{\delta G}{\delta y}\bigg|_{x=x_2} \neq 0. \qquad \ldots(12)$$

The point of this type exists because the curve $y = y(x)$ is not extremals of the functional $K\left[y\right]$. The condition on the point x_2 given in (12) is the condition mentioned earlier. Due to this choice of x_2, equations (6) and (10) give

$$\int_a^b \left[\delta_2 y(x)\right] dx = -\left\{ \frac{\left.\frac{\delta G}{\delta y}\right|_{x=x_1}}{\left.\frac{\delta G}{\delta y}\right|_{x=x_2}} \right\} \int_a^b \left[\delta_1 y(x)\right] dx \qquad \ldots(13)$$

where $\varepsilon' \to 0$ as $\int_a^b \left[\delta_1 y(x)\right] dx \to 0$.

Setting $\lambda = -\dfrac{\left.\frac{\delta F}{\delta y}\right|_{x=x_2}}{\left.\frac{\delta G}{\delta y}\right|_{x=x_2}}$ ...(14)

Equations (13), (14) and (8) give

$$\Delta I = \left\{ \frac{\delta F}{\delta y}\bigg|_{x=x_1} + \varepsilon_1 \right\} \int_a^b \left[\delta_1 y(x)\right] dx - \left\{ \frac{\delta F}{\delta y}\bigg|_{x=x_2} + \varepsilon_2 \right\} \left\{ \frac{\left.\frac{\delta G}{\delta y}\right|_{x=x_1}}{\left.\frac{\delta G}{\delta y}\right|_{x=x_2}} + \varepsilon' \right\} \int_a^b \left[\delta_1 y(x)\right] dx$$

$$= \left\{ \frac{\delta F}{\delta y}\bigg|_{x=x_1} + \lambda \frac{\delta G}{\delta y}\bigg|_{x=x_1} \right\} \int_a^b \left[\delta_1 y(x)\right] dx + \varepsilon \int_a^b \left[\delta_1 y(x)\right] dx \qquad \ldots(15)$$

where $\varepsilon \to 0$ *as* $\int_a^b [\delta_1 y(x)]\, dx \to 0.$

The expression in (15) for ΔI explicitly contains variational derivatives evaluated only at $x = x_1$ and the increment is just $\delta_1 y(x)$. The compensating increment $\delta_2 y(x)$ has been taken into account automatically by using the condition $\Delta K = 0$.

Therefore, the first term in the right hand side of (15) is the pı cipal linear part of ΔI. So, the variation δI of the functional I at $x =$ ₁. is given by

$$\delta I = \left\{ \left.\frac{\delta F}{\delta y}\right|_{x=x_1} + \lambda \left.\frac{\delta G}{\delta y}\right|_{x=x_1} \right\} \int_a^b [\delta_1 y(x)] dx. \qquad \text{...(16)}$$

But the necessary condition for an extremum is that $\delta I = 0$

$\Rightarrow \dfrac{\delta F}{\delta y} + \lambda \dfrac{\delta G}{\delta y} = 0$, as $\int_a^b [\delta_1 y(x)] dx$ is non zero and the point x_1 is arbitrary.

$$\Rightarrow \quad \left\{ \frac{\partial F}{\partial y} - \frac{d}{dx}\left(\frac{\partial F}{\partial y'} \right) \right\} + \lambda \left\{ \frac{\partial G}{\partial y} - \frac{d}{dx}\left(\frac{\partial G}{\partial y'} \right) \right\} = 0 \qquad \text{...(17)}$$

This shows that the curve $y = y(x)$ is an extremal of the functional

$I[y] = \int_a^b (F + \lambda G)\, dx$, where λ is determined by (14).

The problems of this type are known as the isoperimetric problems.

Corollary 9.1: (Generalization of above result)

Suppose $I[y_1, y_2, \ldots, y_n] = \int_a^b F[x, y_1, y_2, y_3, \ldots, y_n]\, dx$ ...(18)

is a functional with boundary conditions

$$y_i(a) = a_i,\ y_i(b) = b_i,\quad i = 1, 2, 3, \ldots, n \qquad \text{...(19)}$$

$$\int_a^b G_K(x, y_1, y_2 \ldots, y_n) dx = l_k,\ k = 1, 2, \ldots, m. \qquad \text{...(20)}$$

Then the necessary condition for an extremum is that

$$\frac{\partial}{\partial y}\left(F + \sum_{k=1}^{m} \lambda_k G_k \right) - \frac{d}{dx}\left\{ \frac{\partial}{\partial y_i}\left(F + \sum_{k=1}^{n} \lambda_k G_k \right) \right\} = 0 \qquad \text{...(21)}$$

for $i = 1, 2, 3, \ldots, n$. Here the values of $2n$ arbitrary constants and the values of m parameters $\lambda_1, \lambda_2, \ldots, \lambda_m$ are determined from the boundary conditions (19) and (20). λ_i's are called Lagrange's multipliers.

Example 3: Find the shape of the curve of the given perimeter enclosing maximum area.

Solution: Let the perimeter of the closed curve C is P then

$$P = \int_{x_1}^{x_2} \sqrt{1+y'^2}\, dx. \qquad \ldots(1)$$

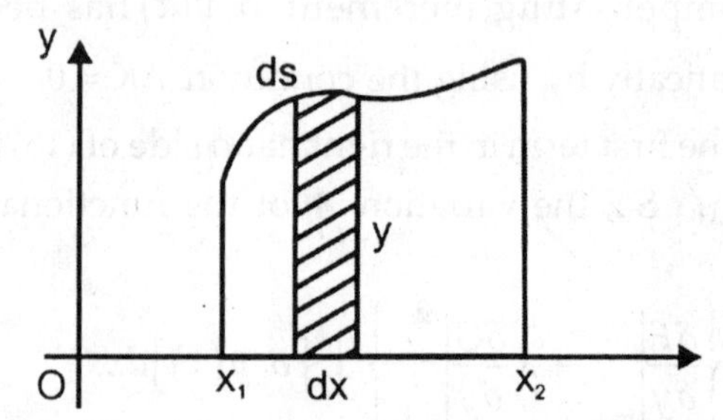

The area A enclosed by the curve, x-axis and two perpendicular lines as shown in the figure is

$$A = \int_{x_1}^{x_2} y\, dx. \qquad \ldots(2)$$

Here we have to find maximum value of (2) under the condition (1) By Lagrange's multiplier method

$$F + \lambda G = y + \lambda\sqrt{1+y'^2}$$

For extremum value of A, the integrated $F + \lambda G$ must satisfy Euler's equation

$$\frac{\partial(F+\lambda G)}{\partial y} - \frac{d}{dx}\left(\frac{\partial(F+\lambda G)}{\partial y'}\right) = 0 \Rightarrow 1 - \lambda\frac{d}{dx}\left[\frac{2y'}{2\sqrt{1+y'^2}}\right] = 0$$

$$\Rightarrow \quad 1 - \lambda\frac{d}{dx}\left[\frac{y'}{\sqrt{1+y'^2}}\right] = 0.$$

On integrating, we obtain

$$x - \lambda\frac{y'}{\sqrt{2+y'^2}} = a \Rightarrow (x-a)^2 = \frac{\lambda^2 y'^2}{1+y'^2}$$

$$\Rightarrow \quad \left[\lambda^2 - (x-a)^2\right]y'^2 = (x-a)^2 \Rightarrow \frac{dy}{dx} = y' = \frac{x-a}{\sqrt{\lambda^2-(x-a)^2}}.$$

Again integrating, we obtain

$$y = -\sqrt{\lambda^2-(x-a)^2} + b \Rightarrow (y-b)^2 = \lambda^2 - (x-a)^2$$

$$\Rightarrow \quad (x-a)^2 + (y-b)^2 = \lambda^2$$

This is an equation of a circle with centre at (a,b) and radius λ.

Example 4: Among all the curves of length l passing through the point $(-a, 0)$ and $(a, 0)$, find the one which encloses maximum area between the curve and x-axis.

Solution: Let $y = g(x)$ be the required curve. The area A enclosed by the curve and x-axis in the interval $[a, -a]$ is given by

$$A = \int_{-a}^{a} y\, dx \qquad \ldots(1)$$

with the conditions

$$y(-a) = 0,\; y(a) = 0 \qquad \ldots(2)$$

and $$\int_{-a}^{a} \sqrt{1+y'^2}\, dx = l. \qquad \ldots(3)$$

Here $F + \lambda G = y + \lambda\sqrt{1+y'^2}$

and the corresponding Euler's equation is

$$\frac{\partial}{\partial y}(F+\lambda G) - \frac{d}{dx}\left(\frac{\partial}{\partial y'}(F+\lambda G)\right) = 0 \Rightarrow 1 - \lambda\frac{d}{dx}\left(\frac{y'}{\sqrt{1+y'^2}}\right) = 0$$

$\Rightarrow \quad (x - C_1)^2 + (y - C_2)^2 = \lambda^2$, as earlier example $\qquad \ldots(4)$

Applying boundary conditions, we obtain

$$y(-a) = 0 \Rightarrow (a + c_1)^2 + c_2^2 = \lambda^2 \qquad \ldots(5)$$

and $$y(a) = 0 \Rightarrow (a - c_1)^2 + c_2^2 = \lambda^2. \qquad \ldots(6)$$

Solving (5) and (6), we get $C_1 = 0,\, C_2 = \sqrt{\lambda^2 - a^2}$

Hence (4) becomes $x^2 + \left(y + \sqrt{\lambda^2 - a^2}\right)^2 = \lambda^2$

This implies $y = \sqrt{\lambda^2 - x^2} - \sqrt{\lambda^2 - a^2} \qquad \ldots(7)$

and $y' = -\dfrac{x}{\sqrt{\lambda^2 - x^2}}$

Now condition (3) gives

$$l = \int_{-a}^{a}\sqrt{1 + \frac{x^2}{\lambda^2 - x^2}}\, dx = \int_{-a}^{a}\frac{\lambda}{\sqrt{\lambda^2 - x^2}}\, dx = 2\lambda \sin^{-1} a/\lambda$$

$$\Rightarrow \quad a = \lambda \sin(l/2\lambda)$$

which is a transcendental equation for λ. Solving it we find certain value $\lambda = \lambda_0$ (say). Thus solution curve becomes

$$y = \sqrt{\lambda_0^{\,2} - x^2} - \sqrt{\lambda_0^{\,2} - a^2}$$

11.14 STURM-LIOUVILLE PROBLEMS

Theorem 10: Suppose $I = \int_a^b (py'^2 - qy^2)\,dx,$ ...(1)

where p, q are functions of the independent variable x is a function subject to the constraint

$$J = \int_a^b ry^2\,dx. \qquad ...(2)$$

Then the extremals of the functional I are the solutions of the Sturm-Liouville equation

$$\frac{d}{dx}[py'] + (q + \lambda r)y = 0 \qquad ...(3)$$

where λ is the Lagrange's multiplier defined by $\lambda = \frac{I}{J}$. ...(4)

and r is function of x.

Proof: If stationary values of the quantity λ are defined by the ratio

$$\lambda = \frac{I}{J} = \frac{\int_a^b (py'^2 - qy^2)\,dx}{\int_a^b ry^2\,dx} \qquad ...(5)$$

The variation in λ is given by

$$\delta\lambda = \frac{J\,\delta I - I\,\delta J}{J^2} = \frac{(\delta I - \lambda\,\delta J)}{J} \qquad ...(6)$$

where

$$\delta I = \left[\frac{\partial F}{\partial y'}\delta y\right]_a^b + \int_a^b \left[\frac{\partial F}{\partial y} - \frac{d}{dx}\frac{\partial F}{\partial y'}\right]\delta y\,dx$$

$$= 2[py'\,\delta y]_a^b - 2\int_a^b \left[\frac{d}{dx}(py') + qy\right]\delta y\,dx \qquad ...(7a)$$

and $\delta J = -2\int_a^b 2ry\,\delta y\,dx.$ ...(7b)

Hence from equation (6), we have

$$\delta\lambda = \frac{2[py'\delta y]_a^b - 2\int_a^b \left[\frac{d}{dx}(py') + qy + \lambda ry\right]\delta y\,dx}{\int_a^b ry^2\,dx}$$

For extremals of I, the condition $\delta I = 0$ leads to $\delta\lambda = 0$.

This gives the relevant Euler equation as

$$\frac{d}{dx}\left(p\frac{dy}{dx}\right) + (q + \lambda r)\,y = 0 \qquad ...(8)$$

and the natural boundary conditions, which require that py' vanishes at an end where y is not pre-assigned:

$y(a)$ prescribed or $[py']_{x=a} = 0$, $y(b)$ prescribed or $[py']_{x=b} = 0$ ...(9)

If the boundary conditions are homogeneous and are of the form

$$y(a) = 0 \text{ or } y'(a) = 0, \; y(b) = 0 \text{ or } y'(b) = 0 \qquad \text{...(10)}$$

Then problem is to find the characteristics values.

The equation (8) is the well known Sturm-Liouville equation, so the extremals of I are the solution of the Sturm-Liouville equation.

If we arbitrarily impose the constraint $J = \int_a^b ry^2\,dx = 1$ ...(11)

in homogeneous cases, then from (5) we follow

$$\delta\lambda = \delta \int_a^b \left(py'^2 - qy^2\right) dx = 0 \qquad \text{...(12)}$$

where y have to satisfy (11) and the relevant end conditions. Again from (6) we follow

$$\delta(I - \lambda J) = 0 \qquad \text{...(13)}$$

with the provision $y \neq 0$, is equivalent to (6). This implies that

$$\frac{d}{dx}(py') + \{(q + \lambda r)y\} = 0.$$

This is the Strum-Liouville equation.

In (13), the constant λ plays the role of a Lagrange multiplier and is to be determined together with y-axis that $I - \lambda J$ is stationary and $y(x) \neq 0$. The condition (11) is known as a normalizing condition and the function $r(x)$, the weight function.

11.15 NATURAL BOUNDARY CONDITIONS

Suppose the functional $I[y] = \int_a^b F(x, y, y')dx$ is to be optimized. The two values $y(a)$ and $y(b)$ at the points may or may not be prescribed. If the values of the unknown $y(x)$ are not pre-assigned at one or both the end points $x = a$ and $x = b$, then we use additional boundary conditions, called the natural boundary conditions. These boundary conditions are to be determined as:

For the extremals of I, the condition $\delta I = 0$ leads

$$\int_a^b \left[\frac{\partial F}{\partial y}\delta y + \frac{\partial F}{\partial y'}\delta y'\right] dx = 0 \Rightarrow \left[\frac{\partial F}{\partial y'}\delta y\right]_a^b + \int_a^b \left(\frac{\partial F}{\partial y} - \frac{d}{dx}\frac{\partial F}{\partial y'}\right)\delta y\,dx = 0$$

which gives Euler's equation

$$\frac{\partial F}{\partial y}-\frac{d}{dx}\left(\frac{\partial F}{\partial y'}\right)=0$$

and the natural boundary conditions

$$\left.\frac{\partial F}{\partial y'}\right|_{x=a}=0 \text{ if } y(a) \text{ is not pre-assigned}$$

$$\left.\frac{\partial F}{\partial y'}\right|_{x=b}=0 \text{ if } y(b) \text{ is not pre-assigned.}$$

Example 5: Find the extremals of $\int_0^1 \left(y'^2-y^2\right)dx$ and the boundary conditions.

i. $y(0)=0,\ y(1)=1$

ii. no boundary condition is given

iii. $y(1)=1$.

Solution: The given functional is $I=\int_0^1 \left(y'^2-y^2\right)dx$.

Here $F=y'^2-y^2$, so the corresponding Euler's equation is

$$\frac{\partial F}{\partial y}-\frac{d}{dx}\left(\frac{\partial F}{\partial y'}\right)=0 \Rightarrow -2y-2y''=0 \qquad \Rightarrow y''+y=0 \qquad \dots(1)$$

The solution of (1) is given by $y=a\cos x+b\sin x$. ...(2)

(i) The boundary conditions $y(0)=0,\ y(1)=1$ imply that a = 0,

$b=\dfrac{1}{\sin 1}$

Hence the extremals are $y=\dfrac{\sin x}{\sin 1}$.

(ii) Here the boundary conditions are not given, so we use the natural boundary conditions

$$y'(0)=0,\ y'(1)=0.$$

Applying these in (2), we obtain $b=0,\ a=0$.

So the required extremals are $y=0$.

(iii) Here, the boundary condition is given at $y(1)=1$, so at $x=0$, we use the natural boundary conditions, which gives

$$\left.\frac{\partial F}{\partial y'}\right|_{x=0} = 0 \Rightarrow y'(0) = 0.$$

Applying these boundary conditions in (2), we obtain $b=0,\ a=\dfrac{1}{\cos 1}$

Hence the required extremals are $y = \dfrac{\cos x}{\cos 1}$.

Example 6: If the functional $I = \int_a^b F(x,y,y')dx + \alpha\, y(a) - \beta\, y(b)$

where α, β are constants and $y(a)$, $y(b)$ are not pre-assigned, then find Euler's equation and the natural boundary conditions.

Solution: For extremals $\delta I = 0$ gives

$$\delta\left[\int_a^b F(x,y,y')dx + \alpha\, y(a) - \beta\, y(b)\right] = 0$$

$$\Rightarrow \quad \int_a^b \left[\frac{\partial F}{\partial y}\delta y + \frac{\partial F}{\partial y'}\delta y'\right] dx + \alpha\,\delta y(a) - \beta\,\delta y(b) = 0$$

$$\Rightarrow \quad \int_a^b \left[\frac{\partial F}{\partial y}\delta y + \frac{\partial F}{\partial y'}\frac{d}{dx}(\delta y)\right] dx + \alpha\,\delta y(a) - \beta\,\delta y(b) = 0,$$

$$\left[\text{since } \frac{d}{dx} \text{ commutes with } \delta\right]$$

$$\Rightarrow \quad \int_a^b \left[\frac{\partial F}{\partial y} - \frac{d}{dx}\left(\frac{\partial F}{\partial y'}\right)\right]\delta y\, dx + \left(\frac{\partial F}{\partial y'}\delta y\right)_{x=a}^{b} + \alpha\,\delta y(a) - \beta\,\delta y(b) = 0$$

$$\Rightarrow \quad \int_a^b \left(\frac{\partial F}{\partial y} - \frac{d}{dx}\frac{\partial F}{\partial y'}\right)\delta y\, dx + \left(\frac{\partial F}{\partial y'} - \beta\right)\left.\delta y\right|_{x=b} - \left(\frac{\partial F}{\partial y'} - \alpha\right)\left.\delta y\right|_{x=a} = 0$$

This leads to the Euler's equation $\dfrac{\partial F}{\partial y} - \dfrac{d}{dx}\left(\dfrac{\partial F}{\partial y'}\right) = 0$

and the natural boundary conditions

$$\left.\frac{\partial F}{\partial y'}\right|_{x=b} = \beta, \qquad \left.\frac{\partial F}{\partial y'}\right|_{x=a} = \alpha.$$

Example 7: Find the solid of maximum volume formed by the revolution of a given surface area.

Solution: Suppose the curve PA passes through origin and is rotated about the x-axis. Then the given surface area

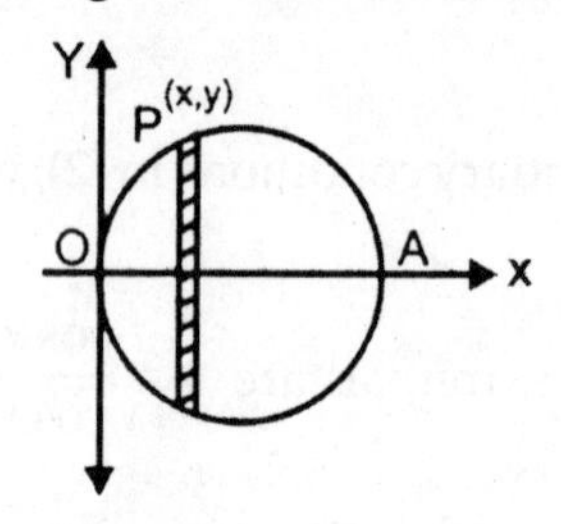

$$S = \int_0^a 2\pi \; y \, ds \;\; S = \int_0^a 2\pi \, y \sqrt{1+y'^2} \, dx$$

and the volume is $V = \int_0^a \pi y^2 \, dx$.

We have to find extremum value of V with the given S.

Here $\quad f = \pi y^2, \; g = 2\pi y \sqrt{(1+y'^2)}$, so $F = f + \lambda g$ gives

$$F = \pi y^2 + \lambda 2\pi y \sqrt{(1+y'^2)}.$$

Form maximum V, F must satisfy Euler's equations. But F does not contain x, we have

$$F - y'\frac{\partial F}{\partial y'} = C \Rightarrow \pi y^2 + \lambda 2\pi y \sqrt{1+y'^2} - y'\frac{1}{2}\frac{2\pi y \; \lambda 2y'}{\sqrt{(1+y'^2)}} = C$$

$$\Rightarrow \quad \pi y^2 + 2\pi y \lambda \sqrt{(1+y'^2)} - \frac{2\pi \lambda y y'^2}{\sqrt{(1+y'^2)}} = C$$

$$\Rightarrow \quad \pi y^2 + \frac{2\pi y \lambda (1+y'^2) - 2\pi y \lambda y'^2}{\sqrt{1+y'^2}} = C$$

$$\Rightarrow \quad \pi y^2 + \frac{2\pi y \lambda}{\sqrt{(1+y'^2)}} = C.$$

As the curve passes through origin $(0, 0)$, so $C = 0$, implies that

$$\pi y^2 + \frac{2\pi y \lambda}{\sqrt{(1+y'^2)}} = 0$$

$$\Rightarrow \quad y + \frac{2\lambda}{\sqrt{(1+y'^2)}} = 0 \Rightarrow y\sqrt{(1+y'^2)} = -2\lambda$$

$$\Rightarrow \quad 1 + y'^2 = \frac{4\lambda^2}{y^2} \Rightarrow y'^2 = \frac{4\lambda^2}{y^2} - 1 = \frac{4\lambda^2 - y^2}{y^2}$$

$$\Rightarrow \quad \frac{dy}{dx} = \frac{\sqrt{4\lambda^2 - y^2}}{y}$$

On integrating, we obtain

$$\int \frac{y\,dy}{\sqrt{(4\lambda^2 - y^2)}} = \int dx + C \Rightarrow -\sqrt{4\lambda^2 - y^2} = x + C$$

$$\Rightarrow \quad \sqrt{4\lambda^2 - y^2} = -x - C. \qquad \ldots(1)$$

The curve passes through $(0,0)$. On putting $x = 0, y = 0$ in (1), we get $\quad -C = 2\lambda.$

Hence equation (1) becomes $\sqrt{4\lambda^2 - y^2} = -x + 2\lambda.$

Squaring it $4\lambda^2 - y^2 = (x - 2\lambda)^2$ *or* $(x - 2\lambda)^2 + y^2 = 4\lambda^2.$

This is the equation of a circle. Hence on revolving the circle about x-axis, the solid of maximum volume is a sphere.

Example 8: Find the extremals of the functional $A = \int_{t_1}^{t_2} \frac{1}{2}(x\dot{y} - y\dot{x})dt$

subject to the integral constant $\int_{t_1}^{t_2} \sqrt{\dot{x}^2 + \dot{y}^2}\, dt = l.$

Solution: Here $f = \frac{1}{2}(x\dot{y} - \dot{x}y)$, $g = \sqrt{\dot{x}^2 + \dot{y}^2}$, hence the integrand is

$$F = f + \lambda g = \frac{1}{2}(x\dot{y} - y\dot{x}) + \lambda\sqrt{\dot{x}^2 + \dot{y}^2}. \qquad \ldots(1)$$

For extremum value of A, F must satisfy the Euler's equations

$$\frac{\partial F}{\partial x} - \frac{d}{dt}\left[\frac{\partial F}{\partial \dot{x}}\right] = 0. \qquad \ldots(2)$$

$$\frac{\partial F}{\partial y} - \frac{d}{dt}\left[\frac{\partial F}{\partial \dot{y}}\right] = 0. \qquad \ldots(3)$$

From (2), we obtain

$$\frac{1}{2}\dot{y} - \frac{d}{dt}\left[-\frac{y}{2} + \frac{\lambda 2\dot{x}}{2\sqrt{\dot{x}^2 + \dot{y}^2}}\right] = 0$$

$$\Rightarrow \quad \frac{d}{dt}\left[y - \frac{\lambda\dot{x}}{\sqrt{\dot{x}^2 + \dot{y}^2}}\right] = 0 \qquad \ldots(4)$$

From (3), we obtain $-\frac{1}{2}\dot{x} - \frac{d}{dt}\left[\frac{x}{2} + \frac{2\lambda\dot{y}}{2\sqrt{\dot{x}^2 + \dot{y}^2}}\right] = 0$

$$\Rightarrow \quad \frac{d}{dt}\left[x - \frac{\lambda \dot{y}}{\sqrt{\dot{x}^2 - \dot{y}^2}}\right] = 0. \qquad \text{...(5)}$$

Integrating (4) and (5), we have

$$y - \frac{\lambda \dot{x}}{\sqrt{\dot{x}^2 + \dot{y}^2}} = c_1 \Rightarrow y - c_1 = \frac{\lambda \dot{x}}{\sqrt{\dot{x}^2 + \dot{y}^2}} \qquad \text{...(6)}$$

$$x - \frac{\lambda \dot{y}}{\sqrt{\dot{x}^2 + \dot{y}^2}} = c_2 \Rightarrow x - c_2 = \frac{\lambda \dot{y}}{\sqrt{\dot{x}^2 + \dot{y}^2}} \qquad \text{...(7)}$$

Squaring (6), (7) and adding, we get

$$(x - c_2)^2 + (y - c_1)^2 = \lambda^2 \left(\frac{\dot{x}^2 + \dot{y}^2}{\dot{x}^2 + \dot{y}^2} \right)$$

$$\Rightarrow (x - c_2)^2 + (y - c_1)^2 = \lambda^2.$$

This is the equation of a circle.

EXERCISE 11.4

1. Find the extremals of the functional $\int_{t_1}^{t_2} \sqrt{\dot{x}^2 + \dot{y}^2 + \dot{z}^2}\, dt$ subject to $x^2 + y^2 + z^2 = a^2$.
2. Find the extremals of the isoperimetric problem $\int_{x_0}^{x_1} y'^2\, dx$ subject to $\int_{x_0}^{x_1} y\, dx = c$.
3. Find extremals of $\int_0^{\pi} y''^2\, dx$ under the conditions $\int_0^{\pi} y^2\, dx = 1$, $y(0) = 0$, $y''(0) = 0$, $y(\pi) = 0$, $y''(\pi) = 0$.
4. Show that an isosceles triangle has the smallest perimeter for a given area and a given base.
5. Find the surface with the smallest area which encloses a given volume.
6. Find the extremals in the isoperimetric problems of the extremum of $\int_0^1 \left(y'^2 + z'^2 - 4xz' - 4z\right) dx$ subject to $\int_0^1 \left(y'^2 - xy' - z'^2\right) dx = 2$, $y(0) = 0$, $z(0) = 0$, $y(1) = 1$, $z(1) = 1$.
7. Obtain the natural boundary conditions for $I = \int_0^1 F[x, y, y']\, dx$, if
   (i) $F = xy'^2 - xy' + y$
   (ii) $F = y'^2 + yy' + y^2$
   (iii) $F = y'^2 + k \cos y$
   (iv) $F = y'^2 - 2\alpha yy' - 2\beta y'$

ANSWERS 11.4

1. Arc of a great circle of a sphere.
2. $4y = \lambda x^2 + 2ax + 2b, \lambda\left(x_1^3 - x_0^3\right) + 3a\left(x_1^2 - x_0^2\right) + 6b\left(x_1 - x_0\right) = 12c.$
3. $y = \sqrt{\left(\frac{2}{\pi}\right)} \sin xx$
5. Sphere
6. $y = \frac{-5x^2}{2} + \frac{7x}{2}, z = x.$
7. (i) $y'(0) = \frac{1}{2}, y'(1) = \frac{1}{2}$
   (ii) $y'(0) = -\frac{y}{2} = y'(1)$
   (iii) $y'(0) = 0 = y'(1)$
   (iv) $y'(0) = \alpha y + \beta = y'(1)$

12 Variational Method of Boundary Value Problems

12.1 INTRODUCTION

The differential equations with boundary conditions are called boundary value problems and their solutions are of immense importance in understanding several physical situations. In this chapter, we first form variational problems from these boundary value problems and then discuss the two important methods namely Rayleigh-Ritz, Galerkin and Kantorovich methods for solving variational problems. These methods can also be used to solve ordinary and partial differential equations.

The procedure to form variational problems from differential equations is as follows:

12.2 VARIATIONAL PROBLEMS FOR DEFORMABLE BODIES

A variational problem can be derived from a differential equation and the associated boundary conditions arise in the theory of elasticity and many other fields.

If we consider a problem to determine the small deflection of a rotating string of length l, the governing differential equation is of the form

$$\frac{d}{dx}\left(F\frac{dy}{dx}\right)+\rho\,\omega^2\,y+p=0 \qquad \ldots(1)$$

where $y(x)$ is the displacement of a point from the axis of rotation, $F(x)$ is the tension, $\rho(x)$ is the linear mass density, ω is the angular velocity of rotation and $p(x)$ is the intensity of a distributed radial load. Suitable boundary conditions are also to be prescribed.

In order to formulate a corresponding variational problem, we first multiply both sides of (1) by a variation δy and integrate the result over the interval $(0, l)$ to obtain

$$\int_0^l \frac{d}{dx}\left(F\frac{dy}{dx}\right)\delta y\,dx+\int_0^l \rho\omega^2 y\,\delta y\,dx+\int_0^l p\,\delta y\,dx=0$$

$$\Rightarrow \quad \left[F\frac{dy}{dx}\delta y\right]_0^l - \int_0^l F\frac{dy}{dx}\frac{d}{dx}(\delta y)dx + \int_0^l \rho\,\omega^2\,\delta\left(\frac{y^2}{2}\right)dx + \int_0^l p\delta y\,dx = 0$$

$$\Rightarrow \quad \int_0^l \delta\left(\frac{1}{2}\rho\,\omega^2\,y^2\right)dx - \int_0^1 \delta\left[\frac{1}{2}F\left(\frac{dy}{dx}\right)^2\right]dx + \int_0^l \delta(py)\,dx + \left[F\left(\frac{dy}{dx}\right)\delta y\right]_0^l = 0$$

$$\Rightarrow \quad \delta\int_0^l\left[\frac{1}{2}\rho\omega^2\,y^2 + py - \frac{1}{2}F\left(\frac{dy}{dx}\right)^2\right]dx + \left[F\left(\frac{dy}{dx}\right)\delta y\right]_0^l = 0. \qquad \ldots(2)$$

If we impose the boundary conditions

$$y(0) = y_0 \text{ or } F\left.\frac{dy}{dx}\right|_{x=0} = 0,\; y(l) = y_l \text{ or } F\left.\frac{dy}{dx}\right|_{x=l} = 0$$

where y_0, y_l are prescribed constants, then the integrated terms in (2) vanish and the equation (2) becomes

$$\delta\int_0^l\left[\frac{1}{2}\rho\,\omega^2\,y^2 + p\,y - \frac{1}{2}F\left(\frac{dy}{dx}\right)^2\right]dx = 0. \qquad \ldots(3)$$

Conversely, (1) is the Euler's equation corresponding to (3) that is, y satisfies (1) and end conditions of the type required in (2), then y provides the integral in (3) stationary.

The boundary conditions in equation (2), that is

$$\left[F\frac{dy}{dx}\delta y\right]_0^l = 0 \qquad \ldots(4)$$

are the so-called natural boundary conditions of the variational problem (3). If we recall that the product $F\frac{dy}{dx}$ is the component of the tensile force normal to the axis of rotation, we see that (4) requires that the end tensions do not work. This situation exists if no end motion is permitted, so that $\delta y = 0$ or if end restraint (normal to the axis of rotation) is present, so that $F\frac{dy}{dx} = 0$.

Clearly the term $\frac{1}{2}\rho(\omega y)^2$ in (3) represents the kinetic energy of the string per unit length, since the speed of an element of the string is given by ωy. Also since $p\,\delta y$ is the work-done by p on the element dx in displacement δy, the term $-p\,y$ is the potential energy per unit length due to radial force distribution $p(x)$. We note that an element of original length dx stretches into an element of length

$$ds = \left[1 + \left(\frac{dy}{dx}\right)^2\right]^{1/2} dx.$$

Then the work per unit length done against the tensile force is

$$F\frac{ds - dx}{dx} = F\left\{\left[1 + \left(\frac{dy}{dx}\right)^2\right]^{1/2} - 1\right\}$$

$$= F\left[1 + \frac{1}{2}\left(\frac{dy}{dx}\right)^2 + \ldots. - 1\right] = \frac{1}{2}F\left(\frac{dy}{dx}\right)^2.$$

if higher powers of slope $\frac{dy}{dx}$ (assumed to be small) are neglected. Thus the term $\frac{1}{2}F\left(\frac{dy}{dx}\right)^2$ represents potential energy per unit length due to the tension in the string.

Case I: *In case of a yielding support at the end $x = 0$, the boundary condition at $x = 0$ would be of the form*

$$\left(F\frac{dy}{dx}\right)_{x=0} = k(y)_{x=0} \qquad \ldots(5)$$

where k is the modulus of the support. Then

$$\left[F\frac{dy}{dx}\delta y\right]_{x=0} = (k\,y\,\delta y)_{x=0} = \delta\left(\frac{1}{2}k\,y^2\right)_{x=0}$$

Since this term would not vanish, (3) would be replaced by

$$\delta\left\{\int_0^l \left[\frac{1}{2}\rho\,\omega^2 y^2 + p\,y - \frac{1}{2}F\left(\frac{dy}{dx}\right)^2\right]dx - \left(\frac{1}{2}k\,y^2\right)_{x=0}\right\} = 0 \qquad \ldots(6)$$

Here the additional term appears corresponding to the energy stored in the support.

Case II: *If the slope of the string at the end $x = 0$ is prescribed as*

$$y'(0) = \alpha \qquad \ldots(7)$$

where α is small, the deflection $y(0)$ then being unknown, there would follow

$$\left(F\frac{dy}{dx}\delta y\right)_{x=0} = (F\,\alpha\,\delta y)_{x=0}$$

Hence (4) becomes

$$\delta\left\{\int_0^l\left[\frac{1}{2}\rho\,\omega^2 y^2+p\,y-\frac{1}{2}F\left(\frac{dy}{dx}\right)^2\right]dx-\alpha\,F(0)y(0)\right\}=0 \qquad \ldots(8)$$

Here the additional term appears corresponding to work done by the component of the tension normal to the x-axis $(F\sin\alpha = F\,\alpha)$ in the end displacement $y(0)$.

12.3 THE RAYLEIGH-RITZ METHOD

The Rayleigh-Ritz method is a procedure for obtaining approximate solutions of problems expressed in variational form. In this method, we determine a function $y(x)$ by assuming that the desired stationary function of a given problem can be approximated by a linear combination of suitably chosen functions as

$$y(x)\approx\phi_0(x)+c_1\,\phi_1(x)+\ldots+c_n\,\phi_n(x) \qquad \ldots(9)$$

where $c_i\,'s(i=1,2,3,\ldots,n)$ are constants to be determined. In general, the functions $\phi_i(x)$ are to be chosen so that this expression satisfies the specified boundary conditions for any choice of $c_i's$.

Thus if y takes on prescribed values at the ends of the interval under consideration, then we require that the function $\phi_0(x)$ takes on the prescribed values at the end points and the each of the remaining functions $\phi_1(x), \phi_2(x), \ldots., \phi_n(x)$ vanish at ends of the interval.

A procedure consists of obtaining a sequence of approximations, in which the first approximation is $\phi_0+c_1\,\phi_1$, the second $\phi_0+c_1\,\phi_1+c_2\,\phi_2$ and so on, the nth assumption being of the form (9). The relevant $c_i's$ are re-determined at each stage of the process as follows.

First the relevant quantity I is expressed as a function of $c_i's$ and then $c_i\,'s$ are so determined that the resultant expression is stationary. Thus in place of using the calculus of variations in attempting to determine that function which renders I stationary with reference to all admissible slightly varied functions, we consider only the family of functions of type (9). Use ordinary differential calculus to seek the member of that family for which I is stationary with reference to slightly modified functions belonging to the same family.

By comparing successive approximations, an estimate of the degree of accuracy attained at any stage of the calculation can be obtained. In order that this process converge as $n \to \infty$, the functions $\phi_0, \phi_1, \phi_2, ..., \phi_n, ...$, should comprise an infinite set of functions such that the unknown function $y(x)$ can with certainty be approximated to any specified degree of accuracy by some linear combination of the ϕ_i's. It is useful to choose polynomials of successively increasing degree, satisfying the specified boundary conditions. In some cases the use of special functions such as sine or cosine harmonics, Bessel functions and so on, may afford computational advantages. The following example explains the procedure.

Example 1: Find variational problem for deformable bodies and its approximate solution subject to the boundary conditions $y(0)=0, y(l)=h$ and subject to a transverse loading (in the y-direction) whose intensity varies linearly, according to the law

$$p(x) = -\frac{q\,x}{l}, \text{ with } w = 0.$$

Solution: First find variational problem (3) as given in article 12.1. Then corresponding variational problem (3) becomes

$$\delta \int_0^1 \left(\frac{1}{2} F\, y'^2 + q \frac{x}{l} y\right) dx = 0 \qquad ...(10)$$

where the integrated terms vanishing by virtue of the boundary conditions

$$y(0)=0, y(l)=h, \qquad ...(11)$$

which require that δy vanishes at the boundary points.

The function ϕ_0 in (9) is to satisfy (11) where as the other co-ordinate functions $\phi_1, \phi_2, ..., \phi_n$, are to vanish when $x=0$ and $x=l$. For convenience, if we use polynomials, then the simplest choices are the functions

$$\phi_0(x) = \frac{h}{l} x, \phi_1(x) = x(x-l), \phi_2(x) = x^2(x-l), ..., \phi_n(x) = x^n(x-l).$$

which correspond to an approximation of the form

$$y(x) \approx \frac{h}{l} x + x(x-l)\left(c_1 + c_2 x + c_3 x^2 + c_3 x^2 + ... + c_n x^{n-1}\right). \qquad ...(12)$$

(I) Consider one-parameter approximation for which $n=1$,

$$y(x) \approx \frac{h}{l} x + c_1 x(x-l) \qquad ...(13)$$

then corresponding to it, (10) becomes

$$\delta \int_0^l \left[\left(\frac{1}{2} F\left\{\frac{h}{l} + c_1(2x-l)\right\}^2\right) + \frac{q\,x}{l}\left(\frac{h\,x}{l} + c_1\left(x^2 - l\,x\right)\right)\right] dx = 0$$

or

$$I=\int_0^1\left[\frac{1}{2}y^2+xy-\frac{1}{2}y'^2\right]dx.$$

On integrating this condition takes the form

$$\delta\left[\frac{1}{2}F\left\{\frac{h^2}{l^2}x+\frac{2hc_1}{l}\left(x^2-lx\right)+c_1^{\,2}\left(\frac{4x^3}{3}-2x^2l+l^2x\right)\right\}+q\left(\frac{hx^3}{3l^2}+\frac{c_1x^4}{4l}-\frac{c_1x^3}{3}\right)\right]_0^l=0$$

or $$\delta\left[\frac{1}{2}F\left\{\frac{h^2}{l}+\frac{c_1^{\,2}}{3}l^3\right\}+q\left(\frac{hl}{3}-\frac{c_1l^3}{12}\right)\right]=0$$

As here only c_1 is varied, we have $\frac{1}{3}Fc_1l^3\,\delta c_1-\frac{1}{12}ql^3\,\delta c_1=0$.

$$\Rightarrow\quad \frac{1}{3}l^3\left(c_1F-\frac{q}{4}\right)\delta c_1=0$$

Since δc_1 is arbitrary, we must follow $c_1=\frac{q}{4F}$,

and hence the desired approximation (13) becomes

$$y(x)\approx\frac{h}{l}x+\frac{q}{4F}x(x-l). \qquad \ldots(14)$$

while the exact solution of this problem is

$$y(x)=\frac{h}{l}x+\frac{q}{6Fl}x\left(x^2-l^2\right) \qquad \ldots(15)$$

which is obtained by elementary methods.

(II) Consider two parameter approximations for which $n=2$

$$y(x)\approx\frac{h}{l}x+c_1x(x-l)+c_2x^2(x-l). \qquad \ldots(16)$$

Then corresponding to it (10) becomes

$$\delta\int_0^l\left[\frac{1}{2}F\left\{\frac{h}{l}+c_1(2x-l)c_2\left(3x^2-2xl\right)\right\}^2+\frac{qx}{l}\left(\frac{h}{l}x+c_1\left(x^2-lx\right)+c_2\left(x^3-x^2l\right)\right)\right]dx=0$$

or

$$\delta\int_0^l\left[\frac{1}{2}F\left\{\frac{h^2}{l^2}+c_1^{\,2}\left(4x^2-4xl+l^2\right)+c_2^{\,2}\left(9x^4-12x^3l+4x^2l^2\right)\right.\right.$$

$$\left.+\frac{2hc_1}{l}(2x-l)+\frac{2hc_2}{l}\left(3x^2-2xl\right)+2c_1c_2\left(6x^3-7x^2l+2xl^2\right)\right\}$$

$$\left.+q\left\{\frac{h}{l^2}x^2+\frac{c_1}{l}\left(x^3-lx^2\right)+\frac{c_2}{l}\left(x^4-x^3\right)\right\}\right]dx=0$$

On integrating this, we obtain

$$\delta\left[\frac{1}{2}F\left\{\frac{h^2}{l}+c_1^{\,2}\left(\frac{4l^3}{3}-2l^3+l^3\right)+c_2^{\,2}\left(\frac{9l^5}{5}-3l^5+\frac{4l^5}{3}\right)+\frac{2hc_1}{l}\left(l^2-l^2\right)\right.\right.$$

$$+\frac{2hc_2}{l}(l^3-l^3)+2c_1c_2\left(\frac{3l^4}{2}-\frac{7l^4}{3}+l^4\right)\Bigg\}$$

$$+q\left\{\frac{1}{3}hl+\frac{c_1}{l}\left(\frac{l^4}{4}-\frac{l^4}{3}\right)+\frac{c_2}{l}\left(\frac{l^5}{5}-\frac{l^5}{4}\right)\right\}\Bigg]=0$$

$$\Rightarrow \delta\left[\frac{1}{2}F\left\{\frac{h^2}{l}+\frac{1}{3}c_1^2 l^3+\frac{2}{15}c_2^2 l^5+\frac{c_1c_2}{3}l^4\right\}+q\left\{\frac{1}{3}hl-\frac{1}{12}c_1 l^3-\frac{1}{20}c_2 l^4\right\}\right]=0.$$

As c_1 and c_2 are varied, we have

$$\frac{1}{3}Fc_1 l^3\,\delta c_1+\frac{2}{15}Fc_2 l^5\,\delta c_2+\frac{c_1}{6}Fl^4\,\delta c_2+\frac{c_2}{6}Fl^4\,\delta c_1-\frac{q}{12}l^3\,\delta c_1-\frac{q}{20}l^4\,\delta c_2=0$$

$$\Rightarrow \left(\frac{1}{3}Fc_1 l^3+\frac{c_2}{6}Fl^4-\frac{q}{12}l^3\right)\delta c_1+\left(\frac{2}{15}Fc_2 l^5+\frac{c_1}{6}Fl^4-\frac{q}{20}l^4\right)\delta c_2=0$$

Since δc_1 and δc_2 are arbitrary and linearly independent, we must have

$$\frac{1}{3}Fc_1 l^3+\frac{c_2}{6}Fl^4-\frac{q}{12}l^3=0 \Rightarrow c_1+\frac{1}{2}c_2 l=\frac{q}{4F} \quad \ldots(17)$$

$$\frac{1}{6}Fc_1^2 l^4+\frac{2}{15}Fc_2 l^5-\frac{q}{20}l^4=0 \Rightarrow c_1+\frac{4}{5}c_2 l=\frac{3q}{10F} \quad \ldots(18)$$

On solving (17) and (18), we have

$$\left(\frac{1}{2}-\frac{4}{5}\right)c_2 l=\left(\frac{1}{4}-\frac{3}{10}\right)\frac{q}{F} \Rightarrow -\frac{3}{10}c_2 l=\frac{(5-6)}{20}\frac{q}{F}$$

$$\Rightarrow \quad c_2=\frac{q}{6Fl}$$

Using it in (17), we have $c_1=\dfrac{q}{4F}-\dfrac{q}{12F}=\dfrac{(3-1)q}{12F}=\dfrac{q}{6F}$

Hence the desired approximation (16) becomes

$$y(x)\approx\frac{h}{l}x+\frac{q}{6F}x(x-1)+\frac{q}{6Fl}x^2(x-l)=\frac{h}{l}x+\frac{qx}{6Fl}(x-l)(x+l)$$

$$\approx\frac{h}{l}x+\frac{qx}{6Fl}(x^2-l^2),$$ which is also an exact solution.

Example 2: Find variational problem for $\frac{d^2y}{dx^2} + xy = -x$ ...(1)

with homogeneous conditions $y(0) = 0 = y(1)$ ...(2)

Also solve it by Rayleigh-Ritz method.

Solution: To find corresponding variational problem, we multiply both sides of (1) by δy and integrate the result over the interval (0, 1) to obtain

$$\int_0^1 (y'' + xy + x)\delta y \, dx = 0 \qquad ...(3)$$

$$\Rightarrow \quad \int_0^1 \frac{d}{dx}\left(\frac{dy}{dx}\right)\delta y \, dx + \int_0^1 xy \, \delta y \, dx + \int_0^1 x \, \delta y \, dx = 0$$

$$\Rightarrow \quad \left[\frac{dy}{dx}\delta y\right]_0^1 - \int_0^1 \frac{dy}{dx}\frac{d}{dx}(\delta y) dx + \int_0^1 \delta\left(\frac{1}{2}xy^2\right) dx + \int_0^1 \delta(xy) dx = 0$$

$$\Rightarrow \quad \delta \int_0^1 \left[\frac{1}{2}xy^2 + xy - \frac{1}{2}\left(\frac{dy}{dx}\right)^2\right] dx = 0. \qquad ...(4)$$

Hence the required variational problem is to optimize the functional

$$I = \int_0^1 \left[\frac{1}{2}xy^2 + xy - \frac{1}{2}\left(\frac{dy}{dx}\right)^2\right] dx. \qquad ...(5)$$

An one-term approximate assumption corresponding to (1) and satisfying (2) be of the form

$$y = c_1 x(1 - x).$$

Substituting in (5), we have

$$I = \int_0^1 \left[\frac{1}{2}c_1^2 x^3 (1-x)^2 + x^2 c_1 (1-x) - \frac{1}{2}c_1^2 (1-2x)^2\right] dx.$$

For extremals, $\delta I = 0$ gives

$$\delta\left\{\int_0^1 \left[\frac{1}{2}c_1^2 x^3 (1 - 2x + x^2) + c_1 (x^2 - x^3) - \frac{1}{2}c_1^2 (1 - 4x + 4x^2)\right]\right\} dx = 0$$

On integrating, we get

$$\delta\left\{\frac{1}{2}c_1^2\left(\frac{1}{4} - \frac{2}{5} + \frac{1}{6}\right) + c_1\left(\frac{1}{3} - \frac{1}{4}\right) - \frac{1}{2}c_1^2\left(1 - 2 + \frac{4}{3}\right)\right\} = 0$$

$$\Rightarrow \quad \delta\left\{\frac{1}{120}c_1^{\,2}+\frac{1}{12}c_1-\frac{1}{6}c_1^{\,2}\right\}=0 \qquad \Rightarrow \delta\left\{\frac{-19}{120}c_1^{\,2}+\frac{1}{12}c_1\right\}=0$$

$$\Rightarrow \quad \left(-\frac{19}{60}c_1+\frac{1}{12}\right)\delta c_1=0.$$

Since δc_1 is arbitrary, we must follow $c_1=\dfrac{5}{19}$

Hence the desired approximation becomes

$$y=\frac{5}{19}x(1-x).$$

Consider the two parameter approximation for which $n=2$, as

$$y=x(1-x)(c_1+c_2\,x)$$

Proceeding as in example 1, we can find $c_1=0.177, c_2=0.173$

Thus a better solution is $y=x(1-x).(0.177+0.173\,x)$.

Here we give working rule to find approximate solution.

Working rule: In order to find approximate solution we can use the following procedure:

Step 1: Substitute the approximate solution

$$y=x(a-x).(c_1+c_2\,x+c_3\,x^2+\ldots+c_n\,x^{n-1})$$

in the functional which is equivalent to the given differential equation if boundary conditions are

$$y(0)=0=y(a)$$

Step 2: After integration we follow that $I=I(c_1,c_2,\ldots,c_n)$

Step 3: Find constants $c_1,c_2,\ldots,c_n$ such that I is extremum. For it necessary conditions are

$$\frac{\partial I}{\partial c_1}=0,\ \frac{\partial I}{\partial c_2}=0,\ \ldots,\ \frac{\partial I}{\partial c_n}=0.$$

Step 4: Solve these simultaneous equations to get $c_1,c_2,\ldots,c_n$.

Step 5: Put these values of $c_1,c_2,\ldots,c_n$ to get best approximate solution of our boundary value problem.

Example 3: Find variational problem for $\dfrac{d^2y}{dx^2}+y=-x$ ...(1)

with boundary conditions $y(0)=0=y(1)$. Also solve it by Rayleigh-Ritz method.

Solution: To find corresponding variational problem, we multiply both sides of (1) by δy and integrate the result over the interval (0, 1) to obtain

$$\int_0^1 (y'' + y + x)\delta y\, dx = 0$$

$$\Rightarrow \quad \int_0^1 \frac{d}{dx}\left(\frac{dy}{dx}\right)\delta y\, dx + \int_0^1 y\,\delta y\, dx + \int_0^1 x\,\delta y\, dx = 0$$

$$\Rightarrow \quad \left[\frac{dy}{dx}\delta y\right]_0^1 - \int_0^1 \frac{dy}{dx}\frac{d}{dx}(\delta y)dx + \int_0^1 \delta\left(\frac{1}{2}y^2\right)dx$$

$$+ \int_0^1 \delta(xy)dx = 0$$

$$\Rightarrow \quad \delta \int_0^1 \left[\frac{1}{2}y^2 + xy - \frac{1}{2}y'^2\right]dx = 0.$$

Hence, the required variation problem is to optimize the functional

$$I = \int_0^1 \left[\frac{1}{2}y^2 + x\,y - \frac{1}{2}y'^2\right]d\,x. \qquad \ldots(2)$$

Consider one-parameter approximate solution

$y(x) = c_1\, x(1-x)$, then $y'(x) = c_1(1-2\,x)$.

Substituting then in (2), we obtain

$$I = \int_0^1 \left[\frac{1}{2}c_1^2\, x^2(1-x)^2 + x^2\, c_1(1-x) - \frac{1}{2}c_1^2(1-2\,x)^2\right]dx.$$

Therefore

$$\frac{\partial I}{\partial c_1} = \int_0^1 \frac{\partial}{\partial c_1}\left[\frac{1}{2}c_1^2\, x^2(1-x)^2 + x^2\, c_1(1-x) - \frac{1}{2}c_1^2(1-2\,x)^2\right]dx$$

$$= \int_0^1 \left[c_1\left(x^2 - 2x^3 + x^4\right) + \left(x^2 - x^3\right) - c_1\left(1 - 4\,x + 4\,x^2\right)\right]dx = \frac{-3c_1}{10} - \frac{1}{12}$$

Thus, for extremum $\frac{\partial I}{\partial c_1} = 0$ gives $c_1 = \frac{5}{18}$.

Therefore required approximate solution is $y(x) = \frac{5}{18}x(1-x)$.

Example 4: Find the first eigen value of the problem

$$y'' + \lambda(1 + x^2)y = 0,\ y(-1) = 0 = y(1).$$

Solution: Here the equivalent variational problem is

$$I = \int_{-1}^1 \left[(y')^2 - \lambda(1 + x^2)y^2\right]dx \qquad \ldots(1)$$

whose Euler's equation coincides with the above second order differential equation. Hence we extremize the functional given by (1).

Consider one parameter approximate solution $y = c(1 - x^2)$, then $y' = -2\,c\,x$.

Substituting these in (1), we obtain

$$I = \int_{-1}^{1}\left[4c^2x^2 - \lambda(1+x^2)c^2(1-x^2)^2\right]dx.$$

Now on differentiating with respect to c, we have

$$\frac{dI}{dc} = \int_{-1}^{1}\left[8cx^2 - 2\lambda c(1+x^2)(1-2x^2+x^4)\right]dx.$$

The condition $\frac{dI}{dc} = 0$, implies that

$$\int_{-1}^{1} 4x^2\,dx = \lambda\int_{-1}^{1}(1+x^2)(1-2x^2+x^4)dx$$

$$\Rightarrow \quad 2\int_{0}^{1} 4x^2\,dx = 2\lambda\int_{0}^{1}(1-x^2-x^4+x^6)dx$$

$$\Rightarrow \quad \frac{8}{3} = \lambda\frac{128}{105} \Rightarrow \lambda = \frac{35}{16} = 2.1875.$$

Example 5: Find the first eigen value λ of the problem

$$(1+x)y'' + y' + \lambda y = 0,\ y(0) = y(1) = 0.$$

Solution: The equivalent variational problem is

$$I = \int_{0}^{1}\left[(1+x)y'^2 - \lambda y^2\right]dx \qquad \text{...(1)}$$

whose Euler's equation coincides with the above second order differential equation. Hence we extremize the functional given by (1).

Consider one parameter appropriate solution $y = cx(1-x)$ then $y' = c(1-2x)$.

Substituting these values in (1), we obtain

$$I = \int_{0}^{1}\left[(1+x)c^2(1-2x)^2 - \lambda c^2x^2(1-x)^2\right]dx.$$

Now on differentiating with respect to c, we have

$$\frac{dI}{dc} = \int_{0}^{1}\left[2c(1+x)^2(1-2x)^2 - 2c\lambda x^2(1-x)^2\right]dx.$$

The condition $\frac{dI}{dc} = 0$ implies that

$$\int_{0}^{1}(1+x)(1-4x+4x^2)dx = \lambda\int_{0}^{1}x^2(1-2x+x^2)dx$$

$$\Rightarrow \quad \int_{0}^{1}(1-3x+4x^3)dx = \lambda\int_{0}^{1}(x^2-2x^3+x^4)dx$$

$$\Rightarrow \quad \frac{1}{2} = \frac{\lambda}{30} \Rightarrow \lambda = 15.$$

Example 6: Find approximately the smallest eigen value λ of

$$y'' + \lambda y = 0,\ y(0) = 0 = y(1).$$

Solution: The equivalent variation problem is $I = \int_0^1 \left[(y')^2 - \lambda y^2\right] dx$ ...(1)

whose Euler's equation coincides with the above second order differential equation. Hence we extremize the functional given by (1).

Consider one parameter approximate solution $y = c\,x(1-x)$, then $y' = c(1-2x)$.

Substituting these values in (1), we obtain

$$I = \int_0^1 \left[c^2(1-2x)^2 - \lambda c^2 x^2 (1-x)^2\right] dx .$$

Now differentiating with respect to c, we have

$$\frac{dI}{dc} = \int_0^1 \left[2c(1-2x)^2 - 2\lambda c x^2 (1-x)^2\right] dx .$$

The condition $\frac{dI}{dc} = 0$ implies that

$$\int_0^1 \left(1 - 4x + 4x^2\right) dx = \lambda \int_0^1 \left(x^2 - 2x^3 + x^4\right) dx$$

$$\Rightarrow \qquad \left(1 - 2 + \tfrac{4}{3}\right) = \lambda\left(\tfrac{1}{3} - \tfrac{1}{2} + \tfrac{1}{5}\right) \Rightarrow \lambda = 10.$$

On the other hand the exact eigen values are $\lambda_n = \pi^2 n^2$.

Therefore the least eigen value is $\lambda_1 = \pi^2 . 1^2 = 9.87$ which is equivalent to its approximate least eigen value 10.

Example 7: Using Rayleigh-Ritz method solve

$$(1 - x^2) y'' - 2xy' + 2y = 0$$

subject to boundary conditions $y(0) = 0, y(1) = 1$.

Solution: The equivalent variational problem becomes

$$I = \int_0^1 \left[(1 - x^2) y'^2 - 2y^2\right] dx \quad ...(1)$$

whose Euler's equation coincides with the above second order differential equation. Here the boundary conditions are not homogeneous, so consider the one parameter approximate solution

$$y = x + c\,x(1-x) = (1+c)x - c\,x^2 \text{, then } y' = 1 + c - 2cx .$$

Substituting these values in (1), we obtain

$$I = \int_0^1 \left[\left(1 - x^2\right)\left(1 + c - 2cx\right)^2 - 2\left\{x(1+c) - cx^2\right\}^2\right] dx$$

$$= \int_0^1 \left(1 - x^2\right)\left\{(1+c)^2 - 4cs(1+c) + 4c^2 x^2\right\} dx$$

$$- \int_0^1 2\left[x^2 (1+c)^2 + c^2 x^4 - 2cx^3 (1+c)\right] dx$$

The condition $\dfrac{dI}{dc} = 0$ implies that

$$\int_0^1 (1-x^2)\{2(1+c) - 4x(1+2c) + 8cx^2\}\,dx$$

$$- \int_0^1 2\{2x^2(1+c) + 2cx^4 - 2x^3(1+2c)\}\,dx$$

$\Rightarrow \quad c = 0.$

Hence the appropriate solution is $y(x) = x$, which is also the exact solution of the given differential equation.

12.4 NECESSARY CONDITION FOR $z = z(x, y)$ TO BE THE EXTREMAL OF THE FUNCTIONAL $\iint_D F(x, y, z, z_k\, z_y)\,dx\,dy$

Let $z = z(x, y)$ be the equation of a surface in Monge's form bounded by a closed curve C whose projection on xy-plane is C′ which forms the boundary of region D. Here we have to find the necessary condition for $z = z(x, y)$ to be the extremals of the functional

$$\iint_D F(x, y, z, z_x, z_y)\,dx\,dy. \qquad \ldots(1)$$

Let $\quad I = \iint_D F(x, y, z, z_x, z_y)\,dx\,dy$, then

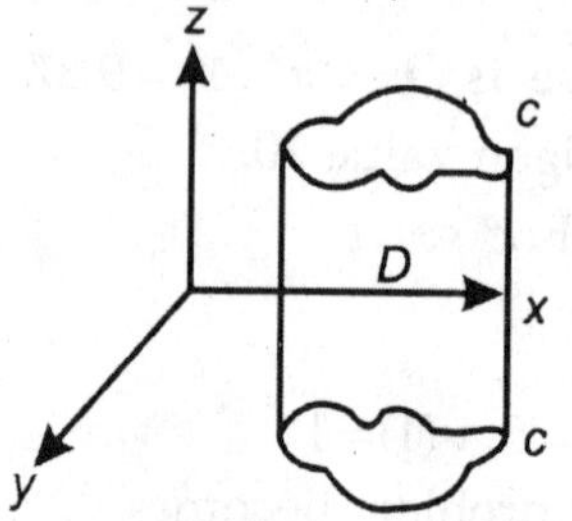

$$\delta I = \iint_D \left[\frac{\partial F}{\partial z}\delta z + \frac{\partial F}{\partial z_x}\delta z_x + \frac{\partial F}{\partial z_y}\delta z_y\right] dx\,dy \qquad \ldots(2)$$

Now $\quad \dfrac{\partial}{\partial x}\left[\dfrac{\partial F}{\partial z_x}\delta z\right] = \dfrac{\partial F}{\partial z_x}\dfrac{\partial}{\partial x}(\delta z) + \delta z \dfrac{\partial}{\partial x}\left[\dfrac{\partial F}{\partial z_x}\right]$

$= \dfrac{\partial F}{\partial z_x}\delta z_x + \delta z \dfrac{\partial}{\partial x}\left(\dfrac{\partial F}{\partial z_x}\right)$, since δ and $\dfrac{\partial}{\partial x}$ commute

$\Rightarrow \quad \dfrac{\partial F}{\partial z_x}\delta z_x = \dfrac{\partial}{\partial x}\left[\dfrac{\partial F}{\partial z_x}\delta z\right] - \delta z \dfrac{\partial}{\partial x}\left[\dfrac{\partial F}{\partial z_x}\right]$

$\Rightarrow \quad \iint_D \left(\dfrac{\partial F}{\partial z_x}\delta z_x\right) dx\,dy = \iint_D \dfrac{\partial}{\partial x}\left(\dfrac{\partial F}{\partial z_x}\delta z\right) dx\,dy - \iint_D \delta z \dfrac{\partial}{\partial x}\left(\dfrac{\partial F}{\partial z_x}\right) dx\,dy$

$$= -\int \frac{\partial F}{\partial z_x}\,\delta z\, dy - \iint_D \delta z \frac{\partial}{\partial x}\left(\frac{\partial F}{\partial z_x}\right) dx\, dy \qquad \ldots(3)$$

[Since by applying Green's theorem $\iint_D \frac{\partial F}{\partial x}\, dx\, dy = -\int F\, dy$ in first term]

Similarly

$$\iint_D \left(\frac{\partial F}{\partial z_y}\,\delta z_y\right) dx\, dy = -\int \frac{\partial F}{\partial z_y}\,\delta z\, dx - \iint_D \delta \frac{\partial}{\partial y}\left(\frac{\partial F}{\partial z_y}\right) dx\, dy. \qquad \ldots(4)$$

Using (3) and (4) in (2), we have

$$\delta I = -\int \left(\frac{\partial F}{\partial z_x}\, dy + \frac{\partial F}{\partial z_y}\, dx\right)\delta z + \iint_D \left[\frac{\partial F}{\partial z} - \frac{\partial}{\partial x}\left(\frac{\partial F}{\partial z_x}\right) - \frac{\partial}{\partial y}\left(\frac{\partial F}{\partial z_y}\right)\right] dx\, dy.$$

For extremum, applying the condition $\delta I = 0$, we obtain

$$\frac{\partial F}{\partial z} - \frac{\partial}{\partial x}\left(\frac{\partial F}{\partial z_x}\right) - \frac{\partial}{\partial y}\left(\frac{\partial F}{\partial z_y}\right) = 0 \qquad \ldots(5)$$

which is the Euler's equation, with the natural boundary condition

$$\frac{\partial F}{\partial z_x}\, dy + \frac{\partial F}{\partial z_y}\, dx = 0. \qquad \ldots(6)$$

Equation (5) and (6) are the required necessary conditions.

12.5 RAYLEIGH-RITZ METHOD FOR SOLVING PDES

Suppose a boundary value problem is

$$-[p\, z_x]_x - [p\, z_y]_y + q\, z = r \qquad \ldots(1)$$

in the region D bounded by a closed curve C and z is supposed to be prescribed on C.

Here the equivalent variational problem becomes

$$I = \iint_D \left[p(z_x^2 + z_y^2) + q\, z^2 - 2\, r\, z\right] dx\, dy \qquad \ldots(2)$$

whose Euler's equation coincides with PDE (1).

Hence we have to optimize the functional

$$\iint_D \left[p\left(z_x^2 + z_y^2\right) + q\, z^2 - 2\, r\, z\right] dx\, dy.$$

We have to find the extremals such that it satisfies the given boundary conditions.

Example 8: Apply Rayleigh-Ritz method to solve the Poisson's equation $z_{xx} + z_{yy} = -1$ in a square $-1 \le x \le 1, -1 \le y \le 1$, with boundary conditions $z(\pm 1, \pm 1) = 0$.

Solution: Hence the corresponding variational problem is to optimize the functional

$$I = \iint_D \left(z_x^2 + z_y^2 - 2z\right) dx\, dy \qquad \ldots(1)$$

whose Euler's equation coincides with PDE

$$z_{xx} + z_{yy} = -1$$

Consider a one-parameter approximate solution

$z(x, y) = x(1-x^2)(1-y^2)$, then

$z_x = -2cx(1-y^2),\ z_y = -2cy(1-x^2)$.

Substituting these in (1), we have

$$I = \int_{-1}^{1}\int_{-1}^{1}\left[4c^2x^2(1-y^2)^2 + 4c^2y^2(1-x^2)^2 - 2c(1-x^2)(1-y^2)\right]dx\,dy$$

$$= 4c^2\left[2\int_0^1 x^2\,dx\, 2\int_0^1 (1-y^2)^2\,dy + 2\int_0^1 (1-x^2)^2\,dx\, 2\int_0^1 y^2\,dy\right]$$

$$- 2c\, 2\int_0^1 (1-x^2) 2\int_0^1 (1-y^2)\,dy$$

$$= 4c^2\left[\frac{2}{3}\frac{16}{15} + \frac{16}{15}\frac{2}{3}\right] - \frac{32}{9}c = \frac{32}{45}\left(8c^2 - 5c\right)$$

For extremals, the condition $\frac{dI}{dc} = 0$ gives $16c - 5 = 0 \Rightarrow c = \frac{5}{16}$

Hence, the appropriate solution is $z(x, y) = \frac{5}{16}(1-x^2)(1-y^2)$

Example 9: Solve the Laplace's equation $z_{xx} + z_{yy} = 0$ is a square $D: -1 \le x \le 1, -1 \le y \le 1$ with the conditions $z(\pm 1, y) = 0, z(x, \pm 1) = 1 - x^2$.

Solution: Here the corresponding variational problem is to optimize the functional

$$I = \iint_D \left[z_x^2 + z_y^2\right] dx\, dy \qquad \ldots(1)$$

whose Euler's equation is $z_{xx} + z_{yy} = 0$.

Consider a one-parameter approximate solution

$z(x, y) = (1-x^2)\left[1 + c(1-y^2)\right]$, then

$z_x = -2x[1 + c(1-y^2)],\ z_y = -2cy(1-x^2)$.

Substituting these values in (1), we have

$$I = \int_{-1}^{1}\int_{-1}^{1}\left[4x^2(1 + c(1-y^2))^2 + 4c^2y^2(1-x^2)^2\right]dx\,dy$$

$$= \int_{-1}^{1}\int_{-1}^{1}\left[4x^2 + 8cx^2(1-y^2) + 4c^2x^2(1-y^2)^2 + 4c^2y^2(1-x^2)^2\right]dx\,dy$$

$$= 2\int_0^1 4x^2\,dx\,2\int_0^1 dy + 8c\,2\int_0^1 x^2\,dx\,2\int_0^1 (1-y^2)\,dy$$

$$+ 4c^2\left[2\int_0^1 x^2\,dx\,2\int_0^1 (1-y^2)^2\,dy + 2\int_0^1 y^2\,dy\,2\int_0^1 (1-x^2)\,dx\right]$$

$$= \frac{16}{3} + 32c\frac{1}{3}\frac{2}{3} + 4c^2\left[\frac{32}{45} + \frac{32}{45}\right] = \frac{16}{3} + \frac{64}{9}c + \frac{256}{45}c^2.$$

For extremals, the condition $\frac{dI}{dc} = 0$ gives

$$\frac{64}{9} + \frac{256}{45}2c = 0 \Rightarrow c = -\frac{5}{8}.$$

Hence the approximate solution is

$$z(x,y) = (1-x^2)\left[1 - \tfrac{5}{8}(1-y^2)\right] = (1-x^2) - \tfrac{5}{8}(1-x^2)(1-y^2).$$

Example 10: Solve the Poisson's equation $z_{xx} + z_{yy} = -1$ in a ellipse $\frac{x^2}{a^2} + \frac{y^2}{b^2} \le 1$ with boundary condition $z = 0$ on $\frac{x^2}{a^2} + \frac{y^2}{b^2} = 1$.

Solution: Here the corresponding variational problem is to optimize the functional

$$I = \iint_D \left[z_x^2 + z_y^2 - 2z\right]dx\,dy \qquad \ldots(1)$$

where D is the region $\frac{x^2}{a^2} + \frac{y^2}{b^2} \le 1$. Its Euler's equation coincides with PDE $z_{xx} + z_{yy} = -1$.

Consider one-parameter approximate solution $z(x,y) = c\left(1 - \frac{x^2}{a^2} - \frac{y^2}{b^2}\right)$

which satisfies the boundary condition $z = 0$ on $\frac{x^2}{a^2} + \frac{y^2}{b^2} = 1$.

Then $z_x = -\frac{2cx}{a^2},\ z_y = -\frac{2cy}{b^2}$.

Substituting these values (1), we have

$$I = \iint_D \left[\frac{4c^2x^2}{a^4} + \frac{4c^2y^2}{b^4} - 2c\left(1 - \frac{x^2}{a^2} - \frac{y^2}{b^2}\right)\right]dx\,dy.$$

Now, the condition $\frac{dI}{dc} = 0$, implies that

$$8c\iint_D \left(\frac{x^2}{a^4} + \frac{y^2}{b^4}\right)dx\,dy - 2\iint_D \left(1 - \frac{x^2}{a^2} - \frac{y^2}{b^2}\right)dx\,dy = 0$$

$$\Rightarrow \quad 4c = \frac{\iint_D \left(1 - \frac{x^2}{a^2} - \frac{y^2}{b^2}\right) dx\, dy}{\iint_D \left(\frac{x^2}{a^4} + \frac{y^2}{b^4}\right) dx\, dy}$$

Using transformation $x = a\, r \cos\theta, y = a\, r \sin\theta,$ we have

$$4c = \frac{\int_0^1 (1 - r^2) r\, dr}{\int_0^1 r^3\, dr} \times \frac{2}{\frac{1}{a^2} + \frac{1}{b^2}} \Rightarrow c = \frac{a^2 b^2}{2(a^2 + b^2)}.$$

Hence the approximate solution is $z(x, y) = \frac{1}{2} \frac{a^2 b^2}{(a^2 + b^2)} \left(1 - \frac{x^2}{a^2} - \frac{y^2}{b^2}\right)$

This is identical to the best solution of the problem.

Example 11: Find approximately the smallest eigen value of $z_{xx} + z_{yy} + \lambda z = 0$ in the region D bounded by the circle $x^2 + y^2 = 1$ with the condition $z = 0$ on the boundary.

Solution: Here the corresponding variational problem is to optimize the functional

$$I = \iint_D \left[z_x^2 + z_y^2 - \lambda z^2\right] dx\, dy \qquad \text{...(1)}$$

where D is the region defined $x^2 + y^2 \leq 1$. Its Euler's equation coincides with PDE

$$z_{xx} + z_{yy} + \lambda z = 0.$$

Consider one parameter approximate solution $z(x, y) = c(1 - x^2 - y^2)$ which satisfies the boundary condition $z = 0$ on $x^2 + y^2 = 1$.

Then $z_x = -2cx, z_y = -2cy$. Substituting these values in (1), we have

$$I = \iint_D \left[4c^2 x^2 + 4c^2 y^2 - \lambda c^2 (1 - x^2 - y^2)^2\right] dx\, dy.$$

Now the condition $\frac{dI}{dc} = 0$ implies that

$$8c \iint_D (x^2 + y^2)\, dx\, dy - 2c\lambda \iint_D (1 - x^2 - y^2)^2\, dx\, dy = 0$$

$$\Rightarrow \quad \lambda = \frac{4 \iint_D (x^2 + y^2)\, dx\, dy}{\iint_D (1 - x^2 - y^2)^2\, dx\, dy}$$

$$= \frac{4 \int_0^1 \int_0^{2\pi} r^3\, d\theta\, dr}{\int_0^1 \int_0^{2\pi} (1 - r^2)^2\, r\, d\theta\, dr}; \text{ by putting } x = r\cos\theta, y = \sin\theta$$

$$= \frac{4 \cdot \frac{1}{4} \cdot 2\pi}{\frac{1}{6} \cdot 2\pi} = 6,$$ which is the required eigen value.

Example 12: Find the estimate of the least eigen value of $z_{xx} + z_{yy} + \lambda z = 0$ in the isosceles right-angled triangular region D bounded by the co-ordinate axes and the line $x + y = 1$ with boundary condition $z = 0$ of the boundary.

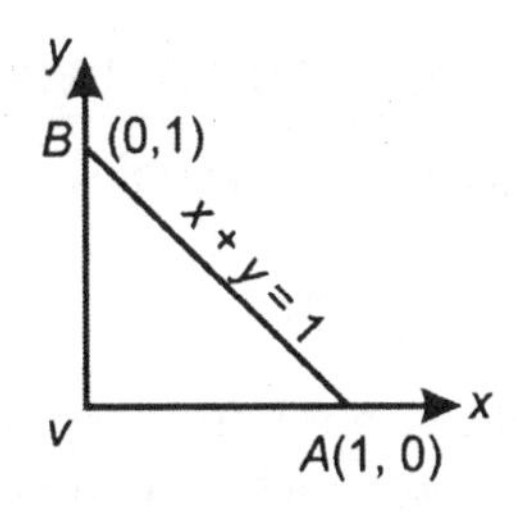

Solution: Here the corresponding variational problem is to optimize the functional

$$I = \iint_D \left[z_x^2 + z_y^2 - \lambda z^2\right] dx\, dy \qquad \dots(1)$$

where D is a isosceles triangular bounded by the co-ordinate axes and the line $x + y = 1$. Its Euler's equation coincides with PDE

$$z_{xx} + z_{yy} + \lambda z = 0.$$

Consider the one parameter solution $z(x, y) = c\,x\,y(1 - x - y)$, which satisfies the boundary condition $z = 0$ on the boundary of isosceles triangular region.

Then $z_x = c\,y(1 - 2x - y), z_y = c\,x(1 - x - 2y)$.

Substituting these values in (1), we have

$$I = \iint_D \left[c^2 y^2 (1 - 2x - y)^2 + c^2 x^2 (1 - x - 2y)^2 - \lambda c^2 x^2 y^2 (1 - x - y)^2\right] dx\, dy$$

Now the condition $\frac{dI}{dc} = 0$ implies that

$$2c \iint_D \left[y^2 (1 - 2x - y)^2 + x^2 (1 - x - 2y)^2\right] dx\, dy - \lambda\, 2c \iint_D x^2 y^2 (1 - x - y)^2\, dx\, dy = 0$$

$$\Rightarrow \lambda = \frac{\iint_D \left[y^2 (1 - 2x - y)^2 + x^2 (1 - x - 2y)^2\right] dx\, dy}{\iint_D x^2 y^2 (1 - x - y)^2\, dx\, dy}$$

$$= \frac{\int_0^1 \int_{y=0}^{1-x} [y^2(1-2x-y)^2 + x^2(1-x-2y)^2]dx\,dy}{\int_0^1 \int_{y=0}^{1-x} x^2 y^2 (1-x-y)^2\,dx\,dy} = 56,$$

on simplification.

This is the required eigen value.

12.6 KANTOROVICH METHOD FOR SOLVING PDES

In Kantorovich method, we choose a trial solution

$$z(x,y) = \sum_{i=1}^{n} \phi_i(x) z_i(x,y)$$

where $z_i(x,y)$ satisfies the given boundary conditions for each i and $\phi_i(x)$ are unknown functions of x. This method is more efficient method in comparison of Rayleigh-Ritz method. The functions $\phi_i(x)$ are than chosen as to extremize the functional I. This is clear from the following illustrative examples.

Example 13: Find an appropriate solution of Poisson's equation $z_{xx} + z_{yy} = -1$, in a square $D - 1 \le x \le 1, -1 \le y \le 1$ with the condition $z = 0$ on the boundary.

Solution: Here the Poisson's equation is the Euler's-Ostrogradsky equation for the extremum of the functional

$$I[z] = \iint_D [z_x^2 + z_y^2 - 2z]dx\,dy \qquad \ldots(1)$$

where D is the square $-1 \le x \le 1; -1 \le y \le 1$.

Consider an appropriate solution of the form $z(x,y) = \phi(x)(1-y^2)$, where $\phi(x)$ is to be determined so that $\phi(\pm 1) = 0$.

Thus clearly on the boundary of the square $D, z = 0$.

Now $\quad z_x = (1-y^2)\phi'(x), z_y = -2y\phi(x)$.

Substituting these values in (1), we obtain

$$I[z] = \iint_D \left[(1-y^2)^2 \phi'^2(x) + 4y^2 \phi^2(x) - 2(1-y^2)\phi(x)\right]dx\,dy$$

$$= \int_{-1}^{1}\int_{-1}^{1} \left[(1-y^2)^2 \phi'^2(x) + 4y^2\phi^2(x) - 2(1-y^2)\phi(x)\right]dx\,dy$$

$$= \int_{-1}^{1} \left[\frac{16}{15}\phi'^2 + \frac{8}{3}\phi^2 - \frac{8}{3}\phi\right]dx.$$

Hence the corresponding Euler equation is

$$-\frac{4}{5}\phi'' + 2\phi - 1 = 0 \Rightarrow \phi'' - \frac{5}{2}\phi = -\frac{5}{4}$$

Its solution is $\phi(x)=\frac{1}{2}+A\cos h\left(\sqrt{\frac{5}{2}}\right)x+B\sin h\left(\sqrt{\frac{5}{2}}\right)x$.

Using the boundary conditions $\phi(\pm 1)=0$, we have

$$\phi(x)=\frac{1}{2}\left[1-\frac{\cos h\left(\sqrt{5/2}\right)x}{\cos h\left(\sqrt{5/2}\right)}\right].$$

Hence the approximate solution is

$$z(x,y)=\frac{\left(1-y^2\right)}{2}\left[1-\frac{\cos h\left(\sqrt{5/2}\right)x}{\cos h\left(\sqrt{5/2}\right)}\right] \qquad \ldots(2)$$

From (2), we have $z(0,0)=0.303$

By Rayleigh-Ritz method, we have $z(0,0)=0.313$

While the exact value of $z(0,0)=0.295$.

These values show the superiority of Kantorovich method over Rayleigh-Ritz method.
More accurate value can be obtained by taking

$$z_2(x,y)=\left(1-y^2\right)\phi_1(x)+\left(1-y^2\right)^2\phi_2\,x.$$

Example 14: Using Kantorovich method estimate the least eigen value of the problem $z_{xx}+z_{yy}+\lambda z=0$ in the square $D|x|\le 1,|y|\le 1$, with the condition $z=0$ on the boundary of square.

Solution: Here the corresponding variational problem is to optimize the functional

$$I[x]=\iint_D\left(z_x^{\,2}+z_y^{\,2}-\lambda z^2\right)dx\,dy \qquad \ldots(1)$$

Here D is a square. Its Euler-Ostrogradsky equation coincides with the PDE $z_{xx}+z_{yy}+\lambda z=0$.

Consider an approximate solution of the form $z(x,y)=\phi(x).(1-y^2)$,

where $\phi(x)$ is to be determined so that $\phi(\pm 1)=0$.

Thus, $z=0$ on the boundary of the square D and

$$z_x=(1-y^2)\phi'(x), z_y=-2y\,\phi(x).$$

Substituting these values in (1), we obtain

$$I[z]=\iint_D\left[(1-y^2)^2\phi'^2+4y^2\phi^2-\lambda\phi^2(1-y^2)^2\right]dx\,dy$$

$$= \int_{-1}^{1}\int_{-1}^{1}\left[(1-2y^2+y^4)\phi'^2 + 4y^2\phi^2 - \lambda\phi^2(1-2y^2+y^4)\right]dx\,dy$$

$$= \int_{-1}^{1}\left[\frac{16}{15}\phi'^2 - \frac{8}{3}\phi^2 - \frac{16}{15}\lambda\phi^2\right]dx = \frac{16}{15}\int_{-1}^{1}\left[\phi'^2 + \left(\lambda - \frac{5}{2}\right)\phi^2\right]dx.$$

Here the corresponding Euler's equation is $\phi'' + \left(\lambda - \frac{5}{2}\right)\phi = 0$

Its solution is $\phi(x) = A\cos\sqrt{(\lambda - 5/2)}\,x + B\sin\sqrt{(\lambda - 5/2)}\,x$.

Using $\phi(\pm 1) = 0$, we have $B = 0$ and $A\cos\sqrt{(\lambda - 5/2)}\,x = 0$.
For non-trivial solution, we have

$$\sqrt{(\lambda - 5/2)} = \frac{\pi}{2} \Rightarrow \lambda = \frac{\pi^2}{4} + \frac{5}{2} = 4.967$$

By using Rayleigh-Ritz method, we have $\lambda = 5$ while the exact value is $\lambda = 4.93$

This shows that the error by Kantorovich method is one half of that by Rayleigh-Ritz method.

Example 15: Find the approximate solution of the Poisson's equation $z_{xx} + z_{yy} = -1$ in an equilateral triangle $x = a$ and $y = \pm\, x/\sqrt{3}$, with $z = 0$ on the boundary of the triangle.

Solution: Here the corresponding variational problem is to optimize the functional

$$I[z] = \iint_D \left[z_x^{\,2} + z_y^{\,2} - 2z\right]dx\,dy \qquad \ldots(1)$$

where D is a equilateral triangle $x = a$ and $y = \pm\, x/\sqrt{3}$

Consider an appropriate solution of the form $z(x, y) = \phi(x)(x^2 - 3y^2)$, where $\phi(a) = 0$ and $\phi(0)$ is finite. This implies

$z_x = (x^2 - 3y^2)\phi'(x) + 2x\,\phi(x), z_y = -6y\,\phi(x)$.

Hence $I[z] = \iint_D [(x^2 - 3y^2)\phi'(x) + 2x\,\phi(x)]^2 + 36y^2\phi^2(x)$

$$-2\phi(x)(x^2 - 3y^2)\,dx\,dy$$

$$= \frac{8\sqrt{3}}{405}\int\left[2x^5\phi'^2 + 10x^4\phi\phi' + 30x^3\phi^2 - 5x^3\phi\right]dx.$$

Here the corresponding Euler equation is

$$-5x^3 + 60x^3\phi + 10x^4\phi' - \frac{d}{dx}\left[4x^5\phi' + 10x^4\phi\right] = 0$$

$$\Rightarrow \quad -5x^3 + 60x^3\phi + 10x^4\phi' - 4x^5\phi'' - 20x^4\phi' - 40x^3\phi = 0$$

$$\Rightarrow \quad -4x^5\phi'' - 20x^4\phi' + 20x^3\phi - 5x^3 = 0$$

$$\Rightarrow \quad x^2 \phi'' + 5 x \phi' - 5 \phi = -\frac{5}{4} \Rightarrow \left[x^2 D^2 + 5 x D - 5\right]\phi = -\frac{5}{4} \quad ...(2)$$

Putting $x = e^z$, we have

$$x D = D_1 \; x^2 D^2 = D_1 (D_1 - 1) \text{where } D \equiv \frac{d}{dx}, D_1 \equiv \frac{d}{dz}$$

Hence the equation (2) becomes

$$\left[D_1 (D_1 - 1) + 5 D_1 - 5\right]\phi = -\frac{5}{4}\tfrac{5}{4} \Rightarrow \left[D_1^2 + 4 D_1 - 5\right]\phi = -\frac{5}{4}$$

Here $C.F. = A c^z + B e^{-5z}$ and $P.I. = \frac{1}{4}$

Hence $\phi(x) = A e^z + B e^{-5z} + \frac{1}{4} = A x + B x^{-5} + \frac{1}{4}$

As $\phi(0)$ is finite, we have $B = 0$

Now $\phi(a) = 0 \Rightarrow A = -\frac{1}{4a}$ implies that $\phi(x) = -\frac{1}{4a} x + \frac{1}{4} = \frac{a - x}{4a}$

Therefore $z(x, y) = \frac{a-x}{4a}(x^2 - 3 y^2)$, this is also the exact solution.

EXERCISE 12.1

1. Find variational problem for the differential equation $y'' - y = 0$, with boundary conditions $y(0) = 0, y(1) = 1$. Also solve it by Rayleigh-Ritz method.
2. Find the approximate solution of equation $y'' = 1$, subject to the boundary conditions $y(0) = 0 = y(1)$.
3. Find the variational problem for deformable bodies.
4. Find the variational problem for $x y'' + y' + y = x$ with boundary conditions $y(0) = 0, y(1) = 1$ and solve it by Rayleigh-Ritz method.
5. Solve $y'' + (1 + x^2) y + 1 = 0$ with boundary conditions $y(1) = 0 = y(-1)$.
6. Find the first eigen value of the problem $y'' + \lambda y = 0, y(\pm 1) = 0$.

   [Hint. Let $y(x) = c(1 - x^2)$]
7. Applying Rayleigh-Ritz method solve the Poisson's equation $z_{xx} + z_{yy} = -1, x^2 + y^2 \le 1$ with condition $z(x, y) = 0$ on $x^2 + y^2 = 1$.
8. Solve the Poisson's equation $z_{xx} + z_{yy} = -1$ in an equilateral triangle bounded by $x = a$ and $y = \pm x/\sqrt{3}$ with the boundary conditions $z = 0$ on the boundary of the triangle.

   [Hint. Let approximate solution $z(x, y) = c(x^2 - 3 y^2)(a - x)$].

9. Find the one term and two term least approximate eigen value λ of the problem $y'' + \lambda x y = 0\ u(0) = 0 = y(1)$.

10. Solve the Laplace's equation $z_{xx} + z_{yy} = 0$ in a square $D: |x| \le 1, |y| \le 1$ with $z(\pm 1, y) = 1 - y^2$ and $z(x, \pm 1) = 0$

11. Applying the Rayleigh-Ritz method, find an appropriate solution of the Poisson's equation $\nabla^2 z = -1$ inside the square $|x| \le a, |y| \le a$ with $z = 0$ on the boundary of the square.

12. Find an appropriate solution of $\nabla^2 z = -1$ in the rectangle $D |x| \le a, |y| \le b$ with $z = 0$ on the boundary of the rectangle.

13. Approximate the first eigen value of the problem $\nabla^2 z + \lambda z = 0$ with $z = 0$ on the boundary of a unit circle D with centre at the origin.

ANSWERS 12 1

1. $y_{1app.} = x - \dfrac{5}{22} x(1 - x)$, $y_{2app.} = 0.854\,x - 0\,59\,x^2 + 0.1628\,x^3$

2. $y(x) = \frac{1}{2}(x^2 - x)$

4. $y(x) = \frac{1}{26}\left(111 - 155\,x + 70\,x^2\right)x$

5. (i) $y_{1app.}(x) = \dfrac{35}{38}\left(1 - x^2\right)$

   (ii) $y_{2app.}(x) = \left(1 - x^2\right)\left(0.933 - 0.054\,x^2\right)$

6. $\lambda = 2.5$.

7. $z(x, y) = \dfrac{1}{4}\left(1 - x^2 - y^2\right)$.

8. $z(x, y) = \dfrac{1}{4a}\left(x^2 - 3y^2\right)(a - x)$.

9. $\lambda_{1app} = 20, \lambda_{2app} = 19.2, \lambda_{Exact} = 18.9$

10. $z(x, y) = \left(1 - y^2\right)\dfrac{\cos h\left(\sqrt{5/2}\right)x}{\cos h\left(\sqrt{5/2}\right)}$.

11. $z = \dfrac{5}{16a^2}\left(x^2 - a^2\right)\left(y^2 - a^2\right)$.

12. $z(x, y) = \dfrac{b^2 - y^2}{2}\left[1 - \dfrac{\cos h\sqrt{5/2}\,\dfrac{x}{b}}{\cos h\sqrt{5/2}\,\dfrac{a}{b}}\right]$

13. $\lambda = 6$ with $z = c\left(1 - x^2 - y^2\right)$.

13

Variational Problems with Moving Boundaries

13.1 INTRODUCTION

So for, we have discussed variational problems with fixed boundary points in the functional. In this chapter, we shall discuss those variational problems in which one or both the boundary points move. This means that the class of admissible curves is extended. In other words, we have to admit curves with variable boundary points in addition to the comparison curves with fixed boundary points. If on a curve $y = y(x)$, an extremum is attained in a problem with moving boundary points, then it is sure that the extremum of all the moving boundaries attained on a restricted class of curves with common (fixed) boundary points. It follows that the curves $y = y(x)$ on which extremum of the above functional is attained in a moving boundary problem must be solutions of the Euler equation

$$F_y - \frac{d}{dx} F_{y'} = 0,$$

so that these curves must be extremals.

In a variational problem with fixed boundary points, the two constants in the solution of Euler's equation are determined from the two boundary conditions at the fixed points (x_1, y_1) and (x_2, y_2). But in a variational problem with moving boundary, one or both of these conditions are missing and so the arbitrary constants in the general solutions of Euler's equation have to be obtained from the necessary condition for extremum; i.e., from $\delta I = 0$.

13.2 FUNCTIONAL DEPENDENT ON ONE FUNCTION

Consider the functional

$$I[y(x)] = \int_{x_1}^{x_2} F(x, y, y')\,dx \qquad \ldots(1)$$

dependent on the function $y(x)$, where one or both the boundary points (x_1, y_1) and (x_2, y_2) can move. For simplicity, suppose that one of the boundary points (x_1, y_1) is fixed while the other boundary point (x_2, y_2)

can move and pass to the point $(x_2 + \delta x_2, y_2 + \delta y_2)$. Since the extremum in a moving boundary problem is attained only on extremals, i.e. on solutions $y = y(x, C_1, C_2)$ of Euler's equations, we shall consider the values of the functional I on such curves. Thus, we see that $I[y(x, C_1, C_2)]$ reduces to a function of the parameters C_1, C_2 and of the limits of integration.

Definition: The admissible curves $y = y(x)$ and $Y = y(x) + \delta y(x)$ are said to close if $|\delta y|$ and $|\delta y'|$ are small. The extremals passing through (x_1, y_1), form a pencil of extremals $y = y(x, C_1)$.

The functional $I[y(x, C_1)]$ on the curves of this pencil becomes a function of C_1 and x.

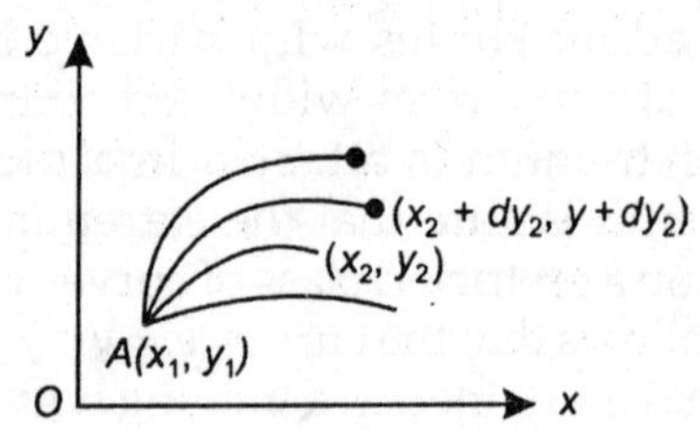

If the curves of the pencil $y = y(x, C_1)$ do not intersect in the neighbourhood of the extremals, then we may consider $I[y(x, C_1)]$ as a single-valued function of (x_2, y_2), because the specification of (x_2, y_2) determines the extremals of the pencil uniquely, and hence determines the value of the functional $I[y(x)]$.

We shall determine the variation of the functional $I[y(x, C_1)]$ when the boundary point moves from (x_2, y_2) to $(x_2 + \delta x_2, y_2 + \delta y_2)$. In such case, the increment ΔI is given by

$$\Delta I = \int_{x_1}^{x_2 + \delta x_2} F(x, y + \delta y, y' + \delta y')dx - \int_{x_1}^{x_2} F(x, y, y')dx$$

$$= \int_{x_1}^{x_2} \left[F(x, y + \delta y, y' + \delta y') - F(x, y, y')\right]dx + \int_{x_2}^{x_2 + \delta x_2} F(x, y + \delta y, y' + \delta y')dx \quad \ldots(2)$$

By mean value theorem, the second term on the R.H.S. of (2) can be written as

$$\int_{x_2}^{x_2 + \delta x_2} F(x, y + \delta y, y' + \delta y')dx = [F]_{x_2 + \theta \delta x_2} \cdot \delta x_2, 0 < \theta < 1. \quad \ldots(3)$$

But by virtue of the continuity of F, we can write

$$[F]_{x_2 + \theta \delta x_2} = [F]_{x_2} + \varepsilon \quad \ldots(4)$$

where ε is an infinitesimal such that $\varepsilon \to 0$ as $\delta x_2 \to 0$ and $\delta y_2 \to 0$. Thus by (3) and (4), we have

$$\int_{x_2}^{x_2+\delta x_2} F(x, y+\delta y, y'+\delta y')dx = [F]_{x=x_2} \cdot \delta x_2 + \varepsilon\, \delta x_2 . \qquad \ldots(5)$$

Now using Taylor's theorem, we may transform the first term on the R.H.S. of (2) as

$$\int_{x_1}^{x_2} \left[F(x, y+\delta y, y'+\delta y') - F(x, y, y')\right]dx$$

$$= \int_{x_1}^{x_2} \left[F_y(x, y, y')\delta y + F_{y'}(x, y, y')\delta y'\right]dx + R \qquad \ldots(6)$$

where R is an infinitesimal such that

$$o(R) > o(\delta y) \text{ or } o(\delta y').$$

Also, using integration by parts the linear part on the R.H.S. of (6) can be written as

$$\left[F_{y'}\, \delta y\right]_{x_1}^{x_2} + \int_{x_1}^{x_2} \left(F_y - \frac{d}{dx} F_{y'}\right)\delta y\, \delta x .$$

Since the values of the functional I are taken only on the extremals, the integral in the second term of the above expression vanishes since $F_y - \dfrac{d}{dx} F_{y'} = 0$. Then the above expression becomes $(F_{y'}, \delta y)_{x=x_2}$, since $(\delta y)_{x=x_1} = 0$. It may be noted that $(\delta y)_{x=x_2} \neq \delta y_2$ since δy_2 is the increment in y_2 when the boundary point is displaced to $(x_2+\delta x_2, y_2+\delta y_2)$ while $(\delta y)_{x=x_2}$ is the increment of the ordinate of the point x_2 when passing from the extremals joining at (x_1, y_1) and (x_2, y_2) to the one joining (x_1, y_1) and $(x_2+\delta x_2, y_2+\delta y_2)$ as shown in the adjoining figure, curves through a given point.

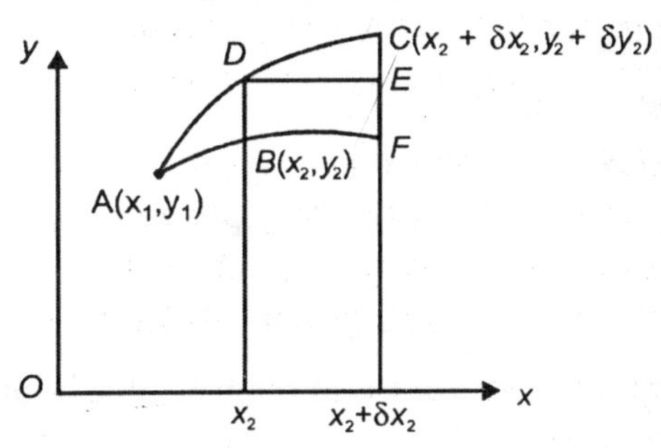

Clearly, $BD = (\delta y)_{x=x_2}$ and $FC = \delta y_2$.

Also, $\quad EC = y'(x_2)\delta x_2$ and hence

$$BD = FC - EC \text{ gives } (\delta y)_{x=x_2} = \delta y_2 - y'(x_2)\delta x_2$$

Evidently this equality is valid within the infinitesimals of higher order.

Thus

$$\int_{x_1}^{x_2}\left[F(x,y+\delta y,y'+\delta y')-F(x,y,y')\right]dx=\left[F_{y'}\right]_{x=x_2}\cdot\left[\delta y_2-y'(x_2)\delta x_2\right].$$

Now using (5) and above relation in (2), we obtain

$$\delta I=[F]_{x=x_2}\,\delta x_2+\left[F_{y'}\right]_{x=x_2}\cdot\left[\delta y_2-y'(x_2)\delta x_2\right]$$

$$=\left[F-y'F_{y'}\right]_{x=x_2}\delta x_2+\left[F_{y'}\right]_{x=x_2}\delta y_2. \quad \ldots(7)$$

Since δx_2 and δy_2 are independent, the necessary condition for the extremum; i.e. $\delta I=0$ then gives

$$\left[F-y'F_{y'}\right]_{x=x_2}=0,\left[F_{y'}\right]_{x=x_2}=0. \quad \ldots(8)$$

13.2.1 Transversality Condition

There may arise the cases when δx_2 and δy_2 are not independent. For example, if the boundary point $(x_2\,,y_2)$ moves along the curve

$$y_2=\phi(x_2) \quad \ldots(9)$$

Then $\delta y_2=\phi'(x_2)\delta x_2$. Thus from (7), we get

$$\left[F+(\phi'-y')F_{y'}\right]_{x=x_2}\delta x_2=0.$$

Since $\delta\,x_2$ is arbitrary, we must have

$$\left[F+(\phi'-y')F_{y'}\right]_{x=x_2}=0. \quad \ldots(10)$$

The equation (1) provides the condition at the free boundary and is known as the *transversality condition*. The conditions (9) and (10) are just sufficient to determine the extremals of the pencil $y=y(x,C_1)$ on which an extremum may be attained. If the boundary point $(x_1\,,y_1)$ also moves along another prescribed curve $y_1=\psi(x_1)$, then we have the condition similar to (10) as

$$\left[F+(\psi'-y')F_{y'}\right]_{x=x_1}=0. \quad \ldots(11)$$

13.2.2 Orthogonality Condition

We now consider the case when the transversality condition (10) reduces to the orthogonality condition. Suppose that

$$F(x,y,y')=A(x,y)(1+y'^2)^{1/2}$$

where $A(x,y)$ does not vanish at the movable boundary point x_2. In this case (10) reduces to

$$A(x,y)(1+\phi'\,y')(1+y'^2)^{1/2}=0 \text{ at } x=x_2 \quad \ldots(12)$$

Since $A(x\,,y)\neq 0$ at $x=x_2$, we have

$$y'=-\frac{1}{\phi'} \text{ at } x=x_2$$

This is *orthogonality condition*.

Example 1: Find the shortest distance between the point $A(1,0)$ and the ellipse $4x^2+9y^2=36$.

Solution: The formula for distance is

$$I[y]=\int_{x_1}^{x_2}\sqrt{1+y'^2}\,dx \text{ where } 4x_1^2+9y_1^2=36 \qquad \ldots(1)$$

The point (x_1, y_1) lies on the curve $y=\phi(x)=\frac{2}{3}\sqrt{9-x^2}$. Here

$$F(x,y,y')=\sqrt{1+y'^2}$$

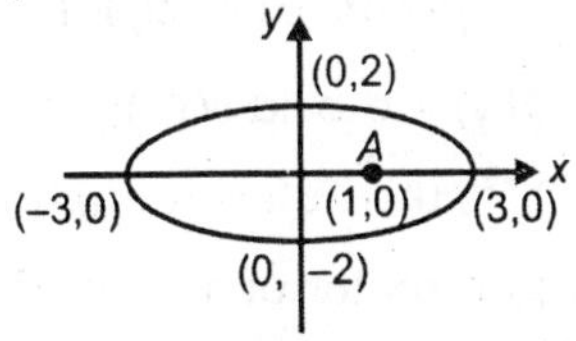

The Euler's equation in this case is

$$F_y - F_{yy'} - F_{yy'}\,y' - F_{y'y'}\,y''=0 \; i.e.\; y''=0$$

Here extremals lie on the straight lines $y=C_1x+C_2$ ...(2)

Since $A(1,0)$ lies on (2), we have $C_1+C_2=0$. ...(3)

Since (x_1, y_1) satisfies (1) and (2), we have

$$y_1=C_1(x_1-1) \qquad \ldots(4)$$

and

$$4x_1^2+9C_1^2(x_1-1)^2=36 \qquad \ldots(5)$$

The transversality condition is

$$\left[F+(\phi'-y')F_{y'}\right]_{x=x_1}=0$$

$$\Rightarrow \quad \left[\sqrt{1+y'^2}+\left(\frac{-2x}{3\sqrt{9-x^2}}-y'\right)\frac{y'}{\sqrt{1-y'^2}}\right]_{x=x_1}=0$$

$$\Rightarrow \quad \sqrt{1+C_1^2}+\left(\frac{-2x_1}{3\sqrt{9-x_1^2}}-C_1\right)\frac{C_1}{\sqrt{1+C_1^2}}=0$$

$$\Rightarrow \quad 3\sqrt{9-x_1^2}=2x_1C_1 \qquad \ldots(6)$$

Equation (5) can be written as $2\sqrt{9-x_1^2}=2C_1(x_1-1)$.

On solving (5) and (7) for x_1 and C_1, we get

$$x_1=\frac{9}{5},\; C_1=2.$$

Also from (4), we have $y_1=\frac{8}{5}$.

Hence the minimum distance $d_{\min}$ is given by

$$d^2_{\min}=(x_1-1)^2+y_1^2=\frac{16}{25}+\frac{64}{25}=\frac{16}{5} \text{ or } d_{\min}=\frac{4}{\sqrt{5}}$$

It lies along the straight line $y=2(x-1)$.

Example 2: Find the shortest distance between the parabola $y = x^2$ and the straight line $x - y = 5$.

Solution: The given problem reduces to finding the extremum the integral

$$I[y(x)] = \int_{x_1}^{x_2} \sqrt{1 + y'^2}\, dx$$

subject to the condition that the point $A(x_1, y_1)$ moves on the parabola $y = x^2$ and the other end point $B(x_2, y_2)$ lies on the straight line $x - y = 5$. Thus we take $\phi(x) = x^2$ and $\psi(x) = x - 5$.

Since $F(x, y, y') = (1 + y'^2)^{1/2}$, the Euler's equation is $y'' = 0$. Hence, the extremals lie on the two-parameter of straight lines $y = C_1 x + C_2$

Since A and B lie on an extremals, we have

$$y_1 = C_1 x_1 + C_2 \qquad \text{...(1)}$$

$$y_2 = C_1 x_2 + C_2 . \qquad \text{...(2)}$$

Since $A(x_1, y_1)$ lies on $y = x^2$ and $B(x_2, y_2)$ on $x - y = 5$, we have

$$y_1 = x_1^2 \qquad \text{...(3)}$$

$$y_2 = x_2 - 5 . \qquad \text{...(4)}$$

The transversality conditions are

$$\left[\sqrt{1 + y'^2} + (2x - y')\frac{y'}{\sqrt{1 - y'^2}}\right]_{x = x_1} = 0$$

and $$\left[\sqrt{1 + y'^2} + (1 - y')\frac{y'}{\sqrt{1 - y'^2}}\right]_{x = x_2} = 0$$

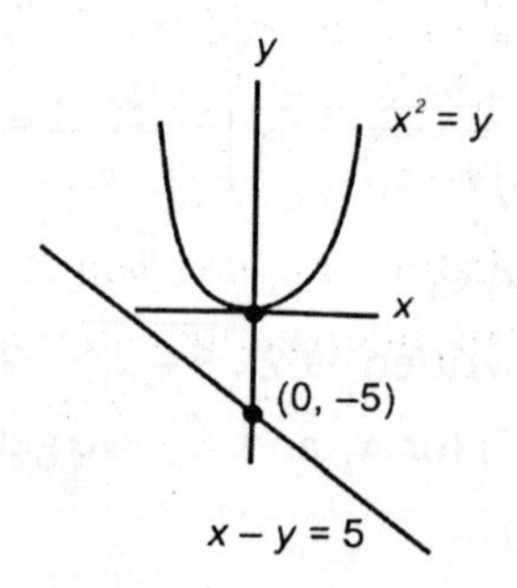

Since $y' = C_1$, these conditions reduce to

$$\sqrt{1 + C_1^2} + (2x_1 - C_1).\frac{C_1}{\sqrt{1 + C_1^2}} = 0 \qquad \text{...(5)}$$

and $$\sqrt{1 + C_1^2} + (1 - C_1).\frac{C_1}{\sqrt{1 + C_1^2}} = 0 . \qquad \text{...(6)}$$

Solving the equations (1) to (6), we obtain

$$x_1 = \frac{1}{2},\ x_2 = \frac{23}{8},\ C_1 = -1,\ C_2 = \frac{3}{4}.$$

Thus the required extremals is $y = -x + \frac{3}{4}$ and the shortest distance between the given parabola and the straight line is

$$L = \int_{1/2}^{23/8} (1+1)^{1/2}\, d\,x = \frac{19\sqrt{2}}{8}.$$

Example 3: Find the shortest distance between the circle $x^2 + y^2 = 1$ and the straight line $x + y = 4$.

Solution: The problem is to find the extremum of the functional

$$I[y(x)] = \int_{x_1}^{x_2} (1 + y'^2)^{1/2}\, dx$$

subject to the condition that the left end of the extremals moves along the circle $x^2 + y^2 = 1$ while the right end moves along the straight line $x + y = 4$. Thus the transversality condition gives

$$\left[\sqrt{1+y'^2} + \left(-\frac{x}{\sqrt{1-x^2}} - y'\right)\frac{y'}{\sqrt{1+y'^2}}\right]_{x=x_1} = 0 \qquad \ldots(1)$$

and

$$\left[\sqrt{1+y'^2} + (-1-y')\frac{y'}{\sqrt{1+y'^2}}\right]_{x=x_2} = 0. \qquad \ldots(2)$$

Since the general solution of Euler's equation $y'' = 0$ is $y = C_1\, x + C_2$ where C_1 and C_2 are constants. It follows that $y' = C_1$. Further, both the end points lie on the extremals $y = C_1\, x + C_2$, hence we must have

$$C_1\, x_1 + C_2 = \sqrt{1 - x_1^{\,2}} \qquad \ldots(3)$$

and $$C_1\, x_2 + C_2 = 4 - x_2 \qquad \ldots(4)$$

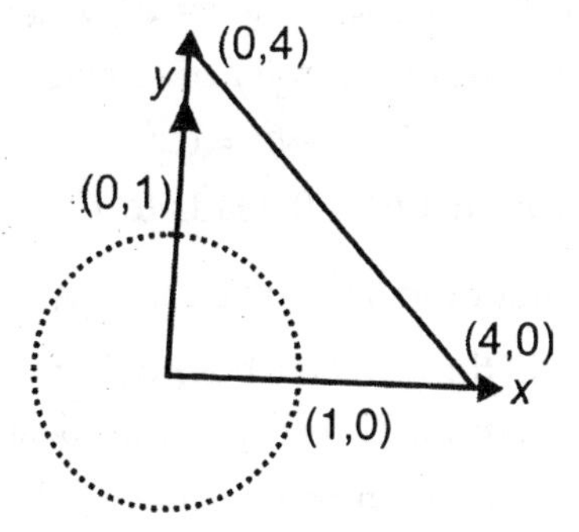

Replacing y' in (1) and (2) by C_1, we have

$$\sqrt{1+C_1^2}+\left(\frac{-x_1}{\sqrt{1-x_1^2}}-C_1\right)\frac{C_1}{\sqrt{1+C_1^2}}=0 \quad ...(5)$$

and $$\sqrt{1+C_1^2}+(-1-C_1)\frac{C_1}{\sqrt{1-C_1^2}}=0. \quad ...(6)$$

Now solving (3), (4), (5) and (6), we get $C_1=1$ and $x_1=\pm\frac{1}{\sqrt{2}}$.

A simple figure shows that the shortest distance occurs in the first quadrant.

Here $x_1=\frac{1}{\sqrt{2}}$ and $y_1=\frac{1}{\sqrt{2}}$

Also $C_2=0,\ x_2=y_2=2$.

Hence the minimum distance $=\int_{1/\sqrt{2}}^{2}\sqrt{2}\,dx=2\sqrt{2}-1$ and lies on the line $y=x$.

Example 4: Using only the basic necessary condition $\delta I=0$ find the curve on which extremum of the functional

$$I[y(x)]=\int_0^{x_1}\frac{(1+y'^2)^{1/2}}{y}dx,\ y(0)=0$$

can be achieved if the second boundary point (x_1, y_1) can move along the circumference

$$(x-9)^2+y^2=9$$

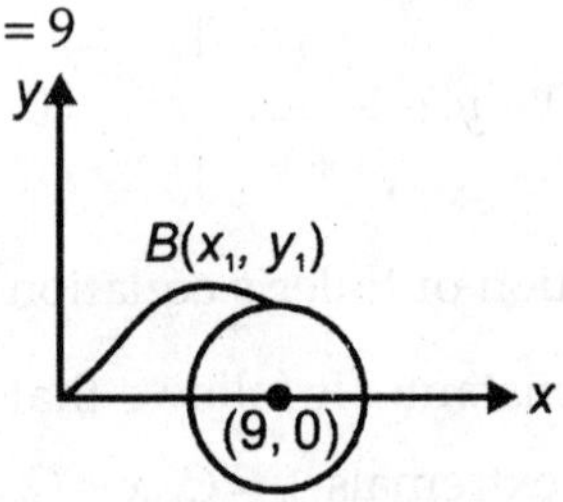

y

B(x₁, y₁)

(9, 0)

x

Solution: Let us denote $F(x,y,y')=\frac{(1+y'^2)^{1/2}}{y}$

Then it is clear that the integral curves $F-y'F_{y'}$ = constant

is the Euler's equation. This gives $1+y'^2=\frac{C_2}{y^2}$, where C_2 is a constant. Its solution is the two-parameter family of circles

$$(x-C_1)^2+y^2=C_2^2 \quad ...(1)$$

The boundary condition $y(0)=0$ leads to $C_1=C_2$.

Again, since the integrand is of the form $A(x,y)(1+y'^2)^{1/2}$, the transversality condition at the movable boundary point (x_1, y_1) reduces to the orthogonality condition. Thus the required extremals will be the arc of a circle belonging to (1) which is orthogonal to $(x-9)^2+y^2=9$.

Since (x_1, y_1) lies on both circles, we must have

$$x_1^2 - 18x_1 + y_1^2 = -72, x_1^2 - 2C_1 x_1 + y_1^2 = 0$$

$$\Rightarrow \quad x_1(C_1 - 9) = -36 \qquad \ldots(2)$$

In view of orthogonality of the two circles at (x_1, y_1), the tangent to (1) at B passes through the centre $(9, 0)$ of the given circle. This yields

$$(9 - C_1)x_1 = 9C_1 \qquad \ldots(3)$$

Solving (2) and (3), we find that $C_1 = 4$ and $x_1 = \frac{36}{5}$ so that the required extremals (2) are the arc of the circle

$$(x-4)^2 + y^2 = 16.$$

13.3 VARIATION PROBLEM WITH A MOVABLE BOUNDARY FOR A FUNCTIONAL DEPENDENT ON TWO FUNCTIONS

Let us consider the functional

$$I[y(x), z(x)] = \int_{x_1}^{x_2} F(x, y(x), z(x), y'(x), z'(x))dx \qquad \ldots(1)$$

Let the point $A(x_1, y_1, z_1)$ corresponding to the lower limit in the above integral be fixed and let the other point $B(x_2, y_2, z_2)$ moves in an arbitrary manner, or along a given curve or surface. Clearly the extremum can be attained only on the integral curves of Euler equations

$$\frac{\partial F}{\partial y} - \frac{d}{dx}\left(\frac{\partial F}{\partial y'}\right) = 0, \frac{\partial F}{\partial z} - \frac{d}{dx}\left(\frac{\partial F}{\partial z'}\right) = 0.$$

The general solution of these equations contains four arbitrary constants. Since the boundary point $A(x_1, y_1, z_1)$ is fixed, it is possible to eliminate two arbitrary constants. The other two constants have to be determined from the necessary condition $\delta I = 0$ for extremum, where δI denotes the variation of I. Now for an extremum $\delta I = 0$ gives

$$\delta I = \left[F - y' F_y - z' F_{z'}\right]_{x=x_2} \delta x_2 + \left[F_{y'}\right]_{x=x_2} \delta y_2 + \left[F_{z'}\right]_{x=x_2} \delta z_2 = 0 \qquad \ldots(2)$$

Suppose the variations $\delta x_2, \delta y_2, \delta z_2$ are independent, then (2) gives

$$\left[F - y' F_{y'} - z' F_{z'}\right]_{x=x_2} = 0, \left[F_{y'}\right]_{x=x_2} = 0, \left[F_{z'}\right]_{x=x_2} = 0.$$

Further, suppose that the boundary point $B(x_2, y_2, z_2)$ moves along some curve $y_2 = \phi(x_2)$, $z_2 = \psi(x_2)$, then

$$\delta y_2 = \phi'(x_2)\delta x_2 \text{ and } \delta z_2 = \psi'(x_2)\delta x_2$$

Thus from (2), we have $\left[F + (\phi' - y')F_{y'} + (\psi' - z')F_{z'}\right]_{x=x_2} = 0 \quad \ldots(3)$

This is the transversality condition in the problem of extremum of (1).

Note that along with the equations $y_2 = \phi(x_2), z_2 = \psi(x_2)$, the condition (3) gives the equations necessary for determining the two arbitrary constants in the general solution of Euler's equations.

On the other hand, if the boundary point $B(x_2, y_2, z_2)$ moves along a given surface $z_2 = \phi(x_2, y_2)$, then

$$\delta z_2 = \phi_{x_2}\, \delta_{x_2} + \phi_{y_2}\, \delta y_2$$

Such that the variations δx_2 and δy_2 are arbitrary. In this case (3) reduces to

$$\left[F - y' F_{y'} + (\phi_x - z')F_{z'}\right]_{x=x_2} \delta x_2 + \left[F_{y'} + \phi_y F_{z'}\right]_{x=x_2} \delta y_2 = 0.$$

Since δx_2 and δy_2 are independent, we find that

$$\left[F - y' F_{y'} + (\phi_x - z')F_{z'}\right]_{x=x_2} = 0 \text{ and} \left[F_{y'} + \phi_y F_{z'}\right]_{x=x_2} = 0 \quad ...(4)$$

These two conditions, together with $z_2 = \phi(x_2, y_2)$ enable us to determine two arbitrary boundaries in the general solution of Euler's equation.

If we consider the function

$$I = \int_{x_1}^{x_2} A(x, y, z)(1 + y'^2 + z'^2)^{1/2}\, dx$$

with the end point (x_2, y_2, z_2) lying on the surface $z = \phi(x, y)$, it may be easily seen that the transversality condition reduces to the orthogonality condition of the extremals to the surfaces $z = L(x, y)$ and $z = M(x, y)$, can only be attained on the straight lines, which are orthogonal to both these surfaces. We can readily extend the above arguments to a functional of the form

$$I = \int_{x_1}^{x_2} F\left(x_1, y_1(x), y_2(x), \ldots\ldots, y_n(x), y_1'(x), y_2'(x), \ldots.\, y_n'(x)\right) dx$$

such that at the moving point x_2, we have

$$\left[F - \sum_{i=1}^{n} y_i' F_{y_i'}\right]_{x=x_2} \delta x_2 + \sum_{i=1}^{n} \left[F_{y'}\right]_{x=x_2} \delta y_i = 0 \quad ...(5)$$

Example 5: Find the extremum of the functional

$$I = \int_{x_1}^{x_2} (y'^2 + z'^2 + 2yz)dx \text{ with } y(0) = 0, z(0) = 0$$

if the point (x_2, y_2, z_2) moves over the fixed plane $x = x_2$.

Solution: In this case

$$F(x, y(x), z(x), y'(x), z'(x)) = y'^2 + z'^2 + 2yz$$

Therefore Euler equations

$$\frac{\partial F}{\partial y} - \frac{d}{dx}\left(\frac{\partial F}{\partial y'}\right) = 0, \frac{\partial F}{\partial z} - \frac{d}{dx}\left(\frac{\partial F}{\partial z'}\right) = 0 \text{ give}$$

$$z'' - y = 0 \text{ and } y'' - z = 0$$

This leads to $y^{iv} - y = 0$ and $z^{iv} - z = 0$.
whose solutions are

$$y = C_1 \cos hx + C_2 \sin hx + C_3 \cos x + C_4 \sin x \quad \text{...(1)}$$

and $$z = C_1 \cos hx + C_2 \sin hx - C_3 \cos x - C_4 \sin x . \quad \text{...(2)}$$

The conditions $y(0) = z(0) = 0$ give $C_1 = C_3 = 0$. Further, the conditions of the moving boundary point (x_2, y_2, z_2) can be derived with $\delta x_2 = 0$ (since x_2 is fixed) as

$$[F_{y'}]_{x=x_2} = 0, [F_{z'}]_{x=x_2} = 0 \Rightarrow y'(x_2) = 0, z'(x_2) = 0 .$$

Thus (1) and (2) lead to

$$C_2 \cos hx_2 + C_4 \cos x_2 = 0 \text{ and } C_2 \cos hx_2 - C_4 \cos x_2 = 0 .$$

If $\cos x_2 \neq 0$, then $C_2 = C_4 = 0$. Therefore, an extremum is attained on $y = 0, z = 0$. But if $\cos x_2 = 0$, then $C_2 = 0$ and C_4 remains arbitrary. In this case, the required extremals is

$$y = C_4 \sin x , z = - C_4 \sin x .$$

13.4 ONE-SIDED VARIATIONS

Consider again the functional $I[y(x)] = \int_{x_1}^{x_2} F(x, y, y')dx \quad \text{...(1)}$

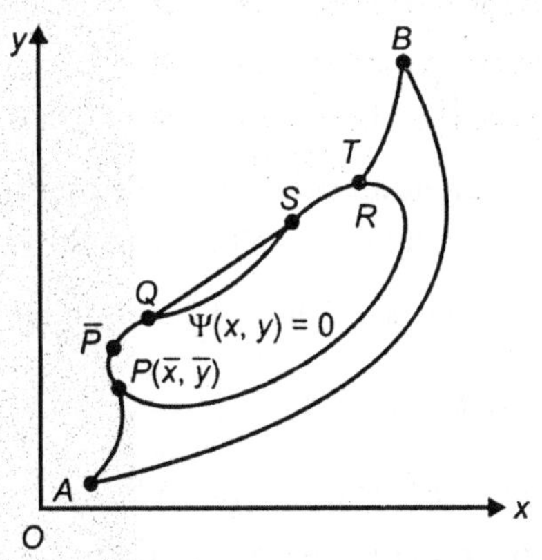

and suppose that a restriction is imposed on the class of permissible curves in such a way that the curves cannot pass through points of a certain region R bounded by the curve $\psi(x, y) = 0$.

In such type of problem the extremizing curve either passes through a region which is complete outside R or C consists of arcs lying outside R and also consists of parts of the boundary of the region R. In the former case, the presence of R does not at all influence the properties of the functional I and its variation δI in the neighbourhood of C. Hence, the extremizing curve must be an extremals. But in the latter case, only one-sided variations of the curve C are possible on parts of the boundary of the region R since the permissible curves are prohibited from entering R.

Clearly parts of the curves C outside the boundary of R are extremals since on these parts two-sided variations are possible.

We now derive conditions at the points of transition P, Q, S and T. We may write

$$I = \int_{x_1}^{x_2} F(x, y, y')dx$$

$$= \int_{x_1}^{\bar{x}} F(x, y, y')dx + \int_{\bar{x}}^{x_2} F(x, y, y')dx = I_1 + I_2, \text{ say} \qquad \dots(2)$$

In order to complete variation of the functional I in (2); i.e., δI, we suppose that the variation is caused solely by the displacement of the point $P(\bar{x}, \bar{y})$ on the curve $\psi(x, y) = 0$. This means that for any positive P on the curve, we may consider that AP is an extremals and the segment $PQSTB$ does not vary.

Also the upper boundary point $\bar{x}$ moves along the boundary of the region R and if $y = \phi(x)$ be the equation of the boundary (as deduced from $\psi(x, y) = 0$), then it follows from transversality condition that

$$\delta I_1 = \left[F + (\phi' - y')F_{y'}\right]_{x=\bar{x}} \delta\bar{x} \qquad \dots(3)$$

The functional $I_2 = \int_{\bar{x}}^{x_2} F(x, y, y')dx$ also has a moving boundary point $(\bar{x}, \bar{y})$. But in the neighbourhood of this point, the curve $y = \phi(x)$ on which an extremum can be achieved does not vary. Thus,

$$\Delta I_2 = \int_{\bar{x}+\Delta\bar{x}}^{x_2} F(x, y, y')dx - \int_{\bar{x}}^{x_2} F(x, y, y')dx$$

$$= -\int_{\bar{x}}^{\bar{x}+\Delta\bar{x}} F(x, y, y')dx = -\int_{\bar{x}}^{\bar{x}+\Delta\bar{x}} F(x, \phi(x), \phi'(x))dx$$

Since $y = \phi(x)$ on the interval $(\bar{x}, \bar{x} + \Delta\bar{x})$.

Now using the mean value theorem and the continuity of F, we get
$\Delta I_2 = -[F(x, \phi, \phi')]_{x=\bar{x}} \Delta\bar{x} + \alpha\Delta\bar{x}$, where $\alpha \to 0$ and $\Delta\bar{x} \to 0$.

Thus we get $\delta I_2 = -[F(x, \phi, \phi')]_{x=\bar{x}} \Delta\bar{x}$ $\qquad \dots(4)$

Combining (3) and (4), we find that

$$\delta I = \delta I_1 + \delta I_2 = \left[F(x, y, y') - F(x, y, \phi') - (y' - \phi')F_{y'}(x, y, y')\right]_{x=\bar{x}} \delta\bar{x}$$

with $y(\bar{x}) = \phi(\bar{x})$.

Since $\delta\bar{x}$ is arbitrary, it follows that the necessary condition $\delta I = 0$ for an extremum reduces to

$$\left[F(x, y, y') - F(x, y, \phi') - (y' - \phi')F_{y'}(x, y, y')\right]_{x=\bar{x}} = 0$$

Again applying the mean value theorem to this equation, we get

$$\left[(y' - \phi')\left\{F_{y'}(x, y, q) - F_{y'}(x, y, y')\right\}\right]_{x=\bar{x}} = 0 \qquad \ldots(5)$$

where $y'(\bar{x}) < q < \phi'(\bar{x})$.

Applying the mean value theorem once more to (5), we finally obtain

$$\left[(y' - \phi')(1 - y')F_{y'y'}(x, y, \bar{q})\right]_{x=\bar{x}} = 0 \text{ where } q < \bar{q} < y'(\bar{x})$$

Let us assume that $F_{y'y'}(x, y, \bar{q}) \neq 0$.

In this case $y'(\bar{x}) = \phi'(\bar{x})$ since $q = y'$ only when $y'(\bar{x}) = \phi'(\bar{x})$.

Hence we conclude that at the point *P*, the extremal *AP* meets the boundary curve *PQ* tangentially.

Example 6: Find the shortest path from the point $A(-2, 3)$ to the point $B(2, 3)$ located in the region $y \leq x^2$.

Solution: In this case our problem is to find the extremum of the functional

$$I[y] = \int_{-2}^{2} \left[1 + y'^2(x)\right]^{1/2} dx$$

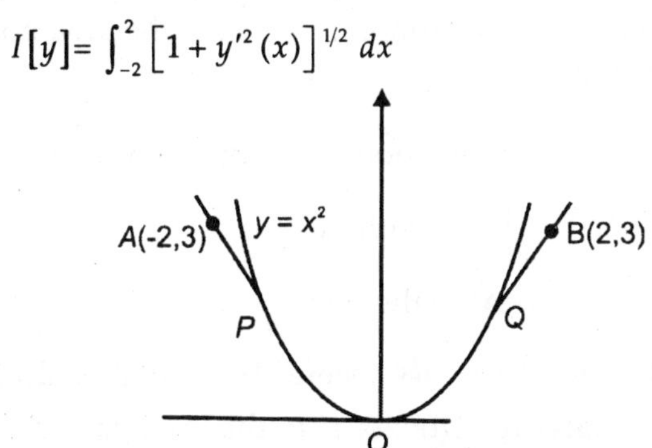

subject to the conditions $y \leq x^2$, $y(-2) = 3$, $y(2) = 3$.

Clearly, the extremals of $I[y]$ are straight lines $y = C_1 x + C_2$

Denoting by F the integrated in $I[y]$, we have

$$F_{y'y'} = \left[1 + y'^2(x)\right]^{3/2} \neq 0.$$

Hence the desired extremals will consist of portions of the straight lines *AP* and *QB* both tangents to the parabola $y = x^2$ and to the portion *POQ* of the parabola.

Let $-\overline{x}$ and $\overline{x}$ denote the abscissa of P and Q, respectively. Then the conditions of tangency of AP and BP at P and Q give

$$C_1\,\overline{x} + C_2 = \overline{x}^{\,2}\,,\, C_1 = 2\,\overline{x} \qquad \ldots(1)$$

Since the tangent QB passes through (2, 3), we have

$$2\,C_1 + C_2 = 3 \qquad \ldots(2)$$

Solving (1) and (2), we get two values for $\overline{x}_1 = 1$ and $\overline{x}_2 = 3$. The second value is clearly inadmissible. Thus, $\overline{x}_1 = 1$. Using it in (1), we get $C_2 = -1\,, C_1 = 2$.

Hence the required extremals is $y = \begin{cases} -2\,x - 1 \; if \; -2 \le x \le -1\,, \\ x^2 \; if \; -1 \le x \le 1\,, \\ 2\,x - 1 \; if \; 1 \le x \le 2\,. \end{cases}$

Clearly, this minimizes the functional.

EXERCISE 13.1

1. Find the function on which the following functional can be extremize $I[y(x)] = \int_0^1 (y''^2 - 2\,xy)dx\,,\, y(0) = y'(0) = 0\;\; y(1) = \frac{1}{120}$ and $y'(1)$ is not given.
2. Find the curve on which the following functional can attain an extremum $I[y] = \int_0^{10} y'^3\,dx\,,\, y(0) = 0 = y(10)$ subject to the condition that the admissible curves cannot pass inside the area bounded by the circle $(x-5)^2 + y^2 = 9$.
3. Determine the stationary function $y(x)$ for the problem $\delta\left\{\int_0^1 y'^2\,dx + [y(1)]^2\,dx\right\} = 0$, with $y(0) = 1$.
4. Prove that for a functional of the form $I[y(x)] = \int_{x_1}^{x^2} A(x\,,y)\sqrt{1+y'^2}\,dx$ where $A(x\,,y) \ne 0$ on the boundary curves, the extremals are orthogonal to the boundary curves.
5. Find the shortest distance between the given points and corresponding straight lines
   (i) $(0\,,2\,,4)$ and $\dfrac{x-1}{1} = \dfrac{y}{3} = \dfrac{z}{4}$
   (ii) $(0\,,2\,,0)$ and $x = y = z$
6. Find the shortest distance between the point $(1\,,1\,,1)$ and the sphere $x^2 + y^2 + z^2 = 1$.

ANSWERS 13.1

1. $y = \dfrac{x^5}{120} + \dfrac{1}{2a}\left(x^2 - x^3\right).$

2. $y(x) = \begin{cases} \pm \frac{3}{4}x, \text{ for } 0 \le x \le \frac{16}{5} \\ y(x) = \pm\sqrt{9-(x-5)^2}, \text{ for } \frac{16}{5} < x \le \frac{34}{5} \\ \pm\frac{3}{4}(x-10), \text{ for } \frac{34}{5} < x \le 10 \end{cases}$

3. $y = 1 - \dfrac{x}{2}.$

5. (i) $\dfrac{2\sqrt{6}}{3}$ units (ii) $\sqrt{\dfrac{105}{26}}$ units

6. $(\sqrt{3} - 1)$ units

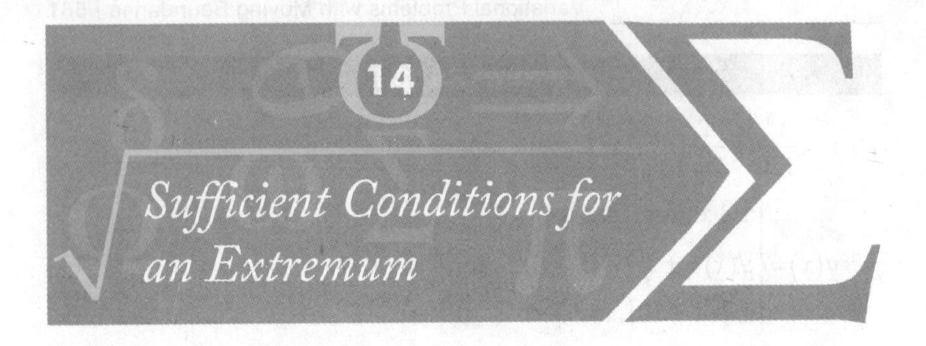

14 Sufficient Conditions for an Extremum

14.1 INTRODUCTION

At an extremum, a functional may have either minimum or maximum value. In the preceding chapters, we have studied necessary conditions for an extremum. At a stationary function, a functional may or may not have an extremum. In the present chapter we shall confine ourselves to study the sufficient conditions for extremum of variational problems with fixed boundaries only.

We first define proper and central fields in order to define field of extremals and then we will discuss sufficient conditions for an extremum of variational problems.

14.1.1 Proper Field

If one and only one curve out of a family of curves $y = y(x, C)$ passes through each point of a given domain D, then this family forms a proper field in the domain D. In other words, if the curves of the family $y = y(x, C)$ cover the entire domain D without self intersecting, then this family forms a proper field in the domain D.

Example 1: Inside the circle $x^2 + y^2 = 1$ the family of curves $y = C\,e^x$, C being arbitrary constant, forms a proper field since through any point of the above circle there passes one and only one curve of the family of curves $y = C\,e^x$.

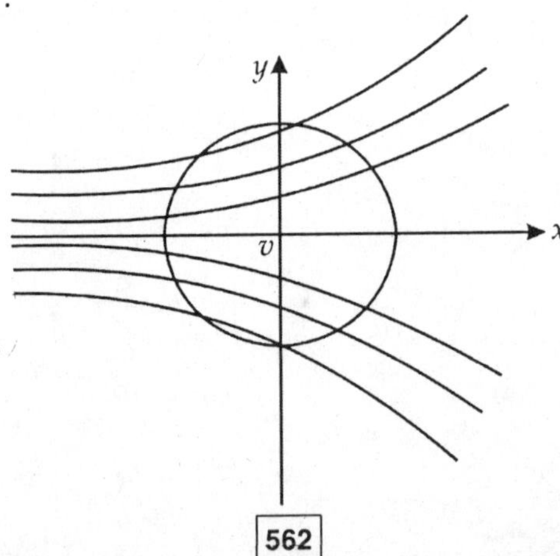

Example 2: The curve of the family of parabolas $y=(x+C)^2$ inside the circle $x^2+y^2=1$ does not constitute a proper field because different curves of the family intersect inside the circle.

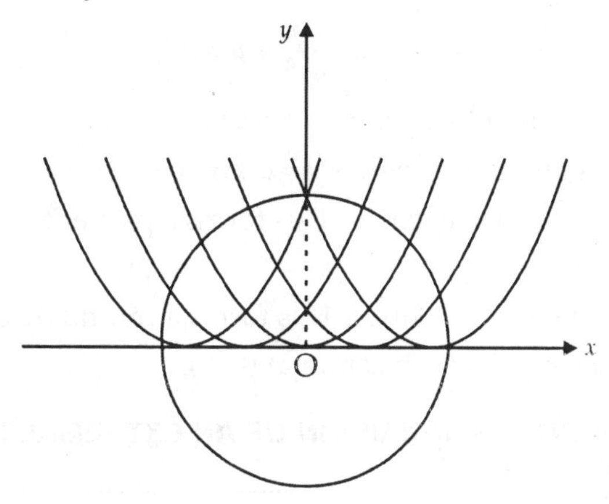

14.1.2 Central Field

If one and only one curve out of a family of curves $y=y(x,C)$ passes through each point of a given domain D and all curves of the family $y=y(x,C)$ pass through a single point (x_0, y_0) outside the domain D without self intersection, then they are said to constitute a central field over a domain D. The point (x_0, y_0) is called the centre of the central field or centre of the pencil of curves.

Example 3: The family of curves $y=Cx$ forms a central field in the domain $x>0$ and the point $(0,0)$ is the centre of the pencil of curves; i.e., straight lines emanating from $O(0,0)$.

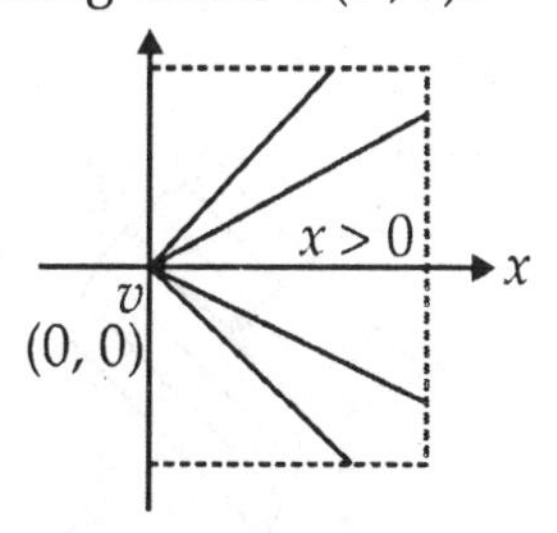

14.1.3 Field of Extremals

A field (proper or central) formed by a family of extremals of some variational problem is called a field of extremals or extremal field.

Example 4: Let the functional be $\int_0^1 (y^2 - y'^2)dx$, then find the proper and central field of extremals.

Solution: Here $F(x, y, y') = y^2 - y'^2$. Then the Euler's equation

$$F_y - \frac{d}{dx} F_{y'} = 0 \text{ gives}$$

$$2y - \frac{d}{dx}[-2y'] = 0 \Rightarrow \frac{d^2y}{dx^2} + y = 0$$

Now the A.E. is $m^2 + 1 = 0 \Rightarrow m = \pm i$.

Hence the general solution is $y = c_1 \cos x + c_2 \sin x$.

For $c_2 = 0$, we have $y = c_1 \cos x$. This forms a proper field of extremals in the domain $0 \le x \le 1$.

For $c_1 = 0$, we have $y = c_2 \sin x$. This forms a central field of extremals in the domain $0 < x \le 1$ with centre at (0, 0).

14.2 EMBEDDING OR INCLUSION OF AN EXTREMAL IN THE FIELD OF EXTREMALS

14.2.1 Embedding in a Proper Field

Let $y = y(x)$ be an extremal of the variational problem

$$I[y(x)] = \int_{x_1}^{x_2} F(x, y, y')dx \qquad \ldots(1)$$

with fixed boundary points $A(x_1, y_1)$ and $B(x_2, y_2)$. We say that the extremal $y = y(x)$ is embedded in a proper field of extremals if a family of extremals $y = y(x, C)$ can be found such that this family forms a proper field with the curve $y = y(x)$ as a member of this field for some value of C, say $C = C_0$ and the extremal $y = y(x)$ does not lie on the boundary of the domain D in which the family $y = y(x, C)$ forms a proper field.

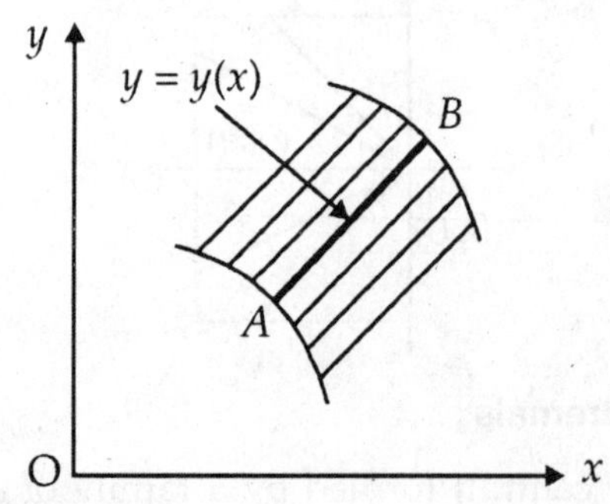

14.2.2 Embedding in a Central Field

We say that the extremal $y = y(x)$ is embedded in a central field of extremals if a family of extremals $y = y(x, C)$ can be found such that this family forms a central field with the curve $y = y(x)$ as a member of this field for some value of C, say $C = C_0$ and the extremal $y = y(x)$ does not lie on the boundary of the domain D in which the family $y = y(x\ C)$ forms a central field, that is, if a pencil of extremals emanating from the point $A(x_1, y_1)$ forms a central field including the extremal $y = y(x)$, then this curve AB is said to be embedded in a central field of extremals. In this case the parameter of family of curves forming the filed can be taken as the slope of the tangent line to the curve at A.

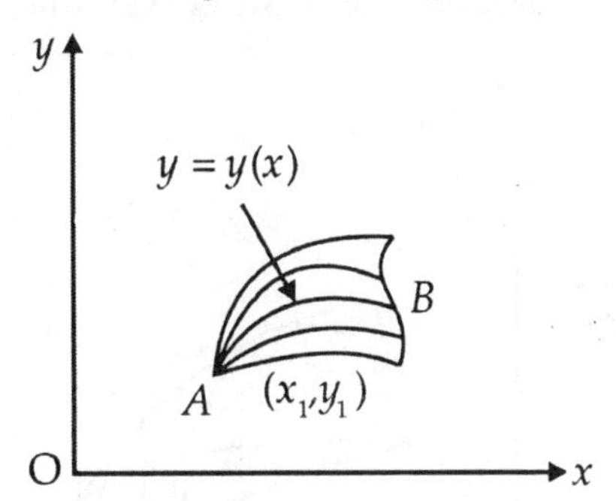

Example 5: Show that the extremal of the variational problem $\int_0^2 (y'^3 + \cos^2 x)dx,\ y(0)=0,\ y(2)=6$ is embedded in a central field of extremals of the given functional.

Solution: Here $F(x, y, y') = y'^3 + \cos^2 x$. The Euler's equation

$$F_y - \frac{d}{dx} F_{y'} = 0 \text{ gives}$$

$$F_{y'} = c, \text{ a constant} \Rightarrow 3y'^2 = c \Rightarrow y' = c_1 \Rightarrow y = c_1 x + c_2 \qquad \ldots(1)$$

This is the equation of extremals. Now applying the boundary conditions $y(0)=0,\ y(2)=6$; we get

$$c_2 = 0 \text{ and } c_1 = 3 \text{ respectively.}$$

Thus from (1), we have $y = 3x$. This is the extremal of the given problem.

If $c_2 = 0$, the equation (1) gives $y = c_1 x$. This is the central field of extremals in the domain $0 < x \le 2$ with centre at (0, 0). For $c_1 = 3$, the equation $y = c_1 x$ gives $y = 3x$. This shows that the extremal $y = 3x$ is embedded in the central field of extremals $y = c_1 x$.

14.3 JACOBI CONDITION

First of all, we shall discuss some necessary concepts before defining Jacobi's conditions.

14.3.1 Envelope

Let us consider one-parameter family of plane curves $\phi(x, y, C) = 0$. A curve which touches each member of family of curves $\phi(x, y, C) = 0$, and at each point is touched by some member of the family, is called the envelope of that family of curves.

14.3.2 C-Discriminant

The C-discriminant of family of plane curves $\phi(x, y, C) = 0$ is the locus of points defined by

$$\phi(x, y, C) = 0, \frac{\partial \phi}{\partial C} = 0. \qquad \ldots(1)$$

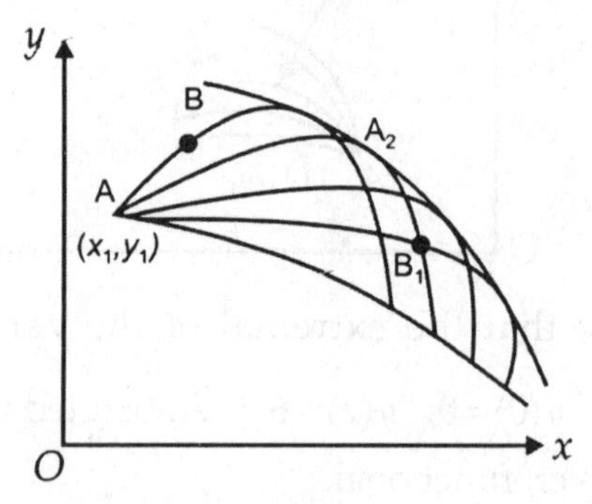

Generally C-discriminant includes the envelope of above family of curves, the locus of nodal points and the locus of cusps. Further, if we have a pencil of curves with centre at $A(x_1, y_1)$, then A belongs to the C-discriminant (1).

14.3.3 Conjugate Point

Consider a pencil of extremals $y = y(x, C)$ passing through $A(x_1, y_1)$ such that its C-discriminant is the curve $\phi(x, y) = 0$. Then it is clear that the envelope Γ of this pencil of curves belongs to $\phi(x, y) = 0$. The point A_1 where the extremal given by $y = y(x)$ touches Γ is called the conjugate point of $A(x_1, y_1)$. It is clear that if the point $B(x_2, y_2)$ lies between A and A_1, then the extremals of the pencil close to AB do not intersect. It follows that the extremals close to AB form a central field. Clearly, this central field includes the arc AB.

Let us consider the extremal AB_1 for which the conjugate point A_2 of A lies between A and B_1. For the extremal AB_1, we observe that the curves of pencil close to AB_1 intersect. Thus in this case the extremal AB_1 cannot be embedded in a central field.

Jacobi condition: From the above discussion we conclude the following:

For an arc AB of an extremal to be included in a central field of extremals with centre at A, it is sufficient that no conjugate point of A lie on the arc AB. In other words, to embed an arc AB of an extremal in a central field of extremals with centre at A, it is sufficient that the conjugate point of A does not lie on the arc AB. This sufficiency condition is known as the Jacobi condition.

The Jacobi condition can be stated mathematically as follows:

Let $y = y(x, C)$ be the equation of a pencil of extremals with centre at A such that the C-discriminant can be regarded as the slope y' of extremals of the pencil at A. The C-discriminant of the pencil is given by $y = y(x, C), \frac{\partial y}{\partial C} = 0$.

We now see that along every fixed curve of the family, $\frac{\partial y(x, C)}{\partial C}$ is a function of x only.

Let $u = \frac{\partial y(x, C)}{\partial C}$, where C is fixed. Since $y = y(x, C)$ is a solution of Euler equation for the extremum of $I[y(x)]$ in (1), we have

$$F_y\left[x, y(x, C), y_x'(x, C)\right] - \frac{d}{dx} F_{y'}\left[x, y(x, C), y_x'(x, C)\right] = 0.$$

Now differentiating this identity w. r. t. 'C', we get

$$\left(F_{yy} - \frac{d}{dx} F_{yy'}\right) u - \frac{d}{dx}\left(F_{y'y'}\, u'\right) = 0 \qquad \ldots(2)$$

This second order equation in u is known as the Jacobi equation.

In equation (2), $F_{yy}(x, y, y'), F_{yy'}$ and $F_{y'y'}$ are known functions of x since the second argument y is a solution of Euler's equation with $C = C_0$ (given) for the extremal AB.

Since the centre of the pencil belongs to the C-discriminant curve, it follows that the solution $u = \frac{\partial y}{\partial C}$ vanishes at $A(x_1, y_1)$.

Hence if the solution $u = \frac{\partial y}{\partial C}$ also vanishes at some point of the interval $x_1 < x < x_2$, then the point conjugate to A defined by $y = y(x, C_0)$ and $\left(\frac{\partial y}{\partial C}\right)_{C = C_0} = 0$ lies on the arc AB of the extremal with

B at the point (x_2, y_2). But if there exists a solution of (2) which vanishes for $x = x_1$ and does not further vanish at any point in $x_1 \le x \le x_2$, then it is clear that there are no points conjugate to A lying on the arc AB. Hence the Jacobi condition is satisfied and the arc AB of the extremal can be embedded in a central field of extremals with centre at A. We have following results.

i. If $u(x) = 0$ for some $x_1 \le x \le x_2$, then a conjugate point of A lies on the arc AB. Therefore by the Jacobi condition the arc AB of the extremal of the variational problem is not included in the central field of extremals with center at A.

ii. If $u(x) \neq 0$ for any $x_1 \le x \le x_2$, then no conjugate point of A lies on the arc AB. Therefore by the Jacobi condition the arc AB of the extremal of the variational problem is included in the central field of extremals with center at A.

Example 6: Find the C-discriminant of the family of curves $(x - c)y = c^2$.

Solution: Here given family of curves is $\phi(x, y, c) = (x - c)y - c^2$.

The equations $\phi(x, y, c) = 0, \dfrac{\partial \phi}{\partial c} = 0$ of C-discriminant imply

$$(x - c)y - c^2 = 0 \text{ and } -y - 2c = 0 \Rightarrow c = -\frac{y}{2}$$

Putting this value of c in (1), we get

$(x + \frac{y}{2})y = \left(-\frac{y}{2}\right)^2 \Rightarrow 4xy + y^2 = 0$. This is equation of the C-discriminant.

Example 7: Find a point conjugate to the point (0, 0) if the given family of curves is $y = c(x^2 + 4x)$.

Solution: Here the family of curves is $\phi(x, y, c) = y - c(x^2 + 4x)$.

This is a pencil of curves with centre at (0, 0). The equations $\phi(x, y, c) = 0, \dfrac{\partial \phi}{\partial c} = 0$ of C-discriminant imply $y - c(x^2 + 4x) = 0$ and

$$(x^2 + 4x) = 0 \Rightarrow x = 0 \text{ or } x = -4$$

For $x = 0$ or $x = -4$, the equation $y - c(x^2 + 4x) = 0$ gives $y = 0$.

Therefore C-discriminant is the set $\{(0, 0), (-4, 0)\}$.

Thus if $y = c_0(x^2 + 4x)$ be a curve belonging to the given pencil of curves, then this curve passes through the C-discrimi. ınt $(-4, 0)$ of the given family of curves. Therefore the point $(-4, 0)$ is the conjugate of the point $(0, 0)$.

Example 8: Show that the Jacobi condition is fulfilled for the extremal of the functional $I[y(x)] = \int_0^a (y'^2 + y^2 + x^2)\,dx$, passing through $A(0,0)$ and $B(a,0)$.

Solution: Here $F = y'^2 + y^2 + x^2$. Thus Euler's equation $F_y - \frac{d}{dx} F_{y'} = 0$ gives $2y - \frac{d}{dx}(2y') = 0 \Rightarrow \frac{d^2y}{dx^2} - 1 = 0$

The general solution is $y = c_1 e^x + c_2 e^{-x}$. The boundary conditions $y(0) = 0, y(a) = 0$ give $c_1 = 0 = c_2$

Hence the extremal of the given variational problem is $y = 0$.

Also $F_{yy} = 2, F_{yy'} = 0, F_{y'y'} = 2$.

Therefore, the Jacobi equation $\left(F_{yy} - \frac{d}{dx} F_{yy'}\right)u - \frac{d}{dx}\left(F_{y'y'}\, u'\right) = 0$ gives

$$\left(2 - \frac{d}{dx}(0)\right)u - \frac{d}{dx}(2u') = 0$$

$$\Rightarrow 2u - 2u'' = 0 \Rightarrow u'' - u = 0 \qquad \ldots(1)$$

which is a linear differential equation in u with constant coefficients. Its solution is

$$u = C_1 \sin hx + C_2 \cos hx \qquad \ldots(2)$$

Using $u(0) = 0$, (2) gives $C_2 = 0$.

Therefore $u = C_1 \sin h\, x$. $\qquad \ldots (3)$

Clearly the curves of pencil $u = C_1 \sin hx$ intersect the x-axis only at $x = 0$ and $u \neq 0$ on $(0, a]$. Therefore no conjugate point lies on the arc AB, i.e. on the curve $y = 0$. Therefore by the Jacobi condition the arc AB of the extremal $y = 0$ of the variational problem is included in the central field of extremals with center at A.

Example 9: Is the Jacobi condition is fulfilled for the extremal of the functional $\int_0^a \left(y'^2 - 4y^2 - e^{-x^2}\right)dx, a \neq \frac{n\pi}{2}$ with fixed boundaries $A(0,0)$ and $B(a, 0)$.

Solution: Here $F(x, y, y') = y'^2 - 4y^2 - e^{-x^2}$. Thus Euler's equation

$F_y - \frac{d}{dx} F_{y'} = 0$ gives

$$-8\,y - \frac{d}{dx} 2\,y' = 0 \Rightarrow \frac{d^2y}{dx^2} + 4 = 0$$

General solution is $y = c_1 \cos 2\,x + c_2 \sin 2\,x$.

This is the equation of extremals. Now the boundary conditions $y(0) = 0,\, y(a) = 0$ gives

$$c_1 \cos 0 + c_2 \sin 0 = 0 \Rightarrow c_1 = 0$$

and $c_2 \sin 2\,a = 0 \Rightarrow c_2 = 0$, since $a \neq \frac{n\,\pi}{2}$ respectively.

Thus $y = 0$ is the extremal of the given variational problem.

Also $F_{y\,y} = -8,\, F_{y\,y'} = 0,\, F_{y'\,y'} = 2$

Therefore, the Jacobi equation $\left(F_{y\,y} - \frac{d}{dx} F_{y\,y'}\right) u - \frac{d}{dx}\left(F_{y'\,y'}\, u'\right) = 0$

gives $\left(-8 - \frac{d}{dx} 0\right) u - \frac{d}{dx}(2\,u') = 0 \Rightarrow \frac{d^2u}{dx^2} + 4\,u = 0$

Here solution is $u = k_1 \cos 2\,x + k_2 \sin 2\,x$. Also $u(0) = 0 \Rightarrow k_1 = 0$.

Now let $u(x) = k_2 \sin 2\,x,\, k_2 \neq 0$, then there two cases arise.

Case I: *For* $a < \pi/2$, *we have* $u(x) = k_2 \sin 2\,x \neq 0,\, \forall\, x \in (0, a]$. *Therefore no conjugate point of A(0, 0) lies on the arc AB, i.e. on the curve* $y = 0$. *Hence by Jacobi condition the curve of the extremal* $y = 0$ *is included in a central field of extremals with centre at A(0, 0).*

Case II: *For* $a > \pi/2$, *we have* $u(x) = k_2 \sin 2\,x = 0$, for $x = \frac{\pi}{2}$ *because* $\frac{\pi}{2} \in (0, a)$. *Therefore conjugate point of A(0, 0) lies on the arc AB, i.e. on the curve* $y = 0$. *Hence by Jacobi condition the curve of the extremal* $y = 0$ *is not included in a central field of extremals with centre at* $A(0, 0)$.

14.4 WEIERSTRASS FUNCTION

Let us consider the extremum of the functional

$$I[y(x)] = \int_{x_1}^{x_2} F(x, y, y')\,dx,\; y(x_1) = y_1,\; y(x_2) = y_2 \qquad \ldots(1)$$

Also, suppose that the extremal C through $A(x_1, y_1)$ and $B(x_2, y_2)$ satisfies the Jacobi condition so that C can be embedded in a central field whose slope is $p(x, y)$.

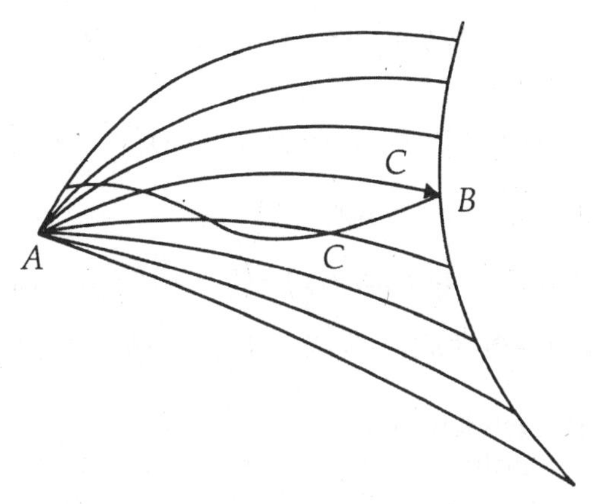

Of course, an extremal can be considered as embedded in a proper field.

In order to determine the sign of the increment ΔI of the functional I while passing from the extremal C of a neighbouring admissible curve $\bar{C}$, we consider the functional

$$I = \int_{\bar{C}} \left[F(x, y, p) + \left(\frac{dy}{dx} - p \right) F_p(x, y, p) \right] dx . \qquad ...(2)$$

Obviously (2) reduces to $\int_C F(x, y, y')dx$ on the extremal C, where $\frac{dy}{dx} = p$.

The functional (2) can be written as

$$I = \int_{\bar{C}} \left[F(x, y, p) - p\, F_p(x, y, p) \right] dx + F_p(x, y, p) dy \qquad ...(3)$$

which is the integral of an exact differential. As a matter of fact, the differential of the functional $\bar{I}(x, y)$ into which the functional $I[y(x)]$ is transformed on the extremals of the field has the form

$$d\,\bar{I} = \left[F(x, y, y') - y'\, F_{y'}(x, y, y') \right] dx + F_{y'}\, dy . \qquad ...(4)$$

Clearly, (4) differs from the integrand in (3) with y' replaced by p, the slope of the tangent line to the extremals.

Since the integrand in (3) is an exact differential, the integral is independent of the path of integration and is known as Hilbert's independence integral. Thus we have

$$\int_{\bar{C}} \left[F(x, y, p) + (y' - p) F_p(x, y, p) \right] dx = \int_C F(x, y, y') dx \qquad ...(5)$$

for any choice of admissible curve $\bar{C}$.

Now using (3), the increment

$$\Delta I = \int_{\bar{C}} F(x, y, y')dx - \int_{C} F(x, y, y')dx$$

can be written as

$$\Delta I = \int_{\bar{C}} \left[F(x, y, y') - F(x, y, p) - (y' - p)F_p(x, y, p)\right]dx$$

$$= \int_{x_1}^{x_2} E(x, y, p, y')dx, \qquad \ldots(6)$$

where $E(x, y, p, y') = F(x, y, p) - (y' - p)F_p(x, y, p)$. ...(7)

The integrand of (6), i.e., $E(x, y, p, y')$ is known as the Weierstrass function.

From (6) it is evident that

(i) a sufficient condition for I to attain a minimum of the extremal C is $E \geq 0$ $\left[\because E \geq 0 \Rightarrow \Delta I \geq 0\right]$

(ii) a sufficient condition for a maximum is $E \leq 0$ $\left[\because E \leq 0 \Rightarrow \Delta I \leq 0\right]$

Further, a sufficient condition for a weak minimum is that $E \geq 0$ is satisfied for values of x, y close to the values of x, y on the extremal and for value of y' close to $p(x, y)$ on the same extremal. But the sufficient condition for a strong minimum is that $E \geq 0$ is satisfied for all values of x, y closed to the corresponding values on the extremal with values of y' arbitrary.

Thus, we may summarize our results on sufficient conditions for an extremum of a functional given by (1) as follows:

Weak Extremum

1. The curve C is an extremal satisfying the boundary conditions in (1).
2. The extremal C must be embeddable in a field of extremals or the Jacobi condition must be satisfied.
3. The Weierstrass function E does not change sign at a point (x, y) close to the curve C and for values of y' close to $p(x, y)$ on the extremal. For a minimum $E \geq 0$ and for a maximum $E \leq 0$.

Strong Extremum

1. The curve C is an extremal satisfying the boundary conditions in (1).
2. It should be possible to embed the extremal C in a field of extremals or Jacobi condition is fulfilled.
3. The function E does not change sign at any point (x, y) close to the curve C and for arbitrary values of y'. For a minimum, $E \geq 0$ and for a maximum $E \leq 0$.

Remark: Every strong extremum at the same time is a weak extremum but converse is not true.

14.4.1 Weierstrass Condition and Elementary Convexity

It is sufficient to observe that the Weierstrass condition $E \geq 0$ (or $E \leq 0$) can be understood in geometrical terms.

To see this, let us take $F(x, y, y') = f(\xi)$ with $y' = \xi$ and (x, y) fixed

Then the condition $E \geq 0$ with E given by (7) implies $f(\xi) \geq l(\xi)$ where $l(\xi)$ is the linear function that agrees with f at $\xi = p$ and whose graph touches the graph of f at that point.

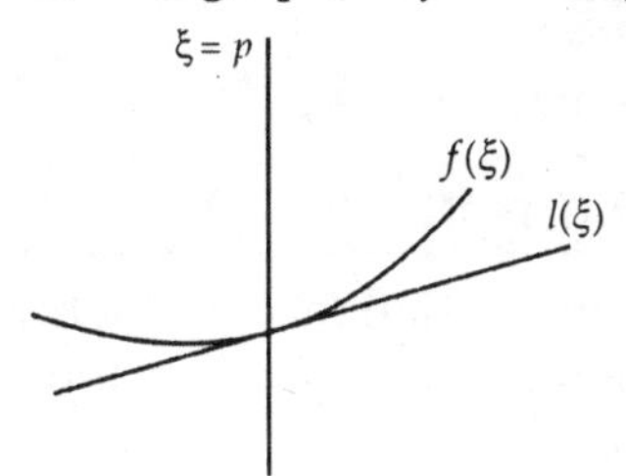

We say that the $f(\xi)$ is convex at $\xi = p$ if a linear function $l(\xi)$ exists such that $l(\xi) \leq f(\xi)$ for all ξ with equality at $\xi = p$. If f is differentiable, then l is necessary tangent at this point.

Example 10: Investigate an extremum of the functional

$$I[y(x)] = \int_0^1 \left(x + 2y + \frac{1}{2} y'^2 \right) dx, \; y(0) = 0, \; y(1) = 0.$$

Solution: Here $F(x, y, y') = x + 2y + \frac{1}{2} y'^2$, $x_1 = 0$, $x_2 = 1$.

Therefore $F_y = 2$, $F_{yy'} = 0$, $F_{y'y'} = 1$.

Hence the Euler's equation $F_y - F_{yy'} - F_{y'y'} y'' = 0$ reduces to $2 - 0 - y'' = 0$ or $y'' = 2$.

whose solution $y = x^2 + C_1 x + C_2$ satisfying the boundary conditions $y(0) = 0, y(1) = 0$ is $y = x^2 - x$.

Also, in this case, the Jacobi equation

$$\left(F_{yy}-\frac{d}{dx}F_{yy'}\right)u-\frac{d}{dx}\left(F_{y'y'},u'\right)=0$$

reduces to $u''=0$ whose solution satisfying $u(0)=0$ is $u(x)=Cx$. Since $u(x)=Cx$ does not vanish (for $C\neq 0$) anywhere in $[0,1]$ except $x=0$, it follows that the Jacobi condition is fulfilled and the extremal is $y=x^2+C_1x$ with the centre at the origin $(0,0)$.

In this case, the Weierstrass function is $E(x,y,p,y')=\frac{1}{2}(y'-p)^2$.

Clearly, $E\geq 0$ for arbitrary y'. Thus a strong minimum $=I\left[x^2-x\right]=\frac{1}{3}$ is attained on the extrema $y=x^2-x$.

14.5 LEGENDRE CONDITION

We observe that in some variational problems, testing the sign of E may pose some difficulties. So, we need an alternative condition which is readily verifiable.

Assume that the function $F(x,y,y')$ is thrice differentiable with respect to y'. Thus by Taylor's formula, we have

$$F(x,y,y')=F(x,y,p)+(y'-p)F_p(x,y,p)+\frac{1}{2}(y'+p)^2F_{y'y'}(x,y,q)\ldots(1)$$

where q lies between y' and p. Thus from (7) of 14.4, we have

$$E(x,y,p,y')=\frac{1}{2}(y'-p)^2F_{y'y'}(x,y,q). \qquad \ldots(2)$$

Since $(y^2-p)^2$ is always positive, it follows that the sign of E is the same as that of $F_{y'y'}(x,y,q)$. Therefore, for a weak extremum $F_{y'y'}(x,y,q)$ must retain its sign for values of x and y close to the points of extremal under consideration and for values of q close to $p(x,y)$. Now if $F_{y'y'}(x,y,y')\neq 0$ at all points of the extremal C, then by virtue of continuity $F_{y'y'}(x,y,y')$ maintains its sign at points close to the curve C and for values of y' close to values of y' on C.

Hence for testing for weak minimum, the condition $E\geq 0$ may be replaced by $F_{y'y'}>0$ on the extremal C and for testing for a weak maximum, the condition $E\leq 0$ may be replaced by $F_{y'y'}<$ on C. These conditions are known as the *Legendre condition*.

On the other hand, if $F_{y'y'}(x, y, q) \geq 0$ at points (x, y) close to points of C and for arbitrary values of q, then (2) implies that a strong minimum is attained on C. Observe that here we have assumed the formula (1) holds good for any y'.

We can extend the above results to a functional of the form

$$I[y_1(x), y_2(x), \ldots, y_n(x)] = \int_{x_1}^{x_2} F(x, y_1, y_2, \ldots, y_n, y_1', \ldots, y_n')\,dx$$

subject to the boundary conditions (BCs) $y_i(x_1) = y_{i1}$, $y_i(x_2) = y_{i2}$.

In this case, the Weierstrass function is

$$E = F(x, y_1, y_2, \ldots, y_n, y_1', y_2', \ldots, y_n') - F(x, y_1, y_2, \ldots, y_n, p_1, p_2, \ldots, p_n)$$
$$- \sum_{i=1}^{n} (y_i' - p_i) F_p(x, y_1, y_2, \ldots, y_n\, p_1, p_2, \ldots, p_n)$$

where p_i are the slope functions of the field. Also, the Legendre condition in this case becomes

$$F_{y_1' y_1'} \geq 0, \begin{vmatrix} F_{y_1' y_1'} & F_{y_1' y_2'} \\ F_{y_2' y_1'} & F_{y_2' y_2'} \end{vmatrix} \geq 0, \ldots, \begin{vmatrix} F_{y_1' y_1'} & F_{y_1' y_2'} & \cdots & F_{y_1' y_n'} \\ F_{y_2' y_1'} & F_{y_2' y_2'} & \cdots & F_{y_2' y_n'} \\ \vdots & \vdots & & \vdots \\ F_{y_n' y_1'} & F_{y_n' y_2'} & \cdots & F_{y_n' y_n'} \end{vmatrix} \geq 0.$$

Example 11: Using Legendre condition test for an extremum of the functional $I[y] = \int_0^2 (e^{y'} + 3)\,dx$, $y(0) = 0$, $y(2) = 1$.

Solution: Denoting the integrand of the functional by $F(x, y, y') = e^{y'} + 3$. As the extremals are the solutions of the Euler's equation

$F_y - \dfrac{d}{dx} F_{y'} = 0$. So we have

$$0 - \frac{d}{dx} e^{y'} = 0 \text{ or } \frac{d}{dx} e^{y'} = 0$$

On integrating, we have $e^{y'} = \text{constant} \Rightarrow y' = C_1$.

Again, integrating we get $y = C_1 x + C_2$

These straight lines are the required extremals.

Applying boundary conditions, we get

$$0 = y(0) = C_1\, 0 + C_2 \Rightarrow C_2 = 0$$

$$\text{and } 1 = y(2) = C_1\, 2 + 0 \Rightarrow C_1 = \frac{1}{2}$$

Therefore $y = \frac{x}{2}$ is the extremal satisfying the given boundary conditions. Certainly it can be embedded in the central field of extremals $y = C\,x$ for $C = \frac{1}{2}$.

Also since $F_{y'y'} = e^{y'} > 0\ \forall\ y'$, a strong minimum is attained on $y = \frac{x}{2}$.

Example 12: Using Legendre condition test for an extremum the functional $I[y] = \int_2^3 \frac{x^3}{y'^2}\,dx,\ y(2) = 4,\ y(3) = 9$.

Solution: Here $F(x, y, y') = \frac{x^3}{y'^2}$. The Euler's equation $F_y - \frac{d}{dx} F_{y'} = 0$ gives

$$\frac{d}{dx}\left[-\frac{2\,x^3}{y'^3}\right] = 0 \Rightarrow -\frac{2\,x^3}{y'^3} = c \Rightarrow y' = c_1\,x \Rightarrow y = c_1\,x^2 + c_2 \qquad (1)$$

This is the equation of the extremals. Now boundary conditions give $y(2) = 4 \Rightarrow c_1\,4 + c_2 = 4$ and $y(3) = 9 \Rightarrow c_1\,9 + c_2 = 9$.

Solving these equations, we get $c_1 = 1, c_2 = 0$.

Thus $y = x^2$ is extremal of the given problem.

For $c_2 = 0$, the equation (1) gives $y = c_1\,x^2$, which for $c_1 = 1$ implies $y = x^2$.

Therefore the extremal $y = x^2$ is included in the central field of extremals $y = c_1\,x^2$.

Now $F_{y'y'} = \frac{\partial}{\partial y'}\left(-\frac{2\,x^3}{y'^3}\right) = \frac{6\,x^3}{y'^4}$, which is a continuous function. For $y' = 0$, the quantity $\frac{6\,x^3}{y'^4}$ is undefined. Therefore, the function $y = x^2$ is neither strong maximum nor strong minimum. Hence, on the extremal $y = x^2$, we have

$$F_{y'y'} = \frac{6\,x^3}{(2\,x)^4} = \frac{3}{8\,x}.$$

14.6 SECOND VARIATION

It may be noted that the sufficient conditions for a weak extremum may be derived by a technique based on the sign of the second variation of the function (1) of 14.4.

$$\Delta I = \int_{x_1}^{x_2} \left[F(x, y + \delta\,y, y' + \delta\,y') - F(x, y, y')\right] dx$$

$$= \int_{x_1}^{x_2} \left[F_y \, \delta y + F_{y'} \, \delta y' \right] dx + \frac{1}{2} \int_{x_1}^{x_2} \left[F_{yy} (\delta y)^2 + 2 F_{yy'} \, \delta y \, \delta y' + F_{y'y'} (\delta y')^2 \right] dx + R \quad \ldots(1)$$

where $o(R) > o\left((\delta y)^2\right)$ and $o(R) > o\left((\delta y')^2\right)$.

Also the first variation represented by the term on R.H.S. of (1) vanishes on the extremal. Hence for weak extremum (for which both δy and $\delta y'$ are small), (1) shows that the sign of ΔI coincides with that $\delta^2 I$, the second variation is given by

$$\delta^2 I = \int_{x_1}^{x_2} \left[F_{yy} (\delta y)^2 + 2 F_{yy'} \, \delta y \, \delta y' + F_{y'y'} (\delta y')^2 \right] dx \quad \ldots(2)$$

Let us consider the integral involving the differentiable function $G(x)$;

$$\int_{x_1}^{x_2} \left[G'(x).(\delta y)^2 + 2 G(x) \delta y \, \delta y' \right] dx \quad \ldots(3)$$

which vanishes due to boundary conditions $y(x_1) = y_1$ and $y(x_2) = y_2$ because the integral is

$$\int_{x_1}^{x_2} d\left(G(x).(\delta y)^2\right) = \left[G(x).(\delta y)^2\right]_{x_1}^{x_2} = 0.$$

Now adding (3) to the second variation above, we get

$$\delta^2 I = \int_{x_1}^{x_2} \left[\left(F_{yy} + G'\right).(\delta y)^2 + 2\left(F_{yy'} + G\right) \delta y \, . \, \delta y' + F_{y'y'} (\delta y')^2 \right] dx \quad (4)$$

Clearly the integrand is a perfect square if

$$F_{y'y'} \left(F_{yy} + G'\right) - \left(F_{yy'} + G\right)^2 = 0 \quad \ldots(5)$$

For such a choice of G, (4) gives

$$\delta^2 I = \int_{x_1}^{x_2} F_{y'y'} \left[\delta y' + \frac{F_{yy'} + G}{F_{y'y'}} \delta y \right]^2 dx$$

From the above it is obvious that the sign of $\delta^2 I$ is the same as that of $F_{y'y'}$. But such a transformation is possible provided (5) admits of a differentiable solution $G(x)$.

Now substituting $G(x) = -F_{yy'} - \left(F_{y'y'} \frac{u'}{u} \right)$ in (5), we get

$$\left(F_{yy'} - \frac{d}{dx} F_{yy'} \right) u - \frac{d}{dx}\left(F_{y'y'} \, u'\right) = 0,$$

This is the Jacobi equation derived earlier. If a solution u of this equation exists that does not vanish in $\left[x_1, x_2\right]$; i.e., if the Jacobi condition is satisfied, then it is clear that there exists a differential function $G(x)$ satisfying (5).

Thus we conclude that fulfillment of both Legendre and Jacobi conditions ensure that the sign of second variation $\delta^2 I$ does not change. Thus they constitute sufficient conditions for a weak minimum $\left(F_{y'y'} > 0\right)$ or weak maximum $\left(F_{y'y'} < 0\right)$.

14.7 VARIATIONAL PRINCIPLE OF LEAST ACTION

Let there be a system of n particles of masses $m_i\ (i = 1, 2, ..., n)$ located at $(x_i, y_i, z_i);\ (i = 1, 2, ..., n)$. According to Hamilton's principle of least action for a system of n particles

$$\int_{t_1}^{t_2} (T - U)\,dt$$

is an extremum for the system for fixed terminal times where the kinetic energy T is given by

$$T = \frac{1}{2}\sum_{i=1}^{n} m_i\left(\dot{x}_i^2 + \dot{y}_i^2 + \dot{z}_i^2\right)$$

and the potential energy U is independent of times t. The Hamiltonian H for this system is

$$H = \sum_{i=1}^{n} m_i\left(\dot{x}_i^2 + \dot{y}_i^2 + \dot{z}_i^2\right) - (T - U,) = T + U,$$

This does not involve t explicitly. Therefore, H remains constant throughout the motion.

Thus we find that for a conservative dynamical system, $T + U$ remains constant during the motion.

In this case, Hamilton's variational principle takes the simple form

$$\delta\int_{t_1}^{t_2} (T - U)\,dt = \delta\int_{t_1}^{t_2} \left[2T - (T + U)\right] dt = 0$$

which, owing to $\delta(T + U) = \delta$ (constant) $=0$, leads to

$$\delta\int_{t_1}^{t_2} 2T\,dt = 0. \qquad ...(1)$$

The condition (1) is known as the variational principle of least action.

Example 13: Find the second variation of the functional

$$I[y] = \int_{x_1}^{x_2} y(x)\,dx$$

Solution: Let $\delta\, y$ be the increment of y, then

$$I[y + \alpha\,\delta y] = \int_{x_1}^{x_2} (y + \alpha\,\delta y)\,dx.$$

This implies $\frac{\partial}{\partial\alpha} I[y + \alpha\,\delta y] = \frac{\partial}{\partial\alpha}\int_{x_1}^{x_2} (y + \alpha\,\delta y)\,dx = \int_{x_1}^{x_2} (0 + \delta y)\,dx$

and $$\frac{\partial^2}{\partial \alpha^2} I[y+\alpha\,\delta y] = \frac{\partial}{\partial \alpha}\int_{x_1}^{x_2} \delta y\,dx = 0.$$

Thus $$\left.\frac{\partial^2}{\partial \alpha^2} I[y+\alpha\,\delta y]\right|_{\alpha=0} = 0 \Rightarrow \delta^2 I = 0.$$

Example 14: Find the second variation of the functional $e^{F(y)}$ where $F(y)$ is a twice differentiable function.

Solution: Let $I[y] = e^{F(y)}$ and δy be the increment of y, then

$$I[y+\alpha\,\delta y] = e^{F(y+\alpha\,\delta y)}.$$

This implies $\frac{\partial}{\partial \alpha} I[y+\alpha\,\delta y] = e^{F(y+\alpha\,\delta y)}\,F'(y+\alpha\,\delta y)\delta y;$

$$\frac{\partial^2}{\partial \alpha^2} I[y+\alpha\,\delta y] = e^{F(y+\alpha\,\delta y)}\,\{F'(y+\alpha\,\delta y)\}^2 (\delta y)^2 + e^{F(y+\alpha\,\delta y)}\,F''(y+\alpha\,\delta y).(\delta y)^2$$

$$= e^{F(y+\alpha\,\delta y)}\left[\{F'(y+\alpha\,\delta y)\}^2 + F''(y+\alpha\,\delta y)\right](\delta y)^2$$

Thus $$\left[\frac{\partial^2}{\partial \alpha^2} I[y+\alpha\,\delta y]\right]_{\alpha=0}$$

$$= e^{F(y+0\,\delta y)}\left[\{F'(y+0\,\delta y)\}^2 + F''(y+0\,\delta y)\right](\delta y)^2$$

$$= e^{F(y)}\left[\{F'(y)\}^2 + F''(y)\right](\delta y)^2 \qquad \ldots(1)$$

Let $I[y] = F(y)$, then $I[y+\alpha\,\delta y] = F(y+\alpha\,\delta y)$. In this case,

$$\frac{\partial}{\partial \alpha} I[y+\alpha\,\delta y] = F'(y+\alpha\,\delta y)\delta y$$

and $$\frac{\partial^2}{\partial \alpha^2} I[y+\alpha\,\delta y] = F''(y+\alpha\,\delta y).(\delta y)^2$$

Thus $\left.\frac{\partial}{\partial \alpha} I[y+\alpha\,\delta y]\right|_{\alpha=0} = F'(y)\delta y$ and

$$\left.\frac{\partial^2}{\partial \alpha^2} I[y+\alpha\,\delta y]\right|_{\alpha=0} = F''(y).(\delta y)^2$$

$$\Rightarrow \delta F = F'(y)\delta y \text{ and } \delta^2 F = F''(y).(\delta y)^2$$

Hence from (1) we have

$$\left.\frac{\partial^2}{\partial \alpha^2} I[y+\alpha\,\delta y]\right|_{\alpha=0} = e^{F(y)}\left[\{\delta F\}^2 + \delta^2 F\right].$$

EXERCISE 14.1

1. Find the proper and central field of extremals for the functional of the variational problem $I[y]=\int_0^4 y'^2\,dx$.
2. Find the proper and central field of extremals for the functional of the variational problem $I[y]=\int_0^{\pi/4}(y'^2-y^2+3x^3+5)\,dx$.
3. Show that the family of parabolas $y=(x-c)^2$ is not a proper field inside the circle $x^2+y^2=9$.
4. Show that the extremal of the variational problem
$I[y]=\int_0^1(y'^2-3xy)\,dx$, $y(0)=0$, $y(1)=0$ is included in a central field of extremals of the given functional.
5. Show that the extremal of the variational problem
$I[y]=\int_0^2(y'^3-\sin^2 x)\,dx$, $y(0)=2$, $y(2)=2$ is included in a proper field of extremals of the given functional.
6. Show that the extremal of the variational problem
$I[y]=\int_0^2(y'^2-x^2)\,dx$, $y(0)=1$, $y(2)=3$ is included in a proper field of extremals of the given functional.
7. Find the minimum of the functional
$I[y]=\int_0^1\left(\frac{1}{2}y'^2+y\,y'+y'+y\right)dx$ if the values at the ends of the interval are not given.
8. Find the extremum of the functional
$$I[y]=\int_1^2\frac{x^3}{y'^2}\,dx,\ y(1)=1,\ y(2)=4.$$
9. (Euler's problem on buckling). If a rod simply supported at one end is compressed by a longitudinal force P acting at the other end, it is then in either stable or unstable equilibrium. This means that after a slight lateral bending, it will either return to its equilibrium position or buckle, depending on whether the magnitude of P is less than or greater than a certain critical value P_0. To determine the bucking force P_0.
10. Find the C-discriminant of the following curves
(i) $y=(x-c)^2$ (ii) $y=c(x+c)$ (iii) $y^2+(x-c)^2=1$
11. Find the conjugate points of the point (0, 0) of the following curves
(i) $y=c(x-1)x$ (ii) $y=c\sin hx$

12. Show that the Jacobi condition is fulfilled for the following functionals

    (i) $\int_0^5 (9y^2 + y'^2 - 3x)dx$ with fixed boundaries $A(0,0)$ and $B(5,0)$.

    (ii) $\int_{-1}^1 (12xy + y'^2 + x^2)dx$ with fixed boundaries $A(-1,-2)$ and $B(1,0)$.

    (iii) $\int_0^a (y'^2 + x^2)dx, a > 0$ with fixed boundaries $A(0,0)$ and $B(a,2)$.

    (iv) $\int_0^1 (1 + y'^2)dx$ with fixed boundaries $A(0,0)$ and $B(1,0)$.

13. Using Legendre condition test the extremal of the functional

$$I[y] = \int_0^2 (4 - e^{y'})dx, y(0) = 0, y(2) = 1.$$

14. Show that the functional $I[y] = y(x), y(x) \in [4, 10]$ is differentiable.

15. Find the variations of the following functionals

    (i) $\int_{x_1}^{x_2} (6y(x) + 7)dx$ (ii) $\int_{x_1}^{x_2} (x + y)dx$

    (iii) $\int_0^x y' \sin y \, dx$ (iv) $\int_3^5 y^2(x)dx$

    (v) $\int_a^b (y^2 - y'^2)dx$

16. Find the variations of the following functional

    (i) $\int_a^b y^2 \, dx$ (ii) $\int_0^1 (xy^2 + y'^3)dx$

ANSWERS 14.1

1. $y = c_2, 0 \le x \le 4$; $y = c_1 x, 0 < x \le 4$ with centre at (0, 0).

2. $y = c_1 \cos x \; 0 \le x \le \pi/4$; $y = c_2 \sin x, 0 < x \le \pi/4$ with centre at (0, 0).

7. A strong minimum is attained on $y = \frac{1}{2}(x^2 - x - 1)$.

8. A weak minimum is attained on $y = x^2$.

9. $P_0 = \frac{\pi^2}{l^2} E I$.

10. (i) $y = 0$ (ii) $y = -\frac{x^2}{4}$ (iii) $y^2 = 1$

11. (i) $(1, 0)$ (ii) No conjugate point exist.

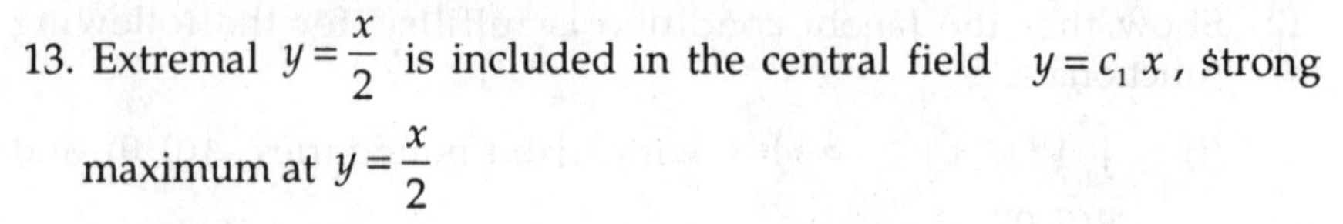

13. Extremal $y=\frac{x}{2}$ is included in the central field $y=c_1 x$, strong maximum at $y=\frac{x}{2}$

15. (i) $6\int_{x_1}^{x_2} \delta y(x)dx$ (ii) $\int_{x_1}^{x_2} \delta y(x)dx$

(iii) $\int_0^x \{(\delta y)' \sin y + y' \cos y\, \delta y\, dx\}$ (iv) $2\int_3^5 y(x)\delta y(x)dx$

(v) $2\int_a^b (y\,\delta y - y'(\delta y)')dx$

16. (i) $2\int_a^b (\delta y)^2\, dx$ (ii) $2\int_0^1 \{x(\delta y)^2 + 3\,y'(\delta y)'^2\}dx$

OBJECTIVE QUESTIONS

1. By a functional we mean a quantity whose values are determined by
   (a) one function only (b) several functions only
   (c) one or several functions (d) all of the above
2. $C[a,b]$ is class of functions which
   (a) is smooth (b) is continuous
   (c) has continuous first derivative (d) none of these
3. The length l between two points (x_0, y_0) and (x_1, y_1) on a curve $y=y(x)$ is a functional and
   (a) $l[y(x)]=\frac{dy}{dx}$
   (b) $l[y(x)]=1+\left(\frac{dy}{dx}\right)^2$
   (c) $\int_0^{x_2}\left[1+\left(\frac{dy}{dx}\right)^2\right]^{1/2} dx = l[y(x)]$
   (d) none of these
4. The shortest curve joining two points is a
   (a) straight line (b) circle
   (c) parabola (d) common catenary
5. Among all the curves joining two gives points (x_0, y_0) and (x_1, y_1) one which generates the surface of minimum area when rotated about x-axis is
   (a) circle (b) straight line
   (c) folium of Descartes (d) catenary
6. The integral curves of Euler's equation are called
   (a) rectifiable (b) functional
   (c) extremals (d) none of these

7. $\int_{t_0}^{t_1} L\,dt$, where L is Lagrangian, is called
   (a) Jacobian (b) action
   (c) Lagrangian (d) least action
8. If two functional have the some extremals, then they are said to be
   (a) equivalent (b) similar
   (c) equal (d) some
9. A family of curves $y = y(x, C)$ forms a proper field in a given domain D of the xy-plane if through any point of D there passes
   (a) all curves of the family
   (b) no curve of the family
   (c) one and only one curve of the family
   (d) those curves which cut the x-axis orthogonally
10. A family of curves $y = f(x, c)$ forms a central field in a given domain D of the xy-plane if
    (a) all the curves pass through a single point $(x_0\ y_0)$
    (b) one and only one curve of the family passes through any point of domain D
    (c) no curve of the family pass through any point of domain D
    (d) the curves which cut the X-axis orthogonally.
11. A field is called a field of extremals if
    (a) a proper filed is formed by a family of extremals
    (b) a central field is formed by a family of extremals
    (c) a proper or central field is formed by a family of extremals
    (d) a field is formed by a family of extremals
12. The necessary condition for $\int_{x_1}^{x_2} f(x, y, y')dx$ to be extremum is
    (a) $\frac{\partial f}{\partial y} = 0$ (b) $\frac{\partial f}{\partial y'} = 0$
    (c) $\frac{\partial f}{\partial y} - \frac{d}{dx}\left(\frac{\partial f}{\partial y'}\right) = 0$ (d) $\frac{\partial f}{\partial y} + \frac{d}{dx}\left(\frac{\partial f}{\partial y'}\right) = 0$

ANSWERS

1.	(c)	2.	(b)	3.	(c)
4.	(a)	5.	(d)	6.	(c)
7.	(b)	8.	(a)	9.	(c)
10.	(a)	11.	(c)	12.	(c)

Index

Reader's Note

Reader's Note

Reader's Note

Reader's Note